Introduction to Sociology

SEVENTH EDITION

Henry L. Tischler
Framingham State College

BUILT-IN STUDY GUIDE AND PRACTICE TESTS BY
Robert Mendelsohn
South Dakota State University

WADSWORTH
THOMSON LEARNING

Australia • Canada • Mexico • Singapore • Spain • United Kingdom • United States

What I know about society could fill a book.
What I don't would fill the world.
Dedicated to my fellow travelers in the journey of life—
Linda, Melissa, and Ben

WADSWORTH

THOMSON LEARNING™

Discipline Editor: Bryan Leake
Marketing Manager: Laura Brennan
Project Editor: Jim Patterson
Production Manager: Suzie Wurzer
Print/Media Buyer: Lisa Kelley
Permissions Editor: Caroline Robbins
Text Designer: Brian Salisbury

Copy Editor: Karen Carriere
Cover Designer: Brian Salisbury
Cover Image: Private Collection/Diana Ong/
 Superstock
Cover Printer: Lehigh Press, Inc.
Compositor: G & S Typesetters, Inc.
Printer: Courier/Kendallville

For more information about our products,
contact us at:
Thomson Learning Academic Resource Center
1-800-423-0563

For permission to use material from this text,
contact us by:

Phone: 1-800-730-2214
Fax: 1-800-730-2215
Web: http://www.thomsonrights.com

Library of Congress Catalog Card Number:
2001086106
ISBN: 0-15-505086-9

Asia
Thomson Learning
60 Albert Street, #15-01
Albert Complex
Singapore 189969

Australia
Nelson Thomson Learning
102 Dodds Street
South Melbourne, Victoria 3205
Australia

Canada
Nelson Thomson Learning
1120 Birchmount Road
Toronto, Ontario M1K 5G4
Canada

Europe/Middle East/Africa
Thomson Learning
Berkshire House
168-173 High Holborn
London WC1 V7AA
United Kingdom

Latin America
Thomson Learning
Seneca, 53
Colonia Polanco
11560 Mexico D.F.
Mexico

Spain
Paraninfo Thomson Learning
Calle/Magallanes, 25
28015 Madrid, Spain

B Process Black

PREFACE

As a freshman at Temple University, my first experience with a college textbook was in my sociology course. I dutifully read the assigned chapter during my first week of class hoping to become familiar with the subject matter of this required course. The only problem was that I had no idea what the author was saying. The writing level was advanced, the style dense, and the book downright threatening, without photos or illustrations. After several hours of reading I felt frustrated and stupid, and I knew no more about sociology than when I started. If this was what college was going to be like, I was not going to make it, I thought. I remember admitting reluctantly that I was probably not what guidance counselors in that day referred to as "college material." I could picture myself dropping out after the first semester and looking for a job selling furniture or driving a cab. My family would be disappointed, but my father was a factory worker, and there was no family history of college attendance to live up to. I continued to struggle with the book and earned a D on the mid-term exam. After much effort, I managed to finish the course with a C, and a burning disinterest in the field of sociology. I did not take another sociology course for two years, and when I did it was "Marriage and the Family," considered the easiest course on campus.

I often wonder how I came from this inauspicious beginning to become a sociology professor, let alone the author of a widely used introductory sociology textbook. Then again, maybe it is not all that unusual, because that experience continues to have an effect on me each day. Those 15 weeks helped to develop my view that little is to be gained by presenting knowledge in an incomprehensible or unnecessarily complicated way, or by making yourself unapproachable. Pompous instructors and intimidating books are a disservice to education. Learning should be an exciting, challenging, and eye-opening experience, not a threatening one.

One of the real benefits of writing seven editions of this textbook is that I have periodically examined every concept and theory presented in an introductory course. In doing so, I have approached the subject matter through a new set of eyes and have consistently tried to find better ways of presenting the material. As instructors, we rarely venture into each other's classrooms and hardly ever do we receive honest, highly detailed, and constructive criticism of how well we are transmitting the subject matter. In the writing of a textbook we receive this type of information, and we can radically restructure or simply fine tune our presentation. It is quite an education for those of us who have devoted our careers to teaching sociology.

Student-Oriented Edition

Prior to revising this edition of *Introduction to Sociology* we surveyed dozens of instructors to find out what they wanted in a textbook and what would assist them in the teaching of sociology, as well as satisfy student needs. This revised text reflects their significant input. In the surveys for this and past editions, we learned that both students and instructors continue to be concerned about the cost of textbooks. Introductory textbooks have become very attractive and expensive during the last decade, as publishers have added hundreds of color photos to the typical volume. This trend has caused the price of textbooks to increase, making them a substantial purchase for the typical student. We did something about the high cost to students—in response to this concern we broke ranks with textbooks with which we typically competed and went back to the basics. A textbook, after all, is meant to be comprehensive, up-to-date, and to serve as an important supplement to a course. It makes no sense to make a book so colorful, and therefore so expensive, that students often forgo purchasing it. To give students the best value for the dollar, we use black-and-white photos instead of color and a soft rather than a hard cover. In this way, students will be getting far greater value because nothing of educational content is sacrificed to produce this saving.

We are not, however, content to merely provide a better value. We also want to provide a better book. We, therefore, include a full study guide with this book that is as extensive, if not more so, than those typically sold separately. By this unusual move, students will be able to purchase the combined textbook and study guide for considerably less than the price of a typical textbook. In fact, the price for our textbook/study guide combination will most likely be lower than the used copy price of a typical hardcover introductory sociology textbook.

Presentation

Even though I began my college career as one of the less-capable students, I was fascinated by what college had to offer. Where else could you be exposed to so much about a world that is so interesting? Belatedly, I began to realize that a great deal of what is interesting falls into the field of sociology. My goal in this book is to demonstrate the vitality, interest, and utility associated with the study of sociology. Examining society and trying to understand how it works is an exciting and absorbing process. I have not set out to make sociologists of my readers (although if that happens I will be delighted), but rather to show how sociology applies to many areas of life and how it is used in day-to-day activities. In meeting this objective I have focused on two basic ideas: that sociology is a rigorous scientific discipline and that a basic knowledge of sociology is essential for understanding social interaction in many different settings, whether they be work or social. In order to understand society, we need to understand how it shapes people and how people in turn shape society.

Each chapter progresses from a specific to a general analysis of society. Each part introduces increasingly more comprehensive factors necessary for a broad-based understanding of social organization.

The material is presented through consistently applied learning aids. Each chapter begins with a **chapter outline.** Then, a thought-provoking **opening vignette** offers a real-life story of the concepts being covered. **Key terms** are presented in boldfaced type in the text. **Key concepts** are presented in italicized type in the text. A **chapter summary** concludes each chapter. An integrated **study guide** follows each chapter. A full **glossary** is in the back of the book for further reference.

Great care has been taken to structure the book in such a way as to permit flexibility in the presentation of the material. **Each chapter is self-contained and, therefore, may be taught in any order.**

It has taken nearly two years to produce this revision. Every aspect of this book has been updated and a great deal has been changed. The information is as current and up-to-date as possible and there are hundreds of 1995 through 2001 **references** throughout the book.

A Comparative and Cross-Cultural Perspective

Sociology is a highly organized discipline shaped by several theoretical perspectives or schools of thought. It is not merely the study of social problems or the random voicing of opinions. In this book no single perspective is given greater emphasis; a balanced presentation of both functionalist theory and conflict theory is supplemented whenever possible by the symbolic interactionist viewpoint.

The book has received a great deal of praise for being cross-cultural in approach and for bringing in examples from a wide variety of societies. Sociology is concerned with the interactions of people wherever and whenever they occur. It would be shortsighted, therefore, to concentrate on only our own society. Often, in fact, the best way to appreciate our own situation is through comparison with other societies. We use our cross-cultural focus as a basis for comparison and contrast with U.S. society.

Features

Opening Vignettes

Each chapter begins with a lively vignette that introduces students to the subject matter of the chapter. Many of these are from real-life events to which students can relate, such as the likelihood of children being abducted by strangers (Chapter 1), the scientific validity of UFO encounters (Chapter 2), socialization during Marine Corps basic training (Chapter 4), education in inner-city schools (Chapter 12), and the personal impact of prenatal screening (Chapter 16). Others deal with unusual circumstances that remind students that there is a wide range of events to which sociology applies. Examples include the eccentric soprano Florence Foster Jenkins (Chapter 6), whites who claim to be black (Chapter 8), a transsexual who believes there are dozens of genders (Chapter 9), the one-child population control policy in China (Chapter 15), and the fear of genital theft in Lagos Nigeria (Chapter 17).

Theme Boxes

Thought-provoking boxed features bring sociological concepts to life for students. This effective learning tool presents sociological concepts in interesting real-life contexts. In this edition, three new themes—Our Diverse Society, Remaking the World, and Society and the Internet—are added. Additionally, new boxes are added for the time-tested themes from previous editions—Controversies in Sociology, Global Sociology, Sociology at Work, and Technology and Society.

Our Diverse Society   NEW

Anyone studying sociology will quickly become aware of the enormous amount of social diversity. The United States with its extensive history of immigration has become one of the most diverse countries in the world. How has this diversity expressed itself in American society? In these boxed features we explore this question when we look at such topics as "How Blacks and Whites Offend Each Other without Realizing It," "The Black Middle Class: Fact or Fiction?" "Should Same-Sex Marriages Be Permitted?" "From an Inner-City High School to the Ivy League," "Disorderly Behavior and Community Decay," and "Why Isn't Life Expectancy in the United States Higher?"

Remaking the World NEW

Traditionally, sociology has often demanded an objective and unbiased approach to social issues. Some people find this position unsatisfactory and respond strongly to social events, particularly when they see a wrong that needs to be corrected. Is there an intersection between objective social research and proactive social action? In these boxed features we explore this question and discuss individuals who have taken strong positions on particular social situations and directed their efforts at making the world a better place. Among the topics we explore are "Should Television Be Used to Teach Values?" "Freeing the Innocent," "The Fight Against Honor Killings," "Jonathan Kozol on Unequal Schooling," "The Campaign to Ban Land Mines," "A World of Child Labor," and "The Lost Art of Healing."

Society and the Internet NEW

We have all witnessed the emergence of the Internet during the last decade and the impact it has had on people's lives. The World Wide Web has greatly enhanced the ability of people with similar interests to find each other. In these boxed features we look at how social interactions and social trends have been influenced by the Internet. Included are such topics as "Hate Sites on the Web," "Is There Gender in Cyberspace?" "Has the Internet Transformed Education?" "Social Movements on the Internet," "Do We Really Need the Information Highway?" and "Religion on the Web."

Controversies in Sociology

The special "Controversies in Sociology" boxed features are designed to show students two sides of an issue. The topics featured will help students realize that most social events require close analysis and that hastily drawn conclusions are often wrong. The students will see that to be a good sociologist, one must be knowledgeable about disparate positions and must be willing to question the validity of all statements and engage in critical thinking. Included in these boxes are such controversies as "Is There a Difference Between Sociology and Journalism?" "Truth in the Courtroom vs. Truth in the Social Sci-

ences," "Is There a Language Instinct?" "Is Day Care Harmful to Children?" "The Continuing Debate Over Capital Punishment: Does It Deter Murderers?" "Is the Income Gap Between the Rich and the Poor a Problem?" "Is the Race and Intelligence Debate Worthwhile?" "Is Transracial Adoption Cultural Genocide?" "Can Gender Identity Be Changed?" "Are College Admission Tests Fair?" "Are Religious Cults Dangerous?" "What Produces Homelessness?" and "Have We Exaggerated the Extent of the Population Problem?"

Global Sociology

To highlight the cross-cultural nature of this book, many chapters include a "Global Sociology" box. These boxed features encourage students to think about sociological issues in a larger context and explore the global diversity present in the world. Included among these boxes are such topics as "Is McDonald's Practicing Cultural Imperialism or Cultural Accommodation?" "An American Success Story Does Not Translate into Japanese," "Cross-Cultural Social Interaction Quiz," "Is Homicide an American Phenomenon?" "Children in Poverty," "Worldwide Racial and Ethnic Prejudice," "Arranged Marriage in India," "College Graduates: A Worldwide Comparison," "When Violence and Politics Equal Democracy," "Worldwide Religious Persecution Is Common," and "HIV/AIDS: Worldwide Facts."

Sociology at Work

These boxed features expand on a concept, theory, or issue discussed in the chapter. They allow instructors and students to examine a specific situation in depth and see its application to sociology. Some of the boxes examine sociologically related research that exposes students to the vibrant nature of the field of sociology. These topics include "If You Are Thinking About Sociology as a Career, Read This," "Suicide in the United States," "Women and the Development of Sociology, 1800 to 1945," "How to Spot a Bogus Poll," "How to Read a Table," "Seymour Martin Lipset on American Exceptionalism," "The Conflict Between Being a Researcher and a Human Being" "Are We Arguing Too Much?" "Does Birth Order Influence Our Social Interactions?" "Public Heroes, Private Felons: Athletes and Sexual Assault," "Serial Murderers and Mass Murderers," "How Easy Is It to Change Social Class?" "Racial Integration in the Military," "Deborah Tannen: Communication Between Women and Men," "Work Is Where the Heart Is: Has the Office Become a Substitute for the Family?" "How Much Are Children Hurt by Their Parent's Divorce?" "The Importance of Presidential Concession Speeches," "Comparing the Political and Moral Values of the

"Comparing the Political and Moral Values of the 1960s with Today," "How to Ruin a City," "Binge Drinking as a Health Problem," "Stereotypes About the Elderly," "Is Vegetarianism a Social Movement?" and "The McDonaldization of Society."

Technology and Society

Social research and technological change often go hand in hand. Technology helps social researchers, at the same time as it produces ethical challenges. Recognizing the importance of the social impact of technology we explore such topics as "Is There Beauty in Research?" "Is Research with Animals Ethical?" "Does Television Reduce Social Interaction?" and "Defining Parenthood: High-Tech Fertility Treatment Versus Adoption."

Built-in Study Guide and Practice Tests

The interactive workbook study guide, by Robert Mendelsohn of South Dakota State University, is fully integrated into the book. Each chapter is followed by a study guide section so students can review the material immediately, without having to search for it elsewhere in the book. This encourages students to see the study guide as an integral part of the learning process.

The study guide provides for ample opportunity to review the material with a variety of styles of review questions. All key terms and key sociologists are reviewed with matching questions. Key concepts are revisited with fill-in questions. Critical Thought Exercises help student contextualize concepts covered in the chapter. Often Web site URLs are provided for students to expand on their exploration of the topic. And, a matching-question answer key is provided to allow students immediate review of their answers.

Practice tests are in the back of the book to provide students with additional preparation for testing. Whereas other practice tests are limited to recognition and recall items, these questions lead students to engage in such higher-level cognitive skills as analysis, application, and synthesis. The tests encourage students to think critically and apply the material to their experiences. Again, an answer key is provided to allow students full review and preparation.

All of these tools will be very useful for students preparing for essay exams and research papers. The textbook also includes the important section, "How to Get the Most Out of Sociology," which discusses how to use the study guide, practice tests, and lecture material in preparing for exams and getting the most out of the introductory sociology course.

The Ancillary Package

The primary objective of a textbook is to provide clear information in a format that promotes learning. In order to assist the instructor in using *Introduction to Sociology* an extensive ancillary package has been developed to accompany the book.

Instructor's Manual and Test Bank

Robert Mendelsohn prepared the revision of the Instructor's Manual and Test Bank, as well as the student study guide and practice tests in the textbook. This provides for unusual consistency and integration among all elements of the teaching and learning package. Both the new and experienced instructor will find plenty of ideas in this Instructor's Manual, which is closely correlated to the textbook and the student study guide. Each chapter of the manual includes teaching objectives, key terms, lecture suggestions, activities, discussion questions, and formatted handouts for many topics. The Instructor's Manual also contains an annotated list of resources for students for reference or as a handout. Instructors will be able to download the Instructor's Manual from the Harcourt Sociology Web site. Consult your sales representative for access information or how to secure the printed version.

The Test Bank contains multiple-choice, true/false, and essay questions keyed to each learning objective. These test items are page referenced to the textbook and include significant numbers of application as well as knowledge questions. Story problems use names drawn from a variety of cultures, reflecting the diversity of U.S. society. Instructors requested that the questions be tied to the practice tests, and we followed that suggestion.

Computerized Test Bank

The computerized version of the Test Bank, available in both Windows and Macintosh formats, allows instructors to modify and add questions as well as to create, scramble, and print tests and answer keys. A telephone hotline is available for anyone who experiences difficulty with the program or its interface with a particular printer. Technical support is available by calling 1-800-447-9457 from 7:00 a.m. to 6:00 p.m. Central time.

PowerPoint Slides

Downloadable from the Harcourt Sociology Web site, this package enhances classroom presentations. Consult your sales representative for access information.

support is available by calling 1-800-447-9457 from 7:00 a.m. to 6:00 p.m. Central time.

PowerPoint Slides

Downloadable from the Harcourt Sociology Web site, this package enhances classroom presentations. Consult your sales representative for access information.

Online Course

Harcourt College Publishers offers a WebCT online course that can be customized to support individual teaching styles and syllabi. Consult your sales representative for additional information.

Sociology on the Web

Harcourt College Publishers provides a distinctive learning tool on its Sociology Web site. The Web site specifically designed to support this textbook includes **Web Resources, Web Links, Glossary, Review Questions, Critical Thinking Exercises, Web Activities by Subject, Student Bulletin Board, Instructor's Resources, Class Act, Syllabus Generator, Instructor's Manual, PowerPoint Slides,** and **WebCT/ Blackboard materials.** Consult your sales representative for additional information and access.

Computer Software

Software packages, which are available to instructors include **Core Concepts in Sociology** (CD-ROM), which combines the expertise of leading sociologists with the power of multimedia presentation to give students an exciting way to explore their sociological imaginations. This CD also includes SocialStat and SimCity software, to let students put ideas to work.

Videos

The instructor has the option of choosing from an extensive collection of videos to enhance the classroom learning experience. These include the following:

- **The *Sociological Imagination* Video Series** of 26-minutes clips from the telecourse by Dallas County Community College, include *Sociological Thinking and Research, Culture, Cities and Populations, The Process of Deviance, Social Class, The Importance of Sex and Gender, Family Religion in America, Political Systems, Science and Technology, Collective Behavior, Social Movements,* and *Social Change.*

- **Social Issues/Social Trends Video Series** includes videos from Films for the Humanities & Sciences and PBS. These videos highlight current social issues, such as *Ethnic Diversity,* and various social trends, including *The Vanishing Father.* Other specialized videos include *Growing Old in a New Age, Marriage and the Family* videos, *The Deadly Deception, Parents and Teenagers,* and *When Families Divorce.* Use of all videos is based on the Harcourt Brace policy. See your publisher's representative for details.

- **Films for the Humanities & Sciences** include *The Death Penalty, The Capital Punishment Industry, Crime and Human Nature, Prisoner on the Run,* and *Bad Cops or Cops Getting a Bad Rap?*

Additional videos are available. Ask your sales representative for a complete listing and access information.

Acknowledgments

The textbook and study guide manuscripts have been written after an extensive survey of faculty at a wide variety of institutions. I am grateful for the thoughtful contributions of the following persons: Froud Stephen Burns, Floyd Junior College; Peter Chroman, College of San Mateo; Mary A. Cook, Vincennes University; William D. Curran II, South Suburban College; Ione Y. Deollos, Ball State University; Brad Elmore, Trinity Valley Community College; Cindy Epperson, St. Louis Community College–Meramac; David A. Gay, University of Central Florida; Daniel T. Gleason, Southern State College; Charlotte K. Gotwald, York College of Pennsylvania; Richard L. Hair, Longview Community College; Selwyn Hollingsworth, University of Alabama; Sharon E. Hogan, Longview Community College; Bill Howard, Lincoln Memorial University; Sidney J. Jackson, Lakewood Community College; Michael C. Kanan, Northern Arizona University; Louis Kontos, Long Island University; Steve Liebowitz, University of Texas, Pan American; Steven Patrick, Boise State University; Thomas Ralph Peters, Floyd College; Kanwal D. Prashar, Rock Valley Community College; Charles A. Pressler, Purdue University, North Central; Stephen Reif, Kilgore College; Richard Rosell, Westchester Community College; Catherine A. Stathakis, Goldey Beacom College; Doris Stevens, McLennan Community College; Gary Stokley, Louisiana Tech University; Elena Stone, Brandeis University; Judith C. Stull, La Salle University; Lorene Taylor, Valencia Community College; Brian S. Vargus, Indiana University–Purdue University Indianapolis; J. Russell Willis, Grambling State

University; and Bobbie Wright, Thomas Nelson Community College.

A project of this magnitude becomes a team effort, with many people devoting enormous amounts of time to ensure that the final product is as good as it can possibly be. At Harcourt College Publishers, Bryan Leake, the acquisitions editor, ushered this project through its many stages along with Christine Caperton, the developmental editor. Jim Patterson, the senior project editor, along with Brian Salisbury, senior art director, and Suzie Wurzer, production manager, made sure that all those things that need to be done between the time the manuscript leaves the author's hands and becomes a book got done. It was a privilege to have the support and assistance of these very capable people. I am also grateful to all those students and instructors who have shared with me their thoughts about this book over the years. Please continue to let me know how you feel about this book.

Henry L. Tischler
htischl@frc.mass.edu
or
txtbks@aol.com

Effective Study: An Introduction

Why should you read this essay? If you think you have an A in your back pocket, perhaps you shouldn't. Maybe you are just not interested in sociology or about learning ways to become a really successful student. Maybe you're just here because an advisor told you that you need a social science course. Maybe you feel, "Hey, a C is good. I'll never need this stuff." If so, you can stop reading now.

But if you want to ace sociology—thereby becoming a more effective participant in society and social life—and if you want to learn some techniques to help you in other classes, too, this is for you. It's filled with the little things no one ever seems to tell you that improve grades, make for better understanding of classes—and may even make classes enjoyable for you. The **choice** is yours: **to read, or not to read.** Be forewarned. These contents may challenge the habits of a lifetime—habits that have gotten you this far but ones that may endanger your future success.

This essay contains ways to help you locate major ideas in your textbook. It contains many techniques that will be of help in reading your other course textbooks. If you learn these techniques early in your college career, you will have a head start on most other college students. You will be able to locate important information, understand lectures better, and probably do better on tests. By understanding the material better, you will not only gain a better understanding of sociology but also find that you are able to enjoy your class more.

The Problem: Passive Reading

Do you believe reading is one-way communication? Do you expect the author's facts will become apparent if you only read hard enough or long enough? (Many students feel this way.) Do you believe the writer has buried critical material in the text somewhere and that you need only find and highlight it to get all that's important? And do you believe that if you can memorize these highlighted details you will do well on tests? If so, then you are probably a passive reader.

The problem with passive reading is that it makes even potentially interesting writing boring. Passive reading reduces a chapter to individual, frequently unrelated facts instead of providing understanding of important concepts. It seldom digs beneath the surface, relying on literal meaning rather than sensing implications. Since most college testing relies on understanding of key concepts rather than simple factual recall, passive reading fails to significantly help students to do well in courses.

The Solution: Active Reading

Active reading is recognizing that a textbook should provide two-way communication. It involves knowing what aids are available to help understand the text and then using them to find the meaning. It involves prereading and questioning. It includes recording of questions, vocabulary learning, and

---

KEY FEATURES OF THE STUDY GUIDE

For each chapter you will find the following:

Key concepts matching exercise
Includes every term defined in the chapter
Promotes association of major thinkers with their key ideas or findings
Provides correct answers

Key thinkers/researchers matching exercise (where relevant)
Includes every important theorist or researcher discussed in the text
Promotes association of major thinkers with their key ideas or findings
Provides correct answers

Critical thinking questions
Promotes depth in reflecting on the material
Encourages creative application of the important concepts to everyday life
Presented in increasing levels of complexity, abstraction, and difficulty
Provides help in preparing for essay exams and papers

Comprehensive practice test
Includes questions on all major points in the chapter
Includes true/false, multiple-choice, and essay questions
Provides correct answers

summarizing. Still, with all these techniques, it frequently takes less time and produces significantly better results than passive reading.

This textbook—especially the Study Guide—is designed to help you become an active reader. For your convenience, the Study Guide material related to each chapter appears right after that chapter. The corners of the Study Guide pages are edged in color for easy reference. In the Study Guide, you will find a variety of learning aids based on the latest research on study skills. If you get into the habit of using the aids presented here, you can apply similar techniques to your other textbooks and become a more successful learner.

Effective Reading: Your Textbook

As an active reader, how should you approach your textbook? Here are some techniques for reading text chapters that you should consider.

GUIDELINES FOR EFFECTIVE READING OF YOUR TEXTBOOK

1. Think first about what you know.
2. Review the learning objectives.
3. Prior to reading the textbook chapter, read the chapter summary as an index to important terms and ideas.
4. Pay attention to your chapter outline.
5. Question as you read.
6. Pay attention to graphic aids.
7. When in doubt, use clues to find main ideas.
8. Do the exercises in the Study Guide.
9. Review right after reading.

1. Think first about what you know. Read the title of your chapter, then ask yourself what experiences you have had that relate to that title. For example, if the title is "Social Interaction and Social Groups," ask yourself, "In what ways have I interacted with others in social situations? Have I ever been part of a social group? If so, what do I remember about the experience?" Answers to these questions personalize the chapter by making it relate to your experiences. They provide a background for the chapter, which experts say improves your chances of understanding the reading. They show that you do know something about the chapter so that its content won't be so alien.

2. Review the learning objectives. Not all textbooks provide learning objectives as this one does, but, where available, they can be a valuable study aid. Learning objectives are stated in behavioral terms—they tell you what you should be able to do when you finish the chapter. Ask yourself questions about the tasks suggested in each learning objective and then read to find the information needed to accomplish that task. For instance, if a learning objective states, "Explain how variations in the size of groups affect what goes on within them," then you'll want to ask yourself something like, "How do groups vary in size?" and "How does each variation affect interaction within the group?"

3. Prior to reading the textbook chapter, read the chapter summary as an index to important terms and ideas. The summary includes all the points you need to find items in the chapter you know already. You may be able to read more quickly through sections covering these items. Some items you may not know anything about. This tells you where to spend your reading time. **A good rule:** Study most what you know least. Wherever it is, the summary is often your best guide to important material.

4. Pay attention to your chapter outline. This textbook, like most other introductory college textbooks, has an outline at the beginning of each chapter. If you do nothing else besides reading the summary and going through this outline before reading the chapter, you will be far ahead of most students because you will be clued in on what is important. The outline indicates the way ideas are organized in the chapter and how those ideas relate to one another. Certain ideas are indented to show that they are subsets or parts of a broader concept or topic. Knowing this can help you organize information as you read.

5. Question as you read. Turn your chapter title into a question, then read up to the first heading to find your answer. The answer to your question will be the main idea for the entire chapter. In forming your question, be sure it contains the chapter title. For example, if the chapter title is "Doing Sociology: Research Method," your question might be "What research methods does sociology use?" or "Why do you need research methods to do sociology?"

As you go through the chapter, turn each heading into a question, and then read to find the answer. Most experts say that turning chapter headings into questions is a most valuable step in focusing reading on important information. You may also want to use the learning objectives as questions, since you know that these objectives will point you toward the most important material in a section. However, it is also a good idea to form your own questions to get into practice for books not containing this helpful aid. A good technique might be to make your own question, then to check it against the appropriate objective before reading. In any case, use a question, then highlight your answer in the text. This will be the most important information under each heading. Don't read as if every word is important; focus on finding answers.

6. **Pay attention to graphic aids.** As you read, note those important vocabulary words appearing in bold type. Find the definitions for these words (in this book, definitions appear in italics right next to key words) and highlight them. These terms will be important to remember. Your Study Guide identifies all these important terms in the section headed "Key Concepts." A "Key Thinkers/Researchers" section, if applicable, identifies the sociologists and other important thinkers in the chapter worth remembering. Both the "Key Concepts" and "Key Thinkers/Researchers" sections are organized as matching exercises. Testing yourself after you read a text chapter (the answer key is at the end of the Study Guide chapter) will let you know whether you recognize the main concepts and researchers.

Pay attention to photos and photo captions. They make reading easier because they provide a visualization of important points in the textbook. If you can visualize what you read, you will ordinarily retain material better than people who don't use this technique. Special boxed sections usually give detailed research information about one or more studies related to a chapter heading. For in-depth knowledge, read these sections, but only after completing the section to which they refer. The main text will provide the background for a better understanding of the research, and the visualization provided by the boxed information will help illuminate the text discussion.

7. **When in doubt, use clues to find main ideas.** It is possible that, even using the questioning technique, there could be places where you are uncertain whether you're getting the important information. You have clues both in the text and in the Study Guide to help you through such places. In the text, it helps to know that main ideas in paragraphs occur more frequently at the beginning and end. Watch for repeated words or ideas—these are clues to important information. Check examples; any point that the author uses examples to document is important. Be alert for key words (such as "first," "second," "clearly," "however," "although," and so on); these also point to important information. Names of researchers (except for those named only within parentheses) will almost always be important. For those chapters in which important social scientists are discussed, you will find a "Key Thinkers/Researchers" section in your Study Guide.

8. **Do the exercises in the Study Guide.** The exercises in the Study Guide are designed as both an encouragement and a model of active learning. The exercises are not about mere regurgitation of material. Rather, you are asked to analyze, evaluate, and apply what you read in the text. By completing these exercises you are following two of the most important principles articulated in this essay: You are actively processing the material, and you are applying it to your own life and relating it to your own experiences. This is a guaranteed recipe for learning.

9. **Review right after reading.** Most forgetting takes place in the first day after reading. A review right after reading is your best way to hold text material in your memory. A strong aid in doing this review is your Study Guide. If a brief review is all you have time for, return to the Learning Objectives at the beginning of the chapter. Can you do the things listed in the objectives? If so, you probably know your material. If not, check the objective and reread the related chapter section to get a better understanding.

An even better review technique is to complete—if you haven't already done so—the exercises. Writing makes for a more active review, and if you do the exercises, you will have the information you need from the chapter. If there are blanks in your knowledge, you can check the appropriate section of text and write the information you find in your Study Guide. This technique is especially valuable in classes requiring essay exams or papers, as it gives you a comprehensive understanding of the material as well as a sense of how it can be applied to real-world situations.

For a slightly longer but more complete review, do the "Key Concepts" and "Key Thinkers/Researchers" matching tests. These will assure you that you have mastered the key vocabulary and know the contributions of the most important researchers mentioned in the chapter. Since a majority of test questions are based on understanding of vocabulary, research findings, and major theories, you will be assuring yourself of a testing benefit during your review.

It is also a good idea to review the "Critical Thinking" questions in the Study Guide. One key objective of sociology—indeed, of all college courses—is to help you develop critical thinking skills. Though basic information may change from year to year as new scientific discoveries are made, the ability to think critically in any field is important. If you get in the habit of going beyond surface knowledge in sociology, you can transfer these skills to other areas. This can be a great benefit not only while you're in school but afterward as well. As with the exercises section, these questions provide the kind of background that is extremely useful for essay exams.

What other methods would an active student use to improve understanding and test scores in sociology? The next several sections present a variety of techniques.

Functioning Effectively in Class

To function effectively in class, you must of course be there. While no one may take attendance or force you to be present, studies show that you have a significantly greater chance of succeeding in your class if you attend regularly. Lecture material is generally important—and it is given only once. If you miss a lecture, in-class discussion, game, or simulation, there is no really effective way to make it up.

> **GUIDELINES FOR EFFECTIVE FUNCTIONING IN CLASS**
>
> 1. Begin each class period with a question.
> 2. Ask questions frequently.
> 3. Join in classroom discussion.

Assuming you are present, there are two ways of participating in your sociology class: actively and passively. Passive participation involves sitting there, not contributing, waiting for the instructor to tell what is important. Passive participation takes little effort, but it is unlikely to result in much learning. Unless you are actively looking for what is significant, the likelihood of finding the important material or of separating if effectively from what is less meaningful is not great. The passive student runs the risk of taking several pages of unneeded notes or of missing key details altogether.

Active students **begin each class period with a question.** "What is this class going to be about today?" They find an answer to that question, usually in the first minute, and use this as the key to important material throughout the lecture or other activity. When there is a point they don't understand, they **ask questions.** Active students know that many other students probably have similar questions but are afraid to ask. Asking questions allows you to help others while helping yourself. Active students also know that what seems a small point today may be critical to understanding a future lecture. Such items also have a way of turning up on tests. **If classroom discussion** is called for, active students are quick to join in. And the funny thing is, they frequently wind up enjoying their sociology class as they learn.

Effective Studying

As you study your sociology text and notes, both the method you use and the time picked for study will have effects on comprehension. Establishing an effective study routine is important. Without a routine, it is easy to put off study—and put it off, and put it off . . . until it is too late. To be most effective, follow the few simple steps listed below.

> **GUIDELINES FOR EFFECTIVE STUDYING**
>
> 1. When possible, study at the same time and place each day.
> 2. Study in half-hour blocks with five-minute breaks.
> 3. Review frequently.
> 4. Don't mix study subjects.
> 5. Reward yourself when you're finished

1. When possible, study at the same time and place each day. Doing this makes use of psychological conditioning to improve study results. "Because it is 7:00 p.m. and I am sitting at my bedroom desk, I realize it is time to begin studying sociology."

2. Study in half-hour blocks with five-minute breaks. Long periods of study without breaks frequently reduce comprehension to the 40% level. That is most inefficient. By using short periods (about 30 minutes) followed by short breaks, you can move that comprehension rate into the 70% range. Note that if 30 minutes end while you are still in the middle of a text section, you should go on to the end of that section before stopping.

3. For even more efficient study, review frequently. Take about a minute at the end of each study session to mentally review what you've studied so far. When you start the next study session, spend the first minute or two rehearsing in your mind what you studied in the previous session. This weaves a tight webbing in which to catch new associations. Long-term retention of material is aided by frequent review, about every two weeks. A 10-minute review planned on a regular basis saves on study time for exams and ensures that you will remember needed material. Another useful way to review is to try to explain difficult concepts or the chapter learning objectives to someone else. One problem students often have is that, while studying and reviewing the material by themselves they think they know it, only to have that knowledge desert them at the time of the exam. Trying to explain something to someone else forces us to be clear about key points and to discover and articulate the relationship among the components of an idea. Ask your friends or family to bear with you as you try to explain the material. After all, they will learn something as well!

4. Don't mix study subjects. Do all of your sociology work before moving on to another course.

Otherwise, your study can result in confusion of ideas and relationships within materials studied.

5. Finally, reward yourself for study well done. Think of something you like to do, and do it when you finish studying for the day. This provides positive reinforcement, which makes for continued good study.

Successfully Taking Tests

Of course, tests are a payoff for you as a student. Tests are where you can demonstrate to yourself and to the instructor that you really know the material. The trouble is, few people have learned how to take tests effectively. And knowing how to take tests effectively makes a significant difference in exam scores. Here are a few tips to improve your test-taking skills.

GUIDELINES FOR SUCCESSFULLY TAKING TESTS

Studying for the Test
1. Think before you study.
2. Begin study a week early.
3. Put notes and related chapters together for study.
4. Take practice tests.

Taking the Test
1. Don't come early; don't come late.
2. Be sure you understand all the directions before you start answering.
3. Read through the test, carefully answering only items you know.
4. Now that you've answered what you know, look carefully at the other questions.
5. If you finish early, stay to check answers.
6. Don't be distracted by other test takers.
7. When you get your test back, use it as a learning experience.

Studying for Tests

1. Think before you study. All material is not of equal value. What did the instructor emphasize in class? What was covered in a week? A day? A few minutes? Were any chapters emphasized more than others? Which learning objectives did your instructor stress? Review the "Key Thinkers/Researchers" and "Key Concepts" sections in your Study Guide for important people and terms. Which of these were given more emphasis by your instructor? Use these clues to decide where to spend most of your study time.

2. Begin study a week early. When you start early, if you encounter material you don't know, you have time to find answers. If you see that you know blocks of material already, you have saved yourself time in future study sessions. You also avoid much of the forgetting that occurs with last-minute cramming.

3. Put notes and related chapters together for study. Integrate the material as much as possible, perhaps by writing it out in a single, comprehensive format. A related technique is to visualize the material on the pages of the text and in your notes. You may even want to think of a visual metaphor for some of the key ideas. This way you can see and remember the connections between similar subjects or similar treatments of the same subject. Grouping the material will also make your studying much more efficient. As you study, don't stop for unknown material. Study what you know. Once you know it, go back and look at what you don't know yet. There is no need to study again what you already know. Put it aside, and concentrate on the unknown.

4. Take practice tests. When you have completed your studying, take the appropriate practice test for each chapter. These tests are grouped together at the back of the book. Tests include true/false and multiple-choice questions, with comprehensive or thematic essays at the end. Each test is divided into sections by major headings in the chapter. Within each section, questions are presented in scrambled order, as they are likely to be on the actual test. Taking the practice test contains a double benefit. First, if you get a good score on this test, you know that you understand the material. Second, the format of the practice test is very similar to that of real tests. For this reason, you should develop confidence in your ability to succeed in course tests from doing well on the practice tests. If your course tests include essay questions, you should, in addition to the practice test essays, use the "Critical Thinking" sections to prepare and practice focused, in-depth answers.

Taking the Test

1. Don't come early; don't come late. Early people tend to develop anxieties; late people lose test time. Studies show that people who discuss test material with others just before a test may forget that material on the test. This is another reason that arriving too early puts students in jeopardy. Get there about two or three minutes early. Relax and visualize yourself doing well on the test. After all, if you followed the study guidelines discussed above, you can't help but do well! Be confident; repeat to yourself as you get ready for the test, "I can do it! I will do it." This will set a positive mental tone.

2. Be sure you understand all the directions before you start answering. Not following directions is the biggest cause of lost points on tests. Ask about whatever you don't understand. The points you save will be your own.

3. Read through the test, carefully answering only items you know. Be sure you read every word and every answer choice as you go. Use a piece of paper or a card to cover the text below the line you are reading. This can help you focus on each line individually—and increase your test score.

Speed creates a serious problem in testing. The mind is moving so fast that it is easy to overlook key words such as *except, but, best example,* and so on. Frequently, multiple-choice questions will contain two close options, one of which is correct, while the other is partly correct. Moving too fast without carefully reading items causes people to make wrong choices in these situations. Slowing your reading speed makes for higher test scores.

The mind tends to work subconsciously on questions you've read but left unanswered. As you're doing questions later in the test, you may suddenly have the answer for an earlier question. In such cases, answer the question right away. These sudden insights quickly disappear and may never come again.

4. Now that you've answered what you know, look carefully at the other questions. Eliminate alternatives you know are wrong, and then guess. Never leave a blank on a test. You may have a 25% chance when you guess on a four-item multiple choice question, but you have a chance. And a chance is better than no chance.

5. If you finish early, stay to check answers. Speed causes many people to give answers that a moment's hesitation would show to be wrong. Read over your choices, especially those for questions that caused you trouble. Don't change answers because you suddenly feel one choice is better than others. Studies show that this is usually a bad strategy. However, if you see a mistake or have genuinely remembered new information, change your answer.

6. Don't be distracted by other test takers. Some people become very anxious because of the noise and movement of other test takers. This is most apparent when several people begin to leave the room after finishing their tests. Try to sit where you will be least apt to see or interact with other test takers. Usually this means sitting toward the front of the room and close to the wall farthest from the door. Turn your chair slightly toward the wall, if possible. The more you insulate yourself from distractions during the test, the better off you will be.

Don't panic when other students finish their exam before you do. Accuracy is always more important than speed. Work at your own pace and budget your time appropriately. For a timed test, always be aware of the time remaining. This means that if a clock is not visible in the classroom, you need to have your own wristwatch. Take as much of the available time as you need to do an accurate and complete job. Remember, your grade will be based upon the answers you give, not on whether you were the first—or the last—to turn in your exam.

7. When you get your test back, use it as a learning experience. Diagnosing a test after it is returned to you is one of the most effective strategies for improving your performance in a course. What kind of material was on the test: theories, problems, straight facts? Where did the material come from: book, lecture, or both? The same kind of material taken from the same source(s) will almost certainly be on future tests.

Look at each item you got wrong. Why is it wrong? If you know why you made mistakes, you are unlikely to make the same ones in the future. Look at the overall pattern of your errors. Did you make most of your mistakes on material from the lectures? Perhaps you need to improve your note-taking technique. Did your errors occur mostly on material from the readings? Perhaps you need to pay more attention to main idea clues and highlight text material more effectively. Were the questions you got wrong evenly distributed between in-class and reading material? Perhaps you need to learn to study more effectively and/or to take steps to reduce test anxiety. Following these steps can make for more efficient use of textbooks, better note-taking, higher test scores, and better course grades.

A Final Word

As you can see, the key to success lies in becoming an active student. Managing time, questioning at the start of lectures, planning effective measures to increase test scores, and using all aids available to make reading and studying easier are all elements in becoming an active student. The Study Guide and Practice Tests for this textbook have been specially designed to help you be that active student. Being passive may seem easier, but it is not. Passive students spend relatively similar amounts of time but learn less. Their review time is likely to be inefficient. Their test scores are more frequently lower—and they usually have less fun in their classes.

Active students are more effective than passive ones. The benefit in becoming an active student is that activity is contagious; if you become an active student in sociology, it is hard not to practice the same active learning techniques in English and math as well. Once you start asking questions in your textbook and using your Study Guide, you may find that you start asking questions in class as well. As you acquire a greater understanding of your subject, you may find that you enjoy your class more—as well as learn more and do better on tests. That is the real benefit in becoming an active learner. It is a challenge we strongly encourage you to meet.

ABOUT THE AUTHOR

©2000 Al Hirschfeld. Drawing reproduced by special arrangement with the Margo Feiden Galleries Ltd., New York.

Henry L. Tischler grew up in Philadelphia and received his bachelor's degree from Temple University and his masters and doctorate degrees from Northeastern University. He pursued post-doctoral studies at Harvard University.

His first venture into textbook publishing took place while he was still a graduate student in sociology when he wrote the fourth edition of *Race and Ethnic Relations* with Brewton Berry. The success of that book led to his authorship of the seven editions of *Introduction to Sociology*.

Tischler has been a professor at Framingham State College in Framingham, Massachusetts, for more than two decades. He has also taught at Northeastern University and Tufts University. He continues to teach introductory sociology every year and has been instrumental in encouraging many students to major in the field. His other areas of interest are race and ethnicity, and crime and deviant behavior.

Professor Tischler has been active in making sociology accessible to the general population and is currently the host of an author interview program on National Public Radio. He has also written a weekly newspaper column called "Society Today" which dealt with a wide variety of sociological topics. Tischler lives outside of Boston with his wife Linda, the managing editor of a national magazine, a daughter Melissa, who is a graduate student, and a son Ben, who is a college student.

CONTENTS IN BRIEF

CONTENTS

FEATURES CONTENTS

TECHNOLOGY AND SOCIETY

The Study of Society

Congratulations. You are about to embark on an exciting voyage of discovery—one in which you will begin to understand society and social interaction. Why should you want to go on this expedition at all? Why should you want to study sociology? What can sociologists tell you about your life that you do not already know?

The first section of this book is designed to answer these questions and more. In Chapter 1 you will encounter the sociological perspective and discover that the sociological point of view causes us to look at the world in a different way. That is, sociology forces us to go beyond our own personal experiences to see the world through the eyes of others and to look for recurring patterns in the behavior of many individuals. This, you will discover, is what makes sociology a social science. There are a number of social sciences, and, of course, all of them deal with people. In Chapter 1 you will learn how sociology is both different from and similar to these other disciplines. You will also discover how the sociological perspective was developed and employed by early sociologists, first in Europe, and then later in the United States. Like other social scientists, however, sociologists do not agree on the best way to organize their understanding of the social world.

Thus you will encounter the major theoretical perspectives used by sociologists. If people and their social behaviors vary widely, and sociologists cannot always agree on the best way to understand and interpret those behaviors, how do you know that sociologists really know anything at all? How can you trust what they tell you about how the social world works? One way is to see if the information is based on research that was done in a scientific manner, which is the subject of Chapter 2. In this chapter you will explore the research process in order to develop criteria for evaluating sociological research, as well as understanding where researchers can go wrong. You will see how bias creeps into social research, and what steps sociologists take to try to avoid it. You will also examine important ethical issues in the collection and use of sociological data.

When you finish this section, you will have a basic understanding of what sociology is and is not, and how sociologists go about studying the social world. You will then be ready to use sociology to understand social behavior.

CHAPTER ONE

The Sociological Perspective

LEARNING OBJECTIVES

After studying this chapter, you should be able to do the following:

- Understand the sociological point of view and how it differs from that of journalists and talk-show hosts.
- Compare and contrast sociology with the other major social sciences.
- Describe the early development of sociology from its origins in nineteenth-century Europe.
- Know the contributions of sociology's pioneers: Comte, Martineau, Spencer, Marx, Durkheim, and Weber.
- Describe the early development of sociology in the United States.
- Understand the functionalism, conflict theory, and the interactionist perspectives.
- Realize the relationship between theory and practice.

At the Shop-Rite supermarket in Wilmington, Delaware, a "Code Adam" message came across the intercom system. The employees knew that Code Adam meant a child was missing in the store. Immediately all doors were monitored, and shoppers were asked to remain while a thorough search of the store was made. Within a few minutes the missing 4-year-old was found hiding in the bakery section.

Code Adam is a program that was started by Wal-Mart, the nation's largest retailer, in 1996. It is named for Adam Walsh, who was abducted from a Sears store in 1981 and was later murdered.

Many other establishments have put programs into effect to protect against child abductions. Home Depot stores as well as other retailers offer parents the option to have their child photographed with identifying information, using a packet called KidCare ID (provided by Polaroid), in the event police need such documentation for a future search. About 7 million children have been so documented to date. Almost 2 million more kids have taken part in a similar camcorder-based program offered annually by Blockbuster Video.

These actions are the result of a fear that has spread among parents of young children, who have come to believe that letting kids out of their sight is extremely dangerous. A study of parents' worries by the Mayo Clinic in Rochester, Minnesota, found that nearly three-quarters of parents said they feared their children might be abducted.

Some groups have suggested that as many as 50,000 children are involved in stranger abduction every year. This number was cited in congressional testimony that eventually led to legislation designed to facilitate recovery of missing kids, in part by centralizing the reporting and tracking of children.

But wait. Let us step back a bit. No doubt we are talking about a social problem with very serious consequences. Is this sociology? Or, for that matter, is this what sociologists do when they study society? The answer would have to be no. A sociologist would want to know how a stranger abduction is defined and how the 50,000 number was obtained.

A sociologist would also point out that the 50,000 stranger abduction number was contradicted by the federal government's Justice Department data. The official definition of a child abduction is as follows: *the child disappears at least overnight, is transported more than 50 miles, is ransomed, or is murdered.* This final category—every parent's worst nightmare—accounts for 200 to 300 cases each year according to the FBI.

This means that, with some 70 million children in the United States, the chances that a child will be abducted and held or murdered by a person outside the immediate family are about 1 in 230,000, a very small risk. In addition, contrary to popular belief, children age 12 and up—particularly teenagers—account for the vast majority of actual long-term child abductions and murders, with numbers dramatically higher among children age 14 to 17 than any other age group (Laliberte, 1998).

Far too infrequently do we realize that people often use data to persuade and that statistics can be used as part of a strategy to promote concern about a social problem. Much of the information we read every day and mistake for sociology is actually an attempt by one group or another to influence social policy. Other information mistaken for sociology is really an attempt to sell a book, or the efforts of television producers to present entertaining programs.

With the constant bombardment of information about social issues, we could come to believe that nearly everyone is engaged in the study of sociology to some extent and that everyone has not only the right but also the ability to put forth valid information about society. This is not the case. Some people have no interest in putting forth objective information and are instead interested in getting us to support their position or point of view. On other occasions the "researchers" do not have the ability or training to disseminate accurate information about drug abuse, homelessness, welfare, high-school dropout rates, white-collar crime, or a host of other sociological topics.

Sociologists have very different goals in mind when they investigate a problem than do journalists or talk-show hosts. A television talk-show host needs to make the program entertaining and maintain high ratings, or the show may be canceled. A journalist is writing for a specific readership. This will certainly limit the choice of topics, as well as the manner in which an issue is investigated. On the other hand, a sociologist must answer to the scientific community as she or he tries to further our understanding of a topic. This means that the goal is not high ratings, but an accurate and scientific approach to the issue being studied.

In this book we will ask you to go beyond popular sociology and investigate society more scientifically than you did before. You will learn to look at major events, as well as everyday occurrences, a little differently and start to notice patterns you may have never seen before. After you are equipped with the tools of sociology, you should be able to evaluate critically popular presentations of sociology. You will see that sociology represents both a body of knowledge and a scientific approach to the study of social issues.

Sociology as a Point of View

Sociology is *the scientific study of human society and social interactions.* As sociologists our main goal is to *understand* social situations and look for repeating patterns in society. We do not use facts selectively to create a lively talk show, sell newspapers, or support one particular point of view. Instead, sociologists are engaged in a rigorous scientific endeavor, which requires objectivity and detachment.

The main focus of sociology is the group, not the individual. Sociologists attempt to understand the forces that operate throughout society—forces that mold individuals, shape their behavior, and thus determine social events.

When you walk into an introductory physics class, you may know very little about the subject and hold very few opinions about the various topics within the field. On the other hand, when you enter your introductory sociology class for the first time, you will feel quite familiar with the subject matter. You have the advantage of coming to sociology with a substantial amount of information, which you have gained simply by being a member of society. Ironically, this knowledge also can leave you at a disadvantage, because these views have not been gathered in a scientific fashion and may not be accurate.

Over the years and through a variety of experiences we develop a set of ideas about the world and how it operates. This point of view influences how we look at the world and guides our attempts to understand the actions and reactions of others. Even though we accept the premise that individuals are unique, we tend to categorize or even stereotype people in order to interpret and predict behavior and events.

Is this personalized approach adequate for bringing about an understanding of ourselves and society? Even though it may serve us quite well in our

Figure 1-1 Levels of Social Understanding: Domestic Violence

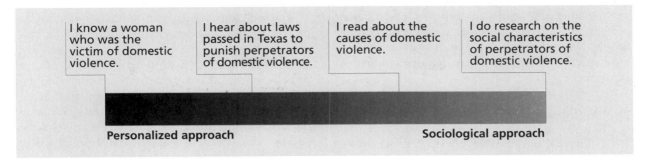

| I know a woman who was the victim of domestic violence. | I hear about laws passed in Texas to punish perpetrators of domestic violence. | I read about the causes of domestic violence. | I do research on the social characteristics of perpetrators of domestic violence. |

Personalized approach   **Sociological approach**

day-to-day lives, a sociologist would answer that it does not give us enough accurate information to develop an understanding of the broader social picture. This picture becomes clear only when we know something about the society in which we live, the social processes that affect us, and the patterns of interaction that characterize our lives.

Let us take the issue of domestic violence. Figure 1-1 shows that we could examine the issue in a variety of ways. If we knew a woman who was the victim of domestic violence, we would have personal information about the experience. If she were willing to discuss her experience with us, we would know more about domestic violence on a specific case level. Although this information is important, it is not yet sociology and is closer to the personalized common-sense approach to understanding society. Sociology tries to move beyond that level of understanding.

If we rely on our own experiences, we are like the blind men of Hindu legend trying to describe an elephant: the first man, feeling its trunk, asserts, "It is like a snake"; the second, trying to reach around the beast's leg, argues, "No, it is like a tree"; and the third, feeling its solid side, disagrees, saying, "It is more like a wall." In a small way, each man is right, but not one of them is able to understand or describe the whole elephant.

If we were to look for recurring patterns in domestic violence, we would now be doing what sociologists do. A sociologist examining the issue might be interested in the age, socioeconomic level, and ethnic characteristics of the victims of domestic violence. A sociologist might want to compare these characteristics with the characteristics of victims of other types of violence: "Are there differences?" they ask. "If so, what kinds and why?"

While studying sociology, you will be asked to look at the world a little differently from the way you usually do. Because you will be looking at the world through other people's eyes—using new points of view—you will start to notice things you may never have noticed before. When you look at life in a middle-class suburb, for instance, what do you see? How does your view differ from that of

a poor inner-city resident? How does the suburb appear to a recent immigrant from Russia or Cuba or India? How does it appear to a burglar? Finally, what does the sociologist see?

Sociology asks you to broaden your perspective on the world. You will start to see that the reason people act in markedly different ways is not because one person is "sane" and another is "crazy." Rather, it is because they all have different ways of making

Vast differences often exist within the same society.

sense out of what is going on in the world around them. These unique perceptions of reality produce varying lifestyles, which in turn produce different perceptions of reality. In order to understand other people, we must stop looking at the world from a perspective based solely on our own individual experiences.

The Sociological Imagination

While most people interpret social events on the basis of their individual experiences, sociologists step back and view society more as an outsider than as a personally involved and possibly biased participant. For example, while we assume that most people in the United States marry because of love, sociologists remind us that the decision to marry—or not to marry—is influenced by a variety of social values taught to us since early childhood. That is, we select our mates based on the social values we internalize from family, peers, neighbors, community leaders, and even our television heroes. As a result, we are less likely to marry someone from a different socioeconomic class, from a different race or religion, or from a markedly different educational background. Thus, as we pair off, we follow somewhat predictable patterns: In most cases the man is older, earns more money, and has a higher occupational status than the woman. These patterns may not be evident to the two people who are in love with each other; indeed, they may not be aware that anything other than romance played a role in their choice of a mate. As sociologists we begin to discern marriage patterns. We may note that marriage rates vary in different parts of the country, that the average age of marriage is related to educational level, and that social class is related to marital stability. These patterns (discussed in Chapter 10) show us that there are forces at work that influence marriage and that may not be evident to the individuals who fall in love and marry.

C. Wright Mills (1959) pointed out different levels on which social events can be perceived and interpreted. He used the term the **sociological imagination** to refer to *this relationship between individual experiences and forces in the larger society that shape our actions.*

The sociological imagination is the process of looking at all types of human behavior patterns and discerning previously unseen connections among them, noting similarities in the actions of individuals with no direct knowledge of one another, and finding subtle forces that mold people's actions. Like a museum-goer who draws back from a painting in order to see how the separate strokes and colors form subtly shaded images, sociologists stand back from individual events in order to see why and how

they occurred. In so doing, they discover patterns that govern our social existence.

The sociological imagination focuses on every aspect of society and every relationship among individuals. It studies the behavior of crowds at ball games and racetracks; shifts in styles of dress and popular music; changing patterns of courtship and marriage; the emergence and fading of different lifestyles, political movements, and religious sects; the distribution of income and access to resources and services; decisions made by the Supreme Court, by congressional committees, and by local zoning boards; and so on. Every detail of social existence is food for sociological thought and relevant to sociological analysis.

The potential for sociology to be put to use—applied to the solution of real-world problems—is enormous. Proponents of applied sociology believe the work of sociologists can and should be used to help bring about an understanding of, and perhaps even guidelines for changing, the complexities of modern society.

The demand for applied sociology is growing, and many sociologists work directly with government agencies or private businesses in an effort to apply sociological knowledge to real-world problems. For example, they might investigate such questions as how the building of a dam will affect the residents of the area, how jury makeup affects the outcome of a case, why voters select one candidate over another, how a company can boost employee morale, and how relationships among administrators, doctors, nurses, and patients affect hospital care. The answers to these questions have practical applications. The growing demand for sociological information provides many new career choices for sociologists (see "Sociology at Work: If You Are Thinking About Sociology as a Career, Read This").

Is Sociology Common Sense?

Common sense is what people develop through everyday life experiences. In a very real sense, it is the set of expectations about society and people's behavior that guides our own behavior. Unfortunately, these expectations are not always reliable or accurate because without further investigation, we tend to believe what we want to believe, to see what we want to see, and to accept as fact whatever appears to be logical. While common sense is often vague, oversimplified, and frequently contradictory, sociology as a science attempts to be specific, to qualify its statements, and to prove its assertions. Upon closer inspection, we find that the proverbial words of wisdom rooted in common sense are often illogical. Why, for example, should you "look before you leap" if "he who hesitates is lost"? How

Sociology at Work

If You Are Thinking About Sociology as a Career, Read This

Speaking from this side of the career-decision hurdle, I can say that being a sociologist has opened many doors for me. It gave me the credentials to teach at the college level and to become an author of a widely used sociology text. It also enabled me to be a newspaper columnist and a talk-show host. Would I recommend this field to anyone else? I would, but not blindly. Realize before you begin that sociology can be an extremely demanding discipline and, at times, an extremely frustrating one.

As in many other fields, the competition for jobs in sociology can be fierce. If you really want this work, do not let the herd stop you. Anyone with motivation, talent, and a determined approach to finding a job will do well. However, be prepared for the long haul: To get ahead in many areas you will need to spend more than 4 years in college. Consider your bachelor's degree as just the beginning. Fields like teaching at the college level and advanced research often require a PhD, which means at least 4 to 6 years of school beyond the BA.

Now for the job possibilities. As you read through these careers, remember that right now your exposure to sociology is limited (you are only on Chapter 1 in your first college sociology text), so do not eliminate any possibilities right at the start. Spend some time thinking about each one as the semester progresses and you learn more about this fascinating discipline.

Most people who go into sociology become teachers. You will need a PhD to teach in college, but often a master's degree will open the door for you at the 2-year college or high-school level.

Second in popularity to teaching are nonacademic research jobs in government agencies, private research institutions, and the research departments of private corporations. Researchers carry on many different functions, including conducting market research, public opinion surveys, and impact assessments. Evaluation research, as the latter field is known, has become more popular in recent years because the federal government now requires environmental impact studies on all large-scale federal projects. For example, before a new interstate highway is built, evaluation researchers attempt to determine the effect the highway will have on communities along the proposed route.

This is only one of many opportunities available in government work. Federal, state, and local governments in policy-making and administrative functions also hire sociologists. For example, a soci-ologist employed by a community hospital provides needed data on the population groups being served and on the health-care needs of the community. Another example: Sociologists working in a prison system can devise plans to deal with the social problems that are inevitable when people are put behind bars. Here are a few additional opportunities in government work: community planner, correction officer, environmental analyst, equal opportunity specialist, probation officer, rehabilitation counselor, resident director, and social worker.

A growing number of opportunities also exist in corporate America, including market researchers, pollsters, human resource managers, affirmative action coordinators, employee assistance program counselors, labor relations specialists, and public information officers, just to name a few. These jobs are available in nearly every field from advertising to banking, from insurance to publishing. Although your corporate title will not be "sociologist," your educational background will give you the tools you need to do the job—and do it well, which, to corporations, is the bottom line.

Whether you choose government or corporate work, you will have the best chance of finding the job you want by specializing in a particular field of sociology while you are still in school. You can become an urban or family specialist or become knowledgeable in organizational behavior before you enter the job market. For example, many demographers, who compile and analyze population data, have specialized in this aspect of sociology. Similarly, human ecologists, who investigate the structure and organization of a community in relation to its environment, have specialized educational backgrounds as well. Keep in mind that many positions require a minor or some course work in other fields such as political science, psychology, ecology, law, or business. By combining sociology with these fields, you will be well prepared for the job market.

What next? Be optimistic and start planning. As the American Sociological Association observed, few fields are as relevant to today and as broadly based as sociology. Yet, ironically, its career potential is just beginning to be tapped. Start planning by reading the *Occupational Outlook Quarterly,* published by the U.S. Bureau of Labor Statistics, as well as academic journals to keep abreast of career trends. Then study hard and choose your specialty. With this preparation, when the time comes to find a job, you will be well prepared.

can "absence make the heart grow fonder" when "out of sight, out of mind"? Why should "opposites attract" when "birds of a feather flock together"?

The "commonsense" approach to sociology is one of the major dangers the new student encounters. Common sense often makes sense after the fact. It is more useful for describing events than for predicting them. It deludes us into thinking we knew the outcome all along (Hawkins & Hastie, 1990).

One researcher (Teigen, 1986) asked students to evaluate actual proverbs and their opposites. When given the actual proverb "Fear is stronger than love," most students agreed that it was true. But so did students who were given the reverse statement, "Love is stronger than fear." The same was true for the statements "Wise men make proverbs and fools repeat them" (actual proverb) and its reversal, "Fools make proverbs and wise men repeat them."

While common sense gleaned from personal experience may help us in certain types of interactions, it will not help us understand why and under what conditions these interactions are taking place. Sociologists as scientists attempt to qualify these statements by specifying, for example, under what conditions do "opposites tend to attract" or "birds of a feather flock together." Sociology as a science is oriented toward gaining knowledge about why and under what conditions events take place in order to understand human interactions better.

Sociology and Science

Sociology is commonly described as one of the social sciences. **Science** refers to *a body of systematically arranged knowledge that shows the operation of general laws.*

Sociology also employs the same general methods of investigation that are used in the natural sciences. Like the natural scientists, sociologists use the **scientific method,** *a process by which a body of scientific knowledge is built through observation, experimentation, generalization, and verification.* The collection of data is an important aspect of the scientific method, but facts alone do not constitute a science. To have any meaning, facts must be ordered in some way, analyzed, generalized, and related to other facts. This is known as theory construction. Theories help organize and interpret facts and relate them to previous findings of other researchers.

Science is only one of the ways in which human beings study the world around them. Take the feeling of joy as an example. A physiologist might describe joy as a biochemical response to certain events. A poet might describe the experience in beautiful language. A theologian might describe joy as the outcome of a relationship with God. Unlike other means of inquiry, science for the most part limits its

investigations to empirical entities, things that can be observed directly or that produce directly observable events. Therefore, one of the basic features of science is **empiricism,** *the view that generalizations are valid only if they rely on evidence that can be observed directly or verified through our senses.* For example, theologians might discuss the role of faith in producing "true happiness"; philosophers might deliberate over what happiness actually encompasses; but sociologists would note, analyze, and predict the consequences of such measurable items as job satisfaction, the relationship between income and education, and the role of social class in the incidence of divorce.

Sociology as a Social Science

The **social sciences** consist of all *those disciplines that apply scientific methods to the study of human behavior.* Although there is some overlap, each of the social sciences has its own area of investigation. It is helpful to understand each of the social sciences and to examine sociology's relationship to them.

Cultural Anthropology The social science most closely related to sociology is *cultural anthropology.* The two have many theories and concepts in common and often overlap. The main difference is in the groups they study and the research methods they use. Sociologists tend to study groups and institutions within large, modern, industrial societies, using research methods that enable them rather quickly to gather specific information about large numbers of people. In contrast, cultural anthropologists often immerse themselves in another society for a long period of time, trying to learn as much as possible about that society and the relationships among its people. Thus, anthropologists tend to focus on the culture of small, preindustrial societies because they are less complex and more manageable using this method of study.

Psychology The study of individual behavior and mental processes is part of *psychology;* the field is concerned with such issues as motivation, perception, cognition, creativity, mental disorders, and personality. More than any other social science, psychology uses laboratory experiments. Psychology and sociology overlap in a subdivision of each field known as *social psychology*—the study of how human behavior is influenced and shaped by various social situations. Social psychologists study such issues as how individuals in a group solve problems and reach a consensus, or what factors might produce nonconformity in a group situation. For the most part, however, psychology studies the individual, and sociology studies groups of individuals, as well as society's institutions.

Sociologists and anthropologists share many theories and concepts. Whereas sociologists tend to study groups and institutions within large, modern, industrial societies, anthropologists tend to focus on the cultures of small, preindustrial societies.

The sociologist's perspective on social issues is broader than that of the psychologist. Take the case of alcoholism, for example. The psychologist might view alcoholism as a personal problem that has the potential to destroy an individual's physical and emotional health, as well as marriage, career, and friendships. The sociologist, on the other hand, would look for patterns in alcoholism. Although each alcoholic makes the decision to take each drink—and each suffers the pain of addiction—the sociologist would remind us to look beyond the personal and to consider the broader aspects of alcoholism, such as its social causes. Sociologists want to know who drinks excessively, when they drink, where they drink, and under what conditions they drink. They are also interested in the social costs of chronic drinking—costs in terms of families torn apart, jobs lost, children severely abused and neglected, costs in terms of highway accidents and deaths, costs in terms of drunken quarrels leading to violence and to murder. Noting the startling increase in heavy alcohol use by adolescents over the past 10 years and the rapid rise of chronic alco-

holism among women, sociologists ask, what forces are at work to account for these patterns?

Economics Economists have developed techniques for measuring such things as prices, supply and demand, money supplies, rates of inflation, and employment. This study of the creation, distribution, and consumption of goods and services is known as *economics*. The economy, however, is just one part of society. It is each individual in society who decides whether to buy an American car or a Japanese import, whether she or he is able to handle the mortgage payment on a dream house, and so on. Whereas economists study price and availability factors, sociologists are interested in the social factors that influence the resulting economic behavior. Is it peer pressure that results in the buying of the large flashy car, or is it concern about gas mileage that leads to the purchase of a small, fuel-efficient, modest vehicle? What social and cultural factors contribute to the differences in the portion of income saved by the average wage earner in different societies? What effect does the unequal allocation of resources have on social interaction? These are examples of the questions sociologists seek to answer.

History Although not exactly a social science, history shares certain attributes with sociology. The study of *history* involves looking at the past in an attempt to learn what happened, when it happened, and why it happened. Sociology also looks at historical events within their social contexts to discover why things happened and, more important, to assess what their social significance was and is. Historians provide a narrative of the sequence of events during a certain period and may use sociological research methods to try to learn how social forces have shaped historical events. Sociologists, on the other hand, examine historical events to see how they influenced later social situations. Historians focus on individual events—the American Revolution or slavery—and sociologists generally focus on phenomena such as revolutions or the patterns of dominance and subordination that exist in slavery. They try to understand the common conditions that contribute to revolutions or slavery wherever they occur.

Let us consider the subject of slavery in the United States. Traditionally, historians might focus on when the first slaves arrived, on how many slaves existed in 1700 or 1850, and the conditions under which they lived. Sociologists and modern social historians would use these data to ask many questions: What social and economic forces shaped the institution of slavery in the United States? How did the Industrial Revolution affect slavery? How has the experience of slavery affected the black family? Although history and sociology have been moving toward each other over the past 20 years, each discipline still retains a somewhat different focus: sociology on the present, history on the past.

Political Science Concentrating on three major areas, *political science* is the study of political theory, the actual operation of government, and, in recent years, political behavior. This emphasis on political behavior overlaps with sociology. The primary distinction between the two disciplines is that sociology focuses on how the political system affects other institutions in society, while political science devotes more attention to the forces that shape political systems and the theories for understanding these forces. However, both disciplines share an interest in why people vote the way they do, why they join political movements, how the mass media are changing political parties and processes, and so on.

Social Work Much of the theory and research methods of social work are drawn from sociology and psychology, but social work focuses to a much greater degree on application and problem solving. The disciplines of sociology and social work are often confused with each other. The main goal of *social work* is to help people solve their problems, while the aim of sociology is to understand why the problems exist. Social workers provide help for individuals and families who have emotional and psychological problems or who experience difficulties that stem from poverty or other ongoing problems rooted in the structure of society. Social workers also organize community groups to tackle local issues such as housing problems and try to influence policy-making bodies and legislation. Sociologists provide many of the theories and ideas used to help others. Although sociology is not social work, it is a useful area of academic concentration for those interested in entering the helping professions.

The Development of Sociology

It is hardly an accident that sociology emerged as a separate field of study in Europe during the nineteenth century. That was a time of turmoil, a period in which the existing social order was being shaken by the growing Industrial Revolution and by violent uprisings against established rulers (the American and French revolutions). People also were affected by the impact of discovering, through world exploration and colonialization, how others lived. At the same time, a void was left by the declining power of the church to impose its views of right and wrong. New social classes of industrialists and business-people emerged to challenge the rule of the feudal aristocracies. Tightly knit communities, held to-

gether by centuries of tradition and well-defined social relationships, were strained by dramatic changes in the social environment. Factory cities began to replace the rural estates of nobles as the centers for society at large. People with different backgrounds were brought together under the same factory roof to work for wages instead of exchanging their services for land and protection. Families now had to protect themselves, to buy food rather than grow it, and to pay rent for their homes. These new living and working conditions led to the development of an industrial, urban lifestyle, which, in turn, produced new social problems.

Many people were frightened by what was going on and wanted to find some way of understanding and dealing with the changes taking place. The need for a systematic analysis of society coupled with acceptance of the scientific method resulted in the emergence of sociology. Henri Saint-Simon (1760–1825) and Auguste Comte (1798–1857) were among the pioneers in the science of sociology.

Auguste Comte (1798–1857)

Born in the French city of Montpellier on January 19, 1798, Auguste Comte grew up in the period of great political turmoil that followed the French Revolution of 1789–1799. In August 1817, Comte met Henri Saint-Simon and became his secretary and eventually his close collaborator. Under

Auguste Comte coined the term *sociology*. He wanted to develop a "science of man" that would reveal the underlying principles of society, much as the sciences of physics and chemistry explained nature and guided industrial progress.

Saint-Simon's influence, Comte converted from an ardent advocate of liberty and equality to a supporter of an elitist conception of society.

Saint-Simon and Comte rejected the lack of empiricism in the social philosophy of the day. Instead they turned for inspiration to the methods and intellectual framework of the natural sciences, which they perceived as having led to the spectacular successes of industrial progress. They set out to develop a "science of man" that would reveal the underlying principles of society much as the sciences of physics and chemistry explained nature and guided industrial progress. During their association the two men collaborated on a number of essays, most of which contained the seeds of Comte's major ideas. Their alliance came to a bitter end in 1824 when Comte broke with Saint-Simon for both financial and intellectual reasons.

Financial problems, lack of academic recognition, and marital difficulties combined to force Comte into a shell. Eventually, for reasons of "cerebral hygiene," he no longer read any scientific work related to the fields about which he was writing. Living in isolation at the periphery of the academic world, Comte concentrated his efforts between 1830 and 1842 on writing his major work, *Cours de Philosophie Positive,* in which he coined the term *sociology.* Comte devoted a great deal of his writing to describing the contributions he expected sociology would make in the future. He was much less concerned with defining sociology's subject matter than with showing how it would improve society. Although Comte was reluctant to specify subdivisions of sociology, he identified two major areas that sociology should concentrate on. These were social statics, the study of how various institutions of society are interrelated, focusing on order, stability, and harmony; and social dynamics, the study of complete societies and how they develop and change over time. Comte believed all societies move through certain fixed stages of development, eventually reaching perfection (as exemplified in his mind by industrial Europe). Sociologists, however, no longer accept the idea of a perfect society.

Harriet Martineau (1802–1876)

Harriet Martineau, an Englishwoman, was an early and significant contributor to the development of sociology. In 1837 she published *Theory and Practice of Society in America,* in which she analyzed the customs and lifestyle present in the nineteenth-century United States. Her book was based on traveling throughout the United States and observing day-to-day life in all its forms, from that which took place in prisons, mental hospitals, and factories to family gatherings, slave auctions, and even proceed-

Harriet Martineau was an early and significant contributor to the development of sociology. She believed that scholars should not simply offer observations but should also use their research to bring about social reform.

ings of the Supreme Court and Senate. The book helped to map out what a sociological work dealt with by examining the impact of immigration, family issues, politics, and religion, as well as race and gender issues. In her book she also compared social stratification systems in Europe with those in the United States. Martineau's work also demonstrated the level of objectivity she thought was necessary for an analysis of society when she noted, "It is hard to tell which is worse, the wide diffusion of things that are not true or the suppression of things that are." Later in her career, she came to the conclusion that scholars should not just offer observations, but should also use their research to bring about social reform for the benefit of society. She asked her readers to "judge for themselves . . . how far the people of the United States lived up to" their stated ideals (Hoecker-Drysdale, 1992).

Martineau's second important contribution to sociology was translating into English August Comte's six-volume *Positive Philosophy.* Her two-volume

Herbert Spencer helped to define the subject matter of sociology. Spencer also became a proponent of a doctrine know as social Darwinism.

edition of this book introduced the field of sociology to England and influenced people such as Herbert Spencer, as well as early American sociologists. (For more on the early contributions of women in sociology see "Sociology at Work: Women and the Development of Sociology, 1800 to 1945" on page 19.)

Herbert Spencer (1820–1903)

A largely self-educated Briton, Herbert Spencer had a talent for synthesizing information. In 1860 he started work toward the goal of organizing human knowledge into one system. The result was his *Principles of Sociology* (1876, 1882), the first sociology textbook. Unlike Comte, Spencer was precise in defining the subject matter of sociology. He declared the field of sociology included the study of the family, politics, religion, social control, work, and stratification.

Spencer believed society was similar to a living organism. Just as the individual organs of the body are interdependent and make their specialized contributions to the living whole, so, too, are the various segments of society interdependent. Every part

of society serves a specialized function necessary to ensure society's survival as a whole.

Spencer became a proponent of a doctrine known as social Darwinism. **Social Darwinism** *applied to society Charles Darwin's notion of "survival of the fittest," in which those species of animals best adapted to the environment survived and prospered, while those poorly adapted died out.* Spencer reasoned that people who could not successfully compete in modern society were poorly adapted to their environment and therefore inferior. Lack of success was viewed as an individual failing, and that failure was in no way related to barriers (such as prejudice or racism) created by society. In this view, to help the poor and needy was to intervene vainly in a natural evolutionary process.

Social Darwinism had a significant effect on those who believed in the inequality of races. They now claimed that those who had difficulty succeeding in the white world were really members of inferior races. The fact that they lost out in the competition for status was proof of their poor adaptability to the environment. The survivors were clearly of superior stock (Berry & Tischler, 1978).

Many whites accepted social Darwinism because it served as a justification for their control over institutions. It enabled them to oppose reforms or social welfare programs, which they viewed as interfering with nature's plan to do away with the unfit. Social Darwinism thus became a justification for the repression and neglect of African Americans following the Civil War. It was also used to justify policies that resulted in the decimation of Native American populations and the complete eradication of the native people of Tasmania (near Australia) between 1803 and 1876 by white settlers (Fredrickson, 1971; Parillo, 1997).

Spencer's ties to social Darwinism have led many scholars to disregard his original contributions to the discipline of sociology. However, Spencer originally formulated many of the standard concepts and terms still current in sociology, and their use derives directly from his works.

During the nineteenth century, sociology developed rapidly under the influence of three scholars of highly divergent temperaments and orientations. Despite their differences, however, Karl Marx, Émile Durkheim, and Max Weber were responsible for shaping sociology into a relatively coherent discipline.

Karl Marx (1818–1883)

Those who are unfamiliar with his writings often think of Karl Marx as a revolutionary proponent of the political and social system we see in countries

Karl Marx's views on class conflict were shaped by the Industrial Revolution. He believed that capitalist societies produce greater conflict because of the deep divisions between the social classes.

today labeled communist. Marx lived in Europe during the early period of industrialization, when the overwhelming majority of people in such societies were poor. The rural poor moved to cities where employment was available in the factories and workshops of the new industrial economies. Those who owned and controlled the factories exploited the masses who worked for them. Even children, some as young as 5 or 6 years old, worked 12-hour days, six and seven days a week (Lipsey & Steiner, 1975), and received only a subsistence wage. "The iron law of wages"—the philosophy that justified paying workers only enough money to keep them alive— prevailed during this early period of industrialization. In this way, the rural poor were converted into an urban poor. Meanwhile, those who owned the means of production possessed great wealth, power, and prestige.

Marx tried to understand the societal forces that produced such inequities and looked for a means to change them in order to improve the human condition. Marx believed the entire history of human societies could be seen as the history of class conflict: the conflict between the bourgeoisie, who own and control the means of production (capitalists), and the proletariat, who make up the mass of workers— the exploiters and the exploited. He believed the capitalists determined the distribution of wealth, power, and even ideas in that society. The wealthy get their power not just from their control of the economy, but also from their control of the political, educational, and religious institutions in their society as well. According to Marx, capitalists make and enforce laws that serve their interests and act against the interests of workers. Their control over all institutions enables them to create common beliefs that make the workers accept their status. These economic, political, and religious ideologies make the masses loyal to the very institutions that are the source of their exploitation and that also are the source of the wealth, power, and prestige of the ruling class. Thus, the prevailing beliefs of any society are those of its dominant group.

Marx predicted that capitalist society eventually would be polarized into two broad classes: the capitalists and the increasingly impoverished workers. Intellectuals like him would show the workers that the capitalist institutions were the source of exploitation and poverty. Gradually, the workers would become unified and organized, and then through revolution they would take over control of the economy. The means of production would then be owned and controlled by the people in a workers' socialist state. Once the capitalist elements of all societies had been eliminated, the governments would wither away. New societies would develop in which people could work according to their abilities and take according to their needs. The seeds of societal conflict and social change would then come to an end as the means of production were no longer privately owned.

In many capitalist societies today, regulatory mechanisms have been introduced to prevent some of the excesses of capitalism. Unions have been integrated into the capitalist economy and the political system, giving workers a legal, legitimate means through which they can benefit from the capitalist system.

Marx was not a sociologist, but his considerable influence on the field can be traced to his contributions to the development of *conflict theory*, which will be discussed more fully in this chapter.

Émile Durkheim (1858–1917)

A student of law, philosophy, and social science, Émile Durkheim was the first professor of sociology at the University of Bordeaux, France. Whereas

Émile Durkheim produced the first true sociological study. Durkheim's work moved sociology fully out of the realm of social philosophy and chartered the discipline's course as a social science.

Spencer wrote the first textbook of sociology, it was Durkheim who produced the first true sociological study. Durkheim's work moved sociology fully out of the realm of social philosophy and helped chart the discipline's course as a social science.

Durkheim believed that individuals were exclusively the products of their social environment and that society shapes people in every possible way. To prove his point, Durkheim studied suicide. He believed that if he could take what was perceived to be a totally personal act and show that it is patterned by social factors rather than exclusively by individual mental disturbances, he would provide support for his point of view.

Durkheim began with the theory that the industrialization of Western society was undermining the social control and support that communities had historically provided for individuals. Industrialization forced or induced individuals to leave rural communities for urban areas, where there were usu-

ally greater economic opportunities. The anonymity and impersonality that they encountered in these urban areas, however, caused many people to become isolated from both family and friends. In addition, in industrial societies people are frequently encouraged to aspire to goals that are difficult to attain. Durkheim refined his theory to state that suicide rates are influenced by group solidarity and societal stability. He believed that low levels of solidarity—which involve more individual choice, more reliance on oneself, and less adherence to group standards—would mean high rates of suicide.

To test his idea, Durkheim decided to study the suicide rates of Catholic versus Protestant countries. He assumed the suicide rate in Catholic countries would be lower than in Protestant countries because Protestantism emphasized the individual's relationship to God over community ties. The comparison of suicide records in Catholic and Protestant countries in Europe supported his theory by showing the probability of suicide was indeed higher in Protestant countries. (See Figure 1-2 for suicide rates in various countries. Also see "Sociology at Work, Suicide in the United States.")

Recognizing the fact that lower suicide rates among Catholics could be based on factors other than group solidarity, Durkheim proceeded to test other groups. Reasoning that married people would be more integrated into a group than single people, or people with children more than people without children, or non-college-educated people more than college-educated people (because college tends to break group ties and encourage individualism), or Jews more than non-Jews, Durkheim tested each of these groups, and in each case his theory held. Then, characteristic of the scientist that he was, Durkheim extended his theory by identifying three types of suicide—egoistic, altruistic, and anomic—that take place under different types of conditions.

Egoistic suicide comes from low group solidarity, an underinvolvement with others. Durkheim argued that loneliness and a commitment to personal beliefs rather than group values can lead to egoistic suicide. He found that single and divorced people had higher suicide rates than did married people and that Protestants, who tend to stress individualism, had higher rates of suicide than did Catholics.

Altruistic suicide derives from a very high level of group solidarity, an overinvolvement with others. The individual is so tied to a certain set of goals that he or she is willing to die for the sake of the community. This type of suicide, Durkheim noted, still exists in the military as well as in societies based on ancient codes of honor and obedience. Perhaps the best-known examples of altruistic suicide come from Japan: the ceremonial rite of *seppuku,* in which

Figure 1-2 Suicide Rates of People 15–24 in Various Countries

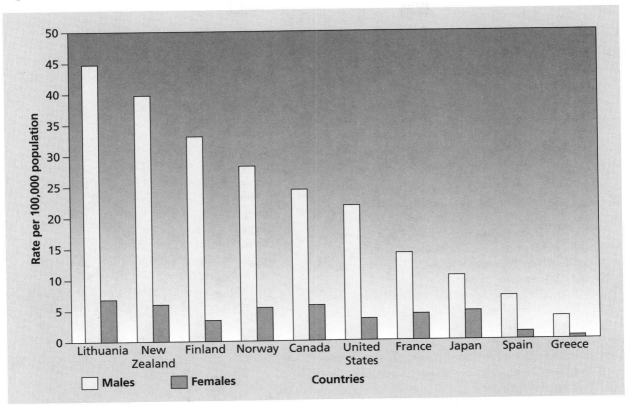

Source: World Health Statistics Annual 1993, 1994, and 1995, by the World Health Organization.

a disgraced person rips open his own belly, and the kamikaze attacks by Japanese pilots toward the end of World War II. Today we often see examples of altruistic suicide with terrorists. These individuals are willing to sacrifice their lives for their cause as they blow up a plane, restaurant, or military installation.

Anomic suicide results from a sense of feeling disconnected from society's values. A person may know what goals to strive for but not have the means of attaining them, or a person may not know what goals to pursue. Durkheim found that times of rapid social change or economic crisis are associated with high rates of anomic suicide.

Durkheim's study was noteworthy not only because it proved that the most personal of all acts, suicide, is in fact a product of social forces, but also because it was one of the first examples of a scientifically conducted sociological study. Durkheim systematically posed theories, tested them, and drew conclusions that led to further theories. He also published his results for everyone to see and criticize.

Durkheim's interests were not limited to suicide. His mind ranged the entire spectrum of social activities. He published studies on *The Division of Labor*

in Society (1893) and *The Elementary Forms of the Religious Life* (1917). In both works he drew on what was known about nonliterate societies, following the lead of Comte and Spencer in viewing them as evolutionary precursors of the contemporary industrial societies of Europe.

Durkheim focused on the forces that hold society together—that is, on the functions of various social structures. This point of view, often called the *functionalist theory* or *functionalist perspective*, remains one of the dominant approaches to the modern study of society.

Max Weber (1864–1920)

Much of Weber's work attempted to clarify, criticize, and modify the works of Marx. For that reason we shall discuss Weber's ideas as they relate to and contrast with those of Marx. Unlike Marx, who was not only an intellectual striving to understand society but also a revolutionary conspiring to overturn the capitalist social system, Weber was essentially a German academic attempting to understand human behavior. Weber believed the role of intellectuals

Sociology at Work

Suicide in the United States

Émile Durkheim studied suicide because he believed people were the product of their social environments. Differences in these environments were responsible for variations in social behavior. The following facts about suicide in the United States indicate that a variety of social factors are involved in suicide. How many can you identify?

- More people die from suicide than from homicide in the United States. In 1997, 30,535 Americans took their own lives. In contrast, 19,491 were homicide victims. On average, 84 Americans commit suicide each day, and there have been more suicides than homicides each year since 1950.

  In 1997, suicide was the eighth leading cause of death in the United States. Suicide was the third leading cause of death for 15- to 24-year-olds. A nationwide survey of high school students found that in the previous year, one-fifth had seriously considered suicide and 1 in 13 had attempted it.

- Suicide rates are especially high among older adults (age 65 and older). Older adults have had the highest suicide rate of all age groups since 1933, the year all states began reporting deaths. Suicide rates tend to rise with age and are highest among white men age 65 and older. Older adults account for almost 20% of suicide deaths, but only 13% of the U.S. population.

- Most suicides are males. In 1997, males accounted for 80% of all suicides in the United States. Among 15- to 19-year-olds, boys were

five times as likely as girls to commit suicide; among 20- to 24-year-olds, males were seven times as likely to commit suicide as females.

- Suicide is only part of the problem of self-directed violence. The number of completed suicides reflects only a small portion of the impact of suicidal behavior. It is estimated that more people are hospitalized because of suicide attempts than are fatally injured. In 1997, an estimated 610,000 visits to U.S. hospital emergency departments were due to self-directed violence.

- Suicide affects the various ethnic populations differently. Suicide rates for Native Americans were 1.5 times the national rates. Asian-American women have the highest suicide rate among women age 65 or older. Hispanic high-school students were significantly more likely than white students to have reported a suicide attempt.

- Most suicides involve firearms. In the United States, nearly three of every five suicides in 1997 (58%) were committed with a firearm. The availability of firearms in the homes of high-risk individuals markedly increases their risk for suicide.

- Many suicides are not reported. Although such reporting may spare the family some emotional distress, it results in the underreporting of suicides in the United States.

Source: Centers for Disease Control and Prevention, *Fact book for the year 2000: Working to prevent and control injury in the United States.* Washington, DC: U.S. Government Printing Office.

was simply to describe and explain truth, whereas Marx believed the scholar should also tell people what to do.

Marx believed that ownership of the means of production resulted in control of wealth, power, and ideas. Weber showed that economic control does not necessarily result in prestige and power. For example, the wealthy president of a chemical company whose toxic wastes have been responsible for the pollution of a local water supply might have little prestige in the community. Moreover, the company's board of directors might deprive the president of any real power.

Although Marx maintained that control of production inevitably results in control of ideologies,

Weber stated that the opposite may happen: Ideologies sometimes influence the economic system. When Marx called religion an "opium of the people," he was referring to the ability of those in control to create an ideology that would justify the exploitation of the masses. Weber, however, showed that religion could be a belief system that contributed to the creation of new economic conditions and institutions. In *The Protestant Ethic and the Spirit of Capitalism* (1904–1905), Weber tried to demonstrate how the Protestant Reformation of the seventeenth century provided an ideology that gave religious justification to the pursuit of economic success through rational, disciplined, hard work. This ideology, called the protestant Ethic, ultimately

Much of Max Weber's work was an attempt to clarify, criticize, and modify the words of Karl Marx. He also showed how religion contributed to the creation of new economic conditions and institutions.

helped transform northern European societies from feudal agricultural communities to industrial capitalist societies.

Yet Weber also predicted that science—the systematic, rational description, explanation, and manipulation of the observable world—would lead to a gradual turning away from religion. The apparent decline in the influence of organized religion in highly industrialized societies seems to support Weber's prediction.

Understanding the development of bureaucracy interested Weber. Whereas Marx saw capitalism as the source of control, exploitation, and alienation of human beings and believed that socialism and communism would ultimately bring an end to this exploitation, Weber believed bureaucracy would characterize both socialist and capitalist societies. He anticipated and feared the domination of individuals by large bureaucratic structures. As he foresaw, bureaucracies now rule our modern industrial world, both capitalist and socialist—economic, political, military, educational, and religious. Given the existing situation, it is easy to appreciate Weber's anxiety. As he put it,

> Each man becomes a little cog in the machine and, aware of this, his one preoccupation is whether he can become a bigger cog. . . . The problem which besets us now is not: how can this evolution be changed?—for that is impossible, but what will become of it? (Quoted in Coser, 1977)

The Development of Sociology in the United States

Sociology had its roots in Europe and did not become widely recognized in the United States until almost the beginning of the twentieth century. The early growth of American sociology took place at the University of Chicago. That setting provided a context in which a large number of scholars and their students could work closely to refine their views of the discipline. It was there that the first graduate department of sociology in the United States was founded in the 1890s. From the 1920s to the 1940s, the so-called Chicago school of sociologists led American sociology in the study of communities, with particular emphasis on urban neighborhoods and ethnic areas. Many of America's leading sociologists from this period were members of the Chicago school, including Robert E. Park, W. I. Thomas, and Ernest W. Burgess. Most of these individuals were Protestant ministers or sons of ministers, and as a group they were deeply concerned with social reform.

W. E. B. DuBois (1868–1963) became the first African-American to receive a PhD from Harvard in 1896 with his dissertation *The Suppression of the African Slave-Trade to the United States*. DuBois then went on to Atlanta University, where he was in charge of the sociology program until 1910. At that point he left to become editor of *The Crisis*, the journal of the National Association for the Advancement of Colored People. By that time DuBois had written dozens of articles and books on the history and sociology of African Americans and was the country's leading African-American sociologist.

At the time when DuBois came of age, racism was very much a part of the American landscape on both a popular and academic level. Writers and politicians were openly preaching that blacks were an inferior and debased race that had contributed nothing to civilization. DuBois believed that doctrines and theories had a powerful effect on social conditions. Slavery and the disenfranchisement of blacks were rooted in the notion of the inferiority of the race. It was important, he felt, to change these beliefs in order to change the status of African Americans. Much of his scholarly work was governed by his view that sociological studies of African Americans would have a positive effect on public opinion (Brotz, 1966; Lewis, 1993).

DuBois argued for the acceptance of African Americans into all areas of society and advocated militant resistance to white racism. He believed that it was not solely the responsibility of blacks, nor was it in their capacity, to alter their collective place in American society, but that it was primarily the responsibility of whites, who held the power to effect such change.

In 1903, he published *The Souls of Black Folk*, a collection of eloquent, well-reasoned essays on race relations. Blending sociology and economics he described the injustices that had scarred the black experience in the United States. "The problem of the 20th century is the problem of the color line," he declared (Lewis, 2000).

W. E. B. DuBois was the first African American to receive a PhD from Harvard University. He wrote dozens of articles and books on the history and sociology of African Americans.

Throughout his life DuBois considered himself torn between being a black man and being an American. This conflict led him to feel like an exile in the United States. As DuBois noted in his autobiography:

Had it not been for the race problem early thrust upon me and enveloping me, I should have probably been an unquestioning worshipper at the shrine of the established social order into which I was born. But just that part of this order which seemed to most of my fellows nearest perfection seemed to me most inequitable and wrong: and starting from that critique, I gradually, as the years went by, found other things to question in my environment. (DuBois, 1919)

DuBois died in 1963 at the age of 95, just before the famous March on Washington took place where Martin Luther King made his "I Have A Dream" speech. It was ironic that America's preeminent black intellectual died on the eve of this great civil rights gathering, which had gained so much energy from his ideas against segregation. DuBois had long ago concluded that the possibility of racial equality was a receding mirage for people of color. At the

time of his death he was leading the life of a political exile in Ghana.

Talcott Parsons (1902–1979) was the sociologist most responsible for developing theories of structural functionalism in the United States. He presided over the Department of Social Relations at Harvard College from the 1930s until he retired in 1973. Parsons's early research was quite empirical, but he later turned to the philosophical and theoretical side of sociology. In *The Structure of Social Action* (1937), Parsons presented English translations of the writings of European thinkers, most notably Weber and Durkheim. In his best-known work, *The Social System* (1951), Parsons portrayed society as a stable system of well-ordered, interrelated parts. His viewpoint elaborated on Durkheim's perspective.

Robert K. Merton also has been an influential proponent of functionalist theory. In his classic work, *Social Theory and Social Structure* (1968), first published in 1949, Merton spelled out the functionalist view of society. One of his main contributions to sociology was to distinguish between two forms of social functions—manifest functions and latent functions. By **social functions** Merton meant *those social processes that contribute to the ongoing operation or maintenance of society.* **Manifest functions** are *the intended and recognized consequences of those processes.* For example, one of the manifest functions of going to college is to obtain knowledge, training, and a degree in a specific area. **Latent functions** are *the unintended or not readily recognized consequences of such processes.* Therefore, college may also offer the opportunity of making lasting friendships and finding potential marriage partners.

Under the leadership of Parsons and Merton, sociology in the United States moved away from a concern with social reform and adopted a so-called value-free perspective. This perspective, which Max Weber advocated, requires description and explanation rather than prescription; it holds that people should be told what is, not what should be. As critics of Parsons and Merton have pointed out, however, interpretations of what exists may differ depending on the perspective from which reality is viewed and on the values of the viewer (Gouldner, 1970; Lee, 1978; Mills, 1959).

Theoretical Perspectives

Scientists need a set of working assumptions to guide them in their work. These assumptions suggest which problems are worth investigating and offer a framework for interpreting the results of studies. Such sets of assumptions are known as paradigms. **Paradigms** are *models or frameworks for questions*

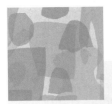

Sociology at Work

Women and the Development of Sociology, 1800 to 1945

In 1982 Shulamit Reinharz, a professor in the sociology department at Brandeis University, taught a social psychology course and was struck and pained by the extensive misogynist imagery in one of the texts she was using. Her first response to this problem was to write about it. Her second was to do something about it. Reinharz's plans for an active response started with the realization that she would be much less likely to encounter misogynist material in sociology texts if she were to use the work of female sociologists in her courses. But who were the female sociologists she could use? As Reinharz pointed out,

> Asking this question made me recognize that my conception of sociology consisted of the work of "great men." Thus, I could not teach a course using the works of female sociologists until I had studied the history of women's contributions to sociology. It did not take me long to realize that I had reached an impasse, because there were no books on this topic and almost no research being done to ameliorate the problem.
>
> I proposed a new course entitled "Women and Intellectual Work" that would focus on women's neglected contributions to sociology and would discuss why women's work is not remembered or incorporated into the canon.
>
> I made the assumption that women had made contributions to sociology in the past and that the fact that I did not know what these contributions were despite my having a Ph.D. in sociology, demonstrated that women's sociological work was not incorporated into the canon for some reason.

The following list is a sampling of the publications or contributions of women in sociology between 1800 and 1945. Works after 1945 are quite extensive and are incorporated into the rest of the chapters in this book.

1818   Francis Wright, *Views of Society and Manners in America.*

1837   Harriet Martineau, *Theory and Practice of Society in America.*

1844   Margaret Fuller, *Women in the Twentieth Century.*

1853   Harriet Martineau publishes translation of Comte's *Positive Philosophy.*

1856   Antoinette Blackwell, *Shadows of Our Social Systems.*

1870   Clémence Royer becomes first female member of Société d'Anthropologie.

1881   Helen Hunt Jackson's *Century of Dishonor* exposes government's treatment of Native Americans.

1883   Helen Campbell, *Prisoners of Poverty.*

1895   Hull House "Maps and Papers" published; helps launch academic studies in urban sociology.

1896   Mary Smith, *Alms House Women.*

1898   Eleanor Touroff Glueck, criminologist, publishes *Evaluative Research in Social Work.*

1901   Frances Kellor, *Experimental Sociology.*

1902   Jane Addams, *Democracy and Social Ethics.*

1906   Elsie Clews Parsons, *The Family.*

1910   Jane Addams, *Twenty Years of Hull House.*

1916   Annie M. MacLean, *Women Workers and Society.*

1919   Joanna Colcord, *Broken Homes: A Study of Family Desertion and Its Social Treatment.*

1925   Dorothy Swaine Thomas, *Social Aspects of the Business Cycle.*
       Hattie Plum Williams, first chair of sociology department, University of Nebraska.

1926   Marianne Weber, Max Weber, *Ein Lebensbild.*

1929   Katherine Davis, *Factors in the Sex Life of 2,200 Women.*

1935   Margaret Mead, *Sex and Temperament in Three Primitive Societies.*

1941   Margaret Lloyd Jarman Hagood, *Statistics for Sociologists.*

1944   Helen Macgill Hughes begins her 17-year term as manager of *The American Journal of Sociology.*

Source: *A Contextualized Chronology of Women's Sociological Work,* 2nd ed., by S. Reinharz, September 1993, Waltham, MA: Brandeis University Women's Studies Program Working Papers Series.

that generate and guide research. Of course, not all paradigms are equally valid, even though at first they seem to be. Sooner or later, some will be found rooted in fact, whereas others will remain abstract and unusable, finally to be discarded. We shall examine those paradigms that have withstood the scrutiny of major sociologists.

Functionalism

Functionalism—or structural functionalism, as it is often called—is rooted in the writings of Spencer and Durkheim and the work of such scholars as Parsons and Merton. **Functionalism** *views society as a system of highly interrelated structures or parts that function or operate together harmoniously.* Functionalists analyze society by asking what each different part contributes to the smooth functioning of the whole. For example, we may assume the education system serves to teach students specific subject matter. However, functionalists might note that it acts as a system for the socialization of the young and as a means for producing conformity. The education system serves as a gatekeeper to the rewards society offers to those who follow its rules.

From the functionalist perspective, society appears quite stable and self-regulating. Much like a biological organism, society is normally in a state of equilibrium or balance. Most members of a society share a value system and know what to expect from one another.

The best-known proponent of the structural-functionalist perspective was Talcott Parsons. His theory centered on the view that interrelated social systems consisted of major areas of social life, such as the family, religion, education, politics, and economics. These systems were analyzed according to functions they performed both for society as a whole and for one another.

Functionalism is a very broad theory in that it attempts to account for the complicated interrelationships of all the elements that make up human societies, including the complex societies of the industrialized (and industrializing) world. In a way it is impossible to be a sociologist and not be a functionalist, because most parts of society serve some stated or unstated purpose. Functionalism is limited in one regard, however: The preconception that societies are normally in balance or harmony makes it difficult for proponents of this view to account for how social change comes about.

A major point of criticism of functionalist theory is its conservative bias. That is, if all the parts of society fit together smoothly, we can assume that the social system is working well. Conflict is then seen as something that disrupts the essential orderliness of the social structure and produces imbalance between the parts and the whole.

Conflict Theory

Conflict theory is rooted in the work of Marx and other social critics of the nineteenth century. **Conflict theory** *sees society as constantly changing in response to social inequality and social conflict.* For the conflict theorists, social change pushed forward by social conflict is the normal state of affairs. Static periods are merely temporary way stations along the road.

Conflict theorists believe social order results from dominant groups making sure that subordinate groups are loyal to the institutions that are the dominant groups' sources of wealth, power, and prestige. The dominant groups will use coercion, constraint, and even force to help control those people who are not voluntarily loyal to the laws and rules they have made. When this order cannot be maintained and the subordinate groups rebel, change comes about.

Conflict theorists are concerned with the issue of who benefits from particular social arrangements and how those in power maintain their positions and continue to reap benefits from them. The ruling class is seen as a group that spreads certain values, beliefs, and social arrangements in order to enhance its power and wealth. The social order then reflects the outcome of a struggle among those with unequal power and resources.

Conflict perspectives are often criticized as concentrating too much on conflict and change and too little on what produces stability in society. They also are criticized for being too ideologically based and making little use of research methods or objective statistical evidence. The conflict theorists counter that the complexities of modern social life cannot be reduced to statistical analysis, and that doing so has caused sociologists to become detached from their object of study and removed from the real causes of human problems.

Both functionalist and conflict theories are descriptive and predictive of social life. Each has its strengths and weaknesses, and each emphasizes an important aspect of society and social life.

The Interactionist Perspective

Functionalism and conflict theory can be thought of as the opposite sides of the same coin. Although quite different from one another, they share certain similarities. Both approaches focus on major structural features of entire societies and attempt to give us an understanding of how societies survive and change. Social life, however, also occurs on an intimate scale between individuals. The **interactionist**

Table 1-1

Major Theoretical Perspectives in Sociology

PERSPECTIVE	SCOPE OF ANALYSIS	POINT OF VIEW	FOCUS OF ANALYSIS
Structural-Functional	Macro level	The various parts of society are interdependent and functionally related. Social systems are highly stable. Social life is governed by consensus and cooperation.	The functional and dysfunctional aspects of institutions and society.
Social-Conflict	Macro level	Society is a system of accommodations among competing interest groups. Social systems are unstable and are likely to change rapidly. Social life involves conflict because of differing goals.	How social inequalities produce conflict. Who benefits from particular social arrangements.
Interactionist	Micro level	Most of what people do has meaning beyond the concrete act. The meanings that people place on their own and one another's behavior can vary.	How people make sense of the world in which they participate.

perspective *focuses on how individuals make sense of—or interpret—the social world in which they participate.* As such, this approach is primarily concerned with human behavior on a person-to-person level. Interactionists criticize functionalists and conflict theorists for implicitly assuming that social processes and social institutions somehow have a life of their own apart from the participants. Interactionists remind us that the educational system, the family, the political system, and indeed all of society's institutions are ultimately created, maintained, and changed by people interacting with one another.

The interactionist perspective includes a number of loosely linked approaches. George Herbert Mead devised a *symbolic interactionist* approach that focuses on signs, gestures, shared rules, and written and spoken language. Harold Garfinkel used *ethnomethodology* to show how people create and share their understandings of social life. Erving Goffman took a *dramaturgical* approach in which he saw social life as a form of theater. (We will discuss ethnomethodology and dramaturgy in Chapter 5.) Of these three approaches, the symbolic interactionist approach has received the widest attention and presents us with a well-formulated theory. (Table 1-1 compares the functionalist, conflict theory, and interactionist approaches to sociology.)

Symbolic Interactionism As developed by George Herbert Mead (1863–1931), **symbolic interaction-ism** is *concerned with the meanings that people place on their own and one another's behavior.* Human beings are unique in that most of what they do with one another has meaning beyond the concrete act. According to Mead, people do not act or react automatically, but carefully consider and even rehearse what they are going to do. They take into account the other people involved and the situation in which they find themselves. The expectations and reactions of other people greatly affect each individual's actions. In addition, people give things meaning and act or react on the basis of these meanings. For example, when the flag of the United States is raised, people stand because they see the flag as representing their country.

Because most human activity takes place in social situations—in the presence of other people—we must fit what we as individuals do with what other people in the same situation are doing. We go about our lives with the assumption that most people share our definitions of basic social situations. This agreement on definitions and meanings is the key to human interactions in general, according to symbolic interactionists. For example, a staff nurse in a mental hospital unlocking a door for an inpatient is doing more than simply enabling the patient to pass from one ward to another. He or she also is communicating a position of social dominance over the patient (within the hospital) and is carrying a powerful symbol of that dominance—the key. The same holds true for a professor writing on a blackboard or a company vice-president dictating to a secretary. Such interactions, therefore, although they appear to be simple social actions, also are laden with highly symbolic social meanings. These symbolic meanings are intimately connected with our understanding of what it is to be and to behave as a human being. This includes our sense of self; how we experience others and their views of us; the joys and pains we feel at home, at school, at work, and among friends and colleagues; and so on.

Symbolic interaction and its various offshoots have been criticized for paying too little attention to the larger elements of society. Interactionists respond that societies and institutions are made up of individuals who interact with one another and do not exist apart from these basic units. They believe that an understanding of the process of social interaction will lead to an understanding of larger social structures. In actual fact, interactionists still must bridge the gap between their studies of social interaction and those of the broader social structures. Nevertheless, symbolic interactionism does complement functionalism and conflict theory in important ways and gives us important insights into how people interact.

Contemporary Sociology

Contemporary sociological theory continues to build on the original ideas proposed in the interactionist perspective, functionalism, and conflict theory. It

Controversies in Sociology

Is There a Difference Between Sociology and Journalism?

It often seems as if sociologists and journalists are engaged in the same activities. Journalists examine and write about social issues. They interview people. They often conduct polls. They make predictions. They offer recommendations for correcting social problems. If journalists do all this, why would someone need to become trained as a sociologist? This is a sociology textbook, so needless to say we are going to make the case that there is a difference between sociologists and journalists.

Daily newspaper and weekly newsmagazines are written for the general public, which wants an overview of local and world events. A fundamental feature of newspapers and magazines is the timely coverage of recent events. Nothing is more stale and uninteresting to readers than an old story. Nothing is more enjoyable for reporters and editors than to scoop the competition with a late-breaking story. Three types of journalists do the writing: reporters, who write stories; editors, who generate ideas for stories and review the copy; and commentators, who interpret events. Journalists usually have a college degree in any of a wide variety of areas, or they may have an advanced degree from a professional journalism program. Professional experts also are invited to write letters to the editor or commentaries for the editorial page. Jargon is kept to a minimum and elaborate explanations must be presented in manageable terms so that the average reader can understand them.

Sociologists engage in the study of society with the primary intent of sharing their work with other sociologists, not necessarily with the general public. They pay special attention to their methods of investigation, their theories of explanation, and their claims of originality. When sociologists publish their results, they are fully aware that other sociologists may dispute the soundness of their findings or the logic of their explanations. The public usually has little interest in sociological disputes about methods, theories, or claims of originality.

The writings of sociologists are typically published as either articles in scholarly journals, chapters in books, or full-length books. These writings are screened by editors and critics hired to evaluate the merits of the work. Sociologists aim to have their colleagues recognize their work as truly significant.

Journalists are always thinking about tomorrow's headlines or next week's cover story. Sociologists never work with such short time frames. A major sociological study may take 3 to 5 years; most take 1 to 2 years. Sociologists also have the freedom to study historical materials. Sociologists are not indifferent to the times in which they live; many hope that their work will be relevant to contemporary debates about current issues.

Essentially, the two fields represent different approaches to social issues. Journalists get a multifaceted overview of an issue, while sociologists have the luxury of exploring a topic in depth and contemplating the ramifications of their findings.

Source: Social Problems: A Critical Thinking Approach, 2nd ed. (pp. 20–22), by P. J. Baker, L. E. Anderson, & D. S. Dorn, 1993, Belmont, CA: Wadsworth.

would be difficult to see contemporary sociological theory as either conflict theory or functionalism in the original sense. Much of it has been modified to include important aspects of each theory. Even symbolic interactionism has not been wholeheartedly embraced and aspects of it have instead been absorbed into general sociological writing.

Very little contemporary sociological theory still can be identified as true functionalism. Part of this is due to the fact sociologists today have abandoned trying to develop all-inclusive theories and instead opt for what Merton (1968) referred to as middle-range theories. **Middle-range theories** *are concerned with explaining specific issues or aspects of society instead of trying to explain how all of society operates.* A middle-range theory might be one that explains why divorce rates rise and fall with certain economic conditions or how crime rates are related to residential patterns.

Modern conflict theory was initially refined by such sociologists as C. Wright Mills (1959), Ralf Dahrendorf (1958), Randall Collins (1975, 1979), and Lewis Coser (1956) to reflect the realities of contemporary society. Mills and Dahrendorf did not see conflict as confined to class struggle. Rather, they viewed it as applicable to the inevitable tensions that arise between groups: parents and children, producers and consumers, professionals and their clients, unions and employers, the poor and the materially comfortable, and minority and majority ethnic groups. Members of these groups have both overlapping and competing interests, and their shared needs keep all parties locked together within one society. At the same time, the groups actively pursue their own ends, thus constantly pushing the society to change in order to accommodate them.

Coser incorporated aspects of both functionalism and conflict theory, seeing conflict as an inevitable element of all societies and as both functional and dysfunctional for society. Conflict between two groups tends to increase their internal cohesion. For example, competition between two divisions of two computer companies to be the first to produce a new product may draw the members closer to one another as they strive to reach the desired goal. This feeling might not have occurred had it not been for the sense of competition, as the conflict itself becomes a form of social interaction. Conflict also could lead to cohesion by causing two or more groups to form alliances against a common enemy. For example, a political contest may cause several groups to unite in order to defeat a common opponent.

During the past 25 years, conflict theory has been influenced by a generation of neo-Marxists. These people have helped produce a more complex and sophisticated version of conflict theory that goes beyond the original emphasis on class conflict and instead shows that conflict exists within almost every aspect of society (Gouldner, 1970, 1980; Skocpol, 1979; Starr, 1982, 1992; Tilly, 1978, 1981; Wallerstein, 1974, 1979, 1980, 1991).

Theory and Practice

Sociological theory gives meaning to sociological practice. Merely assembling countless descriptions of social facts is inadequate for understanding society as a whole. Only when data are collected within the conceptual framework of a theory—in order to answer the specific questions growing out of that theory—is it possible to draw conclusions and make valid generalizations. This pursuit is the ultimate purpose of all science.

Theory without practice (research to test it) is at best poor philosophy and at worst unscientific, and practice uninformed by theory is at best trivial and at worst a tremendous waste of time and resources. Therefore, in the next chapter we shall move from theory to practice—to the methods and techniques of social research.

SUMMARY

We are constantly bombarded with information about social issues. Much of it, however, comes from sources that have an interest in getting people to support a particular point of view (like special interest groups) or that wish to attract attention and maintain high ratings (like journalists and talk-show hosts). By contrast, sociology, which is the scientific study of human society and social interactions, seeks an accurate and scientific understanding of society and social life. The main focus of sociology is the group and not the individual. A sociologist tries to understand the forces that operate throughout the society—forces that mold individuals, shape their behavior, and thus determine social events.

The social sciences consist of all those disciplines that apply scientific methods to the study of human behavior. Though there is some overlap, each of the social sciences has its own area of investigation. Cultural anthropology, psychology, economics, history, political science, and social work all have some things in common with sociology, but each has its own distinct focus, objectives, theories, and methods.

Sociology emerged as a separate field of study in Europe during the nineteenth century. It was a time of turmoil and a period of rapid and dramatic social change. Industrialization, political revolution, urbanization, and the growth of a market economy undermined traditional ways of doing things. The need for a systematic analysis of society coupled with the acceptance of the scientific method resulted in the emergence of sociology.

In the United States, sociology developed in the early twentieth century. Its early growth took place at the University of Chicago, where the first graduate department of sociology in the United States was founded in 1890. The so-called Chicago school of sociology focused on the study of urban neighborhoods and ethnic areas, and included many of America's leading sociologists of the period. As a group, they were deeply concerned with social reform.

Scientists need a set of working assumptions to guide them in their professional activities. These models or frameworks for questions that generate and guide research are known as paradigms. Sociologists have developed several paradigms to help them investigate social processes. Functionalism views society as a system of highly interrelated structures that function or operate together harmoniously. Functionalists analyze society by asking what each part contributes to the smooth functioning of the whole. From the functionalist perspective, society appears quite stable and self-regulating. Critics have attacked the conservative bias inherent in this assumption.

Conflict theory sees society as constantly changing in response to social inequality and social conflict. For these theorists, social conflict is the normal state of affairs. Social order is maintained by coercion. Conflict theorists are concerned with the issue of who benefits from particular social arrangements and how those in power maintain their positions.

The interactionist perspective focuses on how individuals make sense of, or interpret, the social world in which they participate. This perspective consists of a number of loosely linked approaches.

Contemporary sociology has built on and modified the insights of these three theoretical perspectives.

INTERNET RESOURCES

Go to http://web-enhanced.thomsonlearning.com which features Web links relevant to this chapter to expand and enhance the learning experience. This site will be monitored and updated periodically to ensure the links are correct and live.

At Virtual Society: The Wadsworth Sociology Resource Center, http://sociology.wadsworth.com, you will find a Career Center, Sociology in the News, Research Online, InfoTrac College Edition, Virtual Tours in Sociology, and a variety of book-specific student and instructor resources.

CHAPTER ONE STUDY GUIDE

KEY CONCEPTS

Match each concept with its definition, illustration, or explanation presented below.

a. Empiricism
b. Paradigm
c. Dramaturgy
d. Middle-range theories
e. Scientific method
f. Social Darwinism
g. Egoistic suicide
h. Protestant ethic

i. Interaction perspective
j. Social sciences
k. Functionalism
l. Sociology
m. Ethnomethodology
n. Social functions
o. Altruistic suicide
p. Common sense reasoning

q. Conflict theory
r. Science
s. Latent function
t. Symbolic interactionism
u. Sociological imagination
v. Manifest function
w. Anomie suicide
x. Value-free

_____ 1. A theoretical approach that stresses the meanings people place on their own and one another's behavior.

_____ 2. The relationship between individual experiences and forces in the larger society that shape our actions.

_____ 3. A body of systematically arranged knowledge that shows the operation of general laws.

_____ 4. Suicides resulting from a sense of feeling disconnected from society's values.

_____ 5. The view that generalizations are valid only if they rely on evidence that can be observed directly or verified through our senses.

_____ 6. Those disciplines that apply scientific methods to the study of human behavior.

_____ 7. The notion that people who cannot successfully compete in modern society are poorly adapted to their environment and therefore inferior compared to those who successfully adapt.

_____ 8. Processes contributing to the ongoing operation or maintenance of society.

_____ 9. Intended and recognized consequences or outcomes of social processes.

_____ 10. Suicides resulting from low group solidarity, and underinvolvement with others.

_____ 11. A model or framework for organizing the questions that generate and guide research.

_____ 12. Picturing society as a system of highly interrelated parts, operating together in a fairly harmonious manner most of the time.

_____ 13. Picturing society as constantly changing in response to social inequality and resulting conflict.

_____ 14. A perspective seeking to understand how individuals make sense of and interpret their social world.

_____ 15. The scientific study of human society and social interactions.

_____ 16. Studying the sets of rules or guidelines people use in their everyday living practices, especially information about society's unwritten rules for social behavior.

_____ 17. Using the theater as a metaphor for understanding how individuals use roles to create impressions within social settings.

_____ 18. Theories attempting to explain specific issues or selected aspects of society rather than accounting for how the entire society operates.

_____ 19. Describing events and making sense of them after they have happened rather than systematically predicting outcomes in advance.

_____ 20. Suicides resulting from a very high level of group identity, an overinvolvement with others.

_____ 21. Unintended or not readily recognized consequences of social processes.

_____ 22. The approach to research that involves observation, experimentation, generalization, and verification.

_____ 23. An ideology that gave religious justification to the pursuit of economic success through hard work.

_____ 24. An American approach to the study of society that focuses on what is rather than what should be.

KEY THINKERS

Match the thinkers with their main ideas or contributions.

a. Harold Garfinkel
b. Talcott Parsons
c. Karl Marx
d. George Herbert Mead
e. Lewis Coser
f. C. Wright Mills

g. The Chicago School
h. Max Weber
i. Auguste Comte
j. Robert K. Merton
k. Herbert Spencer
l. W.E.B. DuBois

m. Harriet Martineau
n. Émile Durkheim
o. Shulamit Reinharz
p. Henri Saint-Simon
q. K. H. Teigen

_____ 1. Proposed the concept of the sociological imagination.

_____ 2. Coined the term *sociology*.

_____ 3. One of the first sociologists to analyze the customs and lifestyles of nineteenth-century America in a comprehensive, comparative, and objective manner.

_____ 4. An early sociologist who developed many of the standard concepts in the discipline and was a strong advocate of social Darwinism.

_____ 5. Pioneered the conflict theory. Focused on societal forces that produced inequalities, particularly in capitalist societies, which he predicted would eventually be polarized into workers and owners.

_____ 6. Produced the first empirical sociological study in which he explained how such individual acts as suicide could be explained by understanding social integration and other social forces.

_____ 7. Explained how a belief system such as religion could provide an understanding of how new economic systems and institutions could develop. Also pioneered the study of bureaucracy.

_____ 8. A group of early sociologists who developed empirical methods for studying urban neighborhoods.

_____ 9. Early critic of American racial practices. Challenged racist academic theories proposing inferiority of African Americans. Argued that whites, since they held power, had responsibility to promote equality.

_____ 10. A major force in sociology who advocated a value-free approach to functionalist sociology.

_____ 11. A middle-range theorist who developed the concepts of manifest and latent function.

_____ 12. Pioneered the symbolic interaction perspective.

_____ 13. Coined the term *ethnomethodology* as an approach to the sociological study of everyday life.

_____ 14. Integrated functionalism with conflict theory showing different interpretations of social conflict.

_____ **15.** Evaluated commonsense explanations as demonstrated in cultural proverbs and their opposites.

_____ **16.** An early influence on August Comte and the beginnings of the discipline of sociology.

_____ **17.** In 1982, began the focus on women's neglected contributions to sociology and why the early work of women was not remembered or incorporated into the discipline.

CENTRAL IDEA COMPLETIONS

Following the instructions, fill in the appropriate concepts and descriptions for each of the questions posed in the following section.

1. How does using the *sociological imagination* help us to better understand phenomena such as child abductions? What kinds of questions would a sociologist ask when studying a social problem with very serious social consequences? _____

_____

_____

2. What are the major characteristics of suicide in the United States?

   a. _____

   b. _____

   c. _____

3. Considering the issue of eliminating urban poverty in the United States, develop a one-sentence statement representing the views (or solutions) of each of the following pioneer sociologists.

   a. Karl Marx: _____

   b. Émile Durkheim: _____

   c. W. E. B. DuBois: _____

   d. Herbert Spencer: _____

   e. Max Weber: _____

   f. Harriet Martineau: _____

   g. Robert Merton: _____

   h. George Herbert Mead: _____

4. Pursuing this same topic, the elimination of urban poverty in the United States, present a short statement describing how each of the *three perspectives in sociology* might address this social problem.

   a. Structural functionalism: _____

   _____

   b. Conflict theory: _____

   _____

   c. Interactionism: _____

   _____

5. How does sociology, as a point of view, assist one in understanding the phenomenon of domestic violence? _____

   _____

   Regarding domestic violence, elaborate on the differences between a personalized and sociological view.

a. Personalized: _____

b. Sociological: _____

6. You have been asked to represent the Department of Sociology at your college or university in a campus-wide debate on the topic "Resolved: There Is No Difference Between Sociology and Common Sense." What would be the key points of your rebuttal to that resolution? What information would you use to support the idea that sociology as a discipline offers more than common sense to the study of social behavior?

Point 1. _____

Point 2. _____

Point 3. _____

Point 4. _____

7. Assume you were about to begin a term paper on the topic of international terrorism. In the spaces provided state the focus of each of the following social science disciplines (i.e., what would each look at?):

a. Political science: _____

b. Psychology: _____

c. Economics: _____

d. History: _____

e. Cultural anthropology: _____

8. Continuing with the preceding example, how might the use of the sociological imagination enable one to gain insights into the *attitudes, motivations, and behaviors of international terrorists* in a manner different from those used in the social sciences above?

_____

_____

_____

CRITICAL THOUGHT EXERCISES

1. Examine the data on suicide in the box titled "Sociology at Work: Suicide in the United States." Building on your reading of Durkheim's analysis of suicide, what societal factors might account for variations in suicide in the United States?

2. While students enroll in colleges and universities for a variety of reasons ranging from the pursuit of knowledge to a chance to get away from home, most students are hoping their years at school will enable them to enter a meaningful career upon graduation. Unfortunately, students often find it difficult to make the linkage between the subject matter they are studying and the world of work and career. Focusing on the discipline of sociology, create an employment bulletin depicting the variety of jobs for which a major in sociology would be helpful. Select six occupational positions and write relevant corporations, institutions, and agencies for examples of job descriptions, educational requirements, and career benefit packages.

3. Select a contemporary issue in American society (for example, juvenile crime rates, cigarette smoking, or pornography on the Internet). Present your position on how that issue should be addressed or "solved." Focusing on your "solution," discuss the manifest and latent functions that would require consideration in any program seeking the changes you proposed.

4. Building on the box titled "Sociology at Work: If You Are Thinking About Sociology as a Career, Read This", go to the Web site of the American Sociological Association and continue your career exploration as you read the brochures provided.

http://192.231.215.37/asasearch/findit.cfm *

5. W. E. B. DuBois was an early thinker in sociology and spokesperson for the African-American community of his time. Although his writings were produced during the beginnings of this century, the recent Presidential Commission on Racial Relations suggests the issues outlined by DuBois remain relevant today. Check out the following Web sites for information on DuBois and produce an updated primer on his importance for contemporary thought on race relations.

http://www.gms.ocps.k12.fl.us/biopage/ dubois.html *

http://www.cc.columbia.edu/acis/bartleby/ dubois/0.html *

http://www:cc.columbia.edu/acis/bartleby/ dubois/15.html *

*Please note, Web site addresses change frequently. The addresses presented in these exercises should be viewed as guides only, and students should be prepared to utilize appropriate Internet search techniques when necessary should these addresses no longer be accessible.

ANSWERS

KEY CONCEPTS

1. t	5. a	9. v	13. q	17. c	21. s
2. u	6. j	10. g	14. i	18. d	22. e
3. r	7. f	11. b	15. l	19. p	23. h
4. w	8. n	12. k	16. m	20. o	24. x

KEY THINKERS

1. f	4. k	7. h	10. b	13. a	16. p
2. i	5. c	8. g	11. j	14. e	17. o
3. m	6. n	9. l	12. d	15. q	

SOCIAL STATISTICS BRIEFING ROOM

ESBR | Crime | Demography | Education | Health | SSBR

		Previous	Current
CHART: POPclock	**Current Population of the U.S.** Up to the second population estimates. Provided by U.S. Bureau of the Census as of today.		
CHART: POPclock	**Current Population of the World** Up to the second population estimates. Provided by U.S. Bureau of the Census as of today.		
CHART: Median Household Income by Race and Hispanic Origin: 1972 to 1998	**Household Income** Between 1997 and 1998, real median household income increased 3.5 percent to $38,885, the highest median household income level ever recorded. Among the racial groups, non-Hispanic Whites were the only group to have a significant increase in their real median household income between 1997 and 1998–3.0 percent. Hispanic households (who may be of any race) had a 4.8 percent increase in real median income between 1997 and 1998 (not statistically different from the percent increase for non-Hispanic Whites). Provided by U.S. Bureau of the Census / Current Population Survey as of September 30, 1999.	$37,581 1997 (in 1998 dollars)	$38,885 1998 (in 1998 dollars)
CHART: Poverty Rates	**Poverty** The poverty rate for the United States dropped to 12.7 percent in 1998, down from 13.3 percent in 1997. The number of poor dropped significantly also, to 34.5 million people, down from 35.6 million in 1997. *The percentage point change from 1997 to 1998 was -0.5. As a result of rounding, this difference appears to be slightly lower than the difference in the reported rates. Provided by U.S. Bureau of the Census / Current Population Survey as of September 30, 1999.	13.3 1997 percent	12.7* 1998 percent
CHART: Household Wealth	**Household Wealth** Median net worth was $37,587 in 1993. Differences by age group and race were significant. Provided by U.S. Bureau of the Census / Survey of Income and Program Participation as of September 25, 1995.	$38,500 1991 in 1993 dollars	$37,587 1993

CHAPTER TWO

Doing Sociology: Research Methods

LEARNING OBJECTIVES

After studying this chapter, you should be able to do the following:

- Explain the steps in the sociological research process.
- Analyze the strengths and weaknesses of the various research designs.
- Know what independent and dependent variables are.
- Know what sampling is and how to create a representative sample.
- Recognize researcher bias and how it can invalidate a study.
- Explain the strengths and weaknesses of the various measures of central tendency.
- Read and understand the contents of a table.
- Explain the concepts of reliability and validity.
- Understand the problems of objectivity and ethical issues that arise in sociological research.

On Monday, June 19, 1998, the major news outlets were filled with accounts proclaiming that an independent panel of scientists had reexamined UFO data and had concluded that the matter needed to be taken seriously and that important new evidence had been presented. The *Washington Post* wrote of "the first independent scientific review of the controversial topic in almost 30 years" that supposedly found "cases that included intriguing and inexplicable details, such as burns to witnesses, radar detections of mysterious objects, strange lights appearing repeatedly in the skies over certain locales, aberrations in the workings of automobiles, and irradiation and other damage found in vegetation." The Associated Press reported that an international panel of scientists" was convened to conduct "the first independent review of UFO phenomena since 1966." ABC News drew a similar conclusion. All of the media reports noted that the physical evidence in some UFO sightings was dramatic enough to warrant a reevaluation of previous findings.

However, what hardly any of the news organization noted was that this was far from an "independent" review of UFO data. The Society for Scientific Exploration, the organization that made this information available to the media, is not a mainstream scientific organization. Instead, it is a group with a strong bias toward paranormal phenomena. For the past 11 years, the group has published the *Journal of Scientific Exploration,* which includes articles suggesting that some paranormal claims have been verified scientifically.

The group's Web site notes that reincarnation is not only a scientific fact, but that past lives are often indicated by the presence of birthmarks. The Web site also notes that the group's journal will continue to publish supposedly scientific papers on "UFO and related phenomena, clairvoyance, precognition, telepathy, psychokinesis, out-of-body and near-death experiences, astrological claims, ball lightning, crop circles, apparent chemical or biological transmutation (alchemy), etc." The papers published by this organization had been unconvincing to mainstream scientists, and, far from pointing the way to further research, had been ignored.

The media outlets did not know that the UFO panel they were reporting on was at a conference where only pro-UFO researchers had been invited to present "evidence." Skeptical viewpoints were excluded and evidently felt not worth considering. The conference was funded by an individual who had previously provided financial support for other pro-UFO organizations. The conference not only failed to sway the mainstream scientific community, but was also somewhat of an embarrassment to some UFO believers because it included some of the more outlandish claims about UFO abductions.

When the media reported on what was thought to be serious research, they failed to take into account whether any valid study had been done to warrant the findings. As you will see in this chapter, the research process involves a number of specific steps that must be followed in order to produce a valid study. Only when this is done faithfully can we have any confidence in the results of the study. In this chapter we shall examine some of the methods used by scientists in general—and sociologists in particular—to collect data in order to test their ideas.

The Research Process

How should you conduct a research study? After reading Chapter 1, you know that you should not approach a study and draw conclusions on the basis of your personal experience and perceptions; rather, you should approach the study scientifically. In order to approach a study scientifically, you should keep in mind that science has two main goals: (1) to describe in detail particular things or events and (2) to propose and test theories that help us understand these things or events.

There is a great deal of similarity between what a detective does in attempting to solve a crime and what a sociologist does in answering a research problem. In the course of their work, both detectives and sociologists must gather and analyze information. For detectives, the object is to identify and locate criminals and collect enough evidence to ensure their identification is correct. Sociologists, on the other hand, develop hypotheses, collect data, and develop theories to help them understand social behavior. Although their specific goals differ, both sociologists and detectives try to answer two general

questions: Why did it happen, and under what circumstances is it likely to happen again? That is, sociologists seek to explain and predict.

All research problems require their own special emphasis and approach. The research procedure is usually custom tailored to the research problem. Nonetheless, there is a sequence of steps called the research process that is followed when designing a research project. In short, the **research process** *involves defining the problem, reviewing previous research on the topic, developing one or more hypotheses, determining the research design, defining the sample and collecting data, analyzing and interpreting the data, and finally preparing the research report.* The sequence of steps in this process and the typical questions asked at each step are illustrated in Table 2-1. If there are any terms in this table that you are not familiar with, do not become concerned. We will define them as we examine each of the various steps.

Define the Problem

"Love leads to marriage." Suppose you were given this statement as a subject for sociological research. How would you proceed to gather data to prove or disprove it? You must begin by defining love, a task that William Shakespeare himself tried to do in his play *Twelfth Night* when he asked, "What is love?" We would know that we have a problem here, as to this day people are still grappling with the question "How do you know when you are in love?" You could begin by using the definition of love that a researcher used in another study. For example, Hatfield (1988) defined love as "a state of intense longing for union with another." You would now have to find some way of determining whether this condition exists. You also must decide whether both people have to be in love in order for marriage to take place. You may already notice that it may be difficult to achieve the level of precision necessary for a useful research project. Once you accurately define your terms and provide details to clarify your descriptions, you can begin to test the statement we proposed. Even after arriving at a careful definition of your terms and a detailed description of love, you may still have trouble answering the question empirically.

An **empirical question** *can be answered by observing and analyzing the world as it is known.* Examples: How many students in this class have an A average? How many millionaires are there in the United States? Scientists pose empirical questions in order to collect information, to add to what is already known, and to test hypotheses. To turn the statement about love into an empirical question, you must ask: How do we measure the existence of love?

In trying to define and measure love, one researcher (Rubin, 1970; 1973) used an interesting

Table 2-1
The Research Process

STEPS IN THE PROCESS	TYPICAL QUESTIONS
Define the problem	What is the purpose of the study? What information is needed? How can we operationalize the terms? How will the information be used?
Review previous research	What studies have already been done on this topic? Do we need additional information before we begin? From what perspective should we approach this issue?
Develop one or more hypotheses	What are the independent and dependent variables? What is the relationship among the variables? What types of questions do we need to answer?
Determine the research design	Can we use existing data? What will we measure or observe? What research methods should we use?
Define the sample and collect data	Are we interested in a specific population? How large should the sample be? Who will gather the data? How long will it take?
Analyze the data and draw conclusions	What statistical techniques will we use? Have our hypotheses been proved or disproved? Is our information valid and reliable? What are the implications of our study?
Prepare the research report	Who will read the report? What is their level of familiarity with the subject? How should we structure the report?

approach. He prepared a large number of self-descriptive statements that considered various aspects of loving relationships as mentioned by writers, philosophers, and social scientists. After administering these statements to a variety of subjects, he was able to isolate nine items that best reflected feelings of love for another. Three of these items are cited in the following paragraph. In each sentence, the person is to fill in the blank with the name of a particular person and indicate the degree to which the item describes the relationship.

The following statements reflect three components of love. The first is attachment-dependency: "If I were lonely, my first thought would be to seek (blank) out." The second component is caring: "If (blank) were feeling badly, my first duty would be to cheer (him or her) up." The final component is intimacy: "I feel that I can confide in (blank) about virtually everything." These three statements show the strong aspect of mutuality in love relationships.

Using Rubin's scale, you can begin to make some headway toward clarifying an important component of your research problem. In the language of science you have operationalized your definition of love. An **operational definition** is *a definition of an abstract concept in terms of the observable features that describe the thing being investigated.* Attachment-dependency, caring, and intimacy can be three fea-tures of an operational definition of love and can indicate the presence of love in a research study.

Review Previous Research

Which questions are the "right" questions? Although there are no inherently correct questions, some are better suited to investigation than are others. To decide what to ask, researchers must first learn as much as possible about the subject. We would want to familiarize ourselves with as many of the previous studies on the topic as possible, particularly those closely related to what we want to do. By knowing as much as possible about previous research, we avoid duplicating a previous study and are able to build on contributions others have made to our understanding of the topic.

After reviewing the research we might find out that the early anthropologist Ralph Linton thought love was a form of insanity and assuming that it should lead to marriage was absurd. As he noted:

All societies recognize that there are occasional violent emotional attachments between persons of the opposite sex, but our present American culture is practically the only one which has attempted to capitalize these and make them the basis for marriage. Most groups regard them as unfortunate

If you were doing a study of whether love leads to marriage, you would find that it would be quite difficult to define what *love* means.

and point out the victims of such attachments as horrible examples. . . . The percentage of individuals with a capacity for romantic love of the Hollywood type [is] about as large as that of persons able to throw genuine epileptic fits. (Linton, 1936)

Needless to say, Linton would not have thought much of our potential research project.

Develop One or More Hypotheses

Our original statement "Love leads to marriage" is presented in the form of a hypothesis. A **hypothesis** is *a testable statement about the relationships between two or more empirical variables.* A **variable** is *anything that can change (vary).* The number of highway deaths on Labor Day weekends, the number of divorces that occur each year in the United States, the amount of energy the average American family consumes in the course of a year, the daily temperature in Dallas, the number of marathoners in Boston or in Knoxville, Tennessee—all these are variables. The following are not variables: the distance from Los Angeles to Las Vegas, the altitude of Denver, or the number of marriages in Ohio in 2000. These are fixed, unchangeable facts.

As we review the previous research on the topic of love, we find we can develop additional hypothe-

ses that help us investigate the issue further. For example, our reading might show that a common stereotype people hold is the notion that women are more romantic than men. After all, it appears that women enjoy movies about love and romantic novels more than men do.

But wait a minute. We may begin to suspect that common stereotypes may be all wrong, that they are related to traditional gender-role models. We note that in most traditional societies, the male is the breadwinner, while the female is dependent on him for economic support, status, and financial security. Therefore it would seem that when a man marries, he chooses a companion and perhaps a helpmate, while a woman chooses a companion as well as a standard of living. This leads us to hypothesize that in traditional societies men are more likely to marry for love, while women are more likely to marry for economic security.

There is support for this hypothesis. One study designed a scale to measure belief in a romantic ideal in marriage. Males were more likely than females to agree with such statements as "A person should marry whomever he loves regardless of social position" and "As long as they love one another, two people should have no difficulty getting along together in marriage." Men were more likely to disagree than women with the statement "Economic security should be carefully considered before selecting a marriage partner" (Rubin, 1973).

Contrary to popular opinion, men also tend to be more romantic than women (Sprecher & Metts, 1989). This fact should not be that hard to understand when we see that historically men, with their control over resources, have had the luxury to be romantic. Women, who often have not been in charge of their economic destiny, have had to think of men as providers more so than as lovers. We could hypothesize that as gender-role stereotyping declines in the United States and as more and more families come to depend on the income of both spouses, one of two things could happen: Either the importance of romantic love as a basis for marriage will begin to fade, or it will become stronger as the couple now comes together on the basis of mutual attraction as opposed to economic considerations.

Hypotheses involve statements of causality or association. A **statement of causality** says that *something brings about, influences, or changes something else.* "Love between a man and a woman always produces marriage" is a statement of causality.

A **statement of association,** on the other hand, says that *changes in one thing are related to changes in another but that one does not necessarily cause the other.* Therefore, if we propose that "the greater the love relationship between a man and a woman, then the more likely it is they will marry," we are

In one study, males were more likely than females to agree with the statement "A person should marry whomever he or she loves regardless of the person's social position."

making a statement of association. We are noting a connection between love and marriage, but also that one does not necessarily cause the other.

Often hypotheses propose relationships between two different kinds of variables. An **independent variable** *causes or changes another variable*. A **dependent variable** *is influenced by the independent variable*. For example, we might propose the following hypothesis: Men who live in cities are more likely to marry young than are men who live in the country. In this hypothesis the independent variable is the location: Some men live in the city, some live in the country, but presumably their choice of where to live is not influenced by whether they marry young. The age of marriage is the dependent variable because it is possible that the age of marriage depends on where the men live. If research shows that the age of marriage (a dependent variable) is indeed younger among urban men than among rural men, the hypothesis probably is correct. If there is no difference in the age of marriage among urban and rural men—or if it is earlier among rural men—then the hypothesis is not supported by the data. Keep in mind that proving a hypothesis false can be scientifically useful: It eliminates unproductive avenues of thought and suggests other, more productive approaches to understanding a problem.

Even if research shows that a hypothesis is correct, it does not mean the independent variable necessarily produces or causes the dependent variable. For example, if it turns out that we can show that love leads to marriage, we still may not know why. In principle, at least, it is possible to be in love without getting married. However, we still do not know what causes people to take the next step.

Determine the Research Design

Once we have developed our hypotheses, we must design a project in which they can be tested. This is a difficult task that frequently causes researchers a great deal of trouble. If a research design is faulty, it may be impossible to conclude whether the hypotheses are true or false, and the whole project will have been a waste of time, resources, and effort.

A research design must provide for the collection of all necessary and sufficient data to test the stated hypotheses. The important word here is *test*. The researcher must not try to prove a point; rather, the goal is to test the validity of the hypotheses. Although it is important to gather as much information as needed, research designs must guard against the collection of unnecessary information, which can lead to a waste of time and money.

When we design our research project we must also decide which of several research approaches to use. There are four main methods of research used by sociologists: surveys, participant observation, experiments, and secondary analysis. Each has advantages and limitations. Therefore, the choice of methods depends on the questions the researcher hopes to answer.

Surveys A **survey** is *a research method in which a population, or a portion thereof, is questioned in order to reveal specific facts about itself.* Surveys are used to discover the distribution and interrelationship of certain variables among large numbers of people.

The largest survey in the United States takes place every 10 years when the government takes its census. Many of you may have participated in the 2000 census. The U.S. Constitution requires this census in order to determine the apportionment of members to the House of Representatives. In theory, at least, a representative of every family and every unmarried adult responds to a series of questions about his or her circumstances.

From these answers it is possible to construct a picture of the social and economic facts that characterize the American public at one point in time. *Such a study, which cuts across a population at a given time,* is called a **cross-sectional study.** Surveys, by their nature, usually are cross-sectional. If the same population is surveyed two or more times at certain intervals, a comparison of cross-sectional research can give a picture of changes in variables over time. *Research that investigates a population over a period of time is called* **longitudinal research.**

Survey research usually deals with large numbers of subjects in a relatively short time. One of the shortcomings resulting from this method is that investigators are not able to capture the full richness of feelings, attitudes, and motives underlying people's responses. Some surveys are designed to gather this kind of information through interviewing. An **interview** consists of *a conversation between two (or occasionally more) individuals in which one party attempts to gain information from the other(s) by asking a series of questions.*

It would, of course, be ideal to gather exactly the same kinds of information from each research subject. One way researchers attempt to achieve this is through interviews in which all questions are carefully worked out to get at precisely the information wanted (What is your income? How many years of schooling have you had?). Sometimes research participants are forced to choose among a limited number of responses to questions (as in multiple-choice tests). This process results in very uniform data easily subjected to statistical analysis.

A research interview entirely predetermined by a questionnaire (or so-called interview schedule) that is followed rigidly is called a **structured interview.** Structured interviews tend to produce uniform or replicable data that can be elicited time after time by different interviewers.

The use of this method, however, also may allow useful information to slip into "cracks" between the predetermined questions. For example, a questionnaire being administered to married individuals might ask about their age, family background, and what role love played in their reasons for getting married. If, however, we do not ask about social class or ethnicity, we may not find out that these characteristics are very important for our study. If such questions are not built into the questionnaire from the beginning, it is impossible to recover this lost information later in the process when its importance may become apparent.

One technique that can prevent this kind of information loss is the **semistructured,** *or* **open-ended interview,** *in which the investigator asks a list of questions but is free to vary them or even to make up new questions on topics that take on importance in the course of the interview.* This means that each interview will cover those topics important to the research project but, in addition, will yield additional data somewhat different for each subject. Analyzing such diverse and complex data is difficult, but the results are often rewarding.

Interviewing, although it may produce valuable information, is a complex, time-consuming art. Some research studies try to get similar information by distributing questionnaires directly to the respondents and asking them to complete and return them. This is the way the federal government obtains much of its census data. Although it is perhaps the least expensive way of doing social research, it is

One of the shortcomings of survey research is that investigators are not able to capture the full richness of feelings, attitudes, and motives underlying people's responses.

often difficult to assess the quality of data obtained in this manner. For example, people may not answer honestly or seriously for a variety of reasons: They may not understand the questions, they may fear the information will be used against them, and so on. But even data gained from personal interviews may be unreliable. In one study, student interviewers were embarrassed to ask preassigned questions on sexual habits, so they left these questions out of the interviews and filled the answers in themselves afterward. In another study, follow-up research found participants had consistently lied to interviewers.

Participant Observation Researchers entering into a group's activities and observing the group members are engaged in **participant observation.** Unlike sociologists employing survey research, participant observers do not try to make sure they are studying a carefully chosen sample. Rather, they attempt to get to know all members of the group being studied to whatever degree possible.

This research method is generally used to study relatively small groups over an extended period of time. The goal is to obtain a detailed portrait of the group's day-to-day activities, to observe individual and group behavior, and to interview selected informants. Participant observation depends for its success on the relationship that develops between the researchers and research participants. The closer and more trusting the relationship, the more information will be revealed to the researcher—especially the kind of personal information often crucial for successful research.

One of the first and most famous studies employing the technique of participant observation was a study of Cornerville, a lower-class Italian neighborhood in Boston. William Foote Whyte moved into the neighborhood and lived for three years with an Italian family. He published his results in a book called *Street Corner Society* (1943). All the information for the book came from his field notes, which described the behavior and attitudes of the people whom he came to know.

Nearly two decades after Whyte's study, Herbert Gans conducted a participant observation study, published as *The Urban Villagers* (1962), of another Italian neighborhood in Boston. The picture Gans drew of the West End was broader than Whyte's study of Cornerville. Gans included descriptions of the family, work experience, education, medical care, relationships with social workers, and other aspects of life in the West End. Although he covered a wider range of activities than Whyte, his observations were not as detailed.

On rare occasions, participant observers hide their identities while doing research and join groups under false pretenses. Leon Festinger and his stu-

In participant observation, the researcher tries personally to know as many members of the group in question as possible.

dents hid their identities when they studied a religious group preaching the end of the world and the arrival of flying saucers to save the righteous, a group with beliefs similar to the Heaven's Gate group of 1997 (Festinger, Rieken, & Schacter, 1956). However, most sociologists consider this deception unethical. They believe it is better for participant observers to be honest about their intentions and work together with their subjects to create a mutually satisfactory situation. By declaring their positions at the outset, sociologists can then ask appropriate questions, take notes, and carry out research tasks without encountering unnecessary and unethical risks to their study.

Participant observation is a highly subjective research approach. In fact, some scholars reject it outright because the results often cannot be duplicated by another researcher. This method, however, has the benefit of revealing the social life of a group in far more depth and detail than surveys or interviews alone. The participant observer who is able to establish good rapport with the subjects is likely to uncover information that would never be revealed to a survey taker.

The participant observer is in a difficult position, however. He or she will be torn between the need to become trusted (therefore emotionally involved in the group's life) and the need to remain a somewhat detached observer striving for scientific objectivity.

Experiments The most precise research method available to sociologists is the controlled **experiment,** *an investigation in which the variables being studied are controlled and the researcher obtains the results through precise observation and measurement.* Because of their precision, experiments are an attractive means of doing research. Experiments have been used to study patterns of interaction in small groups under a variety of conditions such as stress, fatigue, or limited access to information.

Although experimentation is appropriate for small-group research, most of the issues that interest sociologists cannot be investigated in totally controlled situations. Social events usually cannot be studied in controlled experiments because they simply cannot be controlled. For these reasons, experiments remain the least used research method in sociology.

Secondary Analysis **Secondary analysis** is *the process of making use of data that has been collected by others.* Often the original investigator gathered the data for a specific study. Other times it was merely collected as part of the process of keeping records. The researcher engaged in secondary analysis may use this same data for a new study and a very different purpose. For example, Émile Durkheim in his classic study of suicide in France in the 1890s, engaged in a secondary analysis of official records and developed his theories based on that research.

The enormous amount of material the federal government collects is often used for secondary analysis. The U.S. Bureau of the Census has data on income, birthrates, migration, marriage, divorce, and education levels in the United States that are invaluable for doing social research. Other agencies that provide data that sociologists use for secondary analysis include the Federal Bureau of Investigation, the Department of Labor, the National Centers for Health Statistics, and many others.

An advantage of secondary analysis is that it is useful for collecting or analyzing historical and longitudinal data. It also saves the time and money involved in doing a new study. There are some disadvantages, however. The data may be flawed. You may not know if the original researchers had some biases that are present in the data. Or possibly they were not qualified or knowledgeable enough to collect the data. In addition, the data may not really be suitable for your current study. If you are trying to do a study of economic well-being at different points in history, you may decide to gather that information from certain questions on the U.S. Census of 1950, 1970, and 2000. But which questions should you use? Should you use those on income, those on poverty, those on net worth, those on satisfaction with the political situation, or some other questions? Different types of questions may produce different results and your study may not turn out to be valid because of the choices you make. Therefore the use of secondary analysis requires a thorough understanding of the research process and the problems that can arise from a poorly conceived study. (Table 2-2 compares the advantages and disadvantages of the various research methods.)

Define the Sample and Collect Data

After determining how the needed information will be collected, the researchers must decide what group will be observed or questioned. Depending on the study, this group might be college students, Texans, or baseball players. The particular subset of the population chosen for study is known as a **sample.**

Sampling is *a research technique through which investigators study a manageable number of people, known as the sample, selected from a larger population or group.* If the procedures are carried out correctly, the sample can be called a **representative sample,** or *one that shows, in equivalent proportion, the significant variables that characterize the population as a whole.* In other words, the sample will be representative of the larger population, and the findings from the research will tell us something about the larger group. *The failure to achieve a representative sample is called* **sampling error.**

Suppose you wanted to sample the attitudes of the American public on some issue such as military spending or federal aid for abortions. You could not limit your sample to only New Yorkers or Republicans or Catholics or African Americans or home owners. These groups do not represent the nation as a whole, and any findings you came up with would contain a sampling error.

How do researchers make sure their samples are representative? The basic technique is to use a **random sample**—*to select subjects so that each individual in the population has an equal chance of being chosen.* For example, if we wanted a random sample of all college students in the United States, we might choose every fifth or tenth or hundredth person from a comprehensive list of all registered college students in this country. Or we might assign each student a number and have a computer pick a sample randomly. However, there is a possibility that simply by chance, a small segment of the total college student population would fail to be represented adequately. This might happen with Native-American students, for instance, who make up less than 1% of college students in the United States. For some research

Table 2-2

The Advantages and Disadvantages of Various Research Methods

RESEARCH METHOD	ADVANTAGES	DISADVANTAGES
Social survey	Large numbers of people can be surveyed with questionnaires. Data can be quantified and comparisons among groups can be made. Measures can be taken at different points.	Respondents may give false information or responses they think the researcher wants to hear. Surveys do not leave room for answers that may not fit the standardized categories. There may be a low response rate.
Participant observation	Allows people to be observed in their "natural" environments. Provides a more in-depth understanding of the people being studied. Hypotheses and theories can be developed and changed as the research progresses.	Findings are open to interpretation and subject to researcher bias. The researcher may have an unintended influence on the subjects. It is time consuming. The results may be difficult to replicate.
Experiment	Variables can be isolated and controlled. A cause-and-effect relationship can be found. Easy to replicate	The laboratory setting creates an artificial social environment. The study has to be limited to a few variables. Most things sociologists investigate cannot be studied in a laboratory.
Secondary analysis	Useful for collecting or analyzing historical and longitudinal data. Saves time and money involved in doing a new study.	The data may be flawed. Data may not be suitable for the current study.

purposes this might not matter, but if ethnicity is an important aspect of the research, it would be important to make sure Native-American students were included.

The method used to prevent certain groups from being under- or overrepresented in a sample is to choose a **stratified random sample.** With this technique the population being studied is first divided into two or more groups (or strata) such as age, sex, or ethnicity. A simple random sample is then taken within each group. Finally, the subsamples are combined (in proportion to their numbers in the population) to form a total sample. In our example of college students, the researcher would identify all ethnic groups represented among college students in the United States. Then the researchers would calculate the proportion of the total number of college students represented by each group and draw a random sample separately from each ethnic group. The number chosen from each group should be proportional to its representation in the entire college student population. The sample would still be random, but it would be stratified for ethnicity.

For a study to be accurate, it is crucial that a sample be chosen with care. The most famous example of sampling error occurred in 1936, when *Literary Digest* magazine incorrectly predicted that Alfred E. Landon would win the presidential election. Using telephone directories and automobile registration lists to recruit subjects, *Literary Digest* pollsters sent out more than 10 million straw vote ballots and received 2.3 million completed responses. The survey gave the Republican candidate Landon 55% of the vote and Franklin D. Roosevelt only 41% (the remaining 4% went to a third candidate). Based on this poll, the *Digest* confidently predicted Landon's victory. Instead Landon has become known as the candidate who was buried in a landslide vote for Roosevelt (Squire, 1988).

How could this happen? Two major flaws in the sample accounted for the mistake. First, although the *Literary Digest* sample was large, it was not representative of the nation's voting population because it contained a major sampling error. During the depression years, only the well-to-do could afford telephones and automobiles, and these people were likely to vote Republican.

The second problem with the study was the response rate. Interestingly, of those who claimed to have received a *Literary Digest* ballot, 55% claimed they would have voted for Roosevelt and 44% for Landon. If these people had actually voted in the poll, the *Digest* would have predicted the correct winner. As it turned out, there was a low response rate, and those who did respond were generally better educated, wealthier people who could afford cars and telephones and who tended to be Landon supporters (Squire, 1988).

A stratified random sample is used to prevent certain groups from being underrepresented or overrepresented in a sample.

The outcome of the election was not entirely a surprise to everyone. A young pollster named George Gallup forecast the results accurately. He realized that the majority of Americans supported the New Deal policies proposed by the Democrats. Gallup's sample was much smaller, but far more representative of the American public than that of the *Literary Digest*. This points out that the representativeness of the sample is more important than its size.

A similar problem occurred in January of 2000 when the *Kansas City Star* reported that "Hundreds of Roman Catholic priests across the United States have died of AIDS-related illnesses, and hundreds more are living with HIV, the virus that causes the disease."

With great fanfare, the paper broke the first of a sure-to-be-controversial three-part series on "AIDS in the Priesthood," a nationwide story that challenged the Roman Catholic Church's handling of what the *Star* characterized as the "strikingly high incidence of AIDS-related disease and death among the church's clerics." Heavy nationwide coverage of this story followed.

The study was based on surveys sent to 3,013 priests randomly selected from an alphabetical listing in the *Official Catholic Directory* of 1999, included all 50 states and all Catholic dioceses except one. Eight hundred and one priests (27%) responded

to the survey. This means that nearly three of four priests sent the survey failed to answer it. Normally, when a response rate is this low, follow-up surveys are conducted to increase the returns or at least to learn whether the minority who responded were representative. No follow-up interviews were conducted because of the need for confidentiality.

The *Star* noted that of those who responded, 15% said they were homosexual and 5% bisexual. An additional 8% identified themselves as bisexual or "other," leaving only 77% self-identified heterosexuals. But we have no idea whether the minority who responded were unusually concerned about AIDS, differentially open to questions of personal sexuality, or even more likely to have a homosexual orientation than the 2,212 nonrespondents.

A closer look at the above percentages shows that the *Star* received a total of four responses from priests who said that they definitely had AIDS and three more who feared they might have AIDS. If we simply use this result from four priests as though it were a representative sample projected onto 46,000 priests, the result is roughly 60 priests, nationwide, who would have AIDS and know it. If we further presume that the rate of death from AIDS is going down among priests just as it is in the general population, and if we can further judge that all 60 priests with AIDS today are not likely to die during a single year, we should quickly realize that while 60 priests who will someday die of AIDS is a tragedy, it is also not "hundreds of priests with AIDS," as the *Star* noted. The fact that the sample was not representative means that the *Star* had no solid data on the true number.

A major problem with this survey was that the priests were able to decide to participate or not in the survey. This meant that the sample was unrepresentative, and the *Star* should not project its results onto the nationwide population of priests. Second, the finding of only four priests who say they have AIDS and three more who said they might have AIDS is a slender thread upon which to base national projections.

What is really disturbing, however, is that the *Kansas City Star* presented its study as valid research and that this survey—which has little, if any, real value—is taken seriously by large numbers of people. For a further discussion of deceptive research, see "Sociology at Work: How to Spot a Bogus Poll."

Researcher Bias One of the most serious problems in data collection is **researcher bias,** *the tendency for researchers to select data that support, and to ignore data that seem to go against, their hypotheses.* We see this quite often in mass media publications. They may structure their study to produce the results they wish to obtain, or they may only publicize information that supports their viewpoint.

Researcher bias often takes the form of a self-fulfilling prophecy. A researcher who is strongly inclined toward one point of view may communicate that attitude through questions and reactions in such a way that the subject fulfills the researcher's expectations. For example, a researcher who is trying to prove an association between poverty and antisocial behavior might question low-income subjects in a way that would indicate a low regard for their social attitudes. The subjects, perceiving the researcher's bias, might react with hostility and thus fulfill the researcher's expectations.

Researcher bias might have been the cause of the problems with the data presented by sociologist Lenore Weitzman in her book *The Divorce Revolution* (1985). Weitzman theorized that changes in divorce laws resulting in an increase of no-fault divorces had produced a systematic impoverishment of divorced women and children. Weitzman argued that men usually make more money than women and that in divorce women usually get the children; therefore, treating men and women equally in divorce cheats the wife and children. Weitzman presented data from her study to prove it. She noted that, on the average, divorced women experience a 73% decline in their standard of living in the first

year of divorce, while divorced men experience a 42% rise in their standard of living.

Weitzman's study received a great deal of attention and was cited in 175 newspaper and magazine articles. It was also cited in 348 social science articles, 250 law review articles, 24 state appellate court cases, and one Supreme Court case. The statistic even appeared in President Clinton's 1996 budget. Anne Colby, director of the Murray Research Center at Radcliffe, where the original data are held, called it "one of the most widely quoted statistics in recent history."

Part of the reason the inaccurate data in the study was cited so often may have been that it could be used for both conservative and liberal political purposes. Conservative groups could claim that no-fault divorce should be abolished because it made the breakup of marriage too easy and produced the impoverishment of women and children. Liberal groups could use to data to show that women needed greater protection and more rights after divorce to prevent the inequities in divorce.

Greg Duncan and Saul Hoffman (1988), two researchers who had been tracking the effect of divorce on income for two decades, found the changes following divorce nowhere near as dramatic as Weitzman described. Although they did find a 30% decline in women's living standards and a 10 to 15% improvement in men's living standards, they also found that five years after divorce, the average woman's living standard was actually higher, mainly because many of them had remarried.

What baffled Duncan and Hoffman most was that Weitzman had used their methods to arrive at her numbers. They were not able to duplicate her findings and noted as much. A Census Bureau study eventually confirmed Duncan and Hoffman's study (Bianchi, 1991).

Even though questions continued to be raised about the research, Weitzman refused to allow researchers access to her data, claiming that she wanted to correct some errors in the master computer file before doing so. The refusal to share the data went on for years. Richard R. Peterson of the Social Science Research Council wanted to duplicate the study, but Weitzman would not allow him access to the data either. Peterson appealed to the National Science Foundation (NSF), the organization that had funded Weitzman's research. The NSF threatened to declare Weitzman ineligible for federal grants in the future if she did allow Peterson to examine the data. A year and a half later, Weitzman sent her data to the Murray Research Center at Radcliffe College, which made the research available to others.

After examining the data, Peterson (1996a) came up with figures more in line with other national

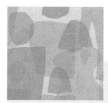

Sociology at Work

How to Spot a Bogus Poll

Opinion surveys can look convincing and be completely worthless. But asking four simple questions of any poll can separate the good numbers from the trash. Politicians use opinion polls as verbal weapons in campaign ads. Journalists use them as props to liven up infotainment shows. Executives are more likely to pay attention to polls when the numbers support their decisions. But this isn't how polls are meant to be used. Opinion polls can be a good way to learn about the views Americans hold on important subjects, but only if you know how to cut through the contradictions and confusion.

Conducting surveys is difficult. It is especially difficult to take a meaningful survey of public opinion, because opinion is a subjective thing that can change rapidly from day to day. Poll questions sometimes produce conflicting or meaningless results, even when they are carefully written and presented by professional interviewers to scientifically chosen samples. That's why the best pollsters sweat the details on the order and wording of questions, and the way data are coded, analyzed, and tabulated. . . . So the next time a poker-faced person tries to give you the latest news about how Americans feel, ask some pointed questions of your own:

Did You Ask the Right People?

Even when you start out with a representative sample, you could end up with a biased one. This is a risk all pollsters take, but some particular methods lend themselves to greater error. For example,

readers of women's magazines are frequently asked to fill out surveys on weighty subjects like crime and sexual behavior. Not only do such polls ignore the opinions of nonreaders, they are biased toward readers who take the trouble to fill out and return the questionnaire, usually at their own expense.

Likewise, many have criticized online polling because Internet users tend to be wealthier, more educated, and more likely to be male than the larger population.

Television news and entertainment shows get into the act by posting toll-free or even toll numbers that viewers can call to "vote" on an issue. These samples are not only biased, they are prone to "ballot-stuffing" by enthusiasts. In other words, viewers who call 12 times get 12 votes.

Conflicts make news. When journalists are trying to liven up a boring political story, they need angry, well-informed citizens like a fish needs water. This is one reason why older men may be quoted more often than other groups. Those aged 50 and older are more likely than younger adults to follow news stories "very closely," according to the Pew Research Center for The People & The Press in Washington, D.C. Men are more likely than women to follow stories about war, business, sports, and politics.

In the last decade, angry white men have dominated media programs designed to give ordinary people a chance to speak out in public. Two-thirds of regular listeners to political talk radio programs are men, according to a 1996 poll taken by Roper Starch Worldwide for the Media Studies Center in

---

studies, with women's standard of living decreasing 27% in the first year after divorce and men's increasing by 10%. This led Peterson to conclude:

> Weitzman's erroneous results have lent support to those seeking to restrict access to no-fault divorce. Whether or not one agrees with those restrictions, . . . policy arguments about no-fault divorce legislation should not be clouded by the use of the erroneous figures reported in *The Divorce Revolution*. While social scientists cannot control how results are reported or used in policy debates, we need to take responsibility for correcting errors that may lead to faulty conclusions or to misguided policies. (Peterson, 1996b)

Felice Levine, executive officer of the American Sociological Association, has pointed out that when

sociologists communicate with the public, they need to make it clear how reliable their results are. It is important to let the public know whether the sociologist's fellow researchers have been able to duplicate the data (Cohen, 2000).

One of the standard means for dealing with research bias is to use **blind investigators,** or *investigators who do not know whether a specific subject belongs to the group of actual cases being investigated or to a comparison group.* For example, in a study on the causes of child abuse, the investigator looking at the children's family background would not be told which children had been abused and which were in the nonabused comparison group. Sometimes double-blind investigators are used. **Double-blind investigators** *are kept uninformed not only of the kinds of subjects (case subjects or comparison group subjects) they are studying but also*

New York City. Republicans outnumber Democrats three to one in the talk-radio audience, and 89% of listeners are white, compared with the national average for voters of 83%. Three in five regular listeners to political talk radio perceive a liberal bias in the mainstream media, compared with one in five nonlisteners.

What is the Margin of Error?

No matter how carefully a survey sample is chosen, there will still be some margin of error. If you selected ten different sets of 1,000 people using the same rules and asked each group the same question, the results would not be identical. The difference between the results is sampling error. Statisticians know that the error is equally likely to be above or below the true mark, and that larger samples have smaller margins of error if they are properly drawn. They are also able to estimate the margin of error, or the amount by which the result could be above or below the truth. Sampling error will always exist unless you survey every member of a population. If you do that, you have conducted a census.

Sampling error is one reason why two professionally conducted polls can show different results and both be correct. . . . Reputable surveys report a margin of error—usually of 3 or 4 percentage points—at a particular confidence level—typically 95%. This means that 5% of the time, or 1 time in 20, the poll's results will not be reliable. The other 95% of the time, it is within 3 or 4 percentage points of the "truth."

Which Came First?

The order in which questions are asked can have a big effect on the results. Most people want to appear consistent to others and to be consistent in their own minds. When a pollster asks a series of related questions, this desire can lead people to take positions they might not have taken if they were asked only one question. Neither way produces an obviously "correct" response, but the results are different. One way to handle this problem is to rotate the order of questions. Then the degree of differences due to question order can be described and interpreted. But not everyone pays heed to such fine distinctions.

What Was the Question?

"Do you want union officials, in effect, to decide how many municipal employees you, the taxpayer, must support?" Well, do you? This question, taken from an actual survey, is obviously biased. The results might make good propaganda for an anti-union group, but they are totally bogus as a poll. So before you pass a survey finding on to others, or even believe it yourself, be sure to look at the actual question.

When you're presented with a new survey or a used car, it helps to ask a few key questions before you buy. But for all their flaws, surveys are essential to the work of politicians, journalists, and businesspeople.

Source: Excerpted from B. Edmondson, "How to Spot a Bogus Poll," *American Demographics, 18*(10) October 1996, pp. 10, 12–15.

of the hypotheses being tested. This eliminates any tendency on their part to find cases that support—or disprove—the research hypothesis.

Analyze the Data and Draw Conclusions

In its most basic sense, **analysis** is *the process through which large and complicated collections of scientific data are organized so that comparisons can be made and conclusions drawn.* It is not unusual for a sociological research project to result in hundreds of thousands of individual pieces of information. By itself this vast array of data has no particular meaning. The analyst must find ways to organize such data into useful categories so that the relationships that exist can be determined. In this way the hypotheses forming the core of the research can be tested, and new hypotheses can be formulated for further

investigation. (One important device to aid in the analysis of data is the table, which is explained in "Sociology at Work: How to Read a Table.")

Sociologists often summarize their data by calculating central tendencies or averages. Actually, sociologists use three different types of averages: the mean, the median, and the mode. Each type is calculated differently, and each can result in a different figure. Suppose you are studying a group of 10 college students whose verbal SAT scores are as follows:

450	690	280	450	760
540	520	450	430	530

Although you can report the information in this form, a more meaningful presentation would give some indication of the central tendency of the 10 SAT scores. The three measures of central tendency,

Sociology at Work

How to Read a Table

Statistical tables are used frequently by sociologists both to present the findings of their own research and to study the data of others. We will use the accompanying table to outline the steps to follow in reading and interpreting a table.

1. *Read the title.* The title tells you the subject of the table. This table presents data on life expectancy at birth in various countries for people of both sexes.
2. *Check the source.* At the bottom of a table you will find its source. In this case, the source is the *Statistical Abstract of the United States: 2000.* Knowing the source of a table can help you decide whether the information it contains is reliable. It also tells you where to look to find the original data and how recent the information is. In our example, the source is both reliable and recent. If the source were the 1958 *Abstract,* it would be of limited value in telling you about life expectancy in those countries today. Improvements in health care, control of epidemic diseases, or national birth control programs all are factors that may have altered life expectancy drastically in several countries since 1958. Likewise, consider a table of data about AIDS cases in Thailand. If its source were a government agency (which might be trying to alleviate the fears of tourists about the rampant spread of the disease in that country), you might well be skeptical about the reliability of the information in the table.
3. *Look for headnotes.* Many tables contain headnotes directly below the title. These may explain how the data were collected, why certain variables (and not others) were studied, why the data are presented in a particular way, whether some data were collected at different times, and so on. In our table on life expectancy, the headnote explains that the numbers in the table refer to the average number of additional years a person can expect to live at birth.
4. *Look for footnotes.* Many tables contain footnotes that explain limitations or unusual circumstances surrounding certain data. For example, in this table, the footnote explains that the data for Serbia is influenced by certain political circumstances.
5. *Read the labels or headings for each row and column.* The labels will tell you exactly what information is contained in the table. It is essential that you understand the labels—both the row headings on the left and the column headings at the top. Here the row headings tell you the names of the countries being compared for life expectancy. For each group, life expectancy is given. Note the units being used in the table. In this case the units are years. Often the figures represent percentages or rates. Many population

like the three different types of averages mentioned earlier, are the mean, median, and mode.

The *mean* is what is commonly called the *average.* To calculate the mean, you add up all the figures and divide by the number of items. In our example, the SAT scores add up to 5,100. Dividing by 10 gives a mean of 510.

The *median* is the figure that falls midway in a series of numbers—there are as many numbers above it as below it. Because we have ten scores—an even number—in our example, the median is the mean (the average) of the fifth and sixth figures, the two numbers in the middle. To calculate the median, rearrange the data in order from the lowest to highest (or vice versa). In our example, you would list the scores as follows: 280, 430, 450, 450, 450, 520, 530, 540, 690, 760. The median is 485—midway between the fifth score (450) and the sixth score (520).

The *mode* is the number that occurs most often in the data. In our example the mode is 450.

The three different measures are used for different reasons, and each has its advantages and disadvantages. The mean is most useful when a narrow range of figures exists, as it has the advantage of including all the data. It can be misleading, however, when one or two scores are much higher or lower than the rest. The median deals with this problem by not allowing extreme figures to distort the central tendency. The mode enables researchers to show which number occurs most often. Its disadvantage is that it does not give any idea of the entire range of data. Realizing the problems inherent in each average, sociologists often state the central tendency in more than one form.

Scientists usually are careful in drawing conclusions from their research. One of the purposes of drawing conclusions from data compiled in the course of research is the ability to apply the information gathered to other, similar situations. Problems thus may develop if there are faults in the

and crime statistics are given in rates per 100,000 people.

6. *Examine the data.* Suppose you want to find the life expectancy of a newborn baby in the United States. First look down the row at the left until you come to "United States." Then look across the columns until you come to the number. Reading across, you discover that on average, a baby born in the United States in 2000 can expect to live another 77.1 years.

7. *Compare the data.* Compare the data in the table both horizontally and vertically. Suppose you want to find in which country people can expect to live longest from birth. Looking down the life expectancy column, we find that people born in Japan can expect to live longest—80.7 years. Among these thirteen nations the United States ranks fourth, behind Japan, Canada and Spain. A baby born in Malawi has the shortest life expectancy—only 37.6 years.

8. *Draw conclusions.* Draw conclusions about the information in the table. After examining the data in the table, you might conclude that a person born in a relatively developed country (Canada, Japan, Spain, United States) is likely to live much longer than is someone born in a poorer nation (Angola, India, Malawi, Sri Lanka, and Zimbabwe).

9. *Pose new questions.* The conclusions you reach might well lead to new questions that could prompt further research. Why, you might want to know, do people in Malawi have shorter life expectancies than people in Japan? What causes the gap in life expectancy between the rich and poor nations?

Table 2-3

Life Expectancy at Birth for Selected Countries

Country	Expectation of Life at Birth, 2000
Angola	38.3
Brazil	62.9
Canada	79.4
Ghana	57.4
India	62.5
Japan	80.7
Malawi	37.6
Poland	73.2
Serbia*	73.5
Spain	78.8
Sri Lanka	71.8
United States	77.1
Zimbabwe	37.8

* Serbia and Montenegro have asserted the formation of a joint independent state, but this entity has not been recognized by the United States. Data in this table are for Serbia alone.

Source: *Statistical Abstract of the United States: 2000,* 120th ed. (p. 826), U.S. Bureau of the Census, Washington, DC, 2001.

research design. For example, the study must show **validity**—that is, *the study must actually test what it was intended to test.* If you want to say one event is the cause of another, you must be able to rule out other explanations to show that your research is valid. Suppose you conclude that marijuana use leads to heroin use. You must show that it is marijuana use and not some other factor, such as peer pressure or emotional problems, that leads to heroin use.

The study must also demonstrate **reliability**—that is, *the findings of the study must be repeatable.* To demonstrate reliability we must show that research can be replicated—repeated for the purpose of determining whether initial results can be duplicated. Suppose you conclude from a study that whites living in racially integrated housing projects, who have contact with African Americans in the same projects, have more favorable attitudes toward blacks than do whites living in racially segregated housing projects. If you or other researchers carry out the same study in housing projects in various cities throughout the country and get the same results, the study is reliable.

It is highly unlikely that any single piece of research will provide all the answers to a given question. In fact, good research frequently leads to the discovery of unanticipated information requiring further research. One of the pleasures of research is that ongoing studies keep opening up new perspectives and posing further questions.

Prepare the Research Report

Research that goes unreported is wasted. Scientific progress is made through the accumulation of research that tests hypotheses and contributes to the ongoing process of bettering our understanding of the world. Therefore, it is usual for agencies that fund research to insist that scientists agree to share their findings.

Scientists generally publish their findings in technical journals. If the information is relevant to the public, many popular and semiscientific publications will report these findings as well. It is especially important that research in sociology and other social sciences be made available to the public, because much of this research has a bearing on social issues and public policies.

Unfortunately, the general public is not always cautious in interpreting research findings. Special interest groups, politicians, and others who have a cause to plead are often too quick to generalize from specific research results, frequently distorting them beyond recognition. This happens most often when the research focuses on something of national or emotional concern. It is therefore important to double-check reports of sociological research appearing in popular magazines with the original research.

For a discussion of the different definitions of truth, see "Controversies in Sociology: Truth in the Courtroom versus Truth in the Social Sciences."

Objectivity in Sociological Research

Max Weber believed that the social scientist should describe and explain what is, rather than prescribe what should be. His goal was a value-free approach to sociology. More and more sociologists today, however, are admitting that completely value-free research may not be possible. In fact, one of the trends in sociology today that could ultimately harm the discipline is that some sociologists who are more interested in social reform than social research have abandoned all claims to objectivity.

Sociology, like any other science, is molded by factors that impose values on research. Gunnar Myrdal (1969) lists three such influential factors: (1) the scientific tradition within which the scientist is educated; (2) the cultural, social, economic, and political environment within which the scientist is trained and engages in research; and (3) the scientist's own temperament, inclinations, interests, concerns, and experiences. These factors are especially strong in sociological research because the researcher usually is part of the society being studied.

Does this mean that all science—sociology in particular—is hopelessly subjective? Is objectivity in sociological research an impossible goal? There are no simple answers to these questions. The best sociologists can do is strive to become aware of the ways in which these factors influence them and to make such biases explicit when sharing the results of their research. We may think of this as disciplined,

or "objective," subjectivity, and it is a reasonable goal for sociological research.

Another problem of bias in sociological research relates to the people being studied rather than to the researchers themselves. The mere presence of investigators or researchers may distort the situation and produce unusual reactions from subjects who now feel special because of their selection for study.

Ethical Issues in Sociological Research

All research projects raise fundamental questions. Whose interests are served by the research? Who will benefit from it? How might people be hurt? To what degree do subjects have the right to be told about the research design, its purposes, and possible applications? Who should have access to and control over research data after a study is completed—the agency that funded the study, the scientists, the subjects? Should research subjects have the right to participate in the planning of projects? Is it ethical to manipulate people without their knowledge in order to control research variables? To what degree do researchers owe it to their subjects not to invade their privacy and to keep secret (and, therefore, not report anywhere) things that were told in strict confidence? What obligations do researchers have to the society in which they are working? What commitments do researchers have in supporting or subverting a political order? Should researchers report to legal authorities any illegal behavior discovered in the course of their investigations? Is it ethical to expose subjects to such risk by asking them to participate in a study?

In the 1960s the federal government began to prescribe regulations for "the protection of human subjects." These regulations are designed to force scientists to consider one central issue: how to judge and balance the intellectual and societal benefits of scientific research against the actual or possible physical and emotional costs to the people being studied. This issue creates three potential dilemmas for the researcher.

The first situation concerns the degree of permissible risk, pain, or harm. Suppose a study that temporarily induces severe emotional distress promises significant benefits. The researcher may justify the study. However, we may wonder whether the benefits will be realized or whether they justify the potential dangers to the subjects, even if the volunteers know what to expect and all possible protective measures have been taken.

A second dilemma is the extent to which subjects should be deceived in a study. It is now necessary for researchers to obtain, in writing, the "informed consent" of the people they study. Questions still

Controversies in Sociology

Truth in the Courtroom versus Truth in the Social Sciences

Lawyers and district attorneys often announce in their opening statements that the upcoming trial will be a search for the "truth." One of the goals of sociology specifically and science in general is to uncover the "truth" and bring about a fuller understanding of an issue. Is there a difference between the search for truth in the courtroom and the search for truth in the social and natural sciences?

The sociologist seeks to discover the truth by interviewing subjects, examining data, and piecing together records. Using this information, generalizations are made about what is likely to occur in the future and new hypotheses are proposed and tested. Even though it may seem objective, this search for the truth is still influenced by personal biases and political orientations.

Sometimes a scientist is looking for one thing and discovers something totally different and unexpected. Other times a discovery may be made by accident or because the original experiment was done incorrectly. None of this invalidates the truth of what was discovered. For example, in 1895 Wilhelm Roentgen accidentally discovered X-rays, a type of radiation that can pass through material that ordinary light cannot. X-rays revolutionized the fields of physics and medicine, and for this discovery Roentgen received the Nobel Prize in 1901. Penicillin, too, was an accidental discovery, as were many others that have changed our lives.

Even if a scientist cheats, exaggerates, or even plagiarizes his or her findings, ultimately the truth or falsity of the information is what remains. In the courtroom there is something known as the exclusionary rule, which throws out any evidence that has not been obtained legally. There is no such thing as an exclusionary rule in science, and truth achieved by unfair means is more likely to endure than falsity achieved by fair means.

In the courtroom, truth is entrusted to a jury of laypeople selected randomly from the population for their lack of information about the facts. Efforts are made to make them representative of the general population in terms of race, gender, or social class. The task of discovering truth in science is entrusted largely to trained experts who have studied the subject for years and are intimately familiar with the relevant facts and theories. They may differ from the general population in many ways.

Finally, the "truths" discovered by science are always subject to reconsideration based on new evidence. There is no such thing as a statute of limitation on gathering evidence. Scientific research involves a continuous process of building and refining on other people's findings. That is not the case in the courtroom. You cannot be tried for the same crime twice even if new evidence suggests guilt. This is known as the prohibition against double jeopardy, and courts strive for a "finality" in a trial and do not want a case to go on indefinitely. The courts do not want cases to be reopened and retried.

Scientific inquiry is basically a search for objective truth. Objective truth is validated by accepted, verifiable, and replicable tests. Truth in a criminal trial is quite different. Truth, although *one* important goal of the criminal trial, is not the *only* goal. If it were, judges would not instruct jurors to acquit a defendant who they believe "probably" did it. The requirement is that guilt must be proved "beyond a reasonable doubt." That is not the same as the search for objective truth, because it prefers one kind of truth over another. The preferred truth is that the defendant did *not* do it, and we require that the jurors err on the side of that truth, even in cases where it is probable that he did do it.

The next time you feel frustrated by a highly publicized trial where a defendant who may seem guilty is acquitted, remember that you are applying a standard of truth that is more typical of science than the courtroom. The courts must be as concerned about Constitutional issues and the fairness of the process as they must be about the search for the truth. Many times, preserving the integrity of the court system requires a decision that goes against what may be the "truth" in the outside world.

Sources: *Reasonable Doubts* (pp. 34–48), by A. M. Dershowitz, 1996, New York: Simon and Schuster; and an interview with Alan M. Dershowitz, March 1996.

arise, however, about whether subjects are informed about the true nature of the study and whether, once informed, they can freely decline to participate.

A third problem in research studies concerns the disclosure of confidential or personally harmful information. Is the researcher entitled to delve into personal lives? What if the researcher uncovers some information that should be brought to the attention of the authorities? Should confidential information be included in a published study (Gans, 1979)?

Rik Scarce, a doctoral student in sociology, had a personal confrontation with the issues of research ethics and confidentiality. He had been doing research on radical environmental groups when a group known as the Animal Liberation Front broke into a research facility at Washington State University, released animals, destroyed computer files, and did general damage to the lab.

A federal grand jury was set up to investigate the break-in and Scarce, although not a suspect, was summoned to appear before the group. Law enforcement officials thought Scarce, while doing his research, might have come across information that would lead to the guilty parties, and they wanted him to testify about his confidential sources.

Scarce answered those questions that he thought would not violate his agreement of confidentiality with his subjects. When he was asked to disclose the names of the actual subjects, he refused. He pointed out that the American Sociological Association's Code of Ethics states that confidentiality *must* be maintained, even at the risk of being jailed. A federal judge did not accept Scarce's ethical obligation and ordered him held in contempt of court until he revealed his confidential reports. Scarce spent 159 days in the Spokane County Jail and was not released until the court realized that he would not reveal his sources. No other researcher has ever been incarcerated for this length of time, and Scarce's commitment to his ethical obligation caused him to pay a substantial price (Monaghan, 1993; Scarce, 1994).

Every sociologist must grapple with these questions and find answers that apply to particular situations. However, two general points are worth noting. The first is that social research rarely benefits the research subject directly. Benefits to subjects tend to be indirect and delayed by many years—as when new government policies are developed to correct problems discovered by researchers. Second, most subjects of sociological research belong to groups with little or no power, because they are easier to find and study. It is hardly an accident that poor people are the most studied, rich people the least. Therefore, research subjects typically have little control over how research findings are used, even though such applications may affect them greatly. This means that sociologists must accept responsibility for the fact that they have recruited research subjects who may be made vulnerable as a result of their cooperation. It is important that researchers establish safeguards limiting the use of their findings, protecting the anonymity of their data, and honoring all commitments to confidentiality made in the course of their research.

The ideal relationship between scientist and research participant is characterized by openness and honesty. Deliberately lying in order to manipulate the participant's perceptions and actions goes directly against this ideal. Yet often researchers must choose between deception and abandoning the research. With few, if any, exceptions, social scientists regard deception of research participants as a questionable practice to be avoided if at all possible. It diminishes the respect due to others and violates the expectations of mutual trust on which organized society is based. When the deceiver is a respected scientist, it may have the undesirable effect of modeling deceit as an acceptable practice. Conceivably, it may contribute to the growing climate of cynicism and mistrust bred by widespread use of deception by important public figures.

It is useful for human beings to seek to understand themselves and the social world in which they live. Sociology has a great contribution to make to this endeavor, both in promoting understanding for its own sake and in providing social planners with scientific information with which well-founded decisions can be made and sound plans for future development adopted. However, sociologists must also shoulder the burden of self-reflection—of seeking to understand the role they play in contemporary social processes while at the same time assessing how these social processes affect their findings.

For a discussion of combining art and science in research, see "Technology and Society: Is There Beauty in Research?"

SUMMARY

To produce a valid study of the social world, we cannot approach it solely on the basis of personal experience and perceptions; rather, we must do so scientifically. Science has two main goals: (1) to describe in detail particular things or events and (2) to propose and test theories that help us understand these things or events. Like detectives, sociologists want to know "Why did it happen?" and "Under what circumstances is it likely to happen again?" The scientific research process involves a sequence of steps that must be followed to produce a valid study.

The research process begins by defining the problem. Next the researcher attempts to find out as much as possible about previous studies on the same topic. The next step is to develop a hypothesis, or a testable statement about the relationship between two or more empirical variables.

Hypotheses are tested by constructing a research design, or a strategy for collecting appropriate data. Sociologists use three main research designs: surveys, participant observation, and experiments.

The particular subset of the population chosen for study is known as a sample. Sampling is a research technique through which investigators study a manageable number of people selected from a larger population. If the procedures are carried out

Technology and Society

Is There Beauty in Research?

Can the research process produce things of beauty? Aesthetics and beauty were probably not the first things that came to mind as you decided to tackle this research methods chapter. Yet take a moment to read these quotes from scientists in various disciplines, as they recount the beauty they encountered in searching for answers to scientific problems. For these people, the art created by science is very much like the art created through music or painting.

■ For an observer, aesthetic pleasure can be found if the artist has conveyed a new way of ordering ideas, or if the artist has supplied the missing part that completes the observer's own creative experience.—William Lipscomb (Nobel Prize winner in chemistry, 1976).

■ The deep satisfaction found in scientific work, akin to the delight derived from genuine art, is one of the fundamental human emotions.—Ilse Rosenthal-Schneider, a student of Albert Einstein.

■ Our science, which we loved above everything, had brought us together. It appeared to us as a flowering garden. In this garden there were well-worn paths where one might look around at leisure and enjoy oneself without effort, especially at the side of a congenial companion.

But we also liked to seek out hidden trails and discovered many an unexpected view, which was pleasing to our eyes; and when the one pointed it out to the other, and we admired it together, our joy was complete.—David Hilbert, a highly acclaimed mathematician.

■ The scientist does not study nature because it is useful to do so. He studies it because he takes pleasure in it; and he takes pleasure in it because it is beautiful. If nature were not beautiful, it would not be worth knowing and life would not be worth living.—Poincare, another well known mathematician and co-discoverer of the special theory of relativity pointed out that:

■ The measure in which science falls short of art is the measure in which it is incomplete as science.—J. W. N. Sullivan, the author of biographies of Ludwig von Beethoven and Sir Issac Newton.

In this chapter then, you are not just studying research methods, you are studying the creative process of answering social science questions.

Source: William N. Lipscomb, "Aesthetic Aspects of Science," in Deane W. Curtin, (ed.) *The Aesthetic Dimension of Science*, New York: Philosophical Library, 1982, pp. 1–24.

correctly, the sample will be a representative sample, or one that shows, in equivalent proportion, the significant variables that characterize the population as a whole. Failure to achieve a representative sample is known as sampling error.

Sociology, like any other science, is molded by factors that impose values on research. Thus completely value-free research may not be possible. Nevertheless, objectivity, or a kind of disciplined subjectivity, is a reasonable goal for sociological research. This means that researchers strive to become aware of the ways in which values influence them and to make such biases explicit when sharing the results of their research.

The central ethical concern in research on human participants is how to judge and balance the intellectual and societal benefits of scientific research against the actual or possible physical and emotional costs to the research participants.

Sociology has a great contribution to make to understanding humans and the social world in which they live, but sociologists must also shoulder the

burden of seeking to understand the role they play in contemporary social processes while simultaneously assessing how these social processes affect their findings.

INTERNET RESOURCES

Go to http://web-enhanced.thomsonlearning.com which features Web links relevant to this chapter to expand and enhance the learning experience. This site will be monitored and updated periodically to ensure the links are correct and live.

At Virtual Society: The Wadsworth Sociology Resource Center, http://sociology.wadsworth.com, you will find a Career Center, Sociology in the News, Research Online, InfoTrac College Edition, Virtual Tours in Sociology, and a variety of book-specific student and instructor resources.

CHAPTER TWO STUDY GUIDE

KEY CONCEPTS

Match each concept with its definition, illustration, or explanation presented below.

a. Longitudinal study
b. Stratified random sample
c. Dependent variable
d. Structured interview
e. Sample
f. Validity
g. Survey
h. Operational definition
i. Semistructured interview
j. Random sample
k. Unrepresentative sample
l. ASA Code of Ethics

m. Statement of causality
n. The order of questions
o. Representative sample
p. Participant observation
q. Review previous research
r. Statement of association
s. Cross-sectional study
t. Research process
u. Reliability
v. Double-blind investigations
w. Sampling error
x. Blind investigators

y. Empirical question
z. Researcher bias
aa. Experiment
bb. Analysis
cc. One goal of science
dd. Hypothesis
ee. Interview
ff. Sampling
gg. Independent variable
hh. Secondary analysis
ii. Variable

_____ 1. Problem with the *Kansas City Star* newspaper report on AIDS among Catholic priests, which affected the validity of the report.

_____ 2. A question that can be answered by observing and analyzing the world as it is known.

_____ 3. A definition of an abstract concept in terms of the observable features that describe the thing being investigated.

_____ 4. A testable statement about the relationships between two or more variables.

_____ 5. Anything that changes or assumes different amounts or values.

_____ 6. An assertion that something brings about, influences, or changes something else.

_____ 7. An assertion that changes in one thing are related to changes in another, but do not necessarily cause the changes in that other.

_____ 8. A variable that changes in response to changes occurring in the independent variable.

_____ 9. A variable that changes for reasons that have nothing to do with the dependent variable.

_____ 10. A research method in which a population, or a portion of it, is questioned in order to reveal specific facts about itself.

_____ 11. A study that cuts across a population at a given time.

_____ 12. An investigation that investigates a population over a period of time.

_____ 13. The general term for a data collection technique where the researcher gains information from another person by asking a series of questions.

_____ 14. A research interview entirely predetermined by a questionnaire (schedule), which is followed rigidly.

_____ 15. The researcher asks a list of questions; however, he or she is free to vary them or even create new questions on topics that become important as the interview unfolds.

_____ 16. Researchers entering into a group's activities and observing the group members.

_____ 17. Research in which the variables being studied are controlled and the researcher obtains results though precise observations and measurements.

_____ 18. The particular subset of the population chosen for the study.

_____ 19. A research technique enabling investigators to study a manageable number of people chosen from a larger population or group.

_____ 20. A sample showing, in equivalent proportion, the significant variables characterizing the whole population.

_____ 21. The failure to achieve a representative sample.

_____ 22. Sampling selecting research participants so that each individual in the population has an equal chance of being selected.

_____ 23. Conditions where researchers are kept uninformed about the hypotheses being tested and the kinds of subjects being studied.

_____ 24. The tendency for researchers to select data that support and ignore data that seem to go against their hypotheses.

_____ 25. Investigators who are not aware of whether a subject belongs to the group under investigation or to a comparison group.

_____ 26. Sampling technique used to prevent certain groups from being under- or overrepresented in the sample.

_____ 27. Procedures for organizing large and complicated sets of scientific data in order that comparisons can be made and conclusions drawn.

_____ 28. The degree to which an investigation actually tested what it was intended to test.

_____ 29. The degree to which a study can demonstrate that its findings are repeatable.

_____ 30. The process of making use of data that have been collected by others.

_____ 31. The sequence of steps that is followed when designing and implementing a research investigation.

_____ 32. To propose and test theories that help us understand things or events.

_____ 33. Method of uncovering what studies have already been done on a topic.

_____ 34. What Rik Scarce was defending when he refused to violate his agreement of confidentiality with his research participants.

_____ 35. During an interview or survey, this factor may affect the results of a study because most people want to appear consistent.

CENTRAL IDEA COMPLETIONS

Following the instructions, fill in the appropriate concepts and descriptions for each of the questions posed in the following section.

1. In the following example, "Research has confirmed that a student's grade point average is directly associated with the amount of time spent preparing for classes," specify:

   a. the independent variable: _____

   b. the dependent variable: _____

2. Compare and contrast the advantages and disadvantages of each of the following four research methods commonly employed by sociologists.

   a. Social survey

   Advantages: _____

   Disadvantages: _____

b. Participant observation

Advantages: _____

Disadvantages: _____

c. Experiment

Advantages: _____

Disadvantages: _____

d. Secondary analysis

Advantages: _____

Disadvantages: _____

3. What is a random sample? _____

4. What were the two major flaws of the *Literary Digest* survey predicting the outcome for the 1936 presidential election? _____

_____

_____

5. How are researcher bias and the self-fulfilling prophecy linked together? _____

_____

_____

6. Given these scores—280, 430, 450, 450, 520, 530, 540, 690, 760, 600, 580—calculate:

a. Mean: _____

b. Median: _____

c. Mode: _____

7. Take the data presented in Table 2-3, "Life Expectancy at Birth for Selected Countries," and present your interpretation following the nine steps in "Sociology at Work: How to Read a Table."

Your interpretation:

_____

_____

8. What is an empirical question? _____

_____

Present an example of an empirical question: _____

_____

9. Assume you just purchased a new bow and arrow set. The set comes with eight arrows. How are the concepts of reliability and validity related to your bow and arrow set? Describe how you might determine the reliability and validity of your equipment.

_____

_____

10. Compare and contrast the techniques for establishing truth in the courtroom and truth in the social sciences.

a. The courtroom: _____

_____

b. The social sciences: _____

_____

11. What was the *main* weakness in the study used as a basis for the *Kansas City Star* report of AIDS among Catholic priests? _____

_____

_____

What is the "really disturbing" outcome of the *Kansas City Star* report? _____

_____

_____

12. How does researcher bias often take the form of a self-fulfilling prophecy? _____

_____

_____

13. Briefly state the ways in which the research process can produce things of beauty. _____

_____

_____

14. What types of precautions would you employ in your effort to spot a bogus poll? _____

_____

_____

_____

15. Briefly indicate why the Weitzman study of divorce, though inaccurate, is "one of the most widely quoted statistics in recent history." _____

_____

_____

16. After reading Chapter 2, what do you see as the central ethical issues connected with sociological research? _____

_____

_____

CRITICAL THOUGHT EXERCISES

1. Select a sociological study from one of the academic journals at your library, then find an example of social research in a popular magazine. Compare and contrast the discussion of the methods of both studies. Consider what steps from the research process were followed in both studies. Evaluate your level of confidence in the accuracy of each study based on the specifications of the methods used in each study.

2. You have been asked to join a presidential task force on "The Attitudes of Average American College Students About Alcohol, Marijuana, and Nicotine". Following the steps of the research process outlined in this chapter, prepare a research outline detailing the manner in which you would guide the conducting of an empirical study by the task force.

3. Compare and contrast the advantages and disadvantages of the four methods of data collection presented in this chapter. Provide one practical research problem that could be appropriately addressed by each data collection method.

4. Among the most important social science data used throughout the United States are the data provided by the U.S. Census. Unfortunately, the census rapidly becomes outdated. To correct this problem, The U.S. Census Bureau conducts the American Community Survey (ACS). Using the Internet, go to the U.S. Census Web site and examine Census Bureau's discussion of the methods it uses to collect these data.

http://www.census.gov/*

5. Renewed interest in UFOs has led to the creation of a new international conference, consisting of a panel of experts reporting the newest evidence on the existence of UFOs. Recalling its experiences with the previous UFO conference, a major media conglomerate has retained you as a sociologist to assist its staff in evaluating the research presented at the conference. What steps would you take to assess the validity of the research claims presented at the conference?

6. Following your reading of "Controversies in Sociology: Truth in the Courtroom versus Truth in the Social Sciences," select a highly publicized criminal trial from the past decade and develop a critical thought essay in which you analyze the available evidence according to both the standards applied in the courtroom and the standards applied in the social sciences. How do the standards for ascertaining truth within each field affect the conclusions of guilt or nonguilt?

7. You are interested in determining why alcohol consumption on college campuses has increased during the past decade. Discuss the pros and cons of each of the four methods of data collection discussed in Chapter 2 for collecting relevant, reliable, and valid information that would assist you in answering this research question.

*Please note, Web site addresses change frequently. The addresses presented in these exercises should be viewed as guides only, and students should be prepared to utilize appropriate Internet search techniques when necessary should these addresses no longer be accessible.

ANSWER KEY

KEY CONCEPTS

1. k	7. r	13. ee	19. ff	25. x	31. t
2. y	8. c	14. d	20. o	26. b	32. cc
3. h	9. gg	15. i	21. w	27. bb	33. q
4. dd	10. g	16. p	22. j	28. f	34. l
5. ii	11. s	17. aa	23. v	29. u	35. n
6. m	12. a	18. e	24. z	30. hh	

PART 2

The Individual in Society

How do sociologists analyze human behavior in groups? We can begin by asking, "What makes us human?" That is, what makes us different from other animal species, and what makes human society unique? Chapter 3 provides the answer: culture. Culture, you will see, is basically a blueprint for living, which we use because we generally lack instinctual programming of our social behavior. Thus, humans are compelled to create themselves and their societies in a symbolic, interpreted fashion. You will discover that it is this very symbolic nature of humans that permits societies to vary so dramatically from one another in their basic beliefs and social practices. Moreover, even within a single society a variety of different ideas, outlooks, and practices may be found. Yet, despite the enormous variation, all human cultures seem to have certain basic requirements in common.

But how do individuals develop and learn to use their symbolic capacities? The answer is through the process of socialization, which you will explore in Chapter 4. You will discover that each human being is a complex mixture of nature, or biological heritage, and nurture, or the entire socialization experience. You will explore the common mechanisms by which each human develops a unique personality and social identity. In Chapter 4 you will encounter some of the major theories about how and why children develop as they do. You will also find out how socialization continues to play a role throughout our lives.

Assuming that individuals have developed the skills through the socialization process, how do humans interact with one another? Are there rules that guide our behavior? If so, how do these rules differ in different situations? How does the context of the interaction (for example, size, setting, level of intimacy) affect the nature, quality, and dynamics of interaction? These issues are explored in Chapter 5, where you will investigate the various elements of human interaction. You also will examine the dynamics of human behavior in groups—groups that range from two people, to large formal associations, to impersonal bureaucracies, to social institutions like the family, the economy, or the political system.

Social interaction is not random; it is clearly organized. But how is that organization achieved? How are the rules for social interaction enforced? What happens when individuals violate social expectations? This is the topic of deviant behavior and social control, which you will explore in Chapter 6. You will discover that what is appropriate and inappropriate behavior is socially defined in each society and subculture and, perhaps somewhat surprisingly, that deviant behavior, or a violation of social expectations, is not only inevitable, but also useful and even necessary in society. You will examine the leading sociological theories for deviant behavior and critically scrutinize the formal system for dealing with deviance in the United States: our system of law and criminal justice.

When you have completed Part 2, you will understand what it means to be a human being in society, how humans construct and alter one another's behavior in various contexts, and how deviant behavior is defined and dealt with.

Culture

LEARNING OBJECTIVES

After studying this chapter, you should be able to do the following:

- Understand how culture makes possible the variation in human societies.
- Distinguish between ethnocentrism and cultural relativism.
- Know the difference between material and nonmaterial culture.
- Understand the importance of language in shaping our perception and classification of the world.
- Discuss whether or not animals have language.
- Understand the roles of innovation, diffusion, and cultural lag in cultural change.
- Explain what subcultures are.
- Describe cultural universals.

Anthropologist Mary Catherine Bateson was worried about her pregnancy, which took place in the Philippines where she had been doing research. After weeks on her back fruitlessly hoping to avert a premature delivery, she gave birth in a Manila hospital to the son she planned to call Martin. The baby died a few hours later. For Bateson, the death of her baby was something that should not have happened. It was unthinkable, unbearable. However, for the gentle Filipino nurses, the loss was sad but part of life, bound to happen from time to time. Their sympathy was firmly mixed with a cheerful certainty that she would be back next year with another baby.

Bateson's fieldwork in the village had allowed her to observe and compare responses to death. On the afternoon of Martin's birth, she described to her husband, Barkev, the way Filipinos would express their sympathy. "Don't expect to be left alone," she said, and "don't expect people tactfully to avoid the subject. Expect friends to seek us out and to show their concern by asking specific factual questions." Rather than a euphemistic handling of the event and a denial of the ordinary course of life, she advised him to be ready for the opposite. An American colleague might shake hands, nod his head sadly, perhaps murmuring, "We were so sorry to hear," and beat a swift retreat; a Filipino friend would say, "It was so sad that your baby died. Did you see him? Who did he look like? Was he baptized? How much did he weigh? How long were you in labor?"

Stereotypes often conceal their opposites. In other contexts, Filipinos describe Americans as "brutally frank," while Americans find Filipinos frustratingly indirect and evasive. Yet in the handling of death, Filipinos behave in a manner that Americans might characterize as "brutally frank" and seem to go out of their way to evoke the expression of emotion, while Americans can only be called euphemistic and indirect, going to great lengths to avoid emotional outbreaks.

If Bateson had not had experience with the culture of the village, the most caring behavior on the part of Filipino friends, genuinely trying to express concern and affection, would have seemed like a violation. To avoid breaking down in the face of sudden reminders of grief, she might have imposed a rigid self-control, which would have reinforced the belief that many Filipinos hold, that Americans don't really grieve; or she might have reacted with anger to the affront, losing valued friends.

Telling of a death, hearing of a death, expressing sympathy in the appropriate way, these are acts in which mutual recognitions of humanness are tested, but there is no single human way of responding. The bereaved is, among other things, a performer in a cultural drama that asserts basic ideas about the nature of life and death and the human heart. Some societies organize their recognitions of bereavement around an effort to help the bereaved regain control and forget, while other societies are geared to support the expression of grief. Some societies rehearse for grief and loss while others deny them.

Filipinos have a worldview that allows them to face the inescapable fact of death, including it in the continuing rhythm of life. Americans treat grief almost like a disease, embarrassing and possibly infectious (Bateson, 1994).

All human societies have complex ways of life that differ greatly from one to the other. These ways have come to be known as *culture*. In 1871, Edward Tylor gave us the first definition of this concept. Culture, he noted, "is that complex whole which includes knowledge, belief, art, law, morals, custom, and other capabilities and habits acquired by man as a member of society" (Tylor, 1958). Robert Bierstadt simplified Tylor's definition by stating, "Culture is the complex whole that consists of all the ways we think and do and everything we have as members of society" (Bierstadt, 1974).

Most definitions of culture emphasize certain features. Namely, culture is shared; it is acquired, not inborn; the elements make up a complex whole; and it is transmitted from one generation to the next.

The Concept of Culture

We will define **culture** as *all that human beings learn to do, to use, to produce, to know, and to believe as they grow to maturity and live out their lives in the social groups to which they belong.* Culture is basically a blueprint for living in a particular society. In common speech, people often refer to a "cultured person" as someone with an interest in the arts, literature, or music, suggesting that the individual has a highly developed sense of style or aesthetic appreciation of the "finer things." To sociologists, however, every human being is "cultured." All human beings participate in a culture, whether they are Harvard educated and upper class, or illiterate and living in a primitive society. Culture is crucial to human existence.

When sociologists speak of culture, they are referring to the general phenomenon that is a characteristic of all human groups. However, when they refer to a culture, they are pointing to the specific culture of a particular group. In other words, all human groups have a culture, but it often varies considerably from one group to the next. Take the concept of time, which we accept as entirely natural. To Westerners, "time marches on" steadily and predictably, with past, present, and future divided into units of precise duration (minutes, hours, days, months, years, and so on). In the Native American culture of the Sioux, however, the concept of time simply does not exist apart from ongoing events: nothing can be early or late—things just happen when they happen. For the Navajo, the future is a meaningless concept—immediate obligations are what count. For natives of the Pacific island of Truk, however, the past has no independent meaning—it is a living part of the present. These examples of cultural differences in the perception of time point to a basic sociological fact: Each culture must be investigated and understood on its own terms before it is possible to make valid cross-cultural comparisons (Hall, 1981).

In every social group, culture is transmitted from one generation to the next. Unlike other creatures, human beings do not pass on many behavioral patterns through their genes. Rather, culture is taught and learned through social interaction.

Culture and Biology

Human beings, like all other creatures, have basic biological needs. We must eat, sleep, protect ourselves from the environment, reproduce, and nurture our young—or else we could not survive as a species. In most other animals, such basic biological needs are met in more or less identical ways by all the members of a species through inherited behavior patterns or instincts. These instincts are specific for a given species as well as universal for all members of that species. Thus, instinctual behaviors, such as the web spinning of specific species of spiders, are constant and do not vary significantly from one individual member of a species to another. This is not true of humans, whose behaviors are highly variable and changeable, both individually and culturally. It is through culture that human beings acquire the means to meet their needs. For example, the young, or larvae, of hornets or yellow jackets are housed in paper-walled, hexagonal chambers that they scrape

Native Americans often wish to follow the traditional practices and customs of their culture. At the same time, the culture of the larger society urges them to adopt the conventions of mainstream American society.

against with their heads when hungry. This is a signal to workers, who immediately feed the young tiny bits of undigested insect parts (Wilson, 1975). Neither the larvae nor the workers learn these patterns of behavior: They are instinctual.

In contrast, although human infants cry when hungry or uncomfortable, the responses to those cries vary from group to group and even from person to person. In some groups, infants are breast-fed; in others, they are fed prepared milk formulas from bottles; and in still others, they are fed according to the mother's preference. Some groups breast-feed children for as long as 5 or 6 years, others for no more than 10 to 12 months. Some mothers feed their infants on demand—whenever they seem to be hungry; other mothers hold their infants to a rigid feeding schedule. In some groups, infants are picked up and soothed when they seem unhappy or uncomfortable. Other groups believe that infants should be left "to cry it out." In the United States, parents differ in their approaches to feeding and handling their infants, but most are influenced by the practices they have observed among members of their families and their social groups. Such habits, shared by the members of each group, express the group's culture. They are learned by group members and are kept more or less uniform by social expectations and pressures.

Culture Shock

Every social group has its own specific culture, its own way of seeing, doing, and making things, its own traditions. Some cultures are quite similar to one another; others are very different. When individuals travel abroad to countries with cultures that are very different from their own, the experience can be quite upsetting. Meals are scheduled at different times of day, "strange" or even "repulsive" foods are eaten, and the traveler never quite knows what to expect from others or what others in turn may expect. Local customs may seem charming or brutal. Sometimes travelers are unable to adjust easily to a foreign culture; they may become anxious, lose their appetites, or even feel sick. Sociologists use the term **culture shock** to describe *the difficulty people have adjusting to a new culture that differs markedly from their own.*

Jonah Blank (1992) experienced culture shock often as he traveled throughout India. One day, he observed three bulls walking in the village:

> [They] ambled lazily by the storefront, leaving three steaming piles of dung in their wake. A few minutes later an old woman waddled along, dropped to her knees, and scooped up the fresh patties with her clapped hands. She slapped them onto an already laden tin plate, and shuffled down the alley. . . . Around the corner from the manure collector, another old woman hung a string of dried cow patties outside her door for luck. A large mound of dung sat at the step, stuck each day with newly plucked flower stems. Had I been rude enough to tell her that the custom was unhygienic, she would assuredly have laughed at my science.

Culture shock can also be experienced within a person's own society. Picture the army recruit having to adapt to a whole new set of behaviors, rules, and expectations in basic training—a new cultural setting.

Ethnocentrism and Cultural Relativism

People often make judgments about other cultures according to the customs and values of their own, a practice sociologists call **ethnocentrism.** Thus, an American might call a Guatemalan peasant's home filthy because the floor is made of packed dirt or believe that the family organization of the Watusi (of East Africa) is immoral because a husband may have several wives. Ethnocentrism can lead to prejudice

The king of Akure in Nigeria appears in the palace court with some of his 156 wives during the festival of Ifa. The mores of our society do not allow this type of behavior. Cultural relativism would require that we not judge this practice from the standpoint of our culture alone.

and discrimination and often results in the repression or domination of one group by another.

Immigrants, for instance, often encounter hostility when their manners, dress, eating habits, or religious beliefs differ markedly from those of their new neighbors. Because of this hostility and because of their own ethnocentrism, immigrants often establish their own communities in their adopted country. Many Cuban Americans, for example, have settled in Miami where they have built a power base through strength in numbers. In Dade County alone, which includes Miami, there are about 850,000 Cuban Americans.

To avoid ethnocentrism in their own research, sociologists are guided by the concept of **cultural relativism,** *the recognition that social groups and cultures must be studied and understood on their own terms before valid comparisons can be made.* Cultural relativism frequently is taken to mean that social scientists never should judge the relative merits of any group or culture. This is not the case. Cultural relativism is an approach to doing objective cross-cultural research. It does not require researchers to abdicate their personal standards. In fact, good social scientists will take the trouble to spell out exactly what their standards are so that both researchers and readers will be alert to possible bias in their studies.

American Moshe Rubinstein encountered the contrasting values between American and Arab culture after a traditional Arabic dinner. Rubinstein was presented with a parable by his host Ahmed.

"Moshe," Ahmed said, putting his fable in the form of a question, "imagine that you, your mother, your wife, and your child are in a boat, and it capsizes. You can save yourself and only one of the remaining three. Whom will you save?" For a moment Moshe froze, while thoughts raced through his mind. No matter what he might say, it would not be right from someone's point of view, and if he refused to answer he might be even worse off. Moshe was stuck. So he tried to answer by thinking aloud as he progressed to a conclusion, hoping for salvation before he said what came to his mind as soon as the question was posed, namely, save the child.

Ahmed was very surprised. Moshe had flunked the test. As Ahmed saw it, there was only one correct answer. "You see," he said, "you can have more than one wife in a lifetime, you can have more than one child, but you have only one mother. You must save your mother!" (Rubinstein, 1975).

This example shows us how the value of individuals such as children, spouses, and mothers can vary greatly from one culture to the next. We can see how what we might consider to be a natural way of thinking is not the case at all in another culture.

Cultural relativism requires that behaviors and customs be viewed and analyzed within the context in which they occur. The packed-dirt floor of the Guatemalan house should be noted in terms of the culture of the Guatemalan peasant, not in terms of suburban America (see "Global Sociology: Is McDonald's Practicing Cultural Imperialism or Cultural Accommodation?"). Researchers, however,

Global Sociology

Is McDonald's Practicing Cultural Imperialism or Cultural Accommodation?

McDonald's may be the most obvious example of American culture abroad. There are approximately 24,500 McDonald's restaurants in more than 115 countries; a new McDonald's opens somewhere in the world every six hours. Like Coke, though, it is easy to denigrate McDonald's as the symbol of the crass, unhealthy, commercial side of American culture. For example, some Japanese critics have blamed junk food for juvenile crime. In France, a few McDonald's have been firebombed.

In order to succeed, McDonald's has had to be responsive to the local cultures and eating habits. The restaurant chain offers *ayran* (a popular chilled yogurt drink) in Turkey; McLaks (a grilled salmon sandwich) in Norway, and teriyaki burgers in Japan. In New Delhi, India, where Hindus shun beef and Muslims refuse pork, the burgers are made of mutton and are called Maharaja Macs. For the many strict Hindus who are vegetarians, there is the McAloo Tikki burger, a spicy vegetarian patty made of potatoes and peas. McDonald's has even figured out how to make a vegetarian mayonnaise without eggs. Workers with green aprons handle only vegetarian food, while those with black aprons handle nonvegetarian food. The restaurants even separate the two menus—realizing that vegetarians do not

want to read about meat dishes. What this has meant is that mixed groups of people, with different tastes and customs, have found a place where they can all eat together.

Compared to American customers, most diners in other countries spend a good deal of time in the restaurants. Far from being a place where they eat and run, in many countries people see McDonald's as a spot where they can linger. In cities where space is at a premium, like Hong Kong, teenagers choose it as an escape from cramped apartments where they can meet their friends and even do their homework. Women in traditional cultures meet at McDonald's because there is no alcohol served, and they see it as a safe, socially acceptable place for women.

The fact that the staff is made up of local people means that the restaurants, though obviously foreign, are not immediately perceived as being American. Even though the golden arches may be conspicuous, whether it is New Delhi, Brazil, or Manila, the food is being cooked and served by people who live nearby and speak the local dialect.

Source: *Golden Arches East: McDonald's In East Asia,* by J. L. Watson, ed., 1997, Berkeley: University of California Press.

---

may find that dirt floors contribute to the incidents of parasites in young children and may therefore judge such construction to be less desirable than wood or tile floors.

Components of Culture

The concept of culture is not easy to understand, perhaps because every aspect of our social lives is an expression of it and also because familiarity produces a kind of nearsightedness toward our own culture, making it difficult for us to take an analytical perspective toward our everyday social lives. Sociologists find it helpful to break down culture into separate components: material culture (objects), nonmaterial culture (rules), cognitive culture (shared beliefs), and language (Hall, 1990).

Material Culture

Material culture *consists of human technology—all the things human beings make and use, from small hand-held tools to skyscrapers.* Without material

culture our species could not long survive, for material culture provides a buffer between humans and their environment. Using it, human beings can protect themselves from environmental stresses, as when they build shelters and wear clothing to protect them from the cold or from strong sunlight. Even more important, humans use material culture to modify and exploit the environment. They build dams and irrigation canals, plant fields and forests, convert coal and oil into energy, and transform ores into versatile metals. Using material culture, our species has learned to cope with the most extreme environments and to survive and even to thrive on all continents and in all climates. Human beings have walked on the floor of the ocean and on the surface of the moon. No other creature can do this: None has our flexibility. Material culture has made human beings the dominant life form on earth.

Nonmaterial Culture

Every society also has a **nonmaterial culture,** which consists of *the totality of knowledge, beliefs, values,*

Housing is an aspect of material culture that can vary widely, as can be seen by comparing this elaborate home in Saudi Arabia and these thatched huts in Nigeria.

and rules for appropriate behavior. The nonmaterial culture is structured by such institutions as the family, religion, education, economy, and government.

While material culture is made up of things that have a physical existence (they can be seen, touched, and so on), the elements of nonmaterial culture are the ideas associated with their use. Although engagement rings and birthday flowers have a material existence, they also reflect attitudes, beliefs, and values that are part of American culture. There are rules for their appropriate use in specified situations.

Norms are central elements of nonmaterial culture. **Norms** are *the rules of behavior that are agreed upon and shared within a culture and that prescribe limits of acceptable behavior.* They define "normal" expected behavior and help people achieve predictability in their lives. For example, one of the few truly universal gestures is the kiss. Anthropologists have speculated that it evolved from the time when mothers would pass food, mouth to mouth, to their infants. Think for a moment how the kiss has permeated our lives. Mothers kiss bruises to "make them better." Athletes kiss their trophies. The French sculptor Auguste Rodin created one—the famous sculpture, *The Kiss.*

Yet most cultures follow unwritten norms concerning kissing in public. In some cultures cheek kissing is a normal way of greeting another person. In Russia, you actually kiss the cheek. In other places, such as France, Italy, or Latin America, you kiss the air—that is, cheeks touch and lips make the sound of kissing, but the lips do not actually press against the cheek. In Latin America, only one cheek

is kissed. In France, each cheek is kissed. In Belguim and Russia, you kiss one cheek, then the other, and back to the first.

In some countries, kissing the hand is the acceptable form of greeting. French etiquette suggests that the man kisses the woman's hand without actually touching it with his lips. Kissing your own fingertips is a European gesture that conveys the message "That's great! That's beautiful." The origin for this gesture probably stems from the custom of ancient Greeks and Romans who, when entering or leaving the temple, threw a kiss to sacred objects. In Mexico a kissing sound is used to summon a waiter in a restaurant. In the Phillipines, street vendors use it to attract the attention of customers.

In the United States kissing between men and women in public is common. Presidential candidates have even given extended mouth-to-mouth kisses to their spouses during prime-time broadcasts. At the other extreme, in certain Asian countries such as Japan, kissing is considered an intimate sexual act and not permissible in public, even as a social greeting (Axtell, 1998).

Mores (pronounced more-ays) are *strongly held norms that usually have a moral connotation and are based on the central values of the culture.* Violations of mores produce strong negative reactions, which are often supported by the law. Desecration of a church or temple, sexual molestation of a child, rape, murder, incest, and child beating all are violations of American mores.

Not all norms command such absolute conformity. Much of day-to-day life is governed by traditions, or **folkways,** which are *norms that permit a wide degree of individual interpretation as long as certain limits are not overstepped.* People who violate folkways are seen as peculiar or possibly eccentric, but rarely do they elicit strong public response. For example, a wide range of dress is now acceptable in most theaters and restaurants. Men and women may wear clothes ranging from business attire to jeans, an open-necked shirt, or a sweater. However, extremes in either direction will cause a reaction. Many establishments limit the extent of informal dress: Signs may specify that no one with bare feet or without a shirt may enter. On the other hand, a person in extremely formal attire might well attract attention and elicit amused comments in a fast-food restaurant.

Good manners in our culture also show a range of acceptable behavior. A man may or may not open a door or hold a coat for a woman, who may also choose to open a door or hold a coat for a man—all four options are acceptable behavior and cause neither comment nor negative reactions from people.

These two examples illustrate another aspect of folkways: They change with time. Not too long ago a man was *always* expected to hold a door open for a woman, and a woman was *never* expected to hold a coat for a man.

Folkways also vary from one culture to another. In the United States, for example, it is customary to thank someone for a gift. To fail to do so is to be ungrateful and ill mannered. Subtle cultural differences can make international gift giving, however, a source of anxiety or embarrassment to well-meaning business travelers. For example, if you give a gift on first meeting an Arab businessman, it may be interpreted as a bribe. If you give a clock in China, it is considered bad luck. In fact, the Mandarin word for *clock* is similar to the one for *death*. In Latin America, you will have a problem if you give knives, letter openers, or handkerchiefs. The first two indicate the end of a friendship; the latter is associated with sadness.

Norms are specific expectations about social behavior, but it is important to add that they are not absolute. Even though we learn what is expected in our culture, there is room for variation in individual interpretations of these norms that deviate from the ideal norm.

Ideal norms are *expectations of what people should do under perfect conditions.* These are the norms we first teach our children. They tend to be simple, making few distinctions and allowing for no exceptions. That drivers should "stop at red lights" is an ideal norm in American society. So is the norm that a marriage will last "until death do us part."

In reality, however, nothing about human beings is ever that dependable. For example, if you interviewed Americans about how drivers respond to red

According to the norms of American culture, a common way for men to greet each other is to shake hands. In France, men greet each other with a kiss on each cheek.

lights, you would get answers like "Ideally, drivers should stop at red lights." In actual fact, however, drivers sometimes run red lights. So even though you can pretty much count on a driver stopping for a red light, it pays to be careful. Also, if it looks like a driver is not going to stop for a light, you had better play it safe and slow down. In other words, people recognize that drivers usually do feel they should stop when a traffic light is red, but they also acknowledge that there are times when a driver will not stop for a red light. The driver may be in a hurry, drunk, upset, or simply not paying attention. **Real norms** are *norms that are expressed with qualifications and allowances for differences in individual behavior.* They specify how people actually behave. They reflect the fact that a person's behavior is a function not only of norm guidance but also of situational elements, as exemplified by the driver who does not always stop at a red light if no car appears to be coming from the other direction. (For an examination of American values and norms, see "Sociology at Work: Seymour Martin Lipset on American Exceptionalism.")

The concepts of ideal and real norms are useful for distinguishing between mores and folkways. For mores, the ideal and the real norms tend to be very close, whereas folkways can be much more loosely connected: Our mores say thou shalt not kill and really mean it, but we might violate a folkway by neglecting to say thank you, for example, without provoking general outrage. More important, the very fact that a culture legitimizes the difference between ideal and real expectations allows us room to interpret norms to a greater or lesser degree according to our own personal dispositions.

Cognitive Culture

Cognitive culture is *the thinking component of culture, which consists of shared beliefs and knowledge*

Sociology at Work

Seymour Martin Lipset on American Exceptionalism

Sociologist Seymour Martin Lipset believes the United States has unique cultural values that set it apart from the rest of the world. However, these values can produce both positive and negative outcomes.

For example, the United States is the most religious, optimistic, productive, well-educated and individualistic country in the world. It is also one of the most crime-ridden and litigious nations, with a wide gap in income distribution, and some of the lowest levels of welfare benefits.

How can widely held cultural values produce both good and bad outcomes? Part of the answer lies in how people attempt to fulfill these values. Take, for example, the American emphasis on individualism and individual happiness. It may produce technological innovation in a wide variety of areas, but Lipset believes it is also responsible for the country's high divorce rate. Americans believe you should be satisfied with your life, be happy with your work, and like your spouse. If that is not the case, you are expected to make changes to correct the situation. Lipset notes that Americans have "an individual happiness" type of reflex, and if there are marital problems we want to know why you are staying in the marriage. If you do not like your job, we expect you to leave it.

The Japanese, in contrast, have a low divorce rate (1.9 divorces per 1,000 people in 1998 vs. 3.6 for the United States) and tend to stay at their jobs for many years. Yet, when the Japanese are asked if they like their jobs or if they like their spouses, they are much more likely to say no than Americans. The Japanese do not automatically

Seymour Martin Lipset.

assume that if you are dissatisfied with your circumstances you must change them.

The United States is a high achievement oriented society. At the same time it also leads the world with many types of crime. Lipset believes these two

---

of what the world is like—what is real and what is not, what is important and what is trivial. The beliefs need not even be true or testable as long as they are shared by a majority of people. Cognitive culture is like a map that guides us through society. Think of a Scout troop on a hike in the wilderness. The troop finds its way by studying a map showing many of the important features of the terrain. The Scouts who use the map share a mental image of the area represented by the map. Yet just as maps differ, each perhaps emphasizing different details of the terrain or using different symbols to represent them, so do cultures differ in the ways in which they represent the world. It is important not to confuse any culture's representation of reality with what ultimately

is real—just as a map is not the actual terrain it charts.

Values are *a culture's general orientations toward life—its notions of what is good and bad, what is desirable and undesirable.* Values themselves are abstractions. They can best be found by looking for the recurring patterns of behavior that express them.

Language and Culture

Language enables humans to organize the world around them into labeled cognitive categories and use these labels to communicate with one another. Language, therefore, makes possible the teaching and sharing of the cognitive and nonmaterial cultures

are also related. In the United States, a lack of success causes the individual to feel much more dissatisfied than in less achievement oriented societies. Hence, people will try to get ahead by whatever means necessary. For some, this may mean crime.

In American society there is a disdain for authority stemming from the country's revolutionary past. The early founders rejected the control of England and produced a sharp break with the authority of the English powers. Americans do not show the kind of deference to authority as is commonly the case in countries such as Canada or Britain that have not had the same kind of revolutionary history.

Lipset notes that the lack of respect for authority carries over into other areas. Some years ago, both Canada and the United States decided it would make sense to adopt the metric system and do away with the system of pounds and inches. With both countries on the same continent, it was decided to go metric as of a joint date 15 years later. Lipset points out that Canadians were told to go metric, and they did. Americans were told to go metric, and they did not. Lipset notes that Americans do not obey if they do not feel like it. Canadians, who do not have a history of rebellion, show much more deference to the state and authority and are more likely to do what they are told.

The same can be seen with respect to voting. The United States has one of the lowest voting rates in the world. Only about 50% of the people vote in presidential elections and fewer still in state and local elections. Seventy-five percent of Canadians vote regularly. In Scandinavian elections, more than 80% vote. Americans vote if they feel like it.

Our lack of respect for authority is also present in our emphasis on first names. We may not have met someone before yet we may call them by their first name. When Lipset taught at Stanford, freshmen who he had never met would walk in and address him by his first name. To Lipset, this is a way of saying, "I'm just as good as you are. You may be my professor, but you are also Marty." A German, Italian, or French professor would throw the student out of the office.

Interestingly, at the same time as there is a substantial amount of social equality in the United States, the country has never displayed any great enthusiasm for socialism. Lipset argues that America's economic system is not socialist, but its culture is. One of the reasons it has been difficult to sell Americans on socialism is because they thought they already had it, not state ownership of property and state planning, but equality of social relationships. They did not really understand what people meant when they said you have to have a revolution to change the system. Thomas Jefferson said, "That government is best that governs least." That principle has been a guiding aspect of American society. Socialism seems to advocate a strong state, and Americans do not want that.

Sources: Based on *American Exceptionalism: A Double-Edged Sword*, by S. M. Lipset, 1996, New York: W.W. Norton and Company; and an interview with Seymour Martin Lipset, January 1996.

we just discussed. It provides the principal means through which culture is transmitted and the foundation on which the complexity of human thought and experience rests.

Language allows humans to transcend the limitations imposed by their environment and biological evolution. It has taken tens of millions of years of biological evolution to produce the human species. On the other hand, in a matter of decades, cultural evolution has made it possible for us to travel to the moon. Biological evolution had to work slowly through genetic changes, but cultural evolution works quickly through the transmission of information from one generation to the next. In terms of knowledge and information, each human generation, because of language, is able to begin where the previous one left off. Each generation does not have to begin anew, as is the case in the animal world.

All people are shaped by the **selectivity** of their culture, *a process by which some aspects of the world are viewed as important while others are virtually neglected.* The language of a culture reflects this selectivity in its vocabulary and even its grammar. Therefore, as children learn a language, they are being molded to think and even to experience the world in terms of one particular cultural perspective.

This view of language and culture, known as the **Sapir-Whorf hypothesis,** *argues that the language a person uses determines his or her perception of reality* (Sapir, 1961; Whorf, 1956). This idea caused

Controversies in Sociology

Is There a Language Instinct?

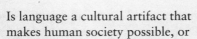

Is language a cultural artifact that makes human society possible, or is it a unique human instinct? Steven Pinker, a proponent of the interdisciplinary field of cognitive science, believes the human capacity to use language is more instinctive than learned. He points out that while human speech is a remarkable ability, it is no more astounding than the bird's ability to fly or the elephant's skills with its trunk.

Over the past 75 years, sociologists and anthropologists have formed a standard view of the interplay between language and culture. It can be summarized as follows:

1. Whereas animals are rigidly controlled by their biology, human behavior is determined by culture and language. Free from biological constraints, human cultures can vary from one another in an infinite number of ways.
2. Human infants are born with nothing more than a few reflexes and an ability to learn. Children learn their culture through their culture's language, socialization, and role models.

This model is a rejection of "biological determinism," which assigns people to fixed slots within a social-political-economic hierarchy based on inborn traits that cannot be changed.

Steven Pinker, a professor at the Massachusetts Institute of Technology, believes that having to choose between the nature versus nurture positions in trying to understand human society reduces our analysis to an overly simplistic level. He believes language is not a cultural artifact that we learn the way we learn to tell time or how the federal government works. Instead it is a distinct piece of the biological makeup of our brains. Language is a complex, specialized skill, which develops in the child spontaneously, without conscious effort or formal instruction. It is deployed without awareness of its underlying logic, and is qualitatively the same in every individual. It is distinct from more general abilities to process information or behave intelligently.

Pinker thinks our brains come equipped with the equivalent of a hard-wired language-learning program. As the brain develops, the program kicks into action and gives children the ability to digest the vocabulary and word order rules supplied by their environment, making them fluent users of spoken language by age 5. All humans possess this unique ability, and the environment merely determines what words we will use. Birds fly, fish swim, and humans do what they do naturally, which is use language.

Pinker believes that essentially there is a language instinct. People know how to talk in more or less the sense that spiders know how to spin webs. Web spinning was not invented by some unsung spider genius and does not depend on having had the right education or on having an aptitude for architecture or the construction trades. Rather, spiders spin webs because they have spider brains, which give them the urge to spin and the competence to succeed.

Thinking of language as an instinct runs contrary to popular wisdom, especially as it is presented in the social sciences and humanities. Pinker believes

some alarm among social scientists at first, for it implied that people from different cultures never quite experience the same reality. Although more recent research has modified this extreme view, it remains true that different languages classify experiences differently—that language is the lens through which we experience the world. The prominent anthropologist Ruth Benedict (1961) pointed out, "We do not see the lens through which we look."

The category corresponding to one word and one thought in language A may be regarded by language B as two or more categories corresponding to two or more words or thoughts. For example, we have only one word for water, but the Hopi Indians have two words—*pahe* (for water in a natural state) and *keyi* (for water in a container). Yet the Hopi have only one word to cover every thing or being that flies, except birds. Strange as it may seem to us, they call a flying insect, an airplane, and a pilot by the same word. Verbs also are treated differently in different cultures. In English we have one verb *to go.* In New Guinea, however, the Manus language has three verbs—depending on direction, distance, and whether the going is up or down.

A little bit closer to home, consider the number of words and expressions pertaining to technology that have entered the English language. These include *cyberspace, virtual reality, hackers, morphing, information surfers, wired,* and *zapped.* These words reflect the preoccupation of American culture with technology. In contrast, many Americans are at a loss for words when they are asked to describe nature:

Stephen Pinker.

language is no more a cultural invention than is upright posture. Though language is a magnificent ability unique to Homo sapiens among living species, many species have unique abilities. Some kinds of bats home in on flying insects using Doppler sonar. Some migratory birds navigate thousands of miles by calibrating the positions of the constellations against the times of day and year. In nature's talent show we are simply a species of primate with our own act, a knack for communicating information about who did what to whom by modulating the sounds we make when we exhale.

Sources: *The Language Instinct* (pp. 18–19, 406), by S. Pinker, 1995, New York: HarperPerennial; and an interview with Steven Pinker, March 1995.

varieties of snow, wind, or rain; kinds of forests; rock formations; earth colors and textures; or vegetation zones. Why? These things are not of great importance in urban American culture.

The translation of one language into another often presents problems. Direct translations are often impossible because (1) words may have a variety of meanings and (2) many words and ideas are culture bound. An extreme example of the first type of these translation problems occurred near the end of World War II. After Germany surrendered, the Allies sent Japan an ultimatum to surrender. Japan's premier responded that his government would *mokusatsu* the surrender ultimatum. *Mokusatsu* has two possible meanings in English: "to consider" or "to take notice of." The premier meant that the government would consider the surrender ultimatum. The English translators, however, used the second interpretation, "to take notice of," and assumed that Japan had rejected the ultimatum. This belief that Japan was unwilling to surrender led to the atomic bombing of Hiroshima and Nagasaki. Most likely the bombing would still have taken place even with the other interpretation, but this example does demonstrate the problems in translating words and ideas from one language into another (Samovar, Porter, & Jain, 1981).

These examples demonstrate the uniqueness of language. No two cultures represent the world in exactly the same manner, and this cultural selectivity, or bias, is expressed in the form and content of a culture's language (see "Controversies in Sociology: Is There a Language Instinct?").

The Symbolic Nature of Culture

All human beings respond to the world around them. They may decorate their bodies, make drawings on cave walls or canvases, or mold likenesses in clay. These all act as symbolic representations of their society. All complex behavior is derived from the ability to use symbols for people, events, or places. Without the ability to use symbols to create language, culture could not exist.

Symbols and Culture

What does it mean to say that culture is symbolic? A **symbol** is *anything that represents something else and carries a particular meaning recognized by members of a culture.* Symbols need not share any quality at all with whatever they represent. Symbols stand for things simply because people agree that they do. Thus, when two or more individuals agree about the things a particular object represents, that object becomes a symbol by virtue of its shared meaning for those individuals. When Betsy Ross sewed the first American flag, she was creating a symbol.

The important point about the meanings of symbols is that they are entirely arbitrary, a matter of cultural convention. Each culture attaches its own meaning to things. Thus, in the United States, mourners wear black to symbolize their sadness at a funeral. In the Far East, people wear white. In this case, the symbol is different but the meaning is the same. On the other hand, the same object can have different meanings in different cultures. Among the Sioux Indians, the swastika (a cross made with ends bent at right angles to its arms) was a religious symbol; in Nazi Germany, its meaning was political.

Looking at culture from this point of view, we would have to say that all aspects of culture—nonmaterial and material—are symbolic. Thus, culture may be said to consist of shared patterns of meanings expressed in symbols (Geertz, 1973). This means that virtually everything we say and do and use as group members has some shared meaning beyond itself. For example, wearing lipstick is more than just coloring one's lips, smoking a cigarette is more than just filling one's lungs with smoke, and wearing high-heeled shoes is more than just trying to be taller. All these actions and artifacts are part of American culture and are symbolic of sexuality and adulthood, among other things. Even a person's clothes and home—material possessions—are not only means of protection from the environment, but are also symbolic of that person's status in the social

The "thumbs up" gesture is appropriate in American society but may be an insult in other countries.

class structure. An automobile, for many people, is more than just a means of transportation—it is symbolic of their socioeconomic status.

Few travelers would think of going abroad without taking along a dictionary or phrase book to help them communicate with the people in the countries they visit. Although most people are aware that symbolic gestures are the most common form of cross-cultural communication, they do not realize that the language of gestures can be just as different, just as regional, and just as likely to cause misunderstanding as the spoken word can.

After a good meal in Naples, a well-meaning American tourist expressed his appreciation to the waiter by making the "A-OK" gesture with his thumb and forefinger. The waiter was shocked. He headed for the manager. The two seriously discussed calling the police and having the hapless tourist arrested for obscene behavior in a public place.

What had happened? How could such a seemingly innocent and flattering gesture have been so misunderstood? In American culture, everyone from astronauts to politicians to signify that everything is fine uses the sign confidently in public. In France and Belgium, however, it means "You're worth zero"; while in Greece and Turkey, it is an insulting or vulgar sexual invitation. In parts of southern Italy, it is an offensive and graphic reference to a part of the anatomy. Small wonder that the waiter was shocked.

In fact, dozens of gestures take on totally different meanings from one country to another. Is thumbs up always a positive gesture? It is in the United States and in most of western Europe. When it was displayed by the emperor of Rome, the upright thumb gesture spared the lives of gladiators in the Coliseum. However, do not try it in Sardinia and northern Greece. There the gesture means the insulting phrase, "Up yours."

The same naiveté that can lead Americans into trouble in foreign countries also may work in reverse. After paying a call on Richard Nixon, Soviet leader Leonid Brazened stood on a balcony at the White House and saluted the American public with his hands clasped together in a gesture many people interpreted as meaning "I am the champ," or "I won." What many Americans perceived as a belligerent gesture was really just the Russian gesture for friendship (Axtell, 1991; Kerman, Friesen, & Bear, 1984).

The communication systems of many animal species consist of more or less complicated systems of signs. For example, a worker bee returning to its hive can communicate accurately to other bees the location of pollen-rich plants by performing a complicated dance. Careful research by the German biologist Karl von Frisch (1967) showed that this is accomplished when the bee, in effect, acts out a miniature version of its flight. The direction of the dance is the direction of the flight, and the duration of the dance is proportional to the distance of the pollen-rich flowers from the hive. Most important, a bee can dance only to communicate where pollen is to be found, not about anything else, such as imminent danger from an approaching bear.

Culture and Adaptation

Over time, cultures adjust to the demands of the environment. Although environmental determinism—the belief that the environment dictates cultural patterns—is no longer accepted, there must be some degree of "fit" between environment and culture. Whereas other species adapt to their environment through the long, slow process of evolution and natural selection, culture has allowed humans to adapt relatively quickly to many different habitats and become the most flexible species on earth.

Culture as an Adaptive Mechanism

Culture probably has been part of human evolution since the time, some 15 million years ago, when our ancestors first began to live on the ground. As we have stressed throughout this chapter, humans are extraordinarily flexible and adaptable. This adaptability, however, is not the result of being biologically fitted to the environment; in fact, human beings are remarkably unspecialized. We do not run very fast, jump very high, climb very well, or swim very far. However, we are specialized in one area: We are culture producing, culture transmitting, and culture dependent. This unique specialization is rooted in the size and structure of the human brain and in our physical ability both to speak and to use tools.

Culture, then, is the primary means by which human beings adapt to the challenges of their environment. Thus, using enormous machines we strip away layers of the earth to extract minerals, and using other machines we transport these minerals to yet more machines, where they are converted to a staggering number of different products. Take away all our technology and American society would cease to exist. Take away all culture and the human species would perish. Culture is as much a part of us as our skin, muscles, bones, and brains.

Adaptation is *the process by which human beings adjust to changes in their environment.* Adaptation can take two different forms: specialization and generalized adaptability. Most cultures make use of both of these means. **Specialization** *involves developing ways of doing things that work extremely well in a particular environment or set of circumstances.* For example, the Inuit (Eskimo) igloo is a specialized way of building a shelter. It works in the Arctic but would fail miserably in the Sahara desert or in a Florida swamp. An American brick apartment building also is specialized. It is fine where the ground is solid and bricks can be delivered by truck or train, but in swamps, deserts, or where people must move around a great deal in order to subsist, the brick apartment building is of no use whatsoever.

Generalized adaptability involves *developing more complicated yet more flexible ways of doing things.* For example, industrial society has very elaborate means of transportation, including trucks, trains, planes, and ships. Industrialized transportation is complex, much more so than, say, the use of camels by desert nomads. At the same time, industrial transportation is a much more flexible transportation system, adaptable to every climate on earth. As such, it displays the quality of generalized adaptability as long as our environment continues to provide enough resources to meet its needs. Should

People who live in Venice, Italy, have adjusted how they build their homes and live their lives to fit their watery environment.

we ever run out of rubber, metals, oil, and the other resources necessary for the operation of this technology, it will have lost its adaptive value. Then the camel might look like a very tempting means of transportation, and our "adaptable" technology would be as overspecialized as the dinosaur, with about the same probability of survival.

Mechanisms of Cultural Change

Cultural change takes place at many different levels within a society. Some of the radical changes that have taken place often become obvious only in hindsight. When the airplane was invented, few people could visualize the changes it would produce. Not only did it markedly decrease the impact of distance on cultural contact, but also it had enormous impact on such areas as economics and warfare.

It is generally assumed that the number of cultural items in a society (including everything from toothpicks to structures as complex as government agencies) has a direct relation to the rate of social change. A society that has few such items will tend

to have few **innovations,** *any new practice or tool that becomes widely accepted in a society.* As the number of cultural items increases, so do the innovations, as well as the rate of social change. For example, an inventory of the cultural items—from tools to religious practices—among the hunting and gathering Shoshone Indians totals a mere 3,000. Modern Americans who also inhabit the same territory in Nevada and Utah are part of a culture with items numbering well into the millions. Social change in American society is proceeding rapidly, while Shoshone culture, as revealed by archaeological excavations, appears to have changed scarcely at all for thousands of years.

Two simple mechanisms are responsible for cultural evolution: innovation and diffusion. Innovation is the source of all **cultural traits**—that is, *items of a culture such as tools, materials used, beliefs, values, and typical ways of doing things.* Innovation takes place in several different ways, including recombining in a new way elements already available to a society (invention), discovering new concepts, finding new solutions to old problems, and devising and making new material objects.

Diffusion is *the movement of cultural traits from one culture to another.* It almost inevitably results when people from one group or society come into contact with another, as when immigrant groups take on the dress or manners of already established groups and in turn contribute new foods or art forms to the dominant culture. Rarely does a trait diffuse directly from one culture into another. Rather, diffusion is marked by **reformulation,** *in which a trait is modified in some way so that it fits better in its new context.* This process of reformulation can be seen in the transformation of black folk-blues into commercial music such as rhythm and blues and rock 'n' roll. Or consider moccasins—the machine-made, chemically waterproofed, soft-soled cowhide shoes—which today differ from the Native-American originals and usually are worn for recreation rather than as part of basic dress, as they originally were. Sociologists would say, therefore, that moccasins are an example of a cultural trait that was reformulated when it diffused from Native-American culture to industrial America.

Cultural Lag

Although the diverse elements of a culture are interrelated, some may change rapidly while others lag behind. William F. Ogburn (1964) coined the term **cultural lag** to describe *the phenomenon through which new patterns of behavior may emerge, even though they conflict with traditional values.* Ogburn observed that technological change (material

When people of one society come in contact with people of another society, cultural diffusion takes place, as evidenced by the Dairy Queen in Japan.

culture) is typically faster than change in nonmaterial culture—a culture's norms and values—and technological change often results in cultural lag. Consequently, stresses and strains among elements of a culture are more or less inevitable. For example, even though the Internet in general and the World Wide Web in particular offer vast educational opportunities, teachers have been slow to incorporate these technologies into the classroom. Traditional school values may be in conflict with use of the Web. Schools often assume that education is best carried out in isolation from the rest of society and that the teacher is the main guide for the students along a path to learning. Education has changed little from 100 years ago and we still expect teachers to talk and groups of students to listen. The Web makes it possible for the student to connect to countless sites outside of the classroom and makes it possible for the student to pursue individual educational goals. The teacher's role and influence becomes less clear with the introduction of this technology. The teacher, instead of being in charge, must now be ready to collaborate with the student and serve as a partner in the exploration of the resources (Maddux, 1997). Traditional teacher/student roles and values are challenged in the process.

Or consider the warning issued to commuters between Kuala Lumpur, the capital of Malaysia, and the town of Port Kelang. Workers building a modern bridge apparently consulted local religious leaders who declared that human heads were needed to appease dangerous spirits during the construction. Commuters were alerted that headhunters had been busy trying to supply them.

Other instances of cultural lag have considerably greater and more widespread negative effects. Advances in medicine have led to lower infant mortality and greater life expectancy, but there has been no corresponding rapid worldwide acceptance of methods of birth control. The result is a potentially disastrous population explosion.

Animals and Culture

Do animals have culture? Not long ago, most scholars would have said no. Language often is cited as the major behavioral difference between humans and animals. Humans possess language, whereas it is said animals do not. Language is the crucial ingredient in the ability to transmit culture from one generation to the next. For years, scientists reported that animals used their calls simply to announce their identity, gender, species, location, and readiness to fight or mate. Some scientists said that this is all animals need to express, as life in the wild is simple. Or is it? Could it be that animals use symbols in other ways that we have overlooked? Is it possible to find the roots of human language in animal language?

For more than three decades, a variety of research has supported this view. For example, Jane van Lawick-Goodall (1971) discovered that chimpanzees living in nature not only use tools but also produce them first and then carry them to where they will use them. These chimps break twigs off trees, strip them of leaves and bark, then carry them to termite mounds where, after wetting them with spit, they poke them into tunnels and pull them out again all covered with delicious termites ready to be licked off. Other animals also use tools. Sea otters, for example, search out flat pieces of rock and, while floating on their backs, place them on their stomachs and crack shellfish open against them.

A number of experiments—the earliest dating back to the mid-1950s—have shown that apes are able to master some of the most fundament aspects of language. Apes, of course, cannot talk. Their mouths and throats simply are not built to produce speech, and no ape has been able to approximate more than four human words. However, efforts to teach apes to communicate by other means have met with a fair amount of success.

The first and most widely known experiment in ape language research began in 1966, under the direction of Allen and Beatrix Gardner of the University of Nevada, with a chimpanzee named Washoe. This experiment consisted of teaching the chimp American Sign Language (ASL), the hand-gesture language used by deaf people. Washoe learned more than 200 distinct signs and was able to ask for food, name objects, and make reference to her environment. The Gardners replicated their results with four other chimpanzees.

Today, Washoe lives with other signing chimpanzees at Central Washington University under the direction of Roger and Debbi Fouts. Washoe now has an adopted son named Loulis. Loulis is not being taught any signs by humans, but has learned at least 54 signs (see "Technology and Society: Is Research With Animals Ethical?").

Another experiment involves a female gorilla named Koko. Francine Patterson has been working with Koko since 1972. Koko uses approximately 400 signs regularly and another 300 occasionally. She also understands several hundred spoken words (so much so that Patterson has to spell such words as *candy* in her presence). In addition, Koko invents signs or creates sign combinations to describe new things. She tells Patterson when she is happy or sad,

refers to past and future events, defines objects, and insults her human companions by calling them such things as "dirty toilet," "nut," and "rotten stink." Once when Patterson was drilling Koko on body parts, the gorilla signed, "Think eye ear nose boring" (Hawes, 1995).

Koko has taken several IQ tests and has recorded scores just below average for a human child—averaging between 70 and 95 points. However, as Patterson has pointed out, the IQ tests have a cultural bias toward humans, and the gorilla may be more intelligent than the test indicates. For example, one item instructs the child, "Point to two things that are good to eat." The choices are a block, an apple, a shoe, a flower, and an ice cream sundae. Reflecting her tastes, Koko pointed to the apple and the flower. She likes to eat flowers and has never seen an ice cream sundae. Although this answer is correct for Koko, it is only half right for humans and therefore was scored incorrect.

Some of the most significant language work with apes has been done by Sue Savage-Rumbaugh and her colleagues, who taught a form of computer language to chimpanzees. Using special symbols, they managed to teach apes to name objects and "converse" with each other.

An unexpected turn of events produced some interesting results with a bonobo, or pygmy chimp, named Kanzi. Rumbaugh was having no luck trying to teach Kanzi's adopted mother Matata to use the keyboard. During the lessons while the mother was trying in vain to figure out what the experimenters wanted, Kanzi was spontaneously picking up on the tasks while crawling about and generally being more of a distraction than a participant. When Rumbaugh finally gave up on Matata and turned her attention to Kanzi assuming that at 2½ he might now be old enough to learn, she was shocked to find that he already knew most of what she wanted to teach him and more. Kanzi was far more adept at the language tasks than any previous chimp and developed these abilities further over the years.

Kanzi may have learned so well because he was immature at the time. It suggests the possibility that there might be in chimps, as in humans, a critical period when some special language-learning mechanism is activated. Yet the problem with the critical learning period approach is that chimps in the wild do not learn a language. Why should a chimp whose ancestors never spoke (and who himself can not speak) demonstrate a critical period for language learning (Deacon, 1997).

Some social scientists have raised serious questions as to whether apes truly are using language (Greenberg, 1980; Terrace et al., 1979). They argue that although apes can acquire large vocabularies,

Technology and Society

Is Research With Animals Ethical?

Roger Fouts has been working with a chimpanzee named Warshoe for more than 30 years as part of an animal language research project. Two other chimps named Moja and Loulis are also part of this family. At a critical point in his study, Fouts started to question the ethics of doing research with animals.

I came to the sobering conclusion that the entire field of ape language research was not as beneficent or benign as I had once thought. One by one the talking chimps were growing up, and by the age of seven they were seen as obsolete by behavioral researchers. A full-grown chimpanzee was still learning; he was just too big, too strong, and too unpredictable to work with in a house or even a caged environment.

Project Warshoe was a cross-fostering experiment in which a chimpanzee became socially and emotionally attached to a human family. But attachment is not a one-way street. From the day I began working on Project Warshoe, I had to break the first commandment of the behavioral sciences. *Thou shalt not love thy research subject.* I was being *paid* to love my research subject so that she could learn language in a natural family setting. Unfortunately, nobody warned me I was supposed to stop loving Warshoe when the experiment was over.

But as I looked around me, I noticed that other scientists did *not* seem so attached to their chim-

panzees. When it came time to choose between keeping the foster family together or advancing science, they chose science.

What would happen to Moja, Pili, Tatu and Dar after they became too big to handle? Now that we'd proven that chimpanzee children could bond with human parents, didn't we have some moral obligation to meet their emotional needs? A person would have to be blind not to see that chimpanzee children were shattered emotionally by the separation from their human mothers. The ultimate proof that cross-fostering between humans and chimpanzees was successful was the fact that the fosterlings were dying when the family bond was broken.

Nobody seemed to notice, much less question, that the chimpanzees were suffering. On the contrary, the scientists around me swore that this shuttling back and forth of chimpanzee babies was a *good* thing because it advanced our own human knowledge.

I don't know why I saw things differently, but I did. All the scientific rationalization in the world could no longer drown out the inner voice of my conscience. I was supposed to take comfort from the fact that my language research with chimpanzees was benefiting autistic children, but it only made me feel worse. The disparity between these loved children and the imprisoned chimps was now unbearable. I didn't want to be part of a system that was breeding more chimpanzees for more suffering.

It was Warshoe who taught me that "human" is only an adjective that describes "being," and that the essence of who I am is not my humanness but my beingness. There are human beings, chimpanzee beings, and cat beings. The distinctions I had once drawn between such things—distinctions that permitted one species to imprison and experiment on another species—were no longer morally defensible to me.

Roger Fouts.

Sources: *Next of Kin: What Chimpanzees Have Taught Me About Who We Are* (pp. 201–204, 325), by R. Fouts, 1997, New York: Morrow; and an interview with Roger Fouts, October 1997.

they cannot produce a sentence and are not using the equivalent of human language. They believe the hand signs are not spontaneous, but dependent on inadvertent cues from the trainers. Terrace believes Kanzi is just "going through a bag of tricks" to get things.

Steven Pinker, a cognitive scientist at MIT (see "Controversies in Sociology: Is There a Language Instinct?"), believes the various animal language experiments are exercises in wishful thinking. He sees the research as analogous to the bears in the Moscow circus that are trained to ride unicycles. He thinks the chimps have not learned anything more sophisticated than how to get the hairless apes on the other side of the glass to give them M&Ms and bananas. Noam Chomsky, whose theory of language acquisition forms the cornerstone of the field, thinks trying to teach linguistic skills to apes is like trying to teach people to flap their wings and fly. The furthest a human can fly is when an athlete does a 30-foot long jump in the Olympics. It does not really compare to what a bird can do. As Chomsky has noted, "If you want to find out about an organism you study what it's good at. If you want to study humans you study language. If you want to study pigeons you study their homing instinct. . . . This research is just some kind of fanaticism" (Johnson, 1995).

Some linguists believe the animal language experiments are motivated by the ideological conviction that humans should be knocked off their self-appointed thrones and that animals are capable of higher-level thinking. As Terrace noted, "I was once stung by the same bug. I really wanted to communicate with a chimpanzee and find out what the world looks like from a chimpanzee's point of view" (Johnson, 1995). For now, the implications of language use by apes remain unresolved, and the evidence continues to mount on both sides.

Language and the production and use of tools are central elements of nonmaterial and material culture. So does it make sense to say that culture is limited to human beings? Although scientists disagree in their answers to this question, they do agree that humans have refined culture to a far greater degree than have any other animals and also that humans depend on culture for their existence much more completely than do any other creatures.

Subcultures

In order to function, every social group must have a culture of its own—its own goals, norms, values, and ways of doing things. As Thomas Lasswell (1965) pointed out, such group culture is not just a "partial or miniature" culture. It is a full-blown, complete culture in its own right. Every family, clique, shop, community, ethnic group, and society has its own culture. Hence, every individual participates in a number of different cultures in the course of a day. Meeting social expectations of various cultures is often a source of considerable stress for individuals in complex, heterogeneous societies like ours. Many college students, for example, find that the culture of the campus varies significantly from the culture of their family or neighborhood. At home they may be criticized for their musical taste, their clothing, their antiestablishment ideas, and for spending too little time with the family. On campus they may be pressured to open up their minds and experiment a little or to reject old-fashioned values.

Sociologists use the term **subculture** to refer to *the distinctive lifestyles, values, norms, and beliefs of certain segments of the population within a society*. The concept of subculture originated in studies of juvenile delinquency and criminality (Sutherland, 1924), and in some contexts the *sub* in *subculture* still has the meaning of inferior. However, sociologists increasingly use *subculture* to refer to the cultures of discrete population segments within a society. The term is primarily applied to the culture of ethnic groups (Italian Americans, Jews, Native Americans, and so on) as well as to social classes (lower or working, middle, upper, and so on). Certain sociologists reserve the term subculture for marginal groups—that is, for groups that differ significantly from the so-called dominant culture.

Types of Subcultures

Several groups have been studied at one time or another by sociologists as examples of subcultures. These can be classified roughly as follows.

Ethnic Subcultures Many immigrant groups have maintained their group identities and sustained their traditions while at the same time adjusting to the demands of the wider society. Though originally distinct and separate cultures, they have become American subcultures. America's newest immigrants, Asians from Vietnam, Korea, Japan, the Philippines, Taiwan, India, and Cambodia, have maintained their values by living together in tight-knit communities in New York, Los Angeles, and other large cities while at the same time encouraging their children to achieve success by American terms.

Occupational Subcultures Certain occupations seem to involve people in a distinctive lifestyle even beyond their work. For example, New York's Wall

Street is not only the financial capital of the world, it is identified with certain values such as materialism, greed, or power. Construction workers, police, entertainers, and many other occupational groups involve people in distinctive subcultures.

Religious Subcultures　Certain religious groups, though continuing to participate in the wider society, nevertheless practice lifestyles that set them apart. These include Christian evangelical groups, Mormons, Muslims, Jews, and many religious splinter groups. Sometimes the lifestyle may separate the group from the culture as a whole as well as the subculture of its immediate community. In a drug-ridden area of Brooklyn, New York, for example, a group of Muslims follows an antidrug creed in a community filled with addicts, drug dealers, and crack houses. Their religious beliefs set them apart from the general society while their attitude toward drugs separates them from many other community members.

Political Subcultures　Small, marginal political groups may so involve their members that their entire way of life is an expression of their political convictions. Often these are so-called left-wing and right-wing groups that reject much of what they see in American society, but remain engaged in society through their constant efforts to change it to their liking.

Geographic Subcultures　Large societies often show regional variations in culture. The United States has several geographical areas known for their distinctive subcultures. For instance, the South is known for its leisurely approach to life, its broad dialect, and its hospitality. The North is noted for "Yankee ingenuity," commercial cunning, and a crusty standoffishness. California is known for its trendy and ultra-relaxed or "laid-back" lifestyle. And New York stands as much for an anxious, elitist, arts-and-literature-oriented subculture as for a city.

Social Class Subcultures　Although social classes cut horizontally across geographical, ethnic, and other subdivisions of society, to some degree it is possible to discern cultural differences among the classes. Sociologists have documented that linguistic styles, family and household forms, and values and norms applied to child rearing are patterned in terms of social class subcultures. (See Chapter 7 for a discussion of social class in the United States.)

Deviant Subcultures　As we mentioned earlier, sociologists first began to study subcultures as a way of explaining juvenile delinquency and criminality. This interest expanded to include the study of a wide variety of groups that are marginal to society in one way or another and whose lifestyles clash with that of the wider society in important ways. Some of the deviant subcultural groups studied by sociologists include prostitutes, strippers, swingers, pool hustlers, pickpockets, drug users, and a variety of criminal groups.

Universals of Culture

In spite of their individual and cultural diversity, their many subcultures and countercultures, human beings are members of one species with a common evolutionary heritage. Therefore, people everywhere must confront and resolve certain common, basic problems such as maintaining group organization and overcoming difficulties originating in their social and natural environments. **Cultural universals** are *certain models or patterns that have developed in all cultures to resolve these problems.*

Among those universals that fulfill basic human needs are the division of labor, the incest taboo, marriage, family organization, rites of passage, and ideology. It is important to keep in mind that although these forms are universal, their specific contents are particular to each culture.

The Division of Labor

Many primates live in social groups in which it is typical for each adult group member to meet most of his or her own needs. The adults find their own food, prepare their own sleeping places, and, with the exception of infant care, mutual grooming, and some defense-related activities, generally fend for themselves.

This is not true of human groups. In all societies—from the simplest bands to the most complex industrial nations—groups divide the responsibility for completing necessary tasks among their members. This means that humans constantly must rely on one another; hence they are the most cooperative of all primates.

The variety of ways in which human groups divide their tasks and choose the kinds of tasks they undertake reflects differences in environment, history, and level of technological development. Yet there are certain commonalities in the division of labor. All cultures distinguish between females and males and between adults and children, and these distinctions are used to organize the division of labor. Thus, in every society there are adult-female tasks, adult-male tasks, and children's tasks. Since the 1960s in the United States, these role distinctions have been changing quite rapidly, a topic we shall explore in Chapter 9.

The Incest Taboo, Marriage, and the Family

All human societies regulate sexual behavior. Sexual mores vary enormously from one culture to another, but all cultures apparently share one basic value: sexual relations between parents and their children are to be avoided. (There is evidence that some primates also avoid sexual relations between males and their mothers.) In most societies it is also wrong for brothers and sisters to have sexual contact (notable exceptions being the brother-sister marriages among royal families in ancient Egypt and Hawaii, and among the Incas of Peru). *Sexual relations between family members* is called **incest,** and because in most cultures very strong feelings of horror and revulsion are attached to incest, it is said to be forbidden by taboo. A **taboo** is *the prohibition of a specific action.*

The presence of the incest taboo means that individuals must seek socially acceptable sexual relationships outside their families. All cultures provide definitions of who is or is not an acceptable candidate for sexual contact. They also provide for institutionalized marriages—ritualized means of publicly legitimizing sexual partnerships and the resulting children. Thus, the presence of the incest taboo and the institution of marriage result in the creation of families. Depending on who is allowed to marry—and how many spouses each person is allowed to have—the family will differ from one culture to another. However, the basic family unit consisting of husband, wife, and children (called the nuclear family) seems to be a recognized unit in almost every culture, and sexual relations among its members (other than between husband and wife) are almost universally taboo. For one thing, this helps keep sexual jealousy under control. For another, it prevents the confusion of authority relationships within the family. Perhaps most important, the incest taboo ensures that family offspring will marry into other families, thus recreating in every generation a network of social bonds among families that knits them together into larger, more stable social groupings.

Rites of Passage

All cultures recognize stages through which individuals pass in the course of their lifetimes. Some of these stages are marked by biological events, such as the start of menstruation in girls. However, most of these stages are quite arbitrary and culturally defined. All such stages—whether or not corresponding to biological events—are meaningful only in terms of each group's culture.

Rarely do individuals drift from one such stage to another; every culture has established **rites of passage,** or *standardized rituals marking major life transitions.* The most widespread—if not universal—rites of passage are those marking the arrival of puberty (often resulting in the individual's taking on adult status), marriage, and death. Typical rites of passage celebrated in American society include baptisms, bar and bat mitzvahs, confirmations, major birthdays, graduation, wedding showers, bachelor parties, wedding ceremonies, major anniversaries, retirement parties, and funerals and wakes. Such rites accomplish several important functions, including helping the individual achieve a sense of social identity, mapping out the individual's life course, and aiding the individual in making appropriate life plans. Finally, rites of passage provide people with a context in which to share common emotions, particularly with regard to events that are sources of stress and intense feelings, such as marriage and death.

Ideology

A central challenge that every group faces is how to maintain its identity as a social unit. One of the most important ways that groups accomplish this is by promoting beliefs and values to which group members are firmly committed. Such **ideologies,** or *strongly held beliefs and values,* are the cement of social structure.

Ideologies are found in every culture. Some are religious, referring to things and events beyond the perception of the human senses. Others are more secular—that is, nonreligious and concerned with the everyday world. In the end, all ideologies rest on untestable ideas rooted in the basic values and assumptions of each culture. There is a story (Geertz, 1973) about a researcher who challenged the assertion of a native of India that the world is supported on the back of an elephant. "Well," he snorted, "what's the elephant standing on?"

"He's standing on the back of a turtle," came the native's confident reply.

"And what's the turtle standing on?"

"From there," the Indian smiled, "it's turtles—turtles all the way down."

Even though ideologies rest on untestable assumptions, their consequences are very real. They give direction and thrust to our social existence and

Certain patterns of behavior, such as marriage, are found in every culture but take various forms. The marriage ceremony, for example, varies greatly among different cultures.

meaning to our lives. The power of ideologies to mold passion and behavior is well known. History is filled with both horrors and noble deeds people have performed in the name of some ideology: thirteenth-century Crusaders, fifteenth-century Inquisitors, pro-states rights and pro-union forces in nineteenth-century America, abolitionists, prohibitionists, trade unionists, Nazis and fascists, communists, segregationists, civil rights activists, feminists, consumer activists, environmentalists. These and countless other groups have marched behind their ideological banners, and in the name of their ideologies they have changed the world, often in major ways.

Culture and Individual Choice

Very little human behavior is instinctual or biologically programmed. In the course of human evolution, genetic programming gradually was replaced by culture as the source of instructions about what to do,

how to do it, and when it should be done. This means that humans have a great deal of individual freedom of action—more than any other creature.

However, as we have seen, individual choices are not entirely free. Simply by being born into a particular society with a particular culture, every human being is presented with a limited number of recognized or socially valued choices. Every society has means of training and of social control that are brought to bear on each person, making it difficult for individuals to act or even think in ways that deviate too far from their culture's norms. To get along in society, people must keep their impulses under some control and express feelings and gratify needs in a socially approved manner at a socially approved time (see "Sociology at Work: The Conflict Between Being a Researcher and a Human Being"). This means that humans inevitably feel somewhat dissatisfied, no matter to which group they belong (Freud, 1930). Coming to terms with this central truth about human existence is one of the great tasks of living.

SUMMARY

All human societies have complex ways of life that differ greatly from one to the other. Each society has its own unique blueprint for living, or culture. Culture consists of all that human beings learn to do, to use, to produce, to know, and to believe as they grow to maturity and live out their lives in the social groups to which they belong.

Humans are remarkably unspecialized; culture allows us to adapt quickly and flexibly to the challenges of our environment. Sociologists view culture as having four major components: material culture, nonmaterial culture, cognitive culture, and language.

Language and the production of tools are central elements of culture. Evidence exists that animals engage, or can be taught to engage, in both of these activities. Does this mean that they have culture? Scientists disagree about how to interpret the evidence. Without question, however, it can be said that humans have refined culture to a far greater degree than other animals and are far more dependent on it for their existence.

Every social group has its own complete culture. Sociologists use the term *subculture* to refer to the distinctive lifestyles, values, norms, and beliefs associated with certain segments of the population within a society. Types of subcultures include ethnic, occupational, religious, political, geographical, social class, and deviant subcultures.

People in all societies must confront and resolve certain common, basic problems. Cultural universals are certain models or patterns that have developed

Sociology at Work

The Conflict Between Being a Researcher and a Human Being

Sociologists and anthropologists are supposed to understand the importance of cultural relativism and realize that cultures must be studied and understood in their own terms before valid comparisons can be made. A researcher should avoid imposing his or her values on a people or interfering with a culture to such an extent that it moves away from its origins. Is this a realistic goal, or are we being overly idealistic when we think this can be accomplished?

Kenneth Good had to deal with these issues often when he studied the Yanomama. The Yanomama are approximately 10,000 South American Indians who live in 125 villages in southern Venezuela and northern Brazil. In terms of their material and technological culture, the Yanomama stand out in their primitiveness. They have no system of numbers, they do not know about the wheel, there is no art or metallurgy. Until recently they made fire by rubbing two sticks together.

Traditionally the Yanomama wear no clothes, but they paint red designs on their bodies. Girls and women adorn their faces by inserting slender sticks through holes in the lower lip at either side of the mouth and in the middle, and then through the nasal septum.

Their lives are characterized by persistent aggression among village members and perpetual warfare with other groups. They engage in club fighting, gang rape, and murder.

Good encountered many events that went against his cultural value system. One day he saw two groups of tribespeople having a tug of war. However, instead of pulling on a thick vine, they were pulling on a woman. Her assailants on one side were three teenage boys. Trying to pull her away from them were three elderly women. The struggle went on for 10 minutes, and Good wanted to stop it. "What are they going to do to her?" he asked a woman. "They're going to rape her," came the answer as casually as if she had said, "They're going to have a picnic."

With a concerted heave, the teenagers pulled her free. Howling in victory, they ran down the trail, yanking her along. As they ran, they were joined by more shouting teenagers. Good followed as the stampede bore her into the jungle.

Kenneth Good and his Yanomama wife, Yarima. (Photo copyright © by Kenneth Good)

in all cultures to resolve those problems. Among them are the division of labor, the incest taboo, marriage, family organization, rites of passage, and ideology. Though the forms are universal, the content is unique to each culture.

By dividing the responsibility for completing necessary tasks among their members, societies create a division of labor. Every culture has established rites of passage, or standardized rituals marking major life transitions. Ideologies, or strongly held beliefs and

I stood there, my heart pounding. I had no doubt I could scare these kids away. On the other hand, I was an anthropologist. I wasn't supposed to take sides and make value judgments. This kind of thing went on. If a woman showed up somewhere unattached, chances were she'd be raped. She knew it, they knew it. It was expected behavior. What was I supposed to do, I thought, try to inject my standards of morality? I hadn't come down here to change these people or because I thought I'd love everything they did: I'd come to study them.

So why was I standing there shaking with anger? Why was I thinking, Come on, Ken, what's wrong with you? Are you going to stand around with your notebook in your hand and observe a gang rape in the name of anthropological science? How could I live with a group of human beings and not be involved with them as a fellow human being?

That afternoon was a turning point for Good. About a month later there was another woman-dragging episode.

After half an hour one of the other men got fed up with the noise. "That's enough," he said, picking up his arrows. "This is really annoying. I'm going to stab her, that will put a stop to it." I watched as he walked up to the three of them with his arrows. He was really going to do it.

When I saw this I yelled, "Don't do that!" He stopped and looked around, surprised. Our eyes met, then he walked back to his hammock and put his arrows down. I knew that I was not the same detached observer I had been before.

Good was scheduled to spend 15 months with the Yanomama. After this time passed, however, he did not leave. He stayed and learned to speak their language and to hunt fish and gather as they did. He learned what it meant to be a nomad. Later he was adopted into the lineage of the village and given a wife, Yarima, according to Yanomama custom and in keeping with the wishes of the tribal leader. Eventually, Yarima had a profound effect on him.

I'm in love. Unbelievable, intense emotion, almost all the time. In the morning when she gets up to start the day, when I see her come in from the gardens with a basket of plantains, especially when we make love. Sure it's universal, except that being in love in Yanomama culture with a Yanomama girl is different, a different game, different rules.

In my wildest dreams it had never occurred to me to marry an Indian woman in the Amazon jungle. I was from suburban Philadelphia. I had no intention of going native. . . . That was what I had come to, after all these years of struggling to fit into the Yanomama world, to speak their language fluently, to grasp their way of life from the inside. My original purpose—to observe and analyze this people . . . had slowly merged with something far more personal.

Eventually, Kenneth Good brought Yarima out of the Amazon jungle to the United States, where she lived from 1988 to 1993. Yarima had never seen flat cleared land, much less houses and cars. She had never worn clothes or walked in shoes. They had three children. On a trip back to the Yanomama as part of a National Geographic documentary, Yarima decided to remain. Good returned to Rutherford, New Jersey, where he lives with their three children and teaches at Jersey City State College. Was Good doing valuable research, or did he impose his values on another culture?

Sources: *Into the Heart*, by K. Good, with D. Chanoff, 1991, New York: Simon & Schuster; and an interview with David Chanoff, July 1994.

values, are the cement of social structure in that they help a group maintain its identity as a social unit.

Due to a lack of instinctual or biological programming, humans have a great deal of flexibility and choice in their activities. Individual freedom of action is limited, however, by the existing culture. Moreover, social pressure to act, think, and feel in socially approved ways inevitably generates individual dissatisfaction. There is thus a tension between the individual and society.

INTERNET RESOURCES

Go to http://web-enhanced.thomsonlearning.com which features Web links relevant to this chapter to expand and enhance the learning experience. This site will be monitored and updated periodically to ensure the links are correct and live.

At Virtual Society: The Wadsworth Sociology Resource Center, http://sociology.wadsworth.com, you will find a Career Center, Sociology in the News, Research Online, InfoTrac College Edition, Virtual Tours in Sociology, and a variety of book-specific student and instructor resources.

CHAPTER THREE STUDY GUIDE

KEY CONCEPTS

Match each concept with its definition, illustration, or explanation presented below.

a. Ideal norms
b. Reformulation
c. Taboo
d. Innovation
e. Cultural relativism
f. Cultural traits
g. Cognitive culture
h. Generalized adaptability
i. Cultural lag
j. McDonald's
k. Ideologies
l. Adaptation

m. Subculture
n. Cultural universals
o. Diffusion
p. Selectivity
q. Mores
r. Nonmaterial culture
s. Norms
t. Culture shock
u. Rites of passage
v. Folkways
w. Symbol
x. Sapir-Whorf hypothesis

y. Ethnocentrism
z. Material culture
aa. Values
bb. Specialization
cc. Incest
dd. Culture
ee. Jane Lawick-Goodall
ff. Biological determinism
gg. Gestures
hh. Washoe
ii. Time
jj. Real norms

_____ 1. Everything humans learn, use, produce, know, and believe as they mature and live out their lives in the social groups to which they belong.

_____ 2. The difficulty people have adjusting to a new culture that differs markedly from their own.

_____ 3. When one makes judgments about other cultures based on the customs and values of one's own.

_____ 4. Recognizing cultures must be understood on their own terms before valid comparisons can be made.

_____ 5. All the things humans make and use, from small hand-held tools to skyscrapers.

_____ 6. The totality of knowledge, beliefs, values, and rules for appropriate behavior.

_____ 7. Agreed upon rules of behavior shared within a culture that prescribe the limits of acceptable behavior.

_____ 8. Strongly held norms that usually have a moral connotation and are based on central values of the culture.

_____ 9. Norms permitting a wide degree of individual interpretation as long as certain limits are not overstepped.

_____ 10. The expectations of what people should do under perfect conditions.

_____ 11. Norms that are expressed with qualifications and allowances for differences in individual behavior.

_____ 12. The thinking component of culture consisting of shared beliefs and knowledge of what is real, what is not real, what is important, and what is trivial.

_____ 13. A culture's general orientation toward life, encompassing what is good, bad, desirable, and undesirable.

_____ 14. A process by which some aspects of the world are viewed as important while others are virtually neglected.

_____ 15. A view of language which argues that the language a person uses determines his or her perceptions of reality.

_____ 16. Anything representing something else, carrying a particular meaning recognized by members of a culture.

_____ **17.** The process by which human beings adjust to changes in their environment.

_____ **18.** Developing ways of doing things that work extremely well in a particular environment or circumstance.

_____ **19.** A society's ability to develop more complicated yet flexible ways of doing things.

_____ **20.** Any new practice or tool that becomes widely accepted by a society.

_____ **21.** Those items of a culture, for example, tools, materials, beliefs, values, and typical ways of doing things.

_____ **22.** The movement of cultural traits from one culture to another culture.

_____ **23.** The way in which a cultural trait is modified in some way so it fits better in its new context.

_____ **24.** The phenomenon through which new patterns of behavior may emerge, even though they conflict with traditional values.

_____ **25.** The distinctive lifestyles, values, norms, and beliefs of certain segments of the population within a society.

_____ **26.** The models or patterns that have developed in all cultures to resolve basic common societal problems.

_____ **27.** Sexual relations between family members.

_____ **28.** The prohibition of a specific action.

_____ **29.** Established or standardized rituals marking major life transitions.

_____ **30.** Strongly held beliefs and values that are the cement of social structure.

_____ **31.** Discovered that chimpanzees living in nature used and produced tools.

_____ **32.** Placing people in fixed slots in a sociopolitical hierarchy based on inborn traits that cannot be changed.

_____ **33.** A famous chimpanzee who was taught American Sign Language.

_____ **34.** A Western example of nonmaterial culture.

_____ **35.** Nonmaterial aspects of a culture that take on totally different meanings from one country to another.

_____ **36.** The most common example of American culture abroad.

CENTRAL IDEA COMPLETIONS

Following the instructions, fill in the appropriate concepts and descriptions for each of the questions posed in the following section.

1. In what ways have McDonald's restaurants taken into account the cultural customs and habits of countries throughout the world? _____

_____

What are some of the differences between customer behaviors and expectations in other countries as contrasted with customers in the United States?

_____

_____

_____

2. Describe how something as apparently spontaneous as a kiss can be analyzed in terms of the norms of a culture.

_____

_____

_____

_____

_____

_____

_____

_____

3. Using the information presented in your text, present the hypothesis that animals have language, then follow with the hypothesis that animals do not have language. Given the available data, which hypothesis would a sociologist most likely support?

 a. Animals have language: _____

 _____

 b. Animals do not have language: _____

 _____

 c. The sociological position: _____

 _____

4. Indicate the importance of the following culture-related concepts:

 a. Innovation: _____

 b. Diffusion: _____

 c. Cultural lag: _____

5. Using an example related to life at your college or university, distinguish between cultural relativism and ethnocentrism.

 a. Example: _____

 b. Cultural relativism: _____

 c. Ethnocentrism: _____

6. List, define, and present an example of the four cultural universals presented in your text.

 a. _____

 b. _____

 c. _____

 d. _____

7. List three important characteristics of material and nonmaterial culture.

Material	Nonmaterial
_____	_____
_____	_____
_____	_____

8. Present a definition of each of the following types of norms and present one example of each norm as it may be found in American society.

 a. Folkway: _____

 b. More: _____

 c. Taboo: _____

9. What is cultural lag? _____

_____

a. What two new forms of technological change do you anticipate will become common in America by the year 2010? _____

_____

b. How might your two forms of technological change be related to behaviors that might conflict with traditional values? _____

_____

10. What are subcultures? _____

a. List five subcultures common to American society in 2001: _____

_____

b. Can you think of five new subcultures that might be present in 2010? _____

_____

11. What is a symbol? _____

_____

How are symbols tied to major values and beliefs in a culture? _____

_____

12. What are the relationships among the incest taboo, marriage, and the family? _____

_____

_____

CRITICAL THOUGHT EXERCISES

1. Your text provides an extensive discussion of the controversies surrounding the question of whether primates other than humans actually can acquire and use language. Pursue these issues by going to the following sites on the Internet.* After reading these additional materials, develop a critical thought essay in which you present your interpretation of the evidence that would answer the question, "Is language common to primates or is it a distinctly human activity?"

a. Do apes use language?
http://whyfiles.news.wisc.edu/
058language/ape_talk.html

b. Language behaviors in human children and chimps
http://pubpages.unh.edu/~jel/
apelang.html

c. The Human-Languages Page—What is human language?
http://www.june29.com/HLP/

d. Chimpanzee and Great Ape Language
http://www.brown.edu/Departments/
Anthropology/apelang.html

2. Since the early days of television, wrestling has been a staple for viewers. After a period of decline, the sport once again has gathered a substantial following. Using a combination of viewing time (where you systematically watch several hours of television wrestling and take notes) and library research on spectator sports, develop a critical thought paper in which you propose and document that this sport and its spectators may be viewed as a subculture. Are there comparable subcultures of spectator sports in other cultures?

3. Explore the Internet for examples of the seven types of subcultures (ethnic, occupational, religious, political, geographic, social class, and deviant) presented in this chapter. What are the characteristics of each type of subculture, and what is distinctive about it relative to the larger normative American culture of which it is a part?

4. One of the major cultural events in the United States takes place on Super Bowl Sunday. Beyond the game itself, there is a week of television precoverage as well as numerous parties around the country that take place on game day. Assume you are a sociologist presenting a paper to a learned society in a non-Western academic setting. Use the concepts of material culture, nonmaterial culture, ideal norms, real norms, and rites of passage in presenting your analysis of Super Bowl Sunday.

5. Using a video camera, collect examples of gestures used by different groups in American society. Interview the people you are videotaping about the meanings they are attaching to the gestures as well as their assumptions about how the gestures are being interpreted by the parties toward whom they are gesturing. Are the gestures examples of group togetherness or ethnocentric putdowns of the parties receiving the gestures?

6. Select a series of formal scientific and colloquial expressions from the United States, Great Britain, and a non-Western society. Discuss the problems of direct translation of these terms across the three cultures. Which types of terms translate easiest and which are most difficult to translate? Support your argument with the discussion of language presented in your text.

7. Humor may be close to a cultural universal as it seems to appear in most societies. What is (are) the function(s) of humor? Some theorists maintain that humor may have a therapeutic function. To examine this possibility, check out the Laughter Therapy Homepage at

http://rampages.onramp.net/~ejunkins/

Develop a presentation on laughter as a form of cultural therapy. Consider how the interpretation of laughter may vary across cultures throughout the world.

* Please note, Web site addresses change frequently. The addresses presented in these exercises should be viewed as guides only, and students should be prepared to utilize appropriate Internet search techniques when necessary should these addresses no longer be accessible.

ANSWER KEY

KEY CONCEPTS

1. dd	7. s	13. aa	19. h	25. m	31. ee
2. t	8. q	14. p	20. d	26. n	32. ff
3. y	9. v	15. x	21. f	27. cc	33. hh
4. e	10. a	16. w	22. o	28. c	34. ii
5. z	11. jj	17. l	23. b	29. u	35. gg
6. r	12. g	18. bb	24. i	30. k	36. j

CHAPTER FOUR

Socialization and Development

LEARNING OBJECTIVES

After studying this chapter, you should be able to do the following:

- Discuss how biology and socialization contribute to the formation of the individual.
- Know what sociobiology is.
- Explain how extreme social deprivation affects early childhood development.
- Identify the stages of cognitive and moral development.
- Explain the views of the self developed by Cooley, Mead, and Freud.
- Understand Erikson's and Levinson's stage models of lifelong socialization.
- Explain how family, schools, peer groups, and the mass media contribute to childhood socialization.
- Know how adult socialization and resocialization differ from primary socialization.

It had been only 11 weeks since the recruits started marine boot camp training at Parris Island, South Carolina, but now, as many revisited civilian society, they felt alienated from their past lives. To Patrick Bayton, "Everything feels different. I can't stand half my friends." Frank Demarco attended a street fair in Bayonne, New Jersey. "It was crowded. Trash everywhere. People were drinking, getting into fights. No politeness whatsoever. This is the way civilian life is—nasty."

As the son of a Wall Street executive, Daniel Keane came from a privileged background. When he came home to spend some time with his family he said, "I didn't know how to act. I didn't know how to carry on a conversation." He found it even more difficult being with his friends. They were "drinking, acting stupid and loud." He was particularly disappointed when two old friends refused to postpone smoking marijuana for a few minutes until he was away from them. "I was disappointed in them doing that. It made me want to be at SOI (the Marines' School of Infantry)."

Keane felt like he had joined a cult or a religion. "People don't understand and I'm not going to waste my breath trying to explain when the only thing that really impresses them is how much beer you can chug down in 30 seconds." Military ideology disapproves of the lack of order and respect for authority that it believes characterizes civilian society. As Sergeant Major James Moore pointed out, "it is difficult to go back into a society of 'What's in it for me?' when a Marine has been taught the opposite for so long" (Ricks, 1997).

During boot camp, the Marine Corps tries to sever a recruit's ties to his or her private life in order to facilitate a process of socialization to the military culture. By the time it is over, the marines come to see themselves as different from society morally and culturally. The boot camp experience has modified their previous years of socialization enough that they now feel most comfortable with others who are also part of that culture.

A marine's first instruction on becoming a member of society began in childhood. This *process of social interaction that teaches the child the intellectual, physical, and social skills needed to function as a member of society* is called **socialization.** It is through socialization experiences that children learn the culture of the society into which they have been born. In the course of this process, each child slowly acquires a **personality**—that is, *the patterns of behavior and ways of thinking and feeling that are distinctive for each individual.* Contrary to popular wisdom, nobody is a born salesperson, criminal, or military officer. These things all are learned and modified as part of the socialization process.

Becoming a Person: Biology and Culture

Every human being is born with a set of **genes,** *inherited units of biological material.* Half are inherited from the mother, half from the father. No two people have exactly the same genes, except for identical twins. Genes are made up of complicated chemical substances, and a full set of genes is found in every body cell. Scientists still do not know how many different genes a human being has, but certainly they number in the tens of thousands.

Genes influence the chemical processes in our bodies and even control some of these processes

The resemblance between this mother and son is quite clear, since half his genes were inherited from her.

completely. For example, such things as blood type, the ability to taste the presence of certain chemicals, and some people's inability to distinguish certain shades of green and red are completely under the control of genes. Most of our body processes, however, are not controlled solely by genes but are the result of the interaction of genes and the environment (physical, social, and cultural). Thus, how tall you are depends on the genes that control the growth of your legs, trunk, neck, and head and also on the amount of protein, vitamins, and minerals in your diet. Genes help determine your blood pressure, but so do the amount of salt in your diet, the frequency with which you exercise, and the amount of stress under which you live.

Nature Versus Nurture: A False Debate

For more than a century, sociologists, educators, and psychologists have argued about which is more important in determining a person's qualities: inherited characteristics (nature) or socialization experiences (nurture). After Charles Darwin (1809–1882) published *On the Origin of Species* in 1859, human beings were seen to be a species similar to all the others in the animal kingdom. Because most animal behavior seemed to the scholars of that time to be governed by inherited factors, they reasoned that human behavior similarly must be determined by instincts—biologically inherited patterns of complex behavior. Instincts were seen to lie at the base of all aspects of human behavior, and eventually researchers cataloged more than 10,000 human instincts (Bernard, 1924).

Then, at the turn of the century, a Russian scientist named Ivan Pavlov (1849–1936) made a startling discovery. He found that if a bell were rung just before dogs were fed, eventually they would begin to salivate at the ringing of the bell itself, even when no food was served. The conclusion was inescapable: So-called instinctive behavior could be molded or, as Pavlov (1927) put it, conditioned, through a series of repeated experiences linking a desired reaction with a particular object or event. Dogs could be taught to salivate. Pavlov's work quickly became the foundation on which a new view of human beings was built—one that stressed their infinite capacity to learn and be molded. The American psychologist John B. Watson (1878–1958) taught a little boy to be afraid of a rabbit by startling him with a loud noise every time he was allowed to see the rabbit. What he had done was to link a certain reaction (fear) with an object (the rabbit) through the repetition of the experience. Watson eventually claimed that if he were given complete control over the environment of a dozen healthy infants, he could train each one to be whatever he wished—doctor, lawyer, artist, merchant, even beggar or thief (Watson, 1925). Among certain psychologists, conditioning became the means through which they explained human behavior.

Sociobiology

The debate over nature versus nurture took a different turn when Edward O. Wilson published his book *Sociobiology: The New Synthesis* (1975). The discipline of sociobiology uses biological principles to explain the behavior of all social beings, both animal and human.

The study of the biological basis of social behavior in animals had long been accepted by biologists. When Wilson published his book, in the 27th and final chapter, he applied the theories of sociobiology to human beings.

According to Edward O. Wilson (1994), human beings inherit a propensity for certain behaviors and social structures. These traits commonly are called human nature and include the division of labor between the sexes, bonding between parents and children, heightened altruism toward closest kin, incest avoidance, some forms of ethical behavior, suspicion of strangers, tribalism, dominance orders within groups, male dominance overall, and territorial aggression over limited resources. Although people have free will, our genes make certain behaviors more likely than others. While cultures may vary, they all begin to lean in certain common directions.

When an especially harsh and prolonged winter leaves an Inuit (Eskimo) family without food supplies, the family must break camp and quickly find a new site in order to survive. Frequently an elderly member of the family, often a grandmother, who may slow down the others and require some of the scarce food, will stay behind and face certain death. Wilson (1975, 1979) saw this as an example of altruism, which ultimately might have a biological component.

Wilson concluded that behavior can be explained in terms of the ways in which individuals act to increase the probability that their genes and the genes of their close relatives will be passed on to the next generation. Proponents of this view, known as sociobiologists, believe that social science one day will be a mere subdivision of biology. Sociobiologists would claim that the grandmother, in sacrificing her own life, is improving her kin's chances of survival. She already has made her productive contribution to the family. Now the younger members of the family must survive to ensure the continuation of the family and its genes into future generations.

Up until the publication of Wilson's book, the prevailing view in the social sciences had been that there was no biologically based human nature, that human behavior is almost entirely sociocultural in origin, and therefore genes play little or no role in social behavior. Wilson took the opposite position, proposing that human behavior cannot be understood without biology (1994).

Critics saw sociobiology as not just intellectually flawed but also morally wrong. If human nature is rooted in heredity, they suggested, then some forms of social behavior probably are intractable. Tribalism and gender differences then might be judged unavoidable and class differences and war in some manner natural. People then also could argue that some racial or ethnic groups also differ irreversibly in personal abilities and emotional attributes. Some people could have inborn mathematical genius, others a bent toward criminal behavior (Wilson, 1994).

The furor over sociobiology was so strong that the American Anthropological Association attempted to pass a resolution to censure sociobiology. Only a passionate speech by noted anthropologist Margaret Mead defeated the motion (Fisher, 1994).

Within Wilson's own department at Harvard University, his chair, Richard Lewontin, and another department member, Stephen Jay Gould, formed the Sociobiology Study Group to publicly denounce Wilson's ideas. Wilson needed bodyguards and was publicly attacked at academic conferences.

Fellow biologist Stephen Jay Gould (1976) proposed another, equally plausible scenario, one that discounts the existence of a particular gene programmed for altruism. He perceived the grandmother's sacrifice as an adaptive cultural trait. (It is widely acknowledged that culture is a major adaptive mechanism for humans.) Gould suggested that the elders remain behind because they have been socially conditioned from earliest childhood to the possibility and appropriateness of this choice. They grew up hearing the songs and stories that praised the elders who stayed behind. Such self-sacrificers were the greatest heroes of the clan. Families whose elders rose to such an occasion survived to celebrate the self-sacrifice, but those families without self-sacrificing elders died out.

Wilson made several major concessions to Gould's viewpoint, acknowledging that among human beings, "the intensity and form of altruistic acts are to a large extent culturally determined" and that "human social evolution is obviously more cultural than genetic." He also left the door open to free will, admitting that even though our genetic coding may have a major influence, we still have the ability to choose an appropriate course of action (Wilson, 1978). However, Wilson insisted, "history did not begin 10,000 years ago. . . . [B]iological history made us what we are no less than culture" (1994).

Gould conceded that human behavior has a biological, or genetic, base. He distinguished, however, between genetic determinism (the sociobiological viewpoint) and genetic potential. What the genes prescribe is not necessarily a particular behavior but rather the capacity to develop certain behaviors. Although the total array of human possibilities is inherited, which of these numerous possibilities a particular person displays depends on his or her experience in the culture.

Although both nature and nurture are important, the debate over the relative contribution of each continues. However, just as a winter snowfall is the result of both the temperature and the moisture in the air, so must the human organism and human behavior be understood in terms of both genetic inheritance and the effects of environment. Nurture—that is, the entire socialization experience—is as essential a part of human nature as are our genes. It is from the interplay between genes and environment that each human being emerges (Fisher, 1994).

Deprivation and Development

Some unusual events and interesting research indicate that human infants need more than just food and shelter if they are to function effectively as social creatures.

Extreme Childhood Deprivation   There are only a few recorded cases of human beings who have grown up without any real contact with other humans. One such case took place in January of 1800, when hunters in Aveyron, in southern France, captured a boy who was running naked through the forest. He seemed to be about 11 years old and apparently had been living alone in the forest for at least 5 or 6 years. He appeared to be thoroughly wild and subsequently was exhibited in a cage, from which he managed to escape several times. Finally, he was examined by "experts" who found him to be an incurable "idiot" and placed him in an institute for deaf-mutes. However, a young doctor, Jean-Marc Itard, thought differently. After close observation, he discovered that the boy was neither deaf nor mute nor an idiot. Itard believed that the boy's wild behavior, lack of speech, highly developed sense of smell, and poor visual attention span all were the result of having been deprived of human contact. It appeared that the crucial socialization provided by a family had been denied him. Though human, the boy had learned little about how to live with other people. Itard took the boy into his house, named him Victor, and tried to socialize him. He had little success. Although Victor slowly learned to wear clothes, to speak and write a few simple words, and to eat with a knife and fork, he ignored human voices unless they were associated with food, developed no relationships with people other than Itard and the housekeeper who cared for him, and died at the age of 40 (Itard, 1932; Shattuck, 1980).

Another sad case concerns a girl named Anna, who grew up in the 1930s and had the misfortune of being born illegitimately to the daughter of an extremely disapproving family. Her mother tried to place Anna with foster parents, was unable to do so, and therefore brought her home. In order to quiet the family's harsh criticisms, the young mother hid Anna away in a room in the attic, where she could be out of sight and even forgotten by the family. Anna remained there for almost 6 years, ignored by the whole family, including her mother, who did the very minimum to keep her alive. Finally, Anna was discovered by social workers. The 6-year-old girl was unable to sit up, to walk, or to talk. In fact, she was so withdrawn from human beings that at first she appeared to be deaf, mute, and brain damaged. However, after she was placed in a special school, Anna did learn to communicate somewhat, to walk (awkwardly), to care for herself, and even to play with other children. Unfortunately, she died at the age of 10 (Davis, 1940).

Another case of extreme childhood isolation was that of a girl named Genie, who came to the attention of authorities in California in 1970. Genie's nearly blind mother went to the California social welfare offices looking for help for herself, not Genie. The social worker was transfixed by the child,

Jean-Marc Itard attempted to establish normal human relationships and communication with the "Wild Boy of Aveyron," whom he nicknamed "Victor."

a small, withered girl with a halting gait and an unnaturally stooped posture. The worker alerted her supervisor to what she thought was an unreported case of autism in a child estimated to be 6 or 7 years old.

Genie was actually 13½ years old, weighed only 59 pounds, and was only 54 inches tall. She could not focus her eyes beyond 12 feet, could not chew solid food, showed no perception of hot and cold, and could not talk. Her condition was due to her father, who throughout her whole life had confined Genie to a small bedroom, harnessed to an infant's potty seat. Genie was left to sit in the harness, unable to move anything except her fingers and hands, feet and toes, hour after hour, day after day, month after month, year after year (Rhymer, 1993).

When Genie was hospitalized she was unsocialized, severely malnourished, unable to speak or even stand upright. After 4 years in a caring environment, Genie had learned some social skills, was able to take a bus to school, had begun to express some feelings toward others, and had achieved the intellectual development of a 9-year-old. There were still, however, serious problems with her language development that could not be corrected, no matter how involved the instruction (Curtiss et al., 1977).

These examples of extreme childhood isolation point to the fact that none of the behavior we think of as typically human arises spontaneously. Humans must be taught to stand up, to walk, to talk, even to think. Human infants must develop **social attachments;** *they must learn to have meaningful interactions and affectionate bonds with others.* This seems to be a basic need of all primates, as the research by Harry F. Harlow shows.

In a series of experiments with rhesus monkeys, Harlow and his coworkers demonstrated the importance of body contact in social development (Harlow, 1959; Harlow & Harlow, 1962). In one experiment, infant monkeys were taken from their mothers and placed in cages, where they were raised in isolation from other monkeys. Each cage contained two substitute mothers: One was made of hard wire and contained a feeding bottle; the other was covered with soft terry cloth but did not have a bottle. Surprisingly, the baby monkeys spent much more time clinging to the cloth mothers than to the wire mothers, even though they received no food at all from the cloth mothers. Apparently, the need to cling to and to cuddle against even this meager substitute for a real mother was more important to them than being fed.

Other experiments with monkeys have confirmed the importance of social contact in behavior. Monkeys raised in isolation *never* learn how to interact with other monkeys or even how to mate. If placed in a cage with other monkeys, those who were raised in isolation either withdraw or become violent and aggressive—threatening, biting, and scratching the others.

Female monkeys who are raised without affection make wretched mothers themselves. After being impregnated artificially and giving birth, such monkeys either ignore their infants or display a pattern of behavior described by Harlow as "ghastly."

As with all animal studies, we must be very cautious in drawing inferences for human behavior. After all, we are not monkeys. Yet Harlow's experiments show that without socialization, monkeys do not develop normal social, emotional, sexual, or maternal behavior. Because human beings rely on learning even more than monkeys do, it is likely that the same is true of us.

It is obvious that the human organism needs to acquire culture to be complete; it is very difficult, if not impossible, for children who have been isolated from other people from infancy onward to catch up. They apparently suffer permanent damage, although human beings do seem to be somewhat more adaptable than were the rhesus monkeys studied by Harlow.

Infants in Institutions   Studies of infants and young children in institutions confirm the view that human beings' developmental needs include more than the mere provision of food and shelter. Psychologist Rene Spitz (1945) visited orphanages in Europe and found that in those dormitories where children were given routine care but otherwise were ignored, they were slow to develop and were withdrawn and sickly. These children's needs for social attachment were not met.

With the fall of the former Soviet Union, people in the West became aware of the conditions in eastern European orphanages. In many of these orphanages, no consistent responsive care giving was provided. Children had multiple caregivers who through understaffing, ignorance, and disinterest tended to the children's needs with as little contact as possible. Rooms were often bare of toys or objects to play with. Malnutrition was common in the culture and worse in the orphanages where meager food supplies were sometimes consumed by malnourished staff. Children were often left in bed without any care or attention for hours. Some children were never spoken to or called by their names. No real attempt was made to provide physical, intellectual, or emotional stimulation. Children from these experiences usually displayed an **attachment disorder**—they were *unable to trust people and to form relationships with others.* Many people who adopted children from these settings and thought that "love was all the children needed" to overcome these early experiences discovered that extensive treatment was necessary for these children to ever become normally functioning adults.

It appears clear that human infants need more than just food and shelter if they are to grow and develop properly. Every human infant needs frequent contact with others who demonstrate affection, who respond to attempts to interact, and who themselves initiate interactions with the child. Infants also need contact with people who find ways to interest the child in his or her surroundings and who teach the child the physical and social skills and knowledge that are needed to function. In addition, in order to develop normally, children need to be taught the culture of their society—to be socialized into the world of social relations and symbols that are the foundation of the human experience.

The Concept of Self

Every individual comes to possess a social identity by occupying **statuses**—*culturally and socially defined positions*—in the course of his or her socialization. This social identity changes as the person moves through the various stages of childhood and adulthood recognized by the society. New statuses are occupied; old ones are abandoned. Picture a teenage girl who volunteers as a candy striper in a community hospital. She leaves that position to attend college, joins a sorority, becomes a premedical major, and graduates. She goes to medical school, completes an internship, becomes engaged, and then enters a program for specialized training in surgery. Perhaps she marries; possibly she has a child. All along the way she is moving through different social identities, often assuming several at once. When, many years later, she returns to the hospital where she was a teenage volunteer, she will have an entirely new social identity: adult woman, surgeon, wife (perhaps), mother (possibly).

This description of the developing girl was the view from the outside, the way that other members of the society experience her social transitions, or what sociologists would call changes in her social identity. A **social identity** is *the total of all the statuses that define an individual.* But what of the person herself? How does this human being who is growing and developing physically, emotionally, intellectually, and socially experience these changes? Is there something constant about a person's experience that allows one to say, "I am that changing person—changing, but yet somehow the same individual?" In other words, do all human beings have personal identities separate from their social identities? Most social scientists believe that the answer is yes. *This changing yet enduring personal identity* is called the **self.**

The self develops when the individual becomes aware of his or her feelings, thoughts, and behaviors as separate and distinct from those of other people. This usually happens at a young age, when children begin to realize that they have their own history, habits, and needs and begin to imagine how these might appear to others. By adulthood, the concept of self is developed fully. Most researchers would agree that the concept of self includes (1) an awareness of the existence, appearance, and boundaries of one's own body (you are walking among the other members of the crowd, dressed appropriately for the occasion, and trying to avoid bumping into people as you chat); (2) the ability to refer to one's own being by using language and other symbols ("Hi, as you can see from my name tag, I'm Harry Hernandez from Gonzales, Texas."); (3) knowledge of one's personal history ("Yup, I grew up in Gonzales. My folks own a small farm there, and since I was a small boy I've wanted to study farm management."); (4) knowledge of one's needs and skills ("I'm good with my hands all right, but I need the intellectual stimulation of doing large-scale planning."); (5) the ability to organize one's knowledge and beliefs ("Let me tell you about planning crop rotation."); (6) the ability to organize one's experiences ("I know what I like and what I don't like."); and (7) the ability to take a step back and look at one's being as others do, to evaluate the impressions one is creating, and to understand the feelings and attitudes one stimulates in others ("It might seem a little funny to you that a farmer like me would want to come to a party for the opening of a new art gallery. Well, as far back as I can remember, I always kinda enjoyed art, and now that I can afford to indulge myself, I thought maybe I'd buy something I like.") (see Cooley, 1909; Erikson, 1964; Gardner, 1978; Mead, 1934).

Dimensions of Human Development

Clearly, the development of the self is a complicated process. It involves many interacting factors, including the acquisition of language and the ability to use symbols. There are three dimensions of human development tied to the emergence of the self: cognitive development, moral development, and gender identity.

Cognitive Development For centuries, most people assumed that a child's mind worked in exactly the same way as an adult's mind. The child was thought of as a miniature adult who simply was lacking information about the world. Swiss philosopher and psychologist Jean Piaget (1896–1980) was instrumental in changing that view, through his studies of the development of intelligence in chil-

The process of socialization involves trying on a variety of roles that will eventually make up the self.

dren. His work has been significant to sociologists because the processes of thought are central to the development of identity and, consequently, to the ability to function in society.

Piaget found that children move through a series of predictable stages on their way to logical thought, and that some never attain the most advanced stages. From birth to age 2, during the *sensorimotor stage,* the infant relies on touch and the manipulation of objects for information about the world, slowly learning about cause and effect. At about age 2, the child begins to learn that words can be symbols for objects. In this, the *preoperational stage* of development, the child cannot see the world from another person's point of view.

The *operational stage* is next and lasts from age 7 to about age 12. During this period, the child begins to think with some logic and can understand and work with numbers, volume, shapes, and spatial relationships. With the onset of adolescence, the child progresses to the most advanced stage of thinking— *formal, logical thought.* People at this stage are capable of abstract, logical thought and can develop ideas about things that have no concrete reference, such as infinity, death, freedom, and justice. In addition, they are able to anticipate possible consequences of their acts and decisions. Achieving this stage is crucial to developing an identity and an ability to enter into mature interpersonal relationships (Piaget & Inhelder, 1969).

Moral Development Every society has a **moral order**—that is, *a shared view of right and wrong.* Without moral order a society soon would fall apart. People would not know what to expect from themselves and one another, and social relationships would be impossible to maintain. Therefore, the

process of socialization must include instruction about the moral order of an individual's society.

The research by Lawrence Kohlberg (1969) suggests that not every person is capable of thinking about morality in the same way. Just as our sense of self and our ability to think logically develop in stages, our moral thinking develops in a progression of steps as well. To illustrate this, Kohlberg asked children from a number of different societies (including Turkey, Mexico, China, and the United States) to resolve moral dilemmas such as the following: A man's wife is dying of cancer. A rare drug might save her, but it costs $2,000. The man tries to raise the money but can come up with only $1,000. He asks the druggist to sell him the drug for $1,000, and the druggist refuses. The desperate husband then breaks into the druggist's store to steal the drug. Should he have done so? Why or why not?

Kohlberg was more interested in the reasoning behind the child's judgment than in the answer itself. Based on his analysis of this reasoning, he concluded that changes in moral thinking progress step by step through six qualitatively distinct stages (although most people never go beyond Stages 3 or 4):

Stage 1. *Orientation toward punishment.* Those who thought the man should steal (pros) said he could get into trouble if he just let his wife die. Those who said he should not steal (cons) stressed that he might be arrested for the crime.

Stage 2. *Orientation toward reward.* The pros said that if the woman lived, the man would have what he wanted. If he got caught in the act of stealing the drug, he could return the drug and probably would be given only a light sentence. The cons said that the man should not blame himself if his wife died; and that if he got caught, she might die before he got out of jail, so he would have lost her anyway. Stealing just would not pay.

Stage 3. *Orientation toward possible disapproval by others.* The pros observed that nobody would think the man was bad if he stole the drug, but that his family never would forgive him if his wife died and he had done nothing to help her. The cons pointed out that not only would the druggist think of the man as a criminal, but the rest of society would, too.

Stage 4. *Orientation toward formal laws and fear of personal dishonor.* The pros said that the man always would feel dishonored if he did nothing and his wife died. The cons said that even if he saved his wife by stealing, he would feel guilty and dishonored for having broken the law.

Stage 5. *Orientation toward peer values and democracy.* The pros said that failure to steal the drug would cost the man his peers' respect, because he would have acted out of fear rather than out of consideration of what was the logical thing to do. The cons countered that the man would lose the respect of the community if he were caught, because he would show himself to be a person who acted out of emotion rather than according to the laws that govern everybody's behavior.

Stage 6. *Orientation toward one's own set of values.* The pros focused on the man's conscience, saying he never would be able to live with himself if his wife died and he had done nothing. The cons argued that although others might not blame the man for stealing the drug, in doing so he would have failed to live up to his own standards of honesty.

Kohlberg found that although these stages of moral development correspond roughly to other aspects of the developing self, most people never progress to Stages 5 and 6. In fact, Kohlberg subsequently dropped Stage 6 from his scheme because it met with widespread criticism that he could not refute. It was felt by critics that Stage 6 was elitist and culturally biased. Kohlberg himself could find no evidence that any of his long-term subjects ever reached this stage (Muson, 1979). At times, people regress from a higher stage to a lower one. For example, when Kohlberg analyzed the explanations that Nazi war criminals of World War II gave for their participation in the systematic murder of mil-

If a child's need for meaningful interaction with another is not met, the development of a social identity will be delayed.

lions of people who happened to possess certain religious (Jewish), ethnic (gypsies), or psychological (mentally retarded) traits, he found that none of the reasons were above Stage 3 and most were at Stage 1—"I did what I was told to do, otherwise I'd have been punished" (Kohlberg, 1967). However, many of these war criminals had been very responsible and successful people in their prewar lives and presumably in those times had reached higher stages of moral development.

Gender Identity One of the most important elements of our sense of self is our gender identity. Certain aspects of gender identity are rooted in biology. Males tend to be larger and stronger than females, but females tend to have better endurance than males. Females also become pregnant and give birth to infants and (usually) can nurse infants with their own milk. However, gender identity is mostly a matter of cultural definition. There is nothing inherently male or female about a teacher, a pilot, a carpenter, or a typist other than what our culture tells us. As we shall see in Chapter 9 gender identity and sex roles are far more a matter of nurture than of nature.

Theories of Development

Among the scholars who have devised theories of development, Charles Horton Cooley, George Herbert Mead, Sigmund Freud, Erik Erikson, and Daniel Levinson stand out because of the contributions they have made to the way sociologists today think about socialization. Cooley and Mead saw the individual and society as partners. They were symbolic interactionists (see Chapter 1) and as such believed that the individual develops a self solely through social relationships—that is, through interaction with others. They believed that all our behaviors, our attitudes, even our ideas of self, arise from our interactions with other people. Hence, they were pure environmentalists, in that they believed that social forces rather than genetic factors shape the individual.

Freud, on the other hand, tended to picture the individual and society as enemies. He saw the individual as constantly having to yield reluctantly to the greater power of society, to keep internal urges (especially sexual and aggressive ones) under strict control.

Erikson and Levinson presented something of a compromise position. They thought of the individual as progressing through a series of stages of development that express internal urges, yet are greatly influenced by societal and cultural factors.

Charles Horton Cooley (1864–1929)

Cooley believed that the self develops through the process of social interaction with others. This process begins early in life and is influenced by such primary groups as the family. Later on, peer groups become very important as we continue to progress as social beings. Cooley used the phrase **looking-glass self** to describe *the three-stage process through which each of us develops a sense of self*. First, we imagine how our actions appear to others. Second, we imagine how other people judge these actions. Finally, we make some sort of self-judgment based on the presumed judgments of others. In effect, other people become a mirror or looking glass for us (1909).

In Cooley's view, therefore, the self is entirely a social product—that is, a product of social interaction. Each individual acquires a sense of self in the course of being socialized and continues to modify it in each new situation throughout life. Cooley believed that the looking-glass self constructed early in life remains fairly stable and that childhood experiences are very important in determining our sense of self throughout our lives.

One of Cooley's principal contributions to sociology was his observation that although our perceptions are not always correct, what we believe is more important in determining our behavior than is what is real. This same idea was also expressed by sociologist W. I. Thomas (1928) when he noted, "If men define situations as real, they are real in their consequences." If we can understand the ways in which people perceive reality, then we can begin to understand their behavior.

George Herbert Mead (1863–1931)

Mead was a philosopher and a well-known social psychologist at the University of Chicago. His work led to the development of the school of thought called symbolic interactionism (described in Chapter 1). As a student of Cooley's, Mead built on Cooley's ideas, tracing the beginning of a person's awareness of self to the relationships between the caregiver (usually the mother) and the child (1934).

According to Mead, the self becomes the sum total of our beliefs and feelings about ourselves. The self is composed of two parts, the "I" and the "me." The "I" *portion of the self wishes to have free expression, to be active and spontaneous.* The "I" wishes to be free of the control of others and to take the initiative in situations. It is the part of the indi-

George Herbert Mead's writings led to the development of the school of thought known as symbolic interactionism.

vidual that is unique and distinctive. The **"me"** *portion of the self is made up of those things learned through the socialization process from family, peers, school, and so on.* The "me" makes normal social interaction possible, while the "I" prevents it from being mechanical and totally predictable.

Mead used the term **significant others** to refer to *those individuals who are most important in our development, such as parents, friends, and teachers.* As we continue to be socialized, we learn to be aware of the views of the generalized others. These **generalized others** are *the viewpoints, attitudes, and expectations of society as a whole, or of a community of people whom we are aware of and who are important to us.* We may believe it is important to go to college, for example, because significant others have instilled this viewpoint in us. While at college we may be influenced by the views of selected generalized others who represent the community of lawyers that we hope to join one day as we progress with our education.

Mead believed that the self develops in three stages (1934). The first or **preparatory stage** is *characterized by the child's imitating the behavior of others, which prepares the child for learning social-role expectations.* In the second or **play stage,** *the child has acquired language and begins not only to imitate behavior, but also to formulate role expectations:* playing house, cops and robbers, and so on. In this stage, the play features many discussions among playmates about the way things "ought" to be. "I'm the boss," a little boy might announce. "The daddy is the boss of the house." "Oh no," his friend might counter, "Mommies are the real bosses." In the third or **game stage,** *the child learns that there are rules that specify the proper and correct relationship among the players.* For example, in a baseball game there are rules that apply to the game in general as well as to a series of expectations about how each position should be played. During the game stage, according to Mead, we learn the expectations, positions, and rules of society at large. Throughout life, in whatever position we occupy, we must learn the expectations of the various positions with which we interact, as well as the expectations of the general audience, if our performance is to go smoothly.

Thus, for Mead the self is rooted in, and begins to take shape through, the social play of children and is well on its way to being formed by the time the child is 8 or 9 years old. Therefore, like Cooley, Mead regarded childhood experience as very important to charting the course of development.

Sigmund Freud (1856–1939)

Freud was a pioneer in the study of human behavior and the human mind. He was a doctor in Vienna, Austria, who gradually became interested in the problem of understanding mental illness.

In Freud's view, the self has three separately functioning parts: the id, the superego, and the ego. The **id** consists of *the drives and instincts that Freud believed every human being inherits, but which for the most part remain unconscious.* Of these instincts, two are most important: the aggressive drive and the erotic or sexual drive (called libido). Every feeling derives from these two drives. The **superego** represents *society's norms and moral values as learned primarily from our parents.* The superego is the internal censor. It is not inherited biologically, like the id, but is learned in the course of a person's socialization. The superego keeps trying to put the brakes on the id's impulsive attempts to satisfy its drives. So, for instance, the superego must hold back the id's unending drive for sexual expression (Freud, 1920, 1923). The id and superego, then, are eternally at war with each other. Fortunately, there is a third functional part of the self called the **ego,** which *tries not only to mediate in the eternal conflict between the id and the superego, but also to find socially acceptable ways for the id's drives to be expressed.* Unlike the id, the ego constantly evaluates

According to Erik Erikson, adolescence is a time when the teenager must develop an identity as well as the ability to establish close personal relationships with others.

Table 4-1

Erikson's Eight Stages of Human Development

STAGE	AGE PERIOD	CHARACTERISTIC TO BE ACHIEVED	MAJOR HAZARDS TO ACHIEVEMENT
Trust vs. mistrust	Birth to 1 year	Sense of trust or security—achieved through parental gratification of needs and affection	Neglect, abuse, or deprivation; inconsistent or inappropriate love in infancy; early or harsh weaning
Autonomy vs. shame and doubt	1 to 4 years	Sense of autonomy—achieved as child begins to see self as individual apart from his/her parents	Conditions that make the child feel inadequate, evil, or dirty
Initiative vs. guilt	4 to 5 years	Sense of initiative—achieved as child begins to imitate adult behavior and extends control of the world around him/her	Guilt produced by overly strict discipline and the internalization of rigid ethical standards that interfere with the child's spontaneity
Industry vs. inferiority	6 to 12 years	Sense of duty and accomplishment—achieved as the child lays aside fantasy and play and begins to undertake tasks and schoolwork	Feelings of inadequacy produced by excessive competition, personal limitations, or other events leading to feelings of inferiority
Indentity vs. role confusion	Adolescence	Sense of identity—achieved as one clarifies sense of self and what he/she believes in	Sense of role confusion resulting from the failure of the family or society to provide clear role models
Intimacy vs. isolation	Young adulthood	Sense of intimacy—the ability to establish close personal relationships with others	Problems with earlier stages that make it difficult to get close to others
Generativity vs. stagnation	30s to 50s	Sense of productivity and creativity—resulting from work and parenting activities	Sense of stagnation produced by feeling inadequate as a parent and stifled at work
Integrity vs. despair	Old age	Sense of ego integrity—achieved by acceptance of the life one has lived	Feelings of despair and dissatisfaction with one's role as a senior member of society

Source: Adapted from *Childhood and Society,* 2nd. ed., by E. H. Erikson, 1963, New York: W. W. Norton. Copyright renewed by the author, 1978.

social realities and looks for ways to adjust to them (Freud, 1920, 1923).

Freud pictured the individual as constantly in conflict: the instinctual drives of the id (essentially sex and aggression) push for expression, while at the same time the demands of society set certain limits on the behavior patterns that will be tolerated. Even though the individual needs society, society's restrictive norms and values are a source of ongoing discontent (Freud, 1930). Freud's theories suggest that society and the individual are enemies, with the latter yielding to the former reluctantly and only out of compulsion.

Erik H. Erikson (1902–1994)

In 1950 Erikson, an artist-turned-psychologist who studied with Freud in Vienna, published an influen-

tial book called *Childhood and Society* (1963). In it he built on Freud's theory of development but added two important elements. First, he stressed that development is a lifelong process and that a person continues to pass through new stages even during adulthood. Second, he paid greater attention to the social and cultural forces operating on the individual at each step along the way.

In Erikson's view, human development is accomplished in eight separate stages (see Table 4-1). Each stage amounts to a crisis of sorts brought on by two factors: biological changes in the developing individual and social expectations and stresses. At each stage the individual is pulled in two opposite directions to resolve the crisis. In normal development, the individual resolves the conflict experienced at each stage somewhere toward the middle of the opposing options. For example, very few people are entirely trusting, and very few trust nobody at all. Most of us are able to trust at least some other people and thereby form enduring relationships, while at the same time staying alert to the possibility of being misled.

Erikson's view of development has proved to be useful to sociologists because it seems to apply to many societies. In a later work (1968), he focused on the social and psychological causes of the "identity crisis" that seems to be so prevalent among American and European youths. Erikson's most valuable contribution to the study of human development has been to show that socialization continues throughout a person's life and does not stop with childhood. There is indeed development after 30 — and after 60 and 70 as well. The task of building the self is lifelong; it can be considered our central task from cradle to grave. We construct the self—our identity—using the materials made available to us by our culture and our society.

Daniel Levinson (1920–1994)

A blending of sociology and psychology has taken place in the area of adult development. Through research in this field, we have come to recognize that there are predictable age-related developmental periods in the adult life cycle, just as there are in the developmental cycles for children and adolescents. These periods are marked by a concerted effort to resolve particular life issues and goals.

Daniel Levinson (1978) did research in this area involving a male population. He recruited 40 men, ages 35 to 45, from four occupational groups: factory workers, novelists, business executives, and academic biologists. Each participant was interviewed several times during a 2- to 3-month period and again, if possible, in a follow-up session 2 years later. Later Levinson (1996) repeated his study with 45 women in three broad categories; academics, corporate-financial careers, and homemakers.

From these studies, Levinson developed the foundation of his theory. He proposed that adults periodically are faced with new but predictable developmental tasks throughout their lives and that working through these challenges is the essence of adulthood. Levinson believed the adult life course is marked by a continual series of building periods, followed by stable periods, and then followed again by periods in which attempts are made to change some of the perceived flaws in the previous design.

Levinson believed that both men and women go through the same periods of adult development, although there are differences due to the external and internal constraints. For example, gender-role expectations in society or the biological demands of childbearing will influence how a woman works on the developmental tasks of each stage.

Levinson's model describes the periods in the adult life cycle:

1. *Early Adult Period* (ages 17 to 22). Ending childhood and leaving the family of origin is the major task of this period. A great deal of energy is expended in trying to reduce dependence on the family for support or authority. Peer support often becomes crucial to this task.

2. *Entering the Adult World* (ages 22 to 28). This period is marked by trying to make a place for oneself in the world. The tasks now are to make some key choices with respect to career, lifestyle, love, marriage, and separation from the family of origin. The individual may also form a dream that serves as a guiding force and provides images of future life structures.

3. *Age 30 Transitional Period* (ages 28 to 33). At this point, the individual begins to reappraise his or her life and explore new possibilities. This is a time of internal instability, in which many aspects of the individual's life are questioned and examined. Divorce and job changes are common during this period.

4. *Settling Down* (ages 33 to 40). Having reworked some of the aspects of his or her life during the previous period, the individual now is ready to seek order and stability. There is a strong desire for achievement and an earnest attempt to make the dream a reality. The individual wants to "sink roots." There is a movement from "junior" to "senior" membership in the adult world.

5. *Midlife Transition* (ages 40 to 45). This is a major transitional period and represents the turning point between young and middle adulthood. The individual starts to see a difference between the dream and the reality of his or her life, leading to a great deal of soul searching. Divorce and career changes once again become real possibilities. The individual may start to give up certain aspects of the dream and become less achievement and advancement oriented.

6. *Beginning of Middle Adulthood* (ages 45 to 50). The previous period of turmoil has produced a greater acceptance of his- or herself. The individual is less dominated by the need to win or to achieve external rewards and more concerned with enjoying his or her life and work. The individual also has a greater concern for other people than before.

7. *Age 50 Transition* (ages 50 to 55). This is a life period similar to the age 30 transition. Life crises are common in this period especially for persons who have made few significant life changes or inappropriate changes during the previous 10 to 15 years.

8. *Culmination of Middle Adulthood* (55 to 60). During this period the individual tries to achieve the remaining major goals and aspirations.

9. *Late Adult Transition* (age 60 to 65). This period requires a profound reappraisal of the past and a shift to a new era.

The model is particularly interesting to sociologists because it appears to show us that there is a close relationship between individual development and one's position in society at a particular time.

There are, however, problems with being too quick to embrace this type of theory. Levinson's theory of adult development is what is commonly referred to as a "stage theory." Stage theories describe a series of changes that follow an orderly sequential pattern. These theories have been criticized as being too rigid, for assuming that the changes are always in one direction, and for assuming that the stages are universal. Critics say that we can apply stages of development to children, but that when we do so with adults, we leave little room for individual differences, social change, and specific cohort experiences (Neugarten, 1979).

Early Socialization in American Society

Children are brought up very differently from one society to another. Each culture has its own child-rearing values, attitudes, and practices. No matter how children are raised, however, each society must provide certain minimal necessities to ensure normal development. The infant's physical needs must, of course, be addressed. More than that is required, however. Children need speaking social partners (some evidence suggests that a child who has received no language stimulation at all in the first 5 to 6 years of life will be unable ever to acquire speech

(Chomsky, 1975). They also need physical stimulation; objects that they can manipulate; space and time to explore, to initiate activity, and to be alone; and finally, limits and prohibitions that organize their options and channel development in certain culturally specified directions (Provence, 1972).

Every society provides this basic minimum care in its own culturally prescribed ways. A variety of agents, which also vary from culture to culture, are used to mold the child to fit into the society. Here we consider some of the most important agents of socialization in American society.

The Family

For young children in most societies—and certainly in American society—the family is the primary world for the first few years of life. Children also are having significant early experiences in day-care centers with no family members. The values, norms, ideals, and standards presented are accepted by the child uncritically as correct—indeed, as the only way things could possibly be. Even though later experiences lead children to modify much of what they have learned within the family, it is not unusual for individuals to carry into the social relationships of adult life the role expectations that characterized the family of their childhood. Many of our gender-role expectations are based on the models of female and male behavior we witnessed in our families.

Every family, therefore, socializes its children to its own particular version of the society's culture. In addition, however, each family exists within certain subcultures of the larger society: It belongs to a geographical region, a social class, one or two ethnic groups, and possibly a religious group or other subculture. Families differ with regard to how important these factors are in determining their lifestyles and their child-rearing practices. For example, some families are very deeply committed to a racial or ethnic identification such as African American, His-

Every family socializes its children to its own particular version of the society's culture. The values and worldview of a boy brought up on a dairy farm in Wisconsin are likely to differ from those of a child born and brought up in a city such as Los Angeles or New York City.

panic, Chinese, Native American, Italian American, Polish American, or Jewish American. Much of family life may revolve around participation in social and religious events of the community and may include speaking a language other than English.

Evidence also shows that social class and parents' occupations influence the ways that children are raised in the United States. Parents who have white-collar occupations are accustomed to dealing with people and solving problems. As a result, white-collar parents value intellectual curiosity and flexibility. Blue-collar parents have jobs that require specific tasks, obeying orders, and being on time. They are likely to reward obedience to authority, punctuality, and physical or mechanical ability in their children (Kohn & Schooler, 1983).

The past three decades have seen major changes in the structure of the American family. High divorce rates, the dramatic increase in the number of single-parent families, and the common phenomenon of two-worker families have meant that the family as the major source of socialization of children is being challenged. Child-care providers have become a major influence in the lives of many young children. (For a discussion of the effects of day care on the socialization of children, see "Controversies in Sociology: Is Day Care Harmful to Children?")

The School

The school is an institution intended to socialize children in selected skills and knowledge. In recent decades, however, the school has been assigned additional tasks. For instance, in poor communities and neighborhoods, school lunch (and breakfast) programs are an important source of balanced nutrition for children. There is also a more basic problem the school must confront. As an institution, the school must resolve the conflicting values of the local community and of the state and regional officials whose job it is to determine what should be taught. For example, in many school systems parents want to reintroduce school prayer or discussions of religion, even though the Supreme Court has ruled that such actions are not permissible. In other instances, education officials make curriculum changes in the classroom despite the complaints of parents, whose objections are dismissed as perhaps uninformed or tradition bound.

AIDS education is a vivid example of schools deciding what issues should or should not be presented to children. Some parents have objected to teaching young children about sexuality, condoms, and homosexuality, despite the health risk that ignorance could pose. Many school boards have taken the position that the schools have a responsibility to provide this information, even when large numbers of parents object.

In coming to grips with their multiple responsibilities, many school systems have established a philosophy of education that encompasses socialization as well as academic instruction. Educators often aim to help students develop to their fullest capacity, not only intellectually, but also emotionally, culturally, morally, socially, and physically. By exposing the student to a variety of ideas, the teachers attempt to guide the development of the whole student in areas of interests and abilities unique to each. Students are expected to learn how to analyze these ideas critically and reach their own conclusions. The ultimate goal of the school is to produce a well-integrated person who will become socially responsible. Two questions arise: Is such an ambitious, all-embracing educational philosophy working? And is it an appropriate goal for our schools?

In a way, the school is a model of much of the adult social world. Interpersonal relationships are not based on individuals' love and affection for one another. Rather, they are impersonal and predefined by the society with little regard for each particular individual who enters into them. Children's process of adjustment to the school's social order is a preview of what will be expected as they mature and attempt to negotiate their way into the institutions of adult society (job, political work, organized recreation, and so on). Of all the socializing functions of the school, this preview of the adult world may be the most important. (The role of the school in socialization will be discussed more extensively in Chapter 12.)

Peer Groups

Peers are *individuals who are social equals.* From early childhood until late adulthood, we encounter a wide variety of peer groups. No one will deny that they play a powerful role in our socialization. Often their influence is greater than that of any other socialization source.

Within the family and the school, children are in socially inferior positions relative to figures of authority (parents, teachers, principals). As long as the child is small and weak, this social inferiority seems natural, but by adolescence a person is almost fully grown, and arbitrary submission to authority is not so easy to accept. Hence, many adolescents withdraw into the comfort of social groups composed of peers.

Parents may play a major role in the teaching of basic values and the desire to achieve long-term goals, but peers have the greatest influence in lifestyle issues, such as appearance, social activities, and dating. Peer groups also provide valuable social

Peer groups provide valuable social support for adolescents.

support for adolescents who are moving toward independence from their parents. As a consequence, their peer-group values often run counter to those of the older generation. New group members quickly are socialized to adopt symbols of group membership such as styles of dress, use and consumption of certain material goods, and stylized patterns of behavior. A number of studies have documented the increasing importance of peer-group socialization in the United States. One reason for this is that parents' life experiences and accumulated wisdom may not be very helpful in preparing young people to meet the requirements of life in a society that is changing constantly. Not infrequently, adolescents are better informed than their parents are about such things as sex, drugs, and technology.

Peer-group influence, for many youths, can lead to wasted lives and violence. For many, gangs—kids banding together for identity, status, petty criminal activity, and mutual protection—often involve drug abuse. In many urban high schools, for example, cocaine is traded, and attempts to emphasize the dangers of drugs fall on deaf ears.

The negative effects of peer pressure are felt on college campuses as well as in poor inner-city neighborhoods. Peer pressure has caused deaths from hazing activities in college fraternities. For example, one month after entering the Massachusetts Institute of Technology, freshman Scott Kruger died after consuming excessive amounts of alcohol as part of a hazing ritual at a fraternity party.

As the authority of the family diminishes under the pressures of social change, peer groups move into the vacuum and substitute their own morality for that of the parents. Peer groups are most effective in molding the behavior of those adolescents whose parents do not provide consistent standards, a principled moral code, guidance, and emotional support. David Elkind (1981, 1994) has expressed the view that the power of the peer group is in direct proportion to the extent that the adolescent feels ignored by the parents. In fact, over a half century ago sociologist David Riesman, in his classic work *The Lonely Crowd* (1950), already thought that the peer group had become the single most powerful molder of many adolescents' behavior and that striving for peer approval had become the dominant concern of an entire American generation—adults as well as adolescents. He coined the term *other directed* to describe those who are overly concerned with finding social approval.

The Mass Media

It is possible that today Riesman would review his thinking somewhat. Since the late 1960s, the mass media—television, radio, magazines, films, newspapers, and the Internet—have become important agents of socialization in the United States. It is almost impossible in our society to escape from the images and sounds of television or radio; in most homes, especially those with children, the media are constantly visible or audible. For the most part, the communication is one-way, creating an audience that is conditioned to receive passively whatever news, messages, programs, or events are brought to them.

Today, 98.2% of all households in the United States have television sets, with an average of two sets per home. Most children become regular watchers of television between ages 3 and 6. Schoolchildren watch an average 2½ hours of television a day on school days and an average of 4 hours and 20 minutes on an average weekend. Their favorite programs are situation comedies, cartoons, music videos, sports, game shows, talk shows, and soap operas. One study concluded that by the time most people reach the age of 18, they will have spent more waking time watching television than doing anything else—talking with parents, spending time with friends, or even going to school (*Statistical Abstract of the United States: 1996*).

Because young children are so impressionable, and because in so many American households the

Controversies in Sociology

Is Day Care Harmful to Children?

People have been concerned about day-care arrangements for children since well before the current period of two-income families. The first day nurseries were established in the mid-1800s during the Industrial Revolution, when women left their homes to work in factories. During the Depression of the 1930s, public funds were used for child-care centers as a way of providing jobs for unemployed teachers, nurses, and cooks. When World War II started and women went to work in industry, child care boomed. But when the war ended so did the child-care boom, not to be rekindled until the 1960s when preschool programs became popular again (Droege, 1995).

According to the Bureau of Labor Statistics (1997), only 13% of families fit the traditional model of husband as wage-earner and wife as homemaker. In 61% of married couple families, both husband and wife work outside the home. Six out of every 10 mothers of children under age 6 are in the labor force. Small wonder that child care is becoming an ever-increasing fact of life.

There are three types of child care. Nearly 32.8% of the children are cared for in their own homes (principally by their fathers or members of the extended family). In the second type of arrangement, usually referred to as "family day care," a woman takes four to six unrelated young children into her home on a regular basis; 31.3% of child care is of this type. Another 29.3% of the children are enrolled in day-care centers, and 5.5% of the children are cared for by a parent (typically the mother) while they are working (see Figure 4-1) (U.S. Bureau of Census, 1997).

With many parents then putting their preschoolers in organized child care, we must ask how well this arrangement serves the children when compared with

traditional child rearing. Sociologists begin to answer this question by noting that just as all parenting is not uniformly good, neither is all day care mediocre. High-quality day care does exist and can be a suitable substitute for home parenting. Unfortunately, poor quality care, which threatens the child with serious physical and psychological harm, is also readily available.

Critics of day care have argued that sensitive and responsive caregiving is vital to developing a secure feeling of attachment between the parent and the child, and long periods in day care threaten this connection. Responsive caregiving causes the child to be more successful with peers and in school, whereas children without this experience tend to be more aggressive and misbehave in school (Belsky, 1988).

As day care has become a common reality for many children, it may be too simplistic to discuss whether day care is good or bad in itself. What is really important is what the quality of the day care is. Studies show that children who spend time in high-quality child care are just as likely to have secure attachment relationships as children who stay at home. In addition, quality child care can also lead to better mother-child interaction (National Institutes of Health, 1997). Children whose home circumstances are less than ideal and who attend high-quality day care have been shown to benefit

Figure 4-1 **Primary Child-Care Arrangements for Preschool Children**

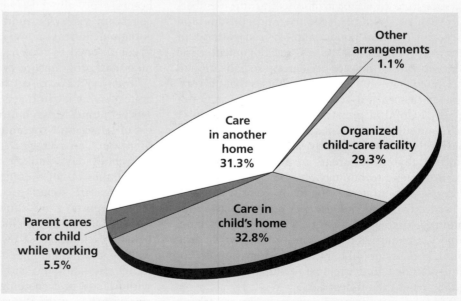

Source: U.S. Bureau of the Census, 1997.

Day-care quality is dependent on the adult/child ratios in the center and the overall size of the group within which the children are cared for.

from the relationship they develop with a sensitive and responsive child-care provider (Howes & Hamilton, 1991).

Children in day care learn cooperation skills better and sooner than children who are cared for at home. These children also have more advanced language skills and tend to be less timid and fearful, more outgoing and cooperative, and function more successfully in social relationships. When contrasted to smaller families with only one or two children, day-care children have a chance to interact with their peers at a younger age. When they get to school, such necessities as sharing and making friends do not come as such a shock (Erbe, 1989).

High-quality day care is dependent on the adult/child ratios in the center and the overall group size in which children are cared for. Staffing guidelines recommend that there be a child-care worker to child ratio of 1:3 for infants under two, 1:4 for toddlers, and 1:7 for preschoolers. Unfortunately, many day-care centers fall short in this area. The National Child Care Staffing Study found that the average adult/child ratio for infants was nearly 1:4, for toddlers 1:6, and for preschoolers over 1:8. The average group size for infants was 7, for toddlers almost 10, and for preschoolers almost 14. Nearly half of all settings exceed optimal guidelines (NCCSS; Whitebook, Phillips, & Howe, 1993).

The stability of the caregiver and peers with whom the child spends the day is another important indicator of quality. Staff turnover in the child-care profession is very high. Child-care workers earn considerably less than others in the workforce with similar levels of education. Many hold college degrees, yet the average salary is below the poverty threshold at less than $12,000 a year (U.S. Bureau of the Census, 1999). Small wonder that there is a 41% turnover rate within the field, since child-care workers see little incentive to remain at the job when other options become available (Whitebook, Phillips, & Howe, 1993).

Problems also arise in centers where there is a high turnover of children. A National Center of Health Statistics Study found that one-fourth of all children switched centers within the previous 12 months. Unfortunately, many children going to day-care centers do not encounter enduring, stable relationships with child-care workers or with the other children (Gill, 1991).

Approximately 20% of working parents place their children in the care of relatives. This is often a less ideal option than a day-care center. One study found that 69% of all the relative care was inadequate (Galinsky et al., 1994). While many parents may feel it is safest and most convenient to leave their child with a family member, day-care centers were more likely to plan activities for the children and the workers had more training in the field than did relatives.

Mothers working outside the home also are providing their children with a different model of what a woman's role in society is. This is particularly important to women who have daughters. Research has shown that the daughters of working women see the world in a less gender-stereotypical fashion than do daughters of nonworking women, are likely to develop a greater sense of independence, are less

Controversies in Sociology continues

Controversies in Sociology

(continued)

fearful and more outgoing and ambitious, and have higher self-esteem (Epstein, 1988).

In addition to the impact that two-worker families have on children and child-care arrangements, there is also a substantial effect on the parents themselves. In the traditional role of mother, women are considered to be responsible for the well-being of the children. If the children have problems, those problems are assumed to be her fault. Mothers are vulnerable to self-blame when their children show signs of distress (Barnett & Baruch, 1987). This issue can be addressed by fathers taking a more active role in the child care.

Parents, particularly mothers, are likely to experience considerable role conflict and role strain as they try to meet the obligations of the job and the family. Yet this may not be bad, because occupying more than one role appears to buffer women from the stress within each role (Crosby, 1991). Women who occupy both family and work roles may be less distressed by problems associated with either role than those more fully invested in only one (Barnett & Marshall, 1991).

Because day care is now a reality for a large segment of the nation's preschoolers, the issue that

needs to be addressed is how to make it a rewarding socialization experience. Clearly, this requires well-trained staff and a properly supervised environment. If we can create that type of environment, then day care can be a positive experience.

Sources: *National Child Care Staffing Study Revisited: Four Years in the Life of Center-Based Care,* by M. Whitebook, D. Phillips, & C. Howe, 1993, Oakland, CA: Child Care Employee Project; *The Study of Children in Family Child Care and Relative Care: Highlights and Findings,* E. Galinsky, C. Howe, S. Koutos, & M. Shinn, 1994, New York: Families and Work Institute; "Results of NICHD Study of Early Child Care Reported at Society for Research in Child Development Meeting," April 3, 1997, National Institutes of Health, *NIH News Release; Juggling: The Unexpected Advantages of Balancing Career and Home for Women and Their Families,* by F. J. Crosby, 1991, New York: The Free Press; "The Relationship Between Women's Work and Family Roles and Their Subjective Well-Being and Psychological Distress," by R. C. Barnett & N. L. Marshall, 1991, in M. Frankenhaeuser, V. Lundberg, & M. Chesney (Eds.), *Women, Work, and Health: Stress and Opportunities,* New York: Plenum; "Child Care: An Educational Perspective," by Kristen Droege, Winter 1995, *MIJCF, Jobs and Capital,* 4, pp. 1–8; "Child Care for Young Children," by C. Howes & C. Hamilton, 1993, in B. Spodek (Ed.), *Handbook of Research in Early Childhood Education,* New York: Macmillan.

---

television is used as an unpaid mechanical baby-sitter, social scientists have become increasingly concerned about the socializing role played by the mass media in our society. By the time the average American child leaves elementary school, she or he has seen 8,000 killings and 100,000 violent acts portrayed on television. By the time that child reaches age 18, this number has increased to 40,000 killings and 200,000 acts of violence (Huston et al., 1992). If children learn from experience, such exposure must certainly have an effect (see "Remaking the World: Should Television Be Used to Teach Values?").

In one study (Centerwall, 1992), an attempt was made to see if there was a connection between the change in violent crime rates following the introduction of television in the United States. It found that murder rates in both the United States and Canada increased, almost 93% in the United States and 92% in Canada (adjusted for population increases) between 1945, when widespread commercial television did not exist, and 1970.

South Africa introduced television many years after its introduction in Canada and the United States, because the apartheid government feared that tele-

vision would be destabilizing. In South Africa, too, the homicide rate rose sharply after the first generation of television children grew up.

These facts caused the researcher (Centerwall) to conclude that "long term childhood exposure to television is a causal factor behind approximately one-half of the homicides committed in the United States, or approximately 10,000 homicides annually." He further estimated that as many as half of all rapes and assaults committed in the United States could be related to television. Others are more conservative in their views and believe that only 10% of all violence is related to television or movies (Comstock, 1992).

In addition to merely seeing violence on television, it is important to evaluate what the message of that violence may be. When 2,693 cable television shows were studied over a 20-week period, a number of answers appeared:

■ Perpetrators of violence were unpunished in 73% of all violent incidents. When violence is unpunished the message is that violence is an acceptable way of resolving problems.

People have become increasingly concerned about the socializing role played by the mass media.

■ Forty-seven percent of all violent interactions showed no harm to the victim, and 58% showed no pain. Fewer still showed any sign of blood. Rarely was there any presence of the long-term negative effects of violence such as psychological, emotional, of financial harm. The cameras quickly cut away from dead or dying bodies. Some have termed this "happy violence" (Gerbner, 1993).

Such information has produced great concern among many. In testimony on behalf of the American Psychological Association before the United States Senate, a researcher (Eron, 1992) noted:

> There can no longer be any doubt that heavy exposure to televised violence is one of the causes of aggressive behavior, crime, and violence in society. . . . The fact that we get the same finding of a relationship between television violence and aggression in children in study after study, in one country after another cannot be ignored.

Lately there have been attempts to warn viewers of violent content in programs. In 1997 the television networks agreed to add content ratings to the existing age-based ratings, using the letter *S* for sex, *L* for Language, *V* for violence, and *D* for suggestive dialogue. The ratings are designed to work in conjunction with the V-chip, a device in all new television sets produced after 1997. The V-chip allows parents to program their sets to screen out any program rated beyond a desired level (Reuters, July 11, 1997).

This may not really help the problem. When boys, especially those aged 10 to 14, saw that a program or movie had a "parental discretion" advisory, they found that show more attractive and wanted to watch it. In addition, no evidence was found that antiviolence public service announcements altered adolescents' attitudes toward the appropriateness of using violence to resolve conflict (Mediascope National Violence Study, 1996).

Proving cause and effect in sociology is never easy, however, and other researchers point out that the relationship between television and violence may be more complex than is generally acknowledged. It is difficult to use television programming as an explanation for race and gender differences in violent behavior. Whites are more likely to subscribe to the violent cable channels. Yet arrests for violence-related crimes are significantly higher for black youths than white youths. Adolescent girls watch nearly as much violent TV as boys. Yet murders by female teens showed stable rates (180 in 1983, 171

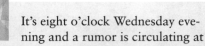

Remaking the World

Should Television Be Used to Teach Values?

It's eight o'clock Wednesday evening and a rumor is circulating at a small-town high school in Massachusetts that a student named Jack is gay. Jack's friends—one of whom is a 15-year-old girl who has been sexually active since she was 13, and another of whom has a mother who has recently committed adultery—assure him it would be okay with them if he were gay, but admit their relief when he says he isn't. An hour later, in San Francisco, a woman named Julia is being beaten by her boyfriend. Meanwhile, in Los Angeles, a young stripper who has given birth out of wedlock learns that her own mother locked her in a basement when she was 3 years old, an experience that she thinks may explain her inability to love her own child.

A typical evening in America? If a visitor from another planet had turned on the television (specifically the WB and Fox networks) on the evening of Wednesday, February 10, 1999, with the aim of learning about our society, he would likely have concluded that it is made up pretty exclusively of photogenic young people with disintegrating nuclear families and liberal attitudes about sex. It's obviously not an accurate picture, but what might our visitor have learned from the programs he watched? Would all the sex, violence, and pathology he saw teach him antisocial behavior? Or might he glean from prime-time dramas and sitcoms the behavior and attitudes that he would do well to adopt if he intended to go native in America?

This is not an idle question—not because aliens might be watching American television, but because people are, particularly impressionable children and teenagers. In a time when 98% of U.S. households own at least one television set—a set which is turned on for an average of nearly seven hours a day. The degree to which people learn from and emulate the behavior of the characters they see on TV is an academic cottage industry.

But if television contributes to poor behavior, might it also be a vehicle for encouraging good behavior? In 1988, Jay Winsten, a professor at the Harvard School of Public Health and the director of the school's Center for Health Communication, conceived a plan to use television to introduce a new social concept—the "designated driver"—to North America. Shows were already dealing with the topic of drinking, Winsten reasoned, so why not add a line of dialogue here and there about not driving drunk? With the assistance of then-NBC chairman Grant Tinker, Winsten met with more than 250

writers, producers, and executives over six months, trying to sell them on his designated driver idea.

Winsten's idea worked; the "designated driver" is now common parlance across all segments of American society and in 1991 won entry into a Webster's dictionary for the first time. An evaluation of the campaign in 1994 revealed that the designated driver "message" had aired on 160 prime-time shows in four seasons and had been the main topic of 25 30-minute or 60-minute episodes. More important, these airings appear to have generated tangible results. In 1989, the year after the "designated driver" was invented, a Gallup poll found that 67% of adults had noted its appearance on network television. What's more, the campaign seems to have influenced adult behavior: polls conducted by the Roper Organization in 1989 and 1991 found significantly increasing awareness and use of designated drivers. By 1991, 37% of all U.S. adults claimed to have refrained from drinking at least once in order to serve as a designated driver, up from 29% in 1989. In 1991, 52% of adults younger than 30 had served as designated drivers, suggesting that the campaign was having greatest success with its target audience.

In 1988 there were 23,626 drunk driving fatalities. By 1997 the number was 16,189. While the Harvard Alcohol Project acknowledges that some of this decline is due to new laws, stricter anti-drunk driving enforcement, and other factors, it claims that many of the 50,000 lives saved by the end of 1998 were saved because of the designated driver campaign. (The television campaign was only a part of the overall campaign; there were strong community-level and public service components as well.)

Of course, making television an explicit vehicle for manipulating behavior has its dangers. My idea of the good may not be yours; if my ideas have access to the airwaves but yours don't, what I'm doing will seem to you like unwanted social engineering. We can all agree that minimizing drunk driving is a good thing—but not everyone agrees on the messages we want to be sending to, say, teenage girls about abstinence versus condoms, about having an abortion, or about whether interfaith marriages are okay. Television's power to mold viewers' understanding of the world is strong enough that we need to be aware that embedding messages about moral values or social behavior can have potent effects—for good or for ill.

Source: "Can TV Improve Us?" by J. Rosenzweig, July/August 1999, *The American Prospect* (45).

in 1991), while they increased sharply for boys (1,476 in 1983, 3,435 in 1991) (Males, 1993). Others suggest that even though heavy television viewing may be associated with violence, it is merely be a symptom of other problems such as parental neglect (Heins, 1994).

Clearly, there are many other factors involved in this relationship that make it difficult to produce a clear cause-and-effect relationship. The relationship between violent acts and antisocial behavior is much more complicated than was originally thought. For both adults and children, the social context, peer influence, values, and attitudes all play at least as important a role in determining behavior as television does.

Adult Socialization

A person's primary socialization is completed when he or she reaches adulthood. **Primary socialization** means that *individuals have mastered the basic information and skills required of members of a society.* He or she has (1) learned a language and can think logically to some degree, (2) accepted the basic norms and values of the culture, (3) developed the ability to pattern behavior in terms of these norms and values, and (4) assumed a culturally appropriate social identity.

There is still much to learn, however, and there are many new social identities to explore. Socialization, therefore, continues during the adult years. **Adult socialization** is *the process by which adults learn new statuses and roles.* It differs from primary socialization in two ways. First, adults are much more aware than young people are of the processes through which they are being socialized. In fact, adults deliberately engage in programs such as advanced education or on-the-job training in which socialization is an explicit goal. Second, adults often have more control over how they wish to be socialized and therefore can generate more enthusiasm for the process. Whether going to business school, taking up a new hobby, or signing up for the Peace Corps, adults can decide to channel their energy into making the most effective use of an opportunity to learn new skills or knowledge.

An important aspect of adult socialization is **resocialization,** which involves *exposure to ideas or values that in one way or another conflict with what was learned in childhood.* This is a common experience for college students who leave their homes for the first time and encounter a new environment, in which many of their family's cherished beliefs and values are held up to critical examination. Changes in religious and political values are not uncommon during the college years, which of-

ten lead to a time of stress for students and their parents.

Erving Goffman (1961a) discussed the major resocialization that takes place in **total institutions**— *environments such as prisons or mental hospitals in which the participants are physically and socially isolated from the outside world.* Goffman noted several factors that produce effective resocialization. These include (1) isolation from the outside world, (2) spending all of one's time in the same place with the same people, (3) shedding individual identity by giving up old clothes and possessions for standard uniforms, (4) a clean break with the past, and (5) loss of freedom of action. Under these circumstances, there usually is a major change in the individual along the lines prescribed by those doing the resocialization.

During the 1970s and 1980s, a number of religious cults gained notoriety because they attracted thousands of followers. Hundreds of cults continue to exist today, but we only notice them when they have trouble with authorities. The methods used by various cults to indoctrinate their members can be seen as a conscious attempt at resocialization.

New members are swept up in the communal spirit of the cult. Group pressure eventually produces major personality changes in the recruits, and new value systems replace the ones learned previously. Consequently, friends and family members no longer recognize the person who had been resocialized. The tactics used by religious cults have been criticized widely. The cults defend their program as simply a means through which they try to get people to rid themselves of old ideas and replace them with new ones.

It is not unusual for those undergoing resocialization experiences to become confused and depressed and to question whether they have chosen the right course. Some drop out; others eventually stop resisting and accept the values of their instructors.

In the following section, we will discuss four events in adult socialization: marriage, parenthood, work, and aging.

Marriage and Responsibility

As Ruth Benedict (1938) noted in a now-classic article on socialization in America, "our culture goes to great extremes in emphasizing contrasts between the child and the adult." We think of childhood as a time without cares, a time for play. Adulthood, by contrast, is marked by work and taking up the burden of responsibility. One of the great adult responsibilities in our society is marriage.

Indeed, many of the traditional role expectations of marriage no longer are accepted uncritically by today's young adults. For both men and women, choices loom large: How much of oneself should one devote to a career? How much to self-improvement and personal growth? How much to a spouse? Ours is a time of uncertainty and experimentation. Even so, marriage still retains its primacy as a life choice for adults. Although divorce has become acceptable in most circles, marriage still is treated seriously as a public statement that both partners are committed to each other and to stability and responsibility. (We will discuss marriage and alternative lifestyles in greater detail in Chapter 10.)

Once married, the new partners must define their relationships to each other and in respect to the demands of society. This is not as easy today as it used to be when these choices largely were determined by tradition. Although friends, parents, and relatives usually are only too ready to instruct the young couple in the "shoulds" and "should nots" of married life, such attempts at socialization are often resented by young people who wish to chart their own courses. One choice they must make is whether or not to become parents.

Parenthood

Once a couple has a child, their responsibilities increase enormously. They must find ways to provide the care and nurturing necessary to the healthy development of their baby, and at the same time they must work hard to keep their own relationship intact, because the arrival of an infant inevitably is accompanied by stress. This requires a reexamination of the role expectations each partner has of the other, both as a parent and as a spouse.

Of course, most parents anticipate some stresses and try to resolve them before the baby is born. Financial plans are made, living space is created, baby care is studied. Friends and relatives are asked for advice, and their future baby-sitting services are secured. However, not all the stresses of parenthood are so obvious. One that is overlooked frequently is the fact that parenthood is itself a new developmental phase.

The psychology of being and becoming a parent is extremely complicated. During the pregnancy, both parents experience intense feelings—some expected, others quite surprising. Some of these feelings may even be very upsetting: for instance, the fear that one will not be an adequate parent or that one might even harm the child. Sometimes such feelings lead people to reconsider their decision to become parents.

The birth of the child brings forth new feelings in the parents, many of which can be traced to the parents' own experiences as infants. As their child grows and passes through all the stages of development we have described, parents relive their own development. In psychological terms, parenthood can

be viewed as a second chance: Adults can bring to bear all that they have learned in order to resolve the conflicts that were not resolved when they were children. For example, it might be possible for some parents to develop a more trusting approach to life while observing their infants grapple with the conflict of basic trust and mistrust (Erikson's first stage).

Career Development: Vocation and Identity

Taking a job involves more than finding a place to work. It means stepping into a new social context with its own statuses and roles, and it requires that a person be socialized to meet the needs of the situation. These may even include learning how to dress appropriately. For example, a young management trainee in a major corporation was criticized for wearing his keys on a ring snapped to his belt. "Janitors wear their keys," his supervisor told him. "Executives keep them in their pockets." The keys disappeared from his belt.

Aspiring climbers of the occupational ladder even may have to adjust their personalities to fit the job. In the 1950s and 1960s, corporations looked for quiet, loyal, tradition-oriented men to fill their management positions—men who would not upset the status quo (Whyte, 1956)—and most certainly not women. Today, especially in the high-tech industry, the trend has been toward recruiting men and women who show drive and initiative and a capacity for creative thinking and problem solving.

Some occupations require extensive resocialization. Individuals wishing to become doctors or nurses, for example, must overcome their squeamishness about blood, body wastes, genitals, and the inside of the body. They also must accept the undemocratic fact that they will receive much of their training while caring for poor patients (usually ethnic minorities). Wealthier patients are more likely to receive care from fully trained personnel.

The armed forces use basic training to socialize recruits to obey orders without hesitating and to accept killing as a necessary part of their work. For many people, such resocialization can be quite confusing and painful. (For an example of career resocialization for an American in Japan, see "Global Sociology: An American Success Story Does Not Translate Into Japanese").

For some individuals, career and identity are so intertwined that job loss can lead to personal crisis. This occurs for many people who are downsized or "encouraged" to retire. For many, losing a job means reevaluation and a new direction. For others, it means spending months looking for a new job and feeling a profound loss of self-identity.

Global Sociology

An American Success Story Does Not Translate Into Japanese

In American society, hard work is seen as admirable and individual accomplishments are rewarded. In other societies, work success depends on the relationships the individual develops with superiors and others in the work group. The following case is an example of what happens when these opposing cultural views clash.

Jeff, a 28-year-old American in Japan went to work for Motorola Nippon, a largely autonomous Japanese subsidiary of the American electronics corporation. Jeff was part of a four-person sales force. He worked under the supervision of Muneo, his Japanese manager who had almost complete authority over him.

Jeff felt from the beginning that Muneo did not like him. At their first meeting he was told his salary was too high. When Jeff met and became engaged to a Japanese woman, Muneo said he did not approve. Muneo also refused Jeff's request for a lower sales quota because of Jeff's difficulty with the Japanese language. However, thanks to his fiancé, Jeff's Japanese language skills improved rapidly.

Jeff believed that he had to prove himself by working harder. He got up at 6:30 a.m. and scheduled numerous sales calls each day.

After 9 months in the job, Jeff was outselling the three other sales executives in his unit, all Japanese. After 18 months he had outsold all of them put together. This was, he felt, a triumph of hard work and persistence. He was out making sales calls most of the time and rarely bothered his boss.

It was therefore a considerable shock when Jeff received an "average" rating from Muneo at his annual performance review. This was the lowest rating possible—short of "unacceptable," which would have led to being fired. It was also the worst appraisal in the department. Jeff was too angry to argue with Muneo, but appealed his "flagrantly unfair" appraisal to Motorola's international human resources office in the United States.

After some months, a ruling came down on Jeff's side and his appraisal was revised upward. Muneo not only refused to speak to Jeff, he refused to look at him. Finally Jeff asked for a meeting. Muneo was so angry he could hardly get any words out. When he did it was "You shoot me, I shoot you." Jeff was outraged at the seeming injustice of the situation.

Given American cultural values, Jeff had performed extremely well. He had worked long hours, made more sales than anyone else, developed a fluency in Japanese, and tried to not bother his boss as he did his job.

But he was living and working in Japan, not the United States, and was being judged by Japanese values, which are quite different from the values to which he was accustomed. Muneo's objections were based on a Japanese management culture.

Muneo was angry with Jeff because he began by asking for favors rather than concentrating on how he could help his boss and the team. When he started to succeed, Jeff did not inform Muneo, solicit his advice, or invite him to share that success. He did not inform other team members about the information and approaches that led to his record sales, so that they could benefit from his knowledge.

Jeff's decision to "not bother" his supervisor was seen by Muneo as a snub, not a favor. As a Japanese boss, he was formally responsible for Jeff's successes and wanted to play a genuine part in them. Muneo felt he was cut out of participating in Jeff's triumphs.

Appealing over Muneo's head to the American owners of the company was seen as an insult to Muneo's authority and undermined the local Japanese management autonomy. When Jeff did not warn Muneo of the appeal or discuss it with him first, it was a rejection of the Japanese ideal of mutual respect. To have a local decision reversed by U.S. headquarters was a matter of shame for Muneo. He lost face before other Japanese colleagues by provoking interference in domestic affairs.

If Jeff had wanted to adjust to the Japanese cultural values, he could have had bimonthly meetings with Muneo and the Japanese sales team in which he would seek their advice and share the background information on his successes. For Muneo to be pleased, the whole team must succeed and Muneo himself must lead that success. Jeff needed to seek advice and support for every move he made. He needed to be as modest as possible, and let the value of shared knowledge speak for him. All of this is not easy when one is used to American workplace cultural values.

Source: Based on *Building Cross-Cultural Competence* (pp. 175–177), by C. Hampden-Turner & F. Trompenaars, 2000, New Haven: Yale University Press.

In many societies, older people are turned to for advice, and their opinions are valued because they are based on life experiences.

Aging and Society

In many societies such as Japan and China, age itself brings respect and honor. Older people are turned to for advice, and their opinions are valued because they reflect a full measure of experience. Often, older people are not required to stop their productive work simply because they have reached a certain age. Rather, they work as long as they are able to, and their tasks may be modified to allow them to continue to work virtually until they die. In this way, people maintain their social identities as they grow old—and their feelings of self-esteem as well.

This is not the case in the United States. Most employers expect their employees to retire before they have reached age 70, and Social Security regulations restrict the amount of nontaxable income that retired people may earn.

Perhaps the biggest concern of the elderly is where they will live and who will take care of them when they get sick. The American nuclear family ordinarily is not prepared to accommodate an aging parent who is sick or whose spouse has recently died. In addition, with the increasing life span many elderly who may be in their late 70s or 80s have sons and daughters who themselves may be in their 50s or even 60s. As a result, those older people who have trouble moving around or caring for themselves often find themselves with no choice but to live in protected environments.

This means that late in life many people are forced to acquire another social identity. Sadly, it is not a valued one, but rather one of social insignificance (de Beauvoir, 1972). This decrease in social significance can be very damaging to older people's self-esteem, and it may even hasten them to their graves. The past decade has seen some attempts at reform to address these issues. Age discrimination in hiring is illegal, and some companies have extended

or eliminated arbitrary retirement ages. However, the problem will not be resolved until elderly people achieve a position of respect and value in American culture equal to that of younger adults.

Even though aging is itself a biological process, becoming old is a social and cultural one: Only society can create a senior citizen. From infancy to old age, both biology and society play important parts in determining how people develop over the course of their lives.

SUMMARY

Unlike other animal species, human offspring have a long period of dependency. During this time, parents and society work together to make children social beings. The process of social interaction that teaches children the intellectual, physical, and social skills and the cultural knowledge needed to function as a member of society is called socialization. In the course of this process each child acquires a personality, or patterns of behavior and ways of thinking that are distinctive for each individual.

Every individual comes to possess a social identity by occupying culturally and socially defined positions. In addition, individuals acquire a changing yet enduring personal identity called the self, which develops when the individual becomes aware of his or her feelings, thoughts, and behaviors as distinct from those of other people. The development of the self is a complex process that has at least three dimensions: cognitive development, moral development, and gender identity.

The agents that mold the child to fit into society vary from culture to culture. In American society, the family is the most important socializing influence in early childhood development. The location of any given family within the social structure has much to do with what its dominant members consider important. However, whatever the family's values, norms, ideals, and standards, they are initially accepted uncritically by the child. As the child grows older and moves into society, other agents of socialization come into play. Schools are increasingly expected to meet a variety of social and emotional needs as well as to pass on knowledge and develop skills. From school age to early adulthood, peers powerfully influence lifestyle orientations and, in some cases, values. The mass media present today's children with an enormous amount of information, both for better and for worse.

Primary socialization ends when individuals reach adulthood. By this time they have (1) learned a language, (2) accepted the basic norms and values of the culture, (3) developed the ability to pattern their behavior in terms of those norms and values, and (4) assumed a culturally appropriate social identity. Although socialization continues throughout one's life, adult socialization differs from primary socialization in that adults are much more aware of the processes through which they are socialized, and they often have more control over the process.

One important form of adult socialization is resocialization, which involves exposure to ideas or values that conflict with what was learned in childhood. Often, resocialization occurs in a total institution, such as a prison or mental hospital, where participants are physically, socially, and psychically isolated from the outside world. Similar techniques are often used by religious cults to indoctrinate recruits. Common events in adult socialization that represent transitions and that precipitate role redefinitions include marriage, parenthood, work, and aging. From infancy to old age, both biology and society play important parts in determining how people develop.

INTERNET RESOURCES

Go to http://web-enhanced.thomsonlearning.com which features Web links relevant to this chapter to expand and enhance the learning experience. This site will be monitored and updated periodically to ensure the links are correct and live.

At Virtual Society: The Wadsworth Sociology Resource Center, http://sociology.wadsworth.com, you will find a Career Center, Sociology in the News, Research Online, InfoTrac College Edition, Virtual Tours in Sociology, and a variety of book-specific student and instructor resources.

CHAPTER FOUR STUDY GUIDE

KEY CONCEPTS

Match each concept with its definition, illustration, or explanation presented below.

a. Sociobiology
b. The "I"
c. Conditioning
d. Superego
e. Gender identity
f. Nurture
g. Self
h. Social attachments
i. Generalized others
j. Preparatory stage
k. Moral order
l. Adult socialization
m. Formal logical stage
n. Significant others

o. Resocialization
p. Mass media
q. Genes
r. Personality
s. Sensorimotor stage
t. Instincts
u. Id
v. Operational stage
w. Looking-glass self
x. Ego
y. Preoperational stage
z. Socialization
aa. Play stage
bb. Libido

cc. The "me"
dd. Game stage
ee. Statuses
ff. Peers
gg. Primary socialization
hh. Total institution
ii. Social identity
jj. Attachment disorder
kk. Age 30 transitional
ll. Early adult period
mm. Late adult transition
nn. Midlife transition
oo. Marriage

_____ 1. The process of social interaction that teaches the child the intellectual, physical, and social skills needed to function as a member of society.

_____ 2. Patterns of behavior and ways of thinking and feeling distinctive to each individual.

_____ 3. Inherited units of biological material.

_____ 4. Biologically inherited patterns of behavior.

_____ 5. A series of repeated experiences linking a desired reaction with a particular object or event.

_____ 6. Discipline using biological principles to explain the behavior of animals and humans.

_____ 7. A person's socialization experiences.

_____ 8. The learning to have meaningful interactions and affectionate bonds with others.

_____ 9. Culturally and socially defined positions.

_____ 10. The total of all the statuses that define an individual.

_____ 11. An individual's changing yet enduring personal identity.

_____ 12. An infant relies on touch and manipulation of objects for information about the world.

_____ 13. A child learns that words can be symbols for objects.

_____ 14. The child begins to think with some logic and can understand and work with numbers.

_____ 15. The period when a child becomes capable of abstract, logical thought and can anticipate possible consequences of acts and decisions.

_____ 16. A society's shared view of right and wrong.

_____ 17. Our sense of self rooted in whether our sex is female or male.

_____ 18. The three-stage process through which each of us develops a sense of self.

_____ 19. Portion of the self, according to Mead, that wishes spontaneous, free expression.

_____ 20. Mead's portion of self composed of things learned through the socialization process.

_____ 21. Those individuals who are most important in our development (e.g., parents, friends).

_____ 22. The viewpoints, attitudes, and expectations of society as a whole or of a community of people whom we are aware of and who are important to us.

_____ 23. Mead's discussion of the period characterized by the child imitating the behavior of others, which prepares the child for learning social-role expectations.

_____ 24. Mead's discussion of the period when a child has acquired language and begins not only to imitate behavior, but also to formulate role expectations.

_____ 25. Mead's discussion of the period where the child learns that there are rules which specify the proper and correct relationship among the players.

_____ 26. The unconscious drives and instincts Freud believed every human being inherits.

_____ 27. Freud's terms for the erotic or sexual drive.

_____ 28. Freud's part of the self that represents society's norms and moral values.

_____ 29. Freud's part of the self that tries not only to mediate conflicts between the id and the superego, but also to find socially acceptable ways for the id to be expressed.

_____ 30. Individuals who are social equals.

_____ 31. A large-scale means of transmitting information through audible, visual, and electronic channels to an audience conditioned to receive passively whatever is brought to them.

_____ 32. When individuals have mastered the basic information and skills required of members of a society.

_____ 33. The process by which adults learn new statuses and roles.

_____ 34. Exposure to ideas or values that in one way or another conflict with what was learned in childhood.

_____ 35. Environments such as prisons or mental hospitals, in which the participants are physically and socially isolated from the rest of the world.

_____ 36. A public statement that both partners are committed to each other and to stability and responsibility.

_____ 37. Levinson's period requiring profound reappraisal of the past and shift to a new era.

_____ 38. Levinson's period of internal instability when divorce and job changes are common.

_____ 39. Levinson's period when great energy is spent trying to reduce family dependence.

_____ 40. Results in children being unable to trust people and to form relationships with others.

_____ 41. Levinson's major transition period between young and middle adulthood wherein the individual sees a difference between the dream and the reality of his or her life.

KEY THINKERS

Match the thinkers with their main ideas or contributions.

a. Harry F. Harlow
b. John B. Watson
c. Lawrence Kohlberg
d. Sigmund Freud
e. Daniel Levinson
f. Stephen Jay Gould

g. Ivan Pavlov
h. Jean Piaget
i. Edward O. Wilson
j. W. I. Thomas
k. Erving Goffman
l. Charles H. Cooley

m. Erik Erikson
n. Charles Darwin
o. George H. Mead
p. Centerwall
q. Jean-Marc Itard
r. Rene Spitz

_____ 1. Author of *On the Origin of Species.* Fostered the view that human beings were a species similar to all others in the animal kingdom.

_____ **2.** Through experiments with dogs, he demonstrated that behavior could be conditioned.

_____ **3.** Linked a certain reaction (fear) with an object (rabbit) through experience repetition thereby demonstrating that humans could be conditioned.

_____ **4.** Coined the term *sociobiology* and was its major advocate as an explanation of human behavior.

_____ **5.** Argued for a cultural interpretation of human behavior as opposed to a sociobiological interpretation.

_____ **6.** Illustrated the harmful effects of social isolation through his experiments with rhesus monkeys.

_____ **7.** In his studies of children, he demonstrated they moved through predictable stages of identity development.

_____ **8.** Maintained that moral thinking developed through five to six distinctive stages.

_____ **9.** Theorized that the self developed in a three-stage process mirroring social interaction with others.

_____ **10.** Author of a theorem that stated if people "define situations as real, they are real in their consequences."

_____ **11.** Founder of symbolic interactionism who maintained that the self develops out of interaction with significant others from whom we learn role expectations for our behaviors.

_____ **12.** Argued the that individual's inner, spontaneous self and society's demands were in continual conflict.

_____ **13.** Developed the view that socialization was a lifelong process through which the person moved through eight distinctive stages, each characterized by a series of stresses and conflicts the individual must resolve.

_____ **14.** Proposed a developmental model for adult socialization consisting of six major stages.

_____ **15.** Proposed the concept of total institution as an environment for the resocialization of persons in adult life.

_____ **16.** Provided one of the early studies of orphanages showing that children need more than routine care to develop.

_____ **17.** A young doctor who worked with the "Wild Boy of Aveyron" and tried, unsuccessfully, to socialize him.

_____ **18.** Conducted major research demonstrating the socialization power of the mass media in the area of children's television viewing habits and later violent behavior.

CENTRAL IDEA COMPLETIONS

Following the instructions, fill in the appropriate concepts and descriptions for each of the questions posed in the following section.

1. Compare and contrast how biology and socialization contribute to the formation of the individual: _____

_____

_____

2. What is sociobiology? _____

_____

_____

3. Drawing on the specific case examples presented in your text, discuss how situations of extreme social isolation and deprivation affect a human's early childhood development:

_____

_____

_____

4. Present a short example of how each of the following variables contributes to the socialization of children in the United States:

a. Family: _____

b. Schools: _____

c. Peer groups: _____

d. Mass media: _____

5. Three giant thinkers in the development of the self are about to give a sound bite of their opinions, theories, and research on the statement, "Among human beings, the self is . . ."

a. Cooley: _____

b. Mead: _____

c. Freud: _____

6. Compare and contrast the essential ideas about moral development as presented by Piaget and Kohlberg:

a. Piaget: _____

_____

b. Kohlberg: _____

_____

7. Define resocialization: _____

8. Define, then compare and contrast how primary socialization and adult socialization differ:

a. Primary socialization characteristics: _____

b. Adult socialization: _____

c. Socialization similarities: _____

d. Socialization contrasts: _____

9. Present the key features of the developmental stage models of Erikson and Levinson:

a. Erikson: _____

_____

_____

b. Levinson: _____

_____

_____

10. Define the concept of *looking-glass self.* Present an example within which you engaged in the *looking-glass self* process:

a. Definition: _____

b. Example: _____

11. What changes in American society are evidenced in the primary child-care arrangements for preschool children? _____

_____

_____

12. What is the connection between television and the teaching of values as exemplified by Winsten's campaign to institute the concept of the "designated driver"?

_____

_____

_____

13. What are the five components involved in Goffman's description of resocialization as it takes place in *total institutions?*

a. _____

b. _____

c. _____

d. _____

e. _____

14. Compare and contrast the differences in socialization toward aging as it takes place in societies such as China and Japan compared with the United States:

_____

_____

_____

CRITICAL THOUGHT EXERCISES

1. Your textbook presents a detailed overview of the arguments regarding the hypothetical relationship between television viewing habits and violent behavior. One of the issues in dispute concerns perceptions and definitions of violence as well as whether any violent behavior when viewed on television is taken as serious or "real." This critical thought activity will involve a series of steps.

   a. Select three television programs that you expect will have violent episodes. Videotape up to a 10-minute segment of each program.

   b. After you have taped the program, watch it again.

   c. What is your assessment of its violent content? Does the violence appear real to you, or do you view it as a make-believe media experience?

   d. Contact three persons (one close to your age and two at least 10 years older than you) whom you know well enough to show the videos. (Remember, this is a nonrandom sample and not necessarily representative of any populations).

Ask their assessments of the level of violence or nonviolence of your selections.

   e. To what extent are their answers related to your age and theirs? Do you agree or disagree with their assessments?

   f. Update this topic with library journal research.

   g. What would one need to do in order to transform this exercise into a complete sociological investigation?

2. What aspects of socialization appear to account for the differences in the definitions of correct office behavior and success as viewed in Japanese society versus American society? Discuss how the concept of resocialization may be relevant to the situation facing Japanese workers coming to the United States or American workers moving to Japan.

3. Although we normally think of newspapers, magazines, and television as the mass media, an emerging form of worldwide media is the Internet. Log on to the Net and select four sites that you feel have the ability to operate as agents of socialization. What socialization messages are being delivered? How effective do you think this medium is as an agent of socialization compared to the family and the schools (junior high or high school)?

4. An enduring controversy in the social sciences has been the so-called nature versus nurture debate. Follow the authors discussed in your text and look up one example of their basic writings (for example, Wilson's *Sociobiology*). Develop a pro and con list of evidence for both positions as one might for a courtroom presentation. What is your position on the debate? In which direction(s) does the weight of the scientific evidence lead? As your text notes, only a passionate plea by Margaret Mead convinced a major association of anthropologists not to condemn (and boycott) sociobiology as an academic explanation for human behavior. In your opinion, why does the issue touch such raw nerves among social scientists? Do you think the debate has the same effect on the average undergraduate student taking an introductory sociology course? Elaborate on your answer.

5. A major force in socialization theory and research are the "developmental" theorists. Several

have been presented in your text. Continue the discussion by analyzing the strengths and weaknesses of developmental theories. Log onto the Internet at

http://www.fishnet/~pparent/johnson1.html*

At this site you will find the work of Diane Clark Johnson, who has constructed an expanded, modified version of Piaget's theory. To what extent does Diane Clark Johnson's developmental model meet the criticisms of earlier developmental theories?

6. Education is a major factor in youth socialization. Log on to the Internet at

http://www.geocities.com/Athens/Forum/2780/social.html*

At this site you will see a discussion of the pros and cons of home schooling as an alternative form of socialization. What do you see as the major issues involved in home schooling versus socialization in traditional school settings?

*Please note, Web site addresses change frequently. The addresses presented in these exercises should be viewed as guides only, and students should be prepared to utilize appropriate Internet search techniques when necessary should these addresses no longer be accessible.

ANSWERS

KEY CONCEPTS

1. z	8. h	15. m	22. i	29. x	36. oo
2. r	9. ee	16. k	23. j	30. ff	37. mm
3. q	10. ii	17. e	24. aa	31. p	38. kk
4. t	11. g	18. w	25. dd	32. gg	39. ll
5. c	12. s	19. b	26. u	33. l	40. jj
6. a	13. y	20. cc	27. bb	34. o	41. nn
7. f	14. v	21. n	28. d	35. hh	

KEY THINKERS

1. n	4. i	7. h	10. j	13. m	16. r
2. g	5. f	8. c	11. o	14. e	17. q
3. b	6. a	9. l	12. d	15. k	18. p

LEARNING OBJECTIVES

After studying this chapter, you should be able to do the following:

- Know what the major types of social interaction are.
- Understand the influence of contexts and norms in social interaction.
- Understand the concepts of status and role.
- Identify the major characteristics of social groups.
- Distinguish between primary and secondary groups.
- Explain the functions of groups.
- Understand the role of reference groups.
- Know the influence of group size.
- Understand the characteristics of bureaucracy.
- Know what Michels's concept of "The Iron Law of Oligarchy" is.
- Understand why social institutions are important.

Wayne Farrell had arrived in New York City only two days ago, but he had already had his fill of strange experiences. It was almost as if panhandlers, con artists, and a host of individuals pushing unusual causes could sense that he was an easy mark. No sooner would he walk down the street and notice someone distributing leaflets than the person would pick him out of the crowd and engage him in conversation. He tried to be polite and listen, but he repeatedly found himself in the midst of uncomfortable situations. Wayne was starting to wonder whether he was doing anything to contribute to these encounters or whether this was just part of living in a large urban environment.

Without realizing it, Wayne was engaging in a pattern of social interaction with these strangers. Coming from a small town in Wyoming, he was unaware of the fact that he was looking at passersby for a second or two longer than the typical person walking down the street. But this was just enough to signal to those looking for such a cue that he was ripe for their approach.

Most Americans use four main distance zones in their business and social relations: Intimate, personal, social, and public. Intimate distance varies from direct physical contact with another person to a distance of 6 to 18 inches and is used for private activities with another close acquaintance. Personal distance varies from 2½ to 4 feet and is the most common spacing used by people in conversation. The third zone—social distance—is employed during business transactions or interactions with a clerk or salesperson. As they work with each other, people tend to use this distance, which is usually between 4 and 12 feet. Public distance, used, for instance, by teachers in classrooms or speakers at public gatherings, can be anywhere from 12 to 25 feet or more.

As Wayne walks down the busy urban street, he is within the fourth zone. Eye behavior in this environment has some specific rules attached to it. For urban whites, once they are within definite recognition distance (usually 16 to 32 feet), there is mutual avoidance of eye contact—unless they want something specific, such as to make a proposition, receive a handout, or obtain information of some kind. In small towns, however, people are much more likely to look at each other, even if they are strangers (Hall, 1966; Hall & Hall, 1990).

Panhandlers and others seeking to establish contact exploit the unwritten, unspoken conventions of eye contact. They take advantage of the fact that once explicit eye contact is established, it is rude to look away, because to do so means brusquely dismissing the other person and his needs. Once having caught the eye of his mark, the individual locks on, not letting go until he moves through the public zone, the social zone, the personal zone, and, finally, the intimate zone, where people are most vulnerable.

Wayne's experiences were not just a coincidence. He actually was engaging in a social interaction with others in his environment that he was not aware of. He started to make a conscious attempt to avoid prolonged eye contact with strangers, and the unwanted encounters began to subside.

As Wayne discovered, there is no way not to interact with others. Social interaction has no opposite. If we accept the fact that all social interaction has a message value, then it follows that no matter how we may try, we cannot not send a message. Activity or inactivity, words or silence all contain a message. They influence others, and others respond to these messages. The mere absence of talking or taking notice of each other is no exception, because there is a message in that behavior also.

When sociologists study human behavior, they are interested primarily in how people affect each other through their actions. They look at the overt behaviors that produce responses from others as well the subtle cues that may result in unintended consequences. Human social interaction is very flexible and quite unlike that of the social animals.

Understanding Social Interaction

Max Weber (1922) was one of the first sociologists to stress the importance of social interaction in the study of sociology. He argued that the main goal of sociology is to explain what he called **social action,** a term he used to refer to *anything people are conscious of doing because of other people.* Weber claimed that in order to interpret social actions, we have to put ourselves in the position of the people we are studying and try to understand their thoughts and motives. The German word Weber used for this is *verstehen,* which can be translated as "sympathetic understanding."

Weber's use of the term *social action* identifies only half of the puzzle, because it deals only with one individual taking others into account before acting. A **social interaction** involves *two or more people taking one another into account.* It is the interplay between the actions of these individuals. In this respect, social interaction is a central concept to understanding the nature of social life.

In this chapter, we shall explain how sociologists investigate social interaction. We shall start with the basic types of social interaction, whether verbal or nonverbal. Next, we shall examine how social interaction affects those involved in it. We then will

Social interaction is a central concept to understanding the nature of social life.

broaden our focus a bit, and move on to groups and social interactions within them. Finally, we will look at the large groupings of people that make social life possible and that ultimately make up the social structure. In other words, we shall start with social behavior at the most basic level and move outward to ever more complicated levels of social interaction.

Contexts

Where a social interaction takes place makes a difference in what it means. Edward T. Hall (1974) identified *three elements that, taken together, define the* **context** *of a social interaction: (1) the physical setting or place; (2) the social environment; and (3) the activities surrounding the interaction—preceding it, happening simultaneously with it, and coming after it.* The context of an interaction consists of many elements. Without knowledge of these elements it is impossible to know the meaning of even the simplest interaction. For example, Germans and Americans treat space very differently. Hall (1969) noted that in many ways, the difference between German and American doors gives us a clue about the space perceptions of these two cultures. In Germany, public and private buildings usually have double doors that create a soundproof environment. Germans feel that American doors, in contrast, are flimsy and light, inadequate for providing the privacy that Germans require. In American offices, doors usually are kept open; in German offices, they are kept closed. In Germany, the closed door does not mean that the individual wants to be left alone or that the people inside are planning something that should not be seen by others. Germans simply think that open doors are sloppy and disorderly. As Hall explained it:

> I was once called in to advise a firm that has operations all over the world. One of the first questions asked was, "How do you get the Germans to keep their doors open?" In this company the open doors were making the Germans feel exposed and gave the whole operation an unusually relaxed and unbusinesslike air. Closed doors, on the other hand, gave the Americans the feeling that there was a conspiratorial air about the place and that they were being left out. The point is that whether the door is open or shut, it is not going to mean the same thing in the two countries. (Hall, 1969)

In Japan there is a third view on the issue. Most Japanese executives prefer to share offices to ensure that information can flow easily and each person knows what is happening in the other's area of responsibility. The Japanese executive does not want to risk being unaware of events as they are developing. Japanese firms have ceremonial rooms for receiving visitors, but few other work areas afford any privacy (Hall & Hall, 1990). (See "Global Sociology: Cross-Cultural Social Interaction Quiz.")

Norms

Human behavior is not random. It is patterned and, for the most part, quite predictable. What makes human beings act predictably in certain situations? For one thing, there is the presence of **norms**—*specific rules of behavior, agreed upon and shared, that prescribe limits of acceptable behavior.* Norms tell us the things we should both do and not do. In fact, our society's norms are so much a part of us that we often are not aware of them until they are violated. Take the unfortunate circumstances that took place in Suzanne Berger's life for example. One day she bent down to pick up her child only to discover she could not straighten up again. In the process of bending she had injured her back so severely that for years thereafter she could neither walk nor sit for more than a few minutes. Traveling anywhere meant that she had to take along a mat and immediately lie down. In effect, she had to violate the norms that assume that you will not stretch out on the floor in a department store, a train station, a classroom, or at public events. As she described it:

> Strangers try not to stare. . . . At airports and train stations people have thought I was a derelict or crazy or maybe homeless; only the dispossessed lie on floors, children lie on floors, dogs lie on floors . . . but adults? *What's that woman doing over there?* A security guard said at the airport. *Dunno, leave her alone. Must be drunk.* With friends inside my house, being down here upsets a balance of conviviality, of the *whereness* that grounds a conversation. I am always looking up, as though younger or subservient. Outside I live down with mother-dirt, grass, the asphalt of the city. Wherever I go, I lie down with my mat. *Hey, lady, what the hell you doing down there?* says a child on a city playground. *You sick? You tired?* (Berger, 1996)

We also have norms that guide us in how we present ourselves to others. We realize that how we dress, how we speak, and the objects we possess relay information about us. In this respect, North Americans are a rather outgoing people. The Japanese have learned that it is a sign of weakness to disclose too much of oneself by overt actions. They are taught very early in life that touching, laughing, crying, or speaking loudly in public are not acceptable ways of interacting.

Global Sociology

Cross-Cultural Social Interaction Quiz

South Africa

After a long voyage you arrive at the airport in Johannesburg, South Africa. A man approaches you with both hands held in a cupped shape. You think the man is begging but you are not sure. What is the man doing?

1. Telling you that he is collecting food for his family.
2. Saying a prayer of thank you for your safe arrival.
3. Offering to be a porter for your baggage.

CORRECT ANSWER: 3

The man is offering to carry your baggage and is merely giving a sign of humbleness that means, "The gift you may give me (for carrying your bags) will mean so much that I will have to carry it with both hands."

The Gambia

You and your roommates decide to invite a foreign exchange student over for an American dinner. The student, Dawda, is from The Gambia in West Africa. Before the meal starts, your roommate Sharon makes a short welcoming statement for the group to Dawda. Immediately thereafter, Dawda proceeds to dip his finger into the main dish and lick it. Everyone is surprised and no one knows what to say. Why did Dawda dip his finger in the food?

1. He was trying to break the ice and be funny.
2. He is checking the quality of the food for contamination before anyone eats it.
3. He did not like the food and would like you to replace it with something more acceptable.

CORRECT ANSWER: 2

Dawda has put his finger in the food to test the quality. The action relates to the Gambian practice

that as a courtesy to others at the table you dip your finger in the food first to test for any potential food poisoning.

France

You are sitting with several other people who are listening to a lecture. One of your companions, Pierre, starts to make believe he is playing an imaginary flute. Why is he doing this?

1. The words coming from the speaker are like music to Pierre's ears.
2. The speaker is talking to much.
3. Pierre needs to go to the bathroom but does not want to be rude.

CORRECT ANSWER: 2

Pierre has heard as much as he wants to hear from this speaker and thinks he is being annoying and long winded.

Mali

Your friend, George, has just returned from his 2-year work assignment in the African nation of Mali. He tells you that when he was there and was introduced to someone he was given the person's first name, but not the last name. Why is that?

1. Last names are only given when people become better acquainted.
2. People do not have last names in Mali.
3. Speaking someone's last name is considered bad luck.

CORRECT ANSWER (1)

The custom in Mali is to let someone know your last name only after you have become better acquainted.

Source: The Web of Culture, http://www.worldculture.com/index.html

Not only can the norms for behavior differ considerably from one culture to another, but they also can differ within our own society. Conflicting interpretations of an action can exist among different ethnic groups. Unfamiliarity with such cultural communication can lead to misinterpretations and even unintended insults. For example, Table 5-1 presents examples of interpretations African-Americans and white Americans may make of particular styles of dress or nonverbal behaviors. Among African-

Americans, listeners are expected to avert their eyes from the speaker and in doing so are showing respect. Among white Americans, looking at the speaker directly is seen as a sign of respect and looking away may indicate disinterest or boredom.

Let us assume that a white teacher is speaking directly to an African-American boy. The youngster may avert his eyes as she is speaking. The teacher may think he is not listening. Worse yet, she may say something like "look at me when I speak to you,"

The location and context in which a social interaction takes place makes a difference in what it means.

and the boy will be even more confused. He may have been led to believe that looking at the teacher for an extended time may be seen as a challenge to her authority.

Sociologists thus need to understand the norms that guide people's behavior, as without this knowledge it is impossible to understand social interaction. (See "Our Diverse Society: How Blacks and Whites Offend Each Other Without Realizing It.")

Ethnomethodology

Many of the social actions we engage in on a day-to-day basis are commonplace events. They tend to be taken for granted and rarely are examined or considered. Harold Garfinkel (1967) has proposed that it is important to study the commonplace. Those things we take for granted have a tremendous hold over us because we accept their demands without question or conscious consideration. **Ethnomethodology** is *the study of the sets of rules or guidelines that individuals use to initiate behavior, respond to behavior, and modify behavior in social settings.* For ethnomethodologists, all social interactions are

equally important because they provide information about a society's unwritten rules for social behavior—the shared knowledge that is basic to social life.

Garfinkel asked his students to participate in a number of experiments in which the researcher would violate some of the basic understandings among people. For example, when a conversation is held between two people, each assumes that certain things are perfectly clear and obvious and do not need further elaboration. Examine the following conversation and notice what happens when one individual violates some of these expectations:

BOB: *That was a very interesting sociology class we had yesterday.*
JOHN: *How was it interesting?*
BOB: *Well, we had a lively discussion about deviant behavior, and everyone seemed to get involved.*
JOHN: *I'm not certain I know what you mean. How was the discussion lively? How were people involved?*
BOB: *You know, they really participated and seemed to get caught up in the discussion.*
JOHN: *Yes, you said that before, but I want to know what you mean by lively and interesting.*
BOB: *What's wrong with you? You know what I mean. The class was interesting. I'll see you later.*

Bob's response is quite revealing. He is puzzled and does not know whether John is being serious. The normal expectations and understandings around which day-to-day forms of expression take place have been challenged. Still, is it not reasonable to ask for further elaboration of certain statements? Obviously not, when it goes beyond a certain point.

Another example of the confusion brought on by the violation of basic understandings was shown when Garfinkel asked his students to act like boarders in their own homes. They were to ask whether they could use the phone, take a drink of water, have a snack, and so on. The results were quite dramatic:

Family members were stupefied. They vigorously sought to make the strange actions intelligible and to restore the situation to normal appearances. Reports were filled with accounts of astonishment, shock, anxiety, embarrassment and anger, and with charges by various family members that the student was mean, inconsiderate, selfish, nasty, or impolite. Family members demanded explanations: What's the matter? What's gotten into you? Did you get fired? Are you sick? What are you being so superior about? Why are you mad? Are you out of your mind? Are you stupid? One student acutely embarrassed his mother in front of her friends by

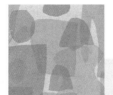

Our Diverse Society

How Blacks and Whites Offend Each Other Without Realizing It

Lena Williams is a black reporter who has worked at the *New York Times* for 25 years. She recently decided to explore how racial prejudices affect even the most casual social interactions. The following section illustrates some of these offenses.

The chic woman threw back her head and ran her fingers through her long flaxen hair. "I hate it when they do that," my brother-in-law whispered. I nodded in agreement.

What may have seemed to be a petty annoyance aroused long simmering memories of racial slights that have yet to heal.

As blacks, we understood instinctively the role hair texture has played in perceptions of beauty and privilege in America. All our lives we've been bombarded with the images of white movie stars, models and other beauty icons with long flowing hair, which has been beyond our reach. For that brief moment in Macy's, we both saw a white woman flaunting a symbol of preference.

Neither my brother-in-law nor I are alone in our perception. When I shared the incident with black friends of various ages, classes and regions, they immediately understood. We knew the woman meant no harm. She was doing what came naturally.

But at a time when the discussion of race has entered new realms of political and scholarly debate and an emerging body of literature is re-examining the significance of race, it is a good time to look at the simple things in life—a look or stare at a store counter, a gesture on a crowded elevator, a maneuver on a sidewalk—that can turn moods red and unleash old emotions that sweep away rationality.

Blacks, as well as whites, are guilty of making false assumptions about the other group's motives. Indeed, resentments over seemingly small slights are felt on both sides of the black-white divide.

So when asked recently by a white male colleague whether there were ordinary things that white people did and said that blacks found offensive, I was more pleasantly surprised than taken aback.

Almost across the board there are things blacks find irritating, if not sometimes downright insulting, in their day-to-day dealings with whites.

The Invisible Black Man or Black Woman

A black person has been standing in line in, say, a restaurant. A white person arrives and cuts the line with not so much as an "excuse me," completely ignores the black person, asks the maitre d' how long the wait will be for a table and then responds to the black's objections by "innocently" saying something like, "Oh, were you waiting for a table?"

Whites might view these incidents simply as someone behaving badly. But many blacks see them as the actions and typical attitudes of a people that still believes, as the Supreme Court declared in its 1857 Dred Scott ruling, that "a black man had no rights that a white man was bound to respect."

Mistaken Identities

We've all done it—call an acquaintance, colleague or classmate by someone else's name. Yet, black Americans are more likely to become agitated than others when whites call them by another's name, because they are convinced the mistake rests on the racial stereotype that blacks "all look alike."

Worse, many blacks will tell you, are whites who tend not to recognize blacks outside their usual surroundings.

As long as you're at your desk in the office, or in the classroom or doing your professional thing, they know you. The minute you're one of the masses, or someplace they think you shouldn't be, you become this faceless blur of blackness.

First-Name Basis?

Among my people, familiarity often breeds contempt. While most Americans enjoy the informality of using first names, many of my black acquaintances, especially men, turn a deaf ear when they are not addressed by their surnames. . . . Historically white people have used first names with blacks to establish superior-inferior relationships.

The Curiosity Factor

It is conveyed through such phrases as these: "Do you get sunburned?" "So what do you do for a living?" And it is sometimes acted out by touching a black [person's] hair or skin.

Is This a Party or What?

Shelby Steele, a black scholar, once talked about parties he'd attended that died from a "lethal injection of the American race issue." To be cornered and peppered with questions about how "the race" feels about this or that is not my idea of a party. . . . Anytime we are asked to speak on behalf of the 15 million to 20 million blacks in America it's off-putting.

"The Look"

It starts with an expressionless stare. The eyes begin to squint. The mouth opens slightly. And you know you're being analyzed or sized up. It is a look whites

Table 5-1

Different Interpretations of Nonverbal Behavior Among African-Americans and White Americans

AFRICAN-AMERICANS	WHITE AMERICANS
Listeners show respect to the speaker by averting their eyes.	Listeners show respect to the speaker by looking at a speaker.
In a conversation the floor is granted to the person who is the most assertive and breaking in during a conversation is accepted.	One person is permitted to speak at a time and you should not interrupt until the other person has made his or her points.
Asking questions of a person met for the first time is seen as intrusive.	Inquiring about job and family of someone met for the first time is seen as friendly.
Touching someone's hair is generally considered offensive.	Touching someone's hair is considered a sign of affection.
Showing emotion during a conflict is seen as honesty and as the first step toward the resolution of the problem.	Showing emotion during a conflict is seen as the beginning of a fight and interferes with the resolution of the conflict.
Men wearing hats and sunglasses indoors is a fashion statement and is similar to women wearing jewelry.	Wearing hats and sunglasses indoors is rude.

often give blacks who don't fit the composite notion of a Negro. If you're black, and you've lived long enough, you know the look when you see it. Often it comes right after you say what you do professionally.

"Really! So how long have you worked for. . . ."

Sometimes "the look" appears when blacks say they have attended a college or university that is not an historically black school.

"You went to Harvard?"

What's Up With That?

Phrases often used by whites that blacks hear as racially loaded: "Isn't she articulate?" "I don't see color." Whenever whites try to talk black slang: "Homie!" "My man!" "Girlfriend."

Why? Blacks often find these remarks patronizing or attempts by whites to show they are not prejudiced or that some of their best friends are black.

If I counted the times a white person, whether a television commentator, journalist or employer, used "articulate" to refer to blacks who use proper diction or English in expressing themselves, I might be left with little else to do.

Over the years, I've learned to lived with perceived racial slights. I refuse to let them define a life of success and achievement. I know better than to judge the actions of all whites by those of a few. Yet the cumulative effect of a lifetime of emotional nicks and scrapes exacts a toll on one's psyche.

"Psychiatrists refer to them as micro-aggressions," Dr. Poussaint explained. "They're the things

you experience every day that then add up and take a toll. They may seem to be minor, but white people don't know how much this society makes blacks people constantly think about being black.

"These many slights and ways of interacting are things whites don't know they're doing," he continued. "It's not like using a racial slur. They're doing it in an intuitive way when they respond to the color of your skin. And they don't know enough about us, as a people, to know that we perceive things differently than they do."

Everyday Slights: The White Take

In no way is this a one-sided issue. I spoke to whites who said there were things blacks often did and said that whites found racially offensive or insensitive. In fact, they were not at a loss for words or examples of ordinary interactions with blacks that left them feeling slighted.

Here are themes that came up often:

The Macho Factor

A white man in his late 40s talked about black men who usurped space on subway cars by sitting wide-legged, causing others to be sardined on their seats. "They sit there defying you to ask them to move over, as if their manhood is at stake," he said.

Another white male said he resented inferences often made by black men that they were somehow more masculine or macho than whites.

Our Diverse Society continues

Our Diverse Society

(continued)

A "Black Thing"

"It's a black thing, you wouldn't understand" was the slogan of a popular T-shirt a few years back. Many whites took offense because the slogan implied that whites lacked a level of understanding about African-Americans.

Being Hopelessly Uncool

The idea that whites don't have rhythm, can't dance or just aren't as hip as blacks has long been a source of irritation to many whites, especially those of a younger generation.

What hurts even more, a few whites indicated, are the suggestions that whites who do appear to be on the cutting edge are "acting black."

The Race Thing

Many whites said they simply hated it when blacks turned innocuous things into a racial guilt trip.

"I remember once, a black friend complained about a white sales clerk in a grocery store putting his change on the counter instead of in his hand," Ms. Lewis said. "When I said, 'She did that to me, too,' he said it wasn't the same thing and stormed off."

Sometimes it has nothing to do with race, but no one believes that, she concluded.

Source: Article based on "It's the Little Things," by L. Williams, December 14, 1997, *The New York Times*. Table based on *Cross-Cultural Communication: An Essential Dimension of Effective Education*, rev. ed., by O. L. Taylor, 1990, Washington DC: The Mid-Atlantic Equity Center.

---

asking if she minded if he had a little snack from the refrigerator. "Mind if you have a little snack? You've been eating little snacks around here for years without asking me. What's gotten into you?" (Garfinkel, 1972)

Ethnomethodology seeks to make us more aware of the subtle devices we use in creating the realities to which we respond. These realities are often intrinsic in human nature, rather than imposed from outside influences. Ethnomethodology addresses questions about the nature of social reality and how we participate in its construction.

Dramaturgy

People create impressions, and others respond with their own impressions. Erving Goffman (1959, 1963, 1971) concluded that a central feature of human interaction is impression formation—the attempt to present oneself to others in a particular way. Goffman believed that much human interaction can be studied and analyzed on the basis of principles derived from the theater. This approach, known as **dramaturgy**, states that *in order to create an impression, people play roles, and their performance is judged by others who are alert to any slips that might reveal the actor's true character.* For example, a job applicant at an interview tries to appear composed, self-confident, and capable of handling the position's responsibilities. The interviewer is seeking to find out whether the applicant is really able to

work under pressure and perform the necessary functions of the job.

Most interactions require some type of play-acting in order to present an image that will bring about the desired behavior from others. Dramaturgy sees these interactions as governed by planned behavior designed to enable an individual to present a particular image to others.

Types of Social Interaction

When two individuals are in each other's presence, they inevitably affect each other. They may do so intentionally, as when one person asks the other for change for a quarter, or they may do so unintentionally, as when two people drift toward opposite sides of the elevator in which they are riding. Whether intentional or unintentional, both behaviors represent types of social interaction.

Nonverbal Behavior

Many researchers have focused our attention on how we communicate with one another by using body movements. This study of body movements, known as *kinesics*, attempts to examine how such things as slight nods, yawns, postural shifts, and other nonverbal cues, whether spontaneous or deliberate, affect communication.

Many of our movements relate to an attitude that our culture, consciously or unconsciously, has taught

The use of hand and arm movements, eye contact, and norms of nonverbal behavior are markedly different for the Arab Bedouins than they are for Americans.

us to express in a specific manner. In the United States, for example, we show a status relationship in a variety of ways. The ritualistic nonverbal movements and gestures in which we engage to see who goes through a door first or who sits or stands first are but a few ways our culture uses movement to communicate status. In the Middle East, status is underscored nonverbally by which individual you turn your back to. In Oriental cultures, bowing and backing out of a room are signs of status relationships. Humility might be shown in the United States by a slight downward bending of the head, but in many European countries this same attitude is manifested by dropping one's arms and sighing. In Samoa, humility is communicated by bending the body forward.

The use of hand and arm movements as a means of communicating also varies among cultures. We all are aware of the different gestures for derision. For some European cultures, it is a closed fist with the thumb protruding between the index and middle fingers. The Russian expresses this same attitude by moving one index finger horizontally across the other.

In the United States, we can indicate that things are okay by making a circle with one's thumb and index finger while extending the others. In Japan, this gesture signifies "money." Among Arabs, if you accompany this gesture with a baring of the teeth, it displays extreme hostility.

In the United States, we say good-bye or farewell by waving the hand and arm up and down. If you

wave in this manner in South America, you may discover that the other person is not leaving but moving toward you. That is because in many countries, what we use as a sign of leaving is a gesture that means "come."

Eye contact is another area in which some interesting findings have been reported. In the United States, the following has been noted:

1. We tend to look at the person we are speaking to more when we are listening than when we are talking. If we are at a loss for a word, we frequently look into space, as if to find the words imprinted somewhere out there.

2. The more rewarding we find the speaker's message to be, the more intensely we will look at that person.

3. How much eye contact we maintain with other people is determined in part by how high we perceive their status to be. When we address someone we regard as having high status, we maintain a modest-to-high degree of eye contact. When we address a person of low status, we make very little effort to maintain eye contact.

4. Eye contact must be regularly broken, because we feel uncomfortable if someone gazes at us for longer than 10 seconds at a time.

These notions of eye contact found in the United States differ from those of other societies. In Japan and China, for example, it is considered rude to look into another person's eyes during conversation. Arabs, in contrast, use personal space very differently; they stand very close to the person they are talking to and stare directly into the eyes. Arabs believe that the eyes are a key to a person's being and that looking deeply into another's eyes allows one to see that person's soul.

The proscribed relationships between males and females in a culture also influence eye behavior. Asian cultures, for example, consider it taboo for women to look straight into the eyes of men. Therefore, most Asian men, out of respect for this cultural characteristic, do not stare directly at women. French men accept staring as a cultural norm and often stare at women in public (Samovar, Porter, & Jain, 1981; Axtell, 1998).

Exchange

When people do something for each other with the express purpose of receiving a reward or return, they are involved in an **exchange interaction.** Most employer-employee relationships are exchange relationships. The employee does the job and is rewarded with a salary. The reward in an exchange interaction, however, need not always be material; it can also be based on emotions such as gratitude. For example, if you visit a sick friend, help someone with a heavy package at the supermarket, or help someone solve a problem, you probably will expect these people to feel grateful to you.

Sociologist Peter Blau (1964) pointed out that exchange is the most basic form of social interaction. He believes social exchange can be observed everywhere once we are sensitized to it.

Cooperation

A **cooperative interaction** occurs *when people act together to promote common interests or achieve shared goals.* The members of a basketball team pass to one another, block off opponents for one another, rebound, and assist one another to achieve a common goal—winning the game. Likewise, family members cooperate to promote their interests as a family—the husband and wife both may hold jobs as well as share in household duties, and the children may help out by mowing the lawn and washing the dishes. College students often cooperate by studying together for tests.

Conflict

In a cooperative interaction, people join forces to achieve a common goal. By contrast, people in conflict struggle with one another for some commonly prized object or value. In most conflict relationships, only one person can gain at someone else's expense. Conflicts arise when people or groups have incompatible values or when the rewards or resources available to a society or its members are limited. Thus, conflict always involves an attempt to gain or use power.

The fact that conflict often leads to unhappiness and violence causes many people to view it negatively. However, conflict appears to be inevitable in human society. A stable society is not a society

Spontaneous cooperation that arises from the needs of a particular situation is the oldest and most natural form of cooperation.

without conflicts, but rather one that has developed methods for resolving its conflicts by justly or brutally suppressing them temporarily. For example, Lewis Coser (1956, 1967) pointed out that conflict can be a positive force in society. The American civil rights movement in the 1950s and 1960s may have seemed threatening and disruptive to many people at the time, but it helped bring about important social changes that may have led to greater social stability.

Coercion involves the use of power that is regarded as illegitimate by those on whom it is exerted. The stronger party can impose its will on the weaker, as in the case of a parent using the threat of punishment to impose a curfew on an adolescent. Coercion rests on force or the threat of force, but usually it operates more subtly. (See "Sociology at Work: Are We Arguing Too Much?")

Competition

The fourth type of social interaction, **competition,** is *a form of conflict in which individuals or groups confine their conflict within agreed-upon rules.* Competition is a common form of interaction in the modern world—not only on the sports field but in the marketplace, the education system, and the political system. American presidential elections, for example, are based on competition. Candidates for each party compete throughout the primaries, and eventually one candidate is selected to represent each of the major parties. The competition grows

Sociology at Work

Are We Arguing Too Much?

From television talk shows to political debates, Deborah Tannen believes we have become an "argument culture." She is referring to a pervasive warlike atmosphere that makes us approach every topic as a fight between two opposing sides. It makes us think that the best way to explore an idea is to set up a debate. Sometimes this approach works well, but often it creates more problems than it solves.

The argument culture urges us to approach the world—and the people in it—in an adversarial frame of mind. It rests on the assumption that opposition is the best way to get anything done. The best way to cover the news is find spokespeople who express the most extreme, polarized views; the best way to settle disputes is litigation that pits one party against the other; and the best way to show you are really thinking is the criticize.

The war on drugs, the war on cancer, the battle of the sexes, political campaigns—in the argument culture, war metaphors pervade our language and shape our thinking. Nearly everything is framed as a battle, or games in which winning or losing is the main concern. These all have their uses and their place, but they are not the only way—and often not the best way—to understand and approach our world. Conflict and opposition are as necessary as cooperation and agreement, but the scale is off balance, with conflict and opposition overweighted.

There are times when it is necessary and right to fight—to defend your country or yourself, to argue for right against wrong or against offensive and dangerous ideas or actions. What Tannen questions is the knee-jerk nature of approaching almost any issue, problem, or public person in an adversarial way. There is a general tendency to use opposition to accomplish *every* goal, even those that do not require fighting but might also be accomplished by other means, such as exploring, expanding, discussing, investigating, and the exchanging of ideas.

When you are arguing with someone, you are usually not trying to understand what the other person is saying or what in their experience leads them to say it. Instead you are readying your response: listening for weaknesses in logic to leap on, points you can distort to make the other person look bad and yourself look good.

Fights have winners and losers. If you are fighting to win, the temptation is great to deny facts that support your opponent's views and to say only what supports your side. In the extreme form, it encourages people to misinterpret or even to lie.

Our determination to pursue truth by setting up a fight between two sides leads us to believe that every issue has two sides—no more, no less. If both sides are given a forum to confront each other, all the relevant information will emerge, and the best case will be made for each side. But opposition does not lead to truth when an issue is not composed of two opposing sides, but is a combination of many sides. Often the truth is in the complex middle, not the oversimplified extremes.

Source: *The Argument Culture: Moving From Debate to Dialogue* (pp. 3–4; 8–10), by D. Tannen, 1998, New York: Random House.

even more intense as the remaining candidates battle directly against each other to persuade a nation of voters they are the best person for the presidency.

One type of relationship may span the entire range of focused interactions: An excellent example is marriage. Husbands and wives cooperate in household chores and responsibilities. They also engage in exchange interactions. Married people often discuss their problems with each other—the partner whose role is listener at one time will expect the spouse to provide a sympathetic ear at another time. Married people also experience conflicts in their relationship. A couple may have a limited amount of money set aside, and each may want to use it for a different purpose. Unless they can agree on a third, mutually desirable use for the money, one spouse will gain at the other's expense, and the marriage may suffer. The husband and wife whose marriage is irreversibly damaged may find themselves in direct competition. If they wish to separate or divorce, their conflict will be regulated according to legal and judicial rules.

Through the course of a lifetime, we constantly are involved in several types of social interaction, because most of our time is spent in some kind of group situation. How we behave in these situations is generally determined by two factors—the statuses we occupy and the roles we play—which together constitute the main components of what sociologists call social organization.

Elements of Social Interaction

People do not interact with one another as anonymous beings. They come together in the context of specific environments and with specific purposes. Their interactions involve behaviors associated with defined statuses and particular roles. These statuses and roles help to pattern our social interactions and provide predictability.

Statuses

Statuses are *socially defined positions that people occupy.* Common statuses may pertain to religion, education, ethnicity, and occupation: for example, Protestant, college graduate, African American, and teacher. Statuses exist independent of the specific people who occupy them (Linton, 1936). For example, our society recognizes the status of politician. Many people occupy that status, including President George W. Bush, Senator Ted Kennedy, Senator Hillary Clinton, and Governor Gray Davis. New politicians appear; others retire or lose popularity or are defeated, but the status, as the culture defines it, remains essentially unchanged. The same is true for all other statuses: occupational statuses such as doctor, computer analyst, bank teller, police officer, butcher, insurance adjuster, thief, and prostitute; and nonoccupational statuses such as son and daughter, jogger, friend, Little League coach, neighbor, gang leader, and mental patient.

It is important to keep in mind that from a sociological point of view, status does not refer—as it does in common usage—to the idea of prestige, even though different statuses often do contain differing degrees of prestige. In the United States, for example, research has shown that the status of Supreme Court justice has more prestige than that of physician, which in turn has more prestige than that of sociologist (Nakao, Keoko, & Treas, 1993).

People generally occupy more than one status at a time. Consider yourself, for example: You are someone's daughter or son, a full-time or part-time college student, perhaps also a worker, a licensed car driver, a member of a church or synagogue, and so forth. Sometimes one of the multiple statuses a person occupies seems to dominate the others in patterning that person's life; such a status is called a master status. For example, George W. Bush has occupied a number of diverse statuses: husband, father, state governor, and presidential candidate. After January 20, 2001, however, his master status was that of president of the United States, as it governed his actions more than did any other status he occupied at the time.

A person's master status will change many times in the course of his or her life cycle. Right now, your master status probably is that of college student. Five years from now it may be graduate student, artist, lawyer, spouse, or parent. Figure 5-1 illustrates the different statuses occupied by a 35-year-old woman who is an executive at a major television network. Although she occupies many statuses at once, her master status is that of vice-president for programming.

In some situations, a person's master status may have a negative influence on the person's life. For example, people who have followed what their culture considers a deviant lifestyle may find that their master status is labeled according to their deviant behavior. Those who have been identified as ex-convicts are likely to be so classified no matter what other statuses they occupy. They will be thought of as ex-convict-painters, ex-convict-machinists, ex-convict-writers, and so on. Their master status has a negative effect on their ability to fulfill the roles of the statuses they would like to occupy. Ex-convicts who are good machinists or house painters may find employers unwilling to hire them because of their police records. Because the label *criminal* can stay with individuals throughout their lives, the criminal justice system is reluctant to label juvenile offenders or to open their records to the courts. Juvenile court

Figure 5-1 **Status and Master Status**

Generally, each individual occupies many statuses at one time. The statuses of a female executive at a major television network include author, wife, mother, pianist, and so on. Other statuses could be added to this list. However, one status—vice-president for programming—is most important in patterning this woman's life. Sociologists call such a status a master status.

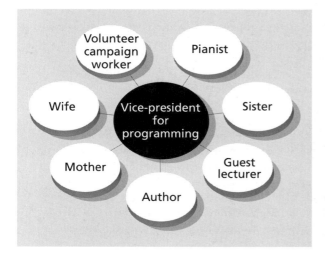

This woman's ascribed status is female; her achieved status is based on her profession.

files are usually kept secret and often permanently sealed when the person reaches age 18.

Some statuses, called **ascribed statuses,** are *conferred upon us by virtue of birth or other significant factors not controlled by our own actions or decisions; people occupy them regardless of their intentions.* Certain family positions, such as that of daughter or son, are typical ascribed statuses, as are one's gender and ethnic or racial identity. Other statuses, called achieved statuses, are occupied as a result of the individual's actions—student, professor, garage mechanic, race car driver, artist, prisoner, bus driver, husband, wife, mother, or father. (See "Sociology at Work: Does Birth Order Influence Our Social Interactions?" for an example of the influence of birth order.)

Roles

Statuses alone are static—nothing more than social categories into which people are put. Roles bring statuses to life, making them dynamic. As Robert Linton (1936) observed, you occupy a status but you play a role. **Roles** are *the culturally defined rules for proper behavior that are associated with every status.* Roles may be thought of as collections of rights and obligations. For example, to be a race car driver you must become well versed in these rights and obligations, as your life might depend on them. Every driver has the right to expect other drivers not to try to pass when the race has been interrupted

by a yellow flag because of danger. Turned around, each driver has the obligation not to pass other drivers under yellow-flag conditions. A driver also has a right to expect race committee members to enforce the rules and spectators to stay off the raceway. On the other hand, a driver has an obligation to the owner of the car to try hard to win.

In the case of our television executive, she has the right to expect to be paid on time, to be provided with good-quality scripts and staff support, and to make decisions about the use of her budget. On the other hand, she has the obligation to act in the best interest of the network, to meet schedules, to stay within her budget, and to treat her employees fairly. What is important is that all these rights and obligations are part of the roles associated with the status of vice-president for programming. They exist without regard to the particular individuals whose behavior they guide (see Figure 5-2).

A status may include a number of roles, and each role will be appropriate to a specific social context. For example, as the child of a military officer, Kay Redfield Jamison found that children had to learn the importance of statuses and roles and the proper

Sociology at Work

Does Birth Order Influence Our Social Interactions?

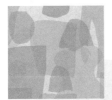

Theodore Kaczynski, the alleged unibomber is a firstborn. Martin Luther King, a peacemaker, was a middle-born. Charles Darwin and Nicholas Copernicus, radical revolutionaries in science, were later-borns.

According to MIT researcher Frank Sulloway, the characteristics of these people are not coincidental, but a result of the birth order. Sulloway has spent nearly 30 years researching the impact of birth order using a huge biographical database of 6,000 people involved in scientific and social revolutions over the past 500 years.

According to Sulloway, later-borns are more open-minded than firstborns. Firstborns tend to support the status quo. They are status conscious and often emerge as leaders. But when faced with revolutionary change, whether it be in politics or science, they resist innovation.

Later-borns are "born to rebel." They take risks and challenge accepted wisdom. Marx, Lenin, Ho Chi Minh, Ralph Nader, Marlon Brando, and Bill Gates were all later-borns.

Why would these differences be present? Siblings within a family compete with each other for the affection of the parents. They can use certain basic strategies. First, they can try to gain parental favoritism directly by helping and obeying parents. Second, they could try to dominate their rivals and hope to eliminate them as competitors. Finally, if they find that they are being dominated they can adopt various countermeasures, including appeasement, rebellion, or a combination of both tactics. Depending on age and size, some of these strategies are more effective than others. This is where birth order enters as a factor. People then continue to use these strategies throughout their lives.

Firstborns seem to be more successful in life. Sulloway believes this is because they act like a surrogate parent when the later-born siblings arrive.

They identify with their parents and try to curry favor with them by being conscientious and supportive of parental authority. There are many rewards for taking that conscientious road in life. It leads to better academic achievement and to moving up the ladder in a conventional way. Therefore, first-borns are overrepresented in business as CEOs and in politics as presidents of the United States. Firstborns are very good at the kind of success that would please your parents.

Later-borns succeed in a very different way. Their successes are fewer, but they win big in the history books as the proponents of radical revolutions. In the history of science, Copernicus was the youngest of four children and developed the theory the earth is not the center of the universe. Charles Darwin, the fifth of six children proposed the theory of evolution. Most of the great revolutionaries in science and social thought have been younger siblings. Sulloway believes that what they are doing in essence is identifying with the underdog, because they were an underdog. They tend to want to pull the rug out from the status quo.

Later-borns are taking greater risks, which is a double-edged sword. On the plus side, they are likely to win big when they take a unique position and end up being right. But they also can be very big losers. Later-borns were very big supporters of phrenology, the idea that bumps on the head could be used as a way of reading personality. This was a radical idea that was totally wrong. Later-borns are likely to take positions before the evidence is fully in. They can seem remarkably early when they are right, but terribly wrong when they are not.

Sources: Born to Rebel: Birth Order, Family Dynamics and Creative Lives, by F. J. Sulloway, 1996, New York: Pantheon Books, and an interview with the author, November 1996.

behavior to be displayed toward those who occupied those positions.

[The] Cotillion was where officers' children were supposed to learn the fine points of manners, dancing, white gloves, and other unrealities of life. It also was where children were supposed to learn, as if the preceding fourteen or fifteen years hadn't already made it painfully clear, that generals outrank colonels, who, in turn, outrank majors and captains and lieutenants, and everyone, but every-

one, outranks children. Within the ranks of children, boys always outrank girls.

One way of grinding this particularly irritating pecking order into the young girls was to teach them the old and ridiculous art of curtsying. It is hard to imagine that anyone in her right mind would find curtsying an even vaguely tolerable thing to do. But having been given the benefits of a liberal education by a father with strongly nonconforming views and behaviors, it was beyond belief to me that I would seriously be expected to

Figure 5-2 Status and Roles

The status of vice-president for programming at a major television network has several roles attached to it, including attending meetings, making programming decisions, and so on.

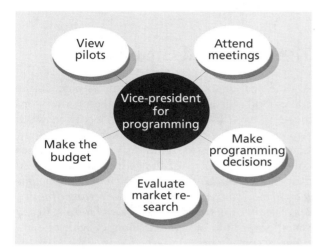

do this. I saw the line of crisply crinolined girls in front of me and watched each of them curtsying neatly. Sheep, I thought, Sheep. Then it was my turn. Something inside of me came to a complete boil. It was one too many times watching one too many girls being expected to acquiesce; far more infuriating, it was one too many times watching girls willingly go along with the rites of submission. I refused. A slight matter, perhaps, in any other world, but within the world of military custom and protocol—where symbols and obedience were everything, and where a child's misbehavior could jeopardize a father's promotion, it was a declaration of war. Refusing to obey an adult, however absurd the request, simply wasn't done. Miss Courtnay, our dancing teacher, glared. I refused again. She said she was very sure that Colonel Jamison would be terribly upset by this. I was, I said, very sure that Colonel Jamison couldn't care less. I was wrong. (Jamison, 1995)

Role Sets

All the roles attached to a single status are known collectively as a **role set.** However, not every role in a particular role set is enacted all the time. An individual's role behaviors depend on the statuses of the other people with whom he or she is interacting. For example, as a college student you behave one way toward other students and another way toward professors. Similarly, professors behave one way toward other professors, another way toward students, and yet a third way toward deans. So the role behavior we expect in any given situation depends on the pairs of statuses occupied by the interacting individuals. This means that role behavior really is defined

Figure 5-3 Role Sets

People's role behaviors change according to the statuses of the other people with whom they interact. The female vice-president for programming will adopt somewhat different roles depending on the statuses of the various people with whom she interacts at the station: a writer, a journalist, her assistants, and so on.

by the rights and obligations that are assigned to statuses when they are paired with one another (see Figure 5-3).

It would be difficult to describe the wide-ranging, unorganized assortment of role behaviors associated with the status of television vice-president for programming. Sociologists find it more useful to describe the specific behavior expected of a network television vice-president for programming interacting with different people. Such a role set would include the following:

Vice-president for programming/network president

Vice-president for programming/other vice-presidents

Vice-president for programming/script writer

Vice-president for programming/administrative assistant

Vice-president for programming/television star

Vice-president for programming/journalist

Vice-president for programming/producer

Vice-president for programming/sponsor

The vice-president's role behavior in each case would be different, meshing with the role behavior of the individual(s) occupying the other status in each pairing (Merton, 1968).

Role Strain

Even though most people try to enact their roles as they are expected to, they sometimes find it difficult.

When a single role has conflicting demands attached to it, individuals who play that role experience **role strain** (Goode, 1960). For example, the captain of a freighter is expected to be sure the ship sails only when it is in safe condition, but the captain also is expected to meet the company's delivery schedule, because a day's delay could cost the company thousands of dollars. These two expectations may exert competing pulls on the captain, especially when some defect is reported, such as a malfunction in the ship's radar system. The stress of these competing pulls is not due to the captain's personality, but rather is built into the nature of the role expectations attached to the captain's status. Therefore, sociologists describe the captain's experience of stress as role strain.

Role Conflict

An individual who is occupying more than one status at a time and who is unable to enact the roles of one status without violating those of another status is encountering **role conflict.** Not too long ago, pregnancy was considered "women's work." An expectant father was expected to get his wife to the hospital on time and pace the waiting room anxiously awaiting the nurse's report on the sex of the baby and its health. Today, men are encouraged and even expected to participate fully in the pregnancy period and the birth of the child. A role conflict arises, however, in that while the new father is expected to be involved, his involvement is defined along male gender role lines. He is expected to be helpful, supportive, and essentially a stabilizing force. He really is not allowed to indicate that he is frightened, nervous, or possibly angry about the baby. His role as a male, even in twenty-first century American society, conflicts with his feelings as a new expectant father (Shapiro, 1987).

As society becomes more complex, individuals occupy increasingly larger numbers of statuses. This increases the chances for role conflict, which is one of the major sources of stress in modern society.

Role Playing

The roles we play can have a profound influence on both our attitudes and our behavior. Playing a new social role often feels awkward at first, and we may feel we are just acting—pretending to be something that we are not. However, many sociologists feel that the roles a person plays are the person's only true self. Peter Berger's (1963) explanation of role playing goes further: The roles we play can transform not only our actions but also ourselves.

One feels more ardent by kissing, more humble by kneeling, and more angry by shaking one's fist— that is, the kiss not only expresses ardor but manufactures it. Roles carry with them both certain actions and emotions and attitudes that belong to these actions. The professor putting on an act that pretends to wisdom comes to feel wise.

The Nature of Groups

A good deal of social interaction takes place in the context of a group. In common speech the word *group* is often used for almost any occasion when two or more people come together. In sociology, however, there are several terms we use for various collections of people, not all of which are considered groups. A **social group** consists of *a number of people who have a common identity, some feeling of unity, and certain common goals and shared norms.* In any social group, the individuals interact with one another according to established statuses and roles. The members develop expectations of proper behavior for people occupying different positions in the social group. The people have a sense of identity and realize they are different from others who are not members. Social groups have a set of values and norms that may or may not be similar to those of the larger society.

Our description of a social group contrasts with our definition of a **social aggregate,** which is made up of *people who temporarily happen to be in physical proximity to each other, but share little else.* Consider passengers riding together in one car of a train. They may share a purpose (traveling to Washington, D.C.) but do not interact or even consider their temporary association to have any meaning. It hardly makes sense to call them a group—unless something more happens. If it is a long ride, for instance, and several passengers start a card game, the cardplayers will have formed a social group: They have a purpose, they share certain role expectations, and they attach importance to what they are doing together. Moreover, if the cardplayers continue to meet one another every day (say, on a commuter train), they may begin to feel special in contrast with the rest of the passengers, who are just "riders."

A social group, unlike an aggregate, does not cease to exist when its members are away from one another. Members of social groups carry the fact of their membership with them and see the group as a distinct entity with specific requirements for membership. A social group has a purpose and is therefore important to its members, who know how to tell an "insider" from an "outsider." It is a social entity that exists for its members apart from any other social relationships that some of them might share. Members of a group interact according to established norms and traditional statuses and roles. As new members are recruited to the group, they move into these traditional statuses and adopt the expected role behavior—if not gladly, then as a result of group pressure.

Consider, for example, a tenants' group that consists of the people who rent apartments in a building. Most such groups are founded because tenants feel a need for a strong, unified voice in dealing with the landlord on problems with repairs, heat, hot water, and rent increases. Many members of a tenants' group may never have met one another before; others may be related to one another; and some may also belong to other groups such as a neighborhood church, the PTA, a bowling league, or political associations. The group's existence does not depend on these other relationships, nor does it cease to exist when members leave the building to go to work or to go away on vacation. The group remains, even when some tenants move out of the building and others move in. Newcomers are recruited, told of the group's purpose, and informed of its meetings; they are encouraged to join committees, take leadership responsibilities, and participate in the actions the group has planned. Members who fail to support group actions (such as withholding rent) will be pressured and criticized by the group.

People sometimes are defined as being part of a specific group because they share certain characteristics. If these characteristics are unknown or unimportant to those in the category, it is not a social group. Involvement with other people cannot develop unless one is aware of them. People with similar characteristics do not become a social group unless concrete, dynamic interrelations develop among them (Lewin, 1948). For example, although all left-handed people fit into a group, they are not a social group just because they share this common characteristic. A further interrelationship must also exist. They may, for instance, belong to Left-Handers International, an organization that champions the rights of left-handers by addressing issues of discrimination and analyzing new products for use by left-handers. About 23,000 left-handers belong to this social group.

Even if people are aware of one another, that is still not enough to make them a social group. We may be classified as Democrats, college students, upper class, or suburbanites. Yet for many of us who fall into these categories, there is no group. We may not be involved with the others in any patterned way that is an outgrowth of that classification. In fact, we personally may not even define ourselves as members of the particular category, even if someone else does.

Social groups can be large or small, temporary or long lasting: Your family is a group, as is your ski club, any association to which you belong, or the clique you hang around with. In fact, it is difficult for you to participate in society without belonging to a number of different groups.

In general, social groups, regardless of their nature, have the following characteristics: (1) permanence beyond the meetings of members, that is, even when members are dispersed; (2) means for identifying members; (3) mechanisms for recruiting new members; (4) goals or purposes; (5) social statuses and roles, that is, norms for behavior; and (6) means for controlling members' behavior.

The traits we described are features of many groups. A baseball team, a couple about to be married, a work unit, players in a weekly poker game, members of a family, or a town planning board all may be described as groups. Yet being a member of a family is significantly different from being a member of a work unit. The family is a primary group, whereas most work units are secondary groups.

Primary and Secondary Groups

The difference between primary and secondary groups lies in the kinds of relationships their members have with one another. Charles Horton Cooley (1909) defined primary groups as groups that are characterized by

> intimate face-to-face association and cooperation. They are primary in several senses, but chiefly in that they are fundamental in forming the social nature and ideas of the individual. The result of intimate association, psychologically, is a certain fusion of individualities in a common whole, so that one's very self, for many purposes at least, is the common life and purpose of the group. Perhaps the simplest way of describing this wholeness is by saying that it is a "we"; it involves the sort of sympathy and mutual identification for which "we" is the natural expression. (p. 23)

Cooley called primary groups the nursery of human nature, because they have the earliest and most fundamental impact on the individual's socialization and development. He identified three basic primary groups: the family, children's play groups, and neighborhood or community groups.

Primary groups involve *interaction among members who have an emotional investment in one another and in a situation, who know one another intimately, and interact as total individuals rather than through specialized roles.* For example, members of a family are emotionally involved with one another and know one anther well. In addition, they interact with one another in terms of their total personalities, not just in terms of their social identities

Table 5-2

Relationships in Primary and Secondary Groups

	PRIMARY	SECONDARY
Physical Conditions	Small number	Large number
	Long duration	Shorter duration
Social Characteristics	Indentification of ends	Disparity of ends
	Intrinsic valuation of the relation	Extrinsic valuation of the relation
	Intrinsic valuation of other person	Extrinsic valuation of other person
	Inclusive knowledge of other person	Specialized and limited knowledge of other person
	Feeling of freedom and spontaneity	Feeling of external constraint
	Operation of informal controls	Operation of formal controls
Sample Relationships	Friend–friend	Clerk–customer
	Husband–wife	Announcer–listener
	Parent–child	Performer–spectator
	Teacher–pupil	Officer–subordinate
Sample Groups	Play group	Nation
	Family	Clerical hierarchy
	Village or neighborhood	Professional association
	Work team	Corporation

Source: *Human Society,* by K. Davis, 1949, New York: Macmillan.

or statuses as breadwinner, student, athlete, or community leader.

A **secondary group,** in contrast, *is characterized by much less intimacy among its members. It usually has specific goals, is formally organized, and is impersonal.* Secondary groups tend to be larger than primary groups, and their members do not necessarily interact with all other members. In fact, many members often do not know one another at all; to the extent that they do, rarely do they know more about one another than about their respective social identities. Members' feelings about, and behavior toward, one another are patterned mostly by their statuses and roles rather than by personality characteristics. The chair of the General Motors board of directors, for example, is treated respectfully by all General Motors employees—regardless of the chair's gender, age, intelligence, habits of dress, physical fitness, temperament, or qualities as a parent or spouse. In secondary groups, such as political parties, labor unions, and large corporations, people are very much what they do. Table 5-2 outlines the major differences between primary and secondary groups.

Functions of Groups

To function properly, all groups—both primary and secondary—must (1) define their boundaries, (2) choose leaders, (3) make decisions, (4) set goals, (5) assign tasks, and (6) control members' behavior.

Defining Boundaries

Group members must have ways of knowing who belongs to their group and who does not. Sometimes devices for marking boundaries are obvious symbols, such as the uniforms worn by athletic teams, lapel pins worn by Rotary Club members, rings worn by Masons, and styles of dress. Other ways by which group boundaries are marked include the use of gestures (special handshakes often used by gang members) and language (dialect differences often mark people's regional origin and social class). In some societies (including our own), skin color also is used to mark boundaries between groups. The idea of the British school tie that, by its pattern and colors, signals exclusive group membership has been adopted by businesses ranging from banking to brewing.

Choosing Leaders

All groups must grapple with the issue of leadership. A **leader** is *someone who occupies a central role or position of dominance and influence in a group.* In some groups, such as large corporations, leadership is assigned to individuals by those in positions of authority. In other groups, such as adolescent peer groups, individuals move into positions of leadership through the force of personality or through particular skills such as athletic ability, fighting, or debating. In still other groups, including political organizations, leadership is awarded through the democratic process of nominations and voting. Think of the long primary process the presidential candidates must endure in order to amass enough votes to carry their parties' nominations for the November election.

Leadership need not always be held by the same person within a group. It can shift from one individual to another in response to problems or situations that the group encounters. In a group of factory workers, for instance, leadership may fall on different members depending on what the group plans to do—complain to the supervisor, head to a bar after work, or organize a picnic for all members and their families.

Politicians and athletic coaches often like to talk about individuals who are "natural leaders." Although attempts to account for leadership solely in terms of personality traits have failed again and again, personality factors may determine what kinds of leadership functions a person assumes. Researchers (Bales, 1958; Slater, 1966) have identified two types of leadership roles: (1) **instrumental leadership**, *in which a leader actively proposes tasks and plans to guide the group toward achieving its goals,* and (2) **expressive leadership**, *in which a leader works to keep relations among group members harmonious and morale high.* Both kinds of leadership are crucial to the success of a group.

Sometimes both leadership functions are fulfilled by one person, but when that is not the case, those functions are often distributed among several group members. The individual with knowledge of the terrain who leads a group of airplane crash survivors to safety is providing instrumental leadership. The group members who think of ways to keep the group from giving in to despair are providing expressive leadership. The group needs both kinds of leadership to survive.

Making Decisions

Closely related to the problem of leadership is the way groups make decisions. In many early hunting and food-gathering societies, important group decisions were reached by consensus—talking about an issue until everybody agreed on what to do (Fried, 1967). Today, occasionally, town councils and other small governing bodies operate in this way. Because this consensus-gathering takes a great deal of time and energy, many groups opt for efficiency by taking votes or simply letting one person's decision stand for the group as a whole.

Setting Goals

As we pointed out before, all groups must have a purpose, a goal, or a set of goals. The goal may be very general, such as spreading peace throughout the world, or it may be very specific, such as playing cards on a railroad train. Group goals may change. For example, the card players might discover that they share a concern about the use of nuclear energy and decide to organize a political-action group.

Assigning Tasks

Establishing boundaries, defining leadership, making decisions, and setting goals are not enough to keep a group going. To endure, a group must do something, if nothing more than ensure that its members continue to make contact with one another. Therefore, it is important that group members know what needs to be done and who is going to do it. This assigning of tasks, in itself, can be an important group activity (think of your family discussions about sharing household chores). By taking on group tasks, members not only help the group reach its goals but also show their commitment to one another and to the group as a whole. This leads members to appreciate one another's importance as individuals and the importance of the group in all their lives—a process that injects life and energy into a group.

Controlling Members' Behavior

If a group cannot control its members' behavior, it will cease to exist. For this reason, failure to conform to group norms is seen as dangerous or threatening, whereas conforming to group norms is rewarded, if only by others' friendly attitudes. Groups not only encourage but often depend for survival on conformity of behavior. A member's failure to conform is met with responses ranging from coolness to criticism or even ejection from the group. Anyone who has tried to introduce changes into the constitution of a club or to ignore long-standing conventions—such as ways of dressing, rituals of greeting, or the assumption of designated responsibilities—probably has experienced group hostility.

Primary groups tend to be more tolerant of members' deviant behavior than are secondary groups. For example, families often will conceal the problems of a member who suffers from chronic alcoholism or drug abuse. Even primary groups, however, must draw the line somewhere, and they will invoke negative sanctions (see Chapter 6) if all else fails to get the deviant member to show at least a willingness to try to conform. When primary groups finally do act, their punishments can be far more severe or harsh than those of secondary groups. Thus, an intergenerational conflict in a family can result in the commitment of a teenager to an institution or treatment center. Secondary groups tend to use formal,

Figure 5-4 **Group Pressure**

In Solomon Asch's experiment on conformity to group pressure, groups of eight students were asked to decide which of the comparison lines (right) was the same length as the standard line (left).

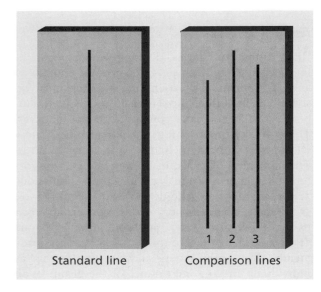

Standard line   Comparison lines

as opposed to informal, sanctions and are much more likely than primary groups simply to expel, or push out, a member who persists in violating strongly held norms: Corporations fire unsatisfactory employees, the army discharges soldiers who violate regulations, and so on. Even though primary groups are more tolerant of their members' behavior, people tend to conform more closely to their norms than to those of secondary groups. This is because people value their membership in a primary group, with its strong interpersonal bonds, for its own sake. Secondary group membership is valued mostly for what it will do for the people in the group, not because of any deep emotional ties. Because primary group membership is so desirable, its members are more reluctant to risk expulsion by indulging in behavior that might violate the group's standards or norms than are secondary group members.

Usually, group members will want to conform as long as the group is experienced as important. Solomon Asch (1955) showed just how far group members will go to promote group solidarity and conformity. In a series of experiments, he formed groups of eight people and then asked each member to match one line against three other lines of varying lengths (see Figure 5-4). Each judgment was announced in the presence of the other group members. The groups were composed of one real subject and seven of Asch's confederates, whose identity was kept secret from the real subject. The confederates had met previously with Asch and had been in-

structed to give a unanimous but incorrect answer at certain points throughout the experiment. Asch was interested in finding out how the individual who had been made a minority of one would respond in the presence of a unanimous majority. The subject was placed in a situation in which a group unanimously contradicted the information of his or her senses. Asch repeated the experiment many times. He found that 32% of the answers by the real subjects were identical with, or in the direction of, the inaccurate estimates of the majority. This was quite remarkable, because there were virtually no incorrect answers in the control groups that lacked Asch's accomplices, which rules out the possibility of optical illusion. What this illustrates is an instance in which individuals are willing to give incorrect answers in order not to appear out of step with the judgments of the other group members. Although groups must fulfill certain functions in order to continue to exist, they serve primarily as a point of reference for their members.

Reference Groups

Groups are more than just bridges between the individual and society as a whole. We spend much of our time in one group or another, and the impact that these groups have on us continues even when we are not actually in contact with the other members. The norms and values of groups we belong to or identify with serve as the basis for evaluating our own and others' behavior.

A **reference group** is *a group or social category that an individual uses to help define beliefs, attitudes, and values and to guide behavior.* They provide a comparison point against which people measure themselves and others. A reference group is often a category we identify with, rather than a specific group we belong to. For example, a communications major may identify with individuals in the media without having any direct contact with them. In this respect, anticipatory socialization is taking place, in that the individual may alter his or her behavior and attitudes toward those he or she perceives to be part of the group the individual plans to join. For example, people who become bankers soon feel themselves part of a group—bankers—and assume ideas and lifestyles that help them identify with that group. They tend to dress in a conservative, "bankerish" fashion, even buying their clothes in shops that other bankers patronize to make sure they have the "right" clothes from the "right" stores. They join organizations such as country clubs and alumni associations so they can mingle with other bankers and clients. Eventually, the norms and values they adopted when they joined the bankers' group become internalized; they see and judge the world around them as bankers.

We can also distinguish between positive and negative reference groups. Positive reference groups are composed of people we want to emulate. Negative reference groups provide a model we do not wish to follow. Therefore, a writer may identify positively with those writers who produce serious fiction, but may think of journalists who write for gossip publications as a negative reference group.

Even though groups are in fact composed of individuals, individuals are also created to a large degree by the groups they belong to through the process of socialization (see Chapter 4). Of these groups, the small group usually has the strongest direct impact on an individual.

Small Groups

The term **small group** refers to *many kinds of social groups, such as families, peer groups, and work groups, that actually meet together and contain few enough members so that all members know one another.* The smallest group possible is a **dyad,** which *contains only two members.* An engaged couple is a dyad, as are the pilot and copilot of an aircraft.

George Simmel (1950) was the first sociologist to emphasize the importance of the size of a group on the interaction process. He suggested that small groups have distinctive qualities and patterns of interaction that disappear when the group grows larger. For example, dyads resist change in their group size: On the one hand, the loss of one member destroys the group, leaving the other member alone; on the other hand, a **triad,** or *the addition of a third member,* creates uncertainty because it introduces the possibility of two-against-one alliances.

Triads are usually unstable groups because the possibility of two-against-one alliances is always present.

Triads are more stable in those situations when one member can help resolve quarrels between the other two. When three diplomats are negotiating off-shore fishing rights, for example, one member of the triad may offer a concession that will break the deadlock between the other two. If that does not work, the third person may try to analyze the arguments of the other two in an effort to bring about a compromise. The formation of shifting pair-offs within triads can help stabilize the group. When it appears that one group member is weakening, one of the two paired members will often break the alliance and form a new one with the individual who had been isolated (Hare, 1976). This is often seen among groups of children engaged in games. In triads in which there is no shifting of alliances and the configuration constantly breaks down into two against one, the group will become unstable and may eventually break up. In Aldous Huxley's novel *Brave New World,* the political organization of the earth was organized into three eternally warring political powers. As one power seemed to be losing, one of the others would come to its aid in a temporary alliance, thereby ensuring worldwide political stability while also making possible endless warfare. No power could risk the total defeat of another because the other surviving power might then become the stronger of the surviving dyad.

As a group grows larger, the number of relationships within it increases, which often leads to the formation of **subgroups**—*splinter groups within the larger group.* Once a group has more than five to seven members, spontaneous conversation becomes difficult for the group as a whole. Then there are two solutions available: The group can split into subgroups (as happens informally at parties), or it can adopt a formal means of controlling communication (use of *Robert's Rules of Order,* for instance). For these reasons, small groups tend to resist the addition of new members because increasing size threatens the nature of the group. In addition, there may be a fear that new members will resist socialization to group norms and thereby undermine group traditions and values. On the whole, small groups are much more vulnerable than large groups to disruption by new members, and the introduction of new members often leads to shifts in patterns of interaction and group norms.

Large Groups: Associations

Although all of us probably would be able to identify and describe the various small groups we belong to, we might find it difficult to follow the same process with the large groups that affect us. As patrons or employees of large organizations and governments, we function as part of large groups all the time. Thus, sociologists must study large groups as well as small groups in order to understand the workings of society.

Much of the activity of a modern society is carried out through large and formally organized groups. Sociologists refer to these groups as **associations,** which are *purposefully created special-interest groups that have clearly defined goals and official ways of doing things.* Associations include such organizations as government departments and agencies, businesses and factories, labor unions, schools and colleges, fraternal and service groups, hospitals and clinics, and clubs for various hobbies from gardening to collecting antiques. Their goals may be very broad and general—such as helping the poor, healing the sick, or making a profit—or quite specific and limited—such as manufacturing automobile tires or teaching people to speak Chinese. Although an enormous variety of associations exist, they all are characterized by some degree of formal structure with an underlying informal structure.

Formal Structure

For associations to function, the work that the association must accomplish is assessed and broken down into manageable tasks that are assigned to specific individuals. In other words, associations are run according to a formal organizational structure that consists of planned, highly institutionalized, and clearly defined statuses and role relationships. The formal organizational structure of large associations in contemporary society is best exemplified by the organizational structure called bureaucracy.

For example, when we consider a college or university, fulfilling its main purpose of educating students requires far more than simply bringing together students and teachers. Funds must be raised, buildings constructed, qualified students and professors recruited, programs and classes organized, materials ordered and distributed, grounds kept up, and buildings maintained. Lectures must be given; seminars must be led; and messages need to be typed, copied, and filed. To accomplish all these tasks, the school must create many different positions: president, deans, department heads, registrar, public-relations staff, groundskeepers, maintenance personnel, purchasing agents, secretaries, faculty, and students. Every member of the school has clearly spelled-out tasks that are organized in relation to one another: Students are taught and evaluated by faculty, faculty members are responsible to department heads or deans, deans to the president, and so on. Underlying these clearly defined assignments are procedures

that are never written down but are worked out and understood by those who have to get the job done.

Informal Structure

Sociologists recognize that formal associations never operate entirely according to their stated rules and procedures. Every association has an informal structure consisting of networks of people who help out one another by "bending" rules and taking procedural shortcuts. No matter how carefully plans are made, no matter how clearly and rationally roles are defined and tasks assigned, every situation and its variants cannot be anticipated. Sooner or later, then, individuals in associations are confronted with situations in which they must improvise and even persuade others to help them do so.

As every student knows, no school ever runs as smoothly as planned. For instance, going by the book—that is, following all the formal rules—often gets students tied up in long lines and red tape. Enterprising students and instructors find shortcuts. A student who wants to change from Section A of Sociology 100 to Section E might find it very difficult or time-consuming to change sections (add and drop classes) officially. However, it might be possible to work out an informal deal—the student stays registered in Section A but attends, and is evaluated in, Section E. The instructor of Section E then turns the grade over to the instructor of Section A, who hands in that grade with all the other Section A grades—as if the student had attended Section A all along. The formal rules have been bent, but the major purposes of the school (educating and evaluating students) have been served.

In addition, human beings have their own individual needs even when they are on company time, and these needs are not always met by attending single-mindedly to assigned tasks. To accommodate these needs, people often try to find extra break time for personal business by getting jobs done faster than would be possible if all the formal rules and procedures were followed. To accomplish these ends, individuals in associations may cover for one another, look the other way at strategic moments, and offer one another useful information about office politics, people, and procedures. Gradually, the reciprocal relationships among members of these informal networks become institutionalized: "unwritten laws" are established, and a fully functioning informal structure evolves.

At the same time as the goals of associations have given rise to an informal structure for job performance, they also have spawned an organizational structure that often increases the formality of procedures. This formal organization structure is called bureaucracy, and it has an impact on the informal structure.

Bureaucracy

Although in ordinary usage the term *bureaucracy* suggests a certain rigidity and red tape, it has a somewhat different meaning to sociologists. Robert K. Merton (1969) defined **bureaucracy** as "*a formal, rationally organized social structure [with] clearly defined patterns of activity in which, ideally, every series of actions is functionally related to the purposes of the organization.*"

Max Weber, the German sociologist, provided the first detailed study of the nature and origins of bureaucracy. Although much has changed in society since he developed his theories, Weber's basic description of bureaucracy remains essentially accurate to this day.

Weber's Model of Bureaucracy: An Ideal Type

Weber viewed bureaucracy as the most efficient—although not necessarily the most desirable—form of social organization for the administration of work.

He studied examples of bureaucracy throughout history and noted the elements that they had in common. Weber's model of bureaucracy is an **ideal type,** which is *a simplified, exaggerated model of reality used to illustrate a concept.* When Weber presented his ideal type of bureaucracy, he combined into one those characteristics that could be found in one form or another in a variety of organizations. It is unlikely that we ever would find a bureaucracy that has all the traits presented in Weber's ideal type. However, his presentation can help us understand what is involved in bureaucratic systems. It is also important to recognize that Weber's ideal type is in no way meant to be ideal in the sense that it presents a desired state of affairs. In short, an ideal type is an exaggeration of a situation that is used to convey a set of ideas. Weber outlined six characteristics of bureaucracies:

1. *A clear-cut division of labor.* The activities of a bureaucracy are broken down into clearly defined, limited tasks, which are attached to formally defined positions (statuses) in the organization. This permits a great deal of specialization and a high degree of expertise. For example, a small-town police department might consist of a chief, a lieutenant, a detective, several sergeants,

Modern bureaucracies have an organizational structure that makes it clear what rights and duties are attached to the various positions.

and a dozen officers. The chief issues orders and assigns tasks; the lieutenant is in charge when the chief is not around; the detective does investigative work; the sergeants handle calls at the desk and do the paperwork required for formal booking procedures; and the officers walk or drive through the community, making arrests and responding to emergencies. Each member of the department has a defined status and duty as well as specialized skills appropriate to his or her position.

2. *Hierarchical delegation of power and responsibility.* Each position in the bureaucracy is given sufficient power so that the individual who occupies it can do assigned work adequately and also compel subordinates to follow instructions. Such power must be limited to what is necessary to meet the requirement of the position. For example, a police chief can order an officer to walk a specific beat but cannot insist that the officer join the Lions Club.

3. *Rules and regulations.* The rights and duties attached to various positions are stated clearly in writing and govern the behavior of all individuals who occupy them. In this way, all members of the organizational structure know what is expected of them, and each person can be held accountable for his or her behavior. For example, the regulations of a police department might state, "No member of the department shall drink intoxicating liquors while on duty." Such rules make the activities of bureaucracies predictable and stable.

4. *Impartiality.* The organization's written rules and regulations apply equally to all its members. No exceptions are made because of social or psychological differences among individuals. Also, people occupy positions in the bureaucracy only because they are assigned according to formal procedures. These positions belong to the organization itself; they cannot become the personal property of those who occupy them. For example, a vice-president of United States Steel Corporation is usually not permitted to pass on that position to his or her children through inheritance.

5. *Employment based on technical qualifications.* People are hired because they have the ability and skills to do the job, not because they have personal contacts within the company. Advancement is based on how well a person does the job. Promotions and job security go to those who are most competent.

6. *Distinction between public and private spheres.* A clear distinction is made between the employees' personal lives and their working lives. It is unusual for employees to be expected to take business calls at home. At the same time, employees' family lives have no place in the work setting.

Although many bureaucracies strive at the organizational level to attain the goals that Weber proposed, most do not achieve them on the practical level.

Bureaucracy Today: The Reality

Just as no building is ever identical to its blueprint, no bureaucratic organization fully embodies all the features of Weber's model. One thing that most bureaucracies do have in common is a structure that separates those whose responsibilities include keeping in mind the overall needs of the entire organization from those whose responsibilities are much more narrow and task oriented. Visualize a modern industrial organization as a pyramid. Management (at the top of the pyramid) plans, organizes, hires, and fires. Workers (in the bottom section) make much smaller decisions limited to carrying out the work assigned to them. A similar division cuts through the hierarchy of the Roman Catholic church. The pope is at the top, followed by the cardinals, archbishops, and the bishops; the clergy are below. Only bishops can ordain new priests, and they plan the church's worldwide activities. The priests administer parishes, schools, and missions; their tasks are quite narrow and confined.

Although employees of bureaucracies may enjoy the privileges of their position and guard them jealously, they may be adversely affected by the system in ways that they do not recognize. Alienation, adherence to unproductive ritual, and acceptance of incompetence are some of the results of a less-than-ideal bureaucracy.

Robert Michels, a colleague of Weber's, also was concerned about the depersonalizing effect of bureaucracy. His views, formulated at the beginning of this century, are still pertinent today.

The Iron Law of Oligarchy

Michels (1911) concluded that the formal organization of bureaucracies inevitably leads to **oligarchy,** *under which organizations that were originally idealistic and democratic eventually come to be dominated by a small self-serving group of people who achieved positions of power and responsibility.* This can occur in large organizations because it becomes physically impossible for everyone to get together every time a decision has to be made. Consequently, a small group is given the responsibility of making decisions. Michels believed that the people in this group would become corrupted by their elite positions and more and more inclined to make decisions to protect their power rather than represent the will of the group they were supposed to serve. In effect, Michels was saying that bureaucracy and democracy do not mix. Despite any protestations and promises that they will not become like all the rest, those placed in positions of responsibility and power often come to believe that they are indispensable to, and more knowledgeable than, those they serve. As time goes on, they become further removed from the rank and file.

The Iron Law of Oligarchy suggests that organizations that wish to avoid oligarchy should take a number of precautionary steps. They should make sure that the rank and file remain active in the organization and that the leaders not be granted absolute control of a centralized administration. As long as there are open lines of communication and shared decision making between the leaders and the rank and file, an oligarchy cannot develop easily.

Clearly, the problems of oligarchy, of the bureaucratic depersonalization described by Weber, and of personal alienation all are interrelated. If individuals are deprived of the power to make decisions that af-fect their lives in many or even most of the areas that are important to them, withdrawal into narrow ritualism and apathy are likely responses.

Institutions and Social Organization

Anyone who has traveled to foreign countries knows that different societies have different ways of doing things. The basic things that get done actually are quite similar—food is produced and distributed; people get married and have children; and children are raised to take on the responsibilities of adulthood. The vehicle for accomplishing the basic needs of any society is the social institution.

Social Institutions

Sociologists usually speak of five areas of society in which basic needs have to be fulfilled: the family sector, the education sector, the economic sector, the religious sector, and the political sector. For each of these areas, there are social groups and associations that carry out the goals and meet the needs of society.

Relatively rapid social change is making less predictable the types of behavior that go along with gender roles.

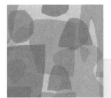

Technology and Society

Does Television Reduce Social Interaction?

People are much less involved in the social life of their communities than in the past. Membership in such diverse organizations as the PTA, the Elks Club, the League of Women Voters, the Red Cross, and labor unions has declined between 25% to 50% in the past three decades. Time spent on informal socializing and visiting is down by 25%.

There have also been sharp declines in political participation, including such things as attending a rally or speech (off 36% between 1973 and 1993), attending a meeting on town or school affairs (off 39%), or working for a political party (off 56%). Even church attendance appears to be off by about 30% since the 1960s.

The usual explanations we hear for this trend include that time pressures are greater now than in the past, suburbanization has caused people to be more spread out, and the movement of women into the paid labor force and the stresses that go along with two-career families have limited free time.

Sociologist Robert Putnam does not accept these reasons and believes the real culprit is television. As television's hold on society became greater, people began to give up more of those activities that would bring them together in social groups. In 1950 barely 10% of the American homes had television sets, but by 1959, 90% did. The viewing hours grew nearly 20% in the 1960s and by an additional 7.5% during the 1970s. By 1995 viewing per TV household was more than 50% higher than it had been in the 1950s.

The average American now watches roughly 4 hours of television a day. This means that television absorbs over 40% of the average person's free time. Each hour spent watching television appears to take away time from social interaction.

What is interesting about this is that other activities do not take from social interaction. For example, people who listen to a good deal of classical music, read a daily newspaper, or pursue other hobbies are more likely, not less likely, to join social groups. Television watching appears to be the only leisure activity that inhibits participation in social gatherings.

Television's negative impact is an outgrowth of both the medium and the message it conveys. Television, with few exceptions, is a one-way medium. The viewer can control which channel to watch, but cannot control the information coming through that channel. Television simply produces a state of passive receptivity.

The message coming from television is one that shows mostly the negative side of people. Violent shows are popular as are talk shows portraying dysfunctional families. Local news focuses on death and destruction. The more television people watch the more they think there is a dangerous world out there that one should stay away from. The message one receives is that it is safer to watch television than to venture out into the world and engage in social activities. This seems to be a destructive force on social interaction.

Source: Adapted from "The Strange Disappearance of Civic America," by R. D. Putnam, Winter 1996, *The American Prospect*, No. 24.

The behavior of people in these groups and associations is organized or patterned by the relevant **social institutions**—*the ordered social relationships that grow out of the values, norms, statuses, and roles that organize those activities that fulfill society's fundamental needs.* Thus, economic institutions organize the ways in which society produces and distributes the goods and services it needs; educational institutions determine what should be learned and how it should be taught; and so forth.

Of all social institutions, the family is perhaps the most basic. A stable family unit is the main ingredient necessary for the smooth functioning of society. For instance, sexual behavior must be regulated and

children must be cared for and raised to fit into society. Hence, the institution of the family provides a system of continuity from one generation to the next.

Using the family as an example, we can see the difference between the concept of group and the concept of institution. A group is a collection of specific, identifiable people. An institution is a system for organizing standardized patterns of social behavior. In other words, a group consists of people, and an institution consists of actions. For example, when sociologists discuss a family (say the Smith family), they are referring to a particular group of people. When they discuss the family, they are referring to the family as an institution—a cluster of statuses, roles, values, and norms that organize the standardized patterns of behavior that we expect to find within family groups. Thus, the family as an American institution typically embodies several master statuses: those of husband, wife, and, possibly, father, mother, and child. It also includes the statuses of son, daughter, brother, and sister. These statuses are organized into well-defined, patterned relationships: Parents have authority over their children, spouses have a sexual relationship with each other (but not with the children), and so on. However, specific family groups may not conform entirely to the ideals of the institution. There are single-parent families, families in which the children appear to be running things, and families in which there is an incestuous parent-child relationship. Although a society's institutions provide what can be thought of as a master plan for human interactions in groups, actual behavior and actual group organization often deviate in varying degrees from this plan.

Social Organization

If we step back from a mosaic, the many multicolored stones are seen to compose a single, coordinated pattern or picture. Similarly, if we step back and look at society, the many actions of all its members fall into a pattern or series of interrelated patterns. These consist of social interactions and relationships expressing individual decisions and choices. These choices, however, are not random; rather, they are an outgrowth of a society's social organization. **Social organization** consists of *the relatively stable pattern of social relationships among individuals and groups in society.* These relationships are based on systems of social roles, norms, and shared meanings that provide regularity and predictability in social interaction. Social organization differs from one society to the next. Thus, Islam allows a man to have up to four wives at once, whereas in American society, with its Judeo-Christian religious tradition, such plural marriage is not an acceptable family form.

Just as statuses and roles exist within ordered relationships to one another, social institutions also exist in patterned relationships with one another in the context of society. All societies have their own patterning for these relationships. For example, a society's economic and political institutions often are closely interrelated. So, too, are the family and religious institutions. Thus, a description of American social organization would indicate the presence of monogamy along with Judeo-Christian values and norms and the institutionalization of economic competition and of democratic political organizations.

A society's social organization tends to be its most stable aspect. The American social organization, however, may not be as static as that of many other societies. American society is experiencing relatively rapid social change because of its complexity and because of the great variety in the types of people that are part of it. This complexity makes life less predictable, because new values and norms being introduced from numerous quarters result in changes in social organization. For example, ideas about the behavior that should go along with female gender roles have changed considerably over the past three decades. Traditionally, married women were expected not to work but to stay home and attend to the rearing of children. Today, the majority of American women are working outside the home, and views on what roles mothers should play in the lives of their children are in flux. (For a look at one of the ways in which technology has affected American social life, see "Technology and Society: Does Television Reduce Social Interaction?")

SUMMARY

Because humans are symbolic creatures, everything we do—or do not do—conveys a message to others. Whether we intend it or not, other people take account of our behavior. For instance, most Americans distinguish among intimate, personal, social, and public distance.

People do not interact with each other as anonymous beings. They come together in the context of specific environments, with specific purposes, and with specific social characteristics. Among the most important of the latter are statuses and roles, which together help to define our social interactions and provide predictability.

Statuses are socially defined positions that people occupy in a group or society that help determine how they interact with one another. Statuses exist independent of the specific people who occupy them. Roles are the culturally defined rules for proper behavior that are associated with every status. A role is basically a collection of rights and obligations.

A social group consists of a number of people who have a common identity, some feeling of unity, and certain common goals and shared norms. Sociologists distinguish between primary groups, which involve intimacy, informality, and emotional investment in one another, and secondary groups, which have specific goals, formal organization, and much less intimacy.

In order to function properly, all groups must define their boundaries, choose leaders, make decisions, set goals, assign tasks, and control members' behavior.

A reference group is a group or social category that an individual uses to help define beliefs, attitudes, and values and to guide behavior. When individuals alter their behavior and attitudes toward those in a group they wish to join, they are engaging in anticipatory socialization.

Associations are purposefully created special-interest groups that have clearly defined goals and official ways of doing things. A modern form of large association is bureaucracy, which is a formal, rationally organized social structure with clearly defined patterns of activity that are functionally related to the purposes of the organization.

Social institutions consist of the ordered relationships that grow out of the values, norms, statuses, and roles that organize those activities that fulfill society's fundamental needs. Institutions are systems for organizing standardized patterns of social behavior. Social organization consists of the relatively stable pattern of social relationships among individuals and groups in society. It differs from one society to the next.

INTERNET RESOURCES

Go to http://web-enhanced.thomsonlearning.com which features Web links relevant to this chapter to expand and enhance the learning experience. This site will be monitored and updated periodically to ensure the links are correct and live.

At Virtual Society: The Wadsworth Sociology Resource Center, http://sociology.wadsworth.com, you will find a Career Center, Sociology in the News, Research Online, InfoTrac College Edition, Virtual Tours in Sociology, and a variety of book-specific student and instructor resources.

CHAPTER FIVE STUDY GUIDE

KEY CONCEPTS

Match each concept with its definition, illustration, or explanation below.

a. Secondary group
b. Conflict
c. Bureaucracy
d. Role conflict
e. Social distance
f. Achieved status
g. Social institutions
h. Group functions
i. Reference group
j. Norms
k. Ethnomethodology
l. Social aggregate
m. Oligarchy
n. Dyad
o. Social interaction
p. The curiosity factor

q. Role strain
r. Subgroup
s. Leader
t. Kinesics
u. Master status
v. Triad
w. Roles
x. Social action
y. Formal structure
z. Ideal type
aa. Public distance
bb. Role playing
cc. Competition
dd. Role set
ee. Social organization
ff. Dramaturgy

gg. Primary group
hh. Informal structures
ii. Ascribed status
jj. Association
kk. Personal distance
ll. Instrumental leadership
mm. Statuses
nn. Coercion
oo. Intimate distance
pp. Expressive leadership
qq. Small group
rr. Exchange interaction
ss. Social group
tt. Social interaction context
uu. Being hopelessly uncool
vv. Cooperative interaction

_____ 1. The social spacing used for intensely personal and private interactions.

_____ 2. At about 2.5 to 4 feet, the spacing used by most Americans in ordinary conversation.

_____ 3. A distance of between 4 and 12 feet, it is used in more impersonal business interactions.

_____ 4. The type of spacing between an individual and a group in a large formal gathering.

_____ 5. Anything people are conscious of doing because of other people.

_____ 6. Two or more people taking each other into account.

_____ 7. The physical setting, social environment, and activities surrounding a social interaction.

_____ 8. Specific rules of behavior that are agreed upon and shared and that prescribe limits of acceptable behavior.

_____ 9. People doing something for each other with the express purpose of receiving a reward or return.

_____ 10. People acting together to promote common interests or achieve shared goals.

_____ 11. The behaviors that have the possibility of transforming our actions and ourselves.

_____ 12. The study of the sets of rules or guidelines that individuals use to initiate behavior, respond to behavior, and modify behavior in social settings.

_____ 13. Networks of people who help out by "bending the rules" in an organizational setting.

_____ 14. The approach to studying how people play roles in order to create an impression and how the impression is judged by others who are alert to any character slips.

_____ 15. People struggling against one another for some commonly prized object or value.

_____ 16. A form of conflict in which one of the parties is much stronger than the other and can impose its will on the weaker party.

_____ 17. A form of conflict confined within agreed-upon rules.

_____ 18. Socially defined positions that people occupy in a group or society and in terms of which they interact with one another.

_____ 19. One of the multiple statuses a person occupies that seems to dominate the others in patterning a person's life.

_____ 20. Statuses conferred upon people by virtue of birth or other significant factors not controlled by their own actions or decisions.

_____ 21. Statuses occupied as a result of an individual's actions.

_____ 22. Culturally defined rules for proper behavior that are associated with every status.

_____ 23. All the roles attached to a single status.

_____ 24. Conflicting demands attached to the same role.

_____ 25. An inability to enact the roles of one status without violating those of another status.

_____ 26. A number of people who have a common identity, some feeling of unity, and certain common goals and shared norms.

_____ 27. People temporarily in physical proximity to one another, but who share little else.

_____ 28. A group in which members have an emotional investment in one another, know one another intimately, and interact as total individuals rather than through specialized roles.

_____ 29. A group that has relatively little intimacy, has specific goals, is formally organized, and is impersonal.

_____ 30. Someone who occupies a central role or position of dominance and influence in a group.

_____ 31. A type of leadership in which the leader actively proposes tasks and plans to guide the group toward achieving its goals.

_____ 32. A type of leadership in which the leader works to keep relations among group members harmonious and group morale high.

_____ 33. A group or social category that an individual uses to help define beliefs, attitudes, and values and to guide behavior.

_____ 34. A group that actually meets together and has few enough members that all members know one another.

_____ 35. The smallest possible group, it contains two members.

_____ 36. A group consisting of three members.

_____ 37. A splinter group within a larger group.

_____ 38. A purposefully created special-interest group that has clearly defined goals and official ways of doing things.

_____ 39. A formal, rationally organized social structure with clearly defined patterns of activity in which, ideally, every series of actions is functionally related to the purposes of the organization.

_____ 40. An exaggerated model of reality used to illustrate a concept.

_____ 41. The domination of an organization by a small, self-serving, self-perpetuating group of people in positions of power and responsibility.

_____ 42. The ordered social relationships that grow out of the values, norms, statuses, and roles that organize the activities that fulfill society's fundamental needs.

_____ 43. The outcome of a type of unrealized behavior by blacks, which is offensive to whites.

_____ 44. Defining boundaries, choosing leaders, making decisions, assigning tasks, and controlling new members' behaviors.

_____ 45. The study of how slight nods, yawns, postural shifts, nonverbal cues, and other body movements affect behavior.

_____ **46.** Planned, highly institutionalized, and clearly defined statuses and role relationships.

_____ **47.** A type of unrealized behavior by whites, which is offensive to blacks.

_____ **48.** The relatively stable pattern of social relationships among individuals and groups in society.

KEY THINKERS/RESEARCHERS

Match the thinkers/researchers with their main ideas or contributions.

a. **George Simmel**
b. **Charles Horton Cooley**
c. **Robert Michels**
d. **Deborah Tannen**

e. **Edward T. Hall**
f. **Max Weber**
g. **Solomon Asch**
h. **Lena Williams**

i. **Erving Goffman**
j. **Kay Redfield Jamison**
k. **Frank Sulloway**
l. **Harold Garfinkel**

_____ **1.** One of the first sociologists to stress the importance of social interaction, he also developed a model of bureaucracy.

_____ **2.** A pioneer in studying the context of social interaction.

_____ **3.** A pioneer in defining and demonstrating the importance of primary groups.

_____ **4.** The first sociologist to emphasize the effect of the size of a group on the interaction process.

_____ **5.** Conducted important research showing that a substantial proportion of individuals were willing to deny the evidence of their senses to conform to the group.

_____ **6.** Proposed that it was important to study the commonplace aspects of everyday life.

_____ **7.** A student of bureaucracy, he developed the Iron Law of Oligarchy.

_____ **8.** Presented a detailed investigation of the effects of birth order on behavior in later life.

_____ **9.** Studied the cotillion as an activity where children learned the importance of statuses and roles and the proper behavior to be displayed toward other persons' positions.

_____ **10.** Developed an approach that focused on how people attempt to create a favorable impression of themselves and the manner in which others judged their performances.

_____ **11.** Suggests that we have developed an argument culture, which urges us to approach the world and people in it in an adversarial frame of mind.

_____ **12.** Provided a glimpse of how blacks and whites offend each other without being aware of it.

CENTRAL IDEA COMPLETIONS

Following the instructions, fill in the appropriate concepts and descriptions for each of the questions posed below.

1. List and define the six major functions of social groups:

a. _____

b. _____

c. _____

d. _____

e. _____

f. _____

2. What are reference groups? _____

_____

What functions do reference groups serve? _____

_____

_____

3. Differentiate between role strain and role conflict: _____

_____

_____

   a. What might be two forms of role strain facing a female college professor? _____

   _____

   _____

   b. What are the sources of these role strains? _____

   _____

   _____

   c. How might the individual resolve these strains? _____

   _____

   _____

4. Present an example of each of the following:
   a. Aggregate: _____
   b. Social group: _____
   c. Primary group: _____
   d. Secondary group: _____

5. Select two locations on your campus. Analyze each in terms of its physical setting, social environment, and activities surrounding the social action taking place:
   a. Physical: _____
   b. Social: _____
   c. Activities: _____

6. Continuing with your work in question 5, answer the following:
   a. What are the distinctive ways in which each of these three variables influence the social interaction taking place? _____

   _____

   _____

   b. What changes might one make to produce a corresponding change in the types and levels of social interaction you have observed? _____

   _____

   _____

7. Explain how the American family may be regarded as a social institution. _____

_____

_____

a. What major functions are assigned to the American family? _____
_____
_____

b. What characteristics of the American family might be dysfunctional for the larger society?
_____
_____

c. What activities in a family illustrate exchange behaviors? _____
_____
_____

d. What activities in a family illustrate conflict behaviors? _____
_____
_____

e. What activities in a family illustrate competitive behaviors? _____
_____
_____

8. Using the American high school as an organization, pick three specific characteristics of bureaucracy outlined by Weber and indicate how informal structures might subvert the three formal characteristics of your example high school:

a. _____

b. _____

c. _____

9. What is Michels' "Iron Law of Oligarchy"? _____
_____
_____

10. List the four major types of personal space outlined in your text:

(1) _____   (2) _____   (3) _____   (4) _____

a. Assuming you were to violate each form of personal space, which violation would you
define as most serious? _____
_____

b. What factors might influence an audience witnessing your violation to regard your behavior as serious? _____
_____

11. Define dramaturgy and present one example of a behavior you might study using this
approach: _____
_____
_____

12. Define ethnomethodology and present one example of a behavior you might study using this
approach: _____
_____
_____

CRITICAL THINKING EXERCISES

1. Develop a list of the top three (in terms of importance to your life) primary groups to which you belong and the top three secondary groups. What, if anything, do these six groups share in common? What features are unique to each? How do the concepts of primary group and secondary group assist you in your examination of the differences among the six groups? Select one of the three secondary groups and discuss what changes would need to occur for that group to become a primary group. Follow the same procedure for one of the three primary groups becoming a secondary group.

2. Following your text's discussion of the four distances most Americans use in business and social relations (intimate, personal, social, and public), conduct your own ethnomethodological study by adopting a personal distance incongruent with the setting in which you are interacting. For example, with a close acquaintance, begin a conversation while standing face to face. As your conversation unfolds, begin backing away and increase your distance from the customary 6 to 18 inches to 2½ to 4 feet, then to more than 4 feet. What happened to the conversation? What was the reaction of your close personal acquaintance? Hypothesize what might have happened if you had elected to use an intimate personal distance within a public distance setting—for example, at a department or grocery store. How might the salesperson or store clerk have reacted if you had begun to move as close as 8 to 18 inches?

3. Attend a meeting of the International Students Organization at your college or university. Observe the public and personal conversational distances engaged in by the students, and attempt to determine whether these distances vary by country of origin of the students. After the meeting, approach several of the international students and ask them if they have noticed any differences in the use of personal space among students in their home country and in the United States. If the opportunity presents itself, inquire whether the students had any initial difficulties adjusting to Americans' use of social space when

they first arrived. Finally, ask them whether they expect (based on the time they are spending in the United States) to have any adjustment problems when they return to their home country.

4. Take the Cross-Cultural Social Interaction Quiz presented in your text. Of the four countries discussed (South Africa, The Gambia, Jamaica, and Jordan), which had behaviors to which you would have had the most difficulty adjusting? Examine the gestures used by yourself and your peers. Which of those might be the most difficult for international visitors to understand? Which might be the easiest? Examining verbal expressions, what expressions are currently in vogue within your group that outsiders might have difficulty understanding? Do you use any common English words or expressions in a unique way that a literal translation would fail to get or perhaps even get a totally opposite meaning from the meaning you have attached?

5. Drawing on the Lena Williams discussion, "How Blacks and Whites Offend Each Other Without Realizing It," keep a journal for one month during which time you attempt to observe and record as many examples as possible of unintended racial prejudice. Assuming you encounter examples of the types of racial prejudice discussed by Williams, what remedies would you propose to reduce or even eliminate them?

6. In order to gain a sense of the importance of group membership across the globe, go to the following Internet site, which provides an alphabetized listing of different campus groups and organizations. What do you see as the commonalities and differences across the groups depicted at these sites? How well do they manage to accomplish the major group tasks discussed in your text?

http://sta.umbc.edu/activities/so.html*

*Please note, Web site addresses change frequently. The addresses presented in these exercises should be viewed as guides only, and students should be prepared to utilize appropriate Internet search techniques when necessary should these addresses no longer be accessible.

ANSWERS

KEY CONCEPTS

1. oo	6. o	11. bb	16. nn	21. f	26. ss
2. kk	7. tt	12. k	17. cc	22. w	27. l
3. e	8. j	13. hh	18. mm	23. dd	28. gg
4. aa	9. rr	14. ff	19. u	24. q	29. a
5. x	10. vv	15. b	20. ii	25. d	30. s

31. ll	34. qq	37. r	40. z	43. uu	46. y
32. pp	35. n	38. jj	41. m	44. h	47. p
33. i	36. v	39. c	42. g	45. t	48. ee

KEY THINKERS/RESEARCHERS

1. f	3. b	5. g	7. c	9. j	11. d
2. e	4. a	6. l	8. k	10. i	12. h

Deviant Behavior and Social Control

LEARNING OBJECTIVES

After studying this chapter, you should be able to:

- Understand deviance as culturally relative.
- Explain the functions and dysfunctions of deviance.
- Distinguish between internal and external means of social control.
- Differentiate among the various types of sanctions.
- Describe and critique biological, psychological, and sociological theories of deviance.
- Discuss the concept of anomie and its role in producing deviance.
- Know how the Uniform Crime Reports and the National Crime Victimization Survey differ as sources of information about crime.
- Describe the major features of the criminal justice system in the United States.

Soprano Florence Foster Jenkins believed she was the "goddess of song." Unfortunately she had no singing talent. She sang wildly out of tune and her voice was quivering and colorless. She was, however, a wealthy New York socialite and did not hesitate to use her money to let the world know that she should be considered a world-class diva.

Several times a year she would rent out the Ritz Carlton Hotel and give stunningly inept renditions of standard opera arias and songs specifically written for her, which she would proceed to mangle with her appalling voice. She would create lavish costumes for these performances and her pianist Cosme McMoon treated her with the utmost respect as he accompanied her with the appropriate music.

Eventually her reputation produced a following and tickets to her bizarre performances, which could only be purchased from her directly, were sold out months in advance and were as difficult to get as those for the Metropolitan Opera. Her following continued to build, and her final performance took place at Carnegie Hall. When she died a month later, she left the following epithet on her grave stone; "Some people say I cannot sing, but no one can say that I didn't sing."

Florence Foster Jenkins was certainly an eccentric, but is that different than being a deviant?

Defining Normal and Deviant Behavior

What determines whether a person's actions end up being seen as eccentric, creative, or deviant? Why will two men walking hand-in-hand in downtown Minneapolis cause raised eyebrows but pass unnoticed in San Francisco or in

Provincetown, Massachusetts? Why do Britons waiting to enter a theater stand patiently in line, whereas people from the Middle East jam together at the turnstile? In other words, what makes a given action—men holding hands, cutting into a line—"normal" in one case but "deviant" in another?

The answer is culture—more specifically, the norms and values of each culture (see Chapter 3). Together, norms and values make up the **moral code** of a culture—*the symbolic system in terms of which behavior takes on the quality of being "good" or "bad," "right" or "wrong."* Therefore, in order to decide whether any specific act is "normal" or "deviant," it is necessary to know more than only what a person did. One also must know who the person is (that is, the person's social identity) and the social and cultural context of the act. For example, if Florence Foster Jenkins held her recitals on a street corner in a seedy neighborhood instead of a stage at the Ritz Carlton Hotel, would people have still been as interested in her events? Of course not.

For sociologists, then, **deviant behavior** is *behavior that fails to conform to the rules or norms of the group in question* (Durkheim, 1960a). Therefore, when we try to assess an act as being normal or deviant, we must identify the group by whose terms the behavior is judged. Moral codes differ widely from one society to another. For that matter, even within a society there exist groups and subcultures whose moral codes differ considerably. Watching television is normal behavior for most Americans, but it would be seen as deviant behavior among the Amish of Pennsylvania.

Making Moral Judgments

As we stated, sociologists take a culturally relative view of normalcy and deviance and evaluate behavior according to the values of the culture in which it takes place. Ideally, they do not use their own values to judge the behavior of people from other cultures. Even though social scientists recognize that there is great variation in normal and deviant behavior and that no science can determine what acts are inherently deviant, there are certain acts that are almost universally accepted as being deviant. For example, parent-child incest is severely disapproved of in nearly every society. Genocide, the willful killing of specific groups of people—as occurred in the Nazi extermination camps during World War II—also is considered to be wrong even if it is sanctioned by the government or an entire society. The Nuremberg trials that were conducted after World War II supported this point. Even though most of the accused individuals tried to claim they were merely following orders when they murdered or arranged for the murder of large numbers of Jews and other groups, many were found guilty. The reasoning was that there is a higher moral order under which certain human actions are wrong, regardless of who endorses them. Thus, despite their desire to view events from a culturally relative standpoint, most sociologists find certain actions wrong, no matter what the context.

The Functions of Deviance

Émile Durkheim observed that deviant behavior is "an integral part of all healthy societies" (1895, 1958). Why is this the case? The answer, Durkheim suggested, is that in the presence of deviant behavior, a social group becomes united in its response. In other words, opposition to deviant behavior creates opportunities for cooperation essential to the survival of any group. For example, let us look at the response to a scandal in a small town as Durkheim described it:

[People] stop each other on the street, they visit each other, they seek to come together to talk of

Sociologists believe that deviant behavior fails to conform to the norms of the group. This man has no shirt on and is lying on the sidewalk in the middle of winter. Would this behavior conform to the standards of any group?

the event and to wax indignant in common. From all the similar impressions which are exchanged, from all the temper that gets itself expressed, there emerges a unique temper . . . which is everybody's without being anybody's in particular. That is the public temper. (1895)

When social life moves along normally, people begin to take for granted one another and the meaning of their social interdependency. A deviant act, however, reawakens their group attachments and loyalties, because it represents a threat to the moral order of the group. The deviant act focuses people's attention on the value of the group. Perceiving itself under pressure, the group marshals its forces to protect itself and preserve its existence.

Deviance also offers society's members an opportunity to rededicate themselves to their social controls. In some cases, deviant behavior actually helps teach society's rules by providing illustrations of violation. Knowing what is wrong is a step toward understanding what is right. Deviance, then, may be functional to a group in that it (1) causes the group's members to close ranks, (2) prompts the group to organize in order to limit future deviant acts, (3) helps clarify for the group what it really does believe in, and (4) teaches normal behavior by providing examples of rule violation. Finally, (5) in some situations, tolerance of deviant behavior acts as a safety valve and actually prevents more serious

instances of nonconformity. For example, the Amish, a religious group that does not believe in using such modern examples of contemporary society as cars, radios, televisions, and fashion-oriented clothing, allows its teenagers a great deal of latitude in their behaviors before they are fully required to follow the dictates of the community. This prevents a confrontation that could result in a major battle of wills.

The Dysfunctions of Deviance

Deviance, of course, has a number of dysfunctions as well, which is why every society attempts to restrain deviant behavior as much as possible. Included among the dysfunctions of deviant behavior are the following: (1) It is a threat to the social order because it makes social life difficult and unpredictable. (2) It causes confusion about the norms and values of that society. People become confused about what is expected, what is right and wrong. The variety of social standards compete with one another, causing tension among the different segments of society. (3) Deviance also undermines trust. Social relationships are based on the premise that people will behave according to certain rules of conduct. When people's actions become unpredictable, the social order is thrown into disarray. (4) Deviance also diverts valuable resources. To control widespread deviance, vast resources must be called upon and shifted from other social needs.

The forms of dress and the types of behavior that are considered deviant depend on who is doing the judging and what the context might be.

Mechanisms of Social Control

In any society or social group, it is necessary to have **mechanisms of social control,** or *a way of directing or influencing members' behavior to conform to the group's values and norms.* Sociologists distinguish between internal and external means of control.

Internal Means of Control

As we already observed in Chapters 3 and 5, people are socialized to accept the norms and values of their culture, especially in the smaller and more personally important social groups to which they belong, such as the family. The word *accept* is important here. Individuals conform to moral standards not just because they know what they are, but also because they have internalized these standards. They experience discomfort, often in the form of guilt, when they violate these norms. In other words, for a group's moral code to work properly, it must be internalized and become part of each individual's emotional life as well as his or her thought processes. As this occurs, individuals begin to pass judgment on their own actions. In this way the moral code of a culture becomes an **internal means of control**—that is, *it operates on the individual even in the absence of reactions by others.*

External Means of Control: Sanctions

External means of control consist of *other people's responses to a person's behavior—that is, rewards and punishments.* They include social forces external to the individual that channel behavior toward the culture's norms and values.

Sanctions are *rewards and penalties used by a group's members to regulate an individual's behavior.* Thus, all external means of control use sanctions of one kind or another. *Actions that encourage the individual to continue acting in a certain way* are called **positive sanctions.** *Actions that discourage the repetition or continuation of the behavior* are **negative sanctions.**

Positive and Negative Sanctions Sanctions take many forms, varying widely from group to group and from society to society. For example, an American audience might clap and whistle enthusiastically to show its appreciation for an excellent artistic or athletic performance, but the same whistling in Europe would be a display of strong disapproval. Or consider the absence of a response. In the United States, a professor would not infer public disapproval because of the absence of applause at the end of a lecture—such applause by students is the rarest of compliments. In many universities in Europe, however, students are expected to applaud after every lecture (if only in a rhythmic, stylized manner). The absence of such applause would be a horrible blow to the professor, a public criticism of the presentation.

Most social sanctions have a symbolic side to them. Such symbolism has a powerful impact on people's self-esteem and sense of identity. Consider the positive feelings experienced by Olympic gold medalists or those elected to Phi Beta Kappa, the national society honoring excellence in undergraduate study. Or imagine the negative experience of being

In this photo from France at the end of World War II, the woman whose head has been shaven is jeered by the crowd as she is escorted out of town. She had been a Nazi collaborator during the war. Émile Durkheim believed that deviant behavior performs an important function by focusing people's attention on the values of the group. The deviance represents a threat to the group and forces it to protect itself and preserve it existence.

given the "silent treatment," such as that imposed on cadets who violate the honor code at the military academy at West Point (to some, this is so painful that they drop out).

Sanctions often have important material qualities as well as symbolic meanings. Nobel Prize winners receive not only public acclaim but also a hefty check. The threat of loss of employment may accompany public disgrace when an individual's deviant behavior becomes known. In isolated, preliterate societies, social ostracism can be the equivalent of a death sentence.

Both positive and negative sanctions work only to the degree that people can be reasonably sure that they actually will take place as a consequence of a given act. In other words, they work on people's expectations. Whenever such expectations are not met, sanctions lose their ability to mold social conformity.

It is important to recognize a crucial difference between positive and negative sanctions. When society applies a positive sanction, it is a sign that social controls are successful: The desired behavior has occurred and is being rewarded. When a negative sanction is applied, it is due to the failure of social controls: The undesired behavior has not been prevented. Therefore, a society that frequently must punish people is failing in its attempts to promote conformity. A school that must expel large numbers of students or a government that frequently must call out troops to quell protests and riots should begin to look for the weaknesses in its own system of internal means of social control to promote conformity.

Formal and Informal Sanctions **Formal sanctions** *are applied in a public ritual—as in the awarding of a prize or an announcement of expulsion—and are usually under the direct or indirect control of authorities.* For example, to enforce certain standards of behavior and protect members of society, our society creates laws. Behavior that violates these laws may be punished through formal negative sanctions. Not all sanctions are formal, however. Many social responses to a person's behavior involve **informal sanctions,** or *actions by group members that arise spontaneously with little or no formal direction.* Gossip is an informal sanction that is used universally. Congratulations are offered to people whose behavior has approval. In teenage peer groups, ridicule is a powerful, informal, negative sanction. The anonymity and impersonality of urban living, however, decreases the influence of these controls except when we are with members of our friendship and kinship groups.

A Typology of Sanctions Figure 6-1 shows the four main types of social sanctions, produced by

Figure 6-1 Types of Social Sanctions

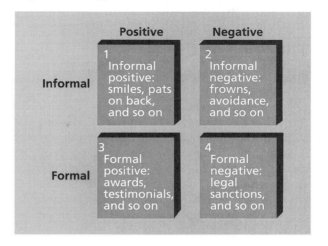

combining the two sets of sanctions we have just discussed: informal and formal, positive and negative. Although formal sanctions might appear to be strong influences on behavior, informal sanctions actually have a greater impact on people's self-images and behavior. This is so because informal sanctions usually occur more frequently and come from close, respected associates.

1. **Informal positive sanctions** are *displays people use spontaneously to express their approval of another's behavior.* Smiles, pats on the back, handshakes, congratulations, and hugs are informal positive sanctions.

2. **Informal negative sanctions** are *spontaneous displays of disapproval or displeasure,* such as frowns, damaging gossip, or impolite treatment directed toward the violator of a group norm.

3. **Formal positive sanctions** are *public affairs, rituals, or ceremonies that express social approval of a person's behavior.* These occasions are planned and organized. In our society, they include such events as parades that take place after a team wins the World Series or the Super Bowl, the presentation of awards or degrees, and public declarations of respect or appreciation (banquets, for example). Awards of money are a form of formal positive sanctions.

4. **Formal negative sanctions** are *actions that express institutionalized disapproval of a person's behavior.* They usually are applied within the context of a society's formal organizations—schools, corporations, the legal system, for example—and include expulsion, dismissal, fines, and imprisonment. They flow directly from decisions made by a person or agency of authority, and frequently there are specialized agencies or

personnel (such as a board of directors, a government agency, or a police force) to enforce them.

Theories of Crime and Deviance

Criminal and deviant behavior has been found throughout history. It has been so troublesome and so persistent that much effort has been devoted to understanding its roots. Many dubious ideas and theories have been developed over the ages. For example, a medieval law specified that "if two persons fell under suspicion of crime, the uglier or more deformed was to be regarded as more probably guilty" (Wilson and Herrnstein, 1985). Modern day approaches to deviant and criminal behavior can be divided into the general categories of biological, psychological, and sociological explanations.

Biological Theories of Deviance

The first attempts to provide "scientific" explanations for deviant and criminal behavior centered around the importance of inherited factors and downplayed the importance of environmental influences. From this point of view, deviant individuals are born, not made.

Cesare Lombroso (1835–1901) was an Italian doctor who believed that too much emphasis was being put on "free will" as an explanation for deviant behavior. While trying to discover the anatomical differences between deviant and insane men, he came upon what he believed was an important insight. As he was examining the skull of a criminal, he noticed a series of features, recalling an apish past rather than a human present:

> At the sight of that skull, I seemed to see all of a sudden, lighted up as a vast plain under a flaming sky, the problem of the nature of the criminal—an atavistic being who reproduces in his person the ferocious instincts of primitive humanity and the inferior animals. (Taylor et al., 1973, p. 4)

According to Lombroso, criminals are evolutionary throwbacks whose behavior is more apelike than human. They are driven by their instincts to engage in deviant behavior. These people can be identified by certain physical signs that betray their savage nature. Lombroso spent much of his life studying and dissecting dead prisoners in Italy's jails and concluded that their criminality was associated with an animal-like body type that revealed an inherited primitiveness (Lombroso-Ferrero, 1972). He also believed that certain criminal types could be identified by their head size, facial characteristics (size

and shape of the nose, for instance), and even hair color. His writings were met with heated criticism from scholars who pointed out that perfectly normal-looking people have committed violent acts. (Modern social scientists would add that by confining his research to the study of prison inmates, Lombroso used a biased sample, thereby limiting the validity of his investigations.)

Shortly before World War II, anthropologist E. A. Hooten argued that the born criminal was a scientific reality. Hooten believed crime was not the product of social conditions but the outgrowth of "organic inferiority."

> [W]hatever the crime may be, it ordinarily arises from a deteriorated organism. . . . You may say that this is tantamount to a declaration that the primary cause of crime is biological inferiority—and that is exactly what I mean. . . . Certainly the penitentiaries of our society are built upon the shifting sands and quaking bogs of inferior human organisms. (Hooten, 1939)

Hooten went to great lengths to analyze the height, weight, shape of the body, nose, and ears of criminals. Hooten was convinced that people betrayed their criminal tendencies by the shape of their bodies:

> The nose of the criminal tends to be higher in the root and in the bridge, and more frequently undulating or concave-convex than in our sample of civilians. . . . (B)ootleggers persistently have broad noses and short faces and flaring jaw angles, while rapists monotonously display narrow foreheads and elongated, pinched noses. (Hooten, 1939)

Following in Hooten's footsteps, William H. Sheldon and his coworkers carried out body measurements of thousands of subjects to determine whether personality traits are associated with particular body types. They found that human shapes could be classified as three particular types: endomorphic (round and soft), ectomorphic (thin and linear), and mesomorphic (ruggedly muscular) (Sheldon & Tucker, 1940). They also claimed that certain psychological orientations are associated with body type. They saw endomorphs as being relaxed creatures of comfort; ectomorphs as being inhibited, secretive, and restrained; and mesomorphs as being assertive, action oriented, and uncaring of others' feelings (Sheldon & Stevens, 1942).

Sheldon did not take a firm position on whether temperamental dispositions are inherited or are the outcome of society's responses to individuals based on their body types. For example, Americans expect heavy people to be good-natured and cheerful, skinny people to be timid, and strongly muscled people to be physically active and inclined toward aggressiveness. Anticipating such behaviors, people often encourage them. In a study of delinquent boys, Sheldon and his colleagues (1949) found that mesomorphs were more likely to become delinquents than were boys with other body types. Their explanation of this finding emphasized inherited factors, although they acknowledged social variables. The mesomorph is quick to anger and lacks the ectomorph's restraint, they claimed. Therefore, in situations of stress, the mesomorph is more likely to get into trouble, especially if the individual is both poor and not very smart. Sheldon's bias toward a mainly biological explanation of delinquency was strong enough for him to have proposed a eugenic program of selective breeding to weed out those types he considered predisposed toward criminal behavior.

In the mid-1960s, further biological explanations of deviance appeared linking a chromosomal anomaly in males, known as XYY, with violent and criminal behavior. Typically, males receive a single X chromosome from their mothers and a Y chromosome from their fathers. Occasionally, a child will receive two Y chromosomes from his father. These individuals will look like normal males; however, based on limited observations, a theory developed that these individuals were prone to commit violent crimes. The simplistic logic behind this theory is that since males are more aggressive than females and possess a Y chromosome that females lack, this Y chromosome must be the cause of aggression, and a double dose means double trouble. One group of researchers noted: "It should come as no surprise that an extra Y chromosome can produce an individual with heightened masculinity, evinced by characteristics such as unusual tallness . . . and powerful aggressive tendencies" (Jarvik, 1972).

Today the XYY chromosome theory has been discounted. It has been estimated (Chorover, 1979; Suzuki & Knudtson, 1989) that 96% of XYY males lead ordinary lives with no criminal involvement. A maximum of 1% of all XYY males in the United States may spend any time in a prison (Pyeritz et al., 1977). No valid theory of deviant and criminal behavior can be devised around such unconvincing data.

Current biological theorists have gone beyond the simplistic notions of Lombroso, Hooten, Sheldon, and the XYY syndrome. Today such theories focus on technical advances in genetics, brain functioning, neurology, and biochemistry. Contemporary biological theories are based on the notion that behavior, whether conforming or deviant, results from the interaction of physical and social environments.

The best-known of these theories is Sarnoff Mednick's theory of inherited criminal tendencies. Mednick proposed that some genetic factors are passed along from parent to child. Criminal behavior is not directly inherited, nor do the genetic factors directly cause the behavior; rather one inherits a greater susceptibility to criminality or to adapt to normal environments in a criminal way (Mednick, Moffitt, & Stacks, 1987).

Mednick believed that certain individuals inherit an autonomous nervous system that is slow to be aroused or to react to stimuli. Such individuals are then slow to learn control of aggressive or antisocial behavior (Mednick, 1977).

The biological basis for deviant behavior continues to be investigated. The conflicting data and conclusions indicate that the existence of biologically, or at least genetically, determined deviant behavior is still far from proven (Liska, 1991)

Psychological Theories of Deviance

Psychological explanations of deviance downplay biological factors and emphasize instead the role of parents and early childhood experiences, or behavioral conditioning, in producing deviant behavior. Although such explanations stress environmental influences, there is a significant distinction between psychological and sociological explanations of deviance. Psychological orientations assume that the seeds of deviance are planted in childhood and that adult behavior is a manifestation of early experiences rather than an expression of ongoing social or cultural factors. The deviant individual therefore is viewed as a "psychologically sick" person who has experienced emotional deprivation or damage during childhood.

Psychoanalytic Theory Psychoanalytic explanations of deviance are based on the work of Sigmund Freud and his followers. Psychoanalytic theorists believe that the unconscious, the part of us consisting of irrational thoughts and feelings of which we are not aware, causes us to commit deviant acts.

According to Freud, our personality has three parts: the id, our irrational drives and instincts; the superego, our conscience and guide as internalized from our parents and other authority figures; and the ego, the balance among the impulsiveness of the id, the restrictions and demands of the superego, and the requirements of society. Because of the id, all of us have deviant tendencies, though through the socialization process we learn to control our behavior, driving many of these tendencies into the unconscious. In this way, most of us are able to function effectively according to our society's norms and values. For some, however, the socialization process is not what it should be. As a result, the individual's behavior is not adequately controlled by either the ego or superego, and the wishes of the id take over. Consider, for example, a situation in which a man has been driving around congested city streets looking for a parking space. Finally he spots a car that is leaving and pulls up to wait for the space. Just as he is ready to park his car, another car whips in and takes the space. Most of us would react to the situation with anger. We might even roll down the car window and direct some angry gestures and strong language at the offending driver. There have been cases, however, in which the angry driver has pulled out a gun and shot the offender. Instead of simply saying, "I'm so mad I could kill that guy," the offended party acted out the threat. Psychoanalytic theorists might hypothesize that in this case, the id's aggressive drive took over, because of an inadequately developed conscience.

Psychoanalytic approaches to deviance have been strongly criticized because the concepts are very abstract and cannot easily be tested. For one thing, the unconscious can be neither seen directly nor measured. Also there is an overemphasis on innate drives at the same time that there is an underemphasis on social and cultural factors that bring about deviant behavior.

Behavioral Theories   According to the behavioral view, people adjust and modify their behavior in response to the rewards and punishments their actions elicit. If we do something that leads to a favorable outcome, we are likely to repeat that action. If our behavior leads to unfavorable consequences, we are not eager to do the same thing again (Bandura, 1969). Those of us who live in a fairly traditional environment are likely to be rewarded for engaging in conformist behavior, such as working hard, dressing in a certain manner, or treating our friends in a certain way. We would receive negative sanctions if our friends found out that we had robbed a liquor store. For some people, however, the situation is reversed. That is, deviant behavior may elicit positive rewards. A 13-year-old who associates with a delinquent gang and is rewarded with praise for shoplifting, stealing, or vandalizing a school is being indoctrinated into a deviant lifestyle. The group may look with contempt at the "straight" kids who study hard, make career plans, and do not go out during the week. According to this approach, deviant behavior is learned by a series of trials and errors. One learns to be a thief in the same way that one learns to be a sociologist.

Crime as Individual Choice   James Q. Wilson and Richard Herrnstein (1985) have devised a theory of criminal behavior that is based on an analysis of individual behavior. Sociologists, almost by definition, are suspicious of explanations that emphasize individual behavior, because they believe such theories neglect the setting in which crime occurs and the broad social forces that determine levels of crime. However, Wilson and Herrnstein have argued that whatever factors contribute to crime—the state of the economy, the competence of the police, the nurturance of the family, the availability of drugs, the quality of the schools—they must affect the behavior of individuals before they affect crime. They believe that if crime rates rise or fall, it must be due to changes that have occurred in areas that affect individual behavior.

Wilson and Herrnstein contend that individual behavior is the result of rational choice. A person will choose to do one thing as opposed to another because it appears that the consequences of doing it are more desirable than the consequences of doing something else. At any given moment, a person can choose between committing a crime and not committing it.

The consequences of committing the crime consist of rewards and punishments. The consequences of not committing the crime also entail gains and losses. Crime becomes likely if the rewards for committing the crime are significantly greater than those for not committing the crime. The net rewards of crime include not only the likely material gain from the crime, but also intangible benefits such as obtaining emotional gratification, receiving the approval of peers, or settling an old score against an enemy. Some of the disadvantages of crime include the pangs of conscience, the disapproval of onlookers, and the retaliation of the victim.

The benefits of not committing a crime include avoiding the risk of being caught and punished and not suffering a loss of reputation or the sense of shame afflicting a person later discovered to have broken the law. All of the benefits of not committing a crime lie in the future, whereas many of the benefits of committing a crime are immediate. The consequences of committing a crime gradually lose their ability to control behavior in proportion to how delayed or improbable they are. For example, millions of cigarette smokers ignore the possibility of fatal consequences of smoking because those consequences are distant and uncertain. If smoking one cigarette caused certain death tomorrow, we would expect cigarette smoking to drop dramatically.

Sociological Theories of Deviance

Sociologists have been interested in the issue of deviant behavior since the pioneering efforts of Émile Durkheim in the late nineteenth century. Indeed, one of the major sociological approaches to understanding this problem derives directly from his work. It is called anomie theory.

Anomie Theory Durkheim published *The Division of Labor in Society* in 1893. In it, he argued that deviant behavior can be understood only in relation to the specific moral code it violates: "We must not say that an action shocks the common conscience because it is criminal, but rather that it is criminal because it shocks the common conscience" (1960a).

Durkheim recognized that the common conscience, or moral code, has an extremely strong hold on the individual in small, isolated societies where there are few social distinctions among people and everybody more or less performs the same tasks. Such *mechanically integrated* societies, he believed, are organized in terms of shared norms and values: All members are equally committed to the moral code. Therefore, deviant behavior that violates the code is felt by all members of the society to be a personal threat. As society becomes more complex—that is, as work is divided into more numerous and increasingly specialized tasks—social organization is maintained by the interdependence of individuals. In other words, as the division of labor becomes more specialized and differentiated, society becomes more *organically integrated*. It is held together less by moral consensus than by economic interdependence. A shared moral code continues to exist, of course, but it tends to be broader and less powerful in determining individual behavior. For example, political leaders among the Cheyenne Indians led their people by persuasion and by setting a moral example (Hoebel, 1960). In contrast with the Cheyenne, few modern Americans actually expect exemplary moral behavior from their leaders, despite the public rhetoric calling for it. We express surprise, but not outrage, when less than honorable behavior is revealed about our political leaders. We recognize that political leadership is exercised through formal institutionalized channels and not through model behavior.

In highly complex, rapidly changing societies such as our own, some individuals come to feel that the moral consensus has weakened. Some people lose their sense of belonging, the feeling of participating in a meaningful social whole. Such individuals feel disoriented, frightened, and alone. Durkheim used the term **anomie** to refer to *the condition of normlessness, in which values and norms have little impact and the culture no longer provides adequate guidelines for behavior.* Durkheim found that anomie was a major cause of suicide, as we discussed in Chapter 1. Robert Merton built on this concept and developed a general theory of deviance in American society.

Strain Theory Robert K. Merton (1938, 1968) believed that American society pushes individuals toward deviance by overemphasizing the importance of monetary success while failing to emphasize the importance of using legitimate means to achieve that success. Those individuals who occupy favorable positions in the social-class structure have many legitimate means at their disposal to achieve success. However, those who occupy unfavorable positions lack such means. Thus, the goal of financial success combined with the unequal access to important environmental resources creates deviance.

As Figure 6-2 shows, Merton identified four types of deviance that emerge from this strain. Each type represents a mode of adaptation on the part of the deviant individual. That is, the form of deviance a person engages in depends greatly on the position he or she occupies in the social structure. Specifically, it depends on the availability to the individual of legitimate, institutionalized means for achieving success. Thus, some individuals, called **innovators,** *accept the culturally validated goal of success but find deviant ways of going about reaching it.* Con artists, embezzlers, bank robbers, fraudulent advertisers, drug dealers, corporate criminals, crooked politicians,

Figure 6-2 **Merton's Typology of Individual Modes of Adaptation**

Mode of adaption		Culture's goals	Institutionalized means
Conformists		Accept	Accept
Deviants	Innovators	Accept	Reject
	Ritualists	Reject	Accept
	Retreatists	Reject	Reject
	Rebels	Reject/Accept	Reject/Accept

Conformists accept both (a) the goals of the culture and (b) the institutionalized means of achieving them. Deviants reject either or both. Rebels are defiants who may reject the goals or the institutions of the current social order and seek to replace them with new ones that they would then embrace.

cops on the take—each is trying to "get ahead" using whatever means are available.

Ritualists are *individuals who reject or deemphasize the importance of success once they realize they will never achieve it and instead concentrate on following and enforcing rules more precisely than was ever intended.* Because they have a stable job with a predictable income, they remain within the labor force but refuse to take risks that might jeopardize their occupational security. Many ritualists are often tucked away in large institutions such as governmental bureaucracies.

Another group of people also lacks the means to attain success but does not have the institutional security of the ritualists. **Retreatists** are *people who pull back from society altogether and cease to pursue culturally legitimate goals.* They are the drug and alcohol addicts who can no longer function—the panhandlers and street people who live on the fringes of society.

Finally, there are the rebels. **Rebels** *reject both the goals of what to them is an unfair social order and the institutionalized means of achieving them.* Rebels seek to tear down the old social order and build a new one with goals and institutions they can support and accept.

Merton's theory has become quite influential among sociologists. It is useful because it emphasizes external causes of deviant behavior that are within the power of society to correct. The theory's weakness is its inability to account for the presence of certain kinds of deviance that occur among all social strata and within almost all social groups in American society: for example, juvenile alcoholism, drug dependence, and family violence (spouse beating and child abuse).

Control Theory In control theory, social ties among people are important in determining their behavior. Instead of asking what causes deviance, control theorists ask: What causes conformity? They believe that what causes deviance is the absence of what causes conformity. In their view, conformity is a direct result of control over the individual. Therefore, the absence of social control causes deviance. According to this theory, people are free to violate norms if they lack intimate attachments with parents, teachers, and peers. These attachments help them establish values linked to a conventional lifestyle. Without these attachments and acceptance of conventional norms, the opinions of other people do not matter and the individual is free to violate norms without fear of social disapproval. This theory assumes that the disapproval of others plays a major role in preventing deviant acts and crimes.

According to Travis Hirschi (1969), one of the main proponents of control theory, we all have the potential to commit deviant acts. Most of us never commit these acts because of our strong bond to society. Hirschi's view is that there are four ways in which individuals become bonded to society and conventional behavior:

1. *Attachment to others.* People form intimate attachments to parents, teachers, and peers who display conventional attitudes and behavior.

2. *A commitment to conformity.* Individuals invest their time and energies in conventional types of activities, such as getting an education, holding a job, or developing occupational skills. At the same time, people show a commitment to achievement through these activities.

3. *Involvement in conventional activities.* People spend so much time engaged in conventional activities that they have no time to commit or even think about deviant activities.

4. *A belief in the moral validity of social rules.* Individuals have a strong moral belief that they should obey the rules of conventional society.

If these four elements are strongly developed, the individual is likely to display conventional behavior. If these elements are weak, deviant behavior is likely.

More recently Hirschi and Gottfredson (1993) have proposed a theory of crime based on one type of control only—self-control. They have suggested that people with high self-control will be less likely during all periods of life to engage in criminal acts. Those with low self-control are more likely to commit crime than those with high self-control. The source of low self-control is ineffective parenting. Parents who do not take an active interest in their children and do not socialize them properly produce children with low self-control. Once established in childhood, the level of social control a person has acquired will guide them throughout the rest of their lives.

Techniques of Neutralization Most of us think we act logically and rationally most of the time. In order to violate the norms and moral values of society, we must have **techniques of neutralization,** *a process that makes it possible for us to justify illegal or deviant behavior* (Sykes & Matza, 1957). In the language of control theory, these techniques provide a mechanism by which people can break the ties to the conventional society that would inhibit them from violating the rules. Techniques of neutralization are learned through the socialization process. They can take several forms:

1. *Denial of responsibility.* These individuals argue that they are not responsible for their actions; forces beyond their control drove them to commit the act, such as a troubled family life, poverty, or being drunk at the time of the incident. In any event, the responsibility for what they did lies elsewhere.

Sociology at Work

Public Heroes, Private Felons: Athletes and Sexual Assault

Thousand of hours of media attention are devoted to broadcasting sports events or reporting on the professional accomplishments of athletes. Athletes are presented as heroes and role models for the nation's youth. Yet at the same time, stories detailing the violence in the private lives of these public figures also make front page news.

Does being a celebrity provide an athlete charged with sexual assault certain built-in advantages in the legal system? Jeff Benedict, the former director of research at the Center for the Study of Sport in Society at Northeastern University, and sociologist Alan Klein wanted to answer this question.

They examined 217 criminal complaints of sexual assault against women by athletes filed with the police between 1986 and 1995. They found that when an athlete was arrested for these complaints, only 43 cases resulted in a plea bargain and only 10 resulted in a jury conviction (see Figure 6-3).

Why are these numbers so low? Jeff Benedict believes the courts are trying very hard to avoid any accusation of favoritism toward the athletes. Athletes do have other advantages, however. First, when an athlete is charged with a crime, he usually has the ability to pay for first-rate legal representation. Even college athletes, who are supposedly unemployed and not earning any income, get excellent representation. Second, the athlete often has a favorable public image whereas the accuser is unknown. Juries have a hard time believing that the well-dressed, popular athlete sitting at the defense table committed a brutal crime. The defense attorney only has to convince one person on the jury that the athlete did not commit the crime and the case will not result in a conviction. Athletes also have coaches, agents, alumni if they are college athletes, and other respected people who

Figure 6-3 Athletes Accused of Sexual Assault

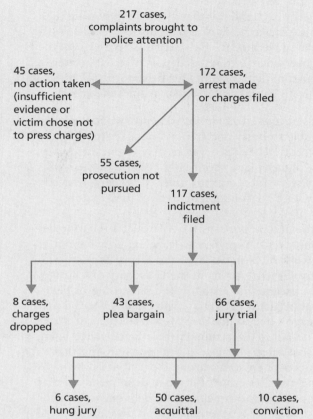

will speak on their behalf. Community support carries a great deal of weight in court. The prosecution often has to settle for a plea bargain or a reduced sentence because getting a conviction in court is usually difficult.

Sources: Based on "Arrest and Conviction Rates for Athletes Accused of Sexual Assault," by J. Benedict & A. Klein, 1997, *Sociology of Sport Journal, 14,* pp. 86–94; *Public Heroes, Private Felons: Athletes and Crimes Against Women,* by J. Benedict, 1997, Boston: Northeastern University Press; and an interview with the author, September 1997.

2. *Denying the injury.* The individual argues that the action did not really cause any harm. Who really got hurt when the individual illegally copied some computer software and sold it to friends? Who is really hurt in illegal betting on a football game?

3. *Denial of the victim.* The victim is seen as someone who "deserves what he or she got." The man who made an obscene gesture to us on the highway deserved to be assaulted when we caught up with him at the next traffic light. Some athletes, when accused of sexual assault on a woman, have claimed the woman consented to sex when she agreed to go to the athlete's hotel room. (See "Sociology at Work: Public Heroes, Private Felons: Athletes and Sexual Assault.")

4. *Condemnation of the authorities.* Deviant or criminal behavior is justified since those who are

in positions of power or are responsible for enforcing the rules are dishonest and corrupt themselves. Political corruption and police dishonesty leave us with little respect for these authority figures, since they are more dishonest than we are.

5. *Appealing to higher principles or authorities.* We claim our behavior is justified since we are adhering to standards that are more important than abstract laws. Acts of civil disobedience against the government are justified because of the government's misguided policy of supporting a corrupt dictatorship. Our behavior may be technically illegal, but the goal justifies the action.

Using these techniques of neutralization, people are able to break the rules without feeling morally unworthy. They may even be able to put themselves on a higher plane specifically because of their willingness to rebel against rules. They are basically redefining the situation in favor of their actions.

Cultural Transmission Theory This theory relies strongly on the concept of learning, growing out of the work of Clifford Shaw and Henry McKay, who received their training at the University of Chicago. They became interested in the patterning of delinquent behavior in that city when they observed that Chicago's high-crime areas remained the same over the decades—even though the ethnic groups living in those areas changed. Further, they found that as members of an ethnic group moved out of the high-crime areas, the rate of juvenile delinquency in that group fell; at the same time, the delinquency rate for the newly arriving ethnic group rose. Shaw and McKay (1931, 1942) discovered that delinquent behavior was taught to newcomers in the context of juvenile peer groups. Also, because such behavior occurred mostly in the context of peer-group activities, youngsters gave up their deviant ways when their families left the high-crime areas.

Edwin H. Sutherland and his student Donald R. Cressey (1978) built a more general theory of juvenile delinquency on the foundation laid by Shaw and McKay. This **theory of differential association** *is based on the central notion that criminal behavior is learned in the context of intimate groups* (see Table 6-1). When criminal behavior is learned, it includes two components: (1) criminal techniques (such as how to break into houses) and (2) criminal attitudes (rationalizations that justify criminal behavior). In this context, people who become criminals are thought to do so when they associate with the rationalizations for breaking the law more than with the arguments for obeying the law. They acquire these attitudes through long-standing interactions with others who hold these views. Thus, among the estimated 70,000 gang members in Los

Table 6-1

Sutherland's Principles of Differential Association

1. Deviant behavior is learned.
2. Deviant behavior is learned in interaction with other persons in a process of communication.
3. The principal part of the learning of criminal behavior occurs within intimate personal groups.
4. When deviant behavior is learned, the learning includes (a) techniques of committing the act, which are sometimes very complicated or sometimes very simple, and (b) the specific direction of motives, drives, rationalizations, and attitudes.
5. The specific direction of motives and drives is learned from definitions of the legal codes as favorable or unfavorable. That is, a person learns reasons for both obeying and violating rules.
6. A person becomes deviant because of an excess of definitions favorable to violating the law over definitions unfavorable to violating the law.
7. Differential associations may vary in frequency, duration, priority, and intensity.
8. The process of learning criminal behavior by association with criminal and anticriminal patterns involves all the mechanisms used in any other learning situation.
9. Although criminal behavior is an expression of general needs and values, it is not explained by those general needs and values, because noncriminal behavior is an expression of the same needs and values.

Source: Adapted from *Criminology,* 10th ed. (pp. 80–82), by E. H. Sutherland & D. R. Cressey, 1978, Philadelphia: Lippincott.

Angeles County, status is often based on criminal activity and drug use. Even arrests and imprisonment are events worthy of respect. A youngster exposed to and immersed in such a value system will identify with it, if only in order to survive. In many respects, differential association theory is quite similar to the behavioral theory we discussed earlier. Both emphasize the learning or socialization aspect of deviance. Both also point out that deviant behavior emerges in the same way that conformist behavior emerges; it is merely the result of different experiences and different associations.

Labeling Theory Under **labeling theory** *the focus shifts from the deviant individual to the social process by which a person comes to be labeled as deviant and the consequences of such labeling for the individual.* This view emerged in the 1950s from the writings of Edwin Lemert (1972). Since then, many other sociologists have elaborated on the labeling approach. Labeling theorists note that although we all break rules from time to time, we do not necessarily think of ourselves as deviant—nor are we so labeled by others. However, some individuals, through a series of circumstances, do come to be defined as deviant by others in society. Paradoxically, this labeling process actually helps bring about more deviant behavior.

Being caught in wrongdoing and branded as deviant has important consequences for one's further social participation and self-image. The most important consequence is a drastic change in the individual's public identity. It places the individual in a new status, and he or she may be revealed as a different kind of person than formerly thought to be. Such people may be labeled as thieves, drug addicts, lunatics, or embezzlers and are treated accordingly.

To be labeled as a criminal, one need commit only a single criminal offense. Yet the word carries a number of connotations of other traits characteristic of anyone bearing the label. A man convicted of breaking into a house and thereby labeled criminal is presumed to be a person likely to break into other houses. Police operate on this premise and round up known offenders for questioning after a crime has been committed. In addition, it is assumed that such an individual is likely to commit other kinds of crimes as well, because he or she has been shown to be a person without "respect for the law." Therefore, apprehension for one deviant act increases the likelihood that this person will be regarded as deviant or undesirable in other respects.

Even if no one else discovers the deviance or endorses the rules against it, the individual who has committed it acts as an enforcer. Such individuals may brand themselves as deviant because of what they did and punish themselves in one way or another for the behavior (Becker, 1963).

There appear to be at least three factors that determine whether a person's behavior will set in motion the process by which he or she will be labeled as deviant: (1) the importance of the norms that are violated, (2) the social identity of the individual who violates them, and (3) the social context of the behavior in question. Let us examine these factors more closely.

1. *The importance of the violated norms.* As we noted in Chapter 3, not all norms are equally important to the people who hold them. The most strongly held norms are mores, and their violation is likely to cause the perpetrator to be labeled deviant. The physical assault of an elderly person is an example. For less strongly held norms, however, much more nonconformity is tolerated, even if the behavior is illegal. For example, running red lights is both illegal and potentially very dangerous, but in some American cities it has become so commonplace that even the police are likely to look the other way rather than pursue violators.

2. *The social identity of the individual.* In all societies there are those whose wealth or power (or even force of personality) enables them to ward off being labeled deviant despite behavior that violates local values and norms. Such individuals are buffered against public judgment and even legal sanction. A rich or famous person caught shoplifting or even using narcotics has a fair chance of being treated indulgently as an eccentric and let off with a lecture by the local chief of police. Conversely, there are those marginal or powerless individuals and groups, such as welfare recipients or the chronically unemployed, toward whom society has little tolerance for nonconformity. Such people quickly are labeled deviant when an opportunity presents itself and are much more likely to face criminal charges.

3. *The social context.* The social context within which an action takes place is important. In a certain situation an action might be considered deviant, whereas in another context it will not. Notice that we say social context, not physical location. The nature of the social context can change even when the physical location remains the same. For example, for most of the year the New Orleans police manage to control open displays of sexual behavior, even in the famous French Quarter. However, during the week of Mardi Gras, throngs of people freely engage in what at other times of the year would be called lewd and indecent behavior. During Mardi Gras, the social context invokes norms for evaluating behavior that do not so quickly lead to the assignment of a deviant label.

Labeling theory has led sociologists to distinguish between primary and secondary deviance. **Primary deviance** is *the original behavior that leads to the application of the label to an individual.* **Secondary deviance** is *the behavior that people develop as a result of having been labeled as deviant* (Lemert, 1972). For example, a teenager who has experimented with illegal drugs for the first time and is arrested for it may face ostracism by peers, family, and school authorities. Such negative treatment may cause this person to turn more frequently to using illegal drugs and to associating with other drug users and sellers, possibly resorting to robberies and muggings to get enough money to buy the drugs. Thus, the primary deviant behavior and the labeling resulting from it lead the teenager to slip into an even more deviant lifestyle. This new lifestyle would be an example of secondary deviance.

Labeling theory has proved useful. It explains why society will label certain individuals deviant but not others, even when their behavior is similar. There are, however, several drawbacks to labeling theory. For one thing, it does not explain primary deviance. That is, even though we may understand how labeling may contribute to future, or secondary, acts of deviance, we do not know why the original, or primary, act of deviance took place. In this respect, labeling theory explains only part of the deviance process. Another problem is that labeling theory ignores the instances when the labeling process may deter a person from engaging in future acts of deviance. It looks at the deviant as a misunderstood individual who really would like to be an accepted, law-abiding citizen. Clearly, this is an overly optimistic view.

It would be unrealistic to expect any single approach to explain deviant behavior fully. In all likelihood, some combination of the various theories discussed is necessary to gain a fuller understanding of the emergence and continuation of deviant behavior.

The Importance of Law

As discussed earlier in this chapter, some interests are so important to a society that folkways and mores are not adequate enough to ensure orderly social interaction. Therefore, laws are passed to give the state the power of enforcement. These laws become a formal system of social control, which is exercised when other informal forms of control are not effective.

It is important not to confuse a society's moral code with its legal code, nor to confuse deviance with crime. Some legal theorists have argued that the legal code is an expression of the moral code, but this is not necessarily the case. For example, although most states and hundreds of municipalities have enacted some sort of antismoking law, smoking is not an offense against morals. Conversely, it is possible to violate American moral sensibilities without breaking the law.

What, then, is the legal code? The **legal code** consists of *the formal rules, called* **laws,** *adopted by a society's political authority.* The code is enforced through the use of formal negative sanctions when rules are broken. Ideally, laws are passed to promote conformity to those rules of conduct that the authorities believe are necessary for the society to function and that will not be followed if left solely to people's internal controls or the use of informal sanctions. Others argue that laws are passed to benefit or protect specific interest groups with political power, rather than society at large (Quinney, 1974; Vago, 1988).

The Emergence of Laws

How is it that laws come into society? How do we reach the point where norms are no longer voluntary and need to be codified and given the power of authority for enforcement? Two major explanatory approaches have been proposed: the consensus approach and the conflict approach.

The **consensus approach** *assumes that laws are merely a formal version of the norms and values of the people.* There is a consensus among the people on these norms and values, and the laws reflect this consensus. For example, people will generally agree that it is wrong to steal from another person. Therefore, laws emerge formally stating this fact and provide penalties for those caught violating the law.

The consensus approach is basically a functionalist model for explaining a society's legal system. It assumes that social cohesion will produce an orderly adjustment in the laws. As the norms and values in society change, so will the laws. Therefore "blue laws," which were enacted in many states during colonial times, and which prohibited people from working or opening shops on Sunday, have been changed, and now vast shopping malls do an enormous amount of business on Sundays.

The conflict approach to explaining the emergence of laws sees dissension and conflict between various groups as a basic aspect of society. The conflict is resolved when the groups in power achieve control. The **conflict approach** to law *assumes that the elite use their power to enact and enforce laws that support their own economic interests and go against the interests of the lower classes.* As William Chambliss (1973) noted:

> Conventional myths notwithstanding, the history of criminal law is not a history of public opinion or public interest. . . . On the contrary, the history of the criminal law is everywhere the history of legislation and appellate-court decisions which in effect (if not in intent) reflect the interests of the economic elites who control the production and distribution of the major resources of the society.

The conflict approach to law was supported by Richard Quinney (1974) when he noted, "Law serves the powerful over the weak . . . moreover, law is used by the state . . . to promote and protect itself."

Chambliss used the development of vagrancy laws as an example of how the conflict approach to law works. He pointed out that the emergence of such laws paralleled the need of landowners for cheap labor in England during a time when the system of serfdom was breaking down. Later, when cheap labor was no longer needed, vagrancy laws were not enforced. Then, in the sixteenth century, the laws were modified to focus on those who were suspected of being involved in criminal activities and interfering with those engaged in the transportation

Conflict theorists believe that laws are used by the state to promote and protect itself.

of goods. Chambliss (1973) noted, "Shifts and changes in the law of vagrancy show a clear pattern of reflecting the interests and needs of the groups who control the economic institutions of the society. The laws change as these institutions change."

Crime in the United States

Crime is *behavior that violates a society's legal code.* In the United States what is criminal is specified in written law, primarily state statutes. Federal, state, and local jurisdictions often vary in their definitions of crimes, though they seldom disagree in their definitions of serious crimes.

A distinction is often made between violent crimes and property crimes. A **violent crime** is *an unlawful event such as homicide, rape, and assault that may result in injury to a person.* Robbery is also a violent crime because it involves the use or threat of force against the person.

A **property crime** is *an unlawful act that is committed with the intent of gaining property but that does not involve the use or threat of force against an individual.* Larceny, burglary, and motor vehicle theft are examples of property crimes.

Criminal offenses are also classified according to how they are handled by the criminal justice system. In this respect, most jurisdictions recognize two classes of offenses: felonies and misdemeanors. Felonies are not distinguished from misdemeanors in the same way in all areas, but most states define **felonies** as *offenses punishable by a year or more in state prison.* Although the same act may be classified as a felony in one jurisdiction and as a misdemeanor in another, the most serious crimes are never misdemeanors, and the most minor offenses are never felonies.

Crime Statistics

It is very difficult to know with any certainty how many crimes are committed in the United States each year. There are two major approaches taken in determining the extent of crime. One measure of crime is provided by the FBI through its *Uniform Crime Reporting* (UCR) program. Since 1929, the FBI has been receiving monthly and annual reports from law enforcement agencies throughout the country, currently representing 96% of the national population. The UCR consists of eight crimes: *homicide, forcible rape, robbery, aggravated assault, burglary, larceny-theft, motor vehicle theft, and arson.* Arrests are reported for 21 additional crime categories also. Not included are federal offenses—political corruption, tax evasion, bribery, or violation of environmental-protection laws, among others.

The UCR has undergone a 5-year redesign effort and will soon be converted to a more comprehensive and detailed program known as the National Incident Based Reporting System (NIBRS).

Sociologists and critics in other fields note that for a variety of reasons, these statistics are not always reliable. For example, each police department compiles its own figures, and definitions of the same crime vary from place to place. Other factors affect the accuracy of the crime figures and rates published in the reports—for example, a law-enforcement agency or a local government may change its method of reporting crimes, so that the new statistics reflect a false increase or decrease in the occurrence of certain crimes. Under some circumstances, UCR data are estimated, because some jurisdictions do not participate or report partial data (Bureau of Justice Statistics, 1997).

A second measure of crime is provided through the *National Crime Victimization Survey* (NCVS), which began in 1973 to collect information on crimes suffered by individuals and households, whether or not these crimes were reported to the police. The UCR only measures reported crimes.

Six crimes are measured in the NCVS: *rape, robbery, assault, household burglary, personal larceny, and motor vehicle theft.* The similarity between these crimes and the UCR categories is obvious and intentional. Some crimes are missing from the NCVS that appear in the UCR. Murder cannot be measured through victim surveys, because obviously the victim is dead. Arson cannot be measured well through such surveys because the victim may in fact have been the criminal. An arson investigator is often needed to determine whether a fire was actually arson. Also, due to problems in questioning the victim, crimes against children under 12 are also excluded.

Whereas the UCR depends on police departments' records of reported crimes, the NCVS attempts to assess the total number of crimes committed. The NCVS obtains its information by asking a nationally representative sample of 49,000 households (about 101,000 people over the age of 12) about their experiences as victims of crime during the previous 6 months. The households stay in the sample for 3 years (Bureau of Justice Statistics, 1997).

Of the 28,800,000 crimes that took place in 1999, the NCVS estimated that only 38% were reported to the police. The specific crimes most likely to be reported were motor vehicle theft (89.7%) and robbery (62%). The crime least likely to be reported was theft of less than $50 dollars (13%).

The particular reason most frequently mentioned for *not* reporting a crime was that it was not important enough. For violent crimes, the reason most often given for not reporting was that it was a private or personal matter. For an example of the reporting rates for a variety of crimes, see Figure 6-4, and see

Figure 6-4 **Percentage of Selected Crimes Reported to the Police**

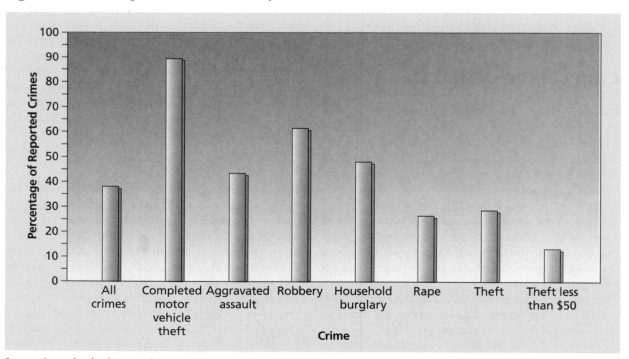

Source: *Sourcebook of Criminal Justice Statistics: 1999, 2000,* U.S. Department of Justice, Bureau of Justice Statistics, Washington, DC: U.S. Government Printing Office.

Figure 6-5 **Likelihood That Someone Will Be Sent to Prison for a Known Crime**

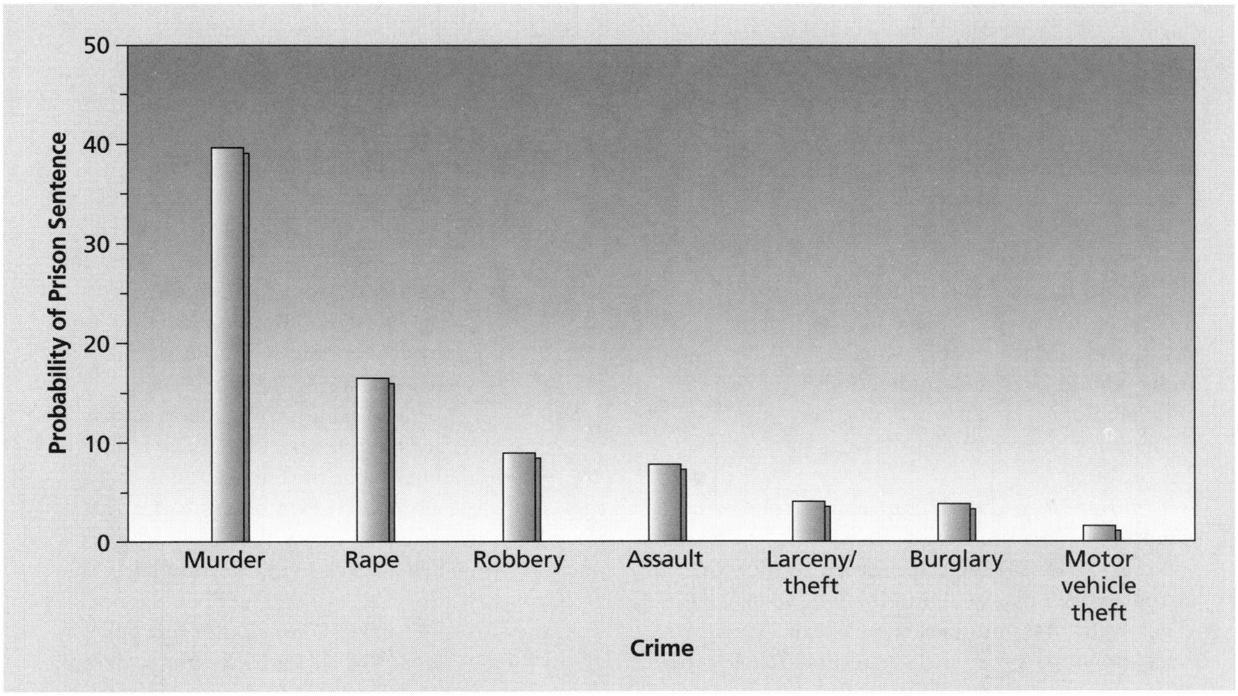

Sources: *Crime in the United States,* 1997, Bureau of Justice Statistics, *Felony Defendants Large Urban Counties,* 1994, January 1998, p. 24.

Figure 6-5 for the likelihood that someone will be sent to prison for a known crime.

Each survey is subject to the kinds of errors and problems typical of its method of data collection. Despite their respective drawbacks, they both are valuable sources of data on nationwide crime.

Kinds of Crime in the United States

The crime committed can vary considerably in terms of the impact it has on the victim and on the self-definition of the perpetrator of the crime. White-collar crime is as different from street crime as organized crime is from juvenile crime. In the next section, we shall examine these differences.

Juvenile Crime

Juvenile crime refers to *the breaking of criminal laws by individuals under the age of 18.* Regardless of the reliability of specific statistics, one thing is clear: Serious crime among our nation's youth is a matter of great concern. Hard-core youthful offenders—perhaps 10% of all juvenile criminals—are responsible, by some estimates, for two-thirds of all serious crimes. Although the vast majority of juvenile delinquents commit only minor violations, the

juvenile justice system is overwhelmed by these hard-core criminals.

Serious juvenile offenders are predominantly male, disproportionately minority group members (compared with their proportion in the population), and typically disadvantaged economically. They are likely to exhibit interpersonal difficulties and behavioral problems, both in school and on the job. They are also likely to come from one-parent families or families with a high degree of conflict, instability, and inadequate supervision.

Arrest records for 1998 show that youths under age 19 accounted for 23.2% of all arrests (*Sourcebook of Criminal Justice Statistics,* 1999, 2000). Arrests, however, are only a general indicator of criminal activity. The greater number of arrests among young people may be due partly to their lack of experience in committing crimes and partly to their involvement in the types of crimes for which apprehension is more likely, for example, theft versus fraud. In addition, because youths often commit crimes in groups, the resolution of a single crime may lead to several arrests. (See Table 6-2 for arrest rates by age.)

Indeed, one of the major differences between juvenile and adult offenders is the importance of gang membership and the tendency of youths to engage in group criminal activities. Gang members are more likely than other young criminals to engage in violent crimes, particularly robbery, rape, assault, and

Sociology at Work

Serial Murderers and Mass Murderers

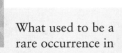

What used to be a rare occurrence in the United States is now becoming increasingly commonplace. There are now two or three mass murders every month in the United States. In fact, 7 of the 10 largest mass killings in American history took place in the past decade. Many people want to know what makes these individuals kill. Sociologist Jack Levin is one of the nation's best-known authorities on this problem.

Jack Levin.

Levin points out that, contrary to popular assumptions, mass murderers do not just "snap" or "go crazy." Their killing sprees are methodical and extremely well planned, and the motive usually is to get even. Mass murderers seek revenge against those individuals they feel are responsible for their problems. Levin notes, "The mass killer may be depressed, disillusioned, despondent, or desperate, but not deranged."

There are two types of multiple homicides. First are the mass killings often in the news. Here the individual kills a number of people within a short period of time. It could take place at the killer's last place of employment, in a restaurant, or at home. Serial killings differ in that instead of killing in a violent outburst, the murderer kills one victim at a time over a period of days, weeks, years, and even decades.

Mass killers tend to be white, middle-class, middle-aged males. Levin believes it takes a prolonged period of frustration to produce the kind of rage necessary for this type of brutal eruption. Mass killers have seen their lives go downhill for decades. Their relationships with others have fallen apart. Many cannot hold a job. They are trying to survive with their many problems, but then a catastrophic event puts them over the edge.

On the surface, serial murderers seem the same in that they are also likely to be white, middle-aged males. The difference is that serial murderers love to kill. Killing becomes a pleasurable end in itself. Most important, killing gives the serial killer a feeling of power. Levin notes that serial murderers have a great need for dominance and control, which they satisfy by taking the last breath from their victims. Very few will use a gun, because they want physical contact with the victim. They are sadistic.

Levin believes it is incorrect to characterize serial killers as insane. These people know what they are doing is wrong; they simply do not care. Levin notes, "They do not have a defect of the mind, they have a defect of character. They are not mad, they are bad. They are not crazy, they are very crafty. They are not sick, they are sickening." They do not feel guilty about their actions. The secret to serial murderers' success lies in the fact that they do not look like the monsters they are. They are thoroughly familiar with the rules of society, but they do not feel the rules apply to them.

When it comes to the issue of whether there should be a death penalty for these killers, Levin points out that in many ways mass murderers are dead already. They want to die and often kill themselves right after their violent binge. For the serial killers, Levin believes we should "lock them up and throw away the key. These people cannot be rehabilitated."

Mass murders and serial murders are a growing but still rare phenomenon. Levin points out that all told about 500 people die a year at the hand of a mass killer or serial killer. This is quite small compared to the 22,000 single victim homicides a year. Levin reminds us that "You are still more likely to contract leprosy or malaria than you are to be murdered by a serial killer or mass killer." We cannot suspect everyone around us of being a killer. Our goal should be to understand the basis for mass killings and serial killings so we may one day be able to prevent them.

---

weapons violations. Gangs that deal in the sale of crack cocaine have become especially violent in the past decade.

There is conflicting evidence on whether juveniles tend to progress from less- to more-serious crimes. It suggests that violent adult offenders began their careers with violent juvenile crimes; thus they began as, and remained, serious offenders. However, minor offenses of youths are often dealt with informally and may not be recorded in crime statistics.

The juvenile courts—traditionally meant to treat, not punish—have had limited success in coping

Juvenile courts have had limited success in dealing with juvenile offenders. The boot camp approach has been tried with some of these individuals, but the success of this method is also questionable.

Table 6-2

Age Distribution of Arrests, 1998

AGE GROUP	PERCENTAGE* OF U.S. POPULATION	PERCENTAGE* OF PERSONS ARRESTED
Age 12 and younger	19.2%	1.6%
13–15	4.2	7.6
16–18	4.0	14.0
19–21	4.1	13.0
22–24	4.4	9.3
25–29	7.4	13.3
30–34	8.5	12.3
35–39	8.4	11.6
40–44	7.6	8.1
45–49	6.4	4.6
50–54	5.1	2.3
55–59	4.2	1.1
60–64	3.9	0.6
Age 65 and older	12.7	0.6

*Percentages do not equal 100% due to rounding.

Source: *Sourcebook of Criminal Justice Statistics—1999* (pp. 348–349), by K. Maguire & A. L. Pastore, eds., 2000, U.S. Department of Justice, Bureau of Justice Statistics, Washington, DC: U.S. Government Printing Office.

with such juvenile offenders (Reid, 1991). Strict rules of confidentiality, aimed at protecting juvenile offenders from being labeled as criminals, make it difficult for the police and judges to know the full extent of a youth's criminal record. The result is that violent youthful offenders who have committed numerous crimes often receive little or no punishment.

Defenders of the juvenile courts contend, nonetheless, that there would be even more juvenile crime without them. Others, arguing from learning and labeling perspectives, contend that the system has such a negative impact on children that it actually encourages **recidivism**—that is, *repeated criminal behavior after punishment.* All who are concerned with this issue agree that the juvenile courts are less than efficient, especially in the treatment of repeat offenders. One reason for this is that perhaps two-thirds of juvenile court time is devoted to processing children guilty of what are called **status offenses**, *behavior that is criminal only because the person involved is a minor* (examples are truancy and running away from home). Recognizing that status offenders clog the courts and add greatly to the terrible overcrowding of juvenile detention homes, states have sought ways to deinstitutionalize status offenders. One approach, known as **diversion**—*steering youthful offenders away from the juvenile justice system to nonofficial social agencies*—has been suggested by Edwin Lemert (1981).

Violent Crime

In 1999, the violent crime rate in the United States reached the lowest level since the Bureau of Justice Statistics (BJS) started measuring it in 1973. There were an estimated 28.8 million violent and property crimes in 1999, compared to 44 million such incidents counted in the first year of the BJS National Crime Victimization survey (*Bureau of Justice Statistics, Criminal Victimization 1999, 2000*).

Yet even with this decline there were 3,047 violent crimes per 100,000 population reported in Atlanta; 2,557 in Tampa; 2,443 in Detroit; and 2,420 in Baltimore. If we keep in mind that only about 44% of all violent crimes are reported to the police, we can see how high the incidence of violent crime really is (*Sourcebook of Criminal, Justice Statistics, 1999*). Bureau of Justice data show that 54% of all violent crime victims knew their attackers. Nearly 70% of the rape and sexual assault victims knew the offender as an acquaintance, friend, relative, or intimate. Twenty-eight percent of the rape and sexual assault victimizations are reported to the police.

The violent crime rate in the United States also includes the highest homicide rate in the industrialized world. There are more homicides in any one of the cities of New York, Detroit, Los Angeles, or Chicago each year than in all of England and Wales combined. (See "Global Sociology: Is Homicide an American Phenomenon?")

In addition to homicide and rape, other violent crimes such as aggravated assault and robbery have an impact on American households. Each year there are about 1.6 million aggravated assaults and about 850,000 robberies (*Bureau of Justice Statistics,* July 1999).

Property Crime

Seventy-five percent of all crime in the United States is what is referred to as crime against property, as opposed to crime against the person. In all instances of crime against property, the victim is not present and is not confronted by the criminal.

The most significant nonviolent crimes are burglary, auto theft, and larceny-theft. In 1997, 4,635,000 households reported a burglary,

1,433,000 reported an auto theft, and 25,817,000 reported a property crime. Keep in mind that only about 32.6% of all household thefts are reported (*Sourcebook of Criminal Justice Statistics,* 1999).

Violent and property crime victimizations disproportionately affected urban residents during 1998. Urbanites accounted for 29% of the United States population and sustained 38% (12 million) of all violent and property crime victimizations.

In comparison, the percentages of suburban and rural residents who were victims of crime were lower than their percentages of the population. Fifty-one percent of the U.S. population were suburban residents who experienced 47% (15 million) of all violent and property victimizations. Rural residents accounted for 20% of the U.S. population but sustained 15% (5 million) of all violent and property crime victimizations, according to National Crime Victimization Survey (NCVS) data (U.S. Department of Justice, Office of Justice Program, Bureau of Justice Statistics Urban, Suburban, and Rural Victimization, 1993–1998, October 2000).

White-Collar Crime

The term **white-collar crime** was coined by Edwin H. Sutherland (1940) to refer to *the acts of individuals who, while occupying positions of social responsibility or high prestige, break the law in the course of their work for the purpose of illegal personal or organizational gain.* White-collar crimes include such illegalities as embezzlement, bribery, fraud, theft of services, kickback schemes, and others in which the violator's position of trust, power, or influence has provided the opportunity to use lawful institutions for unlawful purposes. White-collar offenses frequently involve deception.

Although white-collar offenses are often less visible than crimes such as burglary and robbery, the overall economic impact of crimes committed by such individuals are considerably greater. Not only is white-collar crime very expensive, it is also a threat to the fabric of society, causing some to argue that it causes more harm than street crime (Reiman, 1990). Sutherland (1961) has argued that because white-collar crime involves a violation of public trust, it contributes to a disintegration of social morale and threatens the social structure. This problem is compounded by the fact that in the few cases in which white-collar criminals actually are prosecuted and convicted, punishment usually is relatively light.

New forms of white-collar crime involving political and corporate institutions have emerged since the 1980s. For example, the dramatic growth in high technology has brought with it sensational accounts of computerized heists by sophisticated criminals seated safely behind computer terminals. The possibility of electronic crime has spurred widespread interest in computer security, by business and government alike.

Victimless Crime

Usually we think of crimes as involving culprits and victims—that is, individuals who suffer some loss or injury as a result of a criminal act. However, a number of crimes do not produce victims in any obvious way, and so some scholars have coined the term *victimless crime* to refer to them.

Basically, **victimless crimes** are *acts that violate those laws meant to enforce the moral code.* Usually they involve the use of narcotics, illegal gambling, public drunkenness, the sale of sexual services, or status offenses by minors. If heroin and crack cocaine addicts can support their illegal addictions legitimately, then who is the victim? If a person bets $10 or $20 per week with the local bookmaker, who is the victim? If someone staggers drunkenly through the streets, who is the victim? If a teenager runs away from home because conditions there are intolerable, who is the victim?

Some legal scholars argue that the perpetrators themselves are victims: Their behavior damages their own lives. This is, of course, a value judgment, but then the concept of deviance depends on the existence of values and norms (Schur & Bedau, 1974). Others note that such offenses against the public order do, in fact, contribute to the creation of victims, if only indirectly: Heroin addicts rarely can hold jobs and eventually are forced to steal to support themselves, prostitutes are used to blackmail people and to rob them, chronic gamblers impoverish themselves and bring ruin on their families, drunks drive and get into accidents and may be violent at home, and so on.

Clearly, the problems raised by the existence of victimless crimes are complex. In recent years, American society has begun to recognize that at least some crimes truly are victimless and that they should therefore be decriminalized. Two major activities that have been decriminalized in many states and municipalities are the smoking of marijuana (though not its sale) and sex between unmarried, consenting adults of the same gender.

Victims of Crime

We have been discussing crime statistics, the types of crimes committed, and who commits them. But what about the victims of crime? Is there a pattern? Are some people more apt to become crime victims than others are? It seems that this is true; victims of crime are not spread evenly across society. Although, as we have seen, the available crime data are not always reliable, a pattern of victimization can be seen in the reported statistics. A person's race, gender, age, and socioeconomic status have a great deal to do with whether that individual will become a victim of a serious crime.

Statistics show that, overall, males are much more likely to be victims of serious crimes than females are. When we look at crimes of violence and theft separately, however, a more complex picture emerges. Younger people are much more likely than the elderly to be victims of crime. African Americans are more likely to be victims of violent crime than are whites or members of other racial groups. People with low incomes have the highest violent-crime victimization rates. Theft rates are the highest for people with low incomes (less than $7,5000 per year) and for those with higher incomes (more than $30,000 per year). Students and the unemployed are more likely than homemakers, retirees, or the employed to be victims of crime. Rural residents are less often crime victims than are people living in cities (Bureau of Justice Statistics, 2000).

Despite the growing, albeit unfounded, concern about crimes against the elderly, figures show that young people are most likely to be victims of serious crimes. For example, the violent victimization rates for people aged 16 to 19 are 20.3 times higher than for persons 65 and older. Similarly, 1 in 8 people murdered are under age 18 (Bureau of Justice Statistics, 2000).

The reason the elderly are less likely to be the victims of violent crime than the young is related in part to differences in lifestyle and income. Younger people may more often be in situations that place them at risk. They may frequent neighborhood hangouts, bars, or events that are likely places for an assault to take place. About 22 percent of the elderly reported that they never went out at night for entertainment, shopping, or other activities.

The only crime category that affected the elderly at about the same rate as most others (except those ages 12–24) was personal theft, which includes purse snatching and pocket picking.

Criminals may believe that the elderly are more likely to have large amounts of cash and are less likely to defend themselves. As a result, the elderly are still quite vulnerable to crimes such as robbery, purse snatching, or pickpocketing. (Bureau of Justice Statistics, 2000).

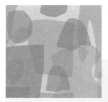

Global Sociology

Is Homicide an American Phenomenon?

The United States has the dubious distinction of having one of the highest homicide rates of all industrialized and non-industrialized countries in the world. The sharp increase in homicides in the late 1980s and much of the subsequent decline are attributable to a rise and fall in gun violence by juveniles and young adults. The nation's murder rate was 6.3 per 100,000 population in 1998 compared to 4.6 per 100,000 population in 1950. This number is two to three times that of most European countries.

Russia and other former Eastern-bloc countries have experienced a great deal of social upheaval since the fall of communism causing their homicides rates to increase dramatically. Today these countries also have homicide rates that have been typical of the United States for the past 30 years (see Figure 6-6).

If we look for answers for this phenomenon, we begin to see that in the United States homicide has become less of a domestic nonstranger event and more of an event that grows out of other criminal

Figure 6-6 **Homicide Rates for Selected Countries**

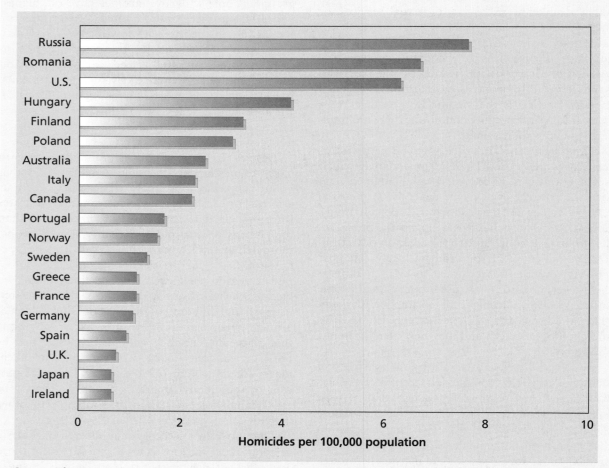

Homicides per 100,000 population

Sources: *The Nature of Homicide: Trends and Changes* (p. 114), by P. K. Lattimore, & C. A. Nahabedian, 1997, Washington, DC: U.S. Department of Justice; *Sourcebook of Criminal Justice Statistics: 1999, 2000,* p. 302, U.S. Department of Justice, Bureau of Justice Statistics, Washington, DC: U.S. Government Printing Office.

Figure 6-7 U.S. Homicide Solution Rates

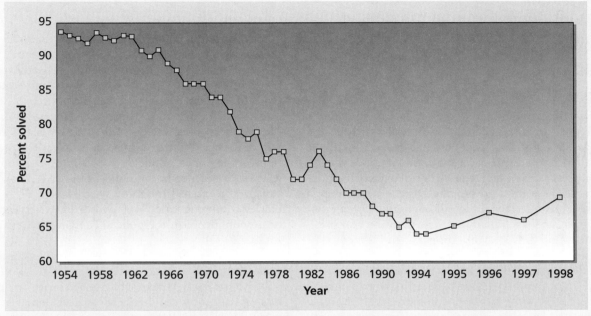

Source: Uniform Crime Reports; FBI, Crime in the United States, 1976 through 1998.

situations. With the advent of cocaine and "crack" in the mid 1980s, homicide has become interconnected with drug trafficking and drug use (Parker 1995).

This change has also made it more difficult to solve homicides. When homicide was more likely to be a domestic or intimate relationship event, nearly all were solved. For example, 94% of all homicides in 1954 were solved. As of 1998, the solution rate had dropped to 69% as homicides are increasingly likely to be perpetrated by a larger group of individuals as they commit a variety of crimes (see Figure 6-7). Homicide has the highest resolution rate of all serious crimes.

Homicide is the 10th leading cause of death in the United States and accounts for 1.1% of all deaths. When compared to the death toll from other major causes such as heart disease and cancer, the percentage attributed to homicide seems quite modest. There is another way of looking at it, however. Homicide disproportionately involves young victims

without any major diseases. The victims are more like those in fatal automobile accidents than those dying from fatal diseases. In any given year, the median age of death of homicide victims is between 25 and 29. The number of years of life lost to a homicide is usually significantly greater than those lost to a disease. Taken as a whole, the years lost to homicide equal nearly 80% of those lost to heart disease and nearly 70% of those lost to cancer (Zimring & Hawkins, 1997).

Sources: *Alcohol and Homicide: A Deadly Combination of Two American Traditions,* by R. N. Parker, 1995, Albany, NY: State University Press; *The Nature of Homicide: Trends and Changes,* U.S. Department of Justice, Office of Justice Programs, 1996, Washington, DC: U.S. Government Printing Office; *Crime Is Not the Problem: Lethal Violence in America,* by F. E. Zimring & G. Hawkins, 1997, New York: Oxford University Press; *Homicide Trends in the United States: 1998 Update,* by J. A. Fox & M. W. Zawitz, Bureau of Justice Statistics, Washington, D. C., March 2000.

Criminal Justice in the United States

Every society that has established a legal code has also set up a **criminal justice system**—*personnel and procedures for arrest, trial, and punishment*—to deal with violations of the law. The three main categories of our criminal justice system are the police, the courts, and the prisons.

The Police

The police system developed in the United States is a highly decentralized one. It exists on three levels: federal, state, and local. On the federal level, the United States does not have a national police system. There are, however, federal laws enacted by Congress. These laws govern the District of Columbia and all states when a federal offense has been committed, such as kidnapping, assassination of a president, mail fraud, bank robbery, and so on. The Federal Bureau of Investigation (FBI) enforces many of these laws and also assists local and state law enforcement authorities in solving local crimes. If a nonfederal crime has been committed, the FBI must be asked by local or state authorities before it can aid in the investigation. If a particular crime is a violation of both state and federal law, state and local police often cooperate with the FBI to avoid unnecessary duplication of effort.

The state police patrol the highways, regulate traffic, and have primary responsibility for the enforcement of some state laws. They provide a variety of other services, such as a system of criminal identification, police training programs, and computer-based records systems to assist local police departments.

The jurisdiction of police officers at the local level is limited to the state, town, or municipality in which the person is a sworn officer of the law. Some problems inevitably result from such a highly decentralized system. Jurisdictional boundaries sometimes result in overlapping, communication problems, and difficulty in obtaining assistance from another law enforcement agency.

Contrary to some expectations, the public has a great deal of confidence in the police. In a year 2000 Gallup Poll, 55% of the public rated the police either "very high" or "high" for honesty and ethics. Only 11% had negative views of the police.

The situation changes, however, once we look at the numbers more closely. African Americans and white Americans differ dramatically on their perceptions of the police. While 63% of whites had a high level of confidence in the police, only 26% of African Americans feel the same way. In fact, 35% of blacks compared to 8% of whites have very little or no confidence in the police (The Gallup Organization, 2000).

People often have unrealistically high expectations of the police. Historically, police in the United States have been young white males with a high school education (or less). Most still come from working-class backgrounds. In the past two decades, attempts have been made to raise the educational levels of the police, as well as produce a more heterogeneous distribution, including women and minorities.

The Courts

The United States has a dual court system consisting of state and federal courts, with state and federal crimes being prosecuted in the respective courts. Some crimes may violate both state and federal

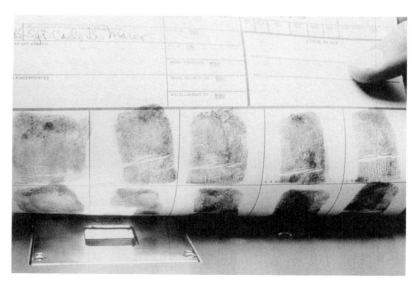

Police use devices that facilitate tracking criminals, such as this finger matrix machine that speeds up the process of analyzing fingerprints.

statutes. About 85% of all criminal cases are tried in the state courts.

The state court system varies from one state to the other. Lower trial courts exist for the most part to try misdemeanors and petty offenses. Higher trial courts can try felonies and serious misdemeanors. All states have appeal courts. Many have only one court of appeal, which is often known as the state supreme court. Some states have intermediate appeal courts.

The federal court system consists of three basic levels, excluding such special courts as the United States Court of Military Appeals. The United States *district courts* are the trial courts. Appeals may be brought from these courts to the *appellate courts.* There are eleven courts at this level, referred to as circuit courts. Finally, the highest court is the *Supreme Court,* which is basically an appeals court, although it has original jurisdiction in some cases.

The lower federal courts and the state courts are separate systems. Cases are not appealed from a state court to a lower federal court. A state court is not bound by the decisions of the lower federal court in its district, but it is bound by decisions of the United States Supreme Court (Reid, 1991).

Prisons

Prisons are a fact of life in the United States. As much as we may wish to conceal them, and no matter how unsatisfactory we think they are, we cannot imagine doing without them. They represent such a fundamental defense against crime and criminals that we now keep a larger portion of our population in prisons than any other nation and for terms that are longer than in many counties. Small wonder that Americans invented the prison.

Before prisons, serious crimes were redressed by corporal or capital punishment. Jails existed, but mainly for pretrial detention. The closest thing to the modern prison was the workhouse. This was a place of hard labor designed almost exclusively for minor offenders, derelicts, and vagrants. The typical convicted felon was either physically punished or fined, but not incarcerated. Today's system of imprisonment for a felony is a historical newcomer.

Goals of Imprisonment Prisons exist to accomplish at least four goals: (1) separation of criminals from society, (2) punishment of criminal behavior, (3) deterrence of criminal behavior, and (4) rehabilitation of criminals.

1. *Separation of criminal from society.* Prisons accomplish this purpose once felons reach the prison gates. Inasmuch as it is important to protect society from individuals who seem bent on repeating destructive behavior, prisons are one logical choice among several others, such as exile and capital punishment (execution). The American criminal justice system relies principally on prisons to segregate convicts from society, and in this regard they are quite efficient.

2. *Punishment of criminal behavior.* There can be no doubt that prisons are extremely unpleasant places in which to spend time. They are crowded, degrading, boring, and dangerous. Not infrequently, prisoners are victims of one another's violence. Inmates are constantly supervised, sometimes harassed by guards, and deprived of normal means of social, emotional, intellectual, and sexual expression. Prison undoubtedly is a severe form of punishment.

3. *Deterrence of criminal behavior.* The general feeling among both the public and the police is that prisons have failed to achieve the goal of deterring criminal behavior. There are good reasons for this. First, by their very nature, prisons are closed to the public. Few people know much about prison life, nor do they often think about it. Inmates who return to society frequently brag to their peers about their prison experiences in order to recover their self-esteem. To use the prison experience as a deterrent, the very unpleasant aspects of prison life would have to be constantly brought to the attention of the population at large. To promote this approach, some prisons have allowed inmates to develop programs introducing high school students to the horrors of prison life. From the scanty evidence available to date, it is unclear whether such programs deter people from committing crimes. Another reason that prisons fail to deter crime is the funnel effect, discussed later. No punishment can deter undesired behavior if the likelihood of being punished is minimal. Thus, the argument regarding the relative merits of different types of punishment is pointless until there is a high probability that whatever forms are used will be applied to all (or most) offenders.

4. *Rehabilitation of criminals.* Many Americans believe that rehabilitation—the resocialization of criminals to conform to society's values and norms and the teaching of usable work habits and skills—should be the most important goal of imprisonment. It is also the stated goal of almost all corrections officials. Yet there can be no doubt that prisons do not come close to achieving this aim. According to the FBI, 63% of all inmates released from prison are arrested again for criminal behavior within 3 years (Bureau of Justice, 1992).

Sociological theory helps explain why rehabilitation is often ineffective. Sutherland's ideas on cultural transmission and differential association point to the fact that inside prisons, the society of inmates has a culture of its own, in which obeying the law is not highly valued. New inmates are socialized quickly to this peer culture and adopt its negative attitudes toward the law. Further, labeling theory tells us that once somebody has been designated as deviant, his or her subsequent behavior often conforms to that label. Prison inmates who are released find it difficult to be accepted in the society at large and to find legitimate work. Hence, former inmates quickly take up with their old acquaintances, many of whom are active criminals. It thus becomes only a matter of time before they are once more engaged in criminal activities.

This does not mean that prisons should be torn down and all prisoners set free. As we have indicated, prisons do accomplish important goals, though certain changes are needed. Certainly it is clear that the entire criminal justice system needs to be made more efficient and that prison terms as well as other forms of punishment must follow predictably the commission of a crime. Another idea, which gained some approval in the late 1960s but seems of late to have declined in popularity, is to create "halfway" houses and other institutions in which the inmate population is not so completely locked away from society. This way, they are less likely to be socialized to the

prison's criminal subculture. Labeling theory suggests that if the process of delabeling former prisoners were made open, formal, and explicit, released inmates might find it easier to win reentry into society. Finally, just as new prisoners are quickly socialized into a prison's inmate culture, released prisoners must be resocialized into society's culture. This can be accomplished only if means are found to bring ex-inmates into frequent, supportive, and structured contact with stable members of the wider society (again, perhaps, through halfway houses). The simple separation of prisoners from society undermines this goal.

To date, no society has been able to come up with an ideal way of confronting, accommodating, or preventing deviant behavior. Although much attention has been focused on the causes of and remedies for deviant behavior, no theory, law, or social-control mechanism has yet provided a fully satisfying solution to the problem. (See "Remaking the World: Freeing the Innocent.")

A Shortage of Prisons

Today's criminal justice system is in a state of crisis over prison crowding. Even though our national prison capacity has expanded, it has not kept up with demands. The National Institute of Justice estimates

Remaking the World
Freeing the Innocent

Merely finishing law school is enough of an accomplishment for most students. Now a few can also claim their studies helped to save someone's life. Having lost their freedom, livelihood, and often their families, many inmates have turned to the Innocence Project for help. Operating out of the Benjamin N. Cardozo School of Law, the Innocence Project provides free legal assistance to inmates who are challenging their convictions. Founded in 1992 by law professors Barry Scheck and Peter Neufeld, the Innocence Project is currently handling more than 200 cases, and more than 1,000 are pending. Clients are spread throughout the country, many serving life sentences and some on death row. The project receives information about hundreds of additional cases every week.

Forensic DNA evidence is often the deciding factor in these cases. DNA testing has become a powerful tool for prosecutors and police in identifying criminals. More recently, DNA testing has also be-

come a key factor in exonerating the wrongfully convicted. Since the advent of forensic DNA testing in the late 1980s, at least 51 people in the United States have been exonerated and set free. Scheck and Neufeld estimate that there may be thousands of innocent people currently incarcerated. Older type blood tests, mistaken eyewitness identification, and unmonitored laboratory practices have contributed to the wrongful conviction of those who are innocent.

Though DNA testing can mean scientific and irrefutable proof of innocence, there is often resistance on the part of police and prosecutors to reopening cases, whether it be for political reasons or for the sake of the finality of the conviction. Reopening the cases of innocent people in prison has often been quite difficult.

Source: *Actual Innocence,* by B. Scheck, P. Neufeld, & J. Dwyer, 2000, New York: Random House.

that we must add 1,000 prison spaces a week just to keep up with the growth in the criminal population.

Compounding the problem is the fact that many states have mandated prison terms for chronic criminals, drunken drivers, and those who commit gun crimes. Yet nearly every community will have an angry uprising if the legislature suggests building a new prison in their neighborhood. Given state financial pressures, community resistance, and soaring construction costs, people face a difficult choice. They must either build more prisons or let most convicted offenders go back to the community.

A key consideration in sending a person to prison is money. The custodial cost of incarceration in a medium-security prison is $15,000 a year. The cost is closer to $25,000 once you add to this the cost of actually building the prison, and additional payments to dependent families. You can see why judges are quick to use probation as an alternative to imprisonment, particularly when the prisons are already overcrowded.

The other side of the question, however, is how much does it cost us *not* to send this person to prison? While it is easy to calculate the cost of an offender's year in prison, it is considerably more difficult to figure the cost to society of letting that individual roam the streets. One study suggests that it is more expensive to release an offender than to incarcerate him when you weigh the value of crime prevented through imprisonment.

An inmate survey revealed that the average inmate had committed 200 crimes in the year before incarceration. Hardened, habitual criminals can be one-person crime waves. These people are committing nearly four crimes a week. Can there be any doubt in anybody's mind that it is cheaper to incarcerate these individuals than to let them pursue their trade on society (Bureau of Justice Statistics, 1988; Chaiken & Chaiken, 1982).

How much does each crime cost the public? The National Institute of Justice has come up with a figure of $2,300 per crime. This number undoubtedly overestimates the value of petty larcenies and underestimates the cost of rapes, murders, and serious assaults. It is an average, however, and it does give us some way of comparing the costs of incarceration with the costs of freedom. Using the $2,300 per-crime cost, we find that a typical inmate committing 200 crimes (the low estimate) is responsible for $460,000 in crime costs per year. Sending 1,000 additional offenders to prison, instead of putting them on probation, would cost an additional $25 million per year. The crimes averted, however, by taking these individuals out of the community would save society more than $460 million.

This approach merely gives us a dollars and cents way of making a comparison. It does not in any way account for the personal anguish and trauma to the victims of crimes that would be averted.

Looking at the issue from this perspective overwhelmingly supports the case for more prison space. It costs communities more in real losses, social damages, and security measures than it does to incarcerate offenders who are crowded out by today's space limitations.

Women in Prison

As prison reform began in this country, the practice was to segregate women into sections of the existing institutions. There were few women inmates, a fact that was used to justify not providing them with a matron. Vocational training and educational programs were not even considered. In 1873, the first separate prison for women, the Indiana Women's Prison, was opened, with its emphasis on rehabilitation, obedience, and religious education.

In contrast with institutions for adult males, institutions for adult women are generally more aesthetic and less secure. This is an outgrowth of the fact that in the past, women inmates were not considered high security risks, nor have they proved to be as violent as male inmates. Women were more likely to commit property crimes, such as larceny, forgery, and fraud. This trend in crimes has changed, however, and women now commit more violent crimes than property crimes. Still, three-quarters of the violent crimes committed by women are the less serious type known as simple assault. Drug offenses by women have also increased dramatically in recent years (Bureau of Justice Statistics, 1997; Sniffen, 1999).

There are some exceptions, but on the whole, women's institutions are built and maintained with the view that their occupants are not great risks to themselves or to others. Women inmates also usually have more privacy than men do while incarcerated, and women usually have individual rooms. With the relatively smaller number of women in prison, there is a greater opportunity for the inmates to have contact with the staff, and there is also a greater chance for innovation in programming (Reid, 1991).

The number of women in state and federal prisons in 1999 reached a record of 90,668. Even though the rate of increase in the number of women going to prison has been greater than that for men, females still make up a relatively small segment of the total prison population—6.6% at the end of 1999. Relative to their number in the U.S. population, men are 15 times more likely than women to be incarcerated. (Bureau of Justice Statistics, "Prisoners in 1999," August 2000). (See Figure 6-8.)

Female, as compared with male, inmates appear to have greater difficulty adjusting to the absence of their families, especially to the absence of their

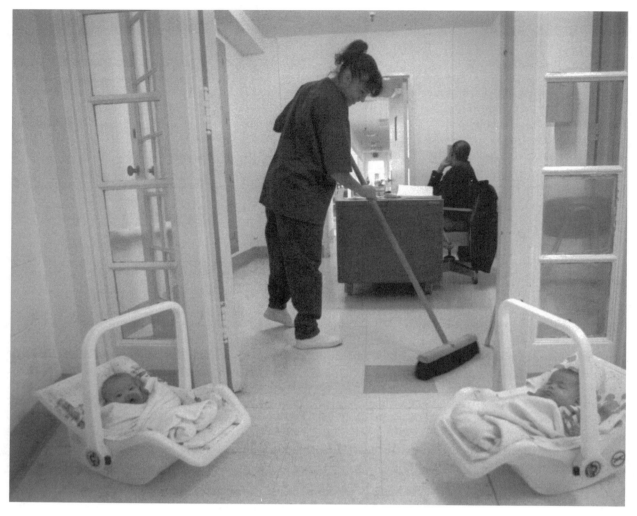

Two-thirds of women in prison are mothers, and the vast majority or their children are under age 18. Some prisons, such as the maximum security women's prison in Bedford Hills, New York, allow inmate mothers to keep their babies with them until the babies are 18 months of age.

children. Two-thirds of women in prison are mothers, and the vast majority (88%) of their children are under 18. In 1996, 96,000 children in the United States had mothers in prison. Only 25% of these children were cared for by the father while the mother was in prison. Most of the children were living with a grandparent (Bureau of Justice Statistics, 1994, 1997).

The Funnel Effect

One complaint voiced by many of those concerned with our criminal justice system is the existence of the funnel effect, in which many crimes are committed, but few people ever seem to be punished. The funnel effect begins with the fact that of all the crimes committed, only 38% are reported to the police (*Sourcebook of Criminal Justice Statistics*, 1999). Only about 26% lead to an arrest. Further,

false arrests, lack of evidence, and plea bargaining (negotiations in which individuals arrested for a crime are allowed to plead guilty to a lesser charge of the crime, thereby saving the criminal justice system the time and money spent in a trial) considerably reduce the number of complaints that actually are brought to trial. To be fair, the situation is not quite as bad as it appears. The number of arrests for serious crimes is considerably higher than it is for crimes in general.

What about punishment? Those who criticize the system's funnel effect seem to regard only a term in prison as an effective punishment. Yet the usual practice is to send to prison only those criminals whose terms of confinement are set at over one year. The number of prisoners in federal and state prisons, after declining through the 1960s, rose sharply through the 1990s, reaching 1,366,721 inmates in 1999 (Bureau of Justice, August 2000).

Figure 6-8 Women Prisoners in State and Federal Institutions, 1925–1999

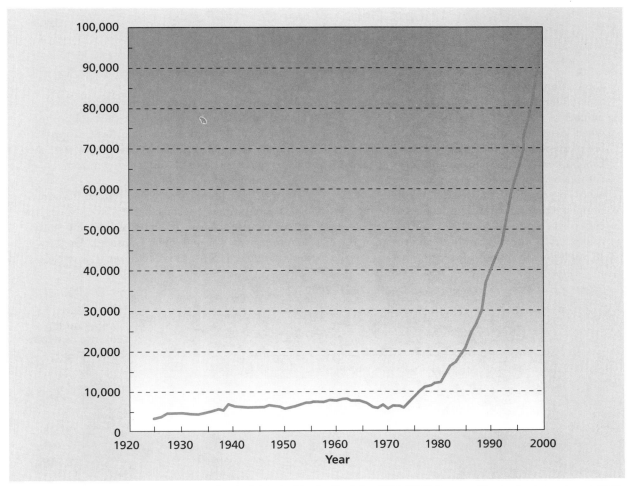

Source: "Prisoners in 1999," (p. 514), *Sourcebook of Criminal Justice Statistics: 1999, 2000;* Bureau of Justice Statistics, August 2000.

Many thousands of other criminals receive shorter sentences and serve them in municipal and county jails. Thus, if the numbers of people sent to local jails as well as to prison are counted, the funnel effect is seen to be less severe than it often is portrayed. The question then becomes one of philosophy: Is a jail term of less than one year an adequate measure for the deterrence of crime? Or should all convicted criminals have to serve longer sentences in federal or state prisons, with jails used primarily for pretrial detention?

Truth in Sentencing

The amount of time offenders serve in prison is almost always shorter than the time they are sentenced to serve by the court. For example, prisoners released in 1996 served an average of 30 months in prison and jail, or 44% of their 85-month sentences. The public has been in favor of longer sentences and

uniform punishments for prisoners. Prison crowding and reductions in prison time for good behavior have often resulted in the release of prisoners well before they served their assigned sentences. In response to complaints that criminals were not paying for their crimes, many states enacted restrictions on the possibility of early release; these laws became known as "truth in sentencing." The truth-in-sentencing laws require offenders to serve a substantial portion of the prison sentence imposed by the court before being eligible for release. The laws are based on the belief that victims and the public are entitled to know exactly what punishments offenders are receiving.

Truth in sentencing gained momentum in the 1990s with the help of the U.S. Congress, which authorized grants to build or expand correctional facilities if states would enact such laws. In order to receive the grants, states had to require persons convicted of violent crimes to serve not less than 85% of their prison sentences.

Controversies in Sociology

The Continuing Debate Over Capital Punishment: Does It Deter Murderers?

Twenty of the 38 states that have capital punishment laws executed 98 prisoners in 1999, 30 more than in 1998. This was the highest number of executions in a single year since 1951, when 105 were put to death bringing the total number of executions to 638 since 1976, the year the U.S. Supreme Court reinstated the death penalty. Although 98 prisoners were executed in 1999, another 75 were given the death sentence that year, bringing the total to 3,527 prisoners awaiting execution (see Figures 6-9 and 6-10). The average person executed in 1999 had been on death row for 11 years and 11 months. It seems obvious that the vast majority of inmates with a death sentence will not be executed.

The public supports the death penalty. The Gallup Poll has been asking Americans about the death penalty for almost 50 years. As of February 2001, two-thirds (67%) of Americans favor the death penalty in cases of murder, down from its high point of 80% in 1994.

Capital punishment has been opposed for many years and for many reasons. In the United States, the Quakers were the first to oppose the death penalty and to provide prison sentences instead. Amnesty International, U.S.A., calls capital punishment a

Figure 6-9 **Persons Sentenced to Death, 1953–2000**

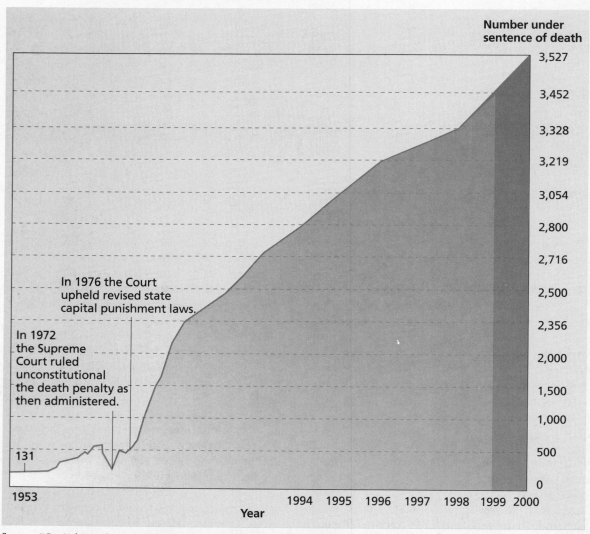

Source: "Capital Punishment, 1999," U.S. Department of Justice, December 2000, Bureau of Justice Statistics.

"horrifying lottery" in which the penalty is death and the odds of escaping are determined more by politics, money, race, and geography than by the crime committed. The group bases its impression on the fact that black men are more likely to be executed than white men; southern states, including Texas, Virginia, Missouri, Louisiana, and Florida, account for the vast majority of executions that have taken place since 1977.

It is also no fluke that nearly all death-row inmates are poor. They often had a public defender who might not have been qualified for the task. Even if the inmate's attorney made errors during the defense, the defendant's appellate attorney must demonstrate that the defense counsel's blunders directly affected the jury's verdict and that without those mistakes, the jury would have returned a different verdict (Prejean, 1993). One study (Radelet, Bedau, & Putnam, 1992), for instance, found that between 1900 and 1991, 416 innocent people were convicted of capital crimes, and 23 actually were executed. The two most frequent causes of errors that produced wrongful convictions were perjury by prosecution witnesses and mistaken eyewitness testimony. At least 48 people have been released from death row since 1973 because of significant

evidence of their innocence (Subcommittee on Civil and Constitutional Rights, 1993).

Yet the arguments for capital punishment continue to mount, centering mainly around the issue of deterrence. As Ernest van den Haag has noted, "If by executing convicted murderers there is any chance, even a mere possibility, of deterring future murderers, I think we should execute them. The life even of a few victims who may be spared seems infinitely precious to me. The life of the convicted murderer has negative value. His crime has forfeited it" (van den Haag, 1991).

Which brings us back to the age-old question: Does the death penalty deter homicide? Until the 1970s, social scientists continued to argue that they could find no evidence that it did. Erlich (1975) has presented information based on sophisticated statistical techniques showing that between 1933 and 1965, every execution may have resulted in seven or eight fewer murders. That is, as many as eight people escape being the victims of future murderers every time an execution takes place. Another study by economist Stephen K. Layson (1985) concluded that every execution of a convicted murderer prevents 18 murders. These results are not widely

Controversies in Sociology continues

Figure 6-10 **Inmates Executed, 1930–2000**

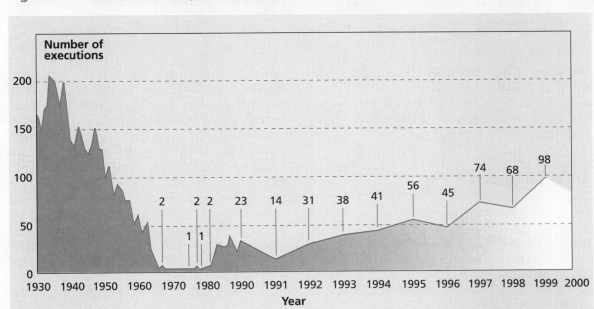

Source: "Capital Punishment, 1999," U.S. Department of Justice, December 2000, *Bureau of Justice Statistics Bulletin.*

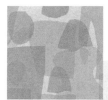

Controversies in Sociology

(continued)

accepted, however, and there is yet no conclusive evidence that the death penalty does or does not deter murder more than other penalties.

There may be more involved in deterrence than we think. Plato believed we are deterred from committing crimes by seeing others punished. He was referring to punishments administered in public, where everyone could see the gory details of torture and execution. Fortunately, today executions are not public and only a small number of people witness them. In place of actually seeing the execution, we now have mass media reports that become our eyes. Therefore, deterrence should be related to how much an execution is publicized.

People also argue that the death penalty is applied in a racially discriminatory fashion. One extensive study (Baldus, Woodworth, & Pulaski, 1990) concluded that the odds of being condemned to death were 4.3 times greater for defendants who killed whites than for defendants who killed blacks. Opponents of the death penalty have used this information to make the case that it should be abolished entirely on the grounds that racial bias is an inevitable part of the administration of capital punishment in the United States and that it would be better to have no death penalty than one influenced by prejudice. Others argue that the remedy to the problem is to do what is known as "leveling up"—increasing the number of people executed for mur-

dering blacks. They point out that if we sentence more murderers of black people to death, we are then eliminating the bias.

A third solution is to impose mandatory death sentences for certain types of crimes. In this way we are eliminating discretionary judgments and the potential for bias (Kennedy 1997).

With public support for the death penalty continuing, and with no broad legal challenges to capital punishment being waged, we can expect executions to continue.

Sources: "Capital Punishment 1999," U.S. Department of Justice, Bureau of September 1994, *Justice Statistics Bulletin;* "The Deterrent Effect of Capital Punishment: A Question of Life and Death," by I. Ehrlich, 1975, *American Economic Review,* pp. 397–417; *Punishing Criminals: Concerning a Very Old and Painful Question,* by E. van den Haag, 1991, New York: University Press Of America; *Dead Man Walking,* by H. Prejean, 1993, New York: Random House, and an interview with the author, September 1993; *In Spite of Innocence,* by M. L. Radelet, H. A. Bedau, & C. E. Putnam, 1992, Boston: Northeastern University Press, and an interview with the authors, January 1992; Staff Report, Subcommittee on Civil and Constitutional Rights, House Judiciary Committee 103rd Congress, 1st Session, October 21, 1993; *Equal Justice and the Death Penalty: A Legal and Empirical Study,* by D. C. Baldus, G. Woodworth, & C. A. Pulaski, Jr., 1990, Boston: Northeastern University Press; *Race Crime and the Law,* by R. Kennedy, 1997, New York: Pantheon Books, and an interview with the author, June 1997; *The Death Penalty in America: Current Controversies,* by H. A. Bedau, ed., 1997, New York: Oxford University Press, and an interview with the author, May 1997.

At this point, two-thirds of the states have established truth-in-sentencing laws. This has limited the powers of parole boards to set release dates, or of prison managers to award good time, earned time, or both (Mackenzie, 2000).

SUMMARY

A culture's norms and values make up its moral code, or the symbolic system by which behavior is viewed as right or wrong, good or bad within that culture. Thus, normal behavior is behavior that conforms to the norms of the group in which it occurs. Deviant behavior is behavior that fails to conform to the group's norms.

Criminal and deviant behavior has been found throughout history. To account for this, scholars have proposed a variety of theories. Biological theo-

ries such as those propounded by Lombroso and Sheldon stressed the importance of inherited factors in producing deviance. Psychological explanations emphasize cognitive or emotional factors within the individual as the cause of deviance. Psychoanalytic theory suggests that criminals act on the irrational impulses of the id because they failed to develop a proper superego, or conscience, in the socialization process. Behaviorists argue that crime is the product of conditioning. Wilson and Herrnstein proposed that criminal activity, like all human behavior, is the product of a rational choice by the individual as a result of weighing the costs and benefits of alternative courses of action.

Sociological theories of deviance rely on patterns of social interaction and the relationship of the individual to the group as explanations. Durkheim argued that, in modern highly differentiated and specialized societies, and particularly under conditions of rapid social change, individuals could become morally disoriented. This condition, which he called anomie, can produce deviance.

Control theorists like Hirschi have argued that everyone is a potential deviant. The issue, for them, is not what causes deviance, but what causes conformity. When individuals have strong bonds to society, their behavior will conform to conventional norms. When any of those bonds is weakened, however, deviance is likely. Other sociologists, like Sykes and Matza, have argued that people become deviant as a result of developing techniques of neutralization or rationalizations that make it possible to justify illegal or deviant behavior. Their view is that these techniques are learned as part of the socialization process.

Cultural transmission theory, pioneered by Shaw and McKay, emphasizes the cultural context in which deviant behavior patterns are learned. Sutherland and Cressey suggested the theory of differential association—that is, that individuals learn criminal techniques and attitudes through intimate contact with deviants. Labeling theory shifts the focus of attention from the deviant individual to the social process by which a person comes to be labeled as deviant and the consequences of such labeling for the individual. In all likelihood some combination of these various theories is necessary to gain a fuller understanding of the emergence and continuance of deviant behavior.

Crime is behavior that violates a society's criminal laws. Violent crime results in injury to a person; property crime is committed with the intent of obtaining property and does not involve the use or threat of force against an individual. While the specific definition varies from state to state, the most serious crimes are termed felonies; less serious crimes are called misdemeanors. The FBI publishes statistics on the frequency of selected crimes in the Uniform Crime Reports. These statistics are not always reliable, however. The National Crime Victimization Survey shows that only a small fraction of all crimes are reported to the authorities.

The U.S. violent crime rate includes the highest homicide rate in the industrialized world. Other violent crimes that have an impact on American households include rape, aggravated assault, and robbery. Ninety percent of all crime in the United States is crime against property, not against a person.

The criminal justice system consists of personnel and procedures for the arrest, trial, and punishment of those who violate the laws. The three main aspects of this system are the police, the courts, and prisons.

The goals of imprisonment include separation of the criminal from society, punishment of criminal behavior, deterrence of criminal behavior, and rehabilitation, or the resocialization of criminals to conform to society's values and norms and the teaching of usable work habits and skills.

 INTERNET RESOURCES

Go to http://web-enhanced.thomsonlearning.com which features Web links relevant to this chapter to expand and enhance the learning experience. This site will be monitored and updated periodically to ensure the links are correct and live.

At Virtual Society: The Wadsworth Sociology Resource Center, http://sociology.wadsworth.com, you will find a Career Center, Sociology in the News, Research Online, InfoTrac College Edition, Virtual Tours in Sociology, and a variety of book-specific student and instructor resources.

CHAPTER SIX STUDY GUIDE

KEY CONCEPTS

Match each concept with its definition, illustration, or explanation below.

a. Property crime
b. Formal positive sanctions
c. Theory of differential association
d. Negative sanctions
e. Laws
f. Formal negative sanctions
g. Victimless crimes
h. Informal sanctions
i. Juvenile crime
j. Normal behavior
k. Crime
l. Mechanisms of social control
m. Informal positive sanctions
n. Primary deviance
o. Formal sanctions
p. Techniques of neutralization

q. Diversion
r. Labeling theory
s. Violent crime
t. Ritualists
u. Consensus approach
v. Status offense
w. Endomorph
x. Mesomorph
y. Positive sanctions
z. Anomie
aa. Retreatists
bb. Moral code
cc. Rebels
dd. Deviant behavior
ee. Conflict approach
ff. Felonies

gg. Innovators
hh. Internal means of control
ii. Sanctions
jj. Funnel effect
kk. Misdemeanors
ll. External means of control
mm. Legal code
nn. Rehabilitation
oo. Secondary deviance
pp. Recidivism
qq. Informal negative sanctions
rr. White-collar crime
ss. Criminal justice system
tt. Nuremberg Trials
uu. Atavistic being
vv. Ectomorph

_____ 1. The symbolic system in terms of which behavior takes on the quality of being good or bad, right or wrong.

_____ 2. Behavior that conforms to the rules or norms of the group in which it occurs.

_____ 3. Behavior that fails to conform to the rules or norms of a group in which it occurs.

_____ 4. Ways of directing or influencing members to conform to the group's values and norms.

_____ 5. The operation of a group's moral code on an individual even in the absence of reactions by others.

_____ 6. The responses of other people to an individual's behavior.

_____ 7. Rewards and penalties by a group's members that are used to regulate an individual's behavior.

_____ 8. Actions that encourage an individual to continue acting in a certain way (i.e., rewards).

_____ 9. Actions that discourage the repetition or continuation of a behavior (i.e., punishments).

_____ 10. Sanctions applied in a public ritual, usually under the direct or indirect control of authorities.

_____ 11. Sanctions applied spontaneously by group members with little or no formal direction.

_____ 12. Displays people use spontaneously to express their approval of another's behavior.

_____ 13. Spontaneous displays of disapproval or displeasure.

_____ 14. Public affairs, rituals, or ceremonies that express social approval of a person's behavior.

_____ 15. Actions that express institutionalized disapproval of a person's behavior.

_____ 16. A state of normlessness, in which values and norms have little impact and the culture no longer provides adequate guidelines for behavior.

_____ 17. Individuals who accept the culturally validated goal of success but find deviant ways of achieving it.

_____ **18.** Individuals who reject or de-emphasize the importance of success and instead concentrate on following and enforcing rules more precisely than was ever intended.

_____ **19.** People who pull back from society altogether and cease to pursue culturally legitimate goals.

_____ **20.** People who reject both the goals and the institutionalized means to achieve them and who wish to build a different social order with alternative goals and means.

_____ **21.** A thought process that makes it possible to justify illegal or deviant behavior.

_____ **22.** The idea that individuals learn criminal techniques and attitudes through intimate contact with deviants.

_____ **23.** The theory that focuses on the social process by which a person comes to be defined as deviant and the consequences of that definition for the individual.

_____ **24.** The original behavior that leads to the application of a label to an individual.

_____ **25.** Behavior that people develop as a result of having been labeled as deviant.

_____ **26.** The body of formal rules adopted by a society's political authority.

_____ **27.** Formal rules.

_____ **28.** Assumes that laws are merely a formal expression of the agreed-upon norms and values of the people.

_____ **29.** Argues that the elite use their power to enact and enforce laws that support their own economic interests to the exclusion of the interests of others.

_____ **30.** Any behavior that violates a society's criminal laws.

_____ **31.** An unlawful event that may result in injury to a person.

_____ **32.** An unlawful act committed with the intent of gaining property, but that does not involve the use or threat of force against an individual.

_____ **33.** The most serious crimes, usually punishable by a year or more in prison.

_____ **34.** Less serious violations of criminal law or minor offenses.

_____ **35.** The breaking of criminal laws by individuals younger than 18.

_____ **36.** Evolutionary throwbacks whose behavior is more apelike than human.

_____ **37.** Behavior that is criminal only because the person involved is a minor.

_____ **38.** Steering offenders away from the justice system to social agencies.

_____ **39.** Acts by individuals who, while occupying positions of social responsibility or high prestige, break the law in the course of their work.

_____ **40.** Acts that violate those laws meant to enforce the moral code.

_____ **41.** Personnel and procedures for arrest, trial, and punishment.

_____ **42.** The resocialization of criminals to conform to society's values and norms and the teaching of usable work habits and skills.

_____ **43.** The fact that, of the many crimes committed, few seem to result in punishment of the offender.

_____ **44.** Argued that there was a higher moral order under which certain human actions were wrong regardless of who endorsed them.

_____ **45.** Criminal behaviors that are repeated even after punishments have occurred.

_____ **46.** A thin and linear body type associated with being inhibited, secretive, and restrained.

_____ **47.** A round and soft body type associated with being a relaxed person.

_____ **48.** A ruggedly muscular body type associated with being assertive and action oriented.

KEY THINKERS/RESEARCHERS

Match the thinkers/researchers with their main idea or contribution.

a. **Gersham Sykes and David Matza**
b. **Robert K. Merton**
c. **Sigmund Freud**
d. **James Q. Wilson and Richard Herrnstein**
e. **Edwin H. Sutherland**
f. **Émile Durkheim**
g. **E. A. Hooten**
h. **Sarnoff Mednick**

i. **William H. Sheldon**
j. **Travis Hirschi**
k. **Clifford Shaw and Henry McKay**
l. **Cesare Lombroso**
m. **Edwin Lemert**
n. **Ernst van den Haag**
o. **Baldus, Woodworth and Prulsli, Jr.**

_____ 1. Argued that deviant behavior is an integral part of all healthy societies; developed the concept of anomie.

_____ 2. Suggested that criminals were evolutionary throwbacks who could be identified by primitive physical features, particularly with regard to the head.

_____ 3. Identified three main body types and suggested that each was responsible for different personality traits.

_____ 4. Argued that crime is produced by the unconscious impulses of the individual.

_____ 5. Argued that crime is the product of a rational choice by an individual as a result of weighing the costs and benefits of alternative courses of action.

_____ 6. Developed a theory of structural strain to explain deviance.

_____ 7. Developed control theory, in which it is hypothesized that the strength of social bonds keep most of us from becoming criminals.

_____ 8. Argued that deviants learn techniques of neutralization to justify their deviance.

_____ 9. Suggested that certain neighborhoods generate a culture of crime that is passed on to residents.

_____ 10. Developed the theory of differential association to explain why some people and not others become deviant; coined the term white-collar crime.

_____ 11. Pioneered the development of labeling theory.

_____ 12. Argued that even if there were only a slim possibility that executing a murderer would deter future murders, society should execute them.

_____ 13. Argued that the born criminal was a scientific reality and that crime was not the product of social conditions but the outgrowth of "organic inferiority."

_____ 14. Proposed that some genetic factors are passed along from parent to child, with a susceptibility to criminality or adaptation to normality being inherited.

CENTRAL IDEA COMPLETIONS

Following the instructions, fill in the appropriate concepts and descriptions for each of the questions posed in the following section.

1. As your text notes, "the amount of time offenders serve in prison is almost always shorter than the time they are sentenced to serve by the court." Respond to (and provide an example for) each of the following questions:

   a. Why is the above statement correct? _____

   _____

   _____

b. What is involved in the "truth-in-sentencing" movement of the 1990s? _____

_____

_____

2. Describe the "Innocence Project." What is the importance of DNA testing for this project?

_____

_____

_____

3. Describe the pattern (in terms of incidence rates) of violent crime in the United States between 1973 and 1999: _____

_____

_____

_____

4. List, define, and provide an example for each of Merton's five categories of individual adaptation to anomie:

a. _____

Example: _____

b. _____

Example: _____

c. _____

Example: _____

d. _____

Example: _____

e. _____

Example: _____

5. Apply Sykes and Matza's five techniques of neutralization to a situation involving cheating in a college or university community:

a. _____

b. _____

c. _____

d. _____

e. _____

6. What five functions of deviance are discussed in your text?

a. _____

Example: _____

b. _____

Example: _____

c. _____

Example: _____

d. _____

Example: _____

e. _____

Example: _____

7. As a sociologist trained in the general area of social deviance, which specific theory would you use to explain to a layperson each of the following forms of social behavior?

a. Heroin addiction: _____

b. Auto theft: _____

c. Embezzlement of stock securities on the Internet: _____

d. Joining a militia group: _____

e. Engaging in forcible rape: _____

8. Compare and contrast the phenomena of serial and mass murders. Be sure to highlight the factor they share in common and the areas in which they differ:

Mass: _____

_____

_____

_____

Serial: _____

_____

_____

_____

9. Discuss and provide examples for each of the four dysfunctions of social deviance:

a. _____

Example: _____

b. _____

Example: _____

c. _____

Example: _____

d. _____

Example: _____

CRITICAL THOUGHT EXERCISES

1. Part A.   Log on to the World Wide Web. Using any available search engine or browser, investigate one or more sites that demonstrate the following:

   a. An example of a behavior you personally would define as both deviant and harmful to society.

   b. An example of a behavior that you would define as deviant, but not socially harmful to society.

   c. One behavior you would define as deviant, yet not illegal.

   d. One behavior you believe many persons older than you might define as deviant, but that you and members of your age cohort would not define as deviant.

   e. One behavior you found so totally, repulsively deviant that you would keep it off the Internet.

Part B.   After you have completed Part A, download an example page for each of the sites you visited and discuss the following:

   a. What are the common features of each form of deviance?

   b. What aspects, if any, of these Web sites made you uncomfortable?

   c. Assuming the sites you defined as deviant are similarly defined by others in a similar manner, what is it about the behavior that leads to these definitions?

d. What role might new technologics, such as the Internet, play in a society's shifting definitions of deviance?

2. Develop a presentation in which you integrate material from two of the supplemental readings in Chapter 6: "Freeing the Innocent" and "The Continuing Debate Over Capital Punishment: Does It Deter Murderers?" Using materials from each reading, discuss the deterrence effects or noneffects of capital punishment on homicide, as well as the relationship between the new technology of DNA testing and the question of an individual's guilt or innocence. As a sociologist, how do you believe these two phenomena are linked together, both theoretically and practically, in American society?

3. Discuss in detail the issues examined in the reading "Public Heroes, Private Felons: Athletes and Sexual Assault." Sociologically, what appears to be the relationship between sports and violent behavior? Apply Sutherland's differential association to the issue of violence and the behavior of athletes.

4. An emerging area of social deviance is road rage. Either through an examination of materials from the Internet or through a review of your library's holdings, describe and then discuss the phenomenon of the 1990s known as road rage. What theory(ies) of social deviance can be used to account for this behavior? How do the ideas of informal and formal social control relate to the emergence of road rage as a contemporary form of American deviance? Are there examples of road rage in other societies throughout the world? If so, what do those examples share with the American varieties of road rage and in what ways are they idiosyncratic to specific countries?

5. While criminologists are likely to focus on deviance that involves the breaking of societal laws, considerable lower level deviance takes place on a daily basis in every society. Prepare a notebook for your field observations, and spend a week observing behaviors on your campus. Conduct your observations relative to these questions:

a. What acts, if any, of deviant behavior did you observe?

b. How did those acts vary by the actor, the place of the activity, and the audiences that witnessed the behavior?

c. Were any of those acts of deviance sanctioned in any way? If yes, how effective were the sanctions in terms of deterring the behavior?

d. How did the deviant individual(s) react, if at all, to the discovery of the deviance by outsiders?

e. After completing your fieldwork, what conclusions are you able to draw about deviance and everyday life on your campus?

ANSWERS

KEY CONCEPTS

1. bb	9. d	17. gg	25. oo	33. ff	41. ss
2. j	10. o	18. t	26. mm	34. kk	42. nn
3. dd	11. h	19. aa	27. e	35. i	43. jj
4. l	12. m	20. cc	28. u	36. uu	44. tt
5. hh	13. qq	21. p	29. ee	37. v	45. pp
6. ll	14. b	22. c	30. k	38. q	46. vv
7. ii	15. f	23. r	31. s	39. rr	47. w
8. y	16. z	24. n	32. a	40. g	48. x

KEY THINKERS/RESEARCHERS

1. f	4. c	7. j	9. k	11. m	13. g
2. l	5. d	8. a	10. e	12. n	14. h
3. i	6. b				

PART 3
Social Inequality

As humans interact in groups and organizations, they do not always do so as equals. In many cases, some people possess more resources than others. Why is this so? On what basis are valued assets distributed unequally? For that matter, who decides what assets or personal characteristics will be valued in a particular society and who gets to establish the basis for distributing these precious goods?

Part 3 addresses the topic of social inequality. In Chapter 7 you will learn that all societies make distinctions among people, resulting in systems of unequal privilege, rewards, opportunities, power, prestige, and influence. When these inequalities are perpetuated by major institutions in society, we speak of a system of social stratification. In Chapter 7 you also will examine the development of different stratification systems and the possibilities people have for changing their position within the social hierarchy, and you will explore the major theories sociologists have developed to explain social stratification. Later in the chapter you will explore the extent of economic inequality in the United States and the world. How are income and wealth distributed? How is poverty determined? Who are the poor? How does the amount of economic inequality in the United States compare to that in other countries? Finally, how does inequality affect the life chances of individuals?

Social class, though, is not the only form of social inequality. Race or ethnic background, gender, and age also have been used historically as a basis for unequally distributing rewards and privileges. Chapter 8 examines the dynamics of racial and ethnic inequality and the position of major ethnic groups and racial minorities in the contemporary U.S. stratification system. Chapter 9 looks at inequality based on gender. How are definitions of gender related to biological differences between the sexes? Why are gender distinctions made at all, and how are they perpetuated? What are the implications of these distinctions, both for individuals and society?

This section of the book will provide you with an understanding of the basic dimensions of inequality in the United States and the world. When you have completed it, you will be prepared to explore the major institutions in society.

CHAPTER SEVEN

Social Stratification and Social Class

LEARNING OBJECTIVES

After studying this chapter you, should be able to do the following:

- Explain the factors that affect one's chances of upward social mobility.
- Compare and contrast the caste, estate, and class systems.
- Describe the distribution of wealth and income in the United States.
- Summarize the functionalist and conflict theory views of stratification.
- Describe the characteristics of each of the social classes in the United States.
- Describe differences in the poverty rate among various groups in American society.
- Compare poverty rates in the United States with those of other industrialized countries.
- Describe some of the personal and social consequences of a person's position in the class structure.

What do you do with a kid who can't read? . . . the kid whose existence became a hallucination at seven, and a catastrophe at fourteen, and a disaster after that. Whose vocation is to be neither a waitress nor a hooker, nor a farmer, nor a janitor, but forever the stepdaughter to a lascivious stepfather and the undefended offspring of a self obsessed mother. The kid who mistrusts everyone, sees the con in everyone, and yet is protected against nothing. Whose capacity to hold on unintimidated is enormous, and yet whose purchase on life is minute. Misfortune's favorite embattled child, the kid to whom everything loathsome that can happen has happened and whose luck shows no sign of changing.

Philip Roth, *The Human Stain*

We wore ties on Sunday, and black wool suits called B-suits, with the school crest on our top pocket. The crest was a dragon, whose head reached out toward the sun. Under that came the motto—*Arduus ad Solem*.

My father had taught me a song to help me do up my tie. It had the tune of "Twinkle, Twinkle, Little Star" and went "Over, under, over, through, pull the little end away from you.". . .

Some Sunday afternoons, the school was like a ghost town. The day before, parents had clogged the school with their Bentleys and Range Rovers and Rollses and driven away their sons. The ones who remained were mostly boys who lived abroad. They weren't foreign. It was just that their parents were working in Singapore or Hong Kong or Bermuda.

Paul Watkins, *Stand Before Your God: A Boarding School Memoir*

The experiment of public housing, which has worked throughout the country to isolate its impoverished and predominantly black tenants from the hearts of their cities, may have succeeded here with even greater efficiency because of Coney Island's utter remoteness. On this peninsula, at the southern tip of Brooklyn, there are almost no stores, no trees . . . nothing, but block after block of gray-cement projects—hulking, prisonlike, and jutting right into the sea. Most summer nights now, an amorphous unease settles over Coney Island, as apartments become stifling and the streets fall prey to the gangs and drug dealers. Officially, Coney Island is part of . . . New York City. But on a night like this, as the dealers set up their drug marts in the streets and alleyways, and the sounds of sirens and gunfire keep pace with the darkening sky, it feels like the end of the world.

Darcy Frey, *The Last Shot*

Americans like to think that social stratification and social class are minor issues. After all, we do not have inherited ranks, titles, or honors. We do not have coats of arms or rigid caste rankings. Besides, equality among men—and women—is an ideal guaranteed by our Constitution and summoned forth regularly in speeches from podiums and lecterns across the land. Yet the vast discrepancy among the environments just described makes it difficult to ignore the life chances of those who grow up in them. What we will begin to see in this chapter is that social stratification is quite complex and open to many subtle variations. It does not always fit neatly into our stereotypes.

The United States is a country characterized by enormous diversity in wealth and power. Once rewards are distributed unequally within a society, then economic, political, and social stratification be-

gin. **Social stratification,** *the division of society into levels, steps, or positions,* is perpetuated by the major institutions of society such as the economy, the family, religion, and education. Even though no formal social stratification policy exists in the United States, there is stratification based on gender, race, education, wealth, and age. Certain groups have greater access than others to better education, medical care, and jobs, and those advantages perpetuate their privileged position in our society. These remain despite legislation, free public education, and political idealism.

The Nature of Social Stratification

Every society makes distinctions among its people. These differences are not always accompanied by qualitative judgments, however. For example, a 5-year-old child is very different from an infant, but is not necessarily viewed as being superior because of the age difference. Although a society may not value a 5-year-old more highly than an infant, it might place a value on old people. Compare, for example, the attitude of respect for the elderly in Japan with the attitude toward senior citizens in the United States.

Values placed on physical characteristics and personality habits also vary from one society to another. For example, among Europeans and Americans, body hair on adult males is considered to be manly and acceptable, but it is seen by the Japanese as ugly. Americans promote competitiveness and individualism; the Kung San of the Kalahari Desert in southern Africa value cooperativeness and modesty (Lee, 1980). Individuals who have characteristics favored

Social inequality involves the uneven distribution of privileges, material rewards, and power.

In some Middle Eastern societies, women are expected to cover themselves in public. Such a situation helps perpetuate inequality between men and women.

societies have a certain amount of mobility. For example, in the United States, an open society, minorities and women continue to struggle against job discrimination. Even in a closed society such as the estate system of medieval Europe, a wealthy merchant whose social position was low could buy his way into the nobility and consolidate his family's new social status by marrying his children to landed aristocracy.

Mobility may come about because of changing one's occupation, marrying into a certain family, and so on. *Movement that involves a change of status with no corresponding change in social class* is known as **horizontal mobility.** For example, Gabriella Zia started her career as an attorney. Within 15 years, she made a number of horizontal career moves, none of which appreciably changed her position in the social hierarchy. She became a financial planner, a stockbroker, a newspaper columnist, an author of a number of successful books on financial planning, and finally a college professor. Although each of these career moves was extremely important to Zia, from a sociological point of view, they are perceived as involving little or no change in prestige, power, or wealth, and hence provide little mobility.

Movement up or down in the hierarchy resulting in a change in social class is known as **vertical mobility.** The United States is filled with success stories of vertical mobility often involving **intergenerational mobility**—that is, *a change in social status that occurs over two or more generations.* The Kennedy family offers a prime example of this type of vertical mobility. Patrick Joseph Kennedy, the grandfather of John F. Kennedy, started life in relative poverty. He had to borrow money from family members to buy a Boston saloon. His son, Joseph P. Kennedy, became an enormously wealthy—and often unscrupulous—business tycoon. John F. Kennedy achieved the pinnacle of success and respectability in this culture by becoming president of the United States.

Another type of vertical mobility is **intragenerational mobility,** which is *a change in social status that occurs during the lifetime of an individual.* There are many examples of men and women who have experienced upward intragenerational mobility. Steven P. Jobs and Stephen G. Wozniak started Apple Computer on approximately $1,200 in Job's garage. William Gates made hundreds of millions of dollars while still in his 20s as head of Microsoft Corporation. George E. Johnson, founder and president of Johnson Products Company, started his empire with $500, half of which was borrowed. Today he is the head of a multimillion-dollar company that manufactures hair-care products and cosmetics for African Americans.

Usually a person's social rank in the stratification hierarchy is consistent and comparatively easy to

by their culture have an advantage over those who do not. It is easier for them to win respect and prestige, to make friends, to find a mate, and to achieve positions of leadership. In all societies, there are some people who are favored, who have more prestige, and who are admired; there are others who are avoided and looked down on. In addition, all groups—even hunters and food gatherers, the most equal-minded of societies—make distinctions on the basis of age and gender. Social evaluations made on the basis of individual characteristics or behaviors lead to **social inequality,** which is *the uneven distribution of privileges, material rewards, opportunities, power, prestige, and influence among individuals and groups.* When social inequality becomes part of the social structure and is transmitted from one generation to the next, social stratification exists.

Social Mobility

Social mobility is *the movement of an individual or group from one social status to another.* The extent of social mobility varies from one society to another. An **open society** *attempts to provide equal opportunity to everyone to compete for desired roles and status regardless of race, religion, gender, or family history.* A **closed society** *fixes at birth the various aspects of people's lives.*

There are no purely open or completely closed societies. Even the most democratic societies assign some roles and statuses, and even the most closed

identify. However, many people do not fit neatly into one social category; their situations are examples of **status inconsistency,** *situations in which people rank differently (higher or lower) on certain stratification characteristics.* A person whose great wealth is known or suspected to have been acquired illegally probably will not become part of the accepted social establishment. An African-American physician, despite the high prestige of the profession, may in certain circles be denied a higher social position because of racial prejudice.

Factors Affecting Social Mobility In the United States, people believe that if individuals work hard enough they can become upwardly mobile—that is, they can become part of the next-higher social class. In fact, several other factors affect social mobility. For example, social structural factors, such as the state of the economy or the age distribution of the population, may either help or hinder social mobility.

During periods of economic expansion, the number of professional and technical jobs increases. Those white-collar jobs can often be filled by upwardly mobile members of other classes. When the supply of jobs increases, one group can no longer determine who will get all the jobs. Consequently, people from lower social classes who have the necessary education, talent, and skills are able to fill some of the positions without having any inside connections. During periods of economic contraction, however, the opposite is true. Getting a job depends on factors that go beyond talent or experience, such as family ties or personal friendships.

Demographic factors also affect upward mobility. When the number of people entering the workforce declines, it is easier for people with the right education, skills, and experience to get a job and advance than it is when the opposite is the case.

Societies also differ in terms of how much they encourage social mobility. The values and norms of American society encourage upward mobility. In fact, Americans are expected to try to succeed and better their status in life. We often look with contempt at those who have no desire to move up the social-class ladder or, worse yet, are downwardly mobile.

What is it that produces mobility? Level of education appears to be an extremely important factor. As would be expected, the greater the level of education attained by the children, the stronger the probability of their upward movement. It can even be said that the impact of education on occupational status is greater than the occupational status of the parents. It is difficult to separate these two factors, however, because the parents' occupations often have an impact on the amount of education received by the children.

The degree of social mobility in a society thus depends in great measure on the type of stratification system that exists.

Stratification Systems

Stratification can come about in two ways: (1) People can be assigned to societal roles, using as a basis for the assignment an ascribed status—some easily identifiable characteristic, such as gender, age, family name, or skin color, over which they have no control. This will produce the caste and estate systems of stratification. Or (2) people's positions in the social hierarchy can be based to some degree on their achieved statuses (see Chapter 5), gained through their individual, direct efforts. This is known as the class system.

The Caste System

The **caste system** is *a rigid form of stratification, based on ascribed characteristics such as skin color or family identity, that determines a person's prestige, occupation, residence, and social relationships.* People are born into and spend their entire lives within a caste, with little chance of leaving it.

Contact between castes is minimal and governed by a set of rules or laws. If interaction must take place, it is impersonal, and examples of the participants' superior or inferior status are abundant. Access to valued resources is extremely unequal. A set of religious beliefs often justifies a caste system. The caste system as it existed for centuries in India before the 1950s is a prime example of how this kind of inflexible stratification works.

The Hindu caste system, in its traditional form in India, consisted of four *varnas* (grades of being), each of which corresponded to a body part of the mythical Purusa, whose dismemberment was believed to have given rise to the human species. Purusa's mouth issued forth priests (Brahmans), and his arms gave rise to warriors (Kshatriyas). His thighs produced artisans and merchants (Vaishyas), and his feet brought forth menial laborers (Shudras). Hindu scripture holds that each person's *varni* is inherited directly from his or her parents and cannot change during the person's life (Gould, 1971).

Each *varna* had clearly defined rights and duties attached to it. Hindus believed in reincarnation of the soul (*karma*) and that, to the extent that an individual followed the norms of behavior of his or her *varna,* the state of the soul increased in purity, and the individual could expect to be born to a higher *varna* in a subsequent life. (The opposite was also true, in that failure to act appropriately according to the *varna* resulted in a person being born to a lower *varna* in the next life.)

This picture of India's caste system is complicated by the presence of thousands of subcastes, or *jatis*. Each of these *jatis* corresponds in name to a particular occupation (leather worker, shoemaker, cattle herder, barber, potter, and so on). Only a minority within each *jati* actually perform the work of that subcaste; the rest find employment when and where they can.

It is important to note that the Hindus have never placidly accepted the caste system. Scholars have frequently noted continuous changes during the centuries of the caste system's development. Even today, changes in the caste system are taking place. *Varnas* are all but nonexistent, and officially the Indian caste system is outlawed, although it still exists informally.

The Estate System

The **estate system** is *a closed system of stratification in which a person's social position is defined by law, and membership is determined primarily by inheritance.* An estate is a segment of a society that has legally established rights and duties. The estate system is similar to a caste system, but not as extreme. Some mobility is present, but by no means as much as exists in a class system.

The estate system of medieval Europe is a good example of how this type of stratification system works. The three major estates in Europe during the Middle Ages were the nobility, the clergy, and, at the bottom of the hierarchy, the peasants. A royal landholding family at the top had authority over a group of priests and the secular nobility, who were quite powerful in their own right. The nobility were the warriors; they were expected to be brave and give military protection to the other two estates. The clergy not only ministered to the spiritual needs of all the people but also were often powerful landowners as well. The peasants were legally tied to the land, which they worked in order to provide the nobles with food and wealth. In return, the nobles were supposed to provide social order, not only with their military strength, but also as the legal authorities who held court and acted as judges in disputes concerning the peasants who belonged to their land. The peasants had little freedom or economic standing, low social status, and almost no power. Just above the peasants was a small but growing group, the merchants and craftsmen. They operated somewhat outside the estate system in that, although they might achieve great wealth and political influence, they had little chance of moving into the estate of nobility or warriors. It is worth noting that it was this marginal group, which was less constricted by norms governing the behavior of the estates, that had the flexibility to gain power when the Industrial Revolution, starting in the eighteenth century, un-dermined the estate system. Individuals were born into one of the estates and remained there throughout their lives. Under unusual circumstances people could change their estate, as, for example, when peasants—using produce or livestock saved from their own meager supply or a promise to turn over a bit of land that by some rare fortune belonged to them outright—could buy a position in the church for a son or daughter. For most, however, social mobility was difficult and extremely limited because wealth was permanently concentrated among the landowners. The only solace for the poor was the promise of a better life in the hereafter (Vanfossen, 1979).

The Class System

A **social class** consists of *a category of people who share similar opportunities, similar economic and vocational positions, similar lifestyles, and similar attitudes and behaviors. A society that contains several different social classes and permits greater social mobility than a caste or estate system* is based on a **class system of stratification.** Class boundaries are maintained by limiting social interaction, intermarriage, and mobility into that class. Some form of class system is usually present in all industrial societies, whether they be capitalist or communist. Mobility is greater in a class system than in either a caste or an estate system. This mobility is often the result of an occupational structure that opens up higher-level jobs to anyone with the education and experience required. A class society encourages striving and achievement. Here in the United States we should find this concept familiar, for ours is basically a class society.

The Dimensions of Social Stratification

If we look closely, we will see that most of the valued attributes that produce stratification fall into three categories: economics, power, and prestige.

Economics

The total economic assets of an individual or a family are known as wealth. For people in the United States, wealth includes income, monetary assets, and various holdings that can be converted into money. These holdings include stocks, bonds, real estate, automobiles, precious metals and jewelry, and trusts.

Information on income and wealth in the United States shows that a high concentration of wealth remains in the hands of a relatively small number of people. This point is highlighted by the fact that the richest 1% of the American population owns about 22% of the nation's wealth. This figure was as high as 36% in 1929, and it illustrates dramatically the extent to which the nation's wealth is controlled by a very few (U.S. Department of Commerce, 1996).

Power

One of the most widely used definitions of power in sociology is a variation of one suggested by Max Weber. **Power** is *the ability to attain goals, control events, and maintain influence over others—even in the face of opposition.* In the United States, ideas about power often have their origins in the struggle for independence. It is a cliché of every Fourth of July speech that the colonists fought the Revolutionary War because of a desire to have a voice in how they were governed, particularly in how they were taxed. The colonists were also making revolutionary political demands on their own leaders by insisting that special conventions be elected to frame constitutions and that the constitutions be ratified by a vote of all free white males without regard to property ownership. In the past, governments had been founded on the power of religious leaders, kings, self-appointed conventions, or parliaments. It was the middle class's resolve for a voice in the decision-making process during the revolutionary period that succeeded in changing our thinking about political representation. The revolutionary period helped develop the doctrine that power in the United States should belong to the people.

Every society has highly valued experiences and material objects. It can be assumed that most people in society would like to have as great a share as possible of these experiences and objects. Those who end up having the most of what people want are then, by inference, the powerful (Domhoff, 1983).

Prestige

Prestige consists of *the approval and respect an individual or group receives from other members of society.* There are two types of prestige. To avoid confusion, we can call the first type esteem, which is potentially open to all. It consists of the appreciation and respect a person wins in his or her daily interpersonal relationships. Thus, for example, among your friends there are some who are looked up to for their outgoing personalities, athletic abilities, reliability in times of need, and so on.

The second form of prestige is much more difficult for many people to achieve. This is the honor that is associated with specific statuses (social positions) in a society. Regardless of personality, athletic ability, or willingness to help others, individuals such as Supreme Court justices, state governors,

Even though she is married to a former president of the United States, Hillary Rodham Clinton has also achieved a high level of prestige in her own right as U.S. Senator for New York.

physicians, physicists, and foreign service diplomats acquire prestige simply because they occupy these statuses. Access to prestigious statuses usually is difficult: Generally speaking, the greater the prestige a status has, the more difficult it is to gain it. For example, few positions carry as much prestige as that of president of the United States—and few positions are as hard to attain.

Occupational Stratification

Occupations are perhaps the most visible statuses to which prestige is attached in an industrial society. Table 7-1 shows the prestige ranking of selected occupations in the United States. These rankings, first undertaken in the 1940s by the National Opinion Research Center, have remained quite stable since then.

Since the 1960s women have had a dramatic impact on the American labor force. As of 1998, about 59.8% of women were working outside the home (*Statistical Abstract of the United States,* 1999, 2000).

The types of jobs held by working women have been changing, although a great deal of occupational segregation still exists. Certain occupations are heavily dominated by women. These include schoolteachers (75.3% female in 1998), cashiers (78.2%), librarians (83.4%), dental hygienists (99.1%), and secretaries (98.4%). In contrast, very few women are firefighters (2.5%), construction workers (2.0%), auto mechanics (0.8%), police officers and detectives (16.3%), and engineers (11.2%).

Even with the persistence of occupational segregation, the representation of women in a number of occupations is growing rapidly. These occupations include lawyers (28.5% in 1998, compared with

Table 7-1

Prestige Ratings of Various Occupations

OCCUPATION	PRESTIGE SCORE
Physician	86
College professor	78
Dentist	74
Architect	73
Lawyer	72
Psychologist	71
Clergy	69
High-school teacher	66
Sociologist	65
Accountant	65
Veterinarian	62
Registered nurse	62
Pharmacist	61
Airline pilot	61
Police officer	60
Dietitian	56
Librarian	54
Social worker	52
Realtor	49
Electrician	49
Secretary	46
Insurance agent	45
Carpenter	43
Bank teller	43
Farmer	40
Brick/stone mason	36
Bus driver	32
Truck driver	30
Sales clerk	29
Bartender	25
Waiter/Waitress	20
Janitor	16
Garbage collector	13

This chart shows how Americans have ranked the prestige of various occupations. Generally, the most prestigious jobs are those that require the greatest number of years of formal education and those that pay the highest income.

Sources: "Occupational Prestige in the United States Revisited: Twenty-five Years of Stability and Change," by K. Nakao & J. Treas, 1990, paper presented at the annual meeting of the American Sociological Association, Washington, DC; *General Social Surveys, 1972–1993: Cumulative Codebook* (pp. 927–945), by the National Opinion Research Center, 1993, Chicago: National Opinion Research Center; Prestige of Occupations, The Harris Poll, May 15–19, 1998.

5% in 1970), doctors (26.6% versus 10%), college and university professors (42.3% versus 29%), mathematical and computer scientists (28.9% versus 14%), and architects (17.5% versus 4%). As of 1998, women made up the majority of professional employees in the United States (*Statistical Abstract of the United States,* 1999, 2000).

One of the key issues determining labor force participation of women is the ability to combine work and family commitments. The United States

has been quite slow in alleviating problems in this area. Sweden, by contrast, has been quite innovative and has encouraged the entry of women into the labor force. Sweden has instituted a network of government-supported day-care centers designed to ease child-care responsibilities. The government has also produced legislation that makes it easier for women to work and have children. The Swedish parent can receive up to nine months maternity or paternity leave at 90% of full pay. They are also guaranteed a job on returning to work. The government further mandates that a worker will receive full-time pay for a shortened workday (6 hours) until a child's eighth birthday. Under these measures, 66% of all women work outside the home. This is the highest female labor force participation rate among the industrialized nations.

Theories of Stratification

Social philosophers have long tried to explain the presence of social inequality—that situation in which the very wealthy and powerful coexist with the poverty-stricken and powerless. In this section, we shall discuss the theories that try to explain this phenomenon.

The Functionalist Theory

Functionalism is based on the assumption that the major social structures contribute to the maintenance of the social system (see Chapter 1). The existence of a specific pattern in society is explained in terms of the benefits that society receives because of that situation. In this sense the function of the family is the socialization of the young, and the function of marriage is to provide a stable family structure.

The functionalist theory of stratification as presented by Kingsley Davis and Wilbert Moore (1945) holds that social stratification is a social necessity. Every society must select individual members to fill a wide variety of social positions (or statuses) and then motivate those people to do what is expected of them in these positions—that is, fulfill their role expectations. For example, our society needs teachers, engineers, janitors, police officers, managers, farmers, crop dusters, assembly-line workers, firefighters, textbook writers, construction workers, sanitation workers, chemists, inventors, artists, bank tellers, athletes, pilots, secretaries, and so on. To attract the most talented individuals to each occupation, society must set up a system of differential rewards based on the skills needed for each position.

If the duties associated with the various positions were all equally pleasant, . . . all equally important to social survival, and all equally in need of the same ability or talent, it would make no difference who got into which positions. . . . But actually it does make a great deal of difference who gets into which positions, not only because [some] positions are inherently more agreeable than others, but also because some require special talents or training and some are functionally more important than others. Also, it is essential that the duties of the positions be performed with the diligence that their importance requires. Inevitably, then, a society must have, first, some kind of rewards that it can use as inducements, and, second, some way of distributing these rewards differentially according to positions. The rewards and their distribution become part of the social order, and thus give rise to stratification. (Davis & Moore, 1945)

According to Davis and Moore, (1) different positions in society make different levels of contributions to the well-being and preservation of society, (2) filling the more complex and important positions in society often requires talent that is scarce and has a long period of training, and (3) providing unequal rewards ensures that the most talented and best-trained individuals will fill the roles of greatest importance. In effect, Davis and Moore believe that those people who are rich and powerful are at the top because they are the best qualified and are making the most significant contributions to the preservation of society (Zeitlin, 1981).

Many scholars, however, disagree with Davis and Moore and their arguments generally take two forms. The first is philosophical and questions the morality of stratification. The second is scientific and questions its functional usefulness. Both criticisms share the belief that social stratification does more harm than good, that it is dysfunctional.

Neither functionalist theory nor conflict theory can fully explain why media people earn very large sums of money.

The Immorality of Social Stratification   On what grounds, one might ask, is it morally justifiable to give widely different rewards to different occupations, when all occupations contribute to society's ongoing functioning? How can we decide which occupations contribute more? After all, without assembly-line workers, mail carriers, janitors, auto mechanics, nurse's aides, construction laborers, truck drivers, secretaries, shelf stockers, sanitation workers, and so on, our society would grind to a halt. How can the multimillion-dollar-a-year incomes of a select few be justified when the earnings of 11.8% of the American population fall below the poverty level, and many others have trouble making ends meet? Why are the enormous resources of our society not more evenly distributed?

Many people find the moral arguments against social stratification convincing enough. However, there are other grounds on which stratification has been attacked—namely, that it is destructive for individuals and society as a whole.

The Neglect of Talent and Merit   Regardless of whether social stratification is morally right or wrong, many critics contend that it undermines the very functions that its defenders claim it promotes. A society divided into social classes (with limited mobility among them) is deprived of the potential contributions of many talented individuals born into the lower classes. From this point of view, it is not necessary to do away with differences in rewards for different occupations. Rather, it is crucial to put aside all the obstacles to achievement that currently handicap the children of the poor.

Barriers to Free Competition   It can also be claimed that access to important positions in society is not really open. That is, those members of society who occupy privileged positions allow only a small number of people to enter their circle. Thereby, shortages are created artificially. This, in turn, increases the perceived worth of those who are in the important positions. For example, the American Medical Association (AMA) is a wealthy and powerful group that exercises great control over the quality and quantity of physicians available to the American public. Historically, the AMA has had a direct influence on the number of medical schools in the United States and thereby the number of doctors produced each year, effectively creating a scarcity of physicians. A direct result of this influence is that medical-care costs and physicians' salaries have increased more rapidly than has the pace of inflation.

This situation is beginning to change, however. The allure of high earnings that attracts many new doctors, along with changing demographic characteristics, could mean a surplus of physicians in the future. In addition, as more and more doctors fight for the same patient dollars, and as health maintenance organizations (HMOs) try to control costs, earnings may begin to suffer. Thus, while barriers to free competition exist in our society, often the marketplace overrules them in the end.

Functionally Important Jobs When we examine the functional importance of various jobs, we become aware that the rewards attached to jobs do not necessarily reflect the essential nature of the functions. Why should a Hollywood movie star receive an enormous salary for starring in a film and a child-protection worker receive barely a living wage? It is difficult to prove empirically which positions are most important to society or what rewards are necessary to persuade people to want to fill certain positions.

Conflict Theory

As we saw, the functionalist theory of stratification assumes society is a relatively stable system of interdependent parts in which conflict and change are abnormal. Functionalists maintain that stratification is necessary for the smooth functioning of society. Conflict theorists, in contrast, see stratification as the outcome of a struggle for dominance. Current views of the conflict theory of stratification are based on the writings of Karl Marx. Later, Max Weber developed many of his ideas in response to Marx's writings.

Karl Marx In order to understand human societies, Karl Marx believed one must look at the economic conditions centering around producing the necessities of life. Stratification emerged from the power struggles for scarce resources.

> The history of all hitherto existing society is the history of class struggles. [There always has been conflict between] freeman and slave, patrician and plebeian, lord and serf, guild-master and journeyman, in a word, oppressor and oppressed. (Marx & Engels, 1961)

Those groups who own or control the means of production within a society obtain the power to shape or maintain aspects of society that favor their interests. They are determined to maintain their advantage. They do this by setting up political structures and value systems that support their position. In this way the legal system, the schools, and the churches are shaped in ways that benefit the ruling class. As Marx and his collaborator Friedrich Engels put it, "The ruling ideas of each age have always

been the ideas of its ruling class" (1961). Thus the pharaohs of ancient Egypt ruled because they claimed to be gods. In the first third of the twentieth century, America's capitalist class justified its position by misusing Charles Darwin's theory of evolution: The capitalists adhered to the view—called social Darwinism (see Chapter 1)—that those who rule do so because they are the most "fit" to rule, having won the evolutionary struggles that promote the "survival of the fittest."

Marx was most interested in the social impact of the capitalist society that was based on industrial production. In a capitalist society, there are two great classes: the bourgeoisie, the owners of the means of production or capital, and the proletariat, or working class. Those in the working class have no resources other than their labor, which they sell to the capitalists. In all class societies, there is exploitation of one class by another.

Marx believed the moving force of history was class struggle, or class conflict. This conflict grows out of differing class interests. As capitalism develops, two conflicting trends emerge. On the one hand, the capitalists try to maintain and strengthen their position. The exploitative nature of capitalism is seen when the capitalists pay the workers a bare minimum wage, below the value of what the workers actually produce. The remainder is taken by the capitalists as profit and adds to their capital. This capital, which Marx said rightfully belongs to the workers, is then used to build more factories, machines, or anything else to produce more goods. As Marx saw it, "Capital is dead labor, that vampire-like, only lives by sucking living labor, and lives the more, the more labor it sucks" (Marx, 1906).

Eventually, in the face of continuing exploitation, the working classes find it in their interest to overthrow the dominant class and establish a social order more favorable to their interests. Marx believed that with the proletariat in power, class conflict would finally end. The proletariat would have no class below it to exploit. The final stage of advanced communism would include an industrial society of plenty, where all could live in comfort.

Marx was basically a materialist. He believed that people's lives are centered on how they deal with the material world. The key issue is how wealth is distributed among the people. Wealth can be distributed in at least four ways:

1. *To each according to need.* In this kind of system, the basic economic needs of all the people are satisfied. These needs include food, housing, medical care, and education. Extravagant material possessions are not basic needs and have no place in this system.

2. *To each according to want.* Here, wealth will be distributed according to what people desire and request. Material possessions beyond the basic needs are included.

3. *To each according to what is earned.* People who live according to this system become the source of their own wealth. If they earn a great deal of money, they can lavish extravagant possessions upon themselves. If they earn little, they must do without.

4. *To each according to what can be obtained—by whatever means.* Under this system, everyone ruthlessly attempts to acquire as much wealth as possible without regard for the hardships that might be brought on others because of these actions. Those who are best at exploiting others become wealthy and powerful, and the others become the exploited and poor (Cuzzort & King, 1980).

In Marxist terms, the first of these four possibilities is what would happen in a socialist society. Although many readers will believe that the third possibility describes U.S. society (according to what is earned), Marxists would say that a capitalist society is characterized by the last choice—the capitalists obtain whatever they can get in any possible way.

Max Weber Weber expanded Marx's ideas about class into a multidimensional view of stratification. Weber agreed with Marx on many issues related to stratification, including the following:

1. Group conflict is a basic ingredient of society.

2. People are motivated by self-interest.

3. Those who do not have property can defend their interests less well than those who have property.

4. Economic institutions are of fundamental importance in shaping the rest of society.

5. Those in power promote ideas and values that help them maintain their dominance.

6. Only when exploitation becomes extremely obvious will the powerless object.

From those areas of agreement, Weber went on to add to and modify many of Marx's basic premises. Weber's view of stratification went beyond the material or economic perspective of Marx. He included status and power as important aspects of stratification as well.

Weber was not interested in society as a whole but in the groups formed by self-interested individuals who compete with one another for power and privilege. He rejected the notion that conflict be-

tween the bourgeoisie and proletariat was the only, or even the most important, conflict relationship in society.

Weber believed there were three sources of stratification: economic class, social status, and political power. Economic classes arise out of the unequal distribution of economic power, a point on which Marx and Weber agreed. Weber went further, however, maintaining that social status is based on prestige or esteem—that is, status groups are shaped by lifestyle, which is in turn affected by income, value system, and education. People recognize others who share a similar lifestyle and develop social bonds with those who are most like themselves. From this inclination comes an attitude of exclusivity: Others are defined as being not as good as those who are a part of the status group. Weber recognized that there is a relationship between economic stratification and social-status stratification. Typically, those who have a high social status also have great economic power.

Inequality in political power exists when groups are able to influence events in their favor. For example, representatives of large industries lobby at the state and federal levels of government for legislation favorable to their interests and against laws that are unfavorable. Thus, the petroleum industry pushed for lifting restrictions on gasoline prices and the auto industry lobbied for quotas on imported cars. In exchange for "correct" votes, a politician is often promised substantial campaign contributions from wealthy corporate leaders or endorsement and funding by a large labor union whose members' jobs will be affected by the government's decisions. The individual consumer who will pay the price for such political arrangements is powerless to exert any influence over these decisions.

Class, status, and power, though related, are not the same. One can exist without the others. To Weber, they are not always connected in some predictable fashion, nor are they always tied to the economic mode of production. An "aristocratic" southern family may be in a state that is often labeled genteel poverty, but the family name still elicits respect in the community. This kind of status sometimes is denied to the rich, powerful labor leader whose family connections and school ties are not acceptable to the social elite. In addition, status and power are often accorded to those who have no relationship to the mode of production. Mother Theresa, for example, controlled no industry, nor did she have any great personal wealth; yet her influence was felt by heads of state the world over.

Whereas Marx was somewhat of an optimist in that he believed that conflict, inequality, and exploitation eventually could be eliminated in future societies, Weber was much more pessimistic about the potential for a more just and humane society.

Modern Conflict Theory

Conflict theorists assume that people act in their own self-interest in a material world in which exploitation and power struggles are prevalent. There are five aspects of modern conflict theory:

1. Social inequality emerges through the domination of one or more groups by other groups. Stratification is the outgrowth of a struggle for dominance in which people compete for scarce goods and services. Those who control these items gain power and prestige. Dominance can also result from the control of property, as others become dependent on the landowners.

2. Those who are dominated have the potential to express resistance and hostility toward those in power. While the potential for resistance is there, it sometimes lies dormant. Opposition may not be organized, because the subordinated groups may not be aware of their mutual interests. They may also be divided because of racial, religious, or ethnic differences.

3. Conflict will most often center on the distribution of property and political power. The ruling classes will be extremely resistant to any attempts to share their advantages in these areas. Economic and political power are the most important advantages in maintaining a position of dominance.

4. What are thought to be the common values of society are really the values of the dominant groups. The dominant groups establish a value system that justifies their position. They control the systems of socialization, such as education, and impose their values on the general population. In this way, the subordinate groups come to accept a negative evaluation of themselves and to believe those in power have a right to that position.

5. Because those in power are engaged in exploitative relationships, they must find mechanisms of social control to keep the masses in line. The most common mechanism of social control is the threat—or the actual use—of force, which can include physical punishment or the deprivation of certain rights. However, more subtle approaches are preferred. By holding out the possibility of a small amount of social mobility for those who are deprived, the power elite will try to induce them to accept the system's basic assumptions. Thus, the subordinate masses will come to believe that by behaving according to the rules, they will gain a better life (Vanfossen, 1979).

The Need for Synthesis

Any empirical investigation will show that neither the functionalist theory nor the conflict theory of stratification is entirely accurate. This does not mean that both are useless in understanding how stratification operates in society. Ralf Dahrendorf (1959) suggested that the two really are complementary rather than opposed. (Table 7-2 compares the two.) We do not need to choose between the two, but instead should see how each is qualified to explain specific situations. For example, functionalism may help explain why differential rewards are needed to serve as an incentive for a person to spend many years training to become a lawyer. Conflict theory

Table 7-2

Functionalist and Conflict Views of Social Stratification: A Comparison

THE FUNCTIONALIST VIEW	THE CONFLICT VIEW
1. Stratification is universal, necessary, and inevitable.	1. Stratification may be universal without being necessary or inevitable.
2. Social organization (the social system) shapes the stratification system.	2. The stratification system shapes social organizations (the social system).
3. Stratification arises from the societal need for integration, coordination, and cohesion.	3. Stratification arises from group conquest, competition, and conflict.
4. Stratification facilitates the optimal functioning of society and the individual.	4. Stratification impedes the optimal functioning of society and the individual.
5. Stratification is an expression of commonly shared social values.	5. Stratification is an expression of the values of powerful groups.
6. Power usually is distributed legitimately in society.	6. Power usually is distributed illegitimately in society.
7. Tasks and rewards are allocated equitably.	7. Tasks and rewards are allocated inequitably.
8. The economic dimension is subordinate to other dimensions of society.	8. The economic dimension is paramount in society.
9. Stratification systems generally change through evolutionary processes.	9. Stratification systems often change through revolutionary processes.

Source: "Some Empirical Consequences of the Davis-Moore Theory of Stratification," by A. L. Stinchcombe, 1969, in J. L. Roach, L. Gross, & O. R. Gursslin, eds., *Social Stratification in the United States* (p. 55), Englewood Cliffs, NJ: Prentice-Hall.

would help explain why the offspring of members of the upper classes study at elite institutions and end up as members of prestigious law firms, while the sons and daughters of the middle and lower classes study at public institutions and become overworked district attorneys.

Social Class in the United States

There is little agreement among sociologists about how many social classes exist in the United States and what their characteristics may be. However, for our purposes here, we will follow a relatively common approach of assuming that there are five social classes in the United States: upper class, upper-middle class, lower-middle class, working class, and lower class (Rossides, 1990). Table 7-3 presents stratification data for each of these classes. (For a discussion of movement between social classes, see "Sociology at Work: How Easy Is It to Change Social Class?")

The Upper Class

Members of the upper class have great wealth, often going back for many generations. They recognize one another, and are recognized by others, by reputation and lifestyle. They usually have high prestige and a lifestyle that excludes those of other classes. Members of this class often have an influence on so-

ciety's basic economic and political structure. The upper class usually isolates itself from the rest of society by residential segregation, private clubs, and private schools. Historically they have been Protestant, especially Episcopalian or Presbyterian. This is less true today. It is estimated that in the United States, the upper class consists of from 1 to 3% of the population.

Since the 1970s, the upper class also has come to include society's new entrepreneurs—people who have often made many millions, and sometimes billions, in business. In many respects these people do not resemble the upper class of the past. Included are people like William Gates, the 46-year-old chief executive officer of Microsoft Corporation, who has a $60 billion net worth, and Larry Ellison, the founder of Oracle Corporation, whose net worth exceeds $50 billion.

Not all billionaires lead opulent lifestyles, and many in the upper class do not approve of displaying wealth. For many, the money is merely a way of keeping score of how well they are doing at their chosen endeavors.

The Upper-Middle Class

The upper-middle class is made up of successful business and professional people and their families. They are usually just below the top in an organizational hierarchy, but still command a reasonably high income. Many aspects of their lives are dominated by their careers, and continued success in this

Table 7-3

Social Stratification in the United States by Occupation

CLASS	OCCUPATION	EDUCATION	CHILDREN'S EDUCATION
Upper class	Corporate ownership; upper-echelon politics; honorific positions in government and the arts	Liberal arts education at elite schools	College and postcollege
Upper-middle class	Professional and technical fields; managers; officials; proprietors	College and graduate school training	College and graduate school training
Lower-middle class	Clerical and sales positions; small-business semi-professionals; farmers	High school; some college	Option of college
Working class	Skilled and semiskilled manual labor; crafts-people; foremen; non-farm workers	Grade school; some or all of high school	High school; vocational school
Lower class	Unskilled labor and service work; private household work and farm labor	Grade school; semi-illiterate	Little interest in education; high school dropouts

Source: Adapted from *Statistical Abstract of the United States: 1981,* U.S. Bureau of the Census, 1981, Washington, DC: U.S. Government Printing Office.

Some members of the upper class can be identified by their dress or behavior. Others wish to remain inconspicuous.

area is a long-term consideration. These people often have a college education, own property, and have a savings reserve. They usually live in comfortable homes in the more exclusive areas of a community, are active in civic groups, and carefully plan for the future. They are very likely to belong to a church. The most common denominations represented are Presbyterians, Episcopalians, Congregationalists, Jews, and Unitarians. In the United States, 10 to 15% of the population falls into this category.

A large percentage of the new upper-middle class are two-income couples, both of whom are college-educated and employed as corporate executives, high government officials, business owners, or professionals. These relatively affluent individuals are changing the face of many communities. They are gentrifying rundown city neighborhoods with their presence and their money.

The Lower-Middle Class

The lower-middle class shares many characteristics with the upper-middle class, but its members have not been able to achieve the same kind of lifestyle because of economic or educational shortcomings. Usually high school graduates with modest incomes, they are semiprofessionals, clerical and sales workers, and upper-level manual laborers. They emphasize respectability and security, have some savings, and are politically and economically conservative. They often are dissatisfied with their standard of living, jobs, and family incomes. They are likely to be represented among the Protestant denominations such as Baptists, Methodists, and Lutherans, or they might be Catholic or Greek Orthodox. They make up 25 to 30% of the United States population.

The Working Class

The working class is made up of skilled and semi-skilled laborers, factory employees, and other blue-collar workers. These are the people who keep the country's machinery going. They are assembly-line workers, auto mechanics, and repair personnel. They are the most likely to be buffeted by economic downturns. More than half belong to unions.

Working-class people live adequately but with little left over for luxuries. They are less likely to

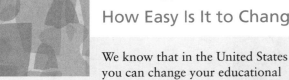

Sociology at Work

How Easy Is It to Change Social Class?

We know that in the United States you can change your educational and financial circumstances through hard work. Is this the same as changing your social class? Or is that a harder job that requires many trade-offs? R. Todd Erkel grew up in a working-class family in Pittsburgh. In the following excerpt, he writes about how the newly middle class often feel uncomfortable, adrift, and profoundly alone as they make their journey to a new culture.

When I decided to go to college, my mother and father offered the only advice they could: "Well," they said, "we hope you know what you're doing." But implicit in their lukewarm endorsement was the obvious truth: I didn't know—any more than they did. So, without a guidance counselor, or parents versed in the calculus of financial aid and college applications, I proceeded arbitrarily, applying to colleges whose names or looks appealed to me. I eventually settled, for no particular reason, on Penn State.

Having applied too late to attend orientation, I arrived in State College, Pennsylvania, with what I soon learned was more than a clothes problem. My credentials—an impressive grade point average and high test scores—gave my application a veneer of promise. But there was no measure for the things I didn't know, nothing to suggest I might have to learn everything about this new world of college from scratch. The system, like me, was blind to the ways in which my working-class background left me unprepared for this new world. It wasn't just poise or spending money that I lacked. Everything from my colloquial speech to my primitive social skills to my wardrobe drew a discreet line between me and my new peer group.

Adrift in this community of 60,000 not-so-kindred souls, I looked to the only place I knew

of to place the blame for my ineptitude—inside, with myself. Overwhelmed by feelings of alienation and worthlessness, I quit. In retrospect, I wonder why nobody—if not my parents, then a teacher or college counselor—could have foreseen my difficult transition to college; not just from one phase of education to another, but from one set of cultural assumptions to another, one entire world to another. I wish, too, that I could have let the full extent of my alienation be known.

Even at the University of Pittsburgh, where I eventually transferred and where the presence of students from working-class backgrounds similar to mine was plain to see, the issue of class remained eerily unspoken. Though part of me knew better, I could not escape the crippling feeling that I remained alone in my bewilderment.

Quiet and hiding in the back corner of a college classroom, I began to make the connection between my feelings of shame and the working-class misgivings toward education I had long ago internalized. For while lip service is paid at every class level in America to the idea that education is the route to a better life, the reality in most working-class homes is that knowledge counts for less than good behavior. The question my parents asked every day when I came home from school was not "What new thing did you learn today?" but "Did you get into any trouble?"

The working-class experience makes the child particularly vulnerable to low self-expectations. Before I could sing the alphabet, I knew something of what it felt like to be my parents: low-achieving, poorly spoken, lacking confidence, afraid to challenge authority, reluctant to ask for help, willing to accept their situation, content to do without.

The message received by children whose parents have battled with the world and come away

vote than the higher classes, and they feel politically powerless. Although they have little time to be involved in civic organizations, they are very much involved with their extended families. The families are likely to be patriarchal with sharply segregated sex roles. They stress obedience and respect for elders. Many of them have not finished high school. The religious makeup is similar to that of the lower-middle class. They represent 25 to 30% of the United States population.

The Lower Class

These are the people at the bottom of the economic ladder. They have little in the way of education or occupational skills and consequently are unemployed or underemployed. Lower-class families often have many problems, including broken homes, illegitimacy, criminal involvement, and alcoholism. Members of the lower class have little knowledge of world events, are not involved with their communities, and

feeling defeated is that they are better off not even trying. A pervasive feeling of helplessness hangs over the working-class house like the second-hand smoke that passes silently from parent to child.

Embracing the promise of an education requires working-class children to construct an inner sense of themselves that is radically different from that of their parents, siblings, and friends, to betray their allegiance to the only source of identity and support they have ever known. At each crossroad, and with every success, I became more aware of the dichotomy—the ways in which my education simultaneously would provide me options and distance me from the life I trusted.

A decade later, the anger I have long felt toward my parents has slowly faded. I realize now that the gifts I so desperately wanted from them (an easy self-confidence and a deep well of optimism) were not theirs to give. Instead, they handed their children a promise, visibly broken, in the hope that we might know better than they did how to make it work. In America, the illusion of free and open passage between classes is preached with religious zeal. But parents who wish for something better for their children must struggle against more than an incomplete education and economic deprivation. They must confront the truth behind the myth of making it in America: The land of opportunity is also the land of persistent class structures and struggles. And though many try, most people never rise very far from the socioeconomic level into which they are born.

Without an awareness of their experience of class discrimination, working-class kids will continue to absorb their parents' shame. They will continue to view the world—through their parents' eyes—as a malevolent force to be approached with rightful caution.

Without some explanation for the feelings present in the home, children internalize their parents' sense of powerlessness. Their unexpressed fears quietly sap the strength of the family. My middle-class friends know enough about the destructive force of class to see me as an exception: a triumph of will over environment. But we sidestep any discussion of a class system in America, afraid maybe of the feelings such awareness might stir. It's easier to assume that we now all occupy one vast, comprehensive middle class.

What I don't mention, and what others don't see, is that I often feel more lost than ever, caught between two widely separated social rungs, never sure whether I should forge ahead or fall back, uncertain whether either option really is mine. I have learned to pose in the middle-class culture, but at a price. I live most of the time on borrowed instincts, afraid to trust that part of the working class I still carry inside.

Today, my family is still wondering what on earth I will one day do for a living and uncomfortable with my willingness to expose the personal details of my life on paper where anyone might read them . . . I am engaged with the world in a way they have never known, and still don't altogether trust. Looking back, the memory of growing apart from the people and habits I know and love stirs a swirl of feelings. I still yearn to believe that my parents know what is best. The adult I've become appreciates why such knowledge eluded them. I understand more clearly the gain of my leaving their world. I'm only now willing to consider the loss.

Source: "The Mighty Wedge of Class" in *Family Therapy Networker*, by R. Todd Erkel, July/August, 1994, pp. 45–47.

---

usually do not identify with other poor people. They have low voting rates. Because of a variety of personal and economic problems, they often have no way of improving their lot in life. For them, life is a matter of surviving from one day to the next. Their dropout rate from school is high, and they also have the highest rates of illiteracy of any of the groups. The lower class is disproportionately African American and Hispanic, but it is not race or even poverty that defines it. Rather, it is a set of characteristics and conditions that are part of a broader lifestyle. Lower-class people often belong to fundamentalist or revivalist sects. About 15 to 20% of the population falls into this class.

Money, power, and prestige are distributed unequally among these classes. However, a desire to advance and achieve success is shared by members of all five classes, which makes them believe that the system is just and that upward mobility is open to all. Therefore, they tend to blame themselves for

Members of the upper class often engage in activities that exclude those of other classes.

lack of success and for material need (Vanfossen, 1979).

Income Distribution

The U.S. Census Bureau has published annual estimates of the distribution of family income since 1947. Those figures show a highly unequal distribution of wealth. In 1999, for example, the richest one-fifth of families earned 49.4% of the total income for the year while the poorest one-fifth earned only 3.6% (U.S. Bureau of the Census, 2000).

Without further elaboration, this information allows us to imagine that the richest one-fifth of families consists of millionaire real estate moguls, Wall Street professionals, and CEOs of major companies. The image is somewhat misleading. In 1999, the richest one-fifth included all families with incomes of $79,375 or more (see Table 7-4). Keep in mind that this is a family income derived from jobs held by husbands, wives, and all other family members. Family incomes for the richest 5% of the population begin at $142,021 (U.S. Bureau of the Census, 2000).

This is not to imply, though, that there is not a significant difference in the distribution of wealth in the United States. Income, however, is only part of the picture. Total wealth—in the form of stocks, bonds, real estate, and other holdings—is even more unequally distributed. The richest 20% of American

families owns more than three-fourths of all the country's wealth. In fact, the richest 5% of all families owns more than half of America's wealth. There is also evidence to support the old adage that "The rich get richer, and the poor get poorer." The number of people in poverty grew from 24.5 million in 1978 to 32.3 million in 1999 (United States Bureau

Table 7-4

Family Income Distribution, 1999

	INCOME AMOUNT	PERCENTAGE OF TOTAL INCOME RECEIVED BY ALL FAMILIES
Richest 5%	$142,021* and above	21.5
Fifth quintile	79,375 and above	49.4
Fourth quintile	50,520– 79,374	23.2
Third quintile	32,000– 50,519	14.9
Second quintile	17,197– 31,999	8.9
First quintile	17,196 or below	3.6

*Richest 5% included in the fifth quintile.

Source: U.S. Bureau of the Census, Current Population Survey, March 2000.

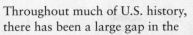

Controversies in Sociology

Is the Income Gap Between the Rich and the Poor a Problem?

Throughout much of U.S. history, there has been a large gap in the amount of wealth held by America's richest and poorest citizens. During the 19th and 20th centuries, the richest 1% of U.S. adults held between 20 and 30% of all private wealth in the country (see Table 7-4).

Around the start of the 1970s, the income gap began to grow—slowly at first, and then more rapidly during the early 1980s. The gap between rich and poor Americans is at the highest level since the late 1970s. The top wage earners have seen large increases in their real incomes, while those at the bottom have actually experienced losses in real income.

One reason for the increasing wage gap was a decline in low-skill jobs based in the United States as manufacturers moved their jobs to other countries where labor was cheaper. It has become harder for people without special skills to land jobs that pay relatively high wages.

Does the U.S. government have an obligation to shrink the income gap? Or is a substantial gap between the highest- and the lowest-paid workers simply a natural part of a capitalist economy? Some people believe the income gap is trapping the lowest-paid workers in poverty. Others downplay the importance of the gap in wages. They say it would be harmful to the economy as a whole to interfere with the distribution of income. People who earn higher incomes have worked hard to attain their positions, they say, and their high wages are a just reward for the innovations and expertise they bring to society.

Others believe the situation is not as dire as some claim because the overall standard of living in the United States has increased in recent decades. Even people at the lowest rung of the wage scale have more material possessions and more money to spend than did their counterparts in earlier generations. In addition, in the United States, they claim, the truly motivated workers can lift themselves into higher income brackets.

It appears the income gap is leading to a society in which people segregate themselves along socioeconomic class lines. People have physically separated themselves from the less fortunate by moving to wealthy areas far away from the problems associated with poverty.

Government approaches to dealing with the income gap include the Earned Income Tax Credit, which provides tax rebates for low-paid workers in nearly 20 million households, and increases in the minimum wage. The next decade will tell us whether these programs are enough to address the problem.

---

of the Census, 2000). (See "Controversies in Sociology: Is the Income Gap Between the Rich and the Poor a Problem?")

Poverty

On a very basic level, poverty refers to a condition in which people do not have enough money to maintain a standard of living that includes the basic necessities of life. Depending on which official or quasi-official approach we use, it is possible to document that anywhere from 14 million to 45 million Americans are living in poverty. The fact is, we really do not have an unequivocal way of determining how many poor people there are in the United States.

Poverty seems to be present among certain groups much more than among others. In 1999, 11.8% of all Americans lived below the poverty level. While 7.7% of all whites were living in poverty, 23.6% of all blacks and 22.8% of those of Hispanic origin fell into this group. People living in certain regions of the United States are much more likely to live in poverty than those living elsewhere. For example, the poverty rates in Louisiana and New Mexico are more than twice those of Maryland (U.S. Bureau of the Census, 2000).

It is also a fact that the level of poverty in rural areas actually is higher than that in our cities. Thirty percent of the nation's poor live in rural America—a reality that is often overlooked by those who focus only on the problems of the urban poor. Even worse, the economic conditions of the rural poor are expected to deteriorate along with the decline of unskilled manufacturing jobs and changes in the mining, agricultural, and oil industries. The problem is especially acute for those with little education or marketable job skills.

The Feminization of Poverty

Different types of families also have different earning potentials. In 1999, a family with both a husband and wife present had a median income of $56,676. For a male householder without a wife the figure was $37,396, and for a female householder without a

Poverty statistics may not adequately count the homeless, who have no permanent address.

husband it was $23,732. This has caused some sociologists to note the "feminization of poverty," a phrase referring to the disproportionate concentration of poverty among female-headed families.

The real impact of these differences becomes even more striking when we look at single women with children. Whereas 11.8% of all people were below the poverty line in 1999, 37.4% of all single women with children were living in poverty. The percentages would be even higher if we just included single African American (47.2%) and Hispanic women (46.4%) with children (Current Population Survey, September 2000).

If present trends continue, 60% of all children born today will spend part of their childhood in a family headed by a mother who is divorced, separated, unwed, or widowed. There is substantial evidence that women in such families are often the victims of poverty. Almost half of all female-headed families with children younger than 18 live below the poverty line.

Not all female-headed families are the same, however. The feminization of poverty is both not as bad as, and much worse than, the previous statement suggests. Families headed by divorced mothers are doing better than the 50% figure, while families headed by never-married mothers are doing much worse.

What accounts for the fact that never-married mothers are so much poorer than their divorced counterparts? Seventy percent of all out-of-wedlock births occur to young women between the ages of 15 and 24. They are, on average, 10 years younger than divorced mothers. Never-married mothers are also, on average, much less educated. The gender gap in poverty rates is greatest at low educational levels. It is much narrower among those with a high school diploma and practically nonexistent among those with a college education. Single mothers without a

high school diploma often have difficulty finding a job that pays enough to cover child-care costs, leading to a dependence on welfare programs (O'Hare, 1996).

How Do We Count the Poor?

To put a dollar amount on what constitutes poverty, the federal government has devised a poverty index of specific income levels, below which people are considered to be living in poverty. Many people use this index to determine how many poor people there are in the United States. According to the index, the poverty level for a family of four in 1999 was $17,029 (see Table 7-5). The poverty-level income figures do not include income received in the form of noncash benefits, such as food stamps, medical care, and subsidized housing.

Table 7-5

Income Levels Below Which Individuals and Families Are Considered to Be Living in Poverty (1999)

SIZE OF UNIT	INCOME
1 person	
Under 65	$ 8,501
Over 65	7,990
2 people	
Under 65	11,214
Over 65	10,075
3 people	13,290
4 people	17,029
5 people	20,127
6 people	22,727
7 people	25,912
8 people	28,967
9 people	34,417

Source: U.S. Bureau of the Census, Current Population Survey, March, 2000.

Seventy percent of all out-of-wedlock births occur to young women between the ages of 15 to 24. Only 53% of them have a high-school diploma. This situation has contributed to the feminization of poverty.

The official definition of poverty used for the index was developed by the Social Security Administration in 1964. It was calculated first by estimating the national average dollar cost of a frugal but adequate diet; then, because a 1955 study found that families of three or more people spent about one-third of their income on food, food costs were multiplied by three to estimate how much total cash income was needed to cover food and other necessities. The poverty index was not originally intended to certify that any individual or family was in need. In fact, the government specifically has warned against using the index for administrative use in any specific program. Despite this warning, people continue to use, or misuse, the poverty index and variations of it for a variety of purposes for which it was not intended. For example, those wanting to show that current government programs are inadequate for the poor will try to inflate the numbers of those living in poverty. Those trying to show that government policies are adequate for meeting the needs of the poor will try to show that the number of poor people is decreasing.

Those who think the poverty index overestimates the poor offer three major criticisms. First, When the federal government developed the poverty index in 1964, about one-quarter of federal welfare benefits were in the form of goods and services. Today, noncash benefits account for about two-thirds of welfare assistance. For example, about 20 million people now receive food stamps, which is not considered income under existing poverty-index rules. Complicating the issue further, the market value of in-kind benefits—such as housing subsidies, school lunch programs, and health-care services, among others—has been multiplied by a factor of 40. Some suggest that if the noncash benefits were counted as income, the poverty rate would be 3 percentage points lower.

Second, the poverty measure looks only at income, not assets. If the value of a home or other assets were included, the poverty rate would also be lower.

Third, food typically accounts for a considerably smaller proportion of family expenses today than it did previously. If we were to try to develop a poverty index today, we would probably have to multiply minimal food costs by a factor of five instead of three.

Those who think the poverty figures underestimate the poor have their criticisms also. First, they point out that money used to pay taxes, alimony, child support, health care, or work related expenses should be excluded when considering assets because these sums cannot be used to buy food or other necessities.

Second, there is no geographic cost-of-living adjustment. The federal government uses the same poverty-level figures for every part of the country. That means the poverty threshold is the same in rural Mississippi as it is in New York City.

Third, many believe the poverty threshold is unrealistically low. Rather than use an absolute number, poverty status should be determined by comparing a person's financial situation with that of the rest of society.

The poverty index has become less and less meaningful. However, its continued existence over all these years has given it somewhat of a sacred character. Few people who cite it know how it is calculated, and they choose to assume it is a fair measure for determining the number of poor in the country. The poverty index has never been a sufficiently precise indicator of need to make it an indisputable test of which individuals and families were poor and which were not.

The number of people living in poverty also is distorted by the fact that the Census Bureau's Current Population Survey is derived from households. It excludes all those people who do not live in traditional housing, specifically the growing numbers of the homeless, estimated at anywhere from 350,000 to 1 million. People in nursing homes and other types of institutions also are not included in the poverty figures because of surveying techniques.

This is not to downplay the number of poor people in the United States. The basic fact is that trying to determine how many poor people there are depends on whom you ask and what type of statistical maneuvering is involved (see Figure 7-1).

Myths About the Poor

We are presented with differing views on poverty and what should be done about it. One side argues that more government aid and the creation of jobs is needed to combat changes in the employment needs of the national economy. The other side contends that government assistance programs launched with the War on Poverty in the mid-1960s have encouraged many of the poor to remain poor and should be eliminated for the able-bodied poor of working age (Murray, 1994).

Our perceptions of the poor shape our views of the various government programs available to help them. It is important that we have a clear understanding of who the poor are in order to direct public policy intelligently. There are a number of common myths about the poor that many Americans believe. Let us try to clear some of them up.

Myth 1: People Are Poor Because They Are Too Lazy to Work Half of the poor are not of working age. About 40% are under 18; another 10% are over 65. Most of the able-bodied poor of working age are working or looking for work. Many of the poor adults who do not work have good reasons for not working. Many of them are ill or disabled, while many others are going to school (mostly those in their late teens from poor families).

Many of the poor work and many work year-round. However, a person working 40 hours a week, every week of the year at minimum wage, will not earn enough to lift a family of three out of poverty. The numbers of the working poor are increasing at

Figure 7-1 Number in Poverty and Poverty Rate, 1959–1999

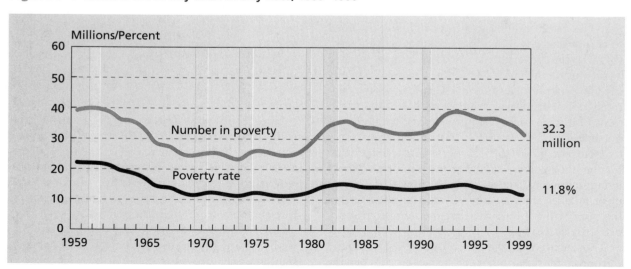

Source: U.S. Bureau of the Census, Current Population Survey, March 2000.

an alarming rate. There are several reasons for this growth. First, although there are more jobs in the economy than ever before, many of these jobs are in low-paying service industries. A janitor or a cook at a fast-food restaurant earns no more than minimum wage. Second, the better jobs the poor used to hold are no longer part of the U.S. economy. Many companies, seeking sources of cheap labor, have set up manufacturing operations overseas to increase their ability to compete in the world market. Finally, many of the working poor are women or young people with few marketable skills. Often, they are forced to settle for poorly paid, part-time work.

In many ways, the working poor are in worse straits than those on the welfare rolls. For example, a mother on welfare may be eligible for public housing and a variety of services that a working poor two-parent family may not be able to receive.

It is easy for the government to ignore the plight of the working poor. Scattered throughout the country and with no collective voice to express protest, they are relatively invisible and, therefore, easily forgotten.

Myth 2: Most Poor People Are Minorities, and Most Minorities Are Poor Neither of these statements is true. Most poor people are white, merely because there are many more whites in the United States than minorities. The poverty rate, however, remains considerably higher for African Americans and Hispanics than whites (23.6 and 22.8%, respectively, versus 7.7% in 1999) (U.S. Bureau of the Census, 2000).

One of the reasons that African Americans are associated with the image of poverty is that they make up more than half of the long-term poor. Another reason is that the War on Poverty was motivated in part, and occurred simultaneously with, the civil rights movement of the 1960s.

Myth 3: Most of the Poor Are Single Mothers With Children While it is true that a disproportionate share of poor households are headed by women, and that the poverty rate for female-headed families is extremely high, the majority of people in poverty live in other family arrangements. Female-headed families with children represent 44.3% of the poor. About 30% live in married couple families, and 22% live alone or with nonrelatives. The remainder live in male-headed families (O'Hare, 1996).

Myth 4: Most People in Poverty Live in the Inner Cities Poverty is growing rapidly in areas not typically identified as poor. Rural communities and areas outside of the central city have seen an increase in the poor in recent years. Approximately 21.5% of the poor live in central cities. About 10.3% of all poor families live outside of the central city, while 17.2% live in nonmetropolitan or rural areas (U.S. Bureau of the Census communication, 1995).

Myth 5: Welfare Programs for the Poor Are Straining the Federal Budget Social assistance programs for low-income people cost the federal government about 14% of the federal budget. A much larger share of the budget (43%) goes to other types of social assistance, such as Social Security and Medicare, which mainly go to middle-class Americans, not the poor. (O'Hare 1996).

Government-Assistance Programs

The public appears to be quite frustrated and upset about what are perceived as soaring welfare costs and growing poverty. Much of this frustration, however, stems from a misperception of what programs are behind the escalating government expenditures, a misunderstanding about who is receiving government assistance, and an exaggerated notion of the amount of assistance going to the typical person in poverty. Most government benefits go to the middle class. Many of the people reading this book will be surprised to know that they or their families actually may receive more benefits than those people typically defined as poor. The value of benefits going to the poor actually has fallen in recent years, while that going to the middle class actually has risen.

Government programs that provide benefits to families or individuals can be divided into two categories: (1) social insurance and cash benefits going to people of all income levels and (2) means-tested programs and cash assistance going only to the poor (see Table 7-6).

Social insurance benefits are not means tested, meaning that you do not have to be poor to receive them. They go primarily to the middle class. Many people receiving payments from social insurance programs, such as Social Security retirement and unemployment insurance, feel they are simply getting back the money they put into these programs. They accuse those receiving benefits from means-tested programs of getting something for nothing. This is not exactly true when we recognize that many social-insurance recipients receive back far more than they put in, and that the poor, the majority of whom work, pay taxes that contribute to their own means-tested benefits.

Social insurance programs account for the overwhelming majority of federal cash assistance

Table 7-6

Expenditures for Major Federal Government Assistance Programs, 2000

SOCIAL INSURANCE AND CASH BENEFITS (RECIPIENTS OF ALL INCOME LEVELS)	BILLIONS
Social Security and Railroad Retirement Established in 1935 to provide cash payments to retired workers and their dependents.	$407.7*
Medicare Covers most medical costs for individuals eligible for Social Security retirement and disability payments.	202.5*
Federal Employees Retirement and Insurance	77.7*
Unemployment Insurance Provides partial wage-replacement payments to workers who lose their jobs.	24.1*
Veterans Health Care	18.6*

MEANS-TESTED PROGRAMS AND CASH ASSISTANCE (RECIPIENTS MUST BE POOR)	
Family Support Payments to States Provides cash payments to needy families.	18
Supplemental Security Income Provides cash payments to needy aged, blind, and disabled people.	29.9
Medicaid Covers most medical costs for families or individuals eligible for Supplemental Security Income.	116
Food Stamps and Nutritional Assistance Distributes coupons redeemable for food to individuals and families with incomes of 130% of the poverty level. Provides free and reduced-price lunches to students from low-income families.	34.2
Housing Assistance Benefits Provides public housing or subsidizes rent in nonpublic housing.	29.2
Earned Income Tax Credit Tax refunds to the poor.	25.7

*Indicates that recipients of benefits are predominantly middle class. The amount going to predominantly middle-income recipients is $730.6 billion. The amount going to low-income recipients is $253 billion.

Source: U.S. Office of Management and Budget, Budget of the United State Government Fiscal Year 2000, Government Printing Office, 2001.

expenditures, and their share has been rising rapidly. Female-headed families in poverty, often portrayed as a heavy drain on the government treasury, account for only 2% of the federal outlays for human resources. In contrast, Social Security for the retired elderly, the vast majority of whom are middle class, accounts for 38%.

The Changing Face of Poverty

It appears that economic rewards are distributed more unequally in the United States than elsewhere in the Western industrialized world. In addition, the United States experiences more poverty than other capitalist countries with similar standards of living. In one international study, the poverty rates for children, working-age adults, and the elderly were tabulated for a variety of countries.

The results showed that the United States has been successful in holding down poverty among the elderly. The American elderly experience far less poverty than the elderly in Great Britain, approximately the same as the old in Norway and Germany, and far more poverty than the elderly in Canada and Sweden.

The United States has been much less successful in keeping children and working-age adults out of poverty. The U.S. child poverty rate is 60% higher than the rate in Great Britain, and more than double the rate in Norway, Sweden, and Germany. (See "Global Sociology: Children in Poverty.")

How has it happened that the United States has made progress in combating poverty among the elderly, but not among other groups? Since 1960, a variety of social policies have been enacted that have improved the standard of living of the elderly relative to that of the younger population. Social Security benefits were increased significantly and protected against the threat of future inflation, Medicare provided the elderly with national health insurance, Supplemental Security Income provided a guaranteed minimum income, special tax benefits for the elderly protected their assets during the later years, and the Older Americans Act supported an array of services specifically for this age group. As a consequence of these measures, poverty among the elderly has declined substantially. While 24.6% of those

The United States had been very successful at lifting the elderly out of poverty. It has not been as successful with children. Not only has child poverty increased over the past two decades, but when compared to other countries in the world the United States appears to do less to improve the living conditions of its poor children than many countries. Even though 30.2% of children in Ireland start out in poverty, the government's assistance expenditures reduce that number to 12% with a variety of programs. In the United States, 25.9% of all children start out in poverty, but the government's programs only reduce that number to 16.9% (see Figure 7-2).

There are a number of reasons why the United States has such high rates of poverty among children. First, there has been an enormous growth in the number of children born to unwed mothers in their teens. These mothers are often poor to begin with, and the birth of the child makes their earnings prospects even more limited. American mothers with limited education or skills are less likely than European mothers to return to work quickly after childbirth, because high-quality child care is comparatively expensive and the jobs will not cover the costs. Second, the United States has more poor immigrants than any other country. Many of the children in poverty are born to poor immigrants who have only been in the country for a few years. A third factor is an out-growth of the realities of the American political system, which depends on advocates supporting programs for special constituencies. Children obviously do not vote and the poor in general have low voting records. The elderly, on the other hand, have high voting records and therefore have substantial political clout. Politicians are usually not voted out of office for cutting benefits to children, whereas they are if they do so for the elderly.

Some have suggested (Besharov, 1995) that other countries have avoided high levels of child poverty by limiting economic growth and lowering the living standard for everyone. Many European countries with generous social assistance programs also have high rates of unemployment and living standards that do not match those in the United States. These critics charge that instead of bringing everyone up, other countries have brought everyone down. They claim that the high rate of American child poverty appears greater because the gap between the rich and the poor is so much larger in the United States and other countries just have gone further in redistributing income.

Despite the arguments over the data, there are substantial numbers of poor children in the United States. Great harm is being done to these children when their poor living conditions are not addressed.

Source: "Low Ranking of Poor American Children," by D. J. Besharov, August 14, 1995, *The New York Times*, p. A9.

Figure 7-2 Worldwide Child Poverty Rates

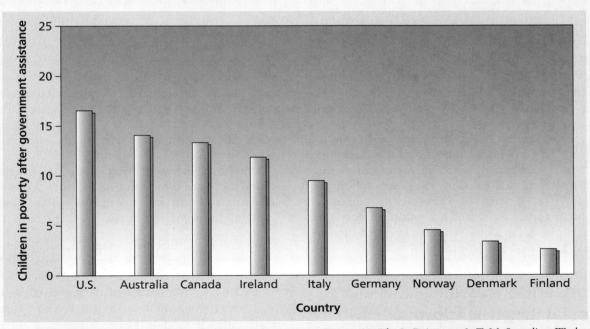

Source: "Doing Poorly: The Real Income of Children in Comparative Perspective," by L. Rainwater & T. M. Smeeding, Working Paper No. 127, Luxembourg Income Study, Maxwell School of Citizenship and Public Affairs, Syracuse, NY: Syracuse University, 1995.

Figure 7-3 Poverty Rates for People Over 65 and Under 18, 1960–1999

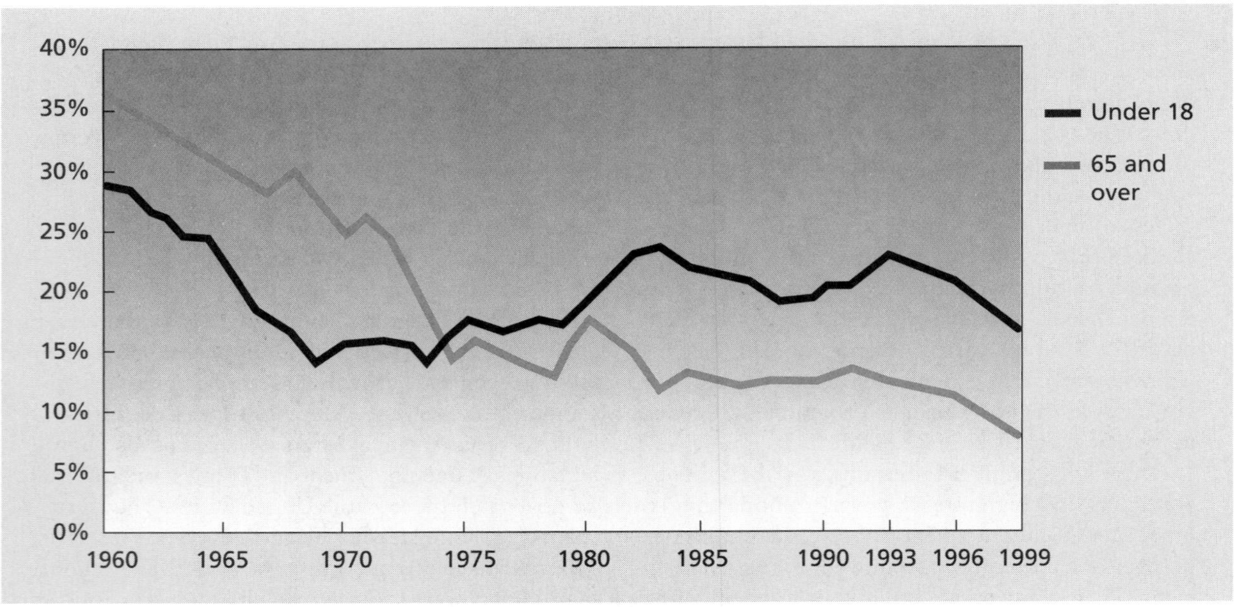

Source: U.S. Bureau of the Census, Current Population Survey, March 2000.

families 65 and older lived below the poverty level in 1970, only 9.7% did so by 1999 (U.S. Bureau of the Census, 2000).

To achieve this dramatic improvement in the conditions of the elderly, it has been necessary to increase greatly the federal money spent on this age group. If these arrangements are maintained, projections show about 60% of the federal budget going to the elderly by the year 2030. A group that has suffered particularly under this shift in expenditures to the elderly is the young. While 14% of children lived in poverty in 1970, 16.9% did so in 1996 (U.S. Bureau of the Census, 2000). (See Figure 7-3.)

It would also surprise many people to learn that not only are the elderly as a group not poor, but that they are actually better off than most Americans. They are more likely than any age group to possess money market accounts, certificates of deposit, U.S. government securities, and municipal and corporate bonds. The median household net worth of those 65 to 69 is the highest of any age group, followed by those 70 to 74 years old. They have the highest rate of home ownership of any age group. Seventy-seven percent of those 65 to 74 own homes, and most of these homes are paid for in full.

Consequences of Social Stratification

Studies of stratification in the United States have shown that social class affects many factors in a person's life. Striking differences in health and life expectancy are apparent among the social classes, especially between the lower-class poor and the other social groups. As might be expected, lower-class people are sick more frequently than are others. For example, in Harlem, a predominantly African-American low-income neighborhood of New York City, the rate of infection with tuberculosis is more than 13 times the national average. The only other area in the Western Hemisphere to exceed this rate is Haiti (Califano, 1994).

Tuberculosis (TB) is especially common among the homeless who are forced to live in urban shelters. Even if tuberculosis is treated properly and promptly, the treatment method makes the problem unlikely to go away. Sufferers are required to take several antibiotics daily over a period of a year—a regimen that is difficult for many of the poor and uneducated to follow. As a result, only about 4 in 10 TB patients in New York City public hospitals get all the medication they need for a complete recovery. As former secretary of Health, Education, and Welfare, Joseph Califano, Jr., noted:

> We have the capacity to prevent TB, but the role of doctors is marginal. The fertile soil in which TB spreads, and the logical focus of attacks on the problem, is decrepit housing and homelessness, unsanitary living conditions, drug and alcohol abuse, and AIDS. (Califano, 1994)

Poverty has a particular impact on the health of the young. Babies born into poverty are significantly more likely to die before their first birthday than those born into families living above the poverty level. One study (Kiely, 1995) found that the infant mortality rate for poor children was more than 50% higher than that for children not born in poverty.

African-American teenagers between 15 and 19 who become pregnant are more than twice as likely as white teenagers to have the baby die. In addition, an African-American mother is more than three times as likely to die giving birth than a white mother (*Statistical Abstract of the United States, 1999, 2000*).

Diet and living conditions also give a distinct advantage to the upper classes, because they have access to better and more sanitary housing and can afford more balanced and nutritious food. A direct consequence of this situation is seen in each social class's life-expectancy pattern. Not surprisingly, lower-class people do not live as long as do those in the upper classes. White males born in 1997 have a life expectancy nearly 7.1 years longer than that for African-American males, many of whom are concentrated in the lower-income brackets (National Vital Statistics Reports, December, 1999). Family,

childbearing, and child-rearing patterns also vary according to social class. Women in the higher-income groups, who have more education, tend to have fewer children than do lower-class women with less schooling. Women more often head the family in the lower class, compared with women in the other groups. Middle-class women discipline their children differently than do working-class mothers. The former punish boys and girls alike for the same infraction, whereas the latter often have different standards for sons and daughters. Also, middle-class mothers judge the misbehaving child's intention, whereas working-class women are more concerned with the effects of the child's action.

There is a direct relationship between a person's social class and the possibility of his or her arrest, conviction, and sentencing if accused of a crime. For the same criminal behavior, the poor are more likely to be arrested; if arrested, they are more likely to be charged; if charged, they are more likely to be convicted; if convicted, they are more likely to be sentenced to prison; and if sentenced, they are more likely to be given longer prison terms than members of the middle and upper classes.

A direct relationship exists between a person's social class and the possibility of his or her arrest, conviction, and sentencing if accused of a crime.

The poor are singled out for harsher treatment at the very beginning of the criminal justice system. Although many surveys show that almost all people admit to having committed a crime for which they could be imprisoned, the police are prone to arrest a poor person and release, with no formal charges, a high-class person for the same offense. A well-to-do teenager who has been accused of a criminal offense is frequently just held by the police at the station house until the youngster can be released to the custody of the parents; poorer teenagers who have committed the same kind of crime more often are automatically charged and referred to juvenile court.

The poor tend to commit violent crimes and crimes against property—they have little opportunity to commit such white-collar crimes as embezzlement, fraud, or large-scale tax evasion—and they are much more severely punished for their crimes than upper-class criminals are for theirs. Yet white-collar crimes are far more damaging and costly to the public than are the crimes more often committed by poor people. The government has estimated that white-collar crimes cost more than $40 billion a year—more than 10 times the total amount of all reported thefts and more than 250 times the amount taken in all bank robberies.

Even the language used to describe the same crime committed by an upper-class criminal and a poor one reflects the disparity in the treatment they receive. The poor thief who takes $2,000 is accused of stealing and usually receives a stiff prison sentence. The corporate executive who embezzles $200,000 merely has "misappropriated" the funds and is given a lighter sentence or none at all, on the promise to make restitution. A corporation often can avoid criminal prosecution by signing a consent decree, which is in essence a statement that it has done nothing wrong and promises never to do it again. Were this ploy available to ordinary burglars, the police would have no need to arrest them; a burglar would merely need to sign a statement promising never to burgle again and file it with the court.

Once charged, the poor are usually dependent on court-appointed lawyers or public defenders to handle their cases. The better-off rely on private lawyers who have more time, resources, and personal interest in defending their cases.

If convicted of the same kind of crime as a well-to-do offender, the poor criminal is more likely to be sentenced and will generally receive a longer prison term. As for prison terms, the sentence for burglary, a crime of the poor, is generally more than twice as long as that for fraud, while a robber will draw an average sentence more than six times longer than that of an embezzler. The result is a prison system heavily populated by the poor.

The mentally ill among the homeless are the least likely to reach out for help and the most likely to remain on the streets.

Another serious consequence of social stratification is mental illness. Studies have also shown that at least one-third of all homeless people suffer from schizophrenia, manic-depressive psychosis, or other mental disorders. Those people are the least likely to reach out for help and the most likely to remain on the streets in utter poverty and despair year after year (Jencks, 1994; Torrey, 1988).

Thus, social class has very real and immediate consequences for individuals. In fact, class membership affects the quality of people's lives more than any other single variable.

SUMMARY

Every society makes distinctions among its members. The outcome of this process of social evaluation is social inequality. It involves the uneven distribution of privileges, material rewards, opportunities, power, prestige, and influence among individuals or groups. When social inequality becomes part of the social structure and is transmitted from one generation to the next, social stratification exists.

Social mobility is the movement of an individual or group from one social status to another within a stratification system. An open society attempts to provide equal opportunity for everyone to compete for desired roles and statuses. In a closed society, various aspects of people's lives are determined at birth and remain fixed. Movement that involves a change of status but no corresponding change in social class is known as horizontal mobility.

Stratification can be based mainly on ascribed or achieved status. The caste system is a rigid form of stratification based on ascribed characteristics such as skin color or family identity. In an estate system a person's position is defined by law and determined primarily by inheritance. Each estate has legally established rights and duties. Class systems are based at least in part on achieved status. A social class consists of a category of people who share similar opportunities, economic and vocational positions, lifestyles, attitudes, and behaviors.

The functionalist theory of stratification as presented by Davis and Moore holds that stratification is socially necessary. They argue that (1) different positions in society make different levels of contributions to the well-being and preservation of society; (2) filling the more complex and important positions in society often requires talent that is scarce and has a long period of training; and (3) providing unequal rewards ensures that the most talented and best-trained individuals will fill the statuses of greatest importance and be motivated to carry out role expectations competently. Critics of this view suggest that stratification is immoral because it creates extremes of wealth and poverty and denigrates the people on the bottom, or that it is dysfunctional in that it neglects the talents and merits of many people who are stuck in the lower classes, ignores the ability of the powerful to limit access to important positions, and overlooks the fact that the level of rewards attached to jobs does not necessarily reflect their functional importance.

Conflict theorists see stratification as the outcome of a struggle for dominance. Karl Marx believed that to understand human societies, one must look at the economic conditions surrounding production of the necessities of life. Those groups who own or control the means of production within a so-ciety also have the power to shape or maintain aspects of society to favor their interests.

Despite the American political ideal of the basic equality of all citizens and the lack of inherited ranks and titles, the United States nonetheless has a class structure that is characterized by extremes of wealth and poverty. Though many people deny their importance, there are class distinctions in the United States based on race, education, family name, career choice, or wealth.

Studies of social stratification have shown that social class affects many aspects of people's lives. Lower-class people get sick more often, and have higher infant mortality rates, shorter life expectancies, and larger families. The poor are more likely to be arrested, charged with a crime, convicted, and sentenced to prison, and are likely to get longer prison terms than middle- and upper-class criminals. Types of mental illness seem to vary with social class, as does the likelihood of spending time in a mental hospital.

In the next chapter we will look at stratification based on race and ethnicity.

 I N T E R N E T   R E S O U R C E S

Go to http://web-enhanced.thomsonlearning.com which features Web links relevant to this chapter to expand and enhance the learning experience. This site will be monitored and updated periodically to ensure the links are correct and live.

At Virtual Society: The Wadsworth Sociology Resource Center, http://sociology.wadsworth.com, you will find a Career Center, Sociology in the News, Research Online, InfoTrac College Edition, Virtual Tours in Sociology, and a variety of book-specific student and instructor resources.

CHAPTER SEVEN STUDY GUIDE

KEY CONCEPTS

Match each concept with its definition, illustration, or explanation below.

a. Intergenerational mobility
b. Estate system
c. Status inconsistency
d. Social inequality
e. Vertical mobility
f. Bourgeoisie
g. Caste system
h. Horizontal mobility
i. Estate
j. Wealth
k. Social stratification

l. Open society
m. Proletariat
n. Power
o. Social class
p. Conflict theory
q. Class system
r. Prestige
s. Closed society
t. Social mobility
u. Functionalist theory
v. Upper class

w. Varnas
x. Upper middle class
y. Lower middle class
z. Poverty
aa. Lower class
bb. Intragenerational mobility
cc. Poverty index
dd. Feminization of poverty
ee. Income distribution
ff. Working class

_____ 1. The uneven distribution of privileges, material rewards, opportunities, power, prestige, and influence among individuals or groups.

_____ 2. The division of society into levels, steps, or positions that is perpetuated by the society's major institutions.

_____ 3. Movement of an individual or a group from one social status to another.

_____ 4. A society characterized by the attempt to provide equal opportunity for everyone to compete for desired statuses.

_____ 5. A society in which the various aspects of people's lives are determined at birth and remain fixed.

_____ 6. A change in status with no corresponding change in social class.

_____ 7. A change in status that results in a change in social class.

_____ 8. A change in social status that takes place over two or more generations.

_____ 9. a change in social status that occurs within the lifetime of one individual.

_____ 10. Situations in which people rank differently on certain stratification characteristics than on others.

_____ 11. A rigid form of stratification based on ascribed characteristics such as skin color or family identity.

_____ 12. A closed system of stratification in which a person's social position is defined by law and membership is determined primarily by inheritance.

_____ 13. A segment of a society that has legally established rights and duties.

_____ 14. A category of people who share similar opportunities, economic and vocational positions, lifestyles, attitudes, and behavior.

_____ 15. A society that contains several different social classes and permits at least some social mobility.

_____ 16. The total economic assets of an individual or family.

_____ 17. The ability of an individual or group to attain goals, control events, and maintain influence over others, even in the face of opposition.

_____ 18. The approval and respect an individual or group received from other members of society.

_____ **19.** The owners of the means of production or capital.

_____ **20.** The working class.

_____ **21.** Hindu grades of being that are the basis for the Hindu caste system.

_____ **22.** U.S. class characterized by corporate ownership, elite schools, upper-echelon politics, and liberal arts educations.

_____ **23.** U.S. class characterized by unskilled labor, little interest in education, grade school completion, service work, and farm labor.

_____ **24.** U.S. class characterized by professional and technical occupations, college and graduate school training, and often managers and proprietors.

_____ **25.** U.S. class characterized by clerical and sales occupations, some college or high-school education, and often small-business owners.

_____ **26.** U.S. occupation characterized by skilled and semi-skilled labor, craftspeople, forepersons, completion of grade school, and completion of some or all of high school.

_____ **27.** The theory stating that in order to attract talented individuals to each occupation, society must set up a system of differential rewards.

_____ **28.** The phrase referring to the disproportionate concentration of poverty among female-headed families.

_____ **29.** The U.S. government's specification of income levels below which people are considered to be living in poverty.

_____ **30.** The U.S. Census estimates of total family income as a percentage of the nation's total income earned within any given year.

_____ **31.** A condition in which people do not have enough money to maintain a standard of living that includes the basic necessities of life.

_____ **32.** Assumes that society is in a constant state of social conflict with only temporary periods of stability.

KEY THINKERS/RESEARCHERS

Match these thinkers/researchers with their main ideas or contributions.

a. Max Weber　　　　　**d. R. Todd Erkel**
b. Ralf Dahrendorf　　　**e. Kingsley Davis and Wilbert Moore**
c. Karl Marx

_____ **1.** Developed the functionalist theory of stratification.

_____ **2.** A pioneering conflict theorist, he developed a critique of capitalist class society and a vision of the transition to socialism.

_____ **3.** Proposed that stratification takes place on the basis of class, status, and power.

_____ **4.** Suggested that functionalist and conflict theories of stratification are really complementary rather than opposed.

_____ **5.** Wrote about how the newly middle class often feel uncomfortable and adrift.

CENTRAL IDEA COMPLETIONS

Following the instructions, fill in the appropriate concepts and descriptions for each of the questions posed in the following section.

1. In the space provided, use the data and information from your text to refute each of the following myths about poverty in America.

a. People are poor because they are too lazy to work:

_____

_____

b. Most poor people are African-American, and most African-Americans are poor:

_____

_____

c. Most of the poor are single mothers with children:

_____

_____

d. Most people in poverty live in inner-city areas:

_____

_____

e. Welfare programs for the poor are straining the federal budget:

_____

_____

2. Based on your reading of this chapter, present a listing of what you believe are the four most detrimental consequences of social stratification in the United States:

a. _____

_____

b. _____

_____

c. _____

_____

d. _____

_____

3. As your text notes, the poor are overrepresented in America's criminal justice system. Yet the poor are not the only Americans committing crime. Compare and contrast the experiences with the criminal justice system for the poor and the rich for each of the areas presented below:

a. Type of crime: _____

_____

b. Legal representation: _____

_____

c. Conviction outcomes: _____

_____

4. What are five key outcomes of the feminization of poverty in the United States?

a. _____

_____

b. _____

_____

c. _____

_____

d. _____

_____

e. _____

_____

5. Present a one-sentence statement about each of the theories of stratification:

a. Functionalist: _____

b. Conflict: _____

c. Modern conflict: _____

6. Present the essential characteristics of each of the stratification systems:

a. Caste: _____

b. Estate: _____

c. Class: _____

7. Describe how different groups in the United States experience poverty:

_____

_____

_____

8. Describe how one's position in the system of social stratification in the United States can affect one's life chances:

_____

_____

_____

9. Define the following:

a. Horizontal mobility: _____

b. Vertical mobility: _____

c. Intergenerational mobility: _____

10. Using Table 7-4, indicate what percentage of total income is received by the richest 5% (fifth quintile) of Americans compared with the first quintile.

a. Fifth quintile: _____

b. First quintile: _____

11. What are three factors which lead many critics to maintain that we are underestimating poverty in the United States?

a. _____

_____

b. _____

_____

c. _____

_____

12. Present four specific outcomes of poverty as it impacts on the health of the young in the United States.

a. _____

b. _____

c. _____

d. _____

CRITICAL THOUGHT EXERCISES

1. Take a position on the following question: Does the U.S. government have an obligation to shrink the income gap, or is a substantial gap between the highest- and lowest-paid workers simply a natural part of the capitalist economy? Support your position with information and data from this chapter as well as any other chapters in your textbook.

2. Your author states that most of you "will be surprised to know that they or their families actually may receive more benefits than those people typically defined as poor." Contact the Department of Social Services closest to your college or university. Ask them for information on the programs and distributions of funds for the poor residents of the county where your school is located. Next, develop a list of the middle- and upper-class governmental benefits common to residents of your community, for example, property tax deductions, home-owner interest deductions, educational assistance programs, and so on. Use your information to create a survey in which you present two columns of governmental benefits: one for the middle and upper classes and the other for the lower classes. Randomly select 10 students from your local campus directory. Ask if they would participate in your "Inquiry into Governmental Assistance" survey, which should take them about 10 minutes. Read them the information from both columns and ask whether they "were aware of" or "were unaware of" each governmental program. What were your results? Were you surprised by the results? Did the results conform (overall) to the thesis put forth by the author of your text? Share your results with your instructor so that he or she may share them with your class.

3. Develop a research paper within which you explore "The Changing Face of Poverty." Address the issue of why "it appears that economic rewards are distributed more unequally in the United States than elsewhere in the Western industrialized world." In which areas has the United States been successful in addressing poverty issues and in which areas has it experienced failure? Assume you were placed at the head of a presidential commission on poverty in America. What central issues would your committee examine on its way to developing a practical proposal to eliminate poverty in America? Which obstacles would be most difficult to overcome, and which obstacles would be most manageable? If your program were to be enacted, what measures of success would you consider?

4. Using your library and the Internet, select two non-Western countries and present a profile of the conditions of poverty in each nation. What poverty features are common to both countries? Are there characteristics of poverty that are unique to either country? What factors may contribute to unique distributions of poverty? What factors are likely to be common across a large range of countries throughout the world?

5. In a now classic argument in sociology, Melvin Tumin challenged the Davis and Moore functionalist explanation for the prevalence of social stratification, particularly as it applied to the position of doctors in American society. Using the functionalist argument, present an explanation for the current salary structure in American sports. Account for why baseball players, for example, are paid more than high-school mathematics teachers. Then logically attack the argument you just presented using the information given in your text that critiques the functionalist approach. Assuming you have presented your best argument for each side of the debate, present your conclusion as to which perspective best accounts for the salaries in major league sports in America.

6. Discuss the impact women have had on the American labor force. Devote special attention to the connection between women entering various occupations and the changing prestige of those occupations. Examine Table 7.1 which depicts the prestige ratings of various occupations. What changes in the prestige ratings of these occupations would you hypothesize are likely to occur in the United States by the year 2025?

ANSWERS

KEY CONCEPTS

1. d	7. e	13. i	19. f	24. x	29. cc
2. k	8. a	14. o	20. m	25. y	30. ee
3. t	9. bb	15. q	21. w	26. ff	31. z
4. l	10. c	16. j	22. v	27. u	32. p
5. s	11. g	17. n	23. aa	28. dd	
6. h	12. b	18. r			

KEY THINKERS/RESEARCHERS

1. e	2. c	3. a	4. b	5. d

Racial and Ethnic Minorities

LEARNING OBJECTIVES

After studying this chapter, you should be able to do the following:

- Describe the genetic, legal and social approaches to defining race.
- Explain the concept of ethnic group.
- Know how the sociological concept of "minority" is used.
- Understand the relationship between prejudice and discrimination.
- Recognize the impact of institutionalized prejudice and discrimination.
- Discuss the history of immigration to the United States.
- Describe the characteristics of the major racial and ethnic groups in the United States.

Paul and Philip Malone, twin brothers, had always dreamed of becoming fire-fighters in the Boston Fire Department. The only problem was that the brothers received scores of 69 and 57, respectively, on the written portion of the entry test, when the passing grade was 82. At the time, Boston was under pressure to increase its minority population within the Fire Department and had separate, and lower, passing grades for nonwhites.

Paul and Philip were aware that they had a great-grandmother who was African-American. The Malones had not given the issue of race much thought during their original application. They decided to apply again, only this time as African-Americans. They took the test again, and this time their scores were considered passing because of the lower passing grades for minority applicants. The Malones were hired and were listed in the records as African-American.

Once they were on the job, the issue of race appeared to be relatively unimportant. The Malones did their jobs well for 10 years, and then decided to take the civil service examinations for lieutenants. They scored exceptionally high and their names were forwarded to the fire commissioner for promotion. As their applications were being reviewed, questions arose about their race because they did not look African-American. They continued to insist that their claim to being African-American was legitimate and produced photos of their great-grandmother. Their protests were to no avail, however, and they were fired for misrepresenting their race on the original application. They appealed the decision in the courts, but eventually the Massachusetts Supreme Judicial Court ruled against them (*The Boston Globe,* September 28, 1999; *The Boston Herald,* November 26, 1988).

If you believe that the Malones misrepresented their race, then what would your view be of Walter White, who was head of the National Association for the Advancement of Colored People from 1931 to 1955? Although he thought of himself as a black man, it has been estimated that he had no more than one sixty-fourth African-American ancestry, considerably less than that of the Malones. White would often go undercover and pass for white while investigating lynchings in the South for the NAACP. In 1923, he even deceived a Ku Klux Klan recruiter into inviting him to Atlanta to advise him on recruitment (Kilker, 1993).

What are we to make of Mark Linton Stebbins's claim of being black? In a close contest for a seat on the Stockton, California, city council, Ralph Lee White, an African-American, lost his spot on the council to Stebbins, a pale-skinned, blue-eyed man with kinky reddish-brown hair. White claimed Stebbins would not represent the minority district because he was white. Stebbins claimed he was African-American. Birth records showed Stebbins' parents and grandparents were white. He has five sisters and one brother, all of whom are white. Yet, when asked to declare his race he noted: "First, I'm a human being, but I'm black." Stebbins did not deny that he was raised as a white person. He began to consider himself black only after he moved to Stockton. "As far as a birth certificate goes, then I'm white; but I am black. There is no question about that."

Stebbins now belongs to an African-American Baptist church and to the NAACP. Most of his friends are African-American. He has been married three times, first to a white woman, and then to two African-American women. He has three children from the first two marriages; two are being reared as whites, the third as African-American. He states he considers himself black—"culturally, socially, genetically."

Ralph Lee White remained unconvinced, especially with his former council seat having gone to Stebbins. "Now, his mama's white and his daddy's white, so how can he be black? If the mama's an elephant and the daddy's an elephant, the baby can't be a lion. He's just a white boy with a permanent."

Stebbins believes the issue of race is tied to identifying with a community in terms of beliefs, aspirations, and concerns. He asserts that a person's racial identity depends on much more than birth records.

These three examples, which on the one hand may seem slightly humorous, are at the same time related to some very serious issues. Throughout history, people have gone to great lengths to determine what race a person belonged to. In many states, laws were devised to determine your race if, like the Malones, you had mixed racial ancestry. Usually these laws existed for the purpose of discriminating against certain minority groups. Our examples also raise the question of whether you can change your race from white to black, even if you have no black ancestors. Mark Stebbins seems to think so. In this chapter, we will explore these and other issues related to race and ethnicity as we try to understand how people come to be identified with certain groups and what that membership means.

The Concept of Race

While the origin of the word is not known, the term *race* has been a highly controversial concept for a long time. Many authorities suspect that it is of Semitic origin, coming from a word that some translations of the Bible render as "race," as in the "race of Abraham" but that is otherwise translated as "seed" or "generation." Other scholars trace the origin to the Czech word *raz*, meaning "artery" or "blood"; others to the Latin *generatio* or the Basque *arraca* or *arraze*, referring to a male stud animal. Some trace it to the Spanish *ras*, itself of Arabic derivation, meaning "head" or "origin." In all these possible sources, the word has a biological significance that implies descent, blood, or relationship.

We shall use the term **race** to refer to *a category of people who are defined as similar because of a number of physical characteristics.* Often the category is based on an arbitrary set of features chosen to suit the labeler's purposes and convenience. As long ago as 1781, German physiologist Johann Blumenbach realized that racial categories did not reflect the actual divisions among human groups. As he put it, "When the matter is thoroughly considered, you see that all [human groups] do so run into one another, and that one variety of mankind does so sensibly pass into the other, that you cannot mark

Throughout history, races have been defined along genetic, legal, and social lines, each presenting its own set of problems.

out the limits between them" (Montagu, 1964). Blumenbach believed racial differences were superficial and changeable, and modern scientific evidence seems to support this view.

Throughout history, races have been defined along genetic, legal, and social lines, each presenting its own set of problems.

Genetic Definitions

Geneticists define race by noting differences in gene frequencies among selected groups. The number of distinct races that can be defined by this method depends on the particular genetic trait under investigation. Differences in traits, such as hair and nose type, have proved to be of no value in making biological classifications of human beings. In fact, the physiological and mental similarities among groups of people appear to be far greater than any superficial differences in skin color and physical characteristics. Also, the various so-called racial criteria appear to be independent of one another. For example, any form of hair may occur with any skin color—a narrow nose gives no clue to an individual's skin pigmentation or hair texture. Thus, Australian aborigines have dark skins, broad noses, and an abundance of curly-to-wavy hair; Asiatic Indians also have dark skins but have narrow noses and straight hair. Likewise, if head form is selected as the major criterion for sorting, an equally diverse collection of physical types will appear in each category thus defined. If people are sorted on the basis of skin color, therefore, all kinds of noses, hair, and head forms will appear in each category.

Legal Definitions

By and large, legal definitions of race have not been devised to determine who was black or of another race, but rather who was not white. The laws were to be used in instances in which separation and different treatments were to be applied to members of certain groups. Segregation laws are an excellent example. If railroad conductors had to assign someone to either the black or white cars, they needed fairly precise guidelines for knowing whom to seat where. Most legal definitions of race were devices to prevent blacks from attending white schools, serving on juries, holding certain jobs, or patronizing certain public places. The official guidelines could then be applied to individual cases. The common assumption that "anyone not white was colored," although imperfect, did minimize ambiguity.

There has been, however, very little consistency among the various legal definitions of race that have been devised. The state of Missouri, for example, made "one-eighth or more Negro blood" the criterion for nonwhite status. Georgia was even more rigid in its definition and noted:

> The term "white person" shall include only persons of the white or Caucasian race, who have no ascertainable trace of either Negro, African, West Indian, Asiatic Indian, Mongolian, Japanese, or Chinese blood in their veins. No person, any of whose ancestors [was] . . . a colored person or person of color, shall be deemed to be a white person.

Virginia had a similar law but made exceptions for individuals with one-fourth or more Indian (Native American) "blood" and less than one-sixteenth Negro "blood." Those Virginians were regarded as Indians as long as they remained on an Indian reservation, but if they moved, they were regarded as blacks (Berry & Tischler, 1978; Novit-Evans & Welch, 1983).

Most of these laws are artifacts of the segregation era. However, if people think that all vestiges of them have disappeared, they are wrong. As recently as 1982, a dispute arose over Louisiana's law requiring anyone of more than one thirty-second African descent to be classified as black. Louisiana's one-thirty-second law is actually of recent vintage, having come into being in 1971. Before this law, racial classification in Louisiana depended on what was referred to as "common repute." The 1971 law was intended to eliminate racial classifications by gossip and inference. In September 1982, Susie Guillory Phipps obtained a copy of her birth certificate so that she could apply for a passport. She was surprised to see that her birth certificate classified her as "black." Phipps, who at the time was 49, had lived her entire life as a white person. She requested that her race be noted as white. The state objected and produced an eleven-generation family tree with ancestors Phipps knew nothing about, including an early eighteenth-century black slave and a white plantation owner. Phipps responded, "My children are white. My grandchildren are white. Mother and Daddy were buried white." Louisiana was not convinced and calculated that she was $3/32$ black, more than enough to make her black under state law (Cose, 1997; Novit-Evans & Welch, 1983).

Social Definitions

The social definition of race, which is the decisive one in most interactions, pays little attention to an individual's hereditary physical features or to whether his or her percentage of "Negro blood" is one-fourth, one-eighth, or one-sixteenth. According to social definitions of race, if a person presents himself or herself as a member of a certain race and others respond to that person as a member of that race,

Table 8-1

Race/Ethnicity Categories in the Census, 1860–2000

CENSUS	1860	1890*	1900	1970	2000
Race	White Black Mulatto	White Black Mulatto Quadroon Octoroon	White Black (Negro Descent)	White Negro or Black	White Black African-American or Negro
		Chinese Japanese Indian	Chinese Japanese Indian	Chinese Japanese Indian (American)	Chinese Japanese American Indian or Alaska Native
				Filipino Hawaiian Korean	Filipino Native Hawaiian Korean Asian Indian Vietnamese Guamanian or Chamorro Samoan Other Asian Other Pacific Islander
				Other	Some Other Race
Hispanic Ethnicity				Mexican	Mexican Mexican American or Chicano
				Puerto Rican Central/South American	Puerto Rican
				Cuban Other Spanish	Cuban Other Spanish/ Hispanic/Latino
				(None of these)	Not Spanish/ Hispanic/Latino

*In 1890, mulatto was defined as a person who was three-eighths to five-eighths black. A quadroon was one-quarter black and an octoroon one-eighth black.

then it makes little sense to say that he or she is not a member of that race.

In Latin American countries, having African ancestry or African features does not automatically define an individual as black. For example, in Brazil many individuals are listed in the census as, and are considered to be, white by their friends and associates even if they had a grandparent who was of pure African descent. It is much the same in Puerto Rico, where anyone who is not obviously of African descent is classified as either mulatto or white.

The U.S. Census relies on a self-definition system of racial classification and does not apply any legal or genetic rules. The 2000 census was the first that allowed people to check more than one category under race. People were able to declare themselves as members of any one or more of five racial categories: American Indian/Alaskan Native, Asian, African-American, Native Hawaiian/Pacific Islander, or

white. Those listing themselves as white and a member of a minority were counted as a minority. (See Table 8-1 for a listing of the various census racial categories that have been used over the years.)

The new census policy was hailed by groups representing parents of mixed-race children who disliked the old rule that forced them to pick one race for their offspring. But civil rights groups expressed concern that the new policy would reduce the number of people who are officially considered black or Asian or Native Indian and would harm minorities when it comes to enforcing civil rights and voting rights laws (U.S. Bureau of the Census, March 11, 2000).

In the 2000 census, 6.8 million people (2.4% of the population) identified themselves as belonging to two or more races. The most common combination was "white" and "some other race" (32%). This was followed by "white" and Native American

(16%); "white" and Asian (13%); and "white and African-American (11%). (U.S. Bureau of the Census, 2001).

Muliracial Ancestry In June of 1999, the Alabama Senate voted to repeal the state's constitutional prohibition against interracial marriage. Even though the United States Supreme Court struck down prohibitions against interracial marriage 32 years earlier, in parts of rural Alabama, probate judges would still refuse to issue marriage licenses to interracial couples (Greenberg, 1999).

The prevalence of interracial marriage is a telling indicator of the social distance between racial groups. About 5% of U.S. married couples include spouses of a different race. This small percentage masks a remarkable growth in the number of interracial marriages. Since 1970, the number of black and white interracial married couples has increased from 300,000 to 1,400,000 million in 1998. Census data also shows that the number of children in interracial families increased from 500,000 in 1970 to 2 million. The National Center for Health Statistics acknowledges that these numbers are probably undercounts, because the father's race is unspecified in a significant number of births each year.

Native Americans are the most likely to marry outside of their group. In fact, they are more likely to marry a white person than another Native American.

People of Asian ancestry are also likely to marry interracially. In about 15% of married couples with an Asian American partner, the other partner is a non-Asian. Racial intermarriage is much higher among native-born Asians, where the percentage who are intermarried reaches 40%.

As more Asian-Americans marry interracially, future generations of Asian-Americans may increasingly blend with other racial and ethnic groups mirroring the experience of European ethnic groups in the United States over the past century.

Interestingly, the future size of the U.S. Asian population depends in part on whether the children of these marriages identify themselves as Asians. If they do, the Asian-American population will increase faster than projected.

African-Americans are much less likely to marry outside their race. About 9% of couples with a black spouse include a spouse of another race. The vast majority of these are black-white unions.

Whites are the least likely to marry outside of their race. Only about 3% of married couples included a white person married to a person of another race. Most of these whites were married to an Asian or Native American.

Interracial births are highest for Native Americans—about one-half of all Indian births are biracial. About 20% of births to Asian women are biracial. Only about 5% of births to white and black women are biracial.

Most people with one white and one black parent, when given the opportunity to label themselves, have historically chosen (or been forced to choose as noted elsewhere in this chapter) one parent's racial identity, and that most often has been and continues to be black. Confusion arises because one can be black and have "white blood," even to the point of having a white parent, but historically one could not be white and have "black blood."

Unlike countries in Latin America and the Caribbean, the United States has no permanent intermediate groups (e.g., mulatto, mestizo) with a separate political, economic, or legal status. The number of children born to interracial couples will continue to increase. The interesting paradox is that at the same time that people are becoming more tolerant of interracial marriage, they are increasingly more likely to celebrate their racial and ethnic heritages.

The Concept of Ethnic Group

An **ethnic group** *has a distinct cultural tradition that its own members identify with and that may or may not be recognized by others* (Glazer & Moynihan, 1975). An ethnic group need not necessarily be a numerical minority within a nation (although the term sometimes is used that way).

Many ethnic groups form subcultures (see Chapter 3): They usually possess a high degree of internal loyalty and adherence to basic customs, maintaining a similarity in family patterns, religion, and cultural values. They often possess distinctive folkways and mores; customs of dress, art, and ornamentation; moral codes and value systems; and patterns of recreation. There is usually something that the whole group is devoted to, such as a monarch, a religion, a language, or a territory. Above all, members of the group have a strong feeling of association. The members are aware of a relationship because of a shared loyalty to a cultural tradition. The folkways may change, the institutions may become radically altered, and the object of allegiance may shift from one trait to another, but loyalty to the group and the consciousness of belonging remain as long as the group exists.

An ethnic group may or may not have its own separate political unit; it may have had one in the past, it may aspire to have one in the future, or its members may be scattered throughout existing countries. Political unification is not an essential feature of this classification. Accordingly, despite the unique

Many ethnic groups form subcultures with a high degree of internal loyalty and adherence to basic customs.

cultural features that set them apart as subcultures, many ethnic groups—Arabs, French Canadians, Flemish, Scots, Jews, and Pennsylvania Dutch, for example—are part of larger political units.

The Concept of Minorities

Whenever race and ethnicity are discussed, it is usually assumed that the object of the discussion is a minority group. Technically this is not always true, as we shall see shortly. A minority is often thought of as being few in number. The concept of minority, rather than implying a small number, should be thought of as implying differential treatment and exclusion from full social participation by the dominant group in a society. In this sense, we shall use Louis Wirth's definition of a **minority** as *a group of people who, because of physical or cultural characteristics, are singled out from others in society for differential and unequal treatment, and who there-*

fore regard themselves as objects of collective discrimination (Linton, 1945).

In his definition, Wirth speaks of physical and cultural characteristics and not of gender, age, disability, or undesirable behavioral patterns. Clearly he is referring to racial and ethnic groups in his definition of minorities. Some writers have suggested, however, that many other groups are in the same position as those more commonly thought of as minorities and endure the same sociological and psychological problems. In this light, women, homosexuals, adolescents, the aged, the handicapped, the radical right or left, and intellectuals can be thought of as minority groups.

Problems in Race and Ethnic Relations

All too often when people with different racial and ethnic identities come together, frictions develop among the various groups. People's suspicions and fears are often aroused by those whom they feel to be different.

Prejudice

There are many definitions of prejudice. People, particularly those with a strong sense of identity, often have feelings of prejudice toward others who are not like themselves. For example, in 1945, 56% of the American public said they opposed a law that would require employees to work alongside persons of any race or color (Gallup Poll, June 14–19, 1945). In 1966, 3 in 10 Americans (31%) thought houses for sale in their neighborhood should not be offered to anyone regardless of race or nationality (Gallup Poll, Mar. 24–29, 1966).

Literally, *prejudice* means a "prejudgment" or "an attitude with an emotional bias" (Wirth, 1944). However, there is a problem with this definition. All of us, through the process of socialization, acquire attitudes, which may not be in response only to racial and ethnic groups but also to many things in our environment. We come to have attitudes toward cats, roses, blue eyes, chocolate cheesecake, television programs, and even ourselves. These attitudes run the gamut from love to hate, from esteem to contempt, from loyalty to indifference. How have we developed these attitudes? Has it been through the scientific evaluation of information, or by other, less logical means? For our purposes we shall define **prejudice** as *an irrationally based negative, or occasionally positive, attitude toward certain groups and their members.*

What is the cause of prejudice? Although pursuing that question is beyond the scope of this book, we can list some of the uses to which prejudice is put and the social functions it serves. First, a prejudice, simply because it is shared, helps draw together those who hold it. It promotes a feeling of "we-ness," of being part of an in-group, and it helps define such group boundaries. Especially in a complex world, belonging to an in-group and consequently feeling special or superior can be an important social identity for many people.

Second, when two or more groups are competing for access to scarce resources (jobs, for example), it makes it easier on the conscience if one can write off his or her competitors as somehow less than human or inherently unworthy. Nations at war consistently characterize each other negatively, using terms that seem to deprive the enemy of any humanity whatsoever.

Third, psychologists suggest that prejudice allows us to project onto others those parts of ourselves that we do not like and therefore try to avoid facing. For example, most of us feel stupid at one time or another. How comforting it is to know that we belong to a group that is inherently more intelligent than another group. Who does not feel lazy sometimes? But how good it is that we do not belong to that group—the one everybody knows is lazy.

Of course, prejudice also has many negative consequences, or dysfunctions, to use the sociological term. For one thing, it limits our vision of the world around us, reducing social complexities and richness to a sterile and empty caricature. Aside from this effect on us as individuals, prejudice also has negative consequences for the whole of society. Most notably, it is the necessary ingredient of discrimination, a problem found in many societies, including our own. (See "Society and the Internet: Hate on the Web" for an example of new outlets for prejudice.)

Discrimination

Prejudice is a subjective feeling, whereas discrimination is an overt action. **Discrimination** refers to *differential treatment, usually unequal and injurious, accorded to individuals who are assumed to belong to a particular category or group.* Discrimination against African Americans and other minorities has occurred throughout U.S. history. At the start of World War II, for example, there were no blacks in the Marine Corps; blacks could be admitted into the Navy only as mess stewards; and the Army had a 10% quota on black enlistments. (For an example of how the military deals with racial issues, see "Our Diverse Society: Racial Integration in the Military.")

Prejudice does not always result in discrimination. Although our attitudes and our overt behavior are closely related, they are neither identical nor dependent on each other. We may have feelings of antipathy without expressing them overtly or even giving the slightest indication of their presence. This simple fact—namely, that attitudes and overt behavior vary independently—has been applied by Robert K. Merton (1966) to the classification of racial prejudice and discrimination. There are, he believes, the following four types of people.

Unprejudiced Nondiscriminators These people are neither prejudiced against the members of other racial and ethnic groups, nor do they practice discrimination. They believe implicitly in the American ideals of justice, freedom, equality of opportunity, and dignity of the individual. Merton recognizes that people of this type are properly motivated to spread the ideals and values of the creed and to fight against those forms of discrimination that make a mockery of them. At the same time, unprejudiced nondiscriminators have their shortcomings. They enjoy talking to one another, engaging in mutual

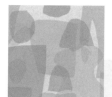

Society and the Internet

Hate Sites on the Web

In recent years we have seen the rise of hate sites on the World Wide Web. Hate sites are generally defined as those that express prejudiced and resentful views about a particular group of people, such as blacks or Jews. The Southern Poverty Law Center (SPLC), a civil-rights group in Montgomery, Alabama, regularly monitors these sites. There may be as many as 800 hate sites.

Some analysts fear that the proliferation of hate sites has allowed racist and bigoted messages to spread faster than ever before. They say the Internet has proven popular with hate groups and prejudiced individuals because it offers an inexpensive and anonymous way of disseminating information to a potentially large audience.

In addition to sites that are specifically anti-black, anti-Jewish, or anti-immigrant, there are hate sites known as "Third Position" sites. They express a mix of "left" and "right" ideas with strong neo-fascist overtones. They can be divided into two groups. The first group includes those sites affiliated with the European Liberation Front and its ally, the Liason Committee for Revolutionary Nationalism. Many of those behind this first group of sites are adherents of Odinism, a neo-Pagan religion popular among Skinheads. The second group of sites are affiliated with a key British group, the International Third Position.

Many people would like to ban hate sites. Yahoo! Inc. is a leading Internet media company that allows users to "chat" and post messages online. Yahoo! users, according to the company's terms of service, may not "promote bigotry, racism, hatred, or harm of any kind against any group or individual."

But groups who champion free speech, such as the American Civil Liberties Union (ACLU), oppose proposals to ban hate sites. "The concept and the principle is that the hate groups have First Amendment rights as they would have in the open market-place of ideas. So if you can display your prejudices on a street corner . . . , then the same principles and rules should apply to the Internet."

In an attempt at a compromise, an antidiscrimination group in New York City has created a computer program called "HateFilter" that blocks access to sites deemed to be bigoted. HateFilter, along with similar products by other companies, is designed for use by parents who want to prevent their children from viewing hate sites.

In addition, some observers fear filters could become subject to manipulation by interest groups who want certain sites blocked. At the request of gay activist groups, CyberNOT software blocks the Internet site of the American Family Association, a group that believes homosexuality is immoral and anti-Christian. The group does not consider itself a hate site and believes its First Amendment rights are being violated.

Some believe our fear of hate sites is overblown. David Goldman, executive director of the Hatewatch Web site, a site that from 1995 to 2001 monitored hate sites, believes hate groups, who once thrived and created fear in the shadows, wither and hide from the public scrutiny on the Internet. What these groups did not count on was that forcing their way into people's homes via the Web would have the effect of mobilizing ordinary people to join in the fight against them. "Far from persuading a supposed 'silent white majority' of angry Aryans to join their ranks, these self proclaimed white warriors made moms and dads into determined anti-hate activists."

Goldman believes the news is much more encouraging than any expected. Hate groups have done an extremely poor job of using the Internet to increase their membership. They have utterly failed to gain widespread acceptance for their belief that bigotry, hate, and violence are viable responses to human diversity. This is not to say that we no longer have cause for concern. The advent of the "lone wolf" gunman whose hatred may be fed by hate group propaganda, bigoted organizations who use e-commerce to support their hateful enterprises, and the newly emerging racist cyberterrorists, all will continue to present great challenges to law enforcement and online civil rights.

Source: "Hate Speech on the Internet," *June 4, 1999, Issues and Controversies on File;* David Goldman, Hatewatch.org.

exhortation and thereby giving psychological support to one another. They believe their own spiritual house is in order, thus they do not feel pangs of guilt and accordingly shrink from any collective effort to set things right.

Unprejudiced Discriminators This type includes those who constantly think of expediency. Though they themselves are free from racial prejudice, they will keep silent when bigots speak out. They will not condemn acts of discrimination but will make

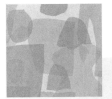

Sociology at Work

Racial Integration in the Military

When one thinks of those institutions in America that have been at the vanguard of social change, the U.S. Army does not spring readily to mind. Yet over the past two decades, the Army has become the most successfully integrated institution in America—from the ranks of the lowliest privates to the highest level of the command. Sociologists Charles Moskos and John Sibley Butler studied the military and found that what has made the Army's experience so striking is that this success was achieved without resort to numerical quotas or manipulation of test scores, nor has the promotion of black officers engendered the racial resentment that has become all too common in business, government, and higher education.

A visitor to an Army dining facility (as the old "mess hall" has been renamed) is likely to see a sight rarely encountered elsewhere in American life: blacks and whites commingling and socializing by choice. This stands in stark contrast to the self-imposed racial segregation in most university dining halls today—not to mention within most other locales in our society. In the Army whites and blacks not only inhabit the same barracks but also patronize equally such nonduty facilities as barber shops, post exchanges, libraries, movie theaters, and snack bars. And, in the course of their military duties, blacks and whites work together with little display of racial animosity. Give or take a surly remark here, a bruised sensibility there, the races get along remarkably well.

Even off duty and off post, far more interracial mingling is noticeable around military bases than in civilian life. Most striking, the racial integration of military life has some carry-over into the civilian sphere. The most racially integrated communities in America are towns with large military installations.

One key difference between the way the Army and many civilian organizations reflect the racial climate is that an officer's failure to maintain a bias-free environment is an absolute impediment to advancement in a military career. Most soldiers we have spoken to could not conceive of an officer who expressed racist views being promoted. We know of many civilian organizations in which this is not true.

Another, perhaps more important, distinction is that the Army does not lower its standards in order to assure an acceptable racial mix. When necessary, the Army makes an effort to compensate for educational or skill deficiencies by providing specialized, remedial training. Affirmative action exists, but without timetables or quotas governing promotions. What goals do exist are pegged to the proportion of blacks in the service promotion pool. Even then, these goals can be bypassed if the candidates do not meet standards.

In this regard, compared to most private organizations, the Army has an obvious advantage. The Army can maintain standards while still promoting African-Americans at all levels because of the large number of black personnel within the organization. The Army's experience with a plenitude of qualified black personnel illuminates an important lesson. When not marginalized, Afro-American cultural patterns can mesh with and add to the effectiveness of mainstream organizations. The overarching point is that the most effective and fairest way to achieve racial equality of opportunity in the United States is to increase the number of qualified Afro-Americans available to fill positions. Doing so is no small task. But as an objective and basic principle, it is infinitely superior to a system under which blacks in visible positions of authority are presumed to have benefited from relaxed standards.

Source: *All That We Can Be: Black Leadership and Racial Integration in the Army* (pp. 2, 9–10), by C. C. Moskos and J. Sibley Butler, 1996, New York: Basic Books.

concessions to the intolerant and will accept discriminatory practices for fear that to do otherwise would hurt their own position.

Prejudiced Nondiscriminators This category is for the timid bigots who do not accept the tenet of equality for all but conform to it and give it lip service when the slightest pressure is applied. Here belong those who hesitate to express their prejudices when in the presence of those who are more tolerant. Among them are the employers who hate certain minorities but hire them rather than run afoul of affirmative-action laws and the labor leaders who suppress their personal racial bias when the majority of their followers demand an end to discrimination.

Prejudiced Discriminators These are the bigots, pure and unashamed. They do not believe in equality, nor do they hesitate to give free expression to their intolerance, both in their speech and in their

Figure 8-1 The Interaction of Prejudice and Discrimination

As this diagram shows, the degree of social pressure being exerted can cause individuals of inherently dissimilar attitudes to exhibit relatively similar behaviors in a given situation.

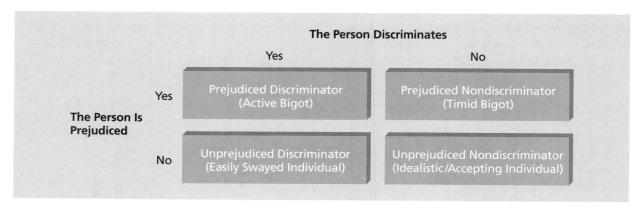

actions. For them there is no conflict between attitudes and behavior. They practice discrimination, believing that it is not only proper but in fact their duty to do so (Berry & Tischler, 1978).

Knowing a person's attitudes does not mean that that person's behavior always can be predicted. Attitudes and behavior are frequently inconsistent because of such factors as the nature and magnitude of the social pressures in a particular situation. The influence of situational factors on behavior can be seen in Figure 8-1. (For a discussion of how prejudice in the United States differs from that in other countries, see "Global Sociology: Worldwide Racial and Ethnic Prejudice").

Institutional Prejudice and Discrimination

Sociologists also tend to distinguish between individual and institutional prejudice and discrimination. When individuals display prejudicial attitudes and discriminatory behavior, it is often based on the assumption of the out-group's genetic inferiority. By contrast, when there is **institutionalized prejudice and discrimination,** we are referring to *complex societal arrangements that restrict the life chances and choices of a specifically defined group, in comparison with those of the dominant group.* In this way, benefits are given to one group and withheld from another. Society is structured in such a way that people's values and experiences are shaped by a prejudiced social order. Discrimination is seen as a by-product of a purposive attempt to maintain social, political, and economic advantage (Davis, 1979).

Some people argue that institutionalized prejudice and discrimination are responsible for the substandard education that many African Americans receive in the United States. Schools that are predominantly black tend to be inferior at every level to schools that are predominantly white. The facilities for blacks are usually of poorer quality than are those for whites. Many blacks also attend unaccredited black colleges, where the teachers are less likely to hold advanced degrees and are poorly paid. The poorer education that blacks receive is one of the reasons they generally are in lower occupational categories than whites are. In this way, institutionalized prejudice and discrimination combine to maintain blacks in a disadvantaged social and economic position (Kozol, 1992).

Patterns of Racial and Ethnic Relations

Relations among racial and ethnic groups seem to include an infinite variety of human experiences. They run the gamut of emotions and appear to be utterly unpredictable, capricious, and irrational. They range from curiosity and hospitality, at the one extreme, to bitter hostility at the other. In this section, we shall show that there is a limited number of outcomes when racial and ethnic groups come into contact. These include assimilation, pluralism, subjugation, segregation, expulsion, and annihilation. In some cases, these categories overlap—for instance, segregation can be considered a form of subjugation—but each has distinct traits that make it worth examining separately. (For a discussion of the value of the race and IQ debate, see "Controversies in Sociology: Is the Race and Intelligence Debate Worthwhile?")

Assimilation

In 1753, 23 years before he signed the Declaration of Independence, Benjamin Franklin wondered about the costs and benefits of German immigration. On

Global Sociology

Worldwide Racial and Ethnic Prejudice

We hear so much about racial and ethnic issues in the United States that we sometimes lose sight of the fact that prejudice and discrimination are worldwide phenomena. Europe has a long history of animosity and discrimination against minority groups. How do European attitudes toward minorities living within their boundaries compare with those of white Americans toward African Americans? According to one study Europeans seem to be quite intolerant of a wide variety of groups. According to Table 8-2, 30% of the people in Lithuania have an unfavorable view of Poles; 42% of Poles have an unfavorable view Ukrainians; and 42% of Ukrainians have an unfavorable view of Azerbaijanis. The most "tolerant" group seems to be the British, with only 21% claiming an unfavorable view of the Irish.

The study also showed that 13% of white people in the United States had an unfavorable view of African Americans. In an ideal world, the number should be zero. Compared to Europeans, however, it appears that white Americans are seemingly less prejudiced. The skeptical response would be that Americans have simply become more adept at hiding their true feelings about race. Yet it would be hard to see why Americans would be more hypocritical on this point. If anything the French

Table 8-2

Percentage of People with Unfavorable Attitudes Toward the Principal Minority Group in Their Country

COUNTRY	MINORITY GROUP	PERCENT WHO VIEW UNFAVORABLY
United States	African Americans	13
Czechoslovakia	Hungarians	49
Germany	Turks	45
France	North Africans	42
Ukraine	Azerbaijanis	42
Poland	Ukrainians	42
Hungary	Romanians	40
Lithuania	Poles	30
Bulgaria	Turks	39
Great Britain	Irish	21

and the British, who pride themselves on their traditions of tolerance, would have more of a reason to disguise their prejudices.

Source: *The Pulse of Europe: A Survey of Political and Social Values and Attitudes* (sec. VIII), by the Times Mirror Center for the People and the Press, 1991, Washington, DC.: Times Mirror Center.

the one hand, he commented that they are "the most stupid" and "great disorders may one day arise among us" because of these immigrants. Yet at the same time he pointed out that they "contribute greatly to the improvement of a country." He finally decided the benefits of German immigration could outweigh the costs if we "distribute them more equally, mix them with the English, establish English schools where they are now too thick settled" (Borjas, 1999).

What Franklin was concerned with was the assimilation of the German immigrants. **Assimilation** *is the process whereby groups with different cultures come to have a common culture.* It refers to more than just dress or language, and it includes less tangible items such as values, sentiments, and attitudes. We are really referring to the fusion of cultural heritages.

Assimilation is the integration of new elements with old ones. The transferring of culture from one group to another is a highly complex process, often

involving the rejection of ancient ideologies, habits, customs, attitudes, and language. It includes, also, the elusive problem of selection. Of the many possibilities presented by the other culture, which ones will be adopted? Why did the Native Americans, for example, when they were confronted with the white civilization, take avidly to guns, horses, rum, knives, and glass beads, while showing no interest in certain other features to which whites themselves attached the highest value?

In the process of assimilation, one society sets the pattern, for the give and take of culture seems never to operate on a 50-50 basis. Invariably, one group has a much larger role in the process than the other does, and various factors interact to make it so. Usually one of the societies enjoys greater power or prestige than the other, giving it an advantage in the assimilation process; one is better suited to the environment than the other; or one has greater numerical strength than the other. Thus, the pattern for the United States was set by the British colonists, and

In their controversial book "The Bell Curve" Richard Hernstein and Charles Murray argued that a transformation was taking place in American society with a cognitive elite becoming the leaders of the country and another group of people with lesser endowments and abilities doomed to labor on the fringes of society. Herrnstein and Murray also noted that American society was experiencing a crisis because certain groups, such as African-Americans, possessed average intelligence levels below those of white Americans. Sociologist Orlando Patterson questions the interest that people have shown in this proposition. He suggests that the discussion would not be important if we were committed to accepting every group as a member of a society that cannot be broken or separated into parts.

It has long been established that significant differences exist in measured average IQ. . . . [Whites] in Tennessee and rural Georgia score significantly lower than [whites] in the Northeast, and, on average, live in far worse conditions than their northern counterparts. No one, however, has ever chosen to call the nation's attention to these differences, except in a sympathetic and appropriately sensitive manner. We do not neglect them, but neither do we make a national issue of them, in the process wantonly insulting and dishonoring these people. No one raises the issue of whether or not these two states in the South are an intellectual drag on the nation; no one bemoans the fact that were they not a part of the nation our ranking in the international IQ parade would be much higher. . . . Why so? Because [whites] in Tennessee and rural Georgia are seen as belonging to the social and moral community that constitutes the American people. Whatever we are incorporates them. If they are different from Northeasterners in one or more respects, the reasoning goes, that is just a fact of our national life, a part of the diverse fabric we take pride in.

The same goes for the elderly. Until very recently, there was a consensus that intellectual functioning declined with age, especially after middle age. . . . According to recent estimates, by 2025 one in five Americans will be over sixty-five. . . . [There] has also been the growing tendency of those over sixty-five remaining in the workplace, reflected most dramatically in the abolition of compulsory ages of retirement. Hence we now face the prospect that one of the fastest growing and most powerful segments of the population—including the working population—is, according to the IQ specialists, an intellectually impaired group with rapidly declining cognitive competence. This surely ought to provoke

Orlando Patterson.

great alarm among those who make it their business to guard the intellectual integrity of the nation. After all, one in five is substantially larger than 13 percent, which is the menace posed by intellectually inferior Afro-Americans; and the elderly are growing a lot faster, from all demographic projections. And yet, we have heard almost nothing on this subject from any of those accustomed to warning the nation about impending psychological and social disasters.

Why not? For the same reason that we do not engage in public handwringing about the intellectual inferiority of rural [white] Georgians and Tennesseeans. The elderly are an integral part of our community; whatever we are as a people, the reasoning goes—and quite rightly, I might add—they help define it. If an important and essential part of what we are is in intellectual decline, then so are we, and so be it.

A principle is at work here. . . . We may call it the principle of infrangibility. It refers to our commitment to a unity that cannot be broken or separated into parts, a commitment to the elements of a moral order and social fabric that is inviolable and cannot be infringed. Afro-Americans too are an infrangible part of the nation.

Source: *The Ordeal of Integration*, (pp. 137–139), by O. Patterson, 1997, Washington: DC: Civitas Counterpoint.

to that pattern the other groups were expected to adapt. This process has often been referred to as **Anglo conformity**—*the renunciation of the ancestral cultures in favor of Anglo-American behavior and values* (Berry & Tischler, 1978; Gordon, 1964).

The Anglo-conformity viewpoint was at its strongest around World War I, as demonstrated by this excerpt from a speech by President Woodrow Wilson:

> You cannot dedicate yourself to America unless you become in every respect and with every purpose of your will thorough Americans. You cannot become thorough Americans if you think of yourselves in groups. America does not consist of groups. A man who thinks of himself as belonging to a particular national group in America has not yet become an American, and the man who goes among you to trade upon your nationality is no worthy son to live under the Stars and Stripes. (Wilson, 1915)

Although assimilation frequently has been a professed political goal in the United States, it has seldom been fully achieved. Consider the case of the Native Americans: In 1924, they were granted full United States citizenship. Nevertheless, the federal government's policies regarding the integration of Native Americans into American society wavered back and forth until the Hoover Commission Report of 1946 became the guideline for all subsequent administrations. The report stated that:

> A program for the Indian peoples must include progressive measures for their complete integration into the mass of the population as full, tax-paying [members of the larger society]. . . . Young employable Indians and the better cultured families should be encouraged and assisted to leave the reservations and set themselves up on the land or in business. (Shepardson, 1963)

However, to this day Native American groups remain largely unassimilated into the mainstream of American life. About 54% live on or near reservations, and most of the rest live in impoverished urban areas. In addition, many Native Americans who left the reservation for greater opportunity in America's cities are returning to the reservation. Despite the economic and lifestyle hardships they face on the reservation, their ethnic pride overrides any desire to assimilate.

Other groups, whether or not by choice, also have not assimilated. The Amish, for instance, have steadfastly maintained their subculture in the face of Anglo-conformity pressures from the larger American society.

China provides an interesting example of what might be called reverse assimilation. Usually it is the defeated minority groups who are assimilated into the culture of the politically dominant group. In the seventeenth century, however, Mongol invaders conquered China and installed themselves as rulers. The Mongols were nomadic pastoralists. They were so impressed with the advanced achievements of the Chinese civilization that they gave up their own ways and took on the trappings of Chinese culture: language, manners, dress, and philosophy. During their rule, the Mongols fully assimilated the Chinese culture.

Pluralism

Pluralism, or *the development and coexistence of separate racial and ethnic group identities within a society,* is a philosophical viewpoint that attempts to produce what is considered to be a desirable social situation. When people use the term *pluralism* today, they believe they are describing a condition that seems to be developing in contemporary American society. They often ignore the ideological foundation of pluralism.

The person principally responsible for the development of the theory of cultural pluralism was Horace Kallen, born in Germany. He came to Boston at the age of 5 and was raised in an Orthodox Jewish home. As he progressed through the Boston public schools, he underwent a common second-generation phenomenon. He started to reject his home environment and religion and developed an uncritical enthusiasm for the United States. As he put it, "It seemed to me that the identity of every human being with every other was the important thing, and that the term 'American' should nullify the meaning of every other term in one's personal makeup."

While Kallen was a student at Harvard, he experienced a number of shocks. Working in a nearby settlement house, he came in contact with liberal and socialist ideas and observed people expressing numerous ethnic goals and aspirations. This exposure caused him to question his definition of what it meant to be an American. This quandary was compounded by his experiences in the American literature class of Barrett Wendell, who believed that Puritan traits and ideals were at the core of the American value structure. The Puritans, in turn, had modeled themselves after the Old Testament prophets. Wendell even suggested that the early Puritans were largely of Jewish descent. These ideas led Kallen to believe that he could be an unassimilated Jew and still belong to the core of the American value system.

After discovering that he could be totally Jewish and still be American, he came to realize that the application could be made to other ethnic groups as well. All ethnic groups, he felt, should preserve their own separate cultures without shame or guilt. As he put it, "Democracy involves not the elimination of differences, but the perfection and conservation of differences."

Pluralism is a reaction against assimilationism and the melting-pot idea. It is a philosophy that not only assumes that minorities have rights but also considers the lifestyle of the minority group to be a legitimate, and even desirable, way of participating in society. The theory of pluralism celebrates the differences among groups of people. The theory also implies a hostility to existing inequalities in the status and treatment of minority groups. Pluralism has provided a means for minorities to resist the pull of assimilation, by allowing them to claim that they constitute the very structure of the social order. From the assimilationist point of view, the minority is seen as a subordinate group that should give up its identity as quickly as possible. Pluralism, on the other hand, assumes that the minority is a primary unit of society and that the unity of the whole depends on the harmony of the various parts.

Switzerland provides an example of balanced pluralism that, so far, has worked out exceptionally well (Kohn, 1956). After a short civil war between the Catholics and the Protestants in 1847, a new constitution—drafted in 1848—established a confederation of cantons (states), and church-state relations were left up to the individual cantons. The three major languages—German, French, and Italian—were declared official languages for the whole nation, and their respective speakers were acknowledged as political equals (Petersen, 1975).

Switzerland's linguistic regions are culturally quite distinctive. Italian-speaking Switzerland has a Mediterranean flavor; in French-speaking Switzerland, one senses the culture of France; and German-speaking Switzerland is distinctly Germanic. However, all three linguistic groups are fiercely pro-Swiss, and the German-Swiss especially have strong anti-German sentiments.

Subjugation

In theory, we could assume that two groups may come together and develop an egalitarian relationship. However, there are few cases in which racial and ethnic groups have established such a relationship. One of the consequences of the interaction of racial and ethnic groups has been **subjugation**—*the subordination of one group and the assumption of a position of authority, power, and domination by the other.* The members of the subordinate group may for a time accept their lower status and even devise ingenious rationalizations for it.

For the most part this is so because there are few instances in which group contact has been based on the complete equality of power. Differences in power will invariably lead to a situation of superior and inferior position. The greater the discrepancy in the power of the groups involved, the greater the extent and scope of the subjugation will be.

By the mid-1870s, the various Native Americans were at the mercy of the white Europeans. Native American traditions, beliefs, and ways of life were condemned as backward and immoral. It was thought that the best way to help Native Americans was to ensure that their tribal cultures were destroyed. The Indian peoples would have to be forced to assimilate into the mainstream culture. President Benjamin Harrison's commissioner of Indian affairs, Thomas Jefferson Morgan, expressed this view in 1889 when he noted:

> The logic of events demands the absorption of the Indians into our national life, not as Indians, but as American citizens. . . . The Indians must conform to "the white man's ways," peaceably if they will, forcibly if they must. . . . This civilization may not be the best possible, but it is the best the Indians can get. They cannot escape it, and must either conform to it or be crushed by it. (Josephy, 1994).

To subjugate and "Americanize" the Native Americans, the government banned their religions, rituals, and sacred ceremonies. Medicine men and shamans were either jailed or exiled. Attempts were even made to stop the speaking of tribal languages. The final step was to send Native American children to boarding schools, where they were taught to become part of the white culture. As a Taos Pueblo youth noted:

> We all wore white man's clothes and ate white man's food and went to white man's churches and spoke white man's talk. And so after a while we also began to say Indians were bad. We laughed at our own people and their blankets and cooking pots and sacred societies and dances. (Embree, 1939/1967)

Why should different levels of power between two groups lead to the domination of one by the other? Gerhard Lenski (1966) proposed that it is because people have a desire to control goods and services. No matter how much they have, they are never satisfied. In addition, high status is often associated with the consumption of goods and services. Therefore demand will exceed supply, and as Lenski asserts, a struggle for rewards will be present in every human society. The outcome of this struggle thus will lead to the subjugation of one group by the other. When a racial or ethnic group is placed in an inferior position, its people often are eliminated as

The subjugation of Native Americans is evident in these before and after photos of three Sioux pupils at the Carlisle Indian School in Pennsylvania. At left are the three boys upon their arrival at the school. At the right we see them 6 months later.

competitors. In addition, their subordinate position may increase the supply of goods and services available to the dominant group.

Segregation

Segregation, *a form of subjugation, refers to the act, process, or state of being set apart.* It is a situation that places limits and restrictions on the contact, communication, and social relations among groups. Many people think of segregation as a negative phenomenon—a form of ostracism imposed on a minority by a dominant group—and this is most often the case. However, for some groups such as the Amish or Chinese, who wish to retain their ethnicity, segregation is voluntary.

The practice of segregating people is as old as the human race itself. There are examples of it in the Bible and in preliterate cultures. American blacks originally were segregated by the institution of slavery and later by both formal sanction and informal discrimination. Although some African-Americans formed groups that preached total segregation from whites as an aid to black cultural development, for most it is an involuntary and degrading experience. The word *ghetto* is derived from the term for the segregated quarter of a city where the Jews in Eu-

rope were often forced to live. Native American tribes were often forced to choose segregation on a reservation in preference to annihilation or assimilation. Segregation has operated in a wide range of circumstances.

Expulsion

Expulsion is *the process of forcing a group to leave the territory in which it resides.* This can be accomplished indirectly by making life increasingly unpleasant for a group, as the Germans did for Jews after Adolf Hitler was appointed chancellor in 1933. Over the following 6 years, Jews were stripped of their citizenship, made ineligible to hold public office, removed from the professions, and forced out of the artistic and intellectual circles to which they had belonged. In 1938, Jewish children were barred from the public schools. At the same time, the government encouraged acts of violence and vandalism against Jewish communities. These actions culminated in Kristallnacht, November 9, 1938, when the windows in synagogues and Jewish homes and businesses across Germany were shattered and individuals were beaten. Under these conditions, Jews left Germany by the thousands. In 1933 there were some 500,000 Jews in Germany; by 1940, before

Hitler began his "final solution"—that is, the murder of all remaining Jews—only 220,000 remained (Robinson, 1976).

Expulsion also can be accomplished through **forced migration**, *the relocation of a group through direct action.* For example, forced migration was a major aspect of the U.S. government's policies toward Native American groups in the nineteenth century. When the army needed to protect its lines of communication to the West Coast, Colonel Christopher "Kit" Carson was ordered to move the Navajos of Arizona and New Mexico out of the way. He was instructed to kill all the men who resisted and to take everybody else captive. He accomplished this in 1864 by destroying their cornfields and slaughtering their herds of sheep, thereby confronting the Navajos with starvation. After a last showdown in Canyon de Chelly, some 8,000 Navajos were rounded up at Fort Defiance. They then were marched on foot 300 miles to Fort Sumner, where they were to be taught the ways of "civilization" (Spicer, 1962).

Although expulsion is an extreme attempt to eliminate a certain minority from an area, annihilation is the most extreme action one group can take against another.

Annihilation

Annihilation refers to *the deliberate extermination of a racial or ethnic group.* In recent years it has also been referred to as *genocide,* a word coined to describe the crimes committed by the Nazis during World War II—crimes that induced the United Nations to draw up a convention on genocide. Annihilation is the denial of the right to live of an entire group of people, in the same way that homicide is the denial of the right to live of one person.

Sometimes annihilation occurs as an unintended result of new contact between two groups. For example, when the Europeans arrived in the Americas, they brought with them a disease, smallpox, new to the people they encountered. Native American groups, including the Blackfeet, the Aztecs, and the Incas, who had no immunity at all against this disease, were nearly wiped out (McNeill, 1976). In most cases, however, the extermination of one group by another has been the result of deliberate action. Thus, the native population of Tasmania, a large island off the coast of Australia, was exterminated by Europeans in the 250 years after the island was discovered in 1642.

The largest, most systematic program of ethnic extermination was the killing of 11 million people, close to 6 million of whom were Jews, by the Nazis before and during World War II. In each country occupied by the Germans, the majority of the Jewish population was killed. Thus, in the mid-1930s, before the war, there were about 3.3 million Jews in Poland, but at the end of the war in 1945 there were only 73,955 Polish Jews left (Baron, 1976). Among them, not a single known family remained intact.

Although there have been attempts to portray this mass murder of Jews as a secret undertaking of the Nazi elite that was not widely supported by the German people, the historical evidence suggests otherwise. For example, during a wave of anti-Semitism (anti-Jewish prejudice, accompanied by violence and repression) in Germany in the 1880s—long before the Nazi regime—only 75 German scholars and other distinguished citizens protested publicly.

These Jews are arriving at the Nazi concentration camp know as Auschwitz. Most of them were murdered and cremated a short time later as part of the German policy of genocide against the Jews.

During the 1930s, the majority of German Protestant churches endorsed the so-called "racial" principles that were used by the Nazis to justify first the disenfranchisement of Jews, then their forced deportation, and finally their extermination. (Jews were blamed for a bewildering combination of "crimes," including "polluting the purity of the Aryan race" and causing the rise of communism while at the same time manipulating capitalist economies through their "secret control" of banks.)

It would seem, then, that the majority of Germans supported the Nazi racial policies or at best were apathetic (Robinson, 1976). Although in 1943 both the Catholic church and the anti-Nazi Confessing church finally condemned the killing of innocent people and pointedly stated that race was no justification for murder, it is fair to say that even this opposition was "mild, vague, and belated" (Robinson, 1976). The fact that such objections were raised points out that the Nazis' plan to exterminate all Jews was not a well-kept military secret. The measure of its success is that some 60% of all Jews in Europe—36% of all Jews in the world—were slaughtered (computed from figures in Baron, 1976).

Another "race" also slated for extermination by the Nazis were the Gypsies, small wandering groups who appear to be the descendants of the Aryan invaders of India, the central Eurasian nomads. For the past thousand years or so, Gypsy bands had spread throughout the continents, largely unassimilated (Ulc, 1975). In Europe, they were widely disliked and constantly accused of small thefts and other criminal behavior.

The sheer magnitude and horror of the Nazi attempt to exterminate the Jews provoked outrage and attempts by the nations of the world to prevent such circumstances from arising again. On December 11, 1946, the General Assembly of the United Nations passed by unanimous vote a resolution affirming that genocide was a crime under international law and that both principals and accomplices alike would be held accountable and would be punished. The assembly called for a convention on genocide that would define the offense more precisely and provide enforcement procedures for its repression and punishment. After 2 years of study and debate, the draft of the convention on genocide was presented to the General Assembly and adopted. Article II of the convention defines genocide as

> any of the following acts committed with intent to destroy, in whole or in part, a national, ethnical, racial or religious group as such:
> (a) Killing members of the group;
> (b) Causing serious bodily or mental harm to members of the group;
> (c) Deliberately inflicting on the group conditions of life calculated to bring about its physical destruction in whole or part;
> (d) Imposing measures intended to prevent births within the group;
> (e) Forcibly transferring children of the group to another group.

The convention further provided that any of the contracting parties could call on the United Nations to take action under its charter for the "prevention and suppression" of acts of genocide. In addition, any of the contracting parties could bring charges before the International Court of Justice.

In the United States, President Harry Truman submitted the resolution to the Senate on June 16, 1949, for ratification. However, the Senate did not act on the measure, and the United States did not sign the document. In 1984, President Ronald Reagan again requested the Senate to hold hearings on the convention so that it could be signed. The United States finally signed the document in 1988.

In the more than 50 years of its existence, the Genocide Convention has never been used to bring charges of genocide against a country. Numerous examples of genocide have occurred during that period. It appears to serve more of a symbolic purpose, by asking nations to go on record as being opposed to genocide, than as an effective means of dealing with actual instances of genocide (Berry & Tischler, 1978).

Racial and Ethnic Immigration to the United States

Since the settlement of Jamestown in 1607, well over 45 million people have immigrated to the United States. Up until 1882, the policy of the United States was almost one of free and unrestricted admittance. The country was regarded as the land of the free, a haven for those oppressed by tyrants, and a place of opportunity. The words of Emma Lazarus, inscribed on the Statue of Liberty, were indeed appropriate:

> Give me your tired, your poor,
> Your huddled masses yearning to breathe free;
> The wretched refuse of your teeming shore.
> Send these, the homeless, tempest-tost to me,
> I lift my lamp beside the golden door!

To be sure, there were those who had misgivings about the immigrants. George Washington wrote to John Adams in 1794, "My opinion with respect to immigration is that except for useful mechanics and some particular descriptions of men or professions, there is no need for encouragement." Thomas

Between 1892 and 1924, some 16 million immigrants came through Ellis Island in New York Harbor on the way to their new life in the United States.

Jefferson was even more emphatic in expressing the wish that there might be "an ocean of fire between this country and Europe, so that it would be impossible for any more immigrants to come hither." Such fears, however, were not widely felt. There was the West to be opened, railroads to be built and canals dug; there was land for the asking. People poured across the mountains, and the young nation was eager for population.

Immigration of white ethnics to the United States can be viewed from the perspective of old migration and new migration. The old migration consisted of people from northern Europe who came before the 1880s. The new migration was much larger in numbers and consisted of people from southern and eastern Europe who came between 1880 and 1920. The ethnic groups that made up the old migration included the English, Dutch, French, Germans, Irish, Scandinavians, Scots, and Welsh. The new migration included Poles, Hungarians, Ukrainians, Russians, Italians, Greeks, Portuguese, and Armenians.

Figure 8-2 shows the number of immigrants who came to the United States in each year from 1820 to 2000. The new migration sent far more immigrants to the United States than the old migration. The earlier immigrants felt threatened by the waves of unskilled and uneducated newcomers, whose ap-

pearance and culture were so different from their own. Public pressure for immigration restriction increased. After 1921, quotas were established limiting the number of people who could arrive from any particular country. The quotas were designed specifically to discriminate against potential immigrants from the southern and eastern European countries. The discriminatory immigration policy remained in effect until 1965, when a new policy was established.

In Table 8-3 you will see a listing of the people who were excluded from immigrating to the United States during each of the periods in its history. As you can see, we were much more lenient during the early days of our history. However, even with our periods of restrictive immigration, the United States has had one of the most open immigration policies in the world, and we continue to take in more legal immigrants than the rest of the world combined (Kotkin & Kishimoto, 1988).

Immigration Today Compared to the Past

The past 35 years have seen a marked shift in United States immigration patterns. From the beginning of the country's birth until the 1960s, most immigrants came primarily from northwestern European

Figure 8-2 **Immigration to the United States, 1820–2000**

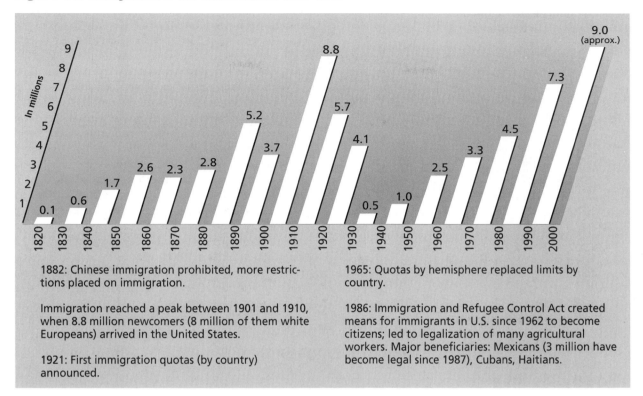

1882: Chinese immigration prohibited, more restrictions placed on immigration.

Immigration reached a peak between 1901 and 1910, when 8.8 million newcomers (8 million of them white Europeans) arrived in the United States.

1921: First immigration quotas (by country) announced.

1965: Quotas by hemisphere replaced limits by country.

1986: Immigration and Refugee Control Act created means for immigrants in U.S. since 1962 to become citizens; led to legalization of many agricultural workers. Major beneficiaries: Mexicans (3 million have become legal since 1987), Cubans, Haitians.

Table 8-3

United States Immigration Restrictions

1769–1875	No restrictions; open-door policy
1875	No convicts; no prostitutes
1882	No idiots; no lunatics; no people likely to need public care
	Start of head tax
1882–1943	No Chinese
1885	No gangs of cheap contract laborers
1891	No immigrants with dangerous contagious diseases; no paupers; no polygamists
	Start of medical inspections
1903	No epileptics; no insane people; no beggars; no anarchists
1907	No feeble-minded; no children under 16 unaccompanied by parents; no immigrants unable to support themselves because of physical or mental defects
	No immigrants from most of Asia or the Pacific islands; no adults unable to read or write
	Start of literacy tests
1921	No more than 3% of foreign-born of each nationality in U.S. in 1910; total about 350,000 annually
1924–1927	National Origins Quota Law; no more than 2% of foreign-born of each nationality already in U.S. in 1890; total about 150,000 annually
1940	Alien Registration Act; all aliens must register and be fingerprinted

1950	Exclusion and deportation of aliens dangerous to national security
1952	Codification, nationalization, and minor alterations of previous immigration laws
1965	National Origins Quota system abolished; no more than 20,000 from any one country outside the Western Hemisphere; total about 170,000 annually
	Start of restrictions on immigrants from other Western Hemisphere countries; no more than 120,000 annually
	Preference to refugees, aliens with relatives here, and workers with skills needed in the United States
1980	Congress passes the Refugee Act of 1980, repealing ideological and geographical preferences that had favored refugees fleeing communism and Middle Eastern countries
1986	The Immigration Reform and Control Act (IRCA) takes effect; grants amnesty to illegal immigrants residing in the United States since 1982; increases sanctions against employers for hiring illegal immigrants
1990	President George Bush increases immigration quotas
1996	Laws enacted to make it easier to deport immigrants who commit crimes
	Tougher restrictions on ability of immigrants to collect welfare

Source: Smithsonian Institution exhibit, Washington, DC.

Table 8-4

**Top Ten Countries of Origin of
U.S. Foreign-Born Residents**

COUNTRY	PERCENTAGE OF FOREIGN BORN
Mexico	27.2
Philippines	4.4
China and Hong Kong	4.3
Cuba	3.5
Vietnam	3.0
India	2.9
Dominican Republic	2.5
El Salvador	2.4
Great Britain	2.4
Korea	2.3

Source: "CPS Publication—*Country of Origin and Year of Entry Into the U.S. of the Foreign Born: March 1997*" U.S. Census Bureau; September 29, 1997; Statistical Abstract of the United States, 2000.

countries—Great Britain, Ireland, Germany, Scandinavia, France—and from Canada.

In 1965 a major change in U.S. immigration policy took place. The national origins quota system, which granted visas mainly to people coming from Western European countries, particularly Great Britain and Germany, was repealed. Under the new immigration system, family ties to persons already living in the United States became the key factor that determined whether a person was admitted into the country. The number of people allowed to come into the country was also increased.

The change has produced such a dramatic shift in the immigrants coming to the United States that by 1999, over half of the foreign-born population originated from Latin America (51%), 27% from Asia, and only 16% from Europe. Canada accounted for 3% of the foreign-born population, and Africa for 2% (see Table 8-4).

These shifts in patterns of immigration have resulted in a much more racially and ethnically diverse foreign-born population. In 1890, only 1.4% of the foreign-born population was nonwhite. By 1970, 27% of this population was nonwhite, and by 1999, 75% of this population was nonwhite.

With the large number of people immigrating to the United States, nearly one in ten U.S. residents (10%) were foreign-born in 2000. This is the highest percentage of foreign-born residents since World War II and double the 1970 level of 4.7% (see Figure 8-3). One-third of the foreign-born population was from Mexico or another Central American country and about one-fourth of this population was from Asia (U.S. Bureau of the Census, 2001).

Many of these people are recent arrivals. Of the 28.4 million foreign-born people living in the United States in March of 2000, 7.5 million had arrived during the previous 7 years. Nearly one-third of these new immigrants live in California, with New York and Florida attracting another 25% (U.S. Bureau of the Census, The Foreign-Born Population in the United States, March 2000).

Today's immigrants are unique in their ethnic origins, education, and skills. The waves of immigration during earlier periods in our history were mostly from European countries. Today's immigrants are primarily from Latin American and Asian countries. Their education levels are at two extremes. While most native-born people have completed high school, immigrants are only half as likely to have done so. At the same time many other immigrants are highly educated and immigrants are more likely than native-born to have advanced college degrees.

The U.S. Census Bureau projects that 880,000 legal immigrants a year will enter the country between 1993 and 2050. In 1999, 51% of the immigrants were from Latin America, 25.5% were from Asian countries, 15.3% were from Europe, and 8.1% were from the rest of the world (see Figure 8-4).

Figure 8-3 U.S. Foreign-Born Population, 1890–2000

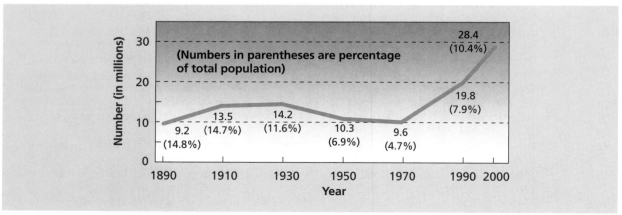

Source: U.S. Bureau of the Census.

Figure 8-4 Racial and Ethnic Makeup of U.S. Population, 2000 and 2050

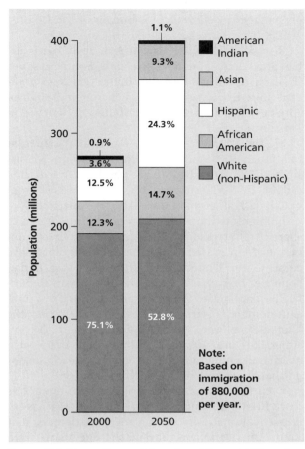

Source: U.S. Census Bureau, "Overview of Race and Hispanic Origin," Census 2000 Brief, March 2001; Population Projections Program, Population Division, U.S. Census Bureau.

Illegal Immigration

Since 1970, illegal immigration has figured prominently in the ethnic makeup of certain regions of the United States. The U.S. Immigration and Naturalization Service (INS) estimates that there are currently 3.5 to 4 million people in the country illegally, and that the number is growing by 200,000 a year. There are two types of illegal immigrants: those who enter the United States legally but overstay their visa limits and those who enter illegally to begin with. Those who migrate over the border each day to work must also be figured into the total.

Mexico, El Salvador, and Guatemala account for 44% of the total number of illegal immigrants. Illegal immigrants tend to settle in certain states. California is home to 40% of the total, with New York, Florida, Texas, and Illinois accounting for most of the rest.

In 1986, the Immigration Reform and Control Act was passed, a law designed to control the flow of illegal immigrants into the United States. The law makes it a crime for employers, even individuals hiring household help, to knowingly employ an illegal immigrant. Stiff fines and criminal penalties can be imposed if they do so. The law also provided legal status to illegal immigrants who entered the United States before 1982 and who have lived here continuously since then. Between 1989 and 1993, 2.7 million people were granted legal resident status under special provisions of the Immigration and Reform Act (IRCA). These people had lived and worked in the country illegally during the 1980s. Interestingly, only 4% of those who applied for amnesty under the 1986 act worked in farming, fishing, or forestry, running counter to the perception that most illegal immigrants are migrant farm workers.

What is America's racial and ethnic composition today? The United States is perhaps the most racially and ethnically diverse country in the world. Unlike many other countries, it has no ethnic group that makes up a numerical majority of the population. In the following discussion, we shall examine the major groups in American society.

White Anglo-Saxon Protestants

About 47 million people claim some English, Scottish, or Welsh ancestors. These Americans of British origin are often grouped together as white Anglo-Saxon Protestants (WASPs). Although in numbers they are a minority within the total American population, they have been in America the longest (aside from the Native Americans) and, as a group, have always had the greatest economic and political power in the country. As a result, WASPs often have acted as if they were the ethnic majority in America, influencing other ethnic groups to assimilate or acculturate to their way of life, the ideal of Anglo conformity.

The Americanization of immigrant groups has been the desired goal of the dominant WASPs during many periods in American history. Contrary to the romantic sentiments expressed on the base of the Statue of Liberty, immigrant groups who came to America after the British Protestants had become established met with considerable hostility and suspicion.

The 1830s and 1840s saw the rise of the "native" American movement, directed against recent immigrant groups (especially Catholics). In 1841, the American Protestant Union was founded in New York City to oppose the "subjugation of our country to control of the Pope of Rome, and his adherents" (Leonard & Parmet, 1971). On the East Coast it was the Irish Catholics who were feared, and in the Midwest it was the German "freethinkers." Protestant religious organizations across America joined forces and urged "native" Americans to organize to offset foreign voting blocs. They also conducted intimidation campaigns against foreigners and attempted to persuade Catholics to renounce their religion for Protestantism (Leonard & Parmet, 1971).

As the twentieth century dawned, American sentiments against immigrants from Southern and Eastern Europe were running high. In Boston, the Immigration Restriction League was formed, which directed its efforts toward keeping out racially "inferior" groups—who were depicted as inherently criminal, mentally defective, and marginally educable. The league achieved its goal in 1924, when the government adopted a new immigration policy that set quotas on the numbers of immigrants to be admitted from various nations. Because the quotas were designed to reflect (and reestablish) the ethnic composition of America in the 1890s, they heavily favored the admission of immigrants from Britain, Ireland, Germany, Holland, and Scandinavia. This new policy was celebrated as a victory for the "Nordic" race (Krause, 1966).

Another expression of Anglo-conformity pressure was the Americanization Movement, which gained strength from the nationalistic passions brought on by World War I. Its stated purpose was to promote the very rapid acculturation of new immigrants. Thus, "federal agencies, state government, municipalities, and a host of private organizations joined in the effort to persuade the immigrant to learn English, take out naturalization papers, buy war bonds, forget his former origins and culture, and give himself over to patriotic hysteria" (Gordon, 1975). From World War II until the early 1960s, Anglo conformity was pretty much an established ideal of the American way of life. Since the upheavals of the 1960s, however, there has been a strong organized reaction among other ethnic groups against Anglo conformity. Strong social-political movements, organized along ethnic group lines, have formed. African-Americans led the way in the late 1960s with the black power movement, and they were joined by Italian Americans, Mexican Americans (Chicanos), Puerto Ricans, Native Americans, and others. America once again is focusing on its ethnic diversity, and the assumptions of Anglo conformity are being questioned.

African-Americans

African-Americans represent the third-largest race/ethnic group in the United States. According to the 2000 census figures there were between 34.7 million and 36.4 million African-Americans in the United States. The exact number is uncertain because

people, for the first time, were able to check off more than one race on the 2000 census form. Blacks now make up 12.3% of the U.S. population, the largest percentage since 1880 (U.S. Bureau of the Census, 2001).

Roughly 86% of all African-Americans live in urban areas, and about 55% live in the Southern states (U.S. Bureau of the Census, 2000). This is a significant shift from the 1940s, when roughly 80% of American blacks lived in the South and worked in agriculture.

In three states, African-Americans represent more than 30% of the population. They are Mississippi (37%), Louisiana (32%), and South Carolina (30%). In the District of Columbia (Washington, D.C.), 62% of population is African-American.

Immigrants have been accounting for a greater share of all blacks in the United States as their numbers have increased. These black immigrants have come to the United States from the West Indian countries of Jamaica, Haiti, the Dominican Republic, Barbados, and Trinidad. In addition, a significant number of African blacks are also entering the United States each year. About 2 million African-Americans are foreign born. In some cities such as Miami, immigrants have become a clearly distinguishable segment of the black population.

Even though the number of African immigrants from Ethiopia, Ghana, Kenya, and Nigeria has increased dramatically during the past decade, it is Latin America in general, and the Caribbean basin specifically, that is the source of most U.S. immigrants of African descent. Of the African-American foreign-born population, about 63% came from Caribbean nations (U.S. Bureau the of Census, 2000).

The African-American economic situation has been improving. Although 9.1 million blacks representing 26.1% of the African-American population fell below the poverty rate in 1998, that was the lowest level since the U.S. Census Bureau began collecting poverty data in 1959. The median African-American household income of $25,351 was an all-time high in 1998. The white family income has been increasing more rapidly, however, and the ratio of African-American to white earnings has actually fallen (U.S. Bureau of the Census, 2000).

Why have African-American families lost ground over the past two decades? A major reason is the growth in female-headed families, who have only one-third the annual income of African-American married-couple families. Only 47% of African-American families are married-couple families. Another factor is that the average African-American family has fewer members in the labor force than white families. Even if African Americans and whites held comparable jobs and earned equal pay,

Immigrants of African descent from the Caribbean basin account for an increasingly greater share of all African-Americans in the United States.

the higher number of wage-earners per family for whites would still keep the average African-American family income below that of whites (U.S. Bureau of the Census, 2000). (See "Our Diverse Society: The Black Middle Class: Fact or Fiction?")

Hispanics (Latinos)

"Hispanic," "Hispano," "Latino," "Latin," "Mejicano," "Spanish," "Spanish-speaking"—which is the right term to use when referring to the Spanish-speaking population in the United States? Political, racial, linguistic, and historical arguments have been advanced for using each of these terms. Historically, usage has gone from Spanish or Spanish-speaking to Latin American, Latino, or Hispanic. Geographically, Hispanic is preferred in the Southeast and much of Texas. New Yorkers use both Hispanic and Latino. Some people in New Mexico prefer Hispano.

Politically, Hispanic has been identified with more conservative viewpoints, while Latino has been associated with more liberal politics. This is partially because Hispanic is an English word meaning "pertaining to ancient Spain." The U.S. Census Bureau has decided to settle on one term—"Hispanic."

The Hispanic population of the United States Increased by 58% between 1990 and 2000. This growth rate was significantly faster than the U.S. population as a whole. The 35.3 million Hispanics in the United States in 2000 constitute 12.5 of the total population and form the nation's second-largest minority. These totals do not include the 3.9 million people living in Puerto Rico. For the next 50 years, the Hispanic population is projected to add more people to the United States than any single race or ethnic group, reaching 98.2 million in 2050. Results from the 2000 census indicate that Hispanics have surpassed non-Hispanic African-Americans

as the nation's largest race/ethnic group behind non-Hispanic whites (U.S. Bureau of the Census, 2001).

Hispanics are not a very well understood segment of the population. First, no one knows exactly how many Hispanics have crossed the border from Mexico as illegal immigrants. Second, many Americans of Hispanic descent do not identify themselves as "Hispanic" on census forms and are not counted as such.

Nearly two-thirds (66.1%) of the nation's Hispanics are of Mexican origin; 14% are of Central and South American origin; 9% are of Puerto Rican origin; and 4% are of Cuban origin.

The states with the highest concentration of Hispanics are New Mexico (where Hispanics constitute 41% of the total population), California (32%), Texas (30%), Arizona (23%), Nevada (17%), and Florida, Colorado, and New York (15% each).

Although people born in Latin America can be found all across the United States, most live in only a few areas. The difference is the place of birth. For example, about three out of four U.S. residents born in the Caribbean lives either in the New York or the Miami metro areas. More than half of the Mexican-born population live in the Los Angeles and Chicago metro areas or the state of Texas (U.S. Bureau of the Census, 2001).

Interestingly, although the vast majority of Hispanics come from rural areas, 90% settle in America's industrial cities and surrounding suburbs. Living together in tightly knit communities, they share a common language and customs.

The Latin migration north is one of the largest mass migrations in U.S. history—about half of the Hispanic residents were born abroad. The adult newcomers begin their U.S. journey with many economic and educational disadvantages compared to the average American: 60% have not graduated

Our Diverse Society

The Black Middle Class: Fact or Fiction?

Fifty years ago the number of African-Americans who could be considered middle class was extremely small. Figures from that era show that 5% of black men and 6% of black women were engaged in white-collar work of any kind. Only 1 out of 20 blacks were either professionals, owners or managers of businesses, salespeople, or secretaries. Six out of 10 African-American women were household workers engaged in cleaning, cooking, or watching children for low wages. The majority of black men were unskilled laborers, sharecroppers, or domestic servants.

Today approximately 40% of all blacks can be considered middle class based on self-identification and income levels. This is about as large as the white middle class was at the end of President Dwight Eisenhower's second term, a time when American society as a whole was described as predominantly middle-class (Thernstrom & Thernstrom, 1997).

Sociologists Melvin Oliver and Thomas Shapiro believe it is premature to celebrate the rise of the black middle class. The glass is both half full and half empty, because at the same time that black wealth has grown, it has fallen even further behind that of whites. Oliver and Shapiro believe that materially whites and blacks constitute two nations with two middle classes. Blacks achieve middle class status because of income not assets, such as property, stocks, and other forms of wealth. In contrast, the white middle class possesses these assets. Even with similar levels of income, middle-class whites have nearly five times as much of their wealth in assets as middle-class blacks. A home, stocks, and savings all add security and stability to a family's lifestyle. The middle-class black position is precarious and fragile (Oliver & Shapiro, 1997).

Orlando Patterson believes that Oliver and Shapiro have discredited "the hard work, intelligence, and industriousness that middle-class [African-Americans] have put into acquiring the very real status and power they possess and the pride that they justly feel in their achievements."

Patterson notes that the extraordinary growth of income inequality in America since the mid-1970s is a national rather than a racial issue. All middle-class Americans have been losing ground to the wealthiest 1% of families, who now own over 42% of all wealth, compared with the 22% they owned in 1975. This is a good example of how a serious nonracial issue is translated into a racial one. If two nations are emerging in America, they are the haves and have-nots, a divide that cuts right across "race." Indeed there is actually greater inequality among the whole of African-Americans than between African-Americans and whites.

Almost all new middle classes in history have had precarious economic starts, including whites, most of whose families rose to middle-class status only after World War II. There is no reason to assume that African-Americans should follow the same paths into middle-class status as whites did 40 or more years ago. In the first place, the American economy has gone through a variety of changes and the avenues of social class mobility of 50 years ago may no longer be available. For many it may make more sense today to continually reinvest in the improvement of skills than to lock up funds in a mortgage. Second, it may well be that the African-American middle class has different lifestyle preferences and asset accumulation may not be one of them. If this is the case, then such choices are entirely their business. Taking a long perspective, the important thing to note is that the children of the new middle class African-Americans will be second- and third-generation members with all the confidence, educational resources, and, most of all, cultural capital to find a more secure niche in the nation's economy.

Sources: *Black Wealth / White Wealth* (pp. 90–125), by M. L. Oliver and T. M. Shapiro, 1997, New York: Routledge; *The Ordeal of Integration* (pp. 24–25), by O. Patterson, 1997, Washington, DC: Civitas; *America in Black and White* (pp. 183–202), by S. Thernstrom and A. Thernstrom, 1997, New York: Simon and Schuster.

from high school; 40% earn less than $20,000 a year; and 33% do not have a driver's license. About 73% of foreign-born Latinos speak mainly Spanish at home. However, most know some English. Only about 20% of those Hispanics who arrived as adults are bilingual in English and Spanish.

One of the most striking indicators of assimilation is the demise of fatalism, the belief, "common among the rural poor of Mexico and certain other Latin countries," that it is pointless to plan for the future because the future is outside of an individual's control (*Migration News*, February 2000).

Mexican Americans   Hispanics of Mexican descent, also known as *Chicanos,* make up the majority of the Spanish-speaking population in the United States. The main reason is proximity. The 1,936-mile border the United States shares with Mexico is the site of millions of legal and illegal crossings each year.

The term *Chicano* is somewhat controversial. It has long been used as a slang word in Mexico to refer to people of low social class. In Texas, it came to be used for Mexicans who illegally crossed over the border in search of work. Many Mexican Americans have taken to using the term themselves to suggest a tough breed of individuals of Mexican ancestry who are committed to achieving success in this country and are willing to fight for it (Madsen, 1973). In his study of Mexican Americans in Texas, William Madsen (1973) noted that among them, three levels of acculturation may be distinguished. Although American technology is appreciated by most Mexican Americans, one group has retained its Mexican peasant culture, at least in regard to values. Another group consists of people torn between the traditional culture of their parents and grandparents on the one hand and Anglo-American culture (learned in American schools) on the other. Many individuals in this group suffer crises of personal and ethnic identity. Finally, there are those Mexican Americans who have acculturated fully and achieved success in Anglo-American society. Some remain proud of and committed to their ethnic origins, but others would just as soon forget them and assimilate fully into the Anglo-American world.

Mexican Americans have been exploited for many years as a source of cheap agricultural labor. Their median family income is $27,883, the lowest of any Spanish-speaking group (U.S. Bureau of the Census, 1999). Thirty-one percent of Mexican Americans live in poverty (U.S. Bureau of the Census, 1999). Also, because of their poverty, they have long been willing to do the stoop labor (literally bending over and working close to the ground, or menial labor in general) that most Anglo-Americans refuse to do. In addition, 21.3% of all Mexican-American families are headed by women, and 37% of all births are out of wedlock (U.S. Bureau of the Census, 1999).

Puerto Ricans   Also included under the category of Hispanics are Puerto Ricans. In 1898, the United States fought a brief war with Spain and as a result took over the former Spanish colonies in the Pacific (the Philippines and Guam) and the Caribbean (Cuba and Puerto Rico). Puerto Ricans were made full citizens of the United States in 1917. Although government programs improved their education and

dramatically lowered the death rate, rapid population growth helped keep the Puerto Rican people poor. American business took advantage of the large supply of cheap, nonunion Puerto Rican labor and built plants there under very favorable tax laws.

There are more than 3 million Puerto Ricans in the United States. Many have migrated to the American mainland seeking better economic opportunities. Most make their homes in the New York City area, but return frequently to the island to visit relatives and friends.

Puerto Ricans living in the United States have a median household income of $28,953, (U.S. Bureau of the Census, 1999). Ironically, the poverty in many Puerto Rican families is due, in part, to the ease of going back and forth between their homeland and the United States. The desire to one day return permanently to Puerto Rico interferes with a total commitment to assimilate into American culture.

Cuban Americans   There were some Cubans in the area now known as Florida as early as the 1500s, but most have come to the United States since the late 1950s. Only in the 1970s did they begin to have a visible cultural and economic impact on the cities where they have settled in sizable numbers.

Many Cubans came to the United States as a result of the 1959 revolution that catapulted Fidel Castro into power on the Cuban island. At that time, Castro's rebel forces overthrew the Fulgencio Batista government. Castro, a Marxist closely aligned with the former Soviet Union, began to restructure the social order, including appropriating for the state privately owned land and property. Professionals and businesspeople who were part of the established Cuban society felt threatened by these changes and fled to the United States. More than 155,000 Cubans immigrated to the United States between 1959 and 1962. As a whole, these immigrants have done extremely well in American society; they had the distinct advantage of coming to the United States with marketable skills and money.

Another large wave of Cuban immigration occurred in 1980, when Castro allowed people to leave Cuba by way of Mariel Harbor. The result was a flotilla of boats bringing 125,000 refugees to the United States. This wave of immigrants was poorer and less well educated than the first. It also included several thousand prisoners and mental patients, many of whom were imprisoned in the United States as soon as they arrived. Others fled into Miami and other cities to lead a life of crime. Serious friction exists between these two waves of immigrants because of differences in background and social class.

Cubans are relatively recent immigrants, and the first-generation foreign-born predominate. They are exiles who came to the United States not so much because they preferred the U.S. way of life, but because they felt compelled to leave their country. For that reason, most Cuban immigrants fiercely attempt to retain the culture and way of life they knew in Cuba. Of all the Hispanic groups, they are the most likely to speak Spanish in the home: 8 out of 10 families do so.

Cubans are largely found in a few major cities. Metropolitan Miami (Dade County, Florida) is the undisputed center, with nearly 65% of all Cubans in the United States living there. In Miami, where Latins make up a larger percentage of the population than Anglos, the city has become distinctively Cuban. Most other Cubans live in New York City, Jersey City and Newark in New Jersey, Los Angeles, and Chicago, all of which are large centers for Hispanics in general.

Acculturation and assimilation have been slow in Miami, in light of the fact that the community is so self-sufficient and has such a large immigrant base. There also appears to be a lack of social and cultural integration between Cubans and other Hispanic groups in U.S. cities with sizable and differentiated Spanish-speaking populations. Of the major Hispanic groups, only the Cubans have come as political exiles, and this has resulted in social, economic, and class differences. In the New York City area, Cubans and Puerto Ricans maintain a distinct social distance. Many Cubans feel or perceive that they have little in common with Puerto Ricans, Mexican Americans, or Dominicans.

The 1.4 million Cubans who live in the United States have fared better than any other Hispanic immigrant group. They are the best educated of any of the Spanish-speaking groups with 23% having a bachelor's degree compared to 6.9% of Mexicans. The median family income of Cubans is $39,530, nearly a third higher than the earnings of the average Hispanic family (U.S. Bureau of the Census, 2001). At the current rate of growth, the Cuban income could surpass the national median income within a few years.

Jews

There is no satisfactory answer to the question "What makes the Jews a people?" other than to say that they see themselves—and are seen by others—as one. Judaism is a religion, of course, but many Jews are nonreligious. Some think of Jews as a race, but their physical diversity makes this notion absurd. For more than 2,000 years, Jews have been dispersed around the world. Reflecting this geographic separation, three major Jewish groups have evolved,

each with its own distinctive culture: the Ashkenazim, the Jews of eastern and western Europe (excluding Spain); the Sephardim, the Jews of Turkey, Spain, and western North Africa; and the Oriental Jews of Egypt, Ethiopia, the Middle East, and central Asia. Nor are Jews united linguistically. In addition to speaking the language of whatever nation they are living in, many Jews speak one or more of three Jewish languages: Hebrew, the language of ancient and modern Israel; Yiddish, a Germanic language spoken by Ashkenazi Jews; and Ladino, an ancient Romance language spoken by the Sephardim.

The first Jews came to America from Brazil in 1654, but it was not until the mid-1800s that large numbers of Jews began to arrive. These were mostly German Jews, refugees from European anti-Semitism. Then, with especially violent anti-Semitism erupting in eastern Europe in the 1880s, there was a massive increase in Jewish immigration to America. It came in two waves: in the last two decades of the nineteenth century and in the first two decades of the twentieth.

Jewish immigration was similar to that of other groups in that it consisted overwhelmingly of young people, though the Jewish immigration also had some unique features. First, it was much more a migration of families than that of other European immigrants, who were mostly single males. Second, Jewish immigrants were much more committed to staying here: Two-thirds of all immigrants to the United States between 1908 and 1924 remained, but 94.8% of the Jewish immigrants settled here permanently. Third, the Jewish immigrant groups contained a higher percentage of skilled and urban workers than did other groups. Fourth, especially after the turn of the century, there were many scholars and intellectuals among Jewish immigrants, which was not true of other immigrant groups (Howe, 1976).

These differences account for the fact that even though Jews encountered at least as much hostility from white Anglo-Saxon Protestants as did other immigrant groups (and also were subject to intense prejudice from Catholics), they have had relatively more success in pulling themselves up the socioeconomic ladder. Of the approximately 6 million Jews in America today, 53% of those working are in the professions and business, versus 25% for the nation on average.

Asian-Americans

According to the 2000 census, there were approximately 10.2 million people of Asian background in the United States, constituting 3.6 % of the total population. An additional 1.7 million people reported that they were Asian and another race. Between 1990 and 2000, the Asian and Pacific Islander population increased more quickly than any other race or ethnic group (U.S. Bureau of the Census, 2001).

Most Asian-Americans are concentrated in the major metropolitan areas. Their percentage of the total population in these cities varies from 29.1% in San Francisco to 0.8% in Detroit. In Honolulu, 70.5% of the population is Asian or Pacific Islander (U.S. Bureau of the Census, 2000).

The first Asians to settle in America in significant numbers were the Chinese. Some 300,000 Chinese migrated here between 1850 and 1880 to escape the famine and warfare that plagued their homeland. Initially they settled on the West Coast, where they took backbreaking jobs mining and building railroads. However, they were far from welcome and were subjected to a great deal of harassment. In 1882, the government limited further Chinese immigration for 10 years. This limitation was extended in 1892 and again in 1904, finally being repealed in 1943. The state of California imposed special taxes on Chinese miners, and most labor unions fought to keep them out of the mines because they took jobs from white workers. In the late 1800s and early 1900s, numerous riots and strikes were directed against the Chinese, who drew back into their "Chinatowns" for protection. The harassment proved successful. In 1880 there were 105,465 Chinese in the United States. By 1900, the figure had dropped to 89,863 and by 1920, to 61,729. The Chinese population in the United States began to rise again only after the 1950s (U.S. Bureau of the Census, 1976). Today, the ethnic Chinese are the largest group of Asian origin in the United States (*Statistical Abstract of the United States,* 1999, 2000).

Japanese immigrants began arriving in the United States shortly after the Chinese and quickly joined them as victims of prejudice and discrimination. Feelings against the Japanese ran especially high in California, where one political movement attempted to have them expelled from the United States. In 1906, the San Francisco Board of Education decreed that all Asian children in that city had to attend a single, segregated school. The Japanese government protested, and after negotiations, the United States and Japan reached what became known as a "gentlemen's agreement." The Japanese agreed to discourage emigration, and President Theodore Roosevelt agreed to prevent the passage of laws discriminating against Japanese in the United States.

Initially, Japanese immigrants were minuscule in number: In 1870 there were only 55 Japanese in

America, and in 1880 there were a mere 148. By 1900 there were 24,326, and subsequently their numbers have grown steadily. By 1970 they had surpassed the Chinese (U.S. Bureau of the Census, 1976), but later figures showed that despite a sharp increase since 1970, the number of ethnic Japanese has been far fewer than the number of Chinese (*Statistical Abstract of the United States,* 1999, 2000).

Japanese Americans were subjected to especially vicious mistreatment during World War II. Fearing espionage and sabotage from the ethnic minorities whose home countries were at war with the United States, President Franklin D. Roosevelt signed Executive Order 9066, empowering the military to "remove any and all persons" from certain regions of the country. Although many German-Americans actively demonstrated on behalf of Germany before the United States entered World War II, no general action was taken against them as a group. Nor was any general action taken against Italian Americans. Nonetheless, General John L. DeWitt ordered that all individuals of Japanese descent be evacuated from three West Coast states and moved inland to relocation camps for the duration of the war. In 1942, 120,000 Japanese, including some 77,000 who were American citizens, were moved and imprisoned solely because of their ethnicity—even though not a single act of espionage or sabotage against the United States ever was attributed to one of their number (Simpson & Yinger, 1972). Many lost their homes and possessions in the process. Included among those who were relocated were members of the 442nd Regimental Combat Team, a fighting group composed solely of Japanese Americans, who fought valiantly in Europe until they were interned.

In 1988, President Reagan signed legislation apologizing for this wartime action. The legislation moved to "right a great wrong" by establishing a $1.25 billion trust fund as reparation for the imprisonment. Each eligible person was to receive a $20,000 tax-free award from the government. The president noted as he signed the legislation: "Yes, the nation was then at war, struggling for its survival. And it's not for us today to pass judgment upon those who may have made mistakes while engaging in the great struggle. Yet we must recognize that the internment of Japanese Americans was just that, a mistake" (*The New York Times,* August 11, 1988).

Compared with the earlier group of Asian immigrants who came primarily from China and Japan, since the 1960s many Asians from Vietnam, the Philippines, Korea, India, Laos, Cambodia, and Singapore have come to the United States. Many of the Asian immigrants from Vietnam, Cambodia, and Laos were involuntary migrants who were forced to leave their homes because they feared

These Japanese Americans are about to be taken to Seattle by a special ferry, which will connect with a train to California, as part of their evacuation and internment during World War II.

persecution after the United States left Southeast Asia. Two distinct waves of refugees came from these countries. The first began in the 1960s and continued until the end of the Vietnam War in 1975. The first group of Southeast Asian immigrants were mainly middle- and upper-class Vietnamese who found ways to get their families and financial assets out of the country when it became clear that military victory for the United States was not going to be swift. This group numbered about 25,000.

The second wave of refugees was very different. Harsh economic conditions, political persecution within Vietnam, and widespread genocide by the Khmer Rouge government in Cambodia created a flood of refugees desperate to leave the area. Many crowded into unsafe boats and hoped to reach Hong Kong, Malaysia, and other neighboring countries. Some of these "boat people" eventually settled in the United States. Between 1975 and 1994, more than 700,000 Vietnamese refugees and 500,000 Cambodians and Laotians resettled in the United States (U.S. Office of Refugee Resettlement, 1995).

The vast majority of Asian immigrants are middle class and highly educated. More than a third have a college degree, twice the rate for Americans born here (immigrants from Vietnam, Cambodia, and other Indochinese countries are the exception). The education, occupations, and income attainments of Asian Americans have been far above the national average. In 1999, the median income of Asian-Americans was $51,205, compared with $44,366 for white Americans. Although Asian-Americans make up 4% of the total U.S. population, they account for nearly 19% of the freshman class at Harvard and other elite universities. They have achieved stunning success in science and business. As a group, they have the highest percentage of college graduates of any group in the United States. For example, 42% of all Asians over 25 have a bachelor's degree. For whites the percentage is 28%. Seven in 10 Asian-Americans between 18 to 21 are attending college, versus half of whites. Asian-Americans received 10% of the doctorates conferred by the nation's colleges and universities. This included 22% of the doctorates in engineering and 21% of those in computer sciences (U.S. Bureau of the Census, 2000).

Native Americans

Early European colonists encountered Native American societies that in many ways were as advanced as their own. Especially impressive were their political institutions. For example, the League of the Iroquois, a confederacy that ensured peace among its five member nations and was remarkably successful in warfare against hostile neighbors, was the model on which Benjamin Franklin drew when he was planning the Federation of States (Kennedy, 1961).

The colonists and their descendants never really questioned the view that the land of the New World was theirs. They took land as they needed it—for agriculture, for mining, and later for industry—and drove off the native groups. Some land was purchased, some acquired through political agreements, some through trickery and deceit, and some through violence. In the end, hundreds of thousands of Native Americans were exterminated by disease, starvation, and deliberate massacre. By 1900, only some 250,000 Indians remained (perhaps one-eighth of their number in precolonial times) (McNeill, 1976). In recent years, however, their numbers have grown dramatically.

There were 2.4 million Native Americans in the country in 2000 (up from 1,479,000 in 1980, representing a 61.6% increase). This increase cannot be attributed to just natural population growth. Other factors that may have contributed to the higher number include improvements in the way the U.S. Census Bureau counts people on reservations and a great propensity for people (especially those of mixed Native American and white parentage) to report themselves as American Indian.

The five states with the largest Native American populations were California (314,000), Oklahoma (263,000), Arizona (261,000), New Mexico (166,000) and Washington (105,000). Overall, roughly one-half of the nation's American Indians and Alaska Natives lived in Western states (U.S. Bureau of the Census, 2000).

More than half of all Native Americans live on or near reservations administered fully or partly by the federal Bureau of Indian Affairs (BIA). Many of the others are living in Native American enclaves in urban areas. Oklahoma, California, Arizona, and New Mexico, respectively, are the states with the largest Native American populations.

To make up for past injustices perpetrated against Native Americans, the federal government pays for a number of programs to assist Native Americans with education, health care, and housing. The government has also granted Native Americans special rights to govern themselves, so that they are subject only to federal rather than state and local laws. These rights, based on hundreds of treaties in the eighteenth and nineteenth centuries, provide for Native American sovereignty and give tribes independence from outside governments in return for land. These rights have made it possible for some tribes to open casinos in states that do not allow gambling establishments.

Interestingly, more than 7 million Americans claim Native American ancestry. That is about 1 in 27 U.S. residents. Yet less than one-third that number identify themselves as Native Americans. Most people who claim Native American ancestry do so in combination with another ancestry group such as the English or Irish. The 69% increase in the Native American population is not due to a baby boom, but rather to the increasing tendency of people to identify with their Native American heritage.

Native Americans have a median household income that is higher than that of African Americans and similar to that of Hispanics.($30,784, $26,608, and $29,110, respectively. Considerable differences exist among the various tribes. The Iroquois and the Creek are much better off economically than the Navajo (U.S. Bureau of the Census, 2000).

Conditions on some reservations are quite bad, particularly for Native American youths. For example, they are only half as likely to have both parents at home as rural white teenagers in Minne-

About one in five Native American households on reservations lack complete plumbing facilities or kitchens.

Controversies in Sociology

Is Transracial Adoption Cultural Genocide?

There are thousands of families waiting to adopt a child at the same time as thousands of children are waiting to be adopted. The issue is an outgrowth of the fact that most of the families who want to adopt are white and of the 100,000 children eligible for adoption 40% are black.

Transracial adoption has been a highly charged topic since 1972, when the National Association of Black Social Workers (NABSW) issued a position paper condemning such adoptions. The members declared that African-American children should be placed only with black families, whether in foster care or for adoption. They believed that African-American children in white homes are cut off from a healthy awareness of themselves as black people. They noted that "we have committed ourselves to go back to our communities and work to end this particular form of genocide" (NABSW, 1972).

The debate over transracial adoption has continued since 1972 and it mirrors a similar debate decades ago over adoptions that crossed religious lines. Many people involved with the adoption process agree that same-race placement should be the first option when placing children in families. The controversy arises over whether that should be considered the only option (as NABSW believes) or whether the next option should be transracial adoption. Critics say they are concerned about the emotional well-being of African-American children and about whether white parents could teach African-American children about their unique cultural heritage. Advocates say such opinions are racist and deny homes to children who desperately need them.

Same-race placement is not always possible for a number of reasons, the primary one being the disproportionate number of African-American children waiting to be adopted. As already noted, the vast majority of people waiting to be adoptive parents are white. Under the NABSW policy of same-race placement, most African-American children are left without homes and put in foster care because there are not enough prospective minority parents to adopt them.

Harvard University Law School professor Elizabeth Bartholet points out that when adoption agencies segregate children waiting for homes on the basis of race "Black children are, as a result, condemned in large numbers to foster limbo, their lives put on hold while they wait for months, years, and often their entire childhoods for color-matched families."

Others point out that the perceived shortage of black adoptive families is due to an adoption system where 75% of the social workers and supervisors are white and in which middle-class values dominate. This produces a system that includes institutional racism, culturally insensitive attitudes on the part of workers, high adoption agency fees, inflexible standards, and poor recruitment techniques. These critics claim that when an agency is sensitive toward the African-American community, it has no trouble finding adoptive parents. Having extended families is common within the African-American community, and many of those families are run by single mothers and grandmothers. White adoption agencies are accused of being unable to appreciate the extended family structures that make up a stable black family.

Critics also point out that transracial adoptions are basically a one-way street. Are we really interested in the "best interests of black children" or "the right of white people to adopt whichever child they choose" (Perry, 1993–1994).

In response to same-race adoption practices, Congress passed the Multiethnic Placement Act. The act advised adoption and foster care agencies that if they receive federal aid, they are prohibited from delaying or denying the placement of a child solely on the basis of race, color, or ethnic origin.

sota, and they are twice as likely to have experienced the death of a parent. One Native American teenager in six has attempted suicide, a rate four times that of other teenagers (*Society*, March/April 1997). The physical environment is poor also. About one in five Native American homes on reservations lack complete plumbing facilities or complete kitchens. More than half of these homes do not have phones (U.S. Bureau of the Census, April 1995).

A Diverse Society

As is evident by now, the many racial and ethnic groups in the United States present a complex and constantly changing picture. Some trends in intergroup relations can be discerned and are likely to continue; new ones may emerge as new groups gain prominence. The resurgence of ethnic-identity movements probably will spread and may

By July of 1996, all states were in compliance with the law. Some legislators believed the law still allowed too much maneuvering by agencies and were trying to pass even more stringent requirements to avoid attempts to block transracial adoptions.

The NABSW adamantly sticks to its position and notes:

> Transracial adoptions should be a last resort only after a documented failure to find an African-American home. (NABSW Position Paper, 1994)

and:

> The lateral transfer of our children to white families is not in our best interest. Having white families raise our children to be white is at least a hostile gesture toward us as a people and at best the ultimate gesture of disrespect for our heritage as African people. (*NABSW Newsletter,* 1988)

Bartholet notes that studies of transracial adoption show that African-American children reared in white homes are doing as well in terms of achievement, adjustment, and self-esteem as children reared in same-race families. "In a society torn by racial conflict . . . these studies show parents and children, brothers and sisters, relating to one another as if race were no barrier to love and commitment. . . . this seems to be a positive good to be celebrated" (Bartholet, 1993).

Another Harvard Law professor, Randall Kennedy, who is himself black, believes that racial matching in adoption reinforces the view that race is destiny. It perpetuates the idea

> that people of different racial backgrounds really are different in some moral, unbridgeable, permanent sense. It affirms the notion that race should be a cage to which people are assigned at birth and from which people should not be allowed to wander. It belies the belief that love and understanding

are boundless and instead instructs us that our affections are and should be bounded by the color line regardless of our efforts. (Kennedy, 1994)

We can also question the idea that black adoptive parents do better than white adoptive parents in raising black children. Those who make this claim rely on the hunch that black adults, as victims of racial oppression, will generally know more than whites about how best to instruct black youngsters on overcoming racial bias. Another hunch could be equally true, that white adults, as insiders to the dominant racial group, will know more than racial minorities about the inner world of whites and how best to maneuver within it to advance one's interests.

Those against interracial adoption say that they are saving the child from living in a setting where the child will be made to feel uncomfortable by a disapproving surrounding community. Yet this view harkens back to the days of segregation when supporters of separate facilities said it protected blacks from the wrath of those whites who would strongly object to integrated schools, housing, and restaurants.

Sources: *Family Bonds: The Politics of Adoptive Parenting,* by E. Bartholet, 1993, Boston: Houghton Mifflin; "Where Do Black Children Belong?: The Politics of Race Matching in Adoption," by E. Bartholet, May 1991, *University of Pennsylvania Law Review;* "National Association of Black Social Workers Position Paper," April 1972; *Transracial Adoptees and Their Families: A Study of Identity and Commitment,* by R. J. Simon & H. Alstein, 1987, New York: Praeger; *National Association of Black Social Workers Newsletter,* Spring 1988, pp. 1–2; National Association of Black Social Workers, "Position Paper: Preserving African-American Families," 1994; "Orphans of Separatism: The Painful Politics of Transracial Adoption," by R. Kennedy, Spring 1994, *The American Prospect* (17), pp. 38–45; "The Transracial Adoption Controversy: An Analysis of Discourse and Subordination," by T. Perry, 1993–1994, *New York University Review of Law and Social Change,* 21, pp. 30–34.

---

be coupled with more collective protest movements among disaffected ethnic and racial minorities who may demand that they be given equal access to the opportunities and benefits of American society.

It is important to realize that the old concept of the United States as a melting pot is both simplistic and idealistic. Many groups have entered the United States. Most encountered prejudice, some severe

discrimination, and others the pressures of Anglo conformity. Contemporary American society is the outcome of all these diverse groups coming together and trying to adjust. Indeed, if these groups are able to interact on the basis of mutual respect, this diversity may offer America strengths and flexibility not available in a homogeneous society. (See "Controversies in Sociology: Is Transracial Adoption Cultural Genocide?")

SUMMARY

Race refers to a category of people who are defined as similar because of a number of physical characteristics. Often these characteristics are arbitrarily chosen to suit the labeler's purposes. Races have historically been defined according to genetic, legal, and social criteria, each with its own problems. While genetic definitions center on inherited traits like hair and nose type, in fact, these traits have been found to vary independently of one another. Moreover, all humans are far more genetically alike than they are different.

An ethnic group is a group with a distinct cultural tradition with which its own members identify and that may or may not be recognized by others. Many ethnic groups form subcultures, with a high degree of internal loyalty and distinctive folkways, mores, values, customs of dress, and patterns of recreation. Above all, members of the group have a strong feeling of association.

In sociological terms, a minority is a group of people who, because of physical or cultural characteristics, are singled out for differential and unequal treatment and who therefore regard themselves as objects of collective discrimination.

Prejudice is an irrationally based negative, or occasionally positive, attitude toward certain groups and their members. Prejudice is a subjective feeling; discrimination is an overt action. Discrimination can be defined as differential treatment, usually unequal and injurious, accorded to individuals who are assumed to belong to a particular category or group.

Assimilation is the process whereby groups with different cultures come to share a common culture. Invariably, one group has a much larger role in the process than the other(s). The particular form of assimilation found in the United States is called Anglo conformity—the renunciation of other ancestral cultures in favor of Anglo-American behavior and values. Pluralism is the development and coexistence of separate racial and ethnic group identities within a society. Pluralism celebrates the differences among groups.

Since the settlement of Jamestown in 1607, more than 45 million people have immigrated to the United States. Historically, the vast majority of these immigrants have been from Europe. The old migration consisted of people from northern Europe who came before 1880. The new migration was far larger and consisted of people who came from southern and eastern Europe. Discriminatory quotas were set up in the beginning of the twentieth century to restrict the immigration of the latter groups. Today, the overwhelming majority of immigrants to the United States come from Latin America, Asia, and the Caribbean.

Legal immigration to the United States has increased in recent decades and is projected to increase even more in the 21st century. It is estimated that there are more than 3.5 to 4 million illegal immigrants in the United States, and the number is growing by 200,000 per year. Forty-four percent are from Mexico, El Salvador, and Guatemala. Illegal immigrants tend to be young and to settle in California, New York, Florida, Texas, and Illinois.

The United States is perhaps the most racially and ethnically diverse country in the world; no single ethnic group makes up a majority of the population.

INTERNET RESOURCES

Go to http://web-enhanced.thomsonlearning.com which features Web links relevant to this chapter to expand and enhance the learning experience. This site will be monitored and updated periodically to ensure the links are correct and live.

At Virtual Society: The Wadsworth Sociology Resource Center, http://sociology.wadsworth.com, you will find a Career Center, Sociology in the News, Research Online, InfoTrac College Edition, Virtual Tours in Sociology, and a variety of book-specific student and instructor resources.

CHAPTER EIGHT STUDY GUIDE

KEY CONCEPTS

Match each concept with its definition, illustration, or explanation below.

a. Expulsion
b. Race
c. Prejudice
d. Genocide
e. Segregation
f. Minority
g. Pluralism
h. Forced migration

i. Discrimination
j. Ethnic group
k. Subjugation
l. Mixed-race children
m. Anglo conformity
n. Executive Order 9066
o. Kristallnacht
p. Annihilation

q. Assimilation
r. Genocide convention
s. Transracial adoption
t. Ghetto
u. Institutionalized prejudice
and discrimination

_____ 1. A category of people who are defined as similar because of a number of physical characteristics.

_____ 2. A group with a distinct cultural tradition with which its own members identify and that may or may not be recognized by others.

_____ 3. A group of people who, because of physical or cultural characteristics, are singled out for differential and unequal treatment, and who therefore regard themselves as objects of collective discrimination.

_____ 4. An irrationally based negative, or occasionally positive, attitude toward certain groups and their members.

_____ 5. Differential treatment, usually unequal and injurious, accorded to individuals who are assumed to belong to a particular category or group.

_____ 6. Complex societal arrangements that restrict the life chances and choices of a specifically defined group, in comparison with those of the dominant group.

_____ 7. The process whereby groups with different cultures come to share a common culture.

_____ 8. The renunciation of other ancestral cultures in favor of Anglo-American behavior and values.

_____ 9. The development and coexistence of separate racial and ethnic group identities within a society.

_____ 10. The subordination of one group and the assumption of a position of authority, power, and domination by the other.

_____ 11. The act, process, or state of being set apart.

_____ 12. Used to describe any kind of segregated living environment.

_____ 13. The process of forcing a group to leave the territory in which it resides.

_____ 14. The relocation of a group through direct action.

_____ 15. The deliberate practice of trying to exterminate a racial or ethnic group.

_____ 16. A synonym for annihilation.

_____ 17. The beginning of Hitler's "final solution" when the windows in synagogues and Jewish homes and businesses across Germany were shattered and individuals beaten.

_____ 18. The UN General Assembly's definition of acts committed with intent to destroy, in whole or in part, a national, ethnic, racial, or religious group.

_____ 19. President Roosevelt's empowering the military "to remove any and all persons" from certain regions of the country; mainly individuals of Japanese descent.

_____ **20.** A hotly debated issue involving the welfare of children, the rights of individuals, and ideas about cultural genocide.

_____ **21.** One outcome of a new feature of the 2000 U.S. Census that allowed people to check more than one category under race.

KEY THINKERS

Match the thinkers with their main ideas or contributions.

a. Robert K. Merton   d. Hernstein and Murray   g. Hoover Commission
b. Johann Blumenbach   e. Orlando Patterson   h. Horace Kallen
c. Gerhard Lenski   f. Benjamin Franklin   i. Oliver and Shapiro

_____ **1.** Eighteenth-century German physiologist who realized that racial categories did not reflect the actual divisions among human groups.

_____ **2.** Developed a model showing the possible attitudinal and behavioral combinations of prejudicial attitudes and discriminatory behaviors.

_____ **3.** Argued that a cognitive elite was constituting the leadership of America and another group with lesser endowments and abilities was doomed to labor on the fringes of society.

_____ **4.** Challenged the notion that certain groups possessed intelligences below white Americans by arguing that the discussion would not be important if Americans were committed to accepting every group as a member of a society that cannot be broken into parts.

_____ **5.** Stated that a program for Native Americans must include progressive measures for their complete integration as tax-paying members of the larger society.

_____ **6.** The person principally responsible for the development of the theory of cultural pluralism.

_____ **7.** Proposed that dominance of one group over another arises because people have a desire to control goods and services.

_____ **8.** Sociologists who believe it is premature to celebrate the rise of the black middle class as, while their incomes have risen, they still do not possess such assets as property, stocks, and other forms of wealth.

_____ **9.** In 1753, he wondered about the costs and benefits of German immigration and concluded they would "contribute greatly to the improvement of a country."

CENTRAL IDEA COMPLETIONS

Following the instructions, fill in the appropriate concepts and descriptions for each of the questions posed in the following section.

1. Compare and contrast the basic features of legal and social definitions of race:

   a. Legal: _____

   b. Social: _____

2. Compare and contrast the concepts of minority group, ethnic group, racial group:

   a. Minority group: _____

   b. Ethnic group: _____

   c. Racial group: _____

3. Compare and contrast the basic characteristics of the following minority groups:

   a. Mexican Americans: _____

   _____

   b. Cuban Americans: _____

   _____

   c. Puerto Ricans: _____

   _____

   d. Latinos: _____

4. What is the principle of infrangibility and how does it relate to the race and intelligence debate?

   _____

   _____

5. What does the data in Table 8-3 suggest about the levels of tolerance of majority Americans toward American minorities relative to prejudice worldwide?

   _____

   _____

6. Explain how the U.S. Army differs from most civilian organizations in the way in which it seeks to create and maintain racial integration in its ranks:

   _____

   _____

7. What key advantage in the area of black leadership does the U.S. Army have over civilian organizations ?

   _____

   _____

8. Provide a college or university situational example relevant for each of Merton's categories of discriminators and nondiscriminators:

   a. Unprejudiced nondiscriminator: _____

   b. Unprejudiced discriminator: _____

   c. Prejudiced nondiscriminator: _____

   d. Prejudiced discriminator: _____

9. Compare and contrast the minority group experience of Asian Americans and Native Americans relative to other minority group experiences in the United States:

   a. Asian Americans: _____

   _____

   b. Native Americans: _____

   _____

10. Define Anglo conformity, showing specific examples of its historical expression:

   1840s: _____

   1920s: _____

   1960s: _____

11. What are the manifest and latent outcomes resulting from the changes made by the census in its *Race/Ethnicity Categories* (Table 8-1)?

_____

_____

_____

12. Discuss the patterns of intermarriage among the following ethnic-racial groups in the United States: Asian Americans, African Americans, Native Americans. What are the manifest and latent outcomes for each group of their intermarriage patterns?

Asian Americans: _____

_____

African Americans: _____

_____

Native Americans: _____

_____

13. In which three states have the largest numbers of immigrants to the United States settled?

_____

14. List and identify the three main causes of *prejudice* presented in your text.

a. _____

b. _____

c. _____

15. After reading the Controversies in Sociology box, what is your conclusion about the issue: Is the race and intelligence debate worthwhile?

_____

_____

_____

CRITICAL THOUGHT EXERCISES

1. Your text discusses the case of the Malone brothers who, although they did not "look" African American, had a great-grandmother who was African American. Recall how they ultimately were fired from the Boston Fire Department for "misrepresenting their race." What was your opinion after reading the evidence? How did you feel about the cases of Walter White and Mark Linton Stebbins? How might the brothers have fared had they been applying for positions within a fire department in Georgia rather than Boston, Massachusetts? Do you believe these are isolated examples or would you guess a number of individuals face the same situation in any given year? Using your college or university library, conduct your own search to see whether the Malones were an isolated case or whether you can uncover other cases where the issue was the "misrepresentation of one's race." After reading your text and conducting your library research, what is your opinion of "what constitutes race?"

2. Transracial adoption has been a highly charged topic since 1972, when the National Association of Black Social Workers (NABSW) issued a position project condemning such adoptions. What are the basic issues presented in the reading in "Controversies in Sociology: Is Transracial Adoption Cultural Genocide?" on pages 260–261? What do you believe are the pros and cons of placing adoptive children into racially and ethnically matched families? What might be the dysfunctions of such policies? In what ways might transracial or transreligious adoptions be viewed as a step along the road of cultural genocide? In the final analysis, what is your position on this sensitive issue?

3. Develop a critical thought paper in which you trace the changes in immigration into the United States since 1965. What factors determined whether a person was admitted into this country? How did changes in the immigration system affect the countries of origin from which the new immigrants have come? What appears to be the pattern of new foreign-born members of the United States? What changes in residence, popular culture, and education do you hypothesize are likely to develop during the next 25 years based on the changing face of U.S. immigration as you have just described it?

4. Compare and contrast the similarities and differences among the major groups making up the Spanish-speaking population of the United States. How does the country of origin influence the distinctive cultural characteristics enjoyed by the various Spanish-speaking members of what is believed to be the largest race-ethnic group in the United States today? How does income differ across the different categories comprising the Hispanic population and what are the consequences of that income distribution?

5. How has the rise of the World Wide Web increased the potential for spreading messages of hate and bigotry? Visit at least six sites that have as their goal the promotion of prejudiced and resentful views about a particular group of people. Based on your research, what are the common features of your six sites? Some persons see these sites as a tremendous threat to American society while others view the entire scene as an overblown issue. After your examination, what is your assessment of the seriousness of these sites? What action, if any, would you propose relative to restricting the dissemination of the information across the Internet?

ANSWERS

KEY CONCEPTS

1. b	5. i	9. g	13. a	16. d	19. n
2. j	6. u	10. k	14. h	17. o	20. s
3. f	7. q	11. e	15. p	18. r	21. l
4. c	8. m	12. t			

KEY THINKERS

1. b	3. d	5. g	7. c	8. i	9. f
2. a	4. e	6. h			

LEARNING OBJECTIVES

After studying this chapter, you should be able to do the following:

- Contrast biological and sociological views of sex and gender.
- Describe the concept of patriarchal ideology.
- Understand the functionalist and conflict theory viewpoints on gender stratification.
- Explain the process of gender-role socialization.
- Describe gender differences in the world of work.
- Be aware of the impact of changes in gender roles in U.S. society.

Kate Bornstein is 6 feet tall; she weighs about 190 pounds, and as you are introduced to her you notice the large hands and the powerful handshake. Kate used to be Al Bornstein before she underwent a sex-change operation. Even though Kate acquired the anatomical features of a woman, she still likes to have physical relationships with women. She considers herself a lesbian transsexual. Why would a man undergo a sex-change operation to become a woman if he/she is physically attracted to women? Kate makes a distinction between gender identity and sexual attraction. She says, "Just because they did a lot of clever surgical manipulation with my genitals, that doesn't change my desire for a romantic partner. I've always been attracted to women. When I grew up in the fifties I knew I wasn't a man, but how could I be a woman if I loved women?"

Kate believes there are many more genders than just male or female. She asserts that her confusion throughout her life stems from being forced to pick one gender over the other. "I know I'm not a man—about that much I'm very clear, and I've come to the conclusion that I'm probably not a woman either" (Bornstein, 1994).

It is difficult for us to imagine that someone could be something other than a man or a woman if we have spent our lives thinking about a world in which there are only men and women. Kate Bornstein forces us to clarify what we mean by gender. What role does our anatomy play? What does it mean to be socialized as a man or a woman? How much of gender is ascribed and how much is achieved?

In this chapter, we will look at some of the differences between the sexes, examine cross-cultural variations in gender roles, and try to understand how a gender identity is acquired. In the process, we will also focus on the changes that are taking place in gender roles in American society.

Are the Sexes Separate and Unequal?

Sociology makes an important distinction between sex and gender. **Sex** refers to *the physical and biological differences between men and women.* In general, sex differences are made evident by physical distinctions in anatomical, chromosomal, hormonal, and physiological characteristics. At birth, the differences are most evident in the male and female genitalia.

We also need to learn how to be a man or a woman. **Gender** refers to *the social, psychological, and cultural attributes of masculinity and femininity that are based on the previous biological distinctions.* Gender pertains to the socially learned patterns of behavior and the psychological or emotional expressions of attitudes that distinguish males from females. Ideas about masculinity and femininity are culturally derived and pattern the ways in which males and females are treated from birth onward. Gender is an important factor in shaping people's self-images and social identities. Sex is thought of as an ascribed status; a person is born either a male or female (although transsexuals such as Kate Bornstein make us realize that sex can be changed). Gender is learned through the socialization process and thus is an achieved status.

In many circles, Kate Bornstein would be dismissed as a "freak of nature" instead of someone who forces us to clarify exactly what we mean by gender. The dominant view in many societies is that gender identities are expressions of what is natural. People tend to assume that acting masculine or feminine is the result of an innate, biologically determined process rather than the result of socialization and social-learning experiences.

To support the view that gender-role differences are innate, people have sought evidence from religion and the biological and social sciences. Whereas most religions tend to support the biological view, biology and the social sciences provide evidence that suggests that what is natural about sex roles expresses both innate and learned characteristics.

Historical Views

We must be careful when we use today's standards to evaluate the statements of people who lived during another time and in another place. Lifting these statements out of context and bringing them into the present has been compared to trying to plant cut flowers (Ellis, 1997). Yet it is important to look at these views and see what has changed since they represented the common thinking.

Aristotle, for example, pointed out:

> It is fitting that a woman of a well-ordered life should consider that her husband's wishes are as laws appointed for her by divine will, along with the marriage state and the fortune she shares. (Aristotle, 1908)

The third-century Chinese scholar Fu Hsuan penned these lines about the status of women in his era:

> Bitter indeed it is to be born a woman,
> It is difficult to imagine anything so low!
> Boys can stand openly at the front gate,
> They are treated like gods as soon as they
> —are born . . .
> But a girl is reared without joy or love,
> And no one in her family really cares for her,
> Grown up, she has to hide in the inner rooms,

Society causes us to expect certain gender-role behaviors from males and females. A changing society, however, produces changes in how these roles are carried out.

From birth, parents all over the world treat boys and girls differently. Although the Amazon Indian boy is the youngest, he wears the chief's headdress.

Cover her head, be afraid to look others in
—the face,
And no one sheds a tear when she is married off.
(Quoted in Bullough, 1973)

In traditional Chinese society, women were subordinate to men. Chinese women were often called *Neiren,* or "inside person." To keep women in shackles, Confucian doctrine created what were known as the three obediences and the four virtues. The three obediences were "obedience to the father when yet unmarried, obedience to the husband when married, and obedience to the sons when widowed." Thus, traditionally, Chinese women were placed under the control of men from cradle to grave.

The four virtues were "women's ethics," meaning a woman must know her place and act in every way in compliance with the old ethical code; "women's speech," meaning a woman must not talk too much, taking care not to bore people; "women's appearance," meaning a woman must pay attention to adorning herself with a view to pleasing men; and "women's chores," meaning a woman must willingly do all the chores in the home.

Thomas Jefferson was an individual who believed in supreme personal liberty and the equal creation of all men. Yet when it came to women, he thought

very differently. He did not think women should be involved in politics or that they should have the right to vote. He felt strongly that women had a single purpose in life, marriage and the subordination to a husband. When his oldest daughter married he wrote her, "The happiness of your life now depends on continuing to please a single person. To this all other objects must be secondary." (Nock, 1996).

In nineteenth-century Europe, attitudes toward women had not improved appreciably. The father of modern sociology, Auguste Comte (1851), in constructing his views of the perfect society, also dealt with questions about women's proper role in society. Comte saw women as the mental and physical inferiors of men. "In all kinds of force, whether physical, intellectual or practical, it is certain that man surpasses women in accordance with the general law prevailing throughout the animal kingdom." Comte did grant women a slight superiority in the realms of emotion, love, and morality.

Comte believed that women should not be allowed to work outside the home, to own property, or to exercise political power. Their gentle nature, he said, required that they remain in the home as mothers tending to their children and as wives tending to their husbands' emotional, domestic, and sexual needs.

Comte viewed equality as a social and moral danger to women. He felt that progress would result only from making the female's life "more and more domestic; to diminish as far as possible the burden of out-door labour." Women, in short, were to be "the pampered slaves of men."

Religious Views

Many religions have overtly declared that men are superior to women. Men are thought by those religions to be spiritually superior to women, who are deemed dangerous and untrustworthy. For centuries of Judeo-Christian history, two short passages in Scripture helped shape the common view of women's character and proper place: the creation of Eve from "Adam's rib," and Eve's transgression and God's subsequent curse on her. The first of these passages presents Eve and women as an afterthought:

And the Lord God said, It is not good that the man should be alone; I will make a help mate for him. . . . And the rib, which the Lord God had taken from man, made he a woman, and brought her unto the man. (Gen. 2:18–23)

The second passage relates how, after God made the world a paradise, Eve disobeyed God and helped to make the world the imperfect place we know:

Unto woman he said, I will greatly multiply thy sorrow and thy conception; in sorrow thou shalt bring forth children; and thy desire shall be to thy husband, and he shall rule over thee. (Gen. 3:16)

War, pestilence, famine, and every sin imaginable were the prices all humanity had to pay for Eve's disobedience. The biblical story of creation presents a God-ordained gender-role hierarchy, with man created in the image of God and woman a subsequent and secondary act of creation. This account has been used as the theological justification for a **patriarchal ideology,** or *the belief that men are superior to women and should control all important aspects of society.* This kind of legitimation of male superiority is displayed in the following passage:

For a man indeed ought not to have his head veiled, forasmuch as he is the image and glory of God: but the woman is the glory of the man: for neither was the man created for the woman but the woman for the man: for this house ought the woman to have a sign of authority on her head. (I Cor. 11:3–10)

In traditional India, the Hindu religion conceived of women as strongly erotic and thus a threat to male asceticism and spirituality. Women were cut off physically from the outside world. They wore veils and voluminous garments and were never seen by men who were not members of the family. Only men were allowed access to and involvement with the outside world.

Women's precarious and inferior position in traditional India is illustrated further by the ancient Manu code, which was drawn up between 200 B.C. and A.D. 200. The code states that if a wife had no children after 8 years of marriage, she would be banished; if all her children were dead after 10 years, she could be dismissed; and if she had produced only girls after 11 years, she could be repudiated.

Stemming from the Hindu patriarchal ideology was the practice of prohibiting women from owning and disposing of property. The prevalent custom in traditional Hindu India was that property acquired by the wife belonged to the husband. Similar restrictions on the ownership of property by women also prevailed in ancient Greece, Rome, Israel, China, and Japan. Such restrictions are still followed by fundamentalist Muslim states like Saudi Arabia and Iran.

Even in Pakistan, a relatively modern Muslim country, women have far less value than men. In a court of law, for example, the testimony of women is given half the weight of the testimony of men. According to a Muslim clergyman, this rule is necessary because of the emotional and irrational nature of women—a nature that makes women intellectu-

ally inferior. From the court's point of view, women have the same value as "the blind, handicapped, lunatics, and children" (*The New York Times,* June 17, 1988).

Biological Views

Supporters of the belief that the basic differences between males and females are biologically determined have sought evidence from two sources: studies of other animal species, including nonhuman primates (monkeys and apes) and studies of the physiological differences between men and women. We shall examine each in turn.

Animal Studies and Sociobiology **Ethology** is *the scientific study of animal behavior.* Ethologists have observed that there are sexual differences in behavior throughout much of the nonhuman animal world. Evidence indicates that these differences are biologically determined—that in a given species, members of the same sex behave in much the same way and perform the same tasks and activities. Popularized versions of these ideas, such as those of Desmond Morris in *The Human Zoo* (1970) or Lionel Tiger and Robin Fox in *The Imperial Animal* (1971), generalize from the behavior of nonhuman primates to that of humans. They maintain that in all primate species, including *Homo sapiens,* there are fundamental differences between males and females. They try to explain human male dominance and the traditional sexual division of labor in all human societies on the basis of inherent male or female capacities. They even have extended their analysis to explain other human phenomena, such as war and territoriality, through evolutionary comparisons with other species. A more sophisticated treatment of this same theme is found in the field of sociobiology (see Chapter 1), through the study of the genetic basis for social behavior (Wilson, 1975, 1978, 1994).

Sociobiologists believe that much of human social behavior has a genetic basis. Patterns of social organization such as family systems, organized aggression, male dominance, defense of territory, fear of strangers, the incest taboo, and even religion are seen to be rooted in the genetic structure of our species. The emphasis in sociobiology is on the inborn structure of social traits.

Critics contend that sociobiologists overlook the important role that learning plays among nonhuman primates in their acquisition of social and sexual behavior patterns (Montagu, 1973). They also assert that by generalizing from animal to human behavior, sociobiologists do not take into account fundamental differences between human and nonhuman primates, such as the human use of a complex language system. While freely acknowledging

Table 9-1

Gender and Disease

Heart Attack—Men are more likely to suffer heart attacks. Heart disease is the leading cause of death for women and a major contributor to their morbidity and disability. More women than men die each year from heart disease, and among those who survive a heart attack, nearly twice as many women as men die within the following year.

Cancer—Cancers are the second leading cause of death for women. Lung cancer (attributed to increased cigarette smoking among women) heads the list with breast cancer coming in second.

HIV/AIDS—HIV incidence is increasing among women, especially in the African-American and Hispanic populations.

Cardiovascular Disease—Men have a greater prevalence of heart disease, but one in nine women ages 45–64 has some cardiovascular disease, rising to one in three at age 65 and older.

Diabetes and Other Chronic Illnesses—Diabetes is a major health problem and is a cause of increased mortality among minority women, especially among middle-aged and older American Indian/Alaska Native, Hispanic, and African-American women.

Osteoporosis—Twenty-four million people, primarily women, suffer from this disease. Eighty percent of people with osteoporosis are women; more than half of women over 65 (mostly white) are afflicted with it.

Immunologic Diseases—Autoimmune thyroid diseases have a 15:1 ratio of women to men. Rheumatoid arthritis has a 3:1 ratio of women to men. Diabetes mellitus and multiple sclerosis occur more often in women.

Mental Disorders—Women are twice as likely as men to be depressed and two to three times more prone to anxiety disorders. In elderly women, the prevalence of depression is 3.64% vs. 1% in elderly men.

Alzheimer's Disease—The incidence of this disease is higher among women, and it increases dramatically after age 85.

Visual and Hearing Impairments—Men have nearly a 50% greater likelihood of experiencing these problems.

Source: National Institutes of Health, Office of Research on Women's Health, 2001.

the biological basis for sex differences, these critics assert that among humans, social and cultural factors overwhelmingly account for the variety in the roles and attitudes of the two sexes. Human expressions of maleness and femaleness, they argue, although influenced by biology, are not determined by it; rather, gender identities acquired through social learning provide the guidelines for appropriate gender-role behavior and expression.

Gender and Physiological Differences Even ardent critics cannot deny that certain genetic and physiological differences exist between the sexes—differences that influence health and physical capacity. Accordingly, the study of gender roles must take those differences into account in such areas as size and muscle development (both usually greater in males), longevity (females, with few exceptions, live longer in nearly every part of the world, sometimes as much as 9 years longer on average), and susceptibility to disease and physical disorders (considerable variation exists between men and women). As you can see from Table 9-1, men and women are afflicted by extremely different chronic conditions.

There has been an explosion of interest in women's health after a highly publicized 1990 government study that raised concerns about the small numbers of women in clinical trials sponsored by the National Institutes of Health (NIH). In 1992, Congress made it illegal to exclude women as subjects in medical research. The National Institutes of Health now has an Office of Research on Women's Health to oversee the representation of women in the insti-

tutes' studies. As a result, women's health research has increased substantially.

The medical treatment of women also is getting more attention in the medical school curriculum after a House Appropriations Committee report noted that U.S. physicians are not trained adequately to address the needs of women. Traditionally, medical schools have taught about disease and treatment in terms of the 70-kilogram male. The NIH women's health research office and the Health Resources and Services Administration have joined forces to find out exactly where the gaps in medical education are and to recommend a model curriculum to Congress. A number of medical schools and teaching hospitals have begun to rethink their curricula to address more medical issues affecting women. The new approaches to teaching medicine require students to consider examples of women in all their subjects, not just in obstetrics and gynecology.

The differences between men and women go far beyond the obvious, and findings from research in the emerging field of gender-based biology could one day lead to treatments that vary depending on the sex of the patient. Gender-based biology, a field less than a decade old, identifies the biological and physiological differences between men and women as well as differences in response to drugs.

A scientific literature review recently presented at the meeting of the Society for the Advancement of Women's Health Research found a growing body of evidence confirming the role of gender on disease. The review identified 10 diseases, including multiple sclerosis, diabetes, lupus and rheumatoid arthritis, that disproportionately affect women.

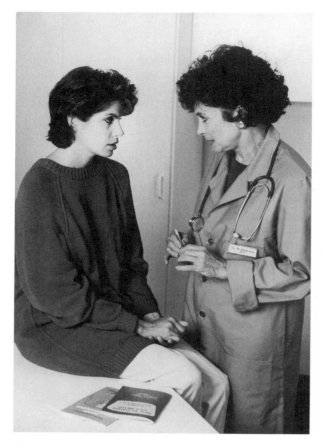

The medical treatment of women is now getting more attention in medical schools, in response to studies that showed that U.S. physicians were not adequately trained in women's health issues.

At menopause, women's brains experience a major decrement in estrogen, while men, paradoxically, continue producing androgen, which metabolizes into an estrogenlike hormone in the brain. As a result, men maintain high levels of estrogen in their brains as they age. Such findings might explain why aging women suffer cognitive decline such as Alzheimer's disease at much higher incidence than men.

Hormone differences between men and women affect how the brain functions. For example, researchers have found that estrogen interacts with both the metabolism and serotonin, a neurotransmitter that helps control moods. Major estrogen receptors are scattered throughout the body but are concentrated in a region of the brain that controls thinking, emotion, sex, and eating.

Researchers will also need to determine the significance of some of the gender differences they are starting to discover (Shelton, 1999).

Responses to Stress   Gender differences also influence the way men and women react to stress—the "fight or flight" reaction that is thought to play a

part in heart disease, stroke, and coronary-artery disease, among other ailments. In earlier days, when primitive man was threatened by wild animals while hunting, testosterone combined with adrenaline enabled him to react quickly to danger. This intense type of reaction is no longer important and may be part of the reason men suffer more heart attacks than women. Women, it appears, react more slowly to stress, putting less pressure on the blood vessels and the heart. While learned behavior may play a role in women's response to stress, biology is no less important.

When women are confronted by stress, they tend to respond by seeking out contact and support from others. The support they seek is usually from other women. This "befriending" behavior has been linked with the hormone oxytocin, which is released by the body during stress. It has been shown to make both rats and humans calmer, less fearful, and more social. While men do secrete oxytocin, male hormones reduce the effect of oxytocin. Female hormones, on the other hand, amplify its effects (Associated Press, May 19, 2000).

Although many differences between males and females have a biological basis, other physical conditions may be tied to cultural influences and variations in environment and activity. Men react differently to psychological stress than women do: Each sex develops severe but dissimilar symptoms. Changing cultural standards and patterns of social behavior have had a pronounced effect on other traits that formerly were thought to be sex linked. For example, the rising incidence of lung cancer among women—a disease historically associated primarily with men—can be traced directly to changes in social behavior and custom, not biology: Women now smoke as freely as men.

In sum, differing learned behaviors contribute to the relative prevalence of certain diseases and disorders in each sex. However, not all male-female differences in disease rate and susceptibility can be attributed to these factors. In addition to genetically linked defects, differences in some basic physiological processes such as metabolic rates and adult secretions of gonadal hormones may make males more vulnerable than females to certain physical problems.

Gender and Sex

We know that men and women think differently about sex, but how differently? Is it just a stereotype that men seek sex more than women? Some researchers have tried to answer this question. In one study (Buss, 1994), men and women were asked, "How many sexual partners would you like to have in a) the next month, b) in the next two years, c) in your lifetime?" Women averaged $8/10$ of a sexual partner in the next month, 1 in the next 2 years, and 4 or 5 over their lifetimes. Men had very different expectations. Men wanted 2 sex partners within the month, 8 in the next 2 years, and 18 over their lifetimes.

Men and women also differed in how well they needed to know the person before having sex. The participants were asked whether they would consider having sex with a desirable partner they had known for (a) one year, (b) one month, or (c) one week. Women preferred to know the person for a year, whereas men had no problem having sex with a woman they knew for a week.

The differences were even more striking in another study (Clark & Hatfield, 1989). Attractive men and women were hired to approach strangers of the opposite sex on college campuses and say to them, "I have been noticing you around campus. I find you very attractive," and then ask one of three questions: (a) "Would you go out with me tonight?" (b) "Would you come over to my apartment tonight?" or (c) "Would you go to bed with me tonight?" Half the men and half the women consented to a date, but that was where the similarity ended. Only 6% of the women consented to go to the confederate's apartment, compared to 69% of the men. None of the women consented to sex. However, 75% of the men consented to sex. Of the remaining 25%, many were apologetic, asking for a rain check or explaining that they could not do it now because they had to meet their girlfriend.

Sociological View: Cross-Cultural Evidence

Most sociologists believe that the way people are socialized has a greater effect on their gender identities than do biological factors. Cross-cultural and historical research offers support for this view, revealing that different societies allocate different tasks and duties to men and women and that males and females have culturally patterned conceptions of themselves and of one another.

Until the pioneering work of anthropologist Margaret Mead (1901–1978) was published in the 1930s, it was widely believed that gender identity (what then was called sex temperament) was a matter of biology alone. It never occurred to Westerners to question their culture's definitions of male and female temperament and behavior, nor did most people doubt that these were innate properties. In 1935, Mead published a refutation of this assumption in *Sex and Temperament*, which has become a classic. While doing research among isolated tribal groups on the island of New Guinea, she found three societies with widely differing expectations of male and female behavior. The Arapesh were characterized as gentle and home loving, with a belief that men and women were of similar temperament. Adult men and women subordinated their needs to those of the younger or weaker members of the society. The Mundugamor, by contrast, assumed a natural hostility between members of the same sex and only slightly less hostility between the sexes. Both sexes were expected to be tough, aggressive, and competitive. The third society, the Tchambuli, believed that the sexes were temperamentally different, but the gender roles were reversed relative to the Western pattern.

> I found . . . in one, both men and women act as we expect women to act—in a mild parental responsive way; in the second, both act as we expect men to act—in a fierce initiative fashion; and in the third, the men act according to our stereotype for women—are catty, wear curls and go shopping, while the women are energetic, managerial, unadorned partners. (Mead, 1935)

Although Mead's findings are interesting and suggestive, anthropologists have cautioned against overinterpreting them. They point out that Mead's research was limited to a matter of months and that her then husband and fellow anthropologist, Reo Fortune, rejected her view that the Arapesh did not distinguish between male and female gender roles.

Most sociologists tend to agree that even in preliterate societies, culture is central to the patterning of gender roles. Nevertheless, biological factors may play a more prominent part in structuring gender roles in societies less technologically developed than our own. Anthropologist Clellan S. Ford (1970) believes that for preindustrial peoples, "the single most important biological fact in determining how men and women live is the differential part they play in reproduction." The woman's life is characterized by a continuing cycle of pregnancy, childbearing, and nursing for periods up to 3 years. By the time the child is weaned, the mother is likely to be pregnant again. Not until menopause, which frequently coincides with the end of the woman's life, is her reproductive role over. In those circumstances, it is not surprising that such activities as hunting, fighting, and forest clearing usually are defined as male tasks; whereas gathering and preparing small game, grains, and vegetables; tending gardens; and building shelters are typically female activities, as is caring for the young.

In an early study, George Murdock (1937) provided data on the division of labor by sex in 224 preliterate societies. Such activities as metalworking,

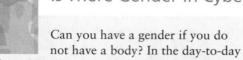

Society and the Internet

Is There Gender in Cyberspace?

Can you have a gender if you do not have a body? In the day-to-day physical world, what you look and sound like relays information about your gender and produces certain reactions. In cyberspace, it is possible to communicate with others without disclosing any of these characteristics. In essence, you have the ability to engage in the totally new experience of communicating without your gender being a key factor in the interaction. For some, this experience may be quite liberating; for others, very disconcerting. Some women may feel that they no longer have to fear the danger that their comments could cause them if they were made in the physical world. Others may feel constrained communicating with someone if they do not know that person's gender. In cyberspace communication groups, men often impersonate women and women impersonate men.

Yet communicating in cyberspace may not really be as devoid of gender as we may think. Women seem to have recognizably different styles of communicating on the Internet. Susan Herring (1994) analyzed male and female participation styles on several electronic lists. She found that the communication was not equal. Men and women had different expectations about what was and what was not appropriate in online interaction. It also appeared that a small male minority dominated the discourse both in terms of sheer amount of postings and with self-promotional and adversarial comments. Moreover, when women do attempt to participate on a more equal basis, they risk being actively censored by the reactions of the men, or they may be ignored. Women tend to be intimidated by the attacks and avoid participating as a result. Herring believes the communication styles appear to be recognizably gendered, and they perpetuate preexisting interaction patterns in society as a whole.

Herring found that 68% of the messages posted by men made use of an adversarial style in which the poster distanced himself from, criticized, or ridiculed other participants, often while promoting his own importance. What is known as "flaming," making offensive or derogatory comments about someone or the person's views, appears to be a male characteristic. The male style is characterized by

"On the Internet, nobody knows you're a dog."

adversarial comments such as put-downs; strong, often contentious assertions; lengthy or frequent postings; self-promotion; and sarcasm. Men are also likely to take an authoritative, self-confident stance in which they represent themselves as experts.

The women, in contrast, displayed such features as hedging, apologizing, asking questions rather than making assertions, and supporting others' thoughts and feelings. The female style includes expressions of appreciation, thanking, and community-building activities that make the participants feel accepted and welcomed.

Far from being gender neutral, online communication patterns tend to show us that gender differences, along with their social consequences, persist in the cyberworld, even when gender is not overtly declared.

Sources: *"Making the Net*Work*: Is There a Z39.50 in Gender Communication?"* by S. Herring, June 1994, paper presented at The American Library Association annual convention, Miami. "Posting in a Different Voice" by S. Herring, 1996, *Philosophical Perspectives on Computer-Mediated-Communication*, C. Ess, ed. New York: State University of New York Press.

making weapons, woodworking and stoneworking, hunting and trapping, building houses and boats, and clearing the land for agriculture were performed by men. Women's activities included grinding grain; gathering and cooking herbs, roots, seeds, fruits,

and nuts; weaving baskets; making pottery; and making and repairing clothing. A review of the cross-cultural literature (D'Andrade, 1966) concluded that the division of labor by sex occurs in all societies. Generally, male tasks require vigorous physical activity or travel, whereas female tasks are less physically strenuous and more sedentary.

The almost universal classification of women to secondary status has had a profound effect on their work and family roles. Anthropologist Sherry Ortner (1974) observed that "Everywhere, in every known culture, women are considered in some degree inferior to men." One important result of this attitude is the exclusion of women from participation in, or contact with, those areas of the particular society believed to be most powerful, whether they be religious or secular. We see these exclusionary patterns in our own society. There are no female priests in the Catholic church or women rabbis in Orthodox Judaism. No woman has ever been the presidential candidate of a major political party and there are very few female chief executive officers of major U.S. corporations.

Another anthropologist, Michelle Rosaldo (1974), believes that "women's status will be lowest in those societies where there is a firm differentiation between domestic and public spheres of activity and where women are isolated from one another and placed under a single man's authority in the home." She believes that the time-consuming and emotionally compelling involvement of a mother with her child is unmatched by any single involvement and commitment by a man. The result is that men are free to form broader associations in the outside world through their involvement in work, politics, and religion. The relative absence of women from these public spheres results in their lack of authority and power. Men's involvements and activities are viewed as important, and the cultural systems accord authority and value to men's activities and roles. Women's work, especially when it is confined to domestic roles and activities, tends to be oppressive and lacking in value and status. Women are seen to gain power and a sense of value only when they can transcend the domestic sphere of activities. This differentiation is most acute in those societies that practice sexual discrimination. Societies in which men value and participate in domestic activities tend to be more egalitarian.

Nevertheless, even though physiological factors tend to play an influential part in gender-role differences, biology does not determine those differences. Rather, people acquire much of their ability to fulfill their gender roles through socialization. (For a discussion of gender in new settings, see "Society and the Internet: Is There Gender in Cyberspace?")

What Produces Gender Inequality?

Sociologists have devoted much thought and research to answering this question. They also have tried to explain why males dominate in most societies. Two theoretical approaches have been used to explain dominance and gender inequality: functionalism and conflict theory.

The Functionalist Viewpoint

From Chapter 1 you may recall functionalists (or structural functionalists, to be more precise) believe that society consists of a system of interrelated parts that work together to maintain the smooth operation of society. Functionalists argue that it was quite useful to have men and women fulfill different roles in preindustrial societies. The society was more efficient when tasks and responsibilities were allocated to particular individuals who were socialized to fulfill specific roles.

The fact that the human infant is helpless for such a long time made it necessary that someone look after the child. It was also logical that the mother who gave birth to the child and nursed it was also the one to take care of it. Because women spent their time near the home, they also took on the duties of preparing the food, cleaning clothes, and attending to the other necessities of daily living. To the male fell the duties of hunting, defending the family, and herding. He also became the one to make economic and other decisions important to the family's survival.

This division of labor created a situation in which the female was largely dependent on the male for protection and food, and so he became the dominant partner. This dominance, in turn, caused his activities to be more highly regarded and rewarded. Over time, this pattern came to be seen as natural and was thought to be tied to biological sex differences. Talcott Parsons and Robert Bales (1955) applied functionalist theory to the modern family. They have argued that the division of labor and role differentiation by sex are universal principles of family organization and are functional to the modern family. They believe the family functions best when the father assumes the *instrumental role,* which focuses on relationships between the family and the outside world. It mainly involves supporting and protecting the family. The mother concentrates her energies on the *expressive role,* which focuses on relationships within the family and requires the mother to provide the love and emotional support needed to sustain the family. The male is required to be dominant and competent, and the female to be passive and nurturant.

As you can imagine, there has been much criticism of the functionalist position. The view that gender roles and gender stratification are inevitable does not fit with cross-cultural evidence and the changing situation in American society (Crano & Aronoff, 1978). Critics contend that industrial society can be quite flexible in assigning tasks to males

Men increasingly are playing expressive roles as well as instrumental roles within the family.

and females. Furthermore, they assert that the functionalist model was developed during the 1950s, an era of very traditional family patterns, and that rather than being predictive of family arrangements, it merely is representative of the era during which it became popular.

The Conflict Theory Viewpoint

While functionalist theory may explain why gender-role differences emerged, it does not explain why they persisted. According to the conflict theory, males dominate females because of their superior power and control over key resources. A major consequence of this domination is the exploitation of women by men. By subordinating women, men gain greater economic, political, and social power. According to conflict theory, as long as the dominant group benefits from the existing relationship, it has little incentive to change it. The resulting inequalities are therefore perpetuated long after they may have served any functional purpose. In this way, gender inequalities resemble race and class inequalities.

Conflict theorists believe the main source of gender inequality is the economic inequality between men and women. Economic advantage leads to power and prestige. If men have an economic advantage in society, that advantage will produce a superior social position in both society and the family.

Friedrich Engels (1942) linked gender inequalities to capitalism, contending that primitive, non-capitalistic hunting and gathering societies without private property were egalitarian. As these societies developed capitalistic institutions of private property, power came to be concentrated in the hands of a minority of men, who used their power to subor-

dinate women and to create political institutions that maintained their power. Engels also believed that to free women from subordination and exploitation, society must abolish private property and other capitalistic institutions. He believed that socialism was the only solution to gender inequality.

Today, many conflict theorists accept the view that gender inequalities may have evolved because they initially were functional. Many functionalists also agree that gender inequalities are becoming more and more dysfunctional. They agree that the origins for gender inequalities are more social than they are biological. (For an example of how gender inequalities lead to violence, see "Remaking the World: The Fight Against Honor Killings.")

Gender-Role Socialization

Gender-role socialization is *a lifelong process whereby people learn the values, attitudes, motivations, and behavior considered appropriate to each sex according to their culture.* In our society, as in all others, males and females are socialized differently. In addition, each culture defines gender roles differently. This process is not limited to childhood but continues through adolescence, adulthood, and old age.

Childhood Socialization

Even before a baby is born, its sex is a subject of speculation, and the different gender-role relationships it will form from birth on already are being decided. A scene from the early musical *Carousel* epitomizes (in somewhat caricatured form) some of the feelings that parents have about bringing up sons

Remaking the World

The Fight Against Honor Killings

Women who are sexually assaulted or raped should be able to expect the police to investigate their cases. In Jordan, women who are raped or are thought to have participated in illicit sexual activity are seen as having compromised their families' honor. Fathers, brothers, and sons see it as their duty to avenge the offense, not by pursuing the perpetrators but by murdering the victims, their own daughters, sisters, and mothers. Honor killings accounted for one-third of the murders of women in Jordan in 1999. Journalist and human rights defender Rana Husseini broke the silence on honor killings when she wrote a series of reports on the killings and launched a campaign to stop them. As a result, Husseini has been threatened and accused of being anti-Islam, antifamily, and anti-Jordan. Yet Queen Noor took up the cause, and later the newly ascended King Abdullah cited the need for protection of women in his opening address to Parliament.

As a reporter at *The Jordan Times* I wrote about thefts, accidents, and fires. Then, after about four or five months on the job, I started to come across crimes of honor. One story really shocked me. In the name of honor, a sixteen-year-old girl was killed by her family because she was raped by her brother. He assaulted her several times and then threatened to kill her if she told anyone. When she discovered that she was pregnant she had to tell her family. After the family arranged an abortion, they married her off to a man fifty years her senior. When he divorced her six months later, her family murdered her.

An honor killing occurs when a male relative decides to take the life of a female relative because, in his opinion, she has dishonored her family's reputation by engaging in an "immoral" act. An immoral act could be that she slept with a man or was simply seen with a strange man. In many cases, women are killed just because of rumors or unfounded suspicions.

I continued to cover stories about women who were killed in an unjust, inhuman way. Most of them did not commit any immoral, much less illegal, act, and even if they did, they still did not deserve to die. But I want to emphasize two things. One is that all women are not threatened in this way in my country. Any woman who speaks to any man will not be killed. These crimes are isolated and limited, although they do cross class and education boundaries. The other thing is a lot of people assume incorrectly that these crimes are mandated by Islam, but they are not. Islam is very strict about killing, and in the rare instances where killing is counseled, it is when adultery is committed within a married couple. In these cases, there must be four eyewitnesses and the punishment must be carried out by the community, not by the family members involved.

Honor killings are part of a culture, not a religion, and occur in Arab communities in the United States and many countries. The killers are treated with leniency, and families assign the task of honor killing to a minor, because under Jordanian Juvenile law, minors who commit crimes are sentenced to a juvenile center where they can learn a profession and continue their education, and then, at eighteen, be released without a criminal record. The average term served for an honor killing is only seven and a half months.

The reason for these killings is that many families tie their reputation to the women. If she does something wrong, the only way to rectify the family's honor is to have a wife, daughter, sister killed. Blood cleanses honor. The killers say, "Yes, she's my sister and I love her, but it is a duty."

A woman who becomes pregnant out of wedlock will turn herself in to the police, to be placed in protective custody. The victim goes to jail. Most of these women are held there indefinitely. They are not charged, and they cannot make bail. If the family bails them out, it is to kill them. So these women remain, wasting their lives in prison.

Since I started reporting on the honor killings, things have started to change for the better. When former King Hussein opened the Thirteenth Parliament, he mentioned women and their rights—the first time a ruler had emphasized women and children. Now King Abdullah is following in his father's footsteps and has asked the prime minister to amend all the laws that discriminate against women.

It is important to realize that people who commit the killings are also victims. Their families put all the burden and pressure on their back. If you don't kill, you are responsible for the family's dishonor. If you do kill, you will be a hero and everyone will be proud.

I hope the day will come when I will no longer need to report of these crimes.

Source: "Honor Killings" by R. Husseini, 2000, in K. K. Cuomo, *Speak Truth to Power* (pp. 31–33), New York: Crown Publishers, and interview with Kerry Kennedy Cuomo, November 2000.

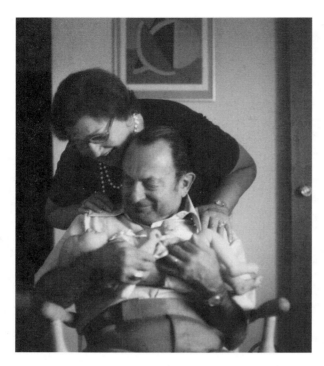

Parents and grandparents respond differently to boys and girls right from the beginning, and they carry in their minds images of what the child should be like, how he or she should behave, and what he or she should be in later life.

as opposed to daughters. A young man discovers he is to be a father. He rhapsodizes about what kind of son he expects to have. The boy will be tall and tough as a tree, and no one will dare to boss him around; it will be all right for his mother to teach him manners but she mustn't make a sissy out of him. He'll be good at wrestling and will be able to herd cattle, run a riverboat, drive spikes, and so on. Then the prospective father realizes, with a start, that the child may be a girl. The music changes to a gentle theme. She will have ribbons in her hair; she will be sweet and petite (just like her mother), and suitors will flock around her. There's a slightly discordant note, introduced for comic relief from sentimentality, when the expectant father brags that she'll be half again as bright as girls are meant to be; but then he returns to the main theme: she must be protected, and he must find enough money to raise her in a setting where she will meet the right kind of man to marry (Maccoby & Jacklin, 1975).

Parents carry in their minds images of what girls and boys are like, how they should behave, and what they should be in later life. Parents respond differently to girls and boys right from the beginning. Girls are caressed more than boys, whereas boys are jostled and roughhoused more. Mothers talk more to their daughters, and fathers interact more with their sons.

These differences are reinforced by other socializing agents—siblings, peers, educational systems, and the mass media. By the first 2 or 3 years of life, core gender identity—the sense of maleness or femaleness—is established as a result of the parents' conviction that their infant's assignment at birth to either the male or female sex is correct. (For a discussion of whether biology determines gender, see "Controversies in Sociology: Can Gender Identity Be Changed?")

Adolescent Socialization

Most societies have different expectations for adolescent girls and boys. Erik Erikson (1968) believed the most important task in adolescence is the establishment of a sense of identity. He believed that during the adolescent stage, both boys and girls undergo severe emotional crises centered on questions of who they are and what they will be. If the adolescent crisis is resolved satisfactorily, a sense of identity will be developed; if not, role confusion will persist. According to Erikson, adolescent boys in our society generally are encouraged to pursue role paths that will prepare them for some occupational commitment, whereas girls generally are encouraged to develop behavior patterns designed to attract a suitable mate. Erikson observed that it is more difficult for girls than for boys to achieve a positive identity in Western society. This is because women are encouraged to be more passive and less achievement oriented than men and to stress the development of interpersonal skills—traits that are not highly valued in our society. Men, by contrast, are encouraged to be competitive, to strive for achievement, and to assert autonomy and independence—characteristics that are held in high esteem in our competitive society.

Nonetheless, female gender roles are changing rapidly, although traditional attitudes toward careers and marriage undoubtedly still remain part of the thinking of many people in our society. Girls are being encouraged not to limit themselves to these stereotyped roles and attitudes. More and more young women expect to pursue careers before and during marriage and child rearing. Marriage no longer is considered the only desirable goal for a woman, nor is it even considered necessary to a woman's success and happiness.

Gender Differences in Social Interaction

There are some interesting differences in how men and women think about the future and solve problems. In one study, researchers (Maines & Hardesty, 1987) asked undergraduates to describe what they

In every society boys and girls are socialized differently, and as they mature, a variety of roles are tried out. In this picture, the 6-year-old boy is experimenting with one version of the male role.

expected would happen to them over the next 10 years. For both male and female undergraduates today, work appears to be a universal expectation. Likewise, the kinds of jobs wanted are nearly identical. Both groups mentioned jobs in business, law, hospital administration, and the computer industry. Further education was anticipated by more than half of the men and women. The vast majority (94%) also saw themselves as eventually married with children.

Thus, at first glance there appears to be a striking similarity in men's and women's future plans. They express the desire for marriage, children, and work, and the desire for higher education is equally present. However, there are significant gender differences in expectations of how family, work, and education will be integrated. Men and women have different assumptions and tactics for achieving the similarly desired events in their lives.

Men operate in what Maines and Hardesty call a linear temporal world. When they try to project what the future might hold for them, they almost always

define it in terms of career accomplishments—lawyer, doctor, college professor, business executive, and so on. Education is seen as something that is pursued to attain the desired career.

Men see a family as desirable and not much of an issue in terms of pursuing career goals. They see little problem in coordinating career and family demands. Many expect to have a traditional division of labor in their families, which will provide a support system for their career pursuit. Mostly, the problems of family living are viewed as being resolved rather easily, and typically there is no mention of career adjustments to address the wife's and children's needs.

Young women, in contrast, operate in contingent temporal worlds. Work, education, and family all are seen as having to be balanced off against one another. Careers are seen as pursuits that may have to be suspended or halted at certain points. The vast majority of women envision problems in their career pursuits, and they see family responsibilities as a major issue that requires adjustments. Nearly half

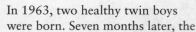

Controversies in Sociology

Can Gender Identity Be Changed?

In 1963, two healthy twin boys were born. Seven months later, the twin named David Reimer lost his entire penis in a botched circumcision. As a result of the irreparable injury, his parents took him to a famed expert in sex research at the renowned Johns Hopkins Hospital in Baltimore. There, Dr. John Money convinced the parents to allow their son to have a surgical sex change. The process involved genital surgery, followed by a 12-year program of social, mental, and hormonal therapy to complete the metamorphosis.

The case was reported in the medical literature as an unqualified success. In 1973, *Time* magazine reported that the case "provides strong support . . . that conventional patterns of masculine and feminine behavior can be altered. It also casts doubt on the theory that major sex differences, psychological as well as anatomical, are immutably set by the genes at conception." For 20 years, this case was cited in numerous sociology, psychology, and human sexuality textbooks. Researchers on these topics used the case as proof that gender is very much the product of socialization and that if there is intervention early enough, a child may be raised in a gender that is different than what he or she was born. As Money noted, "the gender identity gate is open at birth. . . . [I]t stays open at least for something over a year." The case also served as a model of how to treat infants with ambiguous genitalia. Many biological boys had operations that turned them into girls and many girls had procedures that turned them into boys.

The view that gender is the product of socialization is best illustrated by the case of Frankie, who, mistakenly classified as a male at birth, was socialized as a male. At the age of 5, "he" was brought to the hospital for examination and was reclassified as a female whose clitoris had been mistaken for a small penis. Alfred Lindesmith and Anselm Strauss (1956), in a report based on an unpublished document made available to them by one of the nurses assigned to the case, stated that Frankie showed a decided preference for the company of little boys in the children's ward and a disdain for little girls and their "sissy" activities. After the child's real sex had been determined, the nurses were required to treat Frankie as a little girl. One of the nurses observed that this was not easy:

> This didn't sound too difficult—until we tried it. Frankie simply didn't give the right cues. It is amazing how much your response to a child depends on that child's behavior toward you. It was extremely difficult to keep from responding to Frankie's typically little boy behavior in the same way that I responded to other boys in the ward. And to treat Frankie as a girl was jarringly out of key. It was something we all had to continually remind ourselves to do. Yet the doing of it left all of us feeling vaguely uneasy as if we had committed an error. . . . About the same time Frankie became increasingly aware of the change in our attitude toward her. She seemed to realize that behavior which had always before brought forth approval was no longer

say they will quit work for a few years as a solution to family/work demands. Instead of having a clear vision of steps needed to accomplish their career goals, women become much more tentative about their future, since they expect it to entail adjustments and compromises.

Young men seem to take their autonomy for granted. That is, they assume that they will be able to accomplish what they set out to do if they have the necessary education, skill, and good fortune. Young women, however, feel much more limited in the control of their future, even with the necessary education and skill. The problems surrounding the integration of family and career lend an element of uncertainty to their ability to accomplish their goals. Women plan to be flexible to adjust to career and

family needs. This flexibility gives them only partial autonomy in controlling their lives.

This element of tentativeness about the future, the willingness to be flexible and to adjust to the needs of others, and the realization that goals cannot be achieved easily without compromise, evidently produces a difference in how men and women approach issues. This observation has led Carol Gilligan (1982) to believe that men and women think differently when it comes to problem solving.

Men often think that the highest praise they can bestow on a woman is to compliment her for "thinking like a man." That usually means that the woman has been decisive, rational, firm, and clear. To think "like a woman" in our society always has had negative overtones, being characterized as

approved. It must have been far more confusing to her than it was to us and certainly it was bad enough for us. Her reaction was strong and violent. She became extremely belligerent and even less willing to accept crayons, color books and games, which she simply called "sissy" and threw on the floor. (Lindesmith & Strauss, 1956)

The one dissenting voice during this period was Milton Diamond, a young graduate student at the University of Kansas. Diamond was not convinced there was any reason to believe John Money's theory of psychosexual neutrality in children. He published an article on the topic, then he contacted Money and offered to do a joint study with him. Money, a respected researcher, saw no reason to pay any attention to the unknown Diamond.

Yet it now appears that Diamond was the one who was correct and Money woefully misguided. The view that gender can be changed through socialization and surgery appears to be decidedly wrong. After further investigation, it turned out that the twin David never adjusted to being a girl. He steadfastly refused to grow into a woman and stopped taking the estrogen pills that were prescribed for him at age 12. David also refused to undergo additional surgery that the physicians tried to convince him he needed to fully become a woman. At 14 he found other physicians who were sympathetic to his plight. He then underwent a double mastectomy, a phalloplasty, and began a regimen of

male hormones. He refused to ever return to Johns Hopkins University where the original diagnosis and surgery had taken place. After David's return to the male gender identity, he felt his attitudes, behavior, and body were once again united. At 25, David married a woman several years older and adopted her children.

Many of the people who received these gender-changing operations in early life have experienced difficulties. They have formed support groups and some have undergone surgery to reverse the process.

In October of 1998, the American Academy of Pediatrics invited Diamond to address its prestigious annual meeting. Physicians now reject John Money's views on gender identity and no longer treat infants with ambiguous genitalia the way they did 25 years ago.

All of this points to the current view that people are not gender neutral at birth and that they are predisposed to be male or female. Gender is more complex than originally imagined.

Sources: *As Nature Made Him,* by J. Colapinto, 2000, New York: HarperCollins; "Sex Reassignment at Birth: A Long Term Review and Clinical Implications," by M. Diamond & H. K. Sigmundson, March 1997, *Archives of Pediatric and Adolescent Medicine, 151,* pp. 298–304; *Sexual Signatures: On Being a Man or Woman,* by J. Money & P. Tucker, 1975, Boston: Little Brown and Company; *Social Psychology,* by A. R. Lindesmith & A. L. Strauss, 1956, New York: Holt, Rinehart and Winston; "Changing Sex Is Hard to Do," by C. Holden, March 1997, *Science, 275* (5307), p. 1745.

---

fuzzy, indecisive, unpredictable, tentative, and soft-headed.

Gilligan challenged the value judgments made about male versus female styles of reasoning, especially in the area of moral decision making. She argued that a woman's perspective on things is not inferior to a man's; it is just different.

To illustrate, Gilligan described the different responses that 11-year-old boys and girls made to an example used by psychologist Lawrence Kohlberg. In the example, Heinz, a fictional character, is caught up in a complex moral question. Heinz's wife is dying of cancer. The local pharmacist has discovered a drug that might cure her, but it is very expensive. Heinz has done all he can to raise the money necessary to buy it but can come up with

only half the amount, and the druggist demands the full price. The question is: Should Heinz steal the drug?

Boys and girls differ significantly on how they answer this question. Boys often see the problem as the man's individual moral choice, stating that Heinz should steal the drug, as the right to life supersedes the right to property. Case closed.

The girls Gilligan questioned always seemed to get bogged down in peripheral issues. No, they maintained, Heinz should not steal the drug, because stealing is wrong. Heinz should have a long talk with the pharmacist and try to persuade him to do what is right. Besides, they point out, if Heinz steals the drug, he might be caught and go to jail. Then what would happen to his wife? What if there were children?

Sociology at Work

Deborah Tannen: Communication Between Women and Men

Deborah Tannen, a sociolinguist, is the author of the best-seller *You Just Don't Understand: Women and Men in Conversation* (1990) and *Talking From 9 to 5* (1994). Tannen believes that gender differences are widespread in everyday speech, and that in many ways, men and women are living in different worlds when they try to communicate with each other.

Tannen notes that just about every woman she spoke to recalled saying something at a meeting and being ignored, only to have the same idea taken seriously when it was repeated by a man. Tannen believes this is due to the different ways in which men and women use language. Women use language primarily to create intimacy and connections to other people. For women, language is the glue that holds relationships together.

Men, however, use language mainly to convey information. For men, activities are the things that hold people together, and in the absence of doing something with others, talking about an activity is the next best thing. This involves talking about concrete events or facts. This explains the tendency among men to engage in endless discussions about sporting events, batting averages, and so on. Men are acutely aware of the status differences implied by knowledge, or in one way or another of speaking. For men, language and information are used to attain status, not intimacy.

When people speak at a meeting they may speak softly or loudly, with or without a disclaimer, in a self-deprecating or declamatory way, briefly or at length, tentatively or with apparent certainty. The first of each of these choices is more likely to be done by a woman than a man, and this plays a role in how seriously the idea is taken.

Tannen also notes that women engage in "troubles talk," or talking about the day's problems. Confronted with this type of talk, a man is likely to offer advice or solutions and get on with it. Women are not looking for this response. More typically, when one woman mentions a problem, someone else will mention a similar problem. Women are looking for agreement and mutual understanding. When someone offers advice or a solution, it creates "one-up-manship." It is no longer a situation in which two people are in similar circumstances, but one of two people jockeying for status.

Tannen tells a joke about a man whose wife wants to divorce him for not talking to her for 2 years. At the divorce hearing, the judge asks the man why he has not talked to her for all this time. He answers that he did not want to interrupt her. This joke implies that women talk more than men. Upon closer observation, though, we must distinguish between speaking in public situations and speaking at home. Men seem to be more comfortable with the language used in a public setting, while women find the language used in private settings more natural. Essentially, the public language is about information and one person attempting to attain status with knowledge, whereas private language is about personal feelings and sharing them with others. It therefore seems understandable that men talk more in public settings, while women talk more in the home. Men feel they do not need to talk much at home, as their status is established. They use language when they are struggling for status. The home is not such a place.

Should people try to change their conversational styles? In business settings, there may be some advantage to certain ways of presenting information. However, even if no one changes, understanding those gender differences improves relationships. Once people realize that their partners and coworkers have different conversational styles, they are inclined to accept differences without blaming themselves, others, or their relationships.

Sources: Interviews with the author, June 1990 and November 1994; *You Just Don't Understand: Women and Men in Conversation,* by D. Tannen, 1990, New York: William Morrow; *Talking From 9 to 5,* by D. Tannen, 1994, New York: William Morrow.

Instead of labeling this tentativeness as a typical example of women's inability to make firm decisions, Gilligan sees it as an attempt to deal with the consequences of actions rather than simply with "what is right." For women, moral dilemmas involving people have a greater complexity and therefore a greater ambiguity.

If your morality stresses the importance of not hurting others, as seems to be the case with most women, you will often face failure. As one of the men Gilligan interviewed said, making a moral decision is often a matter of "choosing the victim." There is "violence inherent in choice," Gilligan wrote, and "the injunction not to hurt can paralyze women."

If you base your decision on an absolute principle (for example, abortion is murder; therefore it is wrong), you may act with the decisiveness so admired by both men and women. If you base your decision on what you imagine is likely to happen (for example, if this child would be born with no father, if the fetus is likely be seriously defective, if the mother's life would be endangered by the birth, and so on), you often will face uncertainty. Women, whose value systems are focused more on people than on principles, consequently find themselves wrestling with the problems that might result from their decisions.

Gilligan hopes that by ceasing to label a man's perspective as right and a woman's as wrong, we can begin to understand that each may be valuable, though different. For this to happen, according to Gilligan, girls and women must gain confidence in their own ethical perspectives. Indeed, Gilligan feels, if society finally accepted a woman's moral view of the interconnectedness of actions and relationships, enormous consequences could result for everything from scholarship to politics to international rela-tions. (For a further discussion of these issues, see "Sociology at Work: Deborah Tannen: Communication Between Women and Men.")

Gender Inequality and Work

Women's numerical superiority over men has not enabled them as yet to avoid discrimination in many spheres of American society. In this discussion, we shall focus on economic and job-related discrimination, because these data are easily quantified and serve well to highlight the problem. Remember, though, that discrimination against women in America actually is expressed in a far wider range of social contexts and institutions.

Job Discrimination

More than 60% all American women are part of the paid labor force. In 1999, women earned approximately 77% as much as men did (see Figure 9-1).

Figure 9-1 Women's Median Pay as a Percentage of Men's Median Pay, 1890–1999

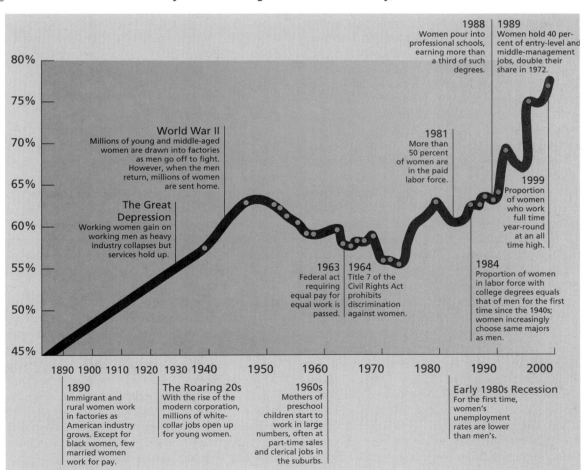

Women and men are concentrated in different occupational groups. Some would argue that the income and earnings differences between them result from pay differences across occupations rather than gender. Within each broad occupational category, women and men tend to work in different specific occupations. In the professional occupations, where women earn the most, they are much less likely than men to be employed in some higher-paying occupations, such as engineers and mathematical and computer scientists.

However, even if we adjust for occupations, we find differences in earnings. Women earn less than men in almost every occupation. For example, male managers and professionals earned $49,304 in 1999, compared with $35,412 for females. Male technical sales workers earned $32,552, while females earned $22,412.

There are differences in educational attainment between men and women. Could it be that these salary differences result from educational disparities? It does not seem as if that is the answer either. In 1999, male high-school graduates who worked full-time year-round had an average salary of $30,160, while women had an average salary of $21,060. For college graduates, the average salaries were $50,804 for men and $38,480 for women.

A striking fact in the above information is that not only are the average earnings for women lower than for men, but male high-school graduates earn not that much below female college graduates. One possible explanation for this disparity is that the less-skilled jobs men hold, such as construction, are typically unionized and therefore pay better. By contrast, women who hold less-skilled jobs, such as waitressing, are not unionized.

Job discrimination against women is a complicated phenomenon. There are three ways in which women experience discrimination in the business world: (1) during the hiring process, when women are given jobs with lower occupational prestige than men who have equivalent qualifications are given; (2) through unequal wage policies, by which women receive less pay than men for equivalent work; and (3) in awarding promotions, as women find it more difficult than men to advance up the career ladder.

Discrimination against women in the economic sector is often quite subtle. Women are more or less channeled away from participation in occupations that are socially defined as appropriate to men. For example, it cannot really be argued that female bank presidents are paid less than men are; instead, there are few women bank presidents. Women and men often do not perform equal work; therefore the phrase equal pay for equal work has little relevance. In some instances, similar work is performed by men and women, but there may be two job titles and two pay scales.

Having painted a somewhat pessimistic picture here, we should note that there has been some improvement in recent years. Women's share of jobs with high earnings has grown. In 1999, women held nearly 47% of all executive, administrative, and managerial jobs, up from 34.2% in 1983. Women now represent 52% of the professional workforce.

Women also constitute a rising share of people being awarded college and postgraduate degrees. In 1996, they represented 55% of people awarded bachelor's degrees, 56% of those awarded master's degrees, 40% of the doctorate recipients, 41% of the of those receiving M.D.s, and 44% of those receiving law degrees.

In addition, young women and men (those under age 25) had fairly similar earnings (young women's earnings were 91% of young men's); however, women's earnings were much lower than men's in older age groups.

Despite this progress, women still dominate low-paying fields. Five of the top 10 occupations employing women are secretaries, bookkeepers, cashiers, sales workers, and typists. The two professional positions that women dominate are relatively low-paying, namely nursing and elementary-school teaching ("Highlights of Women's Earnings in 1999," U.S. Department of Labor, Bureau of Labor Statistics, May 2000).

SUMMARY

Sociology makes a distinction between sex and gender. Sex refers to the physical and biological differences between men and women. Gender refers to the social, psychological, and cultural attributes of masculinity and femininity that are based on biological distinctions. Ideas about masculinity and femininity are culturally derived and are an important factor in shaping people's self-images and social identities. Whereas sex is generally an ascribed status, gender is learned through the socialization process and is thus an achieved status.

Many people, on the other hand, believe that gender identities and masculine or feminine behavior result from an innate, biologically determined process. Historically, societies in both the East and the West viewed women as inherently inferior to men. Not only did intellectuals argue from the point of view of a patriarchal ideology, or a belief that men are superior to women and should control all important aspects of society, but religions in both the East and West supported this view.

Functionalists argue that the sexual division of labor was necessary for the efficient operation of pre-industrial societies. Women, who birthed and nursed infants, remained involved in child care and the necessities of daily living. Men hunted, herded, and defended the family. Because the female was largely dependent on the male for protection and food, he became the dominant partner in the relationship. Functionalists maintain that this role differentiation is functionally necessary and efficient in modern society as well. Critics contend that this view is outdated and that industrial society provides for more flexibility in gender-role assignment.

According to conflict theory, males dominate females because of their superior power and control over key resources. A major consequence of this domination is the exploitation of women by men, and, according to conflict theorists, as long as men benefit from this arrangement, they have little incentive to change it.

Gender-role socialization is a lifelong process whereby people learn the values, attitudes, motivations, and behavior considered appropriate to each sex according to their culture.

INTERNET RESOURCES

Go to http://web-enhanced.thomsonlearning.com which features Web links relevant to this chapter to expand and enhance the learning experience. This site will be monitored and updated periodically to ensure the links are correct and live.

At Virtual Society: The Wadsworth Sociology Resource Center, http://sociology.wadsworth.com, you will find a Career Center, Sociology in the News, Research Online, InfoTrac College Edition, Virtual Tours in Sociology, and a variety of book-specific student and instructor resources.

CHAPTER NINE STUDY GUIDE

KEY CONCEPTS

Match each concept with its definition, illustration, or explanation below.

a. Sex
b. Gender
c. Linear temporal world
d. Eve's disobedience
e. Sociobiology

f. Manu code
g. Achieved gender status
h. Gender identity
i. Ethology

j. Ascribed gender status
k. Contingent temporal world
l. Honor killings
m. Patriarchy

_____ 1. A social science theory focusing on the inborn structure of social traits.

_____ 2. The physical and biological differences between men and women.

_____ 3. Cultural attributes of masculinity and femininity based on biological distinctions.

_____ 4. The term suggesting that a person's sex of either male or female is determined at birth.

_____ 5. The term suggesting that gender is learned through the socialization process.

_____ 6. The way women think about family and career responsibilities as a balancing act.

_____ 7. The Judeo-Christian foundation for the belief in the superiority of men over women.

_____ 8. The individual's sense of maleness of femaleness.

_____ 9. The belief that men are superior to women and should control all important aspects of society.

_____ 10. Ancient law in India stating that if a wife had no children after 8 years of marriage she would be banished.

_____ 11. The scientific study of animal behavior.

_____ 12. The way men view issues of career and family without mention of an adjustment to a wife's career.

_____ 13. The cultural system in Jordan where women who are raped are seen as compromising their families and accordingly are murdered by their fathers, brothers, and sons.

KEY THINKERS AND RESEARCHERS

Match the thinkers with their main ideas or contributions.

a. Erik Erikson
b. Friedrich Engels
c. George Murdock
d. Deborah Tannen
e. Carol Gilligan
f. Auguste Comte

g. Parsons and Bales
h. Ashley Montagu
i. Morris, Fox, Tiger
j. Margaret Mead
k. Michelle Rosaldo
l. John Money

m. Milton Diamond
n. Aristole
o. Thomas Jefferson
p. Maines and Hardesty

_____ 1. The founder of sociology, he believed that women should not be allowed to work outside the home, to own property, or to exercise political power.

_____ 2. A pioneering anthropologist who found that gender roles were not innate properties and could vary widely among societies.

_____ 3. Collected data on the sexual division of labor in more than 200 preliterate societies.

_____ 4. Applied the functionalist theory to the modern family.

_____ **5.** A prominent conflict theorist who linked gender inequalities to capitalism.

_____ **6.** Pointed out that boys and girls are pressured to resolve their adolescent identity crisis in very different ways in our culture.

_____ **7.** Challenged the value judgments made by traditional theorists about male and female styles of reasoning, especially in the area of moral decision making.

_____ **8.** Writers popularizing the work of ethology and generalizing from the behavior of non-human primates to that of humans.

_____ **9.** A critic who argued that sociobiologists do not take into account fundamental differences between human and nonhuman primates such as the human use of complex language structure.

_____ **10.** Studies the status of women and the differentiation between public and private spheres of activity across societies.

_____ **11.** Argued that gender is the product of socialization and if intervention was early enough, a child may be raised in a gender different than what he or she was born.

_____ **12.** Conducted research on the conversational style and content differences among females and males.

_____ **13.** Argued that a woman should consider that her husband's wishes are as divine laws.

_____ **14.** Author of the Declaration of Independence who maintained that women should neither be involved in politics nor be allowed to vote.

_____ **15.** Conducted research demonstrating that men operate in a linear temporal world while women operate in a contingent temporal world.

_____ **16.** Successfully challenged Money's theory of psychosexual neutrality in children.

CENTRAL IDEA COMPLETIONS

Following the instructions, fill in the appropriate concepts and descriptions for each of the questions posed in the following section.

1. Compare and contrast the ideas about women as expressed by Comte with the view of women represented in traditional Hindu religious conceptions:

   a. Comte: _____

   b. Hindu: _____

2. Following the research of Buss (1994), what can be concluded about the differences in how females and males think about sex (in terms of the number of sex partners)?

   a. Females: _____

   b. Males: _____

3. Explain the connection between stress and the releasing of the hormone oxytocin:

   _____

   _____

4. What did the research by Clark and Hatfield (1989) suggest about the differences between males and females in terms of how well they would need to know one another before dating, visiting each other's apartment, or having sex?

   _____

   _____

   _____

5. Compare and contrast the biological and sociological perspectives on sex and gender:

   a. Biological: _____

   b. Sociological: _____

6. A major aspect of gender is gender stratification systems. Compare and contrast the functional perspective with the conflict perspective as explanations for such stratification:

   a. Functionalist: _____

   b. Conflict: _____

7. What did Mead's research on the Arapesh, the Tchambuli, and the Mundugamor suggest about the interaction between culture and gender identity? What were each tribe's gender characteristics?

   a. Mead's overall conclusion: _____

   b. Arapesh: _____

   c. Tchambuli: _____

   d. Mundugamor: _____

8. What were the conclusions from Rosaldo regarding the relationship between the status of women and the degree of differentiation between public and domestic spheres of activity?

   _____

   _____

9. What did the research by Maines and Hardesty (1987) reveal about undergraduates' gender differences in their expectations of how family, work, and education would be integrated in their futures?

   _____

   _____

10. In what ways may gender be related to job discrimination patterns in America?

   _____

   _____

11. Compare and contrast Diamond's and Money's perspectives on the degree to which gender is established at birth:

   _____

   _____

   _____

   _____

CRITICAL THOUGHT EXERCISES

1. Your chapter begins with the very interesting discussion of Kate Bornstein, who used to be Al Bornstein before she underwent a sex-change operation. As noted in the discussion, the decision to change one's sex as well as the resulting sex-other orientations that an individual has after the operation called into question traditional views and assumptions about sex and gender roles. Beginning with your text, then moving into library research and an exploration of personal Web pages on the Internet, develop a research paper within which you explore the following:

a. Where do an individual's ideas about masculinity and femininity originate?

b. In the past three decades, how has American culture shifted its ideas about sex and gender?

c. As any examination of daytime talk-shock television will reveal, the American public has an apparently unending fascination with persons professing an alternative or third gender. What do you see as the source of such interest?

d. Are these individuals, such as Kate Bornstein, the vanguard of a shifting definition of gender and sexuality in America or merely deviants in the limelight of notoriety?

e. Set your search engine to *Transsexual* and visit some of the personal pages of individuals presenting their life stories relative to their gender orientation. What effect, if any, has visiting these pages had on your own interpretations of sex and gender?

2. Continuing with the topic of gender ambiguity and change, select one of Madonna's rock videos and develop an analysis of contemporary representations of sexual pleasure. What images appear on the video and what message, if any, are they presenting to the viewing audience?

3. Karen Walker has conducted research on the friendship patterns of women and men. Her focus has been on what each gender says and does within friendship relationships. Secure a copy of one of her research articles and after reading it along with your text, think about your own friendship patterns. Write a report in which you briefly analyze your own friendship patterns, but also develop a design for a ministudy of the gender friendships of your close associates. What variables and patterns would you be looking for in your research? How would you gather your data? To what extent would you expect your findings to be similar to or distinct from the findings of Karen Walker?

4. After reading the story of Kate Bornstein, visit the Internet and examine several of the sites created by transgendered individuals. You may do a search or visit one of the large transgender-friendly cyberspace centers, for example, youthresource.com, which house individual transgender home pages. What ideas about gender and the sense of the self are presented in those pages? In what ways are the individual home pages similar and in what ways are they distinct from one another? Explain what you have learned about transgendered individuals after visiting a number of their personal home pages.

5. Following the work of Debra Tannen (read her book, *You Just Don't Understand*), select a public place on campus or in the campus community where you and a companion of the opposite gender can sit, listen, and observe 15 minutes of gender conversations. Your task is not to "hide and eavesdrop," but rather to take account of public discourse presented at a decibel level not intended to be private. Observe the conversational styles and content of each of the following:

a. Two males

b. Two females

c. A male and a female

d. A group containing four or more males or females in separate company

e. a group containing four or more males or females in mixed gender company

Prepare a report in which you indicate your findings on the style and content of the conversations. Were your findings consistent with Tannen's research?

6. Discuss in detail the issue of honor killings. Using the resources in your library and those on the World Wide Web, locate some recent examples of honor killings around the world. Are they alike throughout the world, or are there substantial variations by region and culture? If these are a part of another culture, are we in the United States correct or incorrect in taking a stand against honor killings? Assume you have a high position in a company that trades with countries where honor killings occur. Assume further that you personally are against honor killings. How would you go about convincing your company's board of directors to stop doing business with countries where honor killings occur?

ANSWERS

KEY CONCEPTS

1. e	4. j	6. k	8. h	10. f	12. c
2. a	5. g	7. d	9. m	11. i	13. l
3. b					

KEY THINKERS/RESEARCHERS

1. f	4. g	7. e	10. k	13. n	15. p
2. j	5. b	8. i	11. l	14. o	16. m
3. c	6. a	9. h	12. d		

PART 4

Institutions

Every society has standardized ways of fulfilling its members' basic social needs. Sociologists call these patterned interactions institutions. Humans fashion their social institutions in accordance with their society's dominant norms and values. At the same time, social institutions mold and channel individual behavior and values. Social institutions, however, are not monolithic; tensions, defects, and countervailing pressures within them mirror conflicts and controversies in society at large.

In this section of the book, you will explore four major institutions in American society. The first of these is the family. Why do humans live in families? What functions does the family fulfill? What different forms does the family take? How are families formed—that is, what are the rules and practices surrounding marriage? How is the family changing in today's society? All of these questions will be dealt with in Chapter 10.

You also will examine the religious institution. Chapter 11 explores the elements and major types of religious belief. Here you will be introduced to a sociological analysis of religion: competing perspectives on the functions of religious belief for individuals and groups, description and analysis of the social organization of religious life, and examination of religious trends in contemporary American society.

Education is about the acquisition of skills and knowledge. But what else is it? What else goes on in our classrooms and our schools? Indeed, what should be going on there? As you have already seen in examining other issues, functionalists and conflict theorists have different views of why certain features of our society exist and whose interests those features serve. This is also true with regard to education, and in Chapter 12 you will examine these contrasting perspectives on the organization and operation of our educational system. You will also investigate some of the most important social issues facing American education today.

Every society must have mechanisms for making decisions about important issues. These mechanisms are organized into the political institution, which is the subject of Chapter 13. Some of the questions you will explore in this chapter include What is power? How is it exercised legitimately? How and why is government organized the way it is? How does the economic system influence politics and the government, and vice versa? How does political change occur?

When you have completed Part IV of this book, you will have a comprehensive picture of how American society is organized. You will then be ready to explore some important social issues facing our society and our world as we move into the next century.

CHAPTER TEN

Marriage and Alternative Family Lifestyles

LEARNING OBJECTIVES

After studying this chapter, you should be able to do the following:

- Explain the functions of the family.
- Describe the major variations in family structure.
- Define marriage and describe its relationship to the phenomenon of romantic love.
- Describe the various rules governing marriage.
- Explain the ways in which mate selection is not random.
- Summarize recent changes in the family as an institution.
- Explain the impact of changes in divorce and child-custody laws.
- Describe the various alternative lifestyles in contemporary American society.

The American family has changed dramatically since 1970. Some of the changes that have taken place include the following:

The marriage rate has fallen more than 40%.

When men and women marry today they are on average 4 years older than in 1970.

In 1998, single-parent households constituted 27% of family households with children, up from 11% in 1970.

Over the past 30 years, the percentage of persons who have never been married has increased from about 22% to 28%.

The divorce rate has increased by nearly 40%.

Unmarried-couple households have increased nearly fivefold.

One-third of all births in 1999 were to unmarried women, compared with one-tenth in 1970.

Half of all children are expected to spend some part of their childhood in a single-parent home (U.S. Bureau of the Census, 2000).

Men and women who grew up in the 1950s and 1960s watched television programs such as *Leave It to Beaver* and *Father Knows Best,* which presented a glorified version of the family of that era. Even though many families did not fit the model, the typical idealized family was one in which mother stayed at home in the suburbs with a brood of children, baking cookies for them and tending to their needs, while father was off working in the city.

Today, many family forms are common: single-parent families (resulting from either unmarried parenthood or divorce), remarried couples, unmarried couples, stepfamilies, and extended or multigenerational families. Mothers are more likely to be full- or part-time workers than they are to be full-time homemakers.

The United States is not alone in experiencing such changes. Family patterns in the United States reflect broad social and demographic trends that are occurring in most industrialized countries around the world.

Has the American family always been in such flux? Although information about family life during the earliest days of our country is not very precise, it does appear quite clear, beginning with the 1790 Census, that the American family was quite stable. Divorce and family breakup were not common. If a marriage ended because of desertion, death, or divorce, it was seen as a personal and community tragedy.

The American family has always been quite small. There has never been a strong tradition here of the extended family, in which relatives of several generations lived within the same dwelling. Even in the 1700s, most American families consisted of a husband, wife, and approximately three children. This private, inviolate enclave made it possible for the family to endure severe circumstances and to help build the American frontier.

By the 1970s, radical changes were becoming evident. The marriage rate began to fall; the divorce rate, which had been fairly level, began accelerating upward; and fertility began to decline.

The situation today shows that this trend is continuing. In 1999 there were 2,319 million marriages, though there were also 788,000 divorces. This information is even more striking when we realize that divorce statistics do not include desertion or other forms of marital breakup, such as annulment or legal separation.

There also are other signs of change and family instability. Of all children born in 1998, 32.8% were born to single women. Among African Americans, the percentage of children born out of wedlock was 69.1% (*Statistical Abstract of the United States 2000*, 2001).

The inescapable conclusion that can be drawn from all this is that the American family is in a state of transition. But transition to what? In this chapter, we shall study the institution of the family and look more closely at the current trends and what they forecast for the future.

The American family will continue to evolve and change even further during the next decade. Traditional families will account for only two out of three households, and married couples will live in just more than half of all households.

The Nature of Family Life

The U.S. Bureau of the Census distinguishes between a *household* and a *family*. Households consist of people who share the same living space. A household may include one person who lives alone or several people who share a dwelling. A family, on the other hand, has a more precise definition. For example, in his classic study of social organization and the family, anthropologist George P. Murdock (1949) defined the family as:

> a social group characterized by common residence, economic cooperation, and reproduction. It includes adults of both sexes, at least two of whom maintain a socially approved sexual relationship, and one or more children, own or adopted, of the sexually cohabiting adults.

This definition has proved to be too limited. For one thing, it excludes many kinds of social groups that seem, on the basis of the functions they serve, to deserve the label of family. For example, in America single-parent families are widely recognized. Despite Murdock's restricted definition of the family, it was his 1949 study of kinship and family in 250 societies, which showed the institution of the family to be present in every one of them, that led social scientists to believe that some form of the family is found in every known human society.

Perhaps a better understanding of the concept of the family may be gained by examining the various functions it performs in society.

Functions of the Family

Social scientists often assign to the fundamental family unit—married parents and their offspring—a number of the basic functions that it serves in most, if not all, societies. Although in many societies the basic family unit serves some of these functions, in no society does it serve all of them completely or exclusively. For example, among the Nayar of India, a child's biological father is socially irrelevant. Generally, another man takes on the social responsibility of parenthood (Gough, 1952). Or consider the Trobrianders, a people living on a small string of islands north of New Guinea. There, as reported by anthropologist Bronislaw Malinowski (1922), the father's role in parenthood was not recognized, and the responsibility for raising children fell to the family of their mother's brother.

Anthropologist George P. Murdock defined the family as a social group characterized by common residence, economic cooperation, and reproduction.

The basic family is not always the fundamental unit of economic cooperation. Among artisans in preindustrial Europe, the essential economic unit was not the family but rather the household, typically consisting of the artisan's family plus assorted apprentices and even servants (Laslett, 1965). In some societies, members of the basic family group need not necessarily live in the same household. Among the Ashanti of western Africa, for example, husbands and wives live with their own mothers' relatives (Fortes et al., 1947). In the United States, a small but growing number of two-career families is finding it necessary to set up separate households in different communities, sometimes hundreds of miles apart. Husband and wife travel back and forth between these households to be with each other and the children on weekends and during vacations.

In all societies, however, the family does serve the basic social functions discussed next.

Regulating Sexual Behavior No society permits random sexual behavior. All societies have an **incest taboo,** which *forbids intercourse among closely related individuals,* although who is considered to be closely related varies widely. Almost universally, incest rules prohibit sex between parents and their children and between brothers and sisters. However, there are exceptions: The royal families of ancient Egypt, the Inca nation, and early Hawaii did allow sex and marriage between brothers and sisters. In the United States, marriage between parents and children, brothers and sisters, grandparents and grandchildren, aunts and nephews, and uncles and nieces is defined as incest and is forbidden. In addition, approximately 30 states prohibit marriage between first cousins. The incest taboo usually applies to members of one's family (however the family is defined culturally) and thus it promotes marriage—and, consequently, social ties—among members of different families.

Patterning Reproduction Every society must replace its members. By regulating where and with whom individuals may enter into sexual relationships, society also patterns sexual reproduction. By permitting or forbidding certain forms of marriage (multiple wives or multiple husbands, for example) a society can encourage or discourage reproduction.

Organizing Production and Consumption In preindustrial societies, the economic system often depended on each family's producing much of what it consumed. In almost all societies, the family consumes food and other necessities as a social unit. Therefore, a society's economic system and family structures often are closely correlated.

Socializing Children Not only must a society reproduce itself biologically by producing children, it also must ensure that its children are encouraged to accept the lifestyle it favors, to master the skills it values, and to perform the work it requires. In other words, a society must provide predictable social contexts within which its children are to be socialized (see Chapter 4). The family provides such a context almost universally, at least during the period when the infant is dependent on the constant attention of others. The family is ideally suited to this task because its members know the child from birth and are aware of its special abilities and needs.

Providing Care and Protection Every human being needs food and shelter. In addition, we all need
to be among people who care for us emotionally,
who help us with the problems that arise in daily
life, and who back us up when we come into conflict
with others. Although many kinds of social groups
are capable of meeting one or more of these needs,
the family often is the one group in a society that
meets them all.

Providing Social Status Simply by being born
into a family, each individual inherits both material
goods and a socially recognized position defined
by ascribed statuses (see Chapter 5). These statuses
include social class or caste membership and ethnic identity. Our inherited social position, or family
background, probably is the single most important
social factor affecting the predictable course of our
lives.

Thus, we see that Murdock's definition of the
family, based primarily on structure, is indeed too
restrictive. A much more productive way of defining
the family would be to view it as a universal institution that generally serves the functions discussed
previously, although the way it does so may vary
greatly from one society to another. The different
ways in which various social functions are fulfilled
by the institution of the family depend, in some instances, on the form the family takes.

Family Structures

The **nuclear family** is *the most basic family form and
is made up of a married couple and their biological
or adopted children.* The nuclear family is found in
all societies, and it is from this form that all other
(composite) family forms are derived.

Two major composite family forms have been defined: polygamous and extended families. **Polygamous families** are *nuclear families linked together by
multiple marriage bonds, with one central individual
married to several spouses. The family is* **polygynous** *when the central person is male and the multiple spouses are female. The family is* **polyandrous**
*when the central person is female and the multiple
spouses are male.* Polyandry is rare; only one-half of
1% of all societies permit a woman to take several
husbands simultaneously, and the women have to be
wealthy members of those societies to do so. The
Tlingit of southern Alaska are one example. Polyandry also occurs among well-to-do Tibetan families in the highlands of Limi, Nepal (Fisher, 1992).

Extended families *include other relations and
generations in addition to the nuclear family, so that
along with married parents and their offspring, there
may be the parents' parents, siblings of the parents,
the siblings' spouses and children, and in-laws.* All
the members of the extended family live in one
house or in homes close to one another, forming one
cooperative unit.

Families, whether nuclear or extended, trace their
relationships through the generations in several
ways. Under the **patrilineal system,** *the generations
are tied together through the males of a family; all
members trace their kinship through the father's
line.* Under the **matrilineal system,** exactly the opposite is the case: *The generations are tied together
through the females of a family.* Under the **bilateral
system,** *descent passes through both females and
males of a family.* Although in American society descent is bilateral, the majority of the world's societies
are either patrilineal or matrilineal (Murdock, 1949).

In patrilineal societies, social, economic, and political affairs usually are organized around the kinship

Under the patri-
lineal system, the
generations are
tied together
through the males
of the family.

relationships among men, and men tend to dominate public affairs. Polygyny often is permitted, and men also tend to dominate family affairs. Sociologists use the term **patriarchal family** to describe situations in which *most family affairs are dominated by men.* The **matriarchal family,** in which *most family affairs are dominated by women,* is relatively uncommon but does exist. Typically, it emerges in matrilineal societies. The matriarchal family is becoming increasingly common in American society, however, with the rise of single-parent families (most often headed by mothers).

Whatever form the family takes and whatever functions it serves, it generally requires a marriage to exist. Like the family, marriage varies from society to society in its forms.

Defining Marriage

Marriage, an institution found in all societies, is the *socially recognized, legitimized, and supported union of individuals of opposite sexes.* It differs from other unions (such as friendships) in that (1) it takes place in a public (and usually formal) manner; (2) it includes sexual intercourse as an explicit element of the relationship; (3) it provides the essential condition for legitimizing offspring (that is, it provides newborns with socially accepted statuses); and (4) it is intended to be a stable and enduring relationship. Thus, although almost all societies allow for divorce—that is, the breakup of marriage—no society endorses it as an ideal norm.

Romantic Love

In 1900 Elizabeth Cady Stanton, an early advocate for women's rights, was asked why so many marriages were failures. Stanton replied that "People marry for all sorts of reason except the true ones— love and affection" (Crichton, 1998).

Americans are strong believers in the power of love. Over half of American adults (52%) said they believed in "love at first sight." Four in 10 Americans said they have actually fallen in love at first sight and almost three-quarters believed that "one true love" existed (Gallup Poll, February 2001).

We are also strongly wedded to the view that there is a compatibility between romantic love and the institution of marriage. Not only do we believe they are compatible, but we expect them to coexist. Our culture implies that without the prospect of marriage, romance is immoral and that without romance, marriage is empty. This view underlies most romantic fiction and other media presentations.

When college students from 10 countries were asked if they would marry without love, 86% of American college students said they would not. Only 24% of college students from India said no to marriage without love. Cultures are likely to indulge in romantic love if they are wealthy and value individualism over the community (Levine, 1993).

Romantic love can be defined in terms of five dimensions: (1) idealization of the loved one, (2) the notion of a one and only, (3) love at first sight, (4) love winning out over all, and (5) an indulgence of personal emotions (Lantz, 1982).

In some cultures, people expect love to develop after marriage. Hindu children are taught that marital love is the essence of life. Men and women often enter married life enthusiastically expecting a romance to develop. As the Hindu saying goes, "First we marry, then we fall in love" (Fisher 1992). (See "Global Sociology: Arranged Marriage in India.")

In many of the world's other societies, romantic love is unknown or seen as a strange maladjustment. It may exist, but it has nothing to do with marriage. Marriage in these societies is seen as an institution that organizes or patterns the establishment of economic, social, and even political relationships among families. Three families ultimately are involved: **families of origin** or **families of orientation,** *the two families that produced the two spouses* and their **family of procreation,** *the family created by the spouses' union.*

A belief in romantic love helps to make men and women more independent of their relatives. Romantic love weakens the emotional ties that bind people to their families. It makes it natural for partners to be committed to each other and provide mutual support. Romantic love provides a legitimate way to move out of the family without creating tensions and jealousies.

Marriage Rules

In every society, marriage is the binding link that makes possible the existence of the family. All societies have norms or rules governing who may marry whom and where the newlywed couples should live. These rules vary, but certain typical arrangements occur in many societies around the world.

Almost all societies have two kinds of marriage norms or rules: *rules of* **endogamy** *limit the social categories from within which one can choose a marriage partner.* For example, many Americans still attempt to instill in their children the idea that one should marry one's own kind—that is, someone within the ethnic, religious, or economic group of one's family of origin.

Rules of **exogamy,** by contrast, *require an individual to marry someone outside his or her culturally defined group.* For example, in many tribal groups, members must marry outside their lineage. In the

Global Sociology

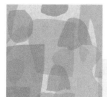

Arranged Marriage in India

Shona Bose's parents gave her all of the freedoms available to progressive, upper-middle-class Indian youngsters of the 1990s. She attended a coeducational college, was encouraged to get a job with a foreign bank in Calcutta, went to discos with her friends, and was told that if she fell in love, she could pick the husband of her choice.

Yet when the time came to get married, Bose—like most other Westernized young urban professionals in India—chose the age-old tradition that had been forced on previous generations—arranged marriage.

"When you fall in love as a teenager, you look for different things: looks, glamour, is he fun to go out with? I met a lot of people I liked, but no one who was suitable for marriage," said Bose. "Love is important, but it is not sufficient."

In contrast to some of their parents, who rebelled in the 1960s and 1970s by opting for "love marriages," members of this generation assess their plans for marriage as they might a business plan. Is it financially viable? Are our lifestyles compatible? Is each side likely to uphold his or her end of the deal?

Young people know the divorce rate in India, long dominated by arranged marriages, is less than 5%, while in the United States roughly half of all marriages break up. While India's low rate can be attributed to cultural disapproval of divorce, advocates of arranged marriage insist it is also because research goes into matching mates with optimum compatibility.

Until a few decades ago, nearly every family married their children to eligible mates based on horoscopes, caste, community standing, and pedigree without consulting the children. The bride and groom were often not allowed to meet before the wedding, and even if they did they were not allowed

to refuse the match. Most children are still expected to marry the spouse of their family's choice. Recently, newspapers carried a story of a young couple who were beheaded for having eloped.

Among educated Indians, some things have changed. Families who once relied on professional neighborhood matchmakers now often opt for computerized marriage bureaus and pages of highly specific matrimonial ads in the Sunday papers.

Young people are commonly allowed to meet alone a few times before the final decision is made, although relationships are still quite restricted compared to the West. Fewer than half of middle-class urban couples had even held hands before the wedding.

Young people support these arranged marriages. In a survey of urban professionals, 81% said their marriages were arranged, and 94% rated these marriages as "very successful." A poll of college students found that two-thirds wanted an arranged marriage.

The young people's attraction to arranged marriage may be a reaction, in part, to a period in the 1970s and 1980s when many marriages that had been based on love ran into trouble. A 1973 study of upper-middle-class north Indian women found 39% said love was essential for marital happiness. Two decades later, that number had fallen to 11%.

Arranged marriages may also persist because they guarantee that almost everyone will have a marriage partner. Perhaps, most of all, they persist because of the family ties they create. There will always be a great deal of family support if the marriage runs into trouble.

Source: "Love No Match for Custom: India's Young Favor Arranged Unions," by I. A. R. Lakshmanan, August 26, 1997, *The Boston Globe*, pp. A1, A10.

United States, there are laws forbidding the marriage of close relatives, although the rules are variable.

These norms vary widely across cultures, but everywhere they serve basic social functions. Rules of exogamy determine the ties and boundaries between social groups, linking them through the institution of marriage and whatever social, economic, and political obligations go along with it. Rules of endogamy, by requiring people to marry within specific groups, reinforce group boundaries and perpetuate them from one generation to the next.

Marriage rules also determine how many spouses a person may have at one time. Among many groups,

Europeans and Americans, for example, marriage is **monogamous**—that is, *each person is allowed only one spouse at a time.* However, many societies allow **multiple marriages,** in which *an individual may have more than one spouse* (polygamy). Polygyny, the most common form of polygamy, is found among such diverse peoples as the Swazi of Africa, the Tiwi of Australia, and, formerly, the Blackfeet Indians of the United States. Polyandry, as noted earlier, is extremely rare (Murdock 1949).

As Marvin Harris (1975) has noted, "Some form of polygamy occurs in 90 percent of the world's cultures." However, within each such society, only a

In India, it is not unusual for girls to be betrothed at an early age; this bride is 4 years old.

minority of people actually can afford it. In addition, the Industrial Revolution favored monogamy, for reasons we shall discuss shortly. Therefore, monogamy is the most common and widespread form of marriage in the world today.

Marital Residence

Once two people are married, they must set up housekeeping. Most societies have strongly held norms that influence where a couple lives. **Marital residence rules** *govern where a couple settles down.* **Patrilocal residence** *calls for the new couple to settle down near or within the husband's father's household*—as among Greek villagers and the Swazi of Africa. **Matrilocal residence** *calls for the new couple to settle down near or within the wife's mother's household*—as among the Hopi Indians of the American Southwest.

Some societies, such as the Blackfeet Indians, allow **bilocal residence,** in which *new couples choose whether to live with the husband's or wife's family of origin.* In modern industrial society, newlyweds typically have even more freedom. With **neolocal residence,** *the couple may choose to live virtually anywhere,* even thousands of miles from their families of origin. In practice, however, it is not unusual for American newlyweds to set up housekeeping near one of their respective families.

Marital residence rules play a major role in determining the compositions of households. With patrilocal residence, groups of men remain in the familiar context of their father's home, and their sisters leave to join their husbands. In other words, after marriage the women leave home to live as strangers among their husband's kinfolk. With matrilocal residence just the opposite is true: Women and their children remain at home, and husbands are the outsiders. In many matrilocal societies this situation often leads to considerable marital stress, with husbands going home to their own mothers' families when domestic conflict becomes intolerable.

Bilocal residence and neolocal residence allow greater flexibility and a wider range of household forms because young couples can move to places in which the social, economic, and political advantages may be greatest. One disadvantage of neolocal residence is that a young couple cannot count on the immediate presence of kinfolk to help out in times of need or with demanding household chores (including the rearing of children). In the United States today, the surrogate or nonkin "family," made up of neighbors, friends, and colleagues at work, may help fill this void. In other societies, polygynous neolocal families help overcome such difficulties, with a number of wives cooperating in household work.

Mate Selection

Like our patterns of family life, America's rules for marriage, which are expressed through mate selection, spring from those of our society's European forebears. Because we have been nourished, through songs and cinema, by the notions of "love at first sight," "love is blind," and "love conquers all," most of us probably believe that in the United States there are no rules for mate selection. Research shows, however, that this is not necessarily true.

If we think statistically about mate selection, we must admit that in no way is it random. Consider for a moment what would happen if it were: Given the population distribution of the United States, blacks would be more likely to marry whites than members of their race; upper-class individuals would have a greater chance of marrying a lower-class person; and various culturally improbable but statistically probable combinations of age, education, and religion would take place. In actual fact, **homogamy**—*the tendency of like to marry like*—is much more the rule.

There are numerous ways in which homogamy can be achieved. One way is to let someone older and wiser, such as a parent or matchmaker, pair up appropriately suited individuals. Throughout history, this has been one of the most common ways by which marriages have taken place. The role of the couple in question can range from having no say about the matter whatsoever to having some sort of veto power. This tradition is quite strong in Islamic countries such as Pakistan. Benazir Bhutto, the two-time Pakistani prime minister, agreed to an arranged marriage despite her Western education and feminist leanings. She submitted to Islamic cultural traditions, under which dating is not considered acceptable behavior for a woman.

In the United States, most people who marry do not use the services of a matchmaker, though the result in terms of similarity of background is so highly patterned that it could seem as if a very conscious homogamous matchmaking effort were involved.

Age   In American society, people generally marry within their own age range. Comparatively few marriages have a wide gap between the ages of the two partners. In addition, only 23.2% of American women marry men who are younger than themselves. On the average, in a first marriage for both the man and the woman, the man tends to be about 2 years older than the woman. This is, however, related to age at the time of marriage. For example, 20-year-old men marry women with a median age only one month younger. Twenty-five-year-old men marry women with a median age of eleven months younger. For 30- to 34-year-old men, the median age for wives is 2.8 years younger; while for men over 65, the median age for wives is nearly 10 years younger (*Statistical Abstract of the United States,* 1991).

In 1890, the estimated median age at first marriage was 26.1 years for men and 22.0 years for women. At that time, a decline in the median age at first marriage began; it did not end until 1956, when the median age reached a low of 20.1 years for women and 22.5 years for men. This 66-year decline has been reversed, and in 1998 the median age at first marriage exceeded the 1890 level, standing at 26.7 years for men and 25.0 for women (U.S. Bureau of the Census, Current Population "Marital Status and Living Arrangements," March 1998).

For remarriages after a divorce, as might be expected, the average age is older. For men it is 37.4, while for women it is 34.2. For widows and widowers, the average age of remarriage is older still, 54.0 for women and 63.1 for men (*Statistical Abstract of the United States,* 1997).

Age homogamy appears to hold for all groups within the population. Studies show that it is true

As late as 1966, 19 states sought to stop interracial marriage through legislation.

for blacks as well as whites and for professionals as well as laborers. Clearly, the norms of our society are very effective in causing people of similar age to marry each other (Leslie, 1979).

Race   Homogamy is most obvious in the area of race. It was only 35 years ago that Thurgood Marshall, who was only months away from appointment to the Supreme Court, suffered an indignity that today seems not just outrageous but almost incomprehensible. He and his wife had found their dream house in a Virginia suburb of Washington, D.C., but could not lawfully live together in that state: he was black and she was Asian. Virginia did not allow such interracial marriages. Fortunately for the Marshalls, in January 1967 the Supreme Court struck down the anti-interracial-marriage laws in Virginia and 18 other states. In 1967 these laws were not mere leftover scraps from an extinct era. Two years before, at the crest of the civil rights revolution, a Gallup poll found that 72% of Southern whites and 42% of Northern whites still wanted to ban interracial marriage (Sailer, 1997).

The laws in 19 states that sought to stop interracial marriage through legislation in the 1960s varied widely, and there was great confusion because of various court interpretations. In Arizona before 1967, it was illegal for a white person to marry a black, Hindu, Malay, or Asian person. The same thing was true in Wyoming, and residents of that state were also prohibited from marrying mulattos.

In 1966, Virginia's Supreme Court of Appeals had to decide on the legality of a marriage that had taken place in Washington, D.C., in 1958 between Richard P. Loving, a white man, and his part-Indian and part-black wife, Mildred Loving. The court unanimously upheld the state's ban on interracial marriages. The couple appealed the case to the United States Supreme Court, which agreed to decide whether state laws prohibiting racial intermarriage were constitutional. Previously, all courts had ruled that the laws were not discriminatory because they applied to both whites and nonwhites. However, on June 12, 1967, the Supreme Court ruled that states could not outlaw racial intermarriage. Today about 5% of U.S. married couples include spouses of a different race. This small percentage masks a remarkable growth in the number of interracial marriages. Since 1970, the number of black and white interracial married couples has increased from 300,000 to 1,400,000 in 1998 (National Center for Health Statistics, 1999).

The most common type of interracial marriage is one between a white husband and a wife of a race other than black. The next most common is between a white woman and a man of a race other than black. These types of interracial marriages have been increasing substantially, while black-white interracial marriages appear to be leveling off.

Interracial marriages are more likely to involve at least one previously married partner than are marriages between spouses of the same race. In addition, brides and grooms who marry interracially tend to be older than the national average.

The degree of education also differs in interracial marriages. White grooms in interracial marriages are more likely to have completed college than white grooms who marry within their race. In all-white couples, 18% of the grooms hold college degrees. In a marriage involving a white groom and a black bride, 24% of the men finished college. Among black men married to women of other races only 5% had completed college, whereas 13% of black men married to white women had done so. This compares with 9% of black men whose spouses were also black who held college degrees.

Although the rate of interracial marriage is growing, the proportions vary widely by state. Hawaii has the distinction of having both the highest number and the greatest proportion of interracial marriage. Almost one-fourth of all marriages there are interracial, the majority of which are Asian-Caucasian unions. Florida runs a distant second. Illinois has the largest number of white brides marrying black grooms of any state. The record for the reverse combination—black brides marrying white grooms—is held by New Jersey. Alaska is second to Hawaii in the proportion of interracial marriages. Thirteen percent of the state's marriages are interracial. None of the other states comes close to these percentages.

Even though their total numbers are still small, racially mixed marriages have become increasingly common in America's melting pot.

Religion Unlike many European nations, none of the American states has ever had legislation restricting interreligious marriage. Religious homogamy is not nearly as widespread as is racial homogamy,

though most marriages still involve people of the same religion.

Attitudes toward religious intermarriage vary somewhat from one religious group to another. Almost all religious bodies try to discourage or control marriage outside the religion, but they vary in the extent of their opposition. Before 1970, the Roman Catholic church would not allow a priest to perform an interreligious marriage ceremony unless the non-Catholic partner promised to rear the children of the union as Catholics, and the Catholic partner promised to encourage the non-Catholic to convert. However, in the late 1960s the Catholic church softened its policy, and since 1970 the pope has allowed local bishops to permit mixed-religion marriages to be performed without a priest and has also eliminated the requirement that the non-Catholic partner promise to rear the children as Catholics. Nevertheless, the Catholic partner must still promise to try to have the children raised as Catholic.

Protestant denominations and sects also differ with regard to the barriers they place on the intermarriage of their members. At one extreme are the Mennonites, who excommunicate any member who marries outside the faith. At the other are numerous denominations that may encourage their members to marry within the faith, but provide no formal penalties for those who do not (Heer, 1980).

Jewish religious bodies also differ in their degree of opposition to religious intermarriage. Orthodox Jews are the most adamantly opposed to intermarriage, while Conservative and Reform Jews, while by no means endorsing it, are more tolerant when it does take place. Jewish intermarriage rates have been increasing dramatically in recent years, with nearly 50% of all Jews marrying someone of another religion.

Religious leaders often are concerned about religious intermarriage, for a variety of reasons. Some claim that one or both intermarrying parties are lost to the religion, and others believe that the potential for marital success is decreased greatly in intermarriage. Complex factors are involved in intermarriages, and simplistic and unequivocal predictions are not warranted.

Most marriages still involve those of the same religion. There is, however, a clear trend of more religious intermarriages in the United States today.

Social Status Level of education and type of occupation are two measures of social status. In these areas, there is usually a great deal of similarity between people who marry each other. Men tend to marry women who are slightly below them in education and social status, though these differences are within a narrow range. Wide-ranging differences between the two people often contain an element of exploitation. One partner may either be trying to make a major leap on the social class ladder or be looking for an easy way of taking advantage of the other partner because of unequal power.

The typical high-school environment often plays a major role in maintaining social status homogamy, as it is in high school that students have to start making plans for their future careers. Some may go to college, and others may plan on going to work directly after graduation. This process causes the students to be divided into two groups: the college bound and the workforce bound. Although the lines separating these two groups are by no means impenetrable, in many high schools these two groups maintain separate social activities. In this way, barriers against dating and future marriage between those of unlike social status are set up. After graduation, those who attend college are more likely to associate with other college students and to choose their mates from that pool. Those who have joined the workforce are more likely to choose their mates from that environment. In this way, similarities in education between marriage partners really are not accidental.

As with several of the items we have discussed already, education, social class, and occupation produce a similarity of experience and values among people. Just as growing up in an Italian-American family may make one feel comfortable with Italian customs and traits, going to college may make one feel comfortable with those who have experienced that environment. Similarities in social status, then, are as much a result of socialization and culture as of conscious choice. We most likely will marry a person we feel comfortable with—a person who has had experiences similar to our own.

Coming to terms with the constraints on our marriage choices can be a sobering experience: What we thought to be freedom of choice in selecting a mate is revealed instead by various studies to be governed by rules and patterns.

The Transformation of the Family

Most scholars agree that the Industrial Revolution had a strong impact on the family. In his influential study of family patterns around the world, William J. Goode (1963) showed that the modern, relatively isolated nuclear family with weak ties to an extensive kinship network is well adapted to the pressures of industrialism.

In 1946 (when this photo was taken), the median family income was $2,800, television was in its infancy, and only 12.5% of people aged 18 to 22 went to college. In the years following World War II, a radical transformation took place in our society and in the family.

First, industrialism demands that workers be geographically mobile so that a workforce is available wherever new industries are built. The modern nuclear family, by having cut many of its ties to extended family networks, is freer to move than its predecessors. It was among laborers' families that extended kinship ties first were weakened. Only in the past few decades have middle- and upper-class families become similarly isolated.

Second, industrialism requires a certain degree of social mobility (see Chapter 7). This is so talented workers may be recruited to positions of greater responsibility (with greater material rewards and increased prestige). A family that is too closely tied to other families in its kinship network will find it difficult to break free and climb into a higher social class. However, if families in the higher social classes are too tightly linked by kinship ties, newly arriving families will find it very difficult to fit into their new social environment. Hence, the isolated nuclear family is well suited to the needed social mobility in an industrial society.

Third, the modern nuclear family allows for inheritance and descent through both sides of the family. Further, material resources and social opportunities are not inherited mainly by the oldest males (or females), as in some societies. This means that all children in a family will have a chance to develop their skills, which in turn means that industry will have a larger, more talented, and flexible labor force from which to hire workers.

By the early twentieth century, then, the nuclear family had evolved fully among the working classes of industrial society. It rested on (1) the child-centered family, (2) **companionate marriage** (that is, *marriage based on romantic love*), (3) increased equality for women, (4) decreased links with extended families or kinship networks, (5) neolocal residence and increased geographical mobility, (6) increased social mobility, and (7) the clear separation between work and leisure. There also was boring and alienating work and an associated expectation that the nuclear family would fulfill the function of providing emotional support for its members.

The World War II years also had a profound effect on the American family, for it was during the war that a process begun in the Great Depression really accelerated. The war made it necessary for hundreds of thousands of women to work outside the home to support their families. They often had to take jobs, vital to the American economy, that had been vacated when their husbands went to fight overseas. After the war, an effort was made to "defeminize" the workforce. Nevertheless, many women remained on the job, and those who left knew what it was like to work for compensation outside the home. Things were never the same again for the American family, and family life began to change.

The initial changes were not all that apparent. Indeed, by the 1950s the United States had entered the most family-oriented period in its history. This was the era of the baby boom, and couples were marrying at the youngest ages in recorded American history. During the 1950s, 96% of women in the child-bearing years married (Blumstein & Schwartz, 1983). The war years' experiences also paved the way for secondary groups and formal organization (see Chapter 5) gradually to take over many of the family's traditional activities and functions. As social historian Christopher Lasch (1977) pointed out, this trend was supported by public policy makers, who came to see the family as an obstacle to social progress. Because the family preserved separatist cultural and religious traditions and other "old-fashioned" ideas that stood in the way of "progress," social reformers sought to diminish the family's hold over its children. Thus, the prime task of socializing the young was shifted from the family to centrally administered schools. Social workers intruded into the home, offering constantly expanding welfare services from outside agencies. The juvenile court system expanded, in the belief that deficiencies in the families of youthful offenders caused crime among children.

Thus, the modern period has seen what sociologists refer to as the transfer of functions from the family to outside institutions. This transfer has had a great effect on the family, and it underlies the trends that currently are troubling many people.

There are several problems in trying to assess the prevailing state of the family. Some feel that the family is deteriorating, and they cite the examples of divorce rates and single-parent families to support their view. Others think of the family as an institution that is in transition but just as stable as ever. Was the family of the past a stable extended-family unit, with everyone working for the betterment of the whole? Or has the family structure changed throughout history in response to the economic and political changes within society? In this next section, we shall explore these views and attempt to clarify the current direction of family life.

Changes in the Marriage Rate

Are fewer people marrying now than in the past? The answer depends on how you evaluate the data. One way to look at it is to ask how many marriages there are. In 1999, there were 2.318 million marriages (see Figure 10-1). A second way to look at it is to calculate the marriage rate, the proportion of the total population marrying. The 2,318 million marriages in 1999, divided by the total population of 273.6 million Americans, yields a marriage rate of 9.9 per 1,000 (National Center for Health Statistics,

Figure 10-1 **Number of Marriages, in Millions, 1960–1999**

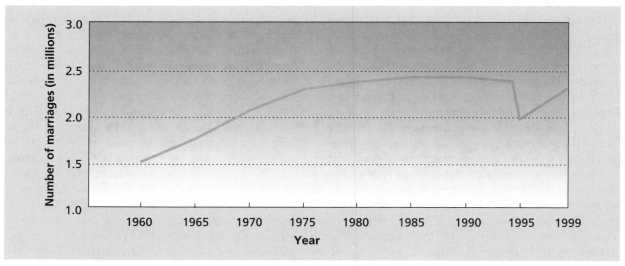

Source: National Center for Health Statistics, 2000.

Figure 10-2 **Number of Marriages per 1,000 Population, 1960–1999**

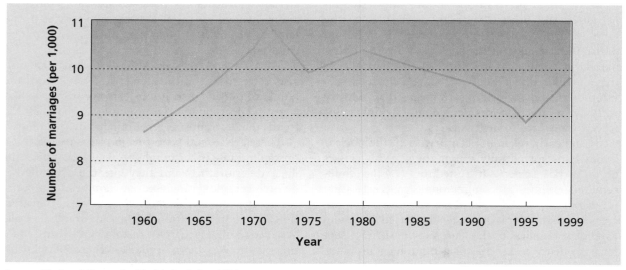

Source: National Center for Health Statistics, 2000.

2000). The number of marriages per 1,000 population reached a peak of 16.4 in 1946, at the end of World War II, and a low of 8.4 in 1958, when an economic recession coincided with a relatively small number of young adults reaching marriageable age in the population. According to the marriage rate, the institution is holding its own with a rate similar to the mid-1970s levels (see Figure 10-2).

The third way to get a true picture of the marriage situation in the United States is to take the marriage rate per 1,000 never-married women ages 15 and older. This is the proportion of eligible women marrying. The marriage rate for unmarried women ages 15 to 44 began to plummet around 1975, and by 1988, it was at an all-time low of 91 per 1,000 (see Figure 10-3).

Since 1960, the number of marriages and the marriage rate have been high because the pool of eligible adults kept expanding. Hidden in these numbers, though, is the fact that a rising share of eligible people are choosing not to marry.

Others (Besharov, 1999) think the declines in marriage are exaggerated. The age 15 cutoff point for comparing marriage rates is wrong, because people now just do not wed at such an early age. Since 1960, the median age of women getting married for the first time has increased by 4.7 years—from 20.3 years to 25.0 in 1998.

Figure 10-3 **Number of Marriages per 1,000 Unmarried Women, Ages 15 to 44, 1940–1995**

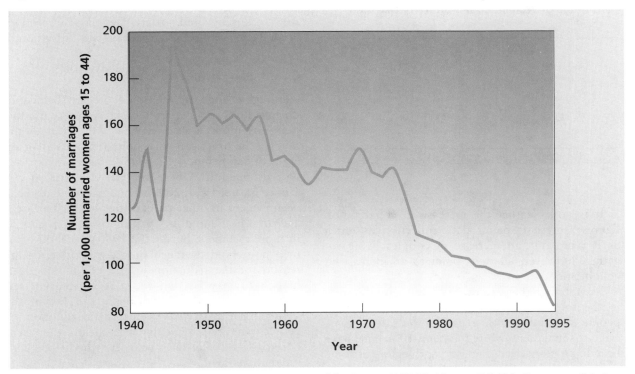

Source: *Statistical Abstract of the United States: 1997;* U.S. Bureau of the Census, 1997, Washington, DC: U.S. Government Printing Office.

To measure marriage rates accurately today, we should look at the percentage of women who are "ever married." Assuming that one should not count teenagers because so few marry, in 1960 the proportion of women twenty and older who had ever been married was 90%. Thirty-six years later, the proportion of women who were then or had at one time been married stood at 83%—only about 7% lower.

Young people today marry later, after they have completed their education and have a better idea of who they are. Because they have seen so much divorce, many want to be very sure before making such a big decision. Studies show they want marriage to be a partnership, with equality between men and women, and to be emotionally satisfying in ways never dreamed of by their parents, let alone grandparents. All of this may begin to explain why cohabitation has emerged as a major change in couple relationships. While the marriage rate may be down, cohabitation is certainly up.

Cohabitation

Cohabitation, or *unmarried couples living together out of wedlock,* has increased dramatically in the past 20 years and is already having a significant im-

pact on the American family. Although we have a great deal of information on marriages, we have very little data on cohabitation in the United States within the past two decades, and almost no information about it before that time.

Most adults in the United States eventually marry. In 2000, 91% of women ages 45 to 54 had been married at least once. The increasing social acceptance of cohabitation has resulted in definitions of being married or being single becoming less clear cut. As the personal lives of unmarried couples start to resemble those of their married counterparts, the meaning and permanence of marriage changes also. The number of unmarried-couple households has grown eightfold since 1970, from 523,000 to 4.2 million in 1999. There were 5 unmarried couples for every 100 married couples in the United States in 1998, compared with only 1 for every 100 in 1970 (*Statistical Abstract of the United States*, 2000). Unmarried-couple households are also likely to include children (43%) (Casper & Cohen, 2000).

An unmarried-couple household, as defined by the U.S. Census Bureau, contains only two adults with or without children under 15 years of age present. The adults must be of opposite sexes and not related to one another (*Statistical Abstract of the United States*, 2000).

Table 10-1

Percentage of Women Cohabitating With a Man

AGE	PERCENTAGE
15 to 19	4.1%
20 to 24	11.2
25 to 29	9.8
30 to 34	7.5
35 to 39	5.2
40 to 44	4.4
All women	7.0

Source: *Statistical Abstract of the United States: 1999, 2000.*

It is possible that the increase in cohabitation figures represents better data collection as much as it represents an increase in couples living together. However, all signs point to an increase in cohabitation.

Even though the percentage of cohabiting couples in the total population is relatively small, many people have lived with a partner outside of marriage at some point. More than 4 in 10 women have been in an unmarried domestic partnership at some point in their lives. The proportion of women in certain age groups cohabitating is quite striking (see Table 10-1).

Childless Couples

Childlessness among married couples has been increasing in recent years. Many women in the childbearing years see postponement of marriage and childbearing as pathway to a good job and economic independence. For many women this temporary postponement becomes permanent either by choice or chance. Women with the highest levels of education, those engaged in managerial and professional occupations, and those with the highest family incomes have the highest levels of childlessness.

Childlessness among women 40 to 44 years old (most women have completed childbearing by this age) increased from 10% in 1980 to 19% in 1998. This is the continuation of a trend that began in 1984. That was the year when there were more childless couples in the United States than couples with children younger than 18. This reversal of the ratio of couples with children to childless couples from what was the case throughout most of our history will continue into the foreseeable future (U.S. Bureau of the Census, May, 1999).

Some of these couples are preparents who plan to have children sometime. Over half of all women younger than 35 who are childless expect to have a child at some point. Others are nonparents either by choice or because of fertility problems. (For a discussion of the issues confronting infertile couples, see "Technology and Society: Defining Parenthood: High-Tech Fertility Treatment Versus Adoption.")

Changes in Household Size

Although changes in household size may be neither positive nor negative, some social scientists use this point to support a negative view of the future of the family. The American household of 1790 had an average of 5.8 members; by 1998, the average number had dropped to 2.6 (*Statistical Abstract of the United States,* 1999, 2000). The same trend also has been evident in other parts of the world. The average rural household in Japan in 1660 often had 20 or more members, but by the 1960s the rural Japanese household averaged only 4.5 members. One reason for the reduction in size of the American household may be that today it is very unusual to house unrelated people (Cohen, 1981). Until the 1940s, for a variety of reasons, it was common for people to have nonkin living with them, either as laborers for the fields or as boarders who helped with the rent payment.

The reduction in the number of nonrelatives living with the family explains only part of the continuing reduction in the average household size. Another reason may be a rapid decrease in the number of aging parents living with grown children and their families. Some point to this as evidence of the fragmentation and loss of intimacy present in the contemporary family. At the turn of the century, more than 60% of those 65 or older lived with one or more of their children; today this figure is less than 10%.

How can we account for so many more old people living apart from their families? We might be tempted to say that the family has become so self-centered and so unable to fulfill the needs of its members that the elderly have become the first and most obvious castoffs. However, this trend of the elderly living away from their children can also be seen as a result of the increasing wealth of the population, including the elderly. In the past, many elderly lived with their children because they could not afford to do otherwise.

Another reason for the change in the size of households is the increasing divorce rate. As more families separate legally and move apart physically, the number of people living under one roof has fallen.

A further explanation for the smaller families of today is the tendency of young people to postpone marriage and the increase in the number of working women. As people marry later, they have fewer

Technology and Society

Defining Parenthood: High-Tech Fertility Treatment Versus Adoption

Elizabeth Bartholet is a professor at the Harvard School of Law. She went through a variety of treatments to deal with her infertility. Later, she encountered many obstacles as she tried to adopt as an unmarried woman. In the following discussion, she questions why our attitudes toward fertility treatments and adoption are so different.

As a society, we define personhood and parenthood in terms of procreation. We push the infertile toward ever more elaborate forms of high-tech treatment. We are moving rapidly in the direction of a new child production market, in which sperm, eggs, embryos, and pregnancy services are for sale so that those who want to parent can produce a child to order. At the same time, we drive prospective parents away from children who already exist and need homes. The claim is that no children are available for adoption, but the fact is that millions of children the world over are in desperate need of nurturing homes. The politics of adoption in today's world prevents these children from being placed in adoptive homes. Our policies make no sense for people interested in parenting, for children in need of homes, or for a world struggling to take care of its population.

The medical profession has a near-monopoly on the information given out as people discover their infertility and explore and exercise their options. When people who have been trying to have a baby realize that something may be wrong, they usually consult their family doctor or their gynecologist, and then, if they can afford it, a fertility specialist. The specialist educates them about the range of treatment possibilities and, if they are willing and financially able, begins to lead them down the same treatment path I traveled: from scheduled sex to fallopian tube surgery to IVF [*in vitro* fertilization].

The adoption world does essentially nothing to reach out to the infertile. Indeed, adoption agency rules push them away and prevent them from obtaining the information they need to consider adoption at an early stage. The accepted ethic among adoption workers is that prospective parents must resolve feelings about infertility before they pursue adoption. The idea behind this makes some sense: they should not enter into adoption thinking of their adopted child as a second-best substitute for the biologic child they still ache to produce. But it may be impossible to know whether the desire to parent will be satisfied by adoption without knowing what adoption is about.

Other factors contribute to the bias toward the medical process. Infertile people are bombarded with messages that reinforce the idea that adoption is an inferior form of parenting and *should* be thought of as a last resort. They are lured into IVF by aggressive advertising.

If infertile people do manage to get accurate information about their parenting options, they find that our society gives vastly preferential treatment to people seeking to produce children. First and foremost, those seeking to reproduce operate in a free market world in which they are able to make their own decisions subject only to financial and physical constraints. Those seeking to adopt operate in a highly regulated world in which the government asserts the right to determine who will be allowed to parent.

The parental screening requirement deters many who might otherwise consider adoption. People don't like to become helpless supplicants, utterly dependent on the grace of social workers, with respect to something as basic as parenting. Screening turns the process of becoming a parent into a bureaucratic nightmare of documents endlessly accumulated and stamped and submitted and copied.

Regulation also sends a powerful message. By subjecting adoptive but not biologic parents to regulation, society suggests that it trusts what goes on when people raise a birth child but profoundly distrusts what goes on when a child is transferred from a birth to an adoptive parent.

Society also discriminates in financial terms. People covered by health insurance are reimbursed for many of the costs involved in infertility treatment, pregnancy, and childbirth. By contrast, those who adopt are generally on their own in paying for the adoption, and only limited subsidies are available for those who adopt children with special needs.

Source: "Blood Knots: Adoption, Reproduction, and the Politics of the Family," by E. Bartholet, Fall 1993, *The American Prospect 15.*

Sociology at Work

Work Is Where the Heart Is: Has the Office Become a Substitute for the Family?

Arlie Russell Hochschild, professor of sociology at the University of California at Berkeley, has been studying the interplay between family and work for many years. She has come to the conclusion that for many employees, work has become a form of "home" and home has become "work." The worlds of home and work have not begun to blur as conventional wisdom goes, but to reverse places. We tend to think that home is where most people feel the most appreciated, the most truly themselves, the most secure and relaxed. We think that work is where most people feel like a "cog in a machine." Yet for many people the work world is becoming a surrogate family.

Nationwide, people are spending more time at work than ever. In 1996, 76.2% of married women with children between 6 and 17 worked outside the home. Nearly 63% of married women with children 6 and under did the same. Meanwhile fathers of small children are not cutting back their hours at work to help out at home (*Statistical Abstract of the United States*, 1997). According to a survey conducted by the Families and Work Institute, American men average 48.8 hours of work a week, and women 41.7 hours, including overtime and commuting. All in all, more men and women are on an economic train that is running faster and faster.

Many companies have "family friendly policies," where the employee can work part time, share a job with another worker, work some hours at home, or take parental leave. Yet few workers use these options. The Families and Work Institute found that fewer than 5% of employees made use of them.

To be sure, some parents have tried to shorten their work hours. Twenty-one percent of the nation's women voluntarily work part time, as do 7%

Arlie Hochschild.

of the men. Yet even though many working parents say they need more time at home, few really try to reduce their time at work.

The most widely held explanation for this situation is that working parents cannot afford to work shorter hours. Certainly this is true for many. But if money is the whole explanation, why is it that the best-paid employees—upper-level managers and professionals—are the least interested in part-time work or job sharing?

Similarly, if money were the answer, we would expect poorer new mothers to return to work more

children. Many couples also are deciding to have no children, as more and more women become involved in work and careers.

All these factors point to the most significant causes of the sharp decline in the average size of the American household: the decrease in the number of children per family and the increase in the number of people living alone (Herbers, 1981). These facts ultimately will have important consequences for the entire structure of society.

Women in the Labor Force

The period since World War II has seen a dramatic change in the labor force participation rates of American women. Nearly 65 million women had paying jobs in 1999, representing more than a 200% increase in 50 years. The number of men in the labor force during this same period increased by only 50%. The change in the labor force is probably the single most important recent change in American society.

quickly after giving birth than better-off mothers. Yet that is not the case either.

A second explanation is that workers do not dare ask for time off because they are afraid it would make them vulnerable to layoffs. With well-paying, secure jobs being replaced by lower-paying, insecure ones, it makes sense to protect your job. But when Arlie Russell Hochschild surveyed employees as to whether they worked long hours for fear of being laid off, virtually everyone said no. Even among a particularly vulnerable group—factory workers who were laid off in the downturn of the early 1980s and were later rehired—most did not cite fear of losing their jobs as the main reason they worked long hours.

Could it be that workers are not aware of their company's family-friendly policies? No. Some even mentioned that they were proud to work for a company that offered such enlightened policies. Workers were not protesting the time bind. They were accommodating to it.

New management techniques have helped transform the workplace into a more appreciative, personal sort of social world. Meanwhile, at home the divorce rate has risen, and the emotional demands have become more baffling and complex. In addition to the normal developments of growing children, the needs of elderly parents are creating more tasks for the modern family, as are former spouses, stepchildren, and blended families. No wonder that many workers feel more confident they can "get the job done" at work than at home.

The one skill still required of family members is the hardest one of all—the emotional work of forging, deepening, or repairing family relationships. It takes time to develop this skill, and even then things can go awry. Family ties are complicated. Yet as broken homes become more common—and as a

sense of belonging to a geographical community grows less and less secure in an age of mobility— the corporate world has created a sense of "neighborhood," of "feminine culture," of family at work. Life at work can be insecure; the company can fire workers. But workers are not so secure at home either. Many workers have been on their job for 20 years, but are on their second or third marriage. The shifting balance between these two "divorce rates" may be the most powerful reason why parents flee a world of unresolved quarrels and unwashed laundry for the orderliness, harmony, and managed cheer of work.

There are several broader historical causes for this reversal also. The past 30 years have witnessed the rapid rise of women in the workplace. At the same time, job mobility has taken families farther from relatives who might lend a hand, making it harder for family members to make close friends of neighbors who could help out. Moreover, as women have acquired more education and have joined men at work, they have absorbed the views of the male-oriented work world. Women have adopted the view that the work world is more honorable and valuable than the world of home and children.

So where do we go from here? There is surely no going back to the mythical 1950s family that confined women to the home. Most women do not wish to return to a full-time role at home—nor can they afford it. But equally troubling is a workaholic culture that strands both men and women outside the home.

Sources: *Time Bind: When Work Becomes Home and Home Becomes Work,* by A. R. Hochschild, 1997, New York: Henry Holt and Company; "There's No Place Like Work," April 20, 1997, *The New York Times Magazine,* pp. 51–55, 81, 84; and an interview with the author, May 1997.

This change has come about because of a number of factors. After World War II, many women who took jobs in record numbers to ease labor shortages during the war remained on the job. Their numbers increased the social acceptability of the working woman. Widespread use of contraceptives is a second important factor. Effective contraception gave women the freedom to decide whether and when to have children. As a result, many women postponed childbearing and continued their educations. Baby

boomers also had different economic expectations when they entered the workforce. Having two incomes became important to ensure the lifestyle and standard of living they had come to expect. (See "Sociology at Work: Work Is Where the Heart Is: Has the Office Become a Substitute for Family?")

Occupational segregation is still present, even though the types of jobs that women are holding have been changing. Certain occupations are heavily dominated by women. These include schoolteachers

Table 10-2

**Percentage of Workforce That Is Female,
Selected Occupations, 1983 and 1998**

	1983	1998
Prekindergarten and kindergarten teachers	98.2%	97.8%
Secretaries	99.0	97.6
Registered nurses	95.8	92.5
Cashiers	84.4	78.2
Social workers	64.3	68.4
Managers, medicine and health	57.0	79.2
Sales-related occupations	58.7	72.7
Accountants and auditors	38.7	58.2
Financial managers	38.6	53.3
College and university faculty	36.3	42.3
Pharmacists	26.7	44.0
Computer programmers	32.5	28.5
Lawyers	15.3	28.5
Physicians	15.8	26.6
Dentists	6.7	19.8
Architects	12.7	17.5
Clergy	5.6	12.0
Airline pilots and navigators	2.1	3.4

Source: *Statistical Abstract of the United States: 1999* (pp. 425–428), U.S. Bureau of the Census, 2000, Washington, DC: U.S. Government Printing Office.

(75.3% female in 1998), cashiers (78.2%), librarians (83.4%), dental hygienists (99.1%), and secretaries (97.6%). In contrast, very few women are firefighters (2.5%), construction workers (2.0%), auto mechanics (0.8%), police officers and detectives (16.3%), and engineers (11.2%) (*Statistical Abstract of the United States,* 1999, 2000) (see Table 10-2).

Family Violence

It appears that we are in greater danger of being the victim of violence when we are with our families than when we are in a public place. Recent years have seen a number of high-profile cases in which spouses have attacked each other, parents have attacked children, and children have attacked and killed parents.

Family violence is greater in poor and minority households, but it occurs in families at all socioeconomic levels. Research has shown the incidence of family violence to be highest among urban lower-class families. It is high among families with more than four children and in those in which the husband is unemployed. Families in which child abuse occurs tend to be socially isolated, living in crowded and otherwise inadequate housing. Research on family violence has tended to focus on lower socio-

economic groups, but scattered data from school counselors and mental health agencies suggest that family violence also is a serious problem among America's more affluent households.

Children who are victims of abuse are more likely to be abusive as adults than children who did not experience family violence. Research shows that about 30% of adults who were abused as children are abusive to their own children (Gelles & Conte, 1990). In essence, they have been socialized to respond to frustrating situations with anger and violent outbursts.

Sociologists are trying to determine whether family violence is rising or decreasing in American society. One national survey (Gelles & Straus, 1988) found that the rate of child abuse had declined by 47% and the rate of spouse abuse had declined by 27% in a decade. The researchers believe that a major reason for the decline is the changing public attitude toward family violence. Behavior that was formerly acceptable is now considered wrong. As a result, people may be less willing to accept violent behavior from a spouse or a parent, and outsiders are more willing to intervene when they suspect abuse.

Changes in society that have affected the family also may have contributed to a decline in family violence. People are getting married later and having fewer children. Many women are in the labor force. These facts lower the risk of family violence. Public policies toward family violence have changed also, and there are many more programs for treating victims and abusers. Shelters for battered women have increased dramatically. Police who used to talk to an abusive husband and try to calm him down are now often required to arrest him. Finally, the media have played a large role in defining family violence as a social problem deserving attention.

Yet other information leads to a less optimistic picture. A Gallup Poll study found that 22% of American women (about 30 million) report having been physically abused at some point in their lives by their spouses or companions. One in five of these women report that the abuse took place within the previous year. In addition, 53% of the people in the study knew of friends or relatives who had been physically abused by their husbands or boyfriends (Gallup Poll, October 1997).

Divorce

Until the past few decades, married couples could divorce only under certain special circumstances. Consequently, the divorce rate (the number of divorces per 1,000 people) was relatively low in the early twentiethth century. In 1900, for example, the divorce rate was only 0.7, or about one-fifth the 1998 divorce rate of 3.6.

Rising divorce rates are not unique to the United States. Most industrialized nations have experienced similar patterns. Divorce rates in France and Great Britain have more than doubled in the past two decades. Divorce, however, is far more prevalent in the United States than elsewhere (*Statistical Abstract of the United States,* 1999, 2000).

All segments of American society have been affected by the rising divorce rate. Among white and Hispanic women ages 15 to 44, more than one-third have experienced a divorce. Among African-American women, the figure is one-half. Religious sanctions against divorce seem to have little influence. Catholics are no less likely to divorce than non-Catholics, despite strong opposition to divorce by the Catholic church (Bumpass, 1990).

The likelihood of divorce varies considerably with several factors. For example, education levels seem to have a strong effect on divorce rates. The likelihood of a first marriage ending in divorce is nearly 60% for those people with some college education but no bachelor's degree. Those who have a college degree but no graduate-school training have nearly a 40% chance of divorce and are the least divorce-prone. We could argue that people with the personality traits and family background that lead them to achieve a college degree are also those most likely to achieve marital stability.

Women who have gone on to graduate school have a greater likelihood of divorce than some less-educated women: Approximately 53% of them will divorce. As more and more women earn graduate degrees and as some of the barriers impeding women in combining professional and personal lives are removed, the higher rate of divorce for these women may also decline.

Although the divorce rate rose sharply during the 1970s, it has dropped off in recent years. In 1998, the divorce rate per 1,000 people was 3.6, the lowest it has been since 1972 (see Figure 10-4).

Even though the rate of increase in divorces has stopped, there is little evidence to suggest that the rate will decline substantially. The United States still has the highest divorce rate in the world. Current divorce rates imply that half of all marriages will end in divorce. Many argue that society cannot tolerate such a high rate of marital disruptions. Thirty years ago, few would have believed society could tolerate even one-third of all married couples divorcing, but that level has already been reached by some marriage cohorts.

The large number of divorces is itself a force that helps to keep the divorce rate high. Divorced people join the pool of available marriage partners, and a large majority remarry. Those remarriages then have a higher overall risk of divorce, and thus an impact on the divorce rate.

Even though divorce rates were lower in 1910, 1930, and 1950 than they are today, can we assume that family life then was happier or more stable? Divorce during those periods was expensive, legally difficult, and socially stigmatized. Many who would have otherwise considered divorce remained married because of those factors. Is it thus accurate to

Figure 10-4 **Annual Divorce Rate per 1,000 Population, 1970–1998**

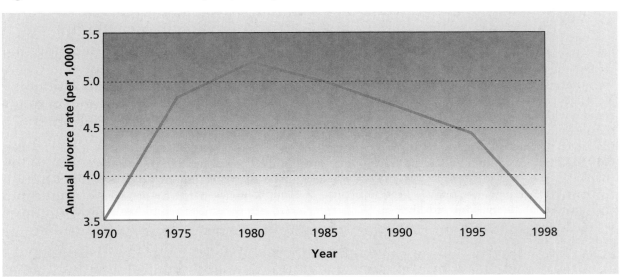

Source: National Center for Health Statistics, 2000.

say that it was better for the children and society for the partners to maintain these marriages?

As divorce becomes more common, it also becomes more visible, and such visibility actually can produce more divorces. Others become a model of how difficult marriages are handled. The model of people suffering in an unhappy marriage is being replaced by one in which people start new lives after dissolving a marriage.

Divorce also may be encouraged by the increasing tendency, mentioned earlier, for outside social institutions to assume traditional family functions that once helped hold the family together. Then again, divorce has become a viable option because people can look forward to living longer today, and they may be less willing to endure a bad marriage if they feel there is time to look for a better way of life.

Another reason for today's high divorce rate is that we have come to expect a great deal from marriage. It is no longer enough, as it might have been at the turn of the century, for the husband to be a good provider and the wife to be a good mother and family caretaker. We now look to marriage as a source of emotional support in which each spouse complements the other in a variety of social, occupational, and psychological endeavors.

Divorce rates have also increased because the possibilities for women in the workforce have improved. During earlier eras, divorced women had great problems contending with the financial realities of survival, and many were discouraged from seeking a divorce because they could not envision a realistic way of supporting themselves. With their greater economic independence, many women can now consider divorce as an option.

Today's high divorce rates can also be traced to a number of legal changes that have taken place to make divorce a more realistic possibility for couples who are experiencing difficulties. Many states have instituted no-fault divorce laws, and many others have liberalized the grounds for divorce to include mental cruelty and incompatibility. These are rather vague terms and can be applied to many problem marriages. Even changes by the American Bar Association, which now allows lawyers to advertise, contribute to the increased divorce rate. Advertisements that state that an uncomplicated divorce will cost only $250 put this option within the reach of many couples.

These legal changes are but a reflection of society's attitudinal changes toward divorce. We are a far cry from a generation ago. Divorce was to be avoided at all costs then, and when it did occur, it became a major source of embarrassment for the entire extended family. The fact that Ronald Reagan was divorced and remarried had little impact on his election to the presidency in 1980 and 1984. With so little public concern being shown, we can be sure that the role of peer-group and public opinion in preventing divorce has been diminished greatly. (See "Sociology at Work: How Much Are Children Hurt by Their Parents' Divorce?")

Divorce Laws

Before 1970, divorces in most American states could be obtained only on fault grounds. Fault grounds are those that assess blame against one of the spouses, and typically consist of adultery, desertion, physical and mental cruelty, long imprisonment for a felony, and drunkenness (Scott, 1990). The fault laws led to wide variations among the states as to how difficult or easy it was to obtain a divorce. New York State had a restrictive law that required one spouse to demonstrate that the other had engaged in adultery. No other grounds for divorce were accepted. Those restrictions led to migratory divorces (getting a divorce in another state) or couples' inventing nonexistent fault grounds.

Those issues led to attempts to change divorce laws. California introduced the Family Law Act of 1969, which allowed divorce if there were irreconcilable differences between the spouses. Thus, California became the first state to enact a no-fault divorce law. Other states quickly followed suit, and by 1985, all 50 states had some form of no-fault divorce law.

No-fault divorce reflects changes in the traditional view of legal marriage. No-fault divorce laws allow couples to dissolve their marriage without either partner having to assume blame for the failure of the marriage. By eliminating the fault-based grounds for divorce and the adversary process, no-fault divorce laws recognize that frequently both parties are responsible for the breakdown of the marriage. Further, these laws recognize that previously, the divorce procedure often worsened the situation by forcing potentially amicable individuals to become antagonists.

The vast majority of states make no-fault divorces possible with virtually no waiting periods. Even with the one-year waiting period that exists in a few states, spouses can divorce much more quickly now than under the previous, fault-based system (Walker & Elrod, 1993). The framers of the no-fault divorce laws did not intend to make divorce easy to obtain, however. Their intent was to permit divorce only after there was considerable evidence that the marriage could not be salvaged (Jacob, 1988).

Critics of no-fault divorce believe it has led to a skyrocketing divorce rate. Between 1969 and 1979, when most no-fault divorce laws were passed, the divorce rate climbed to an all-time high of 5.3, from 3.2. Some claim that no-fault divorce has made it

Sociology at Work

How Much Are Children Hurt by Their Parents' Divorce?

Since 1970, at least a million children a year have seen their parents divorce—building a generation of Americans that is now coming of age. For the past 25 years, Judith Wallerstein has been studying the lives of a specific group of these children of divorce. Now that they have grown to adulthood, they provide important information about the life course of individuals whose parents separated when they were young children.

When I began studying the effects of divorce on children and parents in the early 1970s, I, like everyone else, expected them to rally. But as time progressed, I grew increasingly worried that divorce is a long-term crisis that was affecting the psychological profile of an entire generation. The long-term effects on the children of divorce began to appear in late adolescence and early adulthood, but it was not until the children were fully grown that the whole picture emerged. Divorce is a life-transforming experience. After divorce, childhood is different. Adolescence is different. Adulthood— with the decision to marry or not and have children or not—is different. Whether the final outcome is good or bad, the whole trajectory of an individual's life is profoundly altered by the divorce experience.

Contrary to what we have long thought, the major impact of divorce does not occur during childhood or adolescence. Rather, it rises in adulthood as serious romantic relationships move center stage. When it comes time to choose a life mate and build a new family, the effects of divorce crescendo. A central finding to my research is that children identify not only with their mother and father as separate individuals but with the relationship between them. They carry the template of this relationship into adulthood and use it to seek the image of their new family. The absence of a good image negatively influences their search for love, intimacy, and commitment. Anxiety leads many into making bad choices in relationships, giving up hastily when problems arise, or avoiding relationships altogether.

The divorced family is not a truncated version of the two-parent family. It is a different kind of family in which children feel less protected and less certain about their future than children in reasonably good intact families. Mothers and fathers who share their beds with different people are not the same as mothers and fathers living under the same roof. The divorced family has an entirely new cast of characters and relationships featuring stepparents and stepsiblings, second marriages and second divorces, and often a series of live-in lovers. The child who grows up in a postdivorce family often experiences not one loss—that of the intact family—but a series of losses as people come and go. This new kind of family puts very different demands on each parent, each child, and each of the many new adults who enter the family orbit.

Moreover, divorce brings radical changes to parent-child relationships that run counter to our current understanding. Parenting cut loose from its moorings in the marital contract is often less stable, more volatile, and less protective of children. When that contract dissolves, the perceptions, feelings, and needs of parents and children for one another are transformed. It's not that parents love their children less or worry less about them. It's that they are fully engaged in rebuilding their own lives— economically, socially, and sexually. Parents and children's needs are often out of sync for many years after the breakup. Worried children watch their parents like little hawks, looking for signs of stress that will affect their availability as parents.

Children are not passive vessels but rather active participants who help shape their own destiny and that of their family. They make gallant efforts to fit into the new requirements of the postdivorce family although they hope for many years that their parents will reconcile. Because they are in their formative years, the new roles they assume in the family are built on their character. Some move into the postdivorce vacuum and become principal caregivers of their families. Others learn to hide their true feelings. Some get into trouble hoping that they can bring their parents back together to rescue them. The roles they adopt to adjust to the new circumstances in the divorced family are likely to endure into adulthood and are frequently reinstalled in their adult relationships.

Source: *The Unexpected Legacy of Divorce* (pp. xxvi–xxvii, xxix–xxx), by J. S. Wallerstein, J. M. Lewis, & S. Blakeslee, 2000, New York: Hyperion.

too easy for couples to obtain a divorce and that it encourages couples to not take their marriage vows seriously. The harshest criticism is reserved for so-called unilateral no-fault divorce, which allows one partner to receive a divorce without the other's consent. About 40 states permit unilateral no-fault divorces.

Yet others praise no-fault divorce for making marital breakups less drawn out, expensive, and painful for couples. Defenders of no-fault divorce say it has been essential in helping victims of physical and mental abuse escape damaging relationships without having to prove fault or engage in lengthy court battles. Others point out that divorce was already on the rise before no-fault divorce laws were passed and that the divorce rate has been on the decline since 1981. No-fault divorce, they say, has simply made divorces less agonizing for all parties involved.

Despite the criticism, efforts to repeal no-fault divorce laws have been largely unsuccessful. Legislation that would modify or repeal no-fault divorce has failed in at least 22 states (Issues and Controversies on File, May 7, 1999).

No-fault divorce laws advocate that the financial aspects of marital dissolution be based on equity, equality, and economic need rather than on fault- or gender-based role assignments. Alimony also is to be based on the respective spouses' economic circumstances and on the principle of social equality, not on the basis of guilt or innocence. No longer is alimony automatically awarded to the injured party, regardless of that person's financial needs; no-fault divorce does not recognize an injured party. Instead these laws seek to reflect the changing circumstances of women and their increased participation in the labor force. Under no-fault divorce law, women are encouraged to become self-supporting and husbands are not automatically expected to continue to support their ex-wives throughout their lives.

Some see no-fault divorce legislation as a redefinition of the traditional marital responsibilities of men and women by the institution of a new norm of equality between women and men. Husbands are no longer automatically designated as the head of the household, solely responsible for support, nor are wives alone expected to assume the responsibility of household activities and child rearing. Gender-neutral obligations that fall equally upon the husband and wife have been institutionalized by these new divorce laws. These changes are reflected most clearly in the new considerations for alimony allocation. In addition, property is to be divided equally. Finally, child-support expectations and the standards for child custody also reflect the new egalitarian criteria of no-fault divorce legislation. Today, both father and mother are expected to be equally responsible for the financial support of their children after divorce. In theory, mothers no longer receive custody of the child automatically; rather, a gender-neutral standard instructs judges to award custody in the best interests of the child.

No-fault has worked well for some divorcing couples, yet it has had devastating consequences for many others. Problems often emerge for older homemakers married 35 years or more, lacking any labor-force experience or skills, who may be awarded short-term settlements, ordered to sell the family home, and instructed by the court to pursue job training. Similarly, mothers with toddlers are routinely left with virtually full responsibility for their support.

No-fault divorce laws are based on an idealized picture of women's social, occupational, and economic gains in achieving equality that, in fact, may not reflect their actual conditions and circumstances. This discrepancy between reality and the ideal can have extremely detrimental effects on women's ability to become self-sufficient after divorce.

Child-Custody Laws

The view that children would fare best if they lived with their mothers following divorce was reflected in both the laws and practices of most states until the 1970s. Fathers often were advised by legal counsel of the futility of contesting custody, and the burden of proof was on the father to document the unfitness of the mother or to affirm his ability to be a better parent than the mother. However, there has been an increased recognition of fathers' rights regarding custody, reflecting the changing role of American fathers and the reevaluation of the judicial practice of automatically awarding custody to the mother. Increasing numbers of states are allowing and encouraging joint custody, which involves both parents having decision-making authority for their children even if the child lives predominantly with one parent. California was the pioneer in this area (Fine, 1994). Despite the fact that state laws emphasize joint custody to varying degrees, an overwhelming majority of custody awards are made solely to mothers (Maccoby, Buchanan, Mnookin, & Dornbusch, 1993).

In a legal sense, joint custody means that parental decision-making authority has been given equally to both parents after a divorce. It implies that neither parents' rights will be considered paramount. Both parents will have an equal voice in the children's education, upbringing, and general welfare. Joint legal custody is not a determinant of physical custody or postdivorce living arrangements. It is, however, often confused with complicated situations in which parents share responsibility for the physical day-to-day care of the children. Such arrangements usually require children to alternate between the parents' residences every few days, weeks, or months.

While alternating living environments may accompany joint custody decisions, in most instances they do not. In 90 to 95% of joint custody awards the living arrangements are exactly the same as those under sole-custody orders; namely, the child physically resides with only one parent. However, both parents make decisions regarding the welfare of the child.

Those who believe joint legal custody is a good idea cite a variety of reasons. They assert that sole-custody arrangements, which almost always involve the child's living with the mother, weaken father-child relationships. They create enormous burdens for the mothers and tend to exacerbate hostilities between the custodial parent and the visiting parent. They continue to perpetuate outmoded gender-role stereotyping. Studies also show that sole-custody arrangements are associated with poverty, antisocial behavior in boys, depression in children, lower academic performance, and juvenile delinquency. Advocates of joint custody assume that if fathers are given the opportunity to be available as nurturers, to be accessible, they will begin to participate more in the lives of their children. Furthermore, advocates say, such participation will have beneficial effects on children.

Before we too quickly assume that joint custody alleviates problems and produces benefits, we should note that it is far from being a panacea. If couples had trouble communicating and agreeing on things before the divorce, there is no reason to assume that they will have an easier time of it afterward. Most joint-custody orders are vague and do not decide at what point the joint-custodial parent's rights end and those of the parent with the day-to-day care of the child begin. What sorts of responsibilities can one parent require of the other parent? Issues such as these can easily erupt into disputes, particularly when a history of disagreement and distrust has preceded the joint-custody arrangement.

Joint custody does not give either parent the right to prevail over the other. To solve serious disputes the parents must return to court, where they must engage in litigation to prove that one or the other is unfit—the very process that the original decision of joint custody was to have avoided.

Joint custody appears to work best for those parents who have the capacity, desire, and energy to make it work—and for the children whose characteristics and desires allow them to expend the effort necessary to make it work and to thrive under it.

Remarriage and Stepfamilies

Throughout history, the reputation of stepfamilies has been surprisingly negative. The general thinking has been that the stepchild suffers in such families and that the suffering is caused by the stepmother. The stepmother has come down to us as a figure of cruelty and evil, constantly plotting to harm her stepchildren in a variety of ways. Just think of the Cinderella or Snow White fairy tales. The French word *maratre* means both "stepmother" and a "cruel and harsh mother." In English literature, "stepmother" is often preceded by "wicked." The stepchild was often a child who did not belong or whose status was similar to that of orphan. In fact, "stepchild" originally meant orphan.

Much of this has changed today, and stepfamilies have become a common sight on the American family landscape. The United States has the highest incidence of stepfamilies in the world. It is estimated that about 17% of married couple households involve a stepparent. About one child in six is a stepchild.

The stepfamilies of today are different from those of the past in how they have come into being. The vast majority of stepfamilies now come about because of the marriage or cohabitation of mothers and fathers of children whose other parent is still living. Of these, the largest group by far is composed of families formed by the remarriage of divorced men and women. The high divorce rate is the key factor in the rise of the modern stepfamily.

In the past, stepfamilies resulted from quite different conditions and had different implications. Generally, stepfamilies were the product of remarriages after the death of a spouse. Mortality and the frequency of remarriage by the widowed determined the number of stepfamilies. Moreover, unlike today, children who lived in stepfamilies rarely had more than one living parent. In the past, a stepparent replaced a deceased parent. Today, a stepparent is often an additional parent figure that a child must incorporate.

A stepparent must enter a family system that has been created by the custodial parent and her children. The new marriage partners must establish their own relationship in an existing family structure. They must create new rules for how the family is to run. Such changes may produce disadvantages or tensions among the family members.

Remarriage makes parenthood and kinship an achieved status rather than an ascribed status. Traditionally a person became a father or a mother at the birth of their child. They did not have to do anything else to be a parent, nor could they easily resign from the job, especially in a family system that strongly discouraged divorce.

Remarriage after divorce, though, adds a number of other, potential kinship positions. The new parent must now achieve the status of parent or some other role that may resemble it. Working out these arrangements presents many more problems than

With high divorce rates come high remarriage rates. The resulting blended families produce new relationships that need to develop over time.

are usually encountered in a first marriage family (Cherlin & Furstenberg, 1994).

When people get divorced, they are not rejecting the institution of marriage but rather a specific marriage partner. About two out of three divorced women and three out of four divorced men remarry (Cherlin, 1992). The interval of time between divorce and remarriage is 3.5 to 4 years. In 1970, the median interval between divorce and remarriage was only one year. About 24% of remarriages involve one divorced partner; in 42%, both partners are divorced (Wilson & Clarke, 1992).

A woman's age, race, income, education, and the presence of children all affect her remarriage rates. Men's remarriage rates are less influenced by these factors, and if they are affected, it is in the opposite direction. Women divorced after age 40 have a lower rate of remarriage than younger women. For men, age appears to be a less significant factor in remarriage. Men with higher incomes and higher levels of education are more likely to remarry than those with less education and lower incomes. Women with higher incomes are less likely to remarry, and the education level of white women appears to make little difference in these rates. For African-American women, higher education does increase the likelihood of remarriage (Ahlberg & De Vita, 1992).

The presence of children lowers the probability of remarriage for women, but not for men. Nevertheless, about 80% of remarriages involve children, so it should come as little surprise that with the divorce and remarriage rates so high, stepfamilies are becoming a permanent part of the social landscape. Stepfamilies are changing such businesses as the greeting card industry (we now have birthday wishes to stepmothers and thank-you cards to stepfathers).

Schools must now ask for information on stepparents as well as biological parents.

Stepfamilies, also known as blended families, are transforming basic family relationships. Where there were once two sets of grandparents, there now may be four; an only child may acquire siblings when his mother remarries a man with children.

It used to be that stepfamilies typically followed the death of a spouse. Today's stepfamilies typically follow a divorce. Nearly two-thirds (65%) of children living in stepfamily situations live with their biological mother. Stepfamilies are more likely to be white than black, and they are more often poor than rich. Stepparents are typically younger than other parents, and they are less likely to have a college education (Larson, 1992).

Stepparents are taking on roles formerly held by biological parents. They are staying up nights with the spouse's sick children, attending class recitals, and having heated battles over such essential issues of childhood as curfews, TV, homework, and rights to the family car.

Try as they might, in one study of emotionally healthy middle-class families, only one-half the stepfathers and one-quarter of the stepmothers claimed to have "parental feelings" toward their stepchildren, and fewer still claimed to love them. Stepparenthood is the strongest risk factor for child abuse. A stepparent is 40 to 100 times more likely to kill a young child than a biological parent (Daly & Wilson, 1988). Contributing to this situation is the fact that in the minds of most children, stepparents can never take the place of their real father or mother. Studies have shown that it takes at least 4 years for children to accept a stepparent in the same way they do their biological parents. Researcher James H.

Bray has noted that this acceptance is harder for girls than boys. Although divorce appears harder on elementary school-age boys, remarriage appears harder on girls (Kutner, 1988).

At the heart of most stepfamily relationships are children who, like their parents, are casualties of divorce. In the best stepfamily relationships, all the adults work together to meet the needs of the children, realizing that, all too often, no matter what they do, there still will be problems. In reality, stepfamilies are torn apart by many of the same pressures that divide intact families. Financial problems can be especially acute when parents must support children from different marriages. Resentment builds quickly when stepparents feel they have little power or authority in their own homes. With all the conflicting interests and emotions, it's little wonder that about half of stepfamilies end in divorce (Larson, 1992).

The most difficult stage occurs during the early years, when stepparents want everything to go right. Once stepparents realize that relationships with stepchildren build over time, and that their potential network of allies includes all the other adults in the stepfamily relationship, the adjustment for all is faster and healthier.

Alternative Lifestyles

A number of options are increasingly available to people who, for various reasons, find the traditional form of marriage impractical or incompatible with their lifestyles. More young people are selecting cohabitation as a permanent alternative to marriage (although many consider it a prelude to marriage). In addition, some older men and women are opting to live together in a permanent relationship without getting married. These people choose cohabitation primarily for economic reasons—many would lose sources of income or control of their assets if they entered into a legal marriage. Several other options are discussed next.

The Growing Single Population

Americans traditionally have been the marrying kind. In 1998, nearly 56% of American men and 59% of American women over the age of 18 were married. Younger people, however, may be rejecting this tradition. In 1970, only 10.5% of the women and 19.1% of the men between the ages of 25 and 34 had never been married. In 1998, 35% of women and 48.4% of men that age had never been married (U.S. Bureau of the Census, 1999).

Even though the number of people living alone more than doubled between 1970 and 1998, this trend may mean only that more young people are postponing marriage. It also could mean, however, that a growing proportion of adults is staying single permanently. In fact, as we mentioned earlier in this chapter, some studies show that the marriage rate is declining.

There are a number of reasons why people are choosing not to marry. Working women do not need the financial security that a traditional marriage brings, and sex outside of marriage has become much more widespread. Moreover, many singles view marriage as merely a prelude to divorce and are unwilling to invest in a relationship that is likely to fail. As sociologist Frank Furstenburg notes, "Men who weren't married by their late 20's in the 60s were oddballs. Now they're just successful 29-year-olds."

Clearly, many singles would gladly change their marital status if the right person came along. The ratio of unmarried men to unmarried women varies by age group. There are more unmarried men than women at the younger ages, reflecting the fact that men have a later age at first marriage than women do. After age 40 there are fewer unmarried men than unmarried women because women tend to live longer and are less likely to remarry after divorce and widowhood.

The elderly make up a significant proportion of single-person households. Currently, 39% of one-person households are maintained by people ages 65 or older, and fully 80% of those are elderly women. About 40.9% of women 65 or older live alone, but only 17.4% of men in that age group do so (*Statistical Abstract of the United States,* 1997).

Single-Parent Families

In the early years of the twentieth century, it was not uncommon for children to live with only one parent because of the high death rate. As the death rate started to fall and the number of single widowed parents declined, it was not long before the divorce rate started to increase and single parents because of divorce became more common. Yet even by 1960, nearly one-third of all single mothers with children under 18 were widows. As divorce rates started to rise dramatically in the 1970s, the situation started to change and most single mothers were divorced or separated. By 1980, only 11% of single mothers were widowed and two-thirds were divorced or separated. The path to single motherhood changed again when many women bypassed marriage. By 2000, 40% of single mothers had never been married (Bianchi & Casper, 2000).

There has been a significant increase in the number of single-parent families in the United States. In 1998 they constituted 27% of family households with children, up from 24% in 1990 and 11% in 1970 (National Center for Health Statistics, 2000).

Table 10-3

Percentage of Births to Unmarried Women

Sweden	52%
Denmark	47
France	36
United States	33
United Kingdom	32
Canada	25
Germany	15
Japan	8

Source: *Statistical Abstract of the United States: 1997* (p. 834), U.S. Bureau of the Census, 1997, Washington, DC: U.S. Government Printing Office.

A child in a one-parent family was just as likely to be living with a divorced parent in 2000 as with a never-married parent. A decade ago, a child living with one parent was almost twice as likely to be living with a divorced parent as with a never-married parent.

The increase in the divorce rate is a major reason for the increase in single-parent families. Most divorced parents now set up new households, whereas in earlier times many would have returned to their own parents' households.

People have become alarmed at the high percentage of births to unwed mothers in the United States. After rising dramatically between 1940 and 1990, out-of-wedlock births have leveled off at about 33% of all births. The number of births to unmarried women reached an annual total of 1.3 million in 1999.

The same thing, however, has also been taking place in a number of other industrialized countries. In 1998, half or more of births in Norway and Sweden were out of wedlock. Denmark, France, and the United Kingdom also have high percentages of out-of-wedlock births. Japan is an exception, where nonmarital childbearing is low (*National Vital Statistics Reports,* October 18, 2000). (See Table 10-3.)

Single mothers with children face a multitude of challenges. They are usually the primary breadwinner, disciplinarian, playmate, and caregiver for the children. They must manage the financial and practical aspects of the household.

Most single mothers are not poor, but they tend to be younger, earn lower incomes, and are less educated than married mothers. While the number of single mothers seems to have peaked, the number of single fathers is growing substantially. Men now comprise one-sixth of the nation's 11.9 million single parents. (U.S. Bureau of the Census, 1998)

A key factor contributing to out-of-wedlock births has been the large increase in cohabitation among unmarried couples. About 39% of out-of-wedlock births today are to cohabiting couples.

About one-third of all cohabiting opposite-sex couples have children under 15 present in their homes, but many more cohabiting couples are parents. Many have children living elsewhere with a custodial parent.

Cohabitation is often seen as a prelude to marriage. Single people see it as a way to make sure that the couple is compatible before getting married. At least one of the partners expects the arrangement to result in marriage in 90% of cohabitations. The reality does not bear out these overly optimistic expectations, however. Only 58% of women who cohabitate end up marrying the man (National Center for Health Statistics, 1996). Other research during the 1980s showed that 44% of all couples who ultimately married had lived together (Bumpass, Sweet, & Cherlin, 1991).

It is unlikely that the increase in cohabitation will continue indefinitely. If it did, cohabitation would start to become more common than marriage. However, now that cohabitation does not produce as much disapproval as it once did, it is likely it will become more common and more visible.

Gay and Lesbian Couples

A phenomenon that is not new but one that has become more and more visible is the household consisting of a gay or lesbian couple. Before 1970, almost all gay people wished to avoid the risks that would come with disclosure of their sexual orientation. It was not until 1990 that the U.S. Census attempted to calculate the number of gay and lesbian households. Through a self-identification process, 88,200 gay male couples and 69,200 lesbian couples were recorded (Usdansky, 1993). Other data suggests that there may be 1.7 million gay or lesbian households (*Marital Status and Living Arrangements,* 1994, 1996). These numbers are only a hint at the true numbers, because there are still many couples who did not disclose their arrangements.

Gay and lesbian couples often are not eligible for the same benefits as traditional couples. In 1989, Denmark became the first country to recognize homosexual marriages. In the United States, none of the 50 states permits gay or lesbian marriages. This causes many of the gay and lesbian partners to be ineligible for health care or other benefits that might be provided to the married employees and their families. A number of states and municipalities, however, have passed "domestic partnership" statutes to make such benefits available. In addition, 19 states, the District of Columbia and 119 cities and counties have adopted policies that provide varying degrees of civil rights protection for gays and lesbians (*Harvard Law Review,* 1993). (See "Our Diverse Society: Should Same-Sex Marriages Be Permitted?")

Our Diverse Society

Should Same-Sex Marriages Be Permitted?

The possibility of allowing same-sex marriage first emerged as a real legal prospect in 1993. In May of that year, the Hawaii Supreme Court, somewhat unexpectedly, ruled that a state law barring same-sex marriage violated Hawaii's constitution. Allowing only men to marry women and vice versa, the court filed, ran counter to the state constitution's ban on sex discrimination. The court ruled that unless Hawaii could show a "compelling state interest for banning gay marriage," it must allow same-sex couples to marry.

In anticipation of the impact of the Hawaii decision and under pressure from alarmed members of Congress, then-president Clinton signed a law called the Defense of Marriage Act (DOMA) in September 1996. This law stated that the federal government would not recognize same-sex unions as marriages, meaning that gay married couples could not file joint tax returns or receive any other federal benefits associated with marriage. This act also allowed states to pass laws making clear that they would not recognize same-sex marriages registered in other states. Prior to the act's passage, there had been concern that states would be required to sanction marriages conducted in any state that legalized same-sex marriages. So far, 30 states have passed laws stating that they will not recognize same-sex marriages.

Some have noted that the increased recognition of domestic partnerships in many jurisdictions makes the movement to legalize same-sex marriages unnecessary. Marriage, however is an important issue for lesbians and gay men because if domestic partnership is the only vehicle of legal recognition for same-sex relationships, such relationships are relegated to second-class status. Legal recognition of same-sex marriages, they point out, is an important step closer to full equality.

Lesbians and gays would gain many legal rights from the right to marry. Joint tax returns, insurance policies, social security and pension benefits, property inheritance, and veteran's discounts on medical care and on educational and home loans are all examples of the economic benefits that accompany marriage. Next-of-kin status for hospital visits and medical decisions, as well as adoption and guardianship rights, also go along with the right to marry. Gays and lesbians point out that interracial marriage was once illegal, yet many of the same arguments used to fight the legalization of interracial marriage are used by today's opponents of same-sex marriage.

Supporters of same-sex marriage point out that few would argue that gays and lesbians are capable of the sacrifice, commitment, and responsibilities of marriage. If that is the case and they are denied equal legal standing not because of anything about the relationship itself, but purely because of the involuntary nature of homosexuality itself, then it is a clear example of bias and discrimination.

Same-sex marriage would also provide role models for gay youth. It would bring the gay couple into the heart of the traditional family and do more to heal the gay-straight rift than any gay rights legislation (Sullivan, 1995).

Critics of same-sex marriage note that marriage is a sacred union that unites a man and woman together for life. It is central to every faith and every modern society has embraced this view and rejected same-sex marriages.

Those opposed to same-sex marriage argue that the fundamental purpose of marriage is procreation, and since lesbians and gay men cannot procreate, they should not be allowed to marry. Gays and lesbians counter that if we follow this line of reasoning, then infertile heterosexual couples should not be allowed to marry, nor should those straight couples who choose not to have children for any number of reasons.

Opponents also wonder why same-sex marriage should be recognized when the high divorce rates show that we are having enough trouble maintaining the institution of marriage at all. Critics claim support for same-sex marriage would strike most people as a parody of marriage that could further weaken an already strained institution.

Critics also argue that extending marriage to gay people could lead the way to a further redefinition of marriage. They fear that people might begin to demand to be allowed to marry more than one person at a time (Wilson, 1996).

A ruling by the Vermont Supreme Court in December of 1999 brought the issue to the forefront again when the court ruled that barring same-sex couples from receiving the benefits of marriage violates the state constitution's guarantee of equal protection under the law. Other states will continue to address the issue in future years.

Sources: *Virtually Normal: An Argument about Homosexuality,* by A. Sullivan, 1995, New York: Alfred A. Knopf; "Against Homosexual Marriage," by J. Q. Wilson, March 1996, *Commentary;* "Same-Sex Partnerships," February 18, 2000, *Issues and Controversies on File,* pp. 49–56.

Gay and lesbian couples are highly urban: About 60% of gay couples and 45% of lesbian couples live in only 20 cities in the United States. San Francisco, Washington, D.C., Los Angeles, Atlanta, and New York City are the most common living places for gay and lesbian couples.

Many gay and lesbian families include children (22% of lesbian families and 5% of gay families). Many of these children were part of a mother-father family and continued to live with the parent as she or he made the transition to same-sex relationships. Seventeen percent of gays and 29% of lesbians had previously been in a heterosexual marriage.

Gays and lesbians in couple relationships have higher educational levels than men and women in heterosexual marriages. Cohabiting gay men 25 to 34 years old are twice as likely to have a postgraduate degree as married men. The differences are even greater for lesbian couples in this age group, who are three times as likely to have a postgraduate education compared to married women (Black, 2000).

Traditionally, researchers and the media have concentrated on the ways in which gays and lesbians are different from heterosexuals. Yet gay and lesbian couples are similar to heterosexual couples in many ways. They form long-term relationships and have problems similar to those of heterosexual couples. Gay and lesbian couples are concerned with having family benefits such as health insurance, life insurance, and family leave offered to gay and lesbian couples as well as married couples. Many are also involved in extending adoption rights and same-sex marriage recognition to gay and lesbian couples.

The desire to form a relationship with another person appears to be quite strong among gays and lesbians. By and large, gay men and lesbian women who were not in a relationship reported that they had been in one previously and believed that they would be in one again. There is no doubt that couplehood, as either a reality or as aspiration, is as strong among gay men and lesbian women as it is among heterosexuals (Blumstein & Schwartz, 1983).

The Future: Bright or Dismal?

Given all these changes in the American family, should we be concerned that marriage and family life as we know them will one day disappear? Probably not. The divorce rate is high and will continue to be high during the next decade. It is important, however, to keep things in perspective. Divorce is just as much a social universal as is marriage. Anthropologist Margaret Mead noted that "no matter how free divorce" or "how frequently marriages break up," most societies have assumed that marriages would be permanent. Despite this belief, societies have also recognized that some marriages are incapable of lasting a lifetime and have provided mechanisms for dissolution (Riley, 1991).

Even though the divorce rate is high, the remarriage rate is also very high. The vast majority of people who divorce remarry, usually within a short time after they divorce. The high divorce rate does not necessarily mean that people are giving up on marriage. It just means there is a growing belief that marriage can be better. The high remarriage rate indicates that people are willing to continue trying until they reach their expectations. Obtaining a divorce does not mean the person believes that the idea of marriage is a mistake, only that a particular marriage was a mistake.

Despite claims to the contrary, there is little evidence that the family as an institution is in decline, or any weaker today than a generation ago. Nor is there any indication that people place less value on their own family relationships, or on the role of the family within society, than they once did.

The traditional family is being replaced by family arrangements that better suit today's lifestyles: There are fewer full-time homemakers because more women are in the workforce. Nonfamily households have increased substantially. The typical family with a working dad, homemaker mom, and two or more children is now a distinct minority.

The institutions of marriage and the family have proved to be extremely flexible and durable and have flourished in all human societies under almost every imaginable condition. As we have seen, these institutions take on different forms in differing social and economic contexts, and there is no reason to suspect that they will not continue to do so. Therefore, to make predictions about the future of the American family is equivalent to making predictions about the future of American society in particular and industrial society in general. This is extremely difficult to do, given the social, economic, political, and ecological problems facing us. However, for the foreseeable future, it seems reasonable to assume that the forces of industrialism and public policy that helped shape the current nuclear family in its one-parent and two-parent forms will persist. Therefore, the contemporary nuclear family will continue to provide the basic context within which American society will reproduce itself for several generations to come.

SUMMARY

The American family has changed dramatically in the past 25 years. The marriage rate is down, the divorce rate is up, and more children are being born to single women. The American family is clearly in a state of transition.

A household consists of people who share the same living space. While there is some disagreement over how precisely to define the family, sociologists generally agree that some form of family is found in every known human society, and that the family everywhere serves several important functions: regulating sexual behavior, patterning reproduction, organizing production and consumption, socializing children, providing care and protection, and providing social status.

No society permits random mating; all societies have an incest taboo, which forbids sexual intercourse among closely related individuals. Sex between parents and their children is universally prohibited, but who else is considered to be closely related varies widely among societies.

Every society must replace its members. By regulating where and with whom individuals may enter into sexual relationships, society patterns sexual reproduction. In all societies, the family tends to be the primary unit of consumption; in some societies, the family is a unit of production as well. Every society must also provide predictable social contexts within which its children are socialized. The family is ideally suited to this task because of its intimacy and the special knowledge of the children the adults are likely to have.

In addition to physical needs, humans need to be among people who care for them emotionally, who help with the problems that arise in daily life, and who back them up when there is a conflict with others. Although many kinds of social groups can meet one or more of these needs, the family is often the one group in a society that meets them all. Simply by being born into a family, each individual inherits both material goods and a socially recognized position defined by ascribed statuses. This inherited social position, or family background, is probably the single most important social factor affecting the predictable course of people's lives.

The nuclear family is made up of a married couple and their children. It is the most basic family form and is found in all societies. Polygamous families are nuclear families linked by multiple marriage bonds, with one central individual married to several spouses. The family is polygynous when the central individual is a male and the multiple spouses are females, and polyandrous when the central individual is a female and the multiple spouses are males. Extended families are composed of the nuclear family plus other relations and generations living together and forming one cooperative unit.

Marriage is the socially recognized, legitimized, and supported union of individuals of opposite sexes. It is an institution found in all societies. Marriage differs from other unions in that it takes place in a public (and usually formal) manner, it includes sexual intercourse as an explicit element of the relationship, it provides the essential condition for legitimizing offspring, and it is intended to be a stable and enduring relationship. While almost all societies allow for divorce, or the breakup of marriage, none endorse it as an ideal norm.

American culture is relatively unique in linking romantic love and marriage. Romantic love—which involves the idealization of the loved one, the notion of a one and only, love at first sight, love winning out over all, and an indulgence of personal emotions—has nothing to do with marriage throughout most of the world. Marriage in those societies establishes social, economic, and even political relationships among families. Three families are ultimately involved in a marriage: the two spouses' families of origin or families of orientation—the families in which they were born or raised—and the family of procreation, which is created by the union of the spouses.

INTERNET RESOURCES

Go to http://web-enhanced.thomsonlearning.com which features Web links relevant to this chapter to expand and enhance the learning experience. This site will be monitored and updated periodically to ensure the links are correct and live.

At Virtual Society: The Wadsworth Sociology Resource Center, http://sociology.wadsworth.com, you will find a Career Center, Sociology in the News, Research Online, InfoTrac College Edition, Virtual Tours in Sociology, and a variety of book-specific student and instructor resources.

CHAPTER TEN STUDY GUIDE

KEY CONCEPTS

Match each concept or person with its definition, illustration, or explanation below.

a. Marital residence rules
b. Family of procreation
c. Romantic love
d. Polygamous family
e. Bilateral system
f. Patriarchal family
g. Marriage
h. Cohabitation
i. Patrilineal system
j. Bilocal residence
k. Polyandrous family
l. Extended family
m. Exogamy

n. Trobrianders
o. Monogamy
p. Marriage rate
q. Nuclear family
r. Matrilineal system
s. Incest taboo
t. Patrilocal residence
u. Polygynous family
v. Arlie Hochschild
w. Joint custody
x. Neolocal residence
y. Homogamy

z. Matriarchal family
aa. Divorce
bb. Endogamy
cc. Matrilocal residence
dd. Multiple marriages
ee. Household
ff. Family of orientation
gg. Stepfamilies
hh. No-fault divorce laws
ii. Companionate Marriages
jj. Thurgood Marshall
kk. Elizabeth C. Stanton

_____ 1. People who share the same living space.

_____ 2. A norm that forbids sexual intercourse among closely related individuals.

_____ 3. A married couple and their children.

_____ 4. A set of nuclear families linked by multiple marriage bonds, with one central individual married to several spouses.

_____ 5. A family in which the central individual is a male and the multiple spouses are females.

_____ 6. A family in which the central individual is a female and the multiple spouses are males.

_____ 7. A family that includes other relations and generations in addition to the nuclear family.

_____ 8. A situation in which the generations are tied together through the males of a family.

_____ 9. A situation in which the generations are tied together through the females of a family.

_____ 10. A situation in which descent passes through both females and males of a family.

_____ 11. A situation in which most family affairs are dominated by men.

_____ 12. A situation in which most family affairs are dominated by women.

_____ 13. The socially recognized, legitimized, and supported union of individuals of opposite sexes.

_____ 14. The breakup of a marriage.

_____ 15. A phenomenon characterized by idealization of a loved one, the notion of a one and only, love at first sight, love winning out over all, and an indulgence of personal emotions.

_____ 16. The family in which a person was born and raised.

_____ 17. The family created by marriage.

_____ 18. A situation in which people are directed to marry within certain social groups.

_____ 19. A situation in which individuals are required to marry someone outside their culturally defined group.

_____ 20. A situation in which a person is allowed only one spouse at a time.

_____ **21.** A situation in which an individual may have more than one spouse.

_____ **22.** Norms that govern where a newly married couple settles down and lives.

_____ **23.** A requirement that a new couple settle down near or within the husband's father's household.

_____ **24.** A requirement that a new couple settle down near or within the wife's mother's household.

_____ **25.** A situation in which a newly married couple can choose to live with either the husband's or the wife's family of origin.

_____ **26.** A situation in which a newly married couple may choose to live virtually anywhere.

_____ **27.** The tendency to choose a spouse with a similar social and cultural background.

_____ **28.** Marriage based on romantic love.

_____ **29.** The proportion of the total population marrying.

_____ **30.** A situation in which, in principle, the financial aspects of marital dissolution are to be based on equity, equality, and economic need rather than fault- or gender-based role assignments.

_____ **31.** A legal situation in which parental decision-making authority is given equally to both parents after a divorce.

_____ **32.** Families that include children from parents' previous marriages.

_____ **33.** A situation in which unmarried couples live together.

_____ **34.** Tribe from New Guinea in which the father's role in parenthood was not recognized and responsibility for parenthood fell to the mother's brother's family.

_____ **35.** Argued, "People marry for all sorts of reasons except the true ones—love and affection."

_____ **36.** Supreme Court justice denied residence in a Virginia suburb of Washington, D.C., because he was black and his wife was Asian.

_____ **37.** Wrote *Time Bind,* where she theorized that Americans were developing a workaholic culture where the office was becoming a substitute home.

CENTRAL IDEA COMPLETIONS

Following the instructions, fill in the appropriate concepts and descriptions for each of the questions posed in the following section.

1. Explain each of the major functions of the family and provide one example of how this function is accomplished in a typical, middle-class American family:

  a. Function 1:

    _____

    _____

    Example: _____

  b. Function 2:

    _____

    _____

    Example: _____

c. Function 3:

_____

_____

Example: _____

d. Function 4:

_____

_____

Example: _____

2. Common-sense arguments often suggest that "falling in love" and "selecting a mate" are matters of chance or random events; however, your text demonstrates that mate selection is not at all random. Explain two of the ways in which mate selection is not random:

a. _____

_____

_____

_____

_____

b. _____

_____

_____

_____

3. Contemporary American society abounds with alternative lifestyles. Providing examples, discuss two of these alternative lifestyles as they relate to the family as a social institution:

a. _____

_____

_____

_____

_____

b. _____

_____

_____

_____

4. Your text discusses the phenomenon of romantic love. What are the components of romantic love in America and how is romantic love related to the institution of marriage?

a. Romantic love: _____

_____

_____

_____

b. Its relationship to marriage: _____

_____

_____

_____

5. Patriarchal and matriarchal family structures exist around the globe. State the basic similarities and differences between these two structures. Follow your statement by indicating which structure characterizes your own family, providing specific examples:

a. Partriarchial: _____

_____

_____

_____

b. Matriarchal: _____

_____

_____

_____

c. Your family's structure: _____

_____

_____

_____

_____

6. The American public long has been fascinated with the marital and private family behaviors of its celebrities, not only in Hollywood but also in the realm of government, particularly office holders. To what extent might occupational, celebrity or political position create special strains on marital structures? What role decisions and behaviors might these married couples engage in to minimize such strains? What behaviors might they engage in that would intensify the marital strains?

a. Marital strains: _____

_____

_____

b. Strain-reduction behaviors: _____

_____

_____

c. Strain-enhancing behaviors: _____

_____

_____

7. Family violence is a major concern today. What factors discussed in your text contribute to the incidence of family violence? What sociodemographic variables are associated with family violence?

a. Violence characteristics: _____

_____

_____

b. Sociodemographic variables: _____

_____

_____

8. What praises and criticisms are associated with no-fault divorce in the United States?

   a. Praises of no-fault divorce: _____

   _____

   _____

   b. Criticisms of no-fault divorce: _____

   _____

   _____

9. What are the essential features of child-custody laws in the United States?

   _____

   _____

   _____

10. Profile the condition of marriage in the United States, including the percentages who never marry, the divorce rate, unmarried-couple households, and the birthrate characteristics:

   _____

   _____

   _____

CRITICAL THOUGHT EXERCISES

1. Before reading this chapter, what was your thinking about how parental divorce *hurts* children? Your author devoted considerable attention to the question, how much are children hurt by their parent's divorce? What were his conclusions? What was meant by the statement: The divorced family is not a truncated version of the two-parent family? What factors appear to affect the extent of adjustment or nonadjustment by children to parental divorce?

2. Your text provided an interesting look at arranged marriages in India. Recall Shona Bose's statement, "Love is important, but it is not sufficient." How did you react to that comment relative to your personal life? Given we have entered the next millennium, how is it that such traditional phenomena as arranged marriages continue to exist? What might be the advantages and disadvantages of arranged marriages for the United States? To what extent do you think it would be possible to institute "arranged marriages" in a society like the United States? What do you see as the major structural and cultural barriers to arranged marriages becoming the American norm?

3. Recently, a weekly television program interviewed a newly married couple. What made this couple's situation rather unique was that, upon getting married, the bride did not take the name of the groom, nor did she hyphenate her family of orientation last name with the last name of her husband. In-

stead, the groom gave up his last name and adopted hers. During the interview, both persons indicated that they had discussed this decision and were comfortable with it, although the husband had been somewhat overwhelmed with the public attention his act had generated. What is your opinion about males taking on the last name of the bride at marriage rather than vice versa? If you are male, would you follow this groom's decision? If you are female, would you want your husband to adopt your last name after the ceremony? To what extent might this be a likely or unlikely future trend among married couples in the United States during the next few decades? Base your response on the material on marriage and mate selection presented in your text.

4. Your text notes that "childlessness among married couples has been increasing in recent years." Moreover, "childlessness among women 40 to 44 years old increased from 10% in 1980 to 19% in 1998." What factors do you see as being associated with the childless couple phenomenon? What effect might this trend have on characteristics of American society 20 years from now?

5. In 1996, President Clinton signed a law called the Defense of Marriage Act in which the federal government indicated it would not recognize same-sex unions. What were the major issues presented in the box titled "Our Diverse Society: Should Same-Sex Marriages Be Permitted?" Compare and contrast how this issue might be examined from a sociological perspective as compared with a layperson's nonsociological perspective.

6. The discussion of high-tech fertility treatment versus adoption raised several essential issues relevant to American attitudes about parenthood and procreation. What do you see as the key factors? What might be the consequences of America's moving toward a "new child production market"? Check into the laws and financial costs for adoption in the state where you attend school. Next, write the nearest fertility clinic and see if you can gain information about the qualifications, procedures, time, and fi-

nancial costs associated with providing children through fertility treatment. Would changes in adoption procedures reduce the need for fertility clinics in your area?

7. Discuss interracial marriage in the United States. What are the current patterns and trends in interracial marriage in America and to what extent are they similar or different from patterns in the recent past? Include in your discussion the most common forms of interracial marriage as well as the major variables (for example, education) that are associated with interracial marriage in America.

ANSWERS

KEY CONCEPTS

1. ee	8. i	14. aa	20. o	26. x	32. gg
2. s	9. r	15. c	21. dd	27. y	33. h
3. q	10. e	16. ff	22. a	28. ii	34. n
4. d	11. f	17. b	23. t	29. p	35. kk
5. u	12. z	18. bb	24. cc	30. hh	36. jj
6. k	13. g	19. m	25. j	31. w	37. v
7. l					

Religion

LEARNING OBJECTIVES

After studying this chapter, you should be able to do the following:

- Define the basic elements of religion.
- Differentiate among the major types of religion.
- Describe the functions of religion according to the functionalist perspective.
- Explain the conflict theory perspective on religion.
- Describe the basic types of religious organization.
- Describe important aspects of contemporary American religion.
- Describe the major religions in the United States.

The belief in one God is a basic tenet of Christianity and also the cornerstone of Judaism. Islam also proclaims, "There is no God but Allah, and Muhammad is His Prophet." Hinduism, however, expresses a very different concept: the idea of the "manyness" of God. Hinduism does not have one creed, one founder, one prophet, or one central moment of revelation. It is an expression of 5,000 years of religious and cultural development in Asia. During this development, it has assimilated many ideas and ideologies. In Hinduism, whether there is one god or many gods is unimportant. Rather than worshipping one single god, Hinduism involves worshipping one god at a time. Each Hindu is free to choose his or her own god or goddess.

In his book *All Religions Are True,* Mohandas Karamchand Gandhi wrote:

> It has been a humble but persistent effort on my part to understand the truth of all the religions of the world, and adopt and assimilate in my own thought, word, and deed all that I have found to be best in those religions.

Gandhi's acceptance of diversity, however, has a downside. If all religions are true, then they all must be imperfect. After all, if any one religion were perfect, it would be better than all others.

The idea of the manyness of God is alien not only to Western religion, but also to Western culture. Our monotheism leads us to believe in one Truth. In our minds, the ultimate or best can stand out in many areas, whether it be the best car, the best university, or the ultimate religion (Goldman, 1991).

The important thing to realize is that although religion assumes many different forms, it is a universal human institution. To appreciate the many possible kinds of religious experiences, from the belief in one God to the belief in the manyness of God, requires an understanding of the nature and functions of religion in human life and society.

The Nature of Religion

Religion is *a system of beliefs, practices, and philosophical values shared by a group of people; it defines the sacred, helps explain life, and offers salvation from the problems of human existence.* It is recognized as one of society's important institutions.

In his classic study *The Elementary Form of Religious Life,* first published in 1915, Émile Durkheim observed that all religions divide the universe into two mutually exclusive categories: the profane and the sacred. The **profane** consists of *all empirically observable things*—that is, *things that are knowable through common, everyday experiences.* In contrast, the **sacred** consists of *things that are awe inspiring and knowable only through extraordinary experiences.*

The sacred may consist of almost anything: objects fashioned just for religious purposes (like a cross), a geographical location (Mount Sinai), a place constructed for religious observance (a temple), a word or phrase ("Our Father, who art in heaven . . ."), or even an animal (the cow to Hindus, for example). To devout Muslims, the Sabbath, which falls on Friday, is a sacred day. To Hindus, the cow is holy, not to be killed or eaten. These are not ideas to be debated; they simply exist as unchallengeable truths. Similarly, to Christians, Jesus was the Messiah; to Muslims, Jesus was a prophet; but to sociologists, the person of Jesus is a religious symbol. Religious symbols acquire their particular sacred meanings through the religious belief system of which they are a part.

Durkheim believed that every society must distinguish between the sacred and the profane. This distinction is essentially between the social and nonsocial. What is considered sacred has the capacity to represent shared values, sentiments, power, or beliefs. The profane is not supported in this manner; it may have utility for one or more individuals, but it has little public relevance.

We can look at Babe Ruth's bat as an example of the transformation of the profane to the sacred. At first, it was merely a profane object that had little social value in itself. Today, however, one of Babe Ruth's bats is enshrined in baseball's Hall of Fame. It no longer is used in a profane way but instead is seen as an object that represents the values, sentiments, power, and beliefs of the baseball community. The bat has gained some of the qualities of a sacred object, thus changing from a private object to a public object.

In addition to sacred symbols and a system of beliefs, religion also includes specific rituals. **Rituals** are *patterns of behavior or practices that are related to the sacred.* For example, the Christian ritual of

All religions provide a means for communicating with supernatural beings or forces.

Holy Communion is much more than the eating of wafers and the drinking of wine. To many participants, these substances are the body and blood of Jesus Christ. Similarly, the Sun Dance of the Plains Indians was not merely a group of braves dancing around a pole to which they were attached by leather thongs that pierced their skin and chest muscles. It was a religious ritual in which the participants were seeking a personal communion.

The Elements of Religion

All religions contain certain shared elements, including ritual and prayer, emotion, belief, and organization.

Ritual and Prayer All religions have formalized social rituals, but many also feature private rituals such as prayer. Of course, the particular events that make up rituals vary widely from culture to culture and from religion to religion.

All religions include a belief in the existence of beings or forces that human beings cannot experience. In other words, all religions include a belief in the supernatural. Hence, they also include **prayer,** or *a means for individuals to address or communicate with supernatural beings or forces,* typically by speaking aloud while holding the body in a prescribed posture or making stylized movements or gestures.

Emotion One of the functions of ritual and prayer is to produce an appropriate emotional state. This may be done in many ways. In some religions, participants in rituals deliberately attempt to alter their state of consciousness through the use of drugs, fasting, sleep deprivation, and induction of physical pain. Thus, Scandinavian groups ate mushrooms that caused euphoria, as did many native Siberian tribes. Various Native American religions feature the use of peyote, a buttonlike mushroom that contains a hallucinogenic drug. For the roughly 250,000 members in 100 branches of the Native American Church, peyote ceremonies are a sacred sacrament, the flesh of God put on earth to provide clarity and bring followers closer to the Creator. It is illegal to ingest peyote, in most states, but in 1994, Congress carved out exemptions for "the practice of a traditional Indian religion" by members of federally recognized tribes (Gehrke, 2001). For a while in the late 1960s, a number of counter-cultural groups in America relied on LSD and other drugs to induce religious experiences.

Although not every religion attempts to induce altered states of consciousness in believers, all religions do recognize that such states may happen and believe that they may be the result of divine or sacred intervention in human affairs. Prophets, of course, are thought to receive divine inspiration. Religions differ in the degree of importance they attach to such happenings.

Belief All religions endorse a belief system that usually includes a supernatural order and also often a set of values to be applied to daily life. Belief systems can vary widely. Some religions believe that a valuable quality can flow from a sacred object—animate or inanimate, part or whole—to a lesser object. Numerous Christian sects, for instance, practice the laying on of hands, whereby a healer channels divine energy into afflicted people and thus heals them. Some Christians also believe in the power of relics to work miracles simply because those objects once were associated physically with Jesus or one of the saints. Such beliefs are quite common among the world's religions: Native Australians have their sacred stones, and shamans among African, Asian, and North American societies try to heal through sympathetic touching. In some religions, the source of the valued quality is a personalized deity. In others, it is a reservoir of supernatural force that is tapped.

Just before Christmas of 1996, people were shocked to discover what appeared to be the Virgin Mary on the south window of the Ugly Duck car rental building on Highway 19 in Clearwater, Florida. Soon the parking lot was transformed into

One of the functions of ritual and prayer is to produce an appropriate emotional state.

a shrine as the faithful, many in wheelchairs and on crutches, came to be healed. Many of those who came hoped for a miracle that would cure their ills. Others wanted to receive their own message from Mary, to see visions of Jesus, or to strengthen their faith (Shermer, 2000).

Organization Many religions have an organizational structure through which specialists can be recruited and trained, religious meetings conducted, and interaction facilitated between society and the members of the religion.

The organization also promotes interaction among the members of the religion to foster a sense of unity and group solidarity. Rituals may be performed in the presence of other members. They may be limited to certain locations such as temples, or they may be processions from one place to another. Although some religious behavior may be carried out by individuals in private, all religions demand some public, shared participation. (See "Sociology at Work: Human Societies and the Concept of God.")

Sociology at Work

Human Societies and the Concept of God

Why does God exist? How have the three dominant monotheistic religions—Judaism, Christianity, and Islam—shaped and altered the conception of God? Karen Armstrong, one of Britain's foremost commentators on religious affairs, traced the history of how men and women have perceived and experienced God from the time of Abraham to the present.

The human idea of God has a history, since it has always meant something slightly different to each group of people who have used it at various points of time. The idea of God formed in one generation by one set of human beings could be meaningless in another. Indeed, the statement "I believe in God" has no objective meaning as such, but like any other statement only means something in context, when proclaimed by a particular community.

Consequently, there is no one unchanging idea contained in the word "God"; instead, the word contains a whole spectrum of meanings, some of which are contradictory or even mutually exclusive. Had the notion of God not had this flexibility, it would not have survived to become one of the great human ideas. When one conception of God has ceased to have meaning or relevance, it has been quietly discarded and replaced by a new theology. A fundamentalist would deny this, since fundamentalism is antihistorical: it believes that Abraham, Moses, and the later prophets all experienced their God in exactly the same way as people do today. Yet if we look at our three religions, it becomes clear that there is no objective view of "God": Each generation has to create the image of God that

works for it. The same is true of atheism. The statement "I do not believe in God" has meant something slightly different at each period of history. The people who have been dubbed "atheists" over the years have always denied a particular conception of the divine. Is the "God" who is rejected by atheists today the God of the patriarchs, the God of the prophets, the God of the philosophers, the God of the mystics, or the God of the eighteenth-century deists? All these deities have been venerated as the God of the Bible and the Koran by Jews, Christians, and Muslims at various points of their history. We shall see that they are very different from one another. Atheism has often been a transitional state: thus Jews, Christians, and Muslims were all called "atheists" by their pagan contemporaries because they had adopted a revolutionary notion of divinity and transcendence. Is modern atheism a similar denial of a "God" who is no longer adequate to the problems of our time?

Despite its otherworldliness, religion is highly pragmatic. We shall see that it is far more important for a particular idea of God to work than for it to be logically or scientifically sound. As soon as it ceases to be effective, it will be changed—sometimes for something radically different. This did not disturb most monotheists before our own day because they were quite clear that their ideas about God were not sacrosanct but could only be provisional.

Sources: *A History of God* (pp. xx–xxi), by K. Armstrong, 1994, New York: Ballentine Books; and interviews with the author, October 1994 and 1996.

Magic

In some societies, magic serves some of the functions of religion, though there are essential differences between the two. **Magic** is *an active attempt to coerce spirits or to control supernatural forces. It differs from other types of religious beliefs in that there is no worship of a god or gods.* Magic is used to manipulate and control matters that seem to be beyond human control and that may involve danger and uncertainty. It is usually a means to an end, whereas religion is usually an end in itself, although prayer may be seen as utilitarian when a believer asks for some personal benefit. In most instances, religion serves to unify a group of believers, whereas magic

is designed to help the individual who uses it. Bronislaw Malinowski (1954) explains:

> We find magic wherever the elements of chance and accident, and the emotional play between hope and fear have a wide and extensive range. We do not find magic wherever the pursuit is certain, reliable, and well under the control of rational methods and technological processes. Further we find magic where the element of danger is conspicuous.

During the Middle Ages, when most of the population was illiterate, the belief in magic was quite extensive. Almost everyone believed in sorcery, werewolves, witchcraft, and black magic. If a noblewoman died, her servants ran around the house emptying all containers of water so her soul would not drown. Bloodletting to cure illnesses was popular. Plagues were believed to be the result of an unfortunate conjuncture of the stars and planets. The air was believed to be infested with such soulless spirits as unbaptized infants, ghouls who pulled out cadavers in graveyards and gnawed on their bones, and vampires who sucked the blood of stray children. For the medieval mind, magic provided an understanding of how the world worked. It helped relieve anxiety and allowed people to blame events on bad luck or evil spirits, and it permitted one to cast blame on curses and witchcraft. Astrology was the most popular science of that time. Only religion could rival astrology as an all-embracing explanation for the unpredictability of life (Shermer, 2000).

Rodney Stark and William Bainbridge (1985) have noted that a belief in magic has always been a major part of Christian faith. A common theme throughout the centuries has been the effort of organized religion to prohibit unorthodox practices and practitioners and to monopolize magic. Nonchurch magic was identified as superstition. Serious efforts to root out magic once and for all emerged in the fifteenth century. Eventually, as many as 500,000 people may have been executed for witchcraft. Stark and Bainbridge wrote:

> In order to monopolize religion, a church must monopolize all access to the supernatural. . . . But if the church is to deny others access to the supernatural, it must remain in the magic business. The demand for magic is too great to be ignored. . . . Thus the Catholic Church remained deeply involved in dispensing magic. Immense numbers of magical rites and procedures were developed. . . . Saints and shrines that performed specialized miracles proliferated, and new procedures for seeking saintly intercession abounded. Many forms of illness, especially mental illness, were defined as cases of possession, and legions of official exorcists appeared to treat them.

Stark and Bainbridge noted that magic's respectability has decreased as more scientific attitudes have proliferated. Magic, especially magical healing, is now found mostly among sectarians and cultists. This fact makes the religious beliefs of sects and cults particularly vulnerable to criticism and refutation (Beckwith, 1986).

Major Types of Religions

The earliest evidence of religious practice comes from the Middle East. In Shanidar Cave in Iraq,

archaeologist Ralph Solecki (1971) found remains of burials of Neanderthals—early members of our own species, *Homo sapiens,* once believed to be brutish but now recognized as fully human—dating from between 60,000 and 45,000 years ago (see Chapter 3). Bodies were tied into a fetal position, buried on their sides, provided with morsels of food placed at their heads, and covered with red powder and sometimes with flower petals. Those practices—the food and the ritual care with which the dead were buried—point to a belief in some kind of existence after death.

Using studies of present-day cultures as well as historical records, sociologists have devised a number of ways of classifying religions. One of the simplest and most broadly inclusive schemes recognizes four types of religion: supernaturalism, animism, theism, and abstract ideals.

Supernaturalism

Supernaturalism *postulates the existence of nonpersonalized supernatural forces that can, and often do, influence human events.* These forces are thought to inhabit animate and inanimate objects alike—people, trees, rocks, places, even spirits or ghosts—and to come and go at will. The Melanesian/Polynesian concept of *mana* is a good example of the belief in an impersonal supernatural power.

Mana is *a diffuse, nonpersonalized force that acts through anything that lives or moves,* although inanimate objects such as an unusually shaped rock also may possess *mana.* The proof that a person or thing possesses *mana* lies in its observable effects. A great chief, merely by virtue of his position of power, must possess *mana,* as does the oddly shaped stone placed in a garden plot that then unexpectedly yields huge crops. Although it is considered dangerous because of its power, *mana* is neither harmful nor beneficial in itself, but it may be used by its possessors for either good or evil purposes. An analogy in our culture might be nuclear power, a natural force that intrinsically is neither good nor evil but can be turned to either end by its possessors. We must not carry the analogy too far, however, because we can account for nuclear power according to natural, scientific principles and can predict its effects reliably without resorting to supernatural explanations. A narrower, less comprehensive, but more appropriate analogy in Western society is our idea of luck, which can be good or bad and over which we feel we have very little control.

Although certain objects possess *mana,* taboos may exist in relation to other situations. A **religious taboo** is *a sacred prohibition against touching, mentioning, or looking at certain objects, acts, or people.* Violating a taboo results in some form of pollution.

Taboos may exist in reference to foods not to be eaten, places not to be entered, objects and people not to be touched, and so on. Even a person who becomes a victim of some misfortune may be accused of having violated a taboo and may also become stigmatized.

Taboos exist in a wide variety of religions. Polynesian peoples believed that their chiefs and noble families were imbued with powerful *mana* that could be deadly to commoners. Hence, elaborate precautions were taken to prevent physical contact between commoners and nobles. The families of the nobility intermarried (a chief often would marry his own sister), and chiefs actually were carried everywhere to prevent them from touching the ground and thereby killing the crops. Many religions forbid the eating of selected foods. Jews and Muslims have taboos against eating pork at any time, and until fairly recently, Catholics were forbidden to eat meat on Fridays. Most cultures forbid sexual relations between parents and children and between siblings (the incest taboo).

Supernatural beings fall into two broad categories: those of nonhuman origin, such as gods and spirits, and those of human origin, such as ghosts and ancestral spirits. Chief among those of nonhuman origin are the gods who are believed to have created themselves and may have created or given birth to other gods. Although gods may create, not all peoples attribute the creation of the world to them.

Many of those gods thought to have participated in creation have retired, so to speak. Having set the world in motion, they no longer take part in day-to-day activities. Other creator gods remain involved in ordinary human activities. Whether or not a society has creator gods, many other affairs are left to lesser gods. For example, the Maori of New Zealand have three important gods: a god of the sea, a god of the forest, and a god of agriculture. They call upon each god for help in the appropriate area.

Below the gods in prestige, but often closer to the people, are the unnamed spirits. Some of these can offer constructive assistance, and others take pleasure in deliberately working evil for people.

Ghosts and ancestor spirits represent the supernatural beings of human origin. Many cultures believe that everyone has a soul, or several souls, which survive after death. Some of those souls remain near the living and continue to be interested in the welfare of their kin (Ember & Ember, 1981).

Animism

Animism is *the belief in inanimate, personalized spirits or ghosts of ancestors that take an interest in, and actively work to influence, human affairs.* Spirits may inhabit the bodies of people and animals as well as inanimate phenomena such as winds, rivers,

Only three religions are known to be monotheistic: Judaism, Christianity, and Islam.

or mountains. They are discrete beings with feelings, motives, and a will of their own. Unlike *mana*, spirits may be intrinsically good or evil. Although they are powerful, they are not worshipped as gods, and because of their humanlike qualities, they can be manipulated—wheedled, frightened away, or appeased—by using the proper magic rituals. For example, many Native American and South American Indian societies (as well as many other cultures in the world) think sickness is caused by evil spirits. Shamans, or medicine men or women, supposedly can effect cures because of their special relationships with these spirits and their knowledge of magic rituals. If the shamans are good at their jobs, they supposedly can persuade or force the evil spirit to leave the sick person or to discontinue exerting its harmful influence. In our own culture, there are people who consult mediums, spiritualists, and Ouija boards in an effort to contact the spirits and ghosts of departed loved ones.

Theism

Theism is *the belief in divine beings—gods and goddesses—who shape human affairs.* Gods are seen as powerful beings worthy of being worshipped. Most theistic societies practice **polytheism,** *the belief in a number of gods.* Each god or goddess usually has particular spheres of influence such as childbirth,

Table 11-1

Major Religions of the World

Total World Population 5,929,839,000

		PERCENTAGE OF TOTAL
Christians	1,943,038,000	32.8
Roman Catholic	1,026,501,000	17.3
Protestants	316,445,000	5.3
Orthodox	213,743,000	3.6
Anglican	63,748,000	1.1
Other Christian	373,832,000	6.3
Baha'i	6,764,000	0.1
Buddhists	353,794,000	6.0
Chinese folk-religionists	379,162,000	6.4
Ethnic religionists	248,565,000	4.2
Muslims	1,164,622,000	19.6
Hindus	761,689,000	12.8
Jews	14,111,000	0.2
Sikhs	22,332,000	0.4
Nonreligious	759,655,000	12.8
Atheists	149,913,000	2.5

Sources: *Britannica Book of the Year,* 1999; *Statistical Abstract of the United States: 2000* (p. 831), U.S. Bureau of the Census, Washington, DC: U.S. Government Printing Office, 2000.

rain, or war, and there is generally one who is more powerful than the rest and oversees the others' activities. In the ancient religions of Mexico, Egypt, and Greece, for instance, we find a pantheon, or a host of gods and goddesses.

Monotheism

Monotheism is *the belief in the existence of a single god.* Only three religions are known to be monotheistic: Judaism and its two offshoots, Christianity and Islam. These three religions have the greatest number of believers worldwide (see Table 11-1). Even these faiths are not purely monotheistic, however. Christianity, for example, includes a belief in such divine or semidivine beings as angels, a devil, saints, and the Virgin Mary. Nevertheless, because all three religions contain such a strong belief in the supremacy of one all-powerful being, they are considered to be monotheistic.

Abstract Ideals

Some religions are based on abstract ideals rather than a belief in supernatural forces, spirits, or divine beings. **Abstract ideals** *focus on the achievement of personal awareness and a higher state of consciousness through correct ways of thinking and behaving, rather than by manipulating spirits or worshipping gods.* Such religions promote devotion to religious rituals and practices and adherence to moral codes

of behavior. Buddhism is an example of a religion based on abstract ideals. The Buddhist's ideal is to become "one with the universe," not through worship or magic, but by meditation and correct behavior.

A Sociological Approach to Religion

When sociologists approach the study of religion, they focus on the relationship between religion and society. The functionalist sociologists have examined the functions that religion plays in social life, whereas conflict theorists have viewed religion as a means for justifying the political status quo. In the following section, we will examine these two approaches in detail.

The Functionalist Perspective

Since at least 60,000 years ago, as indicated by the Neanderthal burials at Shanidar Cave, religion has played a role in all known human societies. The question that interests us here is, what universal functions does religion have? Sociologists have identified four categories of religious function: satisfying individual needs, promoting social cohesion, providing a worldview, and helping to adapt to society.

Satisfying Individual Need Religion offers individuals ways to reduce anxiety and to promote emotional integration. Although Sigmund Freud (1918, 1928) thought religion to be irrational, he saw it as helpful to the individual in coming to terms with impulses that induce guilt and anxiety. Freud argued that with a belief in lawgiving, powerful deities could help people reduce their anxieties by providing strong, socially reinforced inducements for controlling dangerous or immoral impulses.

Further, in times of stress, individuals can calm themselves by appealing to deities for guidance or even for outright help, or they can calm their fears by trusting in God. In the face of so many things that are beyond human control and yet may drastically affect human fortunes (such as droughts, floods, or other natural disasters), life can be terrifying. It is comforting to "know" the supernatural causes of both good fortune and bad. Some people attempt to control supernatural forces through magic rituals. Perhaps this is why best-selling author Michael Crichton (a medical doctor and author of *Jurassic Park*) eats the same meal for lunch every day while working on a new novel and why New York Giants football coach Bill Parcells would stop and purchase coffee at two different coffee shops on his way to the stadium before every game.

Social Cohesion Émile Durkheim, one of the earliest functional theorists, noted the ability of religion to bring about group unity and cohesion. According to Durkheim, all societies have a continuing need to reaffirm and uphold their basic sentiments and values. This is accomplished when people come together and communally proclaim their acceptance of the dominant belief system. In this way people are bound to one another, and as a result, the stability of the society is strengthened.

Not only does religion in itself bring about social cohesion, but often the hostility and prejudice directed at its members by outsiders also helps strengthen their bonds. For example, during the 1820s, Joseph Smith, a young farmer from Vermont, claimed that he had received visits from heavenly beings. He said the apparitions enabled him to produce a 600-page history, known as the Book of Mormon, of the ancient inhabitants of the Americas. Shortly after the establishment of the Mormon church, Smith had a revelation that Zion, the place where the Mormons would prepare for the millennium, was to be established in Jackson County, Missouri. Within 2 years, 1,200 Mormons had bought land and settled in Jackson County. The other residents in the area became concerned about the influx, and in 1833 they published their grievances in a doc-

ument that became known as the manifesto, or secret constitution. They charged the Mormons with a variety of transgressions and pledged to remove them from Jackson County. Several episodes of conflict followed, which eventually forced the Mormons to move into an adjoining county. These encounters with a hostile environment produced a sense of collective identity at a time when it was desperately needed. The church was less than 2 years old and included individuals from diverse religious backgrounds. There was a great deal of internal discord, and except for the unity that resulted from the conflict with the townspeople, the group might have disappeared altogether (MacMurray & Cunningham, 1973).

Durkheim's interest in the role of religion in society was aroused by his observation that religion, like the family, seemed to be a universal human institution. This universality meant that religion must serve a vital function in maintaining the social order. Durkheim felt that he could best understand the social role of religion by studying one of the simplest kinds—the totemism of the aboriginal Australian. A **totem** is *an ordinary object such as a plant or animal that has become a sacred symbol to and of a particular group or clan, who not only revere the totem but also identify with it.* Thus, reasoned Durkheim, religious symbols such as totems, as well as religion itself, arose from society itself, not outside it. When people recognize or worship supernatural entities,

Religious rituals fulfill a number of social functions. They bring people together physically, promote social cohesion, and reaffirm a group's beliefs and values.

Figure 11-1 Society, Religion, and the Individual: A Functionalist View

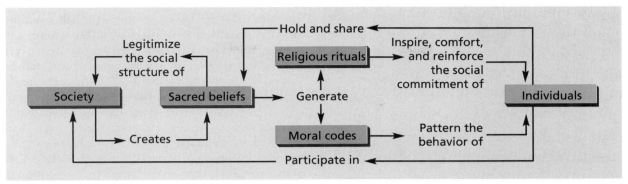

they are really worshipping their own society. They do not realize that their religious feelings are actually the result (a crowd reaction) of the intense emotions aroused when people gather together at a clan meeting, for example. They look for an outside source of this emotional excitement and may settle on a nearby, familiar object as the symbol of both their religion and their society. Thus, society—the clan—is the origin of the clan members' shared religious beliefs, which in turn help cement together their society.

Durkheim saw religious ritual as an important part of this social cement. Religion, through its rituals, fulfills a number of social functions: It brings people together physically, promoting social cohesion; it reaffirms the group's beliefs and values; it helps maintain norms, mores, and prohibitions so that violation of a secular law—murder or incest, for instance—is also a violation of the religious code and may warrant ritual punishment or purification; it transmits a group's cultural heritage from one generation to the next; and it offers emotional support to individuals during times of stress and at important stages in their live cycle, such as puberty, marriage, and death (see Figure 11-1).

In Durkheim's view, these functions are so important that even a society that lacks the idea of the sacred must substitute some system of shared beliefs and rituals. Indeed, some theorists see communism as such a system. Soviet communism had its texts and prophets (Karl Marx, Friedrich Engels), its shrines (V. I. Lenin's tomb), its rituals (May Day parade), and its unique moral code. Durkheim thought that much of the social upheaval of his day could be attributed to the fact that religion and ritual no longer played an important part in people's lives, and that without a shared belief system, the social order was breaking down.

Although many sociologists today take issue with Durkheim's explanation of the origins of religion based on totemism, they nevertheless recognize the value of his functional approach in understanding the vital role of religion in society.

Secular society depends on external rewards and pressures for results, whereas religion depends on the internal acceptance of a moral value structure. Durkheim believed that because religion is effective in bringing about adherence to social norms, society usually presents those norms as an expression of a divine order. For example, in ancient China, as in France until the late eighteenth century, political authority—the right to rule absolutely—rested securely on the notion that emperors and kings ruled because it was divine will that they do so—the divine right of kings. In Egypt, the political authority of pharaohs was unquestioned because they were more than just kings; they were believed to be gods in human form.

Religion serves to legitimize more than just political authority. Although many forms of institutionalized inequality do not operate to the advantage of the subgroups or individuals affected by them, they help perpetuate the larger social order and often are justified by an appeal to sacred authority. In such situations, although religion serves as a legitimator of social inequality, it does function to sustain societal stability. Thus, the Jews in Europe were kept from owning land and were otherwise persecuted because "they had killed Christ" (Trachtenberg, 1961); and even slavery has been defended on religious grounds. In 1700, Judge John Saffin of Boston wrote of the

> Order that God hath set in the world, who hath Ordained different degrees and orders of men, some to the High and Honorable, some to be Low and Despicable . . . yea, some to be born slaves, and so to remain during their lives, as hath been proved (Montagu, 1964a).

Religions do not always legitimize secular authority. In feudal Europe, the church had its own political structure, and there often was tension between church and state. Indeed, just as the church often legitimized monarchs, it also excommunicated those who failed to take its wishes into account. However,

Émile Durkheim believed that when people recognize or worship supernatural entities, they are really worshipping their own society.

the fact remains that religious institutions usually do dovetail quite neatly with other social institutions, legitimizing and helping sustain them. For example, though church and state in medieval Europe were separate structures and often conflicted, the church nonetheless played an important role in supporting the entire feudal system.

Establishing Worldviews   According to Max Weber in his classic book *The Protestant Ethic and the Spirit of Capitalism,* religion responds to the basic human need to understand the purpose of life. In doing so, religion must give meaning to the social world within which life takes place. This means creating a worldview that can have social, political, and economic consequences. For instance, there is the issue of whether salvation can be achieved through active mastery (hard work, for example) or through passive contemplation (meditation). The first approach can be seen in Calvinism, and the second is evident in several of the Eastern religions. Another major issue in creating a worldview is whether salvation means concentrating on a supernatural world, this world, or an inner world.

Using these ideas, Weber theorized that Calvinism fostered the Protestant ethic of hard work and asceticism and that Protestantism was an important influence on the development of capitalism. Calvinism is rooted in the concept of predestination, which holds that before they are born, certain people are selected for heaven and others for hell. Nothing anyone does in this world, Calvinists believe, can change this. The Calvinists consequently were eager to find out whether they were among those chosen for salvation. Worldly success—especially the financial success that grew out of strict discipline, hard work, and self-control—was seen as proof that a person was among the select few. Money was accumulated not to be spent but to be displayed as proof of one's chosenness. Capitalist virtues became Calvinist virtues. It was Weber's view that even though capitalism existed before Calvinist influence, it blossomed only with the advent of Calvinism.

Weber's analysis has been criticized from many standpoints. Calvinist doctrines were not so uniform as Weber pictured them, nor was the work ethic confined to the Protestant value system. Rather, it seems to have been characteristic of the times, promoted by Catholics as well as Protestants. Finally, one could just as well argue the reverse: that the social and economic changes leading to the rise of industrialism and capitalism stimulated the emergence of the new Protestantism—a position that Marxist analysts have taken. Today it is generally agreed that although religious beliefs did indeed affect economic behavior, the tenets of Protestantism and capitalism tended to support each other. However, the lasting value of Weber's work is his demonstration of how religion creates and legitimizes worldviews and how important these views are to human social and political life.

Adaptations to Society   Religion can also be seen as having adaptive consequences for the society in which it exists. For example, many would view the Hindu belief in the sacred cow, which may not be slaughtered, as a strange and not particularly adaptive belief. The cows are permitted to wander freely and defecate along public paths.

Marvin Harris (1966) has suggested that there may be beneficial economic consequences in India from not slaughtering cattle. The cows and their offspring provide a number of resources that could not be provided easily in other ways. For example, a team of oxen is essential to India's many small farms. Oxen could be produced with fewer cows, but food production would have to be devoted to feeding those cows. With the huge supply of sacred cows, although they are not well fed, the oxen are produced at no cost to the economy.

Cow dung is also necessary in India for cooking and as fertilizer. It is estimated that dung equivalent to 45 million tons of coal is burned annually. Alternative sources of fuel, such as wood or oil, are scarce or costly.

Although most Hindus do not eat beef, cattle that die naturally or are slaughtered by non-Hindus are eaten by the lower castes. Without the Hindu taboo against eating beef, these other members of the Indian hierarchy would not have access to this food supply. Therefore, because the sacred cows do not compete with people for limited resources and because they provide a cheap source of labor, fuel, and fertilizer, the taboo against slaughtering cattle may be quite adaptive.

When societies are under great stress or attack, their members sometimes fall into a state of despair analogous, perhaps, to that of a person who becomes depressed. Institutions lose their meaning for people, and the society is threatened with what Durkheim called anomie, or "normlessness." If this continues, the social structure may break down, and the society may be absorbed by another society, unless the culture can regenerate itself. Under these conditions, revitalization movements sometimes emerge. **Revitalization movements** are *powerful religious movements that stress a return to the traditional religious values of the past.* Many of them can be found in the pages of history and are even in existence today.

In the 1880s, the once free and proud plains Indians lived in misery, crowded onto barren reservations by soldiers of the United States government. Cheated out of the pitiful rations that had been promised them, they lived in hunger—and with memories of the past. Then a Paiute by the name of Wovoka had a vision, and he traveled from tribe to tribe to spread the word and demonstrate his Ghost Dance.

> Give up fighting, he told the people. Give up all things of the white man. Give up guns, give up European clothing, give up alcohol, and give up all trade goods. Return to the simple life of the ancestors. Live simply—and dance! Once the Indian people are pure again, the Great Spirit will come, all Indian ancestors will return, and all the game will return. A big flood will come, and after it is gone, only Indians will be left in this good time. (Brown, 1971)

Wovoka's Ghost Dance spread among the defeated tribes. From the Great Plains to California, Native American communities took up the slow, trancelike dance. Some believed that the return of the ancestors would lead to the slaughter of all whites. For others, the dance just rekindled pride in their heritage. For whatever reasons, the Ghost Dance could not be contained, despite the government's attempts to ban it.

On December 28, 1890, the people of a Sioux village camped under federal guard at Wounded Knee, South Dakota, and began to dance. They ignored orders to stop and continued to dance until someone suddenly fired a shot. The soldiers opened fire, and soon more than 200 of the original 350 men, women, and children were killed. The soldiers' losses were 29 dead and 33 wounded, mostly from their own bullets and shrapnel. This slaughter was the last battle between the Indians of the Great Plains and the soldiers of the dominant society (Brown, 1971). (For a discussion of religious tolerance in other countries see "Global Sociology: Worldwide Religious Persecution is Common.")

The Conflict Theory Perspective

Karl Marx asserted that the dominant ideas of each age have always been the ideas of the ruling class (Marx & Engels, 1961), and from this it was a small step to his assertion that the dominant religion of a society is that of the ruling class, an observation that has been borne out by historical evidence. Marxist scholars emphasize religion's role in justifying the political status quo by cloaking political authority with sacred legitimacy and thereby making opposition to it seem immoral.

The concept of alienation is an important part of Marx's thinking, especially in his ideas of the origin and functions of religion. **Alienation** is *the process by which people lose control over the social institutions they themselves invented.* People begin to feel like strangers—aliens—in their own world. Marx further believed that religion is one of the most alienating influences in human society, affecting all other social institutions and contributing to a totally alienated world.

According to Marx, "Man makes religion, religion does not make man" (Marx, 1967). The function of God thus was invented to serve as the model of an ideal human being. People soon lost sight of this fact, however, and began to worship and fear the ideal they had created as if it were a separate, powerful supernatural entity. Thus, religion, because of the fear people feel for the god they themselves have created, serves to alienate people from the real world.

Marx saw religion as the tool that the upper classes used to maintain control of society and to dominate the lower classes. In fact, he referred to it as the "opiate of the masses," believing that through religion, the masses were kept from actions that might change their relationship with those in power. The lower classes were distracted from taking steps toward social change by the promise of happiness through religion. If they followed the rules established by religion, they expected to receive their reward in heaven, and so they had no reason to try to change or improve their condition in this world. These religious beliefs made it easy for the ruling

Global Sociology

Worldwide Religious Persecution Is Common

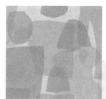

Each year, millions of people throughout the world are persecuted for their religious beliefs. Some governments have passed laws limiting rights for "unapproved religions," while others condone or perpetrate violence against religious minorities. Christians currently represent the most oppressed religious group in the world with 200 million being persecuted. The United States has responded by setting up the Office of International Religious Freedom in the State Department. In 1999, the office issued its first annual report and identified the worst violators of religious freedom. They include the following:

Afghanistan

The country is dominated by an extremist Islamic faction called the Taliban. The Taliban interpretation of Islamic law is extremely harsh. Violators are subject to execution for breaking codes of worship. The Taliban also persecutes women and severely limits their ability to move in society.

China

China, which is officially atheist, is intolerant of religions that have not met with official government approval. Those religions include Tibetan Buddhism, Christianity, and Islam. The Chinese government has given members of these religions long prison sentences and closed their churches and temples.

Iran

The government has persecuted and imprisoned people of the Baha'i faith. The government has desecrated Baha'i places of worship and graveyards and denied its practitioners civil rights, including freedom to assemble and access to higher education. The government has also violated the religious freedom of Jews, Sunni Muslims, and Christians.

Iraq

Iraq's Islamic government has systematically carried out murders, executions, and arbitrary detention of practitioners of the Shiite Muslim faith. The government has also persecuted the country's Assyrian and Chaldean Christians.

Myanmar

Myanmar's military leaders have arrested and imprisoned Buddhist monks who advocate human and political rights. In some areas, the police have tried to force Christians to convert to Buddhism.

Sudan

The Muslim regime has waged war on largely non-Muslim populations in the country's south. Those who deviate from the regime's interpretation of Islamic law have been subject to murder, torture, imprisonment, and forced conversion to Islam.

Sources: "U.S. Report Details World's Religious Persecution," by D. Stout, September 10, 1999, *The New York Times,* p. A14; "Religious Freedom Abroad," January 21, 2000, *Issues and Controversies on File,* pp. 17–24.

classes to continue to exploit the lower classes: Religion served to legitimize upper-class power and authority. Although modern political and social thinkers do not accept all of Marx's ideas, they recognize his contribution to the understanding of the social functions of religion.

Although religion performs a number of vital functions in society—helping maintain social cohesion and control while satisfying the individual's need for emotional comfort, reassurance, and a worldview—it also has negative, or dysfunctional, aspects.

Marx would be quick to point out a major dysfunction of religion: Through its ability to make it seem that the existing social order is the only conceivable and acceptable way of life, it obscures the fact that people construct society and therefore can change society. Religion, by imposing the acceptance of supernatural causes of conditions and events, tends to conceal the natural and human causes of social problems. In fact, in this role of justifying or legitimating the status quo, religion may very well hinder much-needed changes in the social structure. By diverting attention from injustices in the existing social order, religion discourages the individual from taking steps to correct these conditions.

An even more basic and subtle dysfunction of religion is its insistence that only one body of knowledge and only one way of thinking are sacred and correct, thereby limiting independent thinking and the search for further knowledge.

Organization of Religious Life

Several forms or types of organization of religious groups are found in society.

The Universal Church

A **universal church** *includes all the members of a society within one united moral community* (Yinger, 1970). It is fully a part of the social, political, and economic status quo and therefore accepts and supports (more or less) the secular culture. In a preliterate society, in which religion is not really a differentiated institution but rather permeates the entire fabric of social life, a person belongs to the church simply by being a member of the society. In more complex societies, this religious form cuts across divisions of the social structure, such as social classes and ethnic groups, binding all believers into one moral community. A universal church, however, does not seek to change any conditions of social inequality created by the secular society and culture, and indeed, it may even legitimize them. (An example is the Hindu religion of India, which used to perpetuate a rigid caste system.)

The Ecclesia

An **ecclesia** is *a church that shares the same ethical system as the secular society and has come to represent and promote the interest of the society at large.* Like the universal church, an ecclesia extends itself to all members of a society, but because it has so completely adjusted its ethical system to the political structure of the secular society, it comes to represent and promote the interests of the ruling classes. In this process, the ecclesia loses some adherents among the lower social classes, who increasingly reject it for membership in sects, be they sacred or "civil" (Yinger, 1970). An ecclesia is usually the official or national religion. For most people, membership is by birth, rather than conscious decision. Ecclesias have been common throughout human history. Examples include the Catholic Church in Spain and the Roman Empire; the Anglican Church, which is now the official Church of England; Islam in Saudi Arabia; and Confucianism, which was the state religion in China until early in the twentieth century.

The Denomination

A **denomination** *tends to limit its membership to a particular class, ethnic group, or religious group, or at least to have its leadership positions dominated by members of such a group.* It has no official or unofficial connection with the state, and any political involvement is purely a matter of choice by the denomination's leaders, who may either support or oppose any or all of the state's actions and political positions. Denominations do not withdraw themselves from the secular society. Rather, they participate actively in secular affairs and also tend to

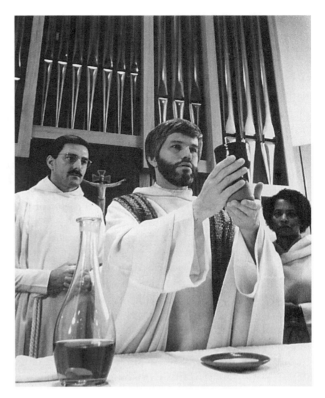

In the United States, Lutheranism and Methodism (and other Protestant groups), Catholicism, and Judaism all embody the characteristics of a denomination.

cooperate with other religious groups. Those two characteristics distinguish them from sects, which are separatist and unlikely to be tolerant of other religious persuasions (Yinger, 1970). (For that matter, universal churches, by their very nature, also typically dismiss other religions.) In America, Lutheranism, Methodism, other Protestant groups, Catholicism, and Judaism embody the characteristics of a denomination.

The Sect

A **sect** is *a small group that adheres strictly to religious doctrine that often includes unconventional beliefs or forms of worship.* Sects generally represent a withdrawal from secular society and an active rejection of secular culture (Yinger, 1970). For example, the Dead Sea Scrolls show clearly that the beliefs of both early Christian and Jewish sects, such as the Essenes, were rooted in a disgust with society's self-indulgent pursuit of worldly pleasures and in a rejection of the corruption perceived in the prevailing religious hierarchy (Wilson, 1969).

Early in their development, sects often are so harsh in their rejection of society that they invite persecution. Some actually thrive on martyrdom, which causes members to intensify their fervent commitment to the faith. Consider, for example, the

Christian martyrs in Rome before the conversion to Christianity of the Emperor Constantine.

Millenarian Movements

Millenarian movements *typically prophesy the end of the world, the destruction of all evil people and their works, and the saving of the just.* Millenarian (from the Latin word for "thousand") prophecies often are linked with the symbolic number 1,000 or multiples thereof.

Throughout human history, religious leaders have emerged in times of stress, foretelling the end of the world and asking everyone to stop whatever they are doing to follow the bearers of the message.

With the advent of the year 1000 A.D. Christendom in Medieval Europe was thrown into a panic by religious doomsday preachers who predicted that the end of the world was imminent. As a result, homes were abandoned, crops were left unharvested, and mobs of the devout took refuge in churches or fled on pilgrimages to the Holy Land.

Similar predictions took place with the advent of the year 2000. Many Christian groups believed the year 2000 would see monumental events predicted in the New Testament Book of Revelations, which predicts a thousand-year kingdom of peace and plenty, and a new heaven and a new Earth.

Why would anyone want to believe that the end of the world is near? Part of the answer is that the apocalypse is not pure destruction. It is the destruction of evil and the victory of good, according to the religious viewpoint. Believers feel they are the center of the universe and the cosmic drama is coming to its ultimate conclusion in their time. Life is filled with meaning.

Apocalyptic beliefs also appeal to people who feel that life has not been fair to them. One of the great appeals of apocalyptic views is that it helps to make sense of the world. All events start to have significance. They also lead to conspiracy theories because every detail is linked to a much larger drama in which good and evil are at odds. There is no such thing as chance.

What happens when the chosen date for the end of the world or other millennial event finally arrives—and nothing happens? Some people may just accept it and move on. More often, they will try to find a way to prove that their prophecy did come true, but not exactly in the way predicted. Others may reset their apocalyptic clocks to another date.

We should not be too quick to dismiss millennial movements. Christopher Columbus believed the world would end in 1650. He considered his discovery of the "New World" part of a divine plan to establish a millennial paradise. "God made me the messenger of the new heaven and the new Earth of which he spoke in the Apocalypse of St. John,"

Columbus wrote in his journal, "and he showed me the spot where to find it" (Sheler, 1997). Many major religious movements received attention because of apocalyptic predictions. Ironically, while the Apocalypse millennialists' prophecy may not come true, they often succeed: the world is a different place after them. It really was the end of the world as we had known it.

Aspects of American Religion

The Pilgrims of 1620 sought to build a sanctuary where they would be free from religious persecution, and the Puritans who followed 10 years later intended to build a community embodying all the virtues of pure Protestantism, a community that would serve as a moral guide to others. Thus, religion pervaded the social and political goals of the early English-speaking settlers and played a major role in shaping colonial society. Today, the three main themes that characterize religion in America are widespread belief, secularism, and ecumenism.

Widespread Belief

Americans generally take religion for granted. Although they differ widely in religious affiliation and degree of church attendance, almost all Americans claim to believe in God. Nine out of every 10 Americans have a religious preference, and say that they attend services on at least some occasions. Two-thirds of Americans maintain an affiliation with a church or synagogue and 6 in 10 consider religion to be of high importance in their personal lives. At the same time, less than 4 out of every 10 American adults attend either church or synagogue each week (Gallup Poll, 2000).

Evidence as to whether America is experiencing a religious revival, as some have claimed, is contradictory. When asked in 2000 if they think religion as a whole is increasing its influence on American life or losing its influence, 35% said it was increasing, while 58% said it was losing influence. This contrasts with 1957 when 69% said religion was increasing in influence and 14% said it was losing influence. Yet about one-third of Americans are devout practitioners of their faith saying they attend church or synagogue at least once a week, and two-thirds believe the Bible answers all or most of the basic questions of life (Gallup Poll, 2000). By and large, and despite dire warnings about the erosion of religiosity because of new ways of thinking and innovative lifestyles, for most Americans religion is a very important part of their lives.

More than half of all religiously affiliated individuals belong to a Protestant denomination, clearly reflecting America's colonial history. However, other denominations are also well represented, especially Catholicism and Judaism. There are more than 200 formally chartered religious organizations in America today. Such pluralism is not typical of other societies and has resulted primarily from the waves of European immigrants who began to arrive in the postcolonial era. Americans' traditional tolerance of religious diversity can be seen as a reflection of the constitutional separation of church and state, so that no one religion is recognized officially as better or more acceptable than any other.

Secularism

Many scholars have noted that modern society is becoming increasingly **secularized**, that is, *less influenced by religion.* Religious institutions are being confined to ever-narrowing spheres of social influence, while people turn to secular sources for moral guidance in their everyday lives (Berger, 1967). This shift is reflected in the reactions of Americans, who for the most part are notoriously indifferent to, and ignorant of, the basic doctrines of their faiths. Rodney Stark and Charles Glock (1968) report that a poll of Americans found that 67% of Protestants and 40% of Catholics could not correctly identify Father, Son, and Holy Spirit as constituting the Holy Trinity; 79% of Protestants and 86% of Catholics could not identify correctly a single prophet from the Old Testament; and 41% of Protestants and 81% of Catholics could not identify the first book of the Bible.

Of course, social and political leaders still rely on religious symbolism to influence secular behavior. The American Pledge of Allegiance tells us that we are "one nation, under God, indivisible," and our currency tells us that "In God We Trust." Since the turn of the century, however, modern society has turned increasingly to science, rather than religion, to point the way. Secular political movements have emerged that attempt to provide most, if not all, of the functions that traditionally have been fulfilled by religion. For example, communism prescribes a belief system and an organization that rival those of any religion. Like religions, communism offers a general concept of the nature of all things and provides symbols that, for its adherents, establish powerful feelings and attitudes and supply motivation toward action. Thus, some political movements lack only a sacred or supernatural component to qualify as religions. In this increasingly secular modern world, however, sacred legitimacy appears to be unnecessary for establishing meaning and value in life.

Ecumenism

Ecumenism refers to *the trend among many religious communities to draw together and project a sense of unity and common direction.* It is partially a response to secularism and is a tendency evident among many religions in the United States.

Unlike religious groups in Europe, where issues of doctrine have fostered sectlike hard-line separatism among denominations, most religious groups in America have focused on ethics—that is, how to live the good and right life. There is less likelihood of disagreement over ethics than over doctrine. Hence, American Protestant denominations typically have had rather loose boundaries, with members of congregations switching denominations rather easily and churches featuring guest appearances by ministers of other denominations. In this context, ecumenism has flourished in the United States far more than in Europe.

Major Religions in the United States

Nowhere is the diversity of the American people more evident than in their religious denominations. There are hundreds of different religious groups in the United States, which vary widely in religious practices, moral views, class structure, family values, and attitudes. A recent survey found surprisingly large and persistent differences among even the major religious groups.

The U.S. Census is prohibited from asking about religion, so the U.S. government generally has little to say on the matter. However, since 1972 the National Opinion Research Center has been conducting the General Social Surveys, which do give us a way of examining American religious attitudes and practices. This group has correlated information on a variety of issues with religious affiliation. Some of its findings are summarized here.

It is useful to think of American Protestant religious denominations as ranked on a scale measuring their degree of traditionalism. Conservative Protestant denominations include the fundamentalists (Pentecostals, Jehovah's Witnesses, etc.), Southern Baptists, and other Baptists. The moderates include Lutherans, Methodists, and inter- or nondenominationalists. Liberal Protestants are represented by Unitarians, Congregationalists, Presbyterians, and Episcopalians. This distinction among Protestants is important because the various branches often differ so markedly in their attitudes, especially toward social issues, that they resemble other religions more than the various denominations of their own. For example, in June of 2000, the Southern Baptist Convention—America's largest Protestant denomination, with 15.8 million members—passed a new Baptist statement of faith saying the Bible is without error and women should not be pastors (Rawls,

Although Americans differ widely in religious affiliation and degree of church attendance, almost all Americans claim to believe in God.

June 15, 2000). This is at a time when 71% of all Americans are in favor of having women pastors, ministers, priests or rabbis (Gallup Poll, 2000).

With respect to the hereafter, 79% of Americans believe there will be a day when God judges whether you go to heaven or hell (Gallup Poll, 2000). Nearly 90% of fundamentalists and Baptists believe in an afterlife. This falls to 80% among the moderate and liberal denominations. Catholics are similar to liberal Protestants in that 75% believe in an afterlife. Among people with no religious affiliation, 46% believe in an afterlife.

A strong belief in sin is typical of fundamentalists and Baptists. This causes them to condemn extramarital and premarital sex, and homosexuality, and to favor outlawing pornography. There is greater sexual permissiveness among the moderate denominations and considerably more among the liberal denominations. Catholics tend to resemble the Protestant moderates, and Jews tend to be more liberal than the liberal Protestant denominations in this area. Attitudes toward drugs and alcohol follow the same pattern, with smoking, drinking, or the frequenting of bars least common among fundamentalists and Baptists.

There are substantial class differences among the major denominations. Jews and Episcopalians have the highest median annual household incomes and Baptists have the lowest (Kosmin & Lachman, 1993). The pattern is the same for occupational prestige and education, with Jews and Episcopalians averaging three more years of education than fundamentalists and Baptists.

Given the wide differences in values and attitudes among religious groups, the relative proportion of the population that belongs to each group helps determine the shape of society. Protestants make up about 57% of the adult population. Among the major Protestant groups, the largest is composed of Baptists, who account for 18% of the adult population. Second are the Methodists with 9% (George W. Bush is the first Methodist to be elected U.S. president since William McKinley, who served from 1897 to 1901), next are the Lutherans with 7%, followed by Episcopalians at 3%, and the Church of Christ at 2% (Gallup Poll, 2000).

Roman Catholics, representing 27% of the population, make up the largest single religious denomination. Jews are 2%, followed by Muslims and a host of religions such as Eastern Orthodox, Hindu, Sufi, and Baha'i.

These percentages are in constant flux, however, because demographic factors such as birthrates and migration patterns may influence the numbers of people in any given religion. Religious conversion can also affect these numbers. Fundamentalism, for example, is gaining among the young and winning converts.

Despite trends toward ecumenicalism, it seems that the magnitude of religious differences, the persistence of established faiths, and the continual development of new faiths will ensure that this pattern of religious diversity will continue (Smith, 1984). (For a discussion of how religions have adapted to the Internet see "Society and the Internet: Religion on the Web.")

Protestantism

Because American Protestantism is so fragmented, many sociologists simply have classified all non-Catholic Christian denominations in the general category of Protestant. However, differences, of greater or lesser significance, exist among the various denominations.

Over the past 30 years there has been a 38% drop in the membership of Methodists, and the Episcopalians have seen a 44% decline. At the same time, the more conservative denominations have increased

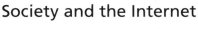

Society and the Internet
Religion on the Web

Religious groups are rushing on-line, setting up church home pages, and establishing theological newsgroups, bulletin boards, and chat rooms. The electronic community of the Internet has come to resemble a high-speed spiritual bazaar, where thousands meet, debate, and swap ideas about faith and religious beliefs.

Every group, it seems, from Lutherans to Tibetan Buddhists, now has a home page, many crammed with technological bells and whistles. Mormon sites offer links to vast genealogical databases, while YaaleVe' Yavo, an Orthodox Jewish site, forwards e-mailed prayers to Jerusalem, where they are affixed to the Western Wall. Two Web sites are devoted to Cao Daiis, the tiny Vietnamese sect that worships French novelist Victor Hugo as a saint, and a handful probe the mysteries of Jainism, an Indian religion in which the truly faithful sweep the ground with a small broom to avoid accidentally stepping on insects or other hapless creatures. Even the famously technophobic Amish are represented online by a Web site run by Ohio State University.

Even holy texts have begun to be adapted to the new technology. The interconnection of religious documents through so-called hyperlinks has produced a new form of scholarship called "hypertheology." Clicking on "Lot" in an electronic Bible, for instance, might connect you to similar stories in the Koran or pertinent twentieth-century moral commentaries. Just as the first illuminated manuscripts exposed readers to early theological debates, these hypertexts open up thousands of interpretations of God's words to anyone curious enough to click a mouse.

For all its seeming newness, however, the marriage between technology and religion is an ancient one. People have always used state-of-the-art communications technology to convey religious views. Five thousand years ago, the Sumerians etched their fears and hopes in cuneiform. Centuries later, Egyptians wrote on papyrus scrolls. The Old Testament was hewed and edited in the first century A.D. when the scrolls were turned into primitive books called codices. The first codices gave upstart Christianity an edge over Roman paganism. While pagan scholars stuck to their scrolls like modern Luddites refusing to embrace e-mail, liberal Christians adopted the efficiencies and portability of books, producing a technological advantage for early Christianity.

For the next millennium cloistered Benedictines toiled for years transcribing copy after copy of the Bible into leather-bound books. A monk would slowly copy a page from a Latin Bible, after which he and his brothers inked and gilded lavishly illustrated pages at the rate of roughly one a day.

As these elaborate Bibles circulated in Europe, they spread more than the word of God. It was also the first time a standard technology for recording and distributing information was used.

Proselytizing via these hand-wrought manuscripts was not an easy task. The Bibles were rare, fragile, and only in Latin. The problems did not go away until a German inventor named Johannes Gutenberg invented his movable-type press. Appropriately enough, the first book Gutenberg printed was the Bible. His simple press revolutionized intellectual commerce. Ideas that once could only be communicated in person took wing across the continent. In 1456, when the first Bible rolled off Gutenberg's press, there were fewer than 30,000 books in Europe. Fifty years later, there were 9 million, most devoted to religious themes.

Now we stand at the start of new movement in the intricate relationship between technology and faith: the marriage between God and the global computer networks. The Internet may become a vast cathedral of the mind, a place where ideas about God and religion can resonate and where faith can be shaped and defined by a collective spirit.

Source: "Finding God on the Web," by J. C. Ramo, December 16, 1996, *Time*, pp. 60–64, 66–67.

dramatically. In the same period, there has been a 211% increase among the Assemblies of God, a 119% increase among Jehovah's Witnesses, a 96% increase among Mormons, and an 8% increase in the number of Baptists (Gallup Poll, 1997).

Since the late 1980s, the fundamentalists and evangelical Christians have become an extremely visible and vocal segment of the Protestant population, and their presence has been felt through the media and through their support of political candidates. Why are the fundamentalist and evangelical churches gaining such popularity? Some of their appeal may lie in the sense of belonging and the comfort they offer through their belief in a well-defined and self-assured religious doctrine—no ambiguities and, hence, few moral choices to be made.

Interfaith marriages have steadily increased as religious prejudices have declined.

The growth of these churches reflects religion's role as a social institution, changing over time and from place to place, partly in response to concurrent social and cultural changes and partly itself acting as an agent of social change.

Catholicism

Twenty-seven percent of Americans are Roman Catholics. One of the most striking things about Catholics in the United States is their youth: Twenty-nine percent are younger than 30, 36% are between 30 and 49, and 35% are over 50. In contrast, 24% of Protestants are younger than 30, and 41% are older than 50. The higher birthrates among Catholics in the baby-boom generation and among Hispanic Catholics account for a large part of this difference. Another part of the explanation is the difficulty that mainline Protestant denominations have had in retaining young people.

American Catholics have long been an immigrant people, and that tradition is continuing. One in five Catholics is a member of a minority group. Hispanics now make up 16% of American Catholics. Another 3% are black, and an additional 3% describe themselves as nonwhite. Very few Hispanics identify themselves as nonwhite, so the data suggest that the influx of Catholic immigrants from Southeast Asia is starting to show in national surveys. Among Protestants, 14% are black and 2% are Hispanic. This means that while the percentage of blacks among Protestants is five times that among Catholics, a higher percentage of Catholics overall come from minority groups.

Since the mid-1960s, Catholics have equaled Protestants in education and income levels. The overall figures for Protestants mask significant differences between denominations. When we compare Catholics with individual Protestant denominations, we find them still ranking behind Presbyterians and Episcopalians, about on a par with Lutherans and Methodists, and well ahead of Baptists on education and income scales. This comparison is striking in light of the fact that large numbers of lower-income minorities are included in the overall Catholic figures.

Catholics remain an urban people, with only one in four living in rural areas. A higher percentage of Catholics (39%) than of any other major denomination live in central cities, and 35% live in suburbs. The vast majority of Catholics are concentrated in the Northeast and Midwest.

Catholics have historically favored larger families than have other Americans, but by 1985 the difference in ideal family size between Catholics and Protestants had disappeared, with both groups considering two children the ideal. Despite the Catholic church's condemnation of artificial means of birth control, American Catholics have favored information about and access to contraceptives in the same proportion as the rest of the population since the 1950s (Gallup & Castelli, 1987).

One of the most important developments in the recent history of Catholicism was the ecumenical council (Vatican II) called by Pope John XXIII, which met from 1962 to 1965 and thoroughly reexamined Catholic doctrine. This ecumenical council led to many changes, often referred to as liberalization and including the substitution of common language for Latin in the Mass. One unintended consequence (or latent function) of Vatican II was that the centralized authority structure of the

Catholic Church was called into question. Lay-people and priests felt free to dispute the doctrinal pronouncements of bishops and even of the pope. What began in the 1960s with a seemingly modest effort of reform has ended with every aspect of Catholic tradition under question. America's biggest single denomination now consists of traditional Catholics and "cafeteria Catholics" who pick and choose what to practice. Surveys show that people who identify as Catholic are more liberal on sexual morals than Protestants as a whole. Birthrates and opinions on abortion are virtually the same among Catholics as among Protestants (Ostling, 1999).

Under the leadership of Pope John Paul II, the Catholic Church took a more conservative turn. It continued to condemn all forms of birth control except the rhythm method and rejected high-technology aids to conception, such as artificial insemination, *in vitro* fertilization, or surrogate motherhood. Calls for a greater role for women in the church or their ordination to the priesthood were rejected. Women were told to seek meaning in their lives through motherhood and giving love to others.

Judaism

There is a strong identification among Jews on both a cultural and a religious level. This sense of connectedness is an important factor in understanding current trends within the religion.

Jews can be divided into three groups, based on the manner in which they approach traditional religious precepts. Orthodox Jews observe traditional religious laws very closely. They maintain strict dietary laws and do not work, drive, or engage in other everyday practices on the Sabbath. Reform Jews, by contrast, allow for major reinterpretations of religious practices and customs, often in response to changes in society. Conservative Jews represent a compromise between the two extremes. They are less traditional than the Orthodox Jews are but not as willing to make major modifications in religious observance as the Reform Jews. In addition, a large secularized segment of the Jewish population still identifies itself as Jewish but refrains from formal synagogue affiliation.

As among Protestants, social-class differences exist among the various Jewish groups. Reform Jews are the best educated and have the highest incomes. For Orthodox Jews, religious rather than secular education is the goal. They have the lowest incomes and the least amount of secular education. As might be expected, Conservative Jews are located between these two poles.

The state of Israel has played a major role in shaping current Jewish thinking. For many Jews, identification with Israel has come to be a secular replacement for religiosity. Support for the country is tied to many deep psychological and emotional responses. To many, Israel and its continued existence represent a way of guaranteeing that never again will millions of Jews perish in a holocaust. The country is seen as a homeland that can help defend world Jewry from the unwarranted attacks that have occurred throughout history. For many Jews, identification with, and support for, Israel is important to the development of cultural or religious ties.

Recently, the Jewish community has had to deal with the issue of ordaining women rabbis. Reform Jews have moved in this direction without a great deal of difficulty, and today women lead some Reform congregations around the country. However, the issue of women rabbis initially produced a bitter fight among Conservative Jews. Today, an increasing number of Conservative women rabbis have become the heads of congregations. Orthodox Jews have not had to address the issue, because for them the existence of a female rabbi would represent too radical a departure from tradition to be contemplated.

At the Wailing Wall in Jerusalem, thousands of Jews gather each day to pray and mourn the destruction of the Second Temple by the Romans in 70 A.D.

According to the Bible, God told the Jewish people to be fruitful and multiply. In the United States, it seems the group is not doing so. Recent surveys estimate that there are 5.98 million Jews in the United States, similar to the number in 1977. Demographers predict that the population will continue to decline. The American Jewish community is thus facing a crisis.

The reasons for the lack of growth in the Jewish community are varied. For one thing, Jews in the United States are not bearing enough children to replace themselves: The Jewish birthrate has been estimated at between 1.3 to 1.7 children per woman per lifetime—well below the replacement rate of 2.1. The Jewish population is also quite old, with about 40% of American Jews having a remaining life expectancy of 20 years or less. Jewish immigration from the former Soviet Union and Israel has increased the population somewhat but is unpredictable. Finally, substantial numbers of young Jews are choosing to marry outside the faith, although this does not necessarily lead to a loss of Jewish identity.

Whether or not there is a population erosion, and what should be done about it, is an ongoing debate in the American Jewish community. Jewish religious groups have attempted to liberalize the definition of who is a Jew, and—in a radical break with tradition—have decided to seek converts.

A common fear is that a further decline in the number of American Jews would lessen their ability to defend their political interests. Others suggest that the United States would lose the contributions of Jewish scientists, artists, and performers. Even though Jews account for 2.3% of the U.S. population (Harris Poll, 2000), they make up 20% of America's Nobel laureates.

While many of the reasons for the shrinking Jewish population are demographic, the issue of how to stem the tide has been the cause of major rifts among the various branches of the religion. Recent statistics, for example, put the rate of Jewish intermarriage at 50%. Jewish opinion is not uniform on the subject, but shows a sharp division between Orthodox Jews the most strictly observant branch of Judaism, whose members are a small minority of the American Jewish population and non-Orthodox. For example, 64% of Orthodox Jews surveyed said they strongly disapproved of interfaith marriages, as opposed to 15% of Conservative Jews, 3% of Reform Jews, and 2% of those who identified themselves as "just Jewish." Interfaith couples have little difficulty today finding rabbis willing to officiate at the marriage ceremony (Niebuhr, 2000).

A subject that has provoked even more controversy is the Reform movement's break with the tradition of matrilineal, or motherly, descent. In 1983, the Reform movement declared that people could be considered Jewish if either the father or mother was Jewish. Before that time, Judaism could be passed on to the child only from the mother.

American Jews also differ with respect to converts. The Reform movement's outreach goes against centuries of Jewish tradition in which proselytizing has been disdained. The outreach program is quite low-key. There is no advertising or airwave sermonizing, more common with Christian evangelical movements. Instead, outreach sessions resemble comparative religion discussion groups, in which introductory classes in Judaism are available to interfaith couples as well as converts.

Islam

A Muslim is someone who has accepted the Islamic creed that "There is no God but Allah, and Mohammed is His Prophet." Like Jews and Christians, Muslims are monotheists, and all three religions share the prophets of the Old Testament.

From its roots in the Arab world, Islam has spread in virtually every direction, as far west as the Atlantic Coast of Africa and as far east as Indonesia. There are an estimated 1 billion Muslims in the world, and almost one-fourth of the world population may be Muslim in the near future if current high levels of fertility continue. Among the five most populous nations, the United States has by far the fewest Muslims. It is, however, the third largest faith in the United States and there are already more Muslims than Presbyterians nationwide.

It is very difficult to ascertain how many Muslims there are in the United States. The census does not collect reliable data on religious preference. Some data do exist on church attendance, but Islam does not require attendance for formal membership in the way that many Christian groups do. Estimates of the number of Muslims in the United States range from 700,000 to 4 million. Despite estimates of millions of U.S. Muslims, there are only a thousand mosques and community centers to serve them. Islam faces the same challenge as Judaism faced with the influx from Eastern Europe toward the end of the nineteenth century. New-style congregations had to be invented, new buildings built, and new schools started to train a new type of rabbi. U.S. Islam is just beginning to create the communal organizations that have served Judaism so well. There is no coherent association of mosques to unite immigrants with native-born blacks, and national organizations of other types are young. Muslims are divided the way Protestants have been, by ethnicity, race, and language. Only in 1996 did U.S. Muslims establish a school for training clergy at the graduate level to parallel the Jewish and Christian seminaries (Ostling, 1999).

About one in four Muslims in the United States is African-American. African-American Muslims

Islamic pilgrims to Mecca must circle the flat-roofed kaaba seven times and at its east corner kiss the Black Stone. The Black Stone is said to have been received by Adam when he fell from paradise.

are of two types—those who follow mainstream Islamic doctrine and those who follow the teachings of the Nation of Islam. This is an important distinction because the Nation of Islam's teachings are very different from those in the Muslim holy book, the Koran.

Muslims tend to be conservative about social issues, favor a close-knit family, and support religious education. The conservatism is based partly on religious beliefs, but also arises from cultural mores. For example, to some Muslims sexual permissiveness is seen as a reflection on the family, rather than on the individual. The family is seen as the ultimate authority and is therefore responsible for the individual's behavior. Many Muslims also believe that the eldest man is the head of the family.

American Muslims sometimes have difficulty reconciling these traditional practices and moral tenets in a non-Islamic society. Immigrant Muslim parents are often at odds with their Americanized offspring about the use of alcohol, which is banned in Islam; dating, which is forbidden; and lack of respect for elders (El-Badry, 1994).

All religions and denominations are affected by the current mood of the country. A heightened social consciousness results in demands for reform, whereas stressful times often produce a movement toward the personalization of religion. In any event, while traditional forms and practices of religion

may be changing in the United States, religion itself is likely to continue to function as a basic social institution.

Social Correlates of Religious Affiliation

Religious affiliation seems to be correlated strongly with many other important aspects of people's lives: Direct relationships can be traced between membership in a particular religious group and a person's politics, professional and economic standing, educational level, family life, social mobility, and attitudes toward controversial social issues. For example, Jews are proportionally the best-educated group; they also have higher incomes than Christians in general; and a greater proportion are represented in business and the professions. Despite their high socioeconomic and educational levels, Jews, like Catholics, occupy relatively few of the highest positions of power in the corporate world and politics: These fields generally are dominated by white Anglo-Saxon Protestants.

Attitudes toward social policy also seem to be correlated, to some extent, with religious affiliations. The fundamentalist and evangelical Protestant sects generally are more conservative on key issues than are the major Protestant sects.

Although it is clear that religious associations show definite correlations with people's political, so-

Controversies in Sociology

Are Religious Cults Dangerous?

Although a sect often develops in response to a rejection of certain religious doctrine or ritual within the larger religious organization, a cult usually introduces totally new religious ideas and principles. Cults generally have charismatic leaders who expect total commitment. Members of cults, who are usually motivated by an intense sense of mission, often must give up individual autonomy and decision making. Many cults require resocialization practices so strong as to make the member seem unrecognizable in personality and behavior to former friends and relatives. Included among the best-known contemporary religious cults are the Unification Church, the Church of Scientology, Church Universal and Triumphant, and The Way International. Because the activities and excesses of some cults have taken over the headlines in recent years, we may tend to forget or even dismiss the fact that many major religions began as cults and sects.

Some contemporary religious cults have been referred to as destructive because of the effect they have on their members. According to Steve Hassan, an ex-cult member who helps people leave such cults, destructive religious cults have the following characteristics:

1. *The doctrine is reality.* The doctrine is the *truth;* it is perfect and absolute. Any flaw in it is viewed as a reflection of the believer's own imperfection. The doctrine becomes a master program for all thoughts, feelings, and actions.
2. *Reality is presented in simple polar terms.* Cult doctrines reduce reality into two basic poles: black versus white, good versus evil, spiritual world versus physical world, us versus them. The cult recognizes no outside group as having any validity, because that would threaten the cult's monopoly on the truth.
3. *An elitist mentality pervades.* Members are made to feel part of an elite group. They consider themselves better, more knowledgeable, and more powerful than anyone else. This produces a heavy burden of responsibility, however, because if

members are told they are not performing their duty fully, they are made to feel that they are failures.

4. *Group goals supersede individual goals.* The self must be subordinated to the will of the group. Absolute obedience to superiors is a common theme in many cults. Conformity to the group is good.
5. *Continued acceptance depends on good performance.* One of the most attractive qualities of a cult is its sense of community. To remain part of this community, however, the cult member learns that he or she must meet group goals, whether they be in recruitment, collecting money, allegiance to the leader, or proper behavior.
6. *Members are manipulated through fear and guilt.* The cult member lives within a narrow corridor of fear, guilt, and shame. Problems are due to the member's weakness, lack of understanding, evil spirits, and so forth. The cult member constantly feels guilty for not being good enough. The devil is seen as always lurking just around the corner, waiting to tempt or seduce the cult member.
7. *The member's time orientation changes.* The past, present, and future are seen in a different light. The cult member looks back on his past life in a distorted way and sees it as totally negative. The present is marked by a sense of urgency in terms of the work that needs to be done. The future is a time when members will be rewarded or punished for what they have done to help the cause.
8. *No reason for leaving is legitimate.* The cult does not recognize a person's right to move on. Members are made to believe that the only reason people leave is because of weakness, insanity, brainwashing (deprogramming), pride, or sin. The cult stresses that if members ever leave, terrible things will happen to themselves, their families, or the world.

Source: *Combating Cult Mind Control* (pp. 78–84), by S. Hassan, 1988, Rochester, VT: Park Street Press.

cial, and economic lives, we must be careful not to ascribe a cause-and-effect relationship to such data, which at most can be considered an indicator of an individual's attitudes and social standing.

The social and political correlates of religious affiliation have had a significant impact on the directions of the various religious denominations and sects in the United States. For a discussion of the differences between traditional religions and cults, see "Controversies in Sociology: Are Religious Cults Dangerous?"

SUMMARY

Religion is a system of beliefs, practices, and philosophical values shared by a group of people that defines the sacred, helps explain life, and offers salvation from the problems of human existence. It is recognized as one of society's most important institutions. Durkheim observed that all religions divide the universe into two mutually exclusive categories. The profane consists of empirically observable things, or things knowable through common, everyday experiences. In contrast, the sacred consists of things that are awe inspiring and knowable only through extraordinary experience. Almost anything may be designated as sacred. Sacred traits or objects symbolize important shared values. Patterns of behavior or practices that are related to the sacred are known as rituals.

All religions have formalized social rituals, but many also feature private rituals such as prayer, which is a means for individuals to address or communicate with supernatural beings or forces. One of the functions of ritual and prayer is to produce an appropriate emotional state. In addition, all religions endorse a belief system that usually includes a supernatural order and a set of values to be applied to daily life. Finally, many religions have an organizational structure through which specialists can be recruited and trained, religious meetings conducted, and interaction facilitated between society and members of the religion. This organization also promotes interaction and solidarity among the members.

Sociologists focus on the relationship between religion and society. Functionalists examine the utility of religion in social life. Religion, they say, offers individuals ways to reduce anxiety and promote emotional integration. In addition, religion provides for group unity and cohesion not only through its own practices, but also sometimes through the hostility and prejudice directed at members of a religious group by outsiders. Durkheim felt that religion serves a vital function in maintaining the social order. He came to this conclusion by studying one of the simplest kinds of religion, totemism. A totem is an ordinary object such as a plant or animal that has become a sacred symbol to and of a particular group or clan, who not only revere the totem but also identify with it. When people recognize or worship supernatural entities, said Durkheim, they are really worshipping their own society.

Religion frequently legitimizes the structure of the society within which it exists. Religion also establishes worldviews that help people to understand the purpose of life. These worldviews can have social, political, and economic consequences. Finally, religion can also help a society adapt to its natural environment or to changing social, economic, and political circumstances. An example of this is that when societies are under great stress or attack, their members sometimes fall into a state of despair. Under this condition, there sometimes emerge revitalization movements, which are powerful religious movements that stress a return to the traditional religious values of the past.

Conflict theorists emphasize religion's role in justifying the political status quo by cloaking political authority with sacred legitimacy and thereby making opposition to it seem immoral. Alienation, the process by which people lose control over the social institutions they themselves have invented, plays an important role in the origin of religion according to this view. Religion tends to conceal the natural and human causes of social problems in the world and discourages people from taking action to correct these problems. In addition, religion limits independent thinking and the search for further knowledge.

INTERNET RESOURCES

Go to http://web-enhanced.thomsonlearning.com which features Web links relevant to this chapter to expand and enhance the learning experience. This site will be monitored and updated periodically to ensure the links are correct and live.

At Virtual Society: The Wadsworth Sociology Resource Center, http://sociology.wadsworth.com, you will find a Career Center, Sociology in the News, Research Online, InfoTrac College Edition, Virtual Tours in Sociology, and a variety of book-specific student and instructor resources.

CHAPTER ELEVEN STUDY GUIDE

KEY CONCEPTS

Match each concept with its definition, illustration, or explanation below.

a. Denomination
b. Magic
c. Polytheism
d. Mana
e. Hypertheology
f. Universal church
g. Theism
h. Sect
i. Rituals

j. Supernaturalism
k. Religion
l. Ecumenism
m. Prayer
n. Abstract ideals
o. Profane
p. Ecclesia
q. Millenarian movements

r. Religious taboo
s. Animism
t. Alienation
u. Revitalization movements
v. Secularized
w. Totem
x. Monotheism
y. Sacred

_____ 1. A system of beliefs, practices, and philosophical values shared by a group of people that defines the sacred, helps explain life, and offers salvation from the problems of human existence.

_____ 2. Things that are knowable through common, everyday experiences.

_____ 3. Things that are awe inspiring and knowable only through extraordinary experience.

_____ 4. Patterns of behavior or practices that are related to the sacred.

_____ 5. A means for individuals to communicate with supernatural beings or forces.

_____ 6. An active attempt to coerce spirits or to control supernatural forces.

_____ 7. Belief in the existence of nonpersonalized supernatural forces that can and often do influence human events.

_____ 8. A diffuse, nonpersonalized force that acts through anything that lives or moves.

_____ 9. A sacred prohibition against touching, mentioning, or looking at certain objects, acts, or people.

_____ 10. Belief in inanimate, personalized spirits or ghosts of ancestors that take an interest in and actively work to influence human affairs.

_____ 11. Belief in divine beings—gods and goddesses—who shape human affairs.

_____ 12. Belief in a number of gods.

_____ 13. Belief in the existence of a single god.

_____ 14. Focus on the achievement of personal awareness and a higher state of consciousness through correct ways of thinking and behaving, rather than manipulating spirits or worshipping gods.

_____ 15. An ordinary object that has become a sacred symbol to or of a particular group or clan.

_____ 16. Religious movements that stress a need to return to the traditional religious values of the past.

_____ 17. The process by which people lose control over the social institutions they themselves invented.

_____ 18. Includes all the members of a society within one united moral community.

_____ 19. A church that shares the same ethical system as the secular society and has come to represent and promote the interest of the society at large.

_____ 20. A religious group that tends to limit its membership to a particular class, ethnic, or religious group.

_____ 21. A small group that adheres strictly to religious doctrine that often includes unconventional beliefs or forms of worship.

_____ 22. Movements that typically prophesy the end of the world, the destruction of all evil people and their works, and the saving of the just.

_____ 23. A society that is less influenced by religion.

_____ 24. The trend among many religious communities to draw together and project a sense of unity and common direction.

_____ 25. The interconnection of religious documents through hyperlinks produced this new form of scholarship.

KEY THINKERS/RESEARCHERS

Match the thinkers/researchers with their main ideas or contributions.

a. **Max Weber**
b. **Karl Marx**
c. **Bronislaw Malinowski**

d. **Émile Durkheim**
e. **Marvin Harris**
f. **Sigmund Freud**

g. **Rodney Stark and William Bainbridge**

_____ 1. The first sociologist who distinguished between the sacred and the profane and discussed religion's role in promoting social cohesion.

_____ 2. Discussed the relationship between magic and Christianity.

_____ 3. Said that religion was irrational, but helpful to the individual in coming to terms with impulses that induce guilt and anxiety.

_____ 4. Discussed, in *The Protestant Ethic and the Spirit of Capitalism,* how the ideology of Calvinism had influenced the development of capitalism.

_____ 5. Showed that the Hindu belief in the sacredness of cows is a positive strategy for adapting to the environment in India and therefore quite rational.

_____ 6. Saw religion is as a tool that the upper classes use to maintain control of society and to dominate the lower classes.

_____ 7. An anthropologist who explained the functional differences between religion and magic, with the former uniting a group of believers while the latter helped the individual who used magic.

CENTRAL IDEA COMPLETIONS

Following the instructions, fill in the appropriate concepts and descriptions for each of the questions posed in the following section.

1. One of the major sociological approaches to religion is the functionalist perspective. Using this perspective, provide an example of how religion fulfills each function below:

   a. Establishing worldviews: _____

   b. Building adaptations to society: _____

   c. Creating social cohesion: _____

   d. Satisfying individual needs: _____

2. Define and provide an example of each of the following basic religious organizations:

   a. Sect: _____

   b. Church: _____

   c. Ecclesia: _____

   d. Denomination: _____

   e. Millenarian movement: _____

3. Define and provide an example of each of the major types of religions:

a. Supernaturalism: _____

b. Animism: _____

c. Theism: _____

d. Abstract ideals: _____

4. Distinguish the similarities and differences between magic and religion:

a. Magic: _____

b. Religion: _____

5. Millennial movements stress the end of the world. What reasons does your text give as to why people want to believe the end of the world is near?

_____

6. What are the characteristics of the major religions in the United States?

_____

_____

7. What does the conflict perspective have to say about religion? How does it differ from the functionalist perspective discussed earlier?

a. Conflict perspective: _____

b. Differences with functionalism: _____

8. What are the relative sizes of each of the following world religions?

a. Protestants: _____

b. Roman Catholics: _____

c. Jews: _____

d. Muslims: _____

e. Hindus: _____

f. Shintoists: _____

g. Buddhists: _____

h. Chinese folk religions: _____

9. List and explain each of the characteristics of destructive cults:

a. _____

b. _____

c. _____

d. _____

e. _____

f. _____

g. _____

h. _____

10. Define and provide a brief example of the sacred and profane distinction:

a. Sacred: _____

b. Profane: _____

11. Discuss how life conditions in Europe during the Middle Ages illustrated your author's point that "we find magic where the element of danger is conspicuous": _____

_____

_____

12. Describe the similarities and differences between Muslims in the United States and the major Christian and Jewish religious communities: _____

_____

_____

13. Reread your text's description of religious persecution around the world. What are the common features of government-condoned or perpetrated violence against religious minorities in these countries?

_____

_____

14. What is the apparent paradox in the United States between people's statements of religious belief and their religious behaviors? _____

_____

CRITICAL THOUGHT EXERCISES

1. Your author discusses the new phenomenon of religion on the World Wide Web. Log on to the Web and conduct a search of any four of the major religious denominations of the United States. Next, search for four smaller sect Web sites. What do all of the sites share in common? How do they differ? Do they provide "hit counters" indicating how many visitors they have monthly? How effective to you think the Web is for the dissemination of religious ideals?

2. Your text outlines the basic features of destructive cults. Log on to the World Wide Web and visit any of the UFO cult sites. What type of information are they presenting at their Web sites? Which of the negative characteristics presented in your text can been uncovered in these cults' Web pages? Which messages are most blatant and which are more subtle? What problems, if any, do you see relative to the movement of religious forms of expression to the World Wide Web?

3. Durkheim provided a detailed discussion of religion, which has formed the core of the functionalist interpretation of religion. In an earlier chapter, you learned about the concepts of dysfunction and latent function. For each of the major functions of religion, provide an analysis of a possible latent dysfunction, complete with an examples.

4. Given each individual has his or her own beliefs (or disbeliefs), how can a sociologist hope to empirically investigate the phenomenon of religion? In what ways can the sociological imagination be a useful tool for studying religion? Apply the sociological imagination to an examination of the major sociological features of your own religion.

5. Your text devoted considerable attention to describing how magic intermixes with religion in many countries and historical periods. Describe the role magic has played in organized religion. Address Stark and Bainbridge's comment, "In order to monopolize religion, a church must monopolize all access to the supernatural. . . . But if the church is to deny others access to the supernatural, it must remain in the magic business."

ANSWERS

KEY CONCEPTS

1. k	6. b	10. s	14. n	18. f	22. q
2. o	7. j	11. g	15. w	19. p	23. v
3. y	8. d	12. c	16. u	20. a	24. l
4. i	9. r	13. x	17. t	21. h	25. e.
5. m					

KEY THINKERS/RESEARCHERS

1. d	3. f	4. a	5. e	6. b	7. c
2. g					

LEARNING OBJECTIVES

After studying this chapter, you should be able to do the following:

- Describe the manifest and latent functions of education.
- Explain the nature of education from the conflict theory view.
- Explain the causes and effects of racial segregation in the public schools.
- Identify issues relating to students who speak English as a second language.
- Discuss the extent to which high-school dropouts are a social problem.
- Discuss the issue of standardized testing.
- Evaluate the idea of special programs for gifted students.

It is June 24, graduation day for Seward Park High School in New York City, and teacher Jessica Siegel has a pocketbook full of bobby pins and tissues. The bobby pins are for her students; the tissues are for herself.

The high school sits empty and silent in the late afternoon, a breeze nudging litter down the street outside, for the commencement is being held 80 blocks north in Manhattan, at Hunter College. There the streets unfold with boutiques, diplomatic mansions, and elegant apartment buildings, with awnings and doormen and gilt doors that glint in the summer sun.

With footsteps both urgent and tentative, the students and their families arrive. Six rise from the Lexington Avenue subway, palms shading their eyes. Five pile out of a Dodge Dart piloted by an uncle. Four emerge from the gypsy cab on which they have splurged, with its smoked windows and air freshener. All of them pause to gape, and the cabbie, too, waits an extra moment before driving on, leaning out the window to admire the promised land. Their world is the Lower East Side, its streets crammed with tenements and bodegas and second-story sweatshops, its belly bursting with immigrants, its class and tongues and customs so foreign at Hunter, only 20 minutes from home.

The seniors have traveled to a border, to a frontier. Some will never cross it. They will have their diplomas and shrink back southward, back to the comforting, familiar things, back to the many menaces. However, others will wield their diplomas as passports and will step into the new, the strange, the almost inconceivable. A few will even attend college (Freedman, 1990).

Jessica Siegel was thrilled to see so many of her students graduate from high school, a major accomplishment that only took place because of her dedication and effort. By helping these students get an education, she had accomplished her goals. What are those goals specifically? It may sound like a simple question, but if you were asked why people should get an education, what would you say? When people across the country are asked just such a question, the most common reason they give is to get a "better job" or "a better paying job" (42%). In fact, 57% believe that schools are not doing enough to develop job skills.

Very few people think the reason to attend school is "to acquire knowledge" (10%), "to learn basic skills" (3%), "to develop an understanding and appreciation for culture" (1%), or "to develop critical thinking skills" (1%) (The Gallup Report, 1994).

Two people responsible for the development of sociology as a field also struggled with trying to decide the purpose of education. Herbert Spencer, the author of the first textbook in sociology and a proponent of social Darwinism, believed education was only important if it had practical value and prepared people for everyday life. Lester Frank Ward, the first president of the American Sociological Society, believed the main purpose of education was to equalize society.

In the 1850s, Herbert Spencer asked, "What knowledge is of most worth?" and concluded that the purpose of education was "to prepare us for complete living." Every study must be judged by whether it had "practical value." The most important knowledge, he believed, was knowledge for gaining a job or trade, being a parent, and carrying out one's civic duties. Spencer believed that the study of science, which was not taught in schools at the time, was the most useful subject that could be studied and should be added to the curriculum. All other subjects had to be judged by how useful they would be in later life.

When Spencer was putting these ideas forth, he enjoyed enormous prestige in the United States because of his views on social Darwinism, which suggested that little could done to help those at the bottom of the social class ladder. His emphasis on practical education was also applauded by a population that was already inclined to doubt the value of book learning.

Lester Frank Ward believed the source of inequality was the unequal distribution of knowledge. The main purpose of education was to equalize society by diffusing knowledge to all. The greatest advances in civilization had been made by people who had the opportunity for an education and the leisure to think. Ward maintained that the potential giants of the intellectual world could be the unskilled workers of today. He wrote, "the number of individuals of exceptional usefulness will be proportionate to the number possessing the opportunity to develop their powers." Ward believed the entire society would benefit if there were more and better educational opportunity. Against both scholarly and popular opinion, he defended "intellectual egalitarianism." The differences between those at the top and the bottom of the social ladder were not due to any difference in intellect, he said, but to differences in knowledge

and education. Ward believed the main job of education was to ensure that the heritage of the past was transmitted to all members of society (Ravitch, 2000).

In this chapter, as we examine the role of education in our society, we will contrast the functionalist and conflict theorist approaches to understanding the American educational system. Functionalists stress the importance of education in socializing the young, transmitting the culture, and developing skills. Conflict theorists, on the other hand, note that education preserves social class distinctions, maintains social control, and promotes inequality. We will also examine the impact of some of the contemporary issues facing education.

Education: A Functionalist View

What social needs does our education system meet? What are its tasks and goals? Education has several *manifest functions* (Chapter 1)—intended and predetermined goals such as the socialization of the young or the teaching of academic skills. There are also some latent functions, which are unintended consequences of the educational process. These may include child care, the transmission of ethnocentric values, and respect for the American class structure.

Socialization

In the broadest sense, all societies must have an educational system. That is, they must have a way of teaching the young the tasks that are likely to be expected of them as they develop and mature into adulthood. If we accept this definition of an educational system, then we must believe that there really is no difference between education and socialization. As Margaret Mead (1943) observed, in many preliterate societies no such distinction is made. Children learn most things informally, almost incidentally, simply by being included in adult activities.

Traditionally, the family has been the main arena for socialization. As societies have become more complex, the family has been unable to fulfill all aspects of its socialization function. Thus, there is a need for the formal educational system to extend the socialization process that starts in the family. In modern industrialized societies, a distinction is made between education and socialization. In ordinary speech, we differentiate between socialization and education by talking of bringing up and educating children as two separate tasks. In modern society, these two aspects of socialization are quite compartmentalized: Whereas rearing children is an informal activity, education or schooling is formal. The role prescriptions that determine interactions between students and teachers are clearly defined, and the curriculum to be taught is explicit. Obviously, the educational process goes far beyond just formalized instruction. In addition, children also learn things in their families and among their peers. In school, children's master status (see Chapter 5) is that of student, and their primary task is to learn.

Schools, as differentiated, formal institutions of education, emerged as part of the evolution of civilization. However, until about 200 years ago, education did not help people become more productive in practical ways, and thus it was a luxury that very few could afford. This changed dramatically with the industrialization of Western culture. Workers with specialized skills were required for production jobs, as were professional, well-trained managers. When the Industrial Revolution moved workers out of their homes and into factories, the labor force consisted not only of adults but also of children. Subsequently, child labor laws were passed to prohibit children from working in factories.

Public schools eventually emerged as agencies dedicated to socializing students, teaching them proper attitudes and behaviors, and encouraging conformity to the norms of social life and the workplace.

Cultural Transmission

The most obvious goal of education is **cultural transmission,** *in which major portions of society's knowledge are passed from one generation to the next.* In relatively small, homogeneous societies, in which almost all members share the culture's norms, values, and perspectives, cultural transmission is a matter of consensus and needs few specialized institutions. In a complex, pluralistic society like ours, with competition among ethnic and other minority groups for economic and political power, the decision about what aspects of the culture will be transmitted is the outgrowth of a complicated process.

If we consider that schools are one of the major means of cultural transmission—both vertically (passing knowledge between generations) and horizontally (disseminating knowledge to adults)—this lack of consensus concerning every important aspect of schooling points to something beyond the educational system itself, to a crisis in the whole of American culture. It is important to understand something of the nature of this cultural crisis before we turn to a narrower discussion of the functions of education in America today. Our pluralistic society contains many cultural differences among its ethnic groups and social classes. Nevertheless, for a society to hold together, there must be certain core values and goals—some common traits of culture—that its constituent social elements share to a greater or lesser degree (see Chapter 3). In America, it seems that this core culture itself is changing rapidly.

A school's curriculum often reflects the ability of organized groups of concerned citizens to impose their views on an educational system, whether local, statewide, or nationwide. Thus, it was a political process that caused African-American history to be introduced into elementary, high school, and college curricula during the 1960s. Similarly, it was political activism that caused the creation of women's studies programs in many colleges. Moreover, even though the concept of evolution is a cornerstone of modern scientific knowledge, it is political pressure (from Christian fundamentalists) that causes some

Until about 200 years ago, education was a luxury that few could afford. Schools, as we know them today, arose out of a need to create a skilled workforce.

textbooks to refer to it as the "theory" of evolution and prevents it from being taught in certain counties.

In recent years, bilingual education has become an educational and political issue. Proponents believe that it is crucial for children whose primary language is not English to be given instruction in their native tongues. They believe that by acknowledging students' native languages, the school system is helping them make the transition into the all-English mainstream—and is also helping to preserve the diversity of American culture.

Others see a danger in these programs. They believe that many bilingual education programs never provide for the transition into English, leaving many youngsters without the basic skills needed to earn a living and participate in our society.

In the end, the debate centers on how closely our sense of who we are as a nation hinges on the language our children speak in school. For the time being, the only agreement between the two sides is that language is the cornerstone for cultural transmission.

Academic Skills

Another crucial function of the schools is to equip children with the academic skills they need to

An important function of education is to equip children with the academic skills that they need in order to function in society.

function as adults—to hold down a job, to balance a checkbook, to evaluate political candidates, to read a newspaper, to analyze the importance of a scientific advance, and so on. Have the schools been successful in this area? Most experts believe they have not.

In 1983, the National Commission on Excellence in Education issued a report titled " Nation at Risk," which bitterly attacked the effectiveness of American education. The message of the report was clear and sobering: "The educational foundations of our society are presently being eroded by a rising tide of mediocrity that threatens our very future."

As a result of this report, reforms were instituted in all 50 states, which stressed the teaching of the "three Rs" and the elimination of frivolous electives that waste valuable student and teacher time. In addition, high-school graduation requirements were raised in 40 states, and in 19 states students were required to pass minimum competency tests before they could receive their high-school diplomas. Forty-eight states also required new teachers to prove their competence by passing a standardized test.

Although the back-to-basics movement has worked, its success has been limited. Five years after "A Nation at Risk" appeared, a follow-up report was issued titled "American Education: Making It Work" (1988). According to the report, "the precipitous downward slide of previous decades has been arrested and we have begun the long climb back to reasonable standards." However, despite this progress, especially among minority groups, the report condemned the performance of American schools as unacceptably low. "Too many students do not graduate from our high schools, and too many of those who do graduate have been poorly educated. . . . Our students know little, and their command of essential skills is too slight," the report stated.

Particularly troublesome is student performance in math and science. According to a study done by the Nation's Report Card (1988), an assessment group that is part of the Educational Testing Service, the math performance of 17-year-olds is "dismal." The study found that although half the nation's 17-year-olds have no trouble with junior high-school math, they flounder when asked to solve multistep, high-school level problems or those involving algebra or geometry. Fewer than 1 in 15 students was able to answer these problems correctly. At fault might be the very back-to-basics movement that was supposed to rescue our educational system from failure in the early 1980s. While rote learning has helped improve the scores of the lowest-level students, it leaves others totally unprepared to analyze complex problems.

The Nation's Report Card also found American students' understanding of science "distressingly low." In a condemnation of the performance of America's schools in the area of science, the report found that most 17-year-olds did not have the skills to handle today's technologically based jobs and that only 7% could cope with college-level science courses.

In a study of students from 21 countries done at the end of the 1994–1995 academic year, American students scored below the international average and only outperformed students from Cyprus, Lithuania, and South Africa (Reuters News Service, February 28, 1998).

A survey by the National Education Goals Panel echoed these findings:

American first-graders may already be academically behind their counterparts in Japan and Taiwan. We know that not only is the content studied by our students less challenging than what students in high-achieving countries are exposed to, but American parents expect less of their children academically.

These attitudes may be changing, however. The past decade has seen dramatic changes in American schooling, including increased public support for school change. Eighty-five percent of those surveyed support standards-based reform (Immerwahr & Johnson. Seventy-one percent of Americans strongly support reforming the existing public school system (Public Attitudes Toward Public Schools, Phi Delta Kappa Gallup Poll, 1997).

In 1994 Congress passed the Education America Act, intended to set a new direction for our schools. The act set eight educational goals that the nation should achieve quickly:

- All children should enter school ready to learn.
- The U.S. high-school graduation rate should reach at least 90%.
- U.S. students should lead the world in mathematics and science performance.
- Every adult should be competent as a citizen and a worker.
- Schools should be disciplined environments free of drugs and violence.
- Students should demonstrate growing academic competencies in specific areas as they progress through school.
- Parents should become more deeply involved in their children's educational welfare.
- Teachers should be helped to expand and perfect their professional skills throughout their careers.

A particularly troubling aspect of American education is that student performance in math and science does not match that of students in other countries.

Some of these goals are no different from those presented by President George Bush at a 1989 education summit and should cause us to be skeptical about the level of progress than can be achieved.

The first goal, that children enter school ready to learn, requires that parents provide a physically and psychologically healthy environment. Many of these environments, however, put the child's educational development at risk. Half of all children are born to mothers who have used alcohol or drugs during pregnancy. Thirty-seven percent of the nation's 2-year-olds do not get all the necessary immunizations against major diseases, and half are not protected against polio (Project Vote Smart, 1994).

The second goal, high-school graduation rates of at least 90%, has been reached and now stands at about 91%. For African Americans the rate is 88.8%, and for the growing Hispanic population, it is 79% (*Statistical Abstract of the United States,* 2000).

Progress on the third goal of leading the world in math and science has been disappointing. Even though the United States spends more for education than most countries, only 4% of American 18-year-olds pass advanced placement tests in subjects like biology and math. Whereas most Japanese high-school students take advanced math courses, their U.S. counterparts are reviewing basic math. Only 7% of American students complete calculus in high school (American Federation of Teachers, 1994).

When it comes to the fourth goal of every adult being a competent citizen and worker, a congressionally funded literacy study found that "nearly half of all adult Americans read and write so poorly that it is difficult for them to hold a decent job." In 1988, the Department of Education was directed to assess the literacy rate of American adults. The result of this effort, the National Adult Literacy Survey

(NALS), represents the most recent and statistically accurate data on literacy in the United States. This research, which included interviews and testing of 26,000 adults, made several important contributions to our understanding and measurement of literacy. First, the researchers broadened the definition of literacy to include not just the ability to read, but skills in problem solving and reasoning. Their study separately measured real-life skills in prose literacy, document literacy, and quantitative literacy. The national survey showed that 21 to 23% of adults in the nation are at the lowest level of literacy. This means that more than 40 million adults cannot locate an intersection on a street map, cannot locate two pieces of information in a sports article, and cannot provide background information on a social security card application. These individuals face difficulty using the skills that are considered necessary for everyday functioning.

The fifth goal, of ensuring that schools are free of drugs and violence, also does not seem to be forthcoming. In 1993, 11% of teachers and 23% of students had been victims of violence in or near their schools. Eighty-two percent of school districts reported that student incidents involving assaults, fistfights, knifings, and shootings had increased during the past 5 years (*The Washington Post,* September 9, 1993).

Although progress has been slow, Goals 2000 has been a driving force in education reform. It has helped 36 states establish content standards in the core academic areas. In addition, most states have some kind of accountability measures for both students and teachers, and many are revising their teacher education and professional development efforts (*Goals 2000: Reforming Education to Improve Student Achievement—April 30, 1998*).

Innovation

A primary task of educational institutions is to transmit society's knowledge, and part of that knowledge consists of the means by which new knowledge is to be sought. Learning how to think independently and creatively is probably one of the most valuable tools the educational institution can transmit. This is especially true of the scientific fields in institutions of higher education. Until well into this century, scientific research was undertaken more as a hobby than a vocation. This was because science was not seen as a socially useful pursuit. Gregor Mendel (1822–1884), who discovered the principles of genetic inheritance by breeding peas, worked alone in the gardens of the Austrian monastery where he lived. Albert Einstein supported himself between 1905 and 1907 as a patent office employee while

Learning how to think independently and creatively is probably one of the most valuable tools that our educational system can transmit.

making several trailblazing discoveries in physics, the most widely known of which is the theory of relativity.

Today, science obviously is no longer the province of part-timers. Modern scientific research typically is undertaken by highly trained professionals, many of whom frequently work as teams; and the technology needed for exploration of this type has become so expensive that most research is possible only with extensive government or corporate funding. In 1998, the United States spent $247 billion on research and development funding (*Statistical Abstract of the United States: 2000,* 2001). In research and development, the areas of national defense, space exploration, and health research receive by far the greatest amount of support. In research alone, the leading three areas are life sciences (biological sciences and agriculture), engineering, and the physical sciences.

The achievements of government and industrial research and development notwithstanding, the importance of the contributions to science by higher academic institutions cannot be overestimated. First, there could be no scientific innovations—no breakthroughs—without the training provided by these schools. In the United States alone, 15,330 doctorates were awarded in 1996 in the health and physical sciences and mathematics, computer science, and engineering (*Statistical Abstract of the United States,* 2000). Second, the universities of the highest caliber continue to generate some of the most significant research in the biological and the physical sciences. (For a further discussion of innovation in education, see "Society and the Internet: Has the Internet Transformed Education?")

It is also worth noting that more women and minorities are earning PhDs than ever before. In 1997, 41% of all PhD recipients were women. Ten years earlier, women received about 33% of the degrees; and in 1967, the total was just 12%. For minorities, the numbers nearly doubled between 1987 and 1997—from 4.6% to about 9% of all doctorates granted (Irvine, 2000).

In addition to their manifest or intended functions, the schools in America have come to fulfill a number of functions that they were not originally designed to serve.

Child Care

One latent function of many public schools is to provide child care outside the nuclear family. This has become increasingly important since World War II, when women began to enter the labor force in large numbers. As of 1998, 72.2% of women with school-age children (ages 6 to 17) were in the labor force. In addition, women with children worked more hours each week on average in 1998 than they did in 1969 (*Monthly Labor Review*, 1999).

A related service of schools is to provide children with at least one nutritious meal per day. In 1975, the number of public school pupils in the United States participating in federally funded school lunch programs was 25,289,000, at a cost of $1.28 billion. By 1999, more than 27 million children each day got their lunch through the National School Lunch Program at a cost of $5.46 billion (U.S. Department of Agriculture, 2000).

Postponing Job Hunting

More and more young American adults are choosing to continue their education after graduating from high school. In 1998, 62.4% of male and 69.8% of female recent high-school graduates were enrolled in college (*Statistical Abstract of the United States:* 2000, 2001). Even though some of these individuals also work at part-time and even full-time jobs, an important latent function of the American educational system is to slow the entry of young adults into the labor market. This helps keep down unemployment, as well as competition for low-paying unskilled jobs.

Originally, two factors pointed to the possibility that college enrollments would not continue to increase. Because of low birthrates, the number of high-school graduates peaked at 3.2 million in 1977 and began a 14-year decline, with only 2.28 million graduating in 1991. That trend has been reversed and 2.50 million graduated in 1999 (*Statistical Abstract of the United States:* 2000, 2001).

Colleges and universities, anticipating enrollment problems, embarked on concerted efforts to ward off disaster. Through hard work and luck, they have succeeded. Total enrollment in 2- and 4-year colleges rose from 11.5 million in 1977 to nearly 15 million in 2000 (*Statistical Abstract of the United States,* 2000).

Colleges have also benefited from the fact that the U.S. economic base has shifted from manufacturing jobs to service jobs. This has caused the demand for professionals and technicians to grow. The salaries for those types of positions are considerably higher than salaries for manufacturing jobs.

Two other trends have also benefited colleges, the first being the increase in the number of women going to college, and the second the increase in older students. Since 1980, the majority of college students have been women. This trend is an outgrowth of changing attitudes about the status of women in our society and the breakdown in gender-role stereotypes (see Chapter 9). The second trend is that people 25 and older represent the most rapidly growing group of college students, accounting for 45% of all undergraduate and graduate students. Women are overrepresented in this group, as are part-time students. Many returning women students are also responding to changing gender-role expectations. (For a discussion of college graduates in other countries see "Global Sociology: College Graduates: A Worldwide Comparison.")

The Conflict Theory View

To the conflict theorist, society is an arena for conflict, not cooperation. In any society, certain groups come to dominate others, and social institutions become the instruments by which those in power control the less powerful. The conflict theorist thus sees the educational system as a means for maintaining the status quo, carrying out this task in a variety of ways. The educational system socializes students into

Society and the Internet

Has the Internet Transformed Education?

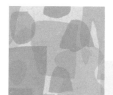

Imagine a party of time travelers from an earlier century, among them one group of surgeons and another of school-teachers, each group eager to see how much things have changed in their profession a hundred or more years into the future. Imagine the bewilderment of the surgeons finding themselves in the operating room of a modern hospital. Although they would know that an operation of some sort was being performed, and might even be able to guess the target organ, they would in almost all cases be unable to figure out what the surgeon was trying to accomplish or what was the purpose of the many strange devices the surgeon and the surgical staff were employing. The rituals of antisepsis and anesthesia, the beeping electronics, and even the bright lights, all so familiar to television audiences, would be utterly unfamiliar to them.

The time-traveling teachers would respond very differently to a modern elementary school classroom. They might be puzzled by a few strange objects. They might notice that some standard techniques had changed—and would likely disagree among themselves about whether the changes they saw were for the better or the worse—but they would fully see the point of most of what was being attempted and could quite easily take over the class.

The time-traveling teachers of our parable who saw nothing in the modern classroom that they did not recognize would have found many surprises had they simply gone home with one or two of the students. There they would have found that with an industriousness and eagerness that school can seldom generate, many of the students had become intensely involved in learning the rules and strategies of what appeared at first glance to be a process much more demanding than any homework assignment—interacting with the software on a modern home computer. The students might describe what they were doing as music file sharing, posting photos to the Internet, communicating in a chat room, or playing a video game, and what they were doing was fun.

While the technology itself might first catch the eye of our visitors, they would in time, being teachers, be struck by the level of intellectual effort that the children were putting into this activity and the level of learning that was taking place, a level that seemed far beyond that which had taken place just a few hours earlier in school. The most open and honest of our time-traveling teachers might well observe that never before had they seen so much being learned in such a confined space and in so short a time.

School would have parents—who honestly don't know how to interpret their children's obvious love affair with computers—believe that children love them and dislike homework because the first is easy and the second hard. In reality, the reverse is more often true. Any adult who thinks these activities are easy need only sit down and try to master them. Many involve complex information, as well as complex techniques that need to be mastered.

The Internet teaches children what computers are beginning to teach adults—that some forms of learning are fast-paced, immensely compelling,

---

values dictated by the powerful majority. Schools are seen as systems that stifle individualism and creativity in the name of maintaining order. To the conflict theorist, the function of school "is to produce the kind of people the system needs, to train people for the jobs the corporations require and to instill in them the proper attitudes and values necessary for the proper fulfillment of one's social role" (Szymanski & Goertzel, 1979).

Social Control

In the United States, schools have been assigned the function of developing personal control and social skills in children. Although the explicit, formally defined school curriculum emphasizes basic skills such as reading and writing, much of what is taught is oriented away from practical concerns. Many critics point out that much of the curriculum (other than in special professional training programs) has little direct, practical application to everyday life. This has led conflict theorists and others to conclude that the most important lessons learned in school are not those listed in the formal curriculum but, rather, involve a hidden curriculum. The **hidden curriculum** refers to *the social attitudes and values taught in school that prepare children to accept the requirements of adult life and to fit into the social, political, and economic statuses the society provides.*

Computer activities often generate in children an industriousness and an eagerness that traditional schooling methods can seldom produce.

and rewarding. The fact that it can be enormously demanding of one's time and require new ways of thinking remains a small price to pay (and is perhaps even an advantage) to be vaulted into the future. Not surprisingly, by comparison school strikes many young people as slow, boring, and frankly out of touch.

Source: *The Children's Machine,* by S. Papert, 1993, New York: Basic Books.

To succeed in school, a student must learn both the official (academic) curriculum and the hidden (social) curriculum. The hidden curriculum is often an outgrowth of the structure within which the student is asked to learn. Within the framework of mass education, it would be impossible to provide instruction on a one-to-one basis or even in very small groups. Consequently, students are usually grouped into relatively larger classes. Because this system obviously demands a great deal of social conformity by the children, those who divert attention and make it difficult for the teacher to proceed are punished. In many respects, the hidden curriculum is a lesson in being docile. For example, an article in *Today's Education,* the journal of the National Education As-

sociation, gives an experienced teacher's advice to new teachers:

> During the first week or two of teaching in an inner-city school, I concentrate on establishing simple routines, such as the procedure for walking downstairs. I line up the children, and . . . have them practice walking up and down the stairs. Each time the group is allowed to move only when quiet and orderly.

Social skills are highly valued in American society, and a mastery of them is widely accepted as an indication of a child's maturity. The school is a

Global Sociology

College Graduates: A Worldwide Comparison

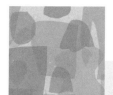

The percentage of young people with a college degree is higher in the United States than anywhere else in the world. One reason for this is that there is virtually no difference between the percentage of male and female degree holders. Japan would have more college degree holders if the same pattern held true there. In Japan however, even though substantially more young men have college degrees than men in the United States, Japanese women obtain degrees at only one-third the rate of Japanese men. Traditional gender-role expectations in Japan are limiting the number of women who pursue higher education.

Source: Organization for Economic Co-operation and Development, INES Project, International Indicators Project, *The Condition of Education 1997,* Indicator 23.

Table 12-1

Percentage of People With a College Degree— Ages 25 to 34

COUNTRY	MEN	WOMEN
Canada	18.0%	18.9%
France	11.9	11.3
Germany	12.7	11.0
Italy	7.7	8.1
Japan	34.2	11.5
United Kingdom	15.7	11.7
United States	23.4	23.5

miniature society, and many individuals fail in school because they are either unable or unwilling to learn or use the values, attitudes, and skills contained in the hidden curriculum. We do a great disservice to these students when we make them feel that they have failed in education, when they have, in fact, only failed to conform to the school's socialization standards.

Screening and Allocation: Tracking

From its beginning, the American school system, in principle, has been opposed to tracking, or the stratification of students by ability, social class, and various other categories. Educators saw in compulsory education a way to diminish the grip of inherited social stratification by providing the means for individuals to rise as high as their achieved skills would allow. In the words of Horace Mann, an influential American educator of the late nineteenth century, public education was to be "the great equalizer of the conditions of men."

Despite the principles on which it is based, the American educational system utilizes tracking, and even though teachers may oppose tracking, the general public favors it by a 68 to 30% margin (Associated Press poll, March 1994). Although tracking is not as formally structured or as completely irreversible in America as in most other industrial so-

cieties, it is influenced by many factors, including socioeconomic status, ethnicity, and place of residence. It is also consistently expressed in the differences between public and private schools as well as in the differences among public schools. (In New York City, for example, there are highly competitive math and science-oriented and arts-oriented high schools, neighborhood high schools, and vocational high schools.) Of course, tracking occurs in higher education in the selection of students by private colleges and universities, state colleges, and junior colleges.

Tracking begins with stratifying students into "fast," "average," and "slow" groups, from first grade through high school. It can be difficult for a student to break out of an assigned category because teachers come to expect a certain level of performance from an individual. The student, sensing this expectation, will often give the level of performance that is expected. In this way, tracking becomes a self-fulfilling prophecy. (See "Remaking the World: Jonathan Kozol on Unequal Schooling.")

In one study of this phenomenon, R. Rosenthal and L. Jacobson (1966) gave IQ tests to 650 lower-class elementary school pupils. Their teachers were told that the test would predict which of the students were the "bloomers" or "spurters." In other words, the tests would identify the superior students. This approach was, in fact, not the one employed. Twenty percent of the students were randomly

As societies have become more complex, a greater need has developed for the formal educational system to extend the socialization process that starts in the family.

selected to be designated as superior, even though there was no measured difference between them and the other 80% of the school population. The point of the study was to determine whether the teacher's expectations would have any effect on the "superior" students.

At the end of the first year, all the students were tested again. There was a significant difference in the gain in IQ scores between the "superior" group and the control group. This gain was most pronounced among students in the first and second grades. Yet the following year, when these students were promoted to another class and assigned to teachers who had not been told that they were "superior," they no longer made the sort of gains they had evidenced during the previous year. Nonetheless, the "superior" students in the upper grades continued to gain during their second year, showing that there had been long-term advantages from positive teacher expectations for them. Apparently, the younger students needed continuous input to benefit from the teachers' expectations, whereas the older students needed less.

Some authors (Bowles & Gintis, 1976) go so far as to assert that the idea of educational success being determined by merit or intelligence is an illusion fostered in the schools. Educational success, they claim, is much more likely to be determined by the possession of appropriate personality traits and by conformity to school norms. Those traits are acquired in the family and home environments of the students. The educational system, while claiming to reward those who demonstrate objective displays of merit, in fact is rewarding behavioral characteristics already possessed by individuals from specific social class backgrounds—that is, the middle class.

The Credentialized Society

Conflict theorists would also argue that we have become a credentialized society (Collins, 1979). A degree or certificate has become necessary to perform a vast variety of jobs. This credential may not necessarily cause the recipient to perform the job better. Even in professions such as medicine, engineering, and law, most knowledge is acquired by performing tasks on the job. However, credentials have become a rite of passage and a sign that a certain process of indoctrination and socialization has taken place. It is recognized that the individual has gone through a process of educational socialization that constitutes adequate preparation to hold the occupational status. Therefore, colleges and universities act as gatekeepers, allowing those who are willing to play by the rules to succeed, while barring those who may disrupt the social order.

At the same time, advanced degrees are undergoing constant change and becoming less specialized. A law degree from Harvard, Yale, or Columbia is less a measure of the training of a particular candidate than a basis on which leading corporations, major public agencies, and important law firms can recruit those who will maintain the status quo. The degree signifies that the candidate has forged links with the established networks and achieved a grade necessary to obtain a degree.

Remaking the World

Jonathan Kozol on Unequal Schooling

Jonathan Kozol.

Jonathan Kozol, author of *Savage Inequalities* and *Death at an Early Age*, contends that five decades after the Supreme Court ordered America to integrate its schools, the country has established two separate and unequal school systems, divided by class and race, with the gap between them growing larger each year.

Kozol studied 30 schools and found that some children were going to schools that were in Third World conditions, while only a short bus ride away there were beautiful, well-equipped school campuses.

"These children have done nothing wrong. They have committed no crime. They are too young to have offended us in any way. One searches for some way to understand why a society so rich would leave them in squalor for so long, and with so little public indignation."

Kozol notes that the gaps in school funding are so consistent from one metropolitan area to another—with rich schools in many areas spending more than twice per pupil as poor schools—that it suggests a deliberate pattern. He points out that students in suburban New Trier High School, outside Chicago, "will compete against each other and against graduates of other schools attended by rich children. They will not compete against the poor, few of whom will graduate from high school; fewer still will go to college; scarcely any will attend good colleges. There will be more spaces for children of New Trier as a consequence."

While their children enjoy richer schools, bolstered by higher property tax bases and federal deductions for property taxes and mortgage payments, suburban parents are lulled by politicians who claim that spending more money on inner-city schools will not make any difference.

Kozol traces the country's retreat from equal public education to two crucial Supreme Court decisions—the Rodriquez decision in 1974, a Texas case in which the court declared that disparities in public education were a constitutional issue, and the Milliken decision a year later, a Michigan case in which the court exempted white suburban schools from urban school desegregation programs.

Kozol points out that "there is a deep-seated reverence for fair play in the United States, but this is not the case in education, health care, or inheritance of wealth. In these elemental areas we want the game to be unfair and we have made it so."

For many poor students, dropping out of school appears a rational strategy, given how little they are getting and how depressing their schools are. If large numbers of dropouts ever did return to school, most urban school systems, already stretched to the breaking point, would have no room for them.

Kozol asserts that "conservatives are often the first to rise to protest an insult to the flag. But they soil the flag in telling us to fly it over ruined children's heads in ugly segregated schools. Flags in these schools hang motionless and gather dust, and are frequently no cleaner than the schools themselves. Children in a dirty school are asked to pledge a dirtied flag."

Sources: *Death at an Early Age,* by J. Kozol, 1989, New York: Plume Books; and *Savage Inequalities,* by J. Kozol, 1992, New York: Harper Perennial.

Colleges and universities are miniature societies more than centers of technical and scientific education. In these environments, students learn to operate within the established order and to accept traditional social hierarchies. In this sense, they provide the power structure with a constantly replenished army of defenders of the established order. At the same time, those who could disrupt the status quo are not permitted to enter positions of power and responsibility.

Issues in American Education

How well have American schools educated the population? The answer depends on the standards one applies. Americans take it for granted that everyone has a basic right to an education and that the state should provide free elementary and high-school classes. The United States pioneered this concept long before similar systems were introduced in Europe.

As we have attempted to provide formal education to everyone, we also have had to contend with a wide variety of problems stemming from the diverse population. In this section, we will examine some of the concerns in contemporary American education.

Unequal Access to Education

American minorities have sought equal access to public schools for two centuries. Tracing those efforts over the generations reveals a pattern of dissatisfaction with integrated as well as segregated schools.

African-American parents attributed the ineffective instruction at the schools attended by their children to one of two causes. If the schools were all-black, failure was attributed to the racially segregated character of those schools. If whites were attending the schools, African-American parents concluded, conditions would be better. This has been the dominant theme in the nineteenth and twentieth centuries.

Discontent also has occurred when African-American children have attended predominantly white schools. In those instances, the racially integrated character of the school was seen as a problem because whites students were thought to be favored by the teachers.

In the 1954 case *Brown v. Board of Education* of Topeka, Kansas, the Supreme Court ruled that school segregation was illegal. The court held that "In the field of public education, the doctrine of separate but equal has no place. Separate educational facilities are inherently unequal." Segregating African-American schoolchildren from white schoolchildren was a violation of the equal-protection clause of the Constitution. However, while the court's verdict banned **de jure segregation,** or *laws prohibiting one racial group from attending school with another,* it had little effect on **de facto segregation,** or *segregation resulting from residential patterns.* For example, minority groups often live in areas of a city where there are few, if any, whites. Consequently, when children attend neighborhood schools, they usually are taught in an environment that is racially segregated.

Ten years after the 1954 ruling, the federal government attempted to document the degree to which equality of education had been achieved. It financed a cross-sectional study of 645,000 children in grades 1, 3, 6, 9, and 12 attending some 4,000 different schools nationwide. The results, appearing in James S. Coleman's now-famous report Equality of Educational Opportunity (1966), supported unequivocally the conclusion that American education remains largely unequal in most parts of the country, including those where "Negroes form any significant proportion of the population." Coleman noted further that on all tests measuring pupils' skills in areas crucial to job performance and career advancement, not only did Native Americans, Mexican Americans, Puerto Ricans, and African Americans score significantly below whites, but also that the gaps widened in the higher grades. Now for a subtle but extremely important point: Although there are acknowledged wide inequalities of educational opportunity throughout the United States, the discrepancies between the skills of minorities and those of their white counterparts could not be accounted for in terms of how much money was spent on education per pupil, quality of school buildings, number of laboratories or libraries, or even class sizes. In spite of good intentions, a school presumably cannot usually outweigh the influence of the family backgrounds of its individual students and of its student population as a whole. The Coleman study thus provided evidence that schools per se do not play as important a role in student achievement as was once thought. It appears that the home environment, the quality of the neighborhood, and the types of friends and associates one has are much more influential in school achievement than is the quality of the school facilities or the skills of teachers. In effect, then, the areas that schools have least control over—social influence and development—are the most important in determining how well an individual will do in school.

The Coleman report pointed out that lower-class nonwhite students showed better school achievement when they went to school with middle-class whites. Racial segregation, therefore, hindered the educational attainments of nonwhites.

A direct outgrowth of the Coleman report pointing to the harm of de facto segregation was the busing of children from one neighborhood to another to achieve racial integration in the schools. The fundamental assumption underlying school busing was that it would bring about improved academic achievement among minority groups. Nationwide, many parents, black and white, responded

The facilities and education available at some elite, private high schools are better than those found at many colleges.

negatively to the idea that their school-age children must leave their neighborhoods. For all practical purposes, busing is no longer a major issue. Those schools that needed to desegregate have mostly done so, and other, less disruptive approaches to desegregation are attempted.

One factor that has increased the difficulty of integrating public schools is white flight, the continuing exodus of white Americans by the hundreds of thousands from the cities to the suburbs. White flight has been prompted partly by the migration of African Americans from the South to the inner cities of the North and Midwest since the 1960s, but some authorities strongly maintain that it also is related closely to school-desegregation efforts in the large cities.

For example, in a later view of desegregation attempts (1977), Coleman vastly revised the position in his 1966 report, stating that urban desegregation has in some instances had the self-defeating effect of emptying the cities of white pupils. Some authorities (Pettigrew & Green, 1975) took exception to the Coleman thesis, and others believe that what may appear to be flight is related more directly to the characteristic tendency of the American middle class to be upwardly mobile and constantly to seek a better lifestyle. Even though there is some evidence of a countertrend, in which middle-class whites are be-

ginning to regentrify inner cities, there seems to be no abating of this migration. Nor have most established communities relinquished the ideal of self-determination as embodied in the right to maintain neighborhood schools.

Since the Coleman report was issued, the integration of our schools has continued to be a crucial problem, especially in urban areas where the vast majority of minority group members live. According to a study by the National School Boards Association, there has been "no significant progress on the desegregation of black students in urban districts since the mid-1970s," and some areas have shown "severe increases in racial isolation" (Fiske, 1988). Schools in Atlanta and Detroit, for example, are more segregated now than they were a decade ago.

Only about 3% of the nation's white students attend central city schools. As a result, these schools have become almost irrelevant to the nation's white population. There are currently many more minority and far fewer white students in our public schools than in the past. Colleges and universities have had mixed results in increasing minority enrollment. Since the mid-1980s, the number of black undergraduates has not increased appreciably. Many qualified students do not apply because their families cannot afford tuition, even though complete aid packages are available. To overcome this problem, many schools have taken an aggressive recruitment stance, believing that once they find qualified candidates, they can persuade them to attend. At Harvard University, 9% of the freshman class of 2001 was African-American—the highest percentage in the school's history. Harvard has the luxury of not turning any student away because of financial need.

Still, thousands of qualified minority students never attend college, and many who do attend fail to graduate. The reasons for low graduation rates include financial problems, poor preparation, and the feeling of being unwelcome. Many cannot afford the loss of income that comes with being a full-time student. For many, family survival depends on the money they contribute. Others are victims of inadequate schools. They simply do not have the skills needed to complete college. Still others drop out because they feel out of place in the predominantly white world of higher education.

Students Who Speak English as a Second Language

The Department of Education reports that 6.3 million children aged 5 to 17, or 14%, speak a language other than English at home. Another 3.2 million elementary and secondary school students are classified as having limited English proficiency. The majority of these students are of Hispanic origin. Virtually all of these students are enrolled in remedial language programs.

Bilingual education has been used to teach academic subjects to immigrant children in their native languages (most often Spanish), while slowly and simultaneously adding English instruction. In theory, the children do not fall behind in other subjects while they are learning English. When they are fluent in English, they can then "transition" to English instruction in academic subjects at the grade level of their peers. Further, the theory goes, teaching immigrants in their native language values their family and community culture and reinforces their sense of self-worth, thus making their academic success more likely.

In contrast, bilingual education's critics tell a quite different story. In the early twentieth century, public schools assimilated immigrants to American culture and imparted workplace skills essential for upward mobility. Children were immersed in English instruction and, when forced to "sink or swim," they swam. Today, however, separatist (usually Hispanic) community leaders and their supporters, opposed to assimilation, want Spanish instruction to preserve native culture and traditions. This is especially dangerous because the proximity of Mexico and the possibility of returning home give

today's immigrants the option of "keeping a foot in both camps"—an option not available to previous immigrants who were forced to assimilate. Today's attempts to preserve immigrants' native languages and cultures will not only balkanize the American melting pot but hurt the children upon whom bilingual education is imposed because their failure to learn English well will leave them unprepared for the workplace. Bilingual education supporters may claim that it aims to teach English, but high dropout rates for immigrant children and low rates of transition to full English instruction prove that, even if educators' intentions are genuine, the program is a failure.

In general, the younger foreign-born children are when they enter school, the better they will perform. Immigrant children's chances for succeeding in school also improve in direct relation to their parents' educational attainment and income. The obstacles are especially great for children whose parents are illiterate and cannot help with homework.

Modern research findings on bilingual education are mixed. As with all educational research, it is so difficult to control for complex background factors that affect academic outcomes that no single study is ultimately satisfying.

English language proficiency is important today because the work world is more complex than during previous decades. Immigrants now come from a broad range of backgrounds and enter a more complex culture. In the past, less-educated people could find well-paid factory jobs. In today's more complex economy, the poorly educated and illiterate are often unemployed.

High-School Dropouts

Dropping out of high school has long been viewed as a serious educational and social problem. By leaving high school before graduation, dropouts risk serious educational deficiencies that severely limit their economic and social well-being. Over the past 90 years, the proportion of people in the adult population who have failed to finish high school has decreased substantially (see Figure 12-1). In 1910, the proportion of the adult population (ages 25 and older) that had completed at least 4 years of high school was 13.5%. It stood at 24.5% in 1940, at 55.2% in 1970, and at 82.8% in 1998. Among young people (ages 25 to 29) the drop is even more striking, with 88% having completed high school in 1999 (U.S. Bureau of the Census, Current Population Survey, October 1999).

Despite these long-term declines in dropout rates, interest in the dropout issue among educators and policy makers has increased substantially in recent

Figure 12-1 **Percentage of People 25 and Older with at Least a High-School Diploma, 1960–1998**

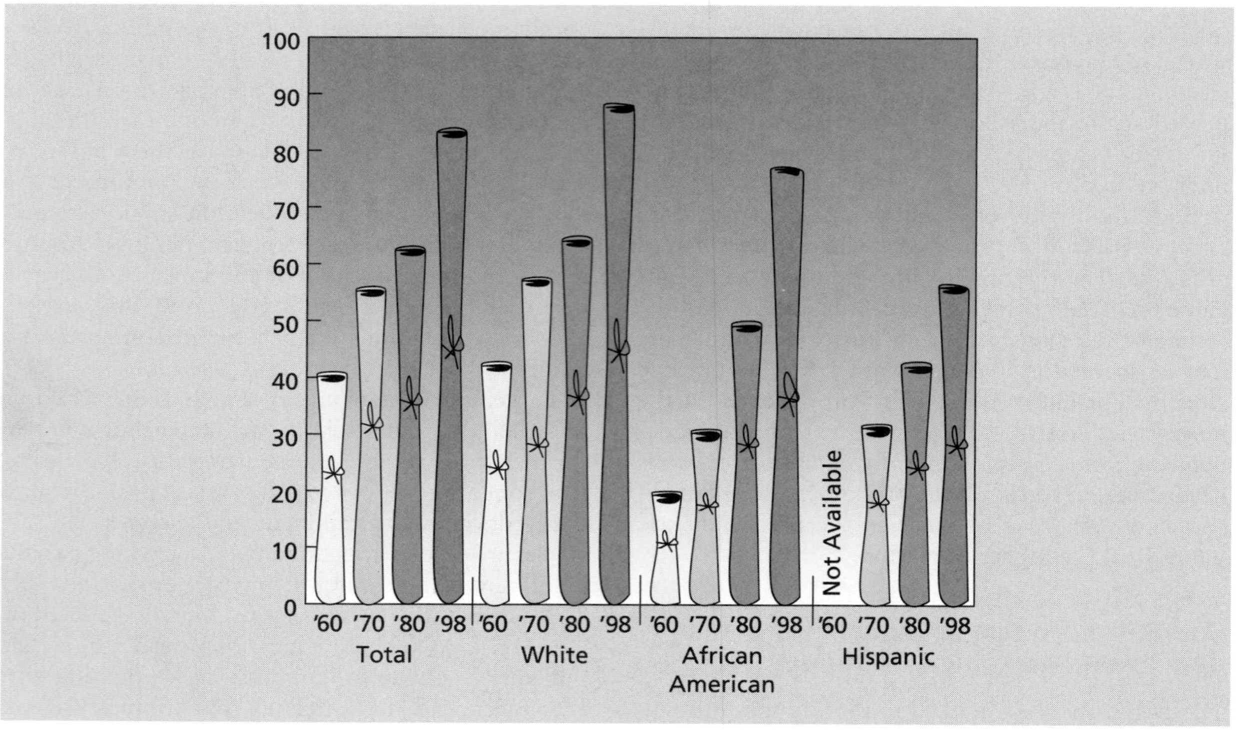

Source: U.S. Bureau of the Census, *Educational Attainment in the United States: March 1999*, pp. 520–528.

years. Legislators and education officials are devoting ever more time and resources to dealing with the issue.

If the long-term trend is that dropout rates are declining, why has the concern for this problem increased lately? First, although the long-term dropout trend has declined, the short-term trend has remained steady and even increased for some groups.

A second reason is that minority populations, who always have had higher dropout rates than whites, are increasing as a proportion of the public high-school population. Racial and ethnic minorities represent the majority of students enrolled in most large U.S. cities and more than 90% of all students in such cities as Newark, N.J.; Atlanta; and San Antonio (*Statistical Abstract of the United States,* 2000). (See "Our Diverse Society: From an Inner-City High School to the Ivy League" for an example of a student who overcame problems in a high school with high dropout rates.)

Dropout rates are higher for members of racial, ethnic, and language minorities; higher for males than females; and higher for people from the lower socioeconomic classes. Hispanics have the highest dropout rates. In 1996, 29.4% of Hispanics aged 18 to 24 had dropped out of high school, compared with 17.0% for African-Americans and 8.5% for

whites. Among Hispanics, Puerto Ricans have the highest dropout rates, followed by Mexican Americans and Cuban Americans. Dropout rates are also particularly high among Native Americans (*Statistical Abstract of the United States,* 1997).

Factors associated with dropping out include low educational and occupational attainment levels of parents, low family income, speaking a language other than English in the home, single-parent families, and poor academic achievement.

The influence of peers is also important, but it has not received much attention in previous research. Many dropouts have friends who are dropouts, but it is not clear to what extent and in what ways a student's friends and peers influence the decision to leave school (Rumberger, 1987).

Dropping out of high school affects not only those who leave school, but also society in general, due to the following reasons:

1. Dropouts pay less in taxes, because of their lower earnings (see Figure 12-2).

2. Dropouts increase the demand for social services including welfare, medical assistance, and unemployment compensation.

3. Dropouts are less likely to vote.

4. Dropouts have poorer health.

Figure 12-2 Median Income by Education Level, 1998

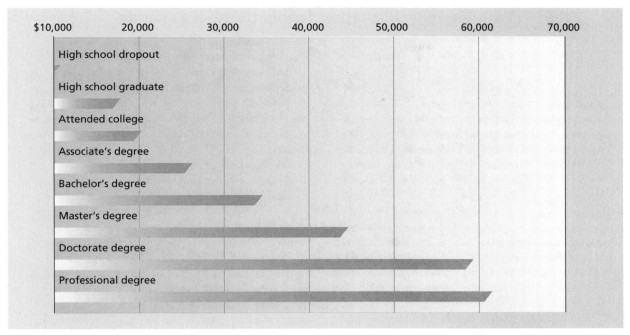

| | $10,000 | 20,000 | 30,000 | 40,000 | 50,000 | 60,000 | 70,000 |

High school dropout

High school graduate

Attended college

Associate's degree

Bachelor's degree

Master's degree

Doctorate degree

Professional degree

Source: U.S. Bureau of the Census, Income in 1998 by Educational Attainment for People 18 Years Old and Over, by Age, Sex, Race, and Hispanic Origin, March 1999.

Given these facts, it is small wonder that the U.S. Department of Education has focused an increasing amount of attention on how to improve high-school completion rates.

Violence in the Schools

Nothing undermines the effectiveness of our educational system more than unsafe schools. Throughout the country, students bring to school drugs, guns, knives, and other paraphernalia of destruction. Gone are books, pencils, and paper—the tools of learning that must be present for education to take place. Many urban school systems screen students with metal detectors when entering school grounds. Nearly 3 million thefts and violent crimes occur on or near school campuses every year (Bureau of Justice Statistics).

Nearly 20% of all students in grades 9 to 12 reported they had carried a weapon at least once during the previous month. Nationwide, 4% of students had missed 1 or more days of school during the 30 days preceding the survey because they had felt unsafe at school or when traveling to or from school (Centers for Disease Control and Prevention, 1999).

In two recent academic years there were 105 cases of school-associated violent deaths. In all, 85 of these cases were homicides and 20 were suicides. The events took place at 101 different schools, in 103 separate incidents.

Nothing undermines the effectiveness of our education system more than unsafe schools.

Our Diverse Society

From an Inner-City High School to the Ivy League

Cedric Jennings was a student at Frank W. Ballou High School, the most troubled and violent school in the blighted southeast corner of Washington, D.C. Cedric was "on a mission to get out of here, to be the one who makes it." He graduated second in his class with a 4.02 grade point average. Pride in such an accomplishment is acceptable behavior for outstanding students at high schools across the land, but at Ballou and other urban schools like it, something else is at work. Educators have even coined a phrase for it. They call it the crab/bucket syndrome: when one crab tries to climb from the bucket, the others pull it back down. The forces dragging students toward failure—especially those who have crawled farthest up the side—flow through every corner of the school. Inside the bucket, there is little chance of escape. Cedric's story is one of somebody who escaped.

With the school's dropout/transfer rate at nearly 50 percent, it's understandable that kids at Ballou act as though they're just passing through. Academics are a low priority, so few stop to read the names of the honor students as they jostle by the Wall of Honor bulletin board. . . .

The wall is a paltry play by administrators to boost the top students' self-esteem—a tired mantra here and at urban schools everywhere. The more practical effect is that the kids listed here become possible targets of violence, which is why some students slated for the Wall of Honor speed off to the principal's office to plead that their names not be listed, that they not be singled out. To replace their fear with confidence, Principal Washington has settled on a new tactic: bribery. Give straight-A students cash and maybe they'll get respect, too. Any student with perfect grades in any of the year's four marking periods receives a $100 check. For a year-long straight-A performance, that's $400. Real money. The catch? Winners have to personally receive their checks at awards assemblies.

At the start, the assemblies were a success. The gymnasium was full, and honor students seemed happy to attend, flushed out by the cash. But after a few such gatherings, the jeering started. It was thunderous. "Nerd!" "Geek!" "Egghead!" And the harshest, "Whitey!" Crew members, sensing a hearts-and-minds struggle, stomped on the bleachers and howled. No longer simply names on the Wall of Honor, the "whiteys" now had faces. The honor students were hazed for months afterward. With each assembly, fewer show up.

Cedric Jennings did not show up at the assembly to receive his cash prize.

Across a labyrinth of empty corridors, an angular, almond-eyed boy is holed up in a deserted chemistry classroom. Cedric Jennings often retreats here. It's his private sanctuary, the one place at Ballou where he feels completely safe, where he can get some peace.

Cedric had this dream of going to an Ivy League university. After fours years of hard work and total ostracism by his peers, Cedric is admitted to Brown University. His high school asks him to give a speech at graduation.

Gangs have also become a major problem in many schools. Fifteen percent of the students said their school had gangs, and 16% claimed that a student had attacked or threatened a teacher at the school (Bureau of Justice Statistics, 2000). In some schools, where gangs freely sell drugs to students within school buildings, principals vainly chain doors in an effort to keep dealers out and students in. In other schools, students are afraid to use the filthy bathrooms because gang members hang out there.

Nationwide, there have been many proposals to limit school violence. Education Secretary Richard Riley has proposed an initiative that would provide $175 million for school security. New York City has opted for stationing uniformed police with nightsticks in public schools. Los Angeles has approved the use of plainclothes officers. In Carter County, Kentucky, opaque backpacks have been banned, because they can be used to hide weapons.

These actions show the desperate conditions of inner-city schools and the measures some are willing to take to improve them. Many inner-city schools are made up primarily of minority students, have high dropout rates, high assault rates, and a staff that expects failure and violence.

Home Schooling

Home schooling is emerging as one of the most significant social trends in education. It is an alternative to traditional schooling in which parents assume the primary responsibility for the education of their children. This trend of what is in fact an

Cedric pushes forward gamely, keeping his voice loud and even:

"When I was asked to deliver the salutatory address I was afraid because it seemed an awesome responsibility. . . ."

"Blah, blah, blah," says a man leaning close to his teenage daughter a few rows from the stage, and the girl laughs. Cedric hears the whole thing.

He tries to remain composed, again rustling pages on the lectern. "In our high school years, we have learned great lessons that will serve us well in the future. Most importantly, we have learned to hold tight to our dreams, although there have been many obstacles on our way to a high school diploma.

"You see," he begins, his voice halting but seeming to sound conversational, "we have learned how to fight off Dreambusters. Yes, Dreambusters. Their favorite lines are 'you cannot' or 'you will not.' Many of us have been called crazy or even laughed at for having big dreams."

Some in the crowd look up, perplexed, as though they aren't sure what he just said.

"I will never forget being laughed at for saying that I wanted to go to the Ivy League. I've been told that I wouldn't make it."

He can hear students mumbling to each other in the middle seats just in front of him, and he imagines what they must be saying: Can it be that the nerd is giving some back? Giving it back to the whole class. Who does he think he is?

"When one of my peers found out that I was going to Brown, he told me I wouldn't last two years."

While they were laughing in the corner and trying to predict my future, I laughed back . . . 'THERE IS *NOTHING* ME AND MY GOD CAN'T HANDLE.'"

The crowd erupts. It's thunderous. A few people are standing. Even the badass kids have to laugh—the human punching bag is finally punching back.

Cedric's mom is up, screaming, "THAT'S MY BOY!" loud enough even Cedric can hear her.

Now the frog prince is flying, up on his toes, preaching, reaching down . . .

"For the race is not given to the swift nor to the strong," he signifies, "but to him who endureth until the end!" The mothers, the powerful church-women, start to cry out from all corners, and distinctions between the cool stage and the surrounding lowlands are gone as the room thumps as one big tent revival.

Cedric goes on to finish the speech, ending nicely with the Langston Hughes poem, but he knows—even as he recites it—that no one will remember much of the end. All they'll remember is that some boy preached today.

Cedric went on to attend Brown University and graduated recently.

Source: *A Hope in the Unseen* (pp. 3–4, 136–138), by R. Susskind, 1998, New York: Broadway Books.

old practice has occurred for a distinctly modern reason: a desire to wrest control from public education and reestablish the family as central to a child's learning.

Home schooling is almost always a matter of choice and not a necessity because of the unavailability of schools. Public schools are there, but home-schooling families choose not to use them. In the past decade there has been an explosive growth in this type of schooling. The numbers are still growing. The number of home schoolers nearly tripled in the 5 years from 1990–1991 to 1995–1996, when there were, according to the best possible estimate, about 700,000 home schoolers. If this growth rate has continued, there should be around 1.5 to 2 million children home-schooled today (about 3 to 4% of school-aged children nationwide) (Lines, 2000).

The contemporary home schooling trend began as a liberal, not a conservative, alternative to the public school. Some families in the late 1950s and early 1960s found schools were too rigid and conservative. They instead wanted to pursue a more liberal philosophy of education as advocated by educators such as John Holt, the author of *Why Children Fail*. Holt suggested the best learning took place when children were allowed to pursue their own interests without an established curriculum.

Conservative and religious families joined the home-schooling trend in the 1980s when they decided public schools were undermining their values. Some believed religious duty required them to teach their own children; others sought to integrate religion, learning, and family life. Joining the liberal and conservative wings of the home-schooling movement

Approximately 750,000 to 1 million students are being educated at home, up from 15,000 in the early 1980s.

are families who simply seek the highest quality education for their child, which they believe public and even private schools can no longer provide.

Home schooling is not a new idea or practice. For centuries, children have learned outside formal school settings. Compulsory schooling is relatively new. Not until the nineteenth century did state legislatures begin requiring local governments to build schools and parents to enroll their children in them. Even then, compulsory requirements extended to only a few months a year. Only recently have we begun to treat schooling as a full-time affair entrusted to professional teachers. Yet in such a short span of time, most of the nation has come to accept classroom schooling as the norm, and so the recent upsurge in home schooling has come to many as a surprise (Lines, 2000).

The typical home-schooling family is religious, conservative, white, middle-income, and better educated than the general population. Home schoolers are more likely to be part of a two-parent family, and the mother typically assumes the largest share of the teaching responsibility, although fathers almost always are involved also.

Most home-schooling children spend time at libraries, museums, or classes offered at a local public school. Normally, parents plan and implement the learning program. The Internet has provided an important resource for these parents and enables them to share information on books and learning opportunities.

How well do home schoolers do? The question is usually related to test scores. Many home-schooling parents reject this criterion, since their mission is to impart not simply skills but a particular set of values. That said, virtually all of the reported data show that home-schooled children score above average, sometimes well above average. Self-selection may be an important factor here because the parents are often highly motivated in bringing a quality education to their children (Lines, 2000).

The success or failure of home schooling depends on the success or failure of the family's interpersonal relationships. Home schooling is a complex issue and represents a tremendous commitment on the part of the parents. In most cases the father is the sole earner in the family and the mother spends her time instructing the children.

More research on home schooling is necessary. Up to now research has been limited to case studies of families or self-reports from participants in home schooling. We need a more accurate and thorough assessment of this growing trend in education.

Standardized Testing

In American schools, the standardized test is the most frequently used means of evaluating students' aptitudes and abilities. Every year, more than 100 million standardized tests are administered, ranking the mental talents of students from nursery to graduate school.

Children encounter standardized tests almost from the first day they go to school. Usually, their first experience with testing is an intelligence test. These are given to more than 2 million youngsters each year. Students are also required to take a number of achievement tests, beginning in elementary school. High school and college seniors take college admissions tests that decide whether they will be accepted at universities and graduate schools.

Much criticism has been leveled at standardized tests. The testing services say the tests merely try to chart, scientifically and objectively, different levels of mental achievement and aptitude. The critics assert that the tests are invalid academically and biased against minorities.

The Educational Testing Service's (ETS) Scholastic Aptitude Test (SAT) is the best-known college admissions test and is required by about 1,200 U.S. colleges and universities. Another 2,800 American colleges require or recommend the American College Test (ACT). Students wishing to go to graduate school are required to take other exams, which measure the ability and skills used in the fields that they wish to enter.

Figure 12-3 Average SAT Scores, 1970–2000

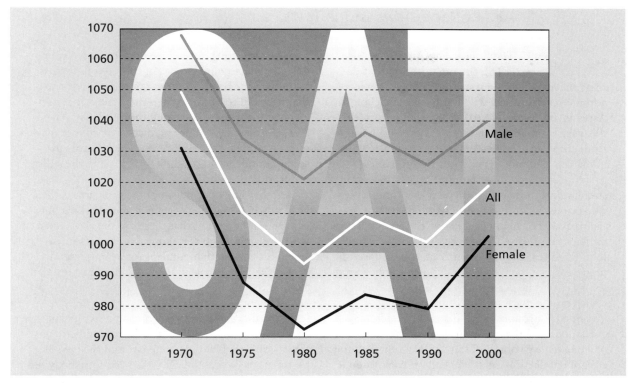

Source: College Entrance Examination Board, 2000.

The ETS professes to be meticulous in its test construction. It hires college students, teachers, and professors to assist its staff in writing questions. Each of the approximately 3,000 questions that are created each year are reviewed by about 15 people for style, content, or racial bias (see "Controversies in Sociology: Are College Admissions Tests Fair?")

The criticism of standardized tests, however, continues to grow. Many assert that all standardized tests are biased against minorities. The average African-American or Hispanic youngster encounters references and vocabulary on a test that are likely to be more familiar to white, middle-class students. Many others oppose the secrecy surrounding the test companies. Groups have pushed for truth in testing, meaning that the test makers must divulge all exam questions and answers shortly after the tests are given. This would enable people to evaluate the tests more closely for cultural bias and possible scoring errors. The testing industry opposes such measures, which would force it to create totally new tests for each administration without the possibility of reusing valid and reliable questions.

No one would contend that standardized tests are perfect measuring instruments. At best, they can provide an objective measure to be used in conjunction with teachers' grades and opinions. At worst, they may discriminate against minorities or not validly measure potential ability. Nonetheless, college admissions officers insist that results from standardized college admissions tests give them a significant tool for evaluating students from a variety of backgrounds and many different parts of the country.

More students are taking the SAT than ever before, a fact that might be expected to result in lower scores. Instead, 2000 scores are now above the 1980 level, reversing a trend that saw scores drop 90 points between 1963 and 1980. In part, the stabilized scores can be attributed to the more rigorous standards of the back-to-basics movement—a movement that has brought about modest gains among poor performers (see Figure 12-3).

Gender Bias in the Classroom

Who is at greater risk of failing in our schools these days—girls or boys? Some researchers (Sadker & Sadker, 1994) believe that sitting in the same classrooms, reading the same textbooks, listening to the same teachers, boys and girls receive very different educations. A body of research, including most prominently a 1992 study sponsored by the American Association of University Women, "How Schools Shortchange Girls," has held that girls face much deeper difficulties in school, while boys get the

Controversies in Sociology

Are College Admissions Tests Fair?

The Educational Testing Service (ETS), the organization that designs the standardized tests that colleges use for admissions decisions, will administer its Scholastic Assessment Test (SAT) to some 9 million students this year. Admissions testing has never been free from criticism. Today, in fact, its critics are more numerous and more vociferous than ever.

The oldest and most familiar accusation against standardized tests is that they are discriminatory. The prime evidence for this charge is the test results themselves. For many years now, the median score of blacks on the SAT has been 200 points below that of whites. Less dramatic, but no less upsetting to groups like the Center for Women's Policy Studies, has been the persistent 35-point gender gap in scores on the math section of the SAT.

The SAT produces such disparate results, say critics, because the questions favor certain kinds of students over others. Thus, fully comprehending a reading selection might depend on background knowledge naturally available to an upper-middle-class white student (by virtue, say, of foreign travel or exposure to the performing arts) but just as naturally unavailable to a lower-class black student from the ghetto.

At the same time, women are said to be put at a disadvantage by the multiple-choice format itself. Singled out for blame are math questions that emphasize abstract reasoning and verbal exercises

based on selecting antonyms, both of which supposedly favor masculine modes of thought. In fact, so biased are the tests, according to their opponents, that they fail to perform even the limited function claimed for them: forecasting future grades. The SAT, consistently "under predicts" the college marks of both women and minorities, which hardly inspires confidence in its ability to measure the skills it purports to identify.

Another line of criticism of the tests grants their accuracy in measuring certain academic skills but challenges the notion that these are the skills most worth having. High test scores, opponents insist, reveal little more than a talent for taking tests. It is also suggested that no mere standardized test can capture the qualities that translate into real-world achievement.

The critics have already won some significant concessions. Faced with both adverse publicity and threats of legal action by activists and the U.S. Department of Education, ETS has tried to remedy differences in group performance. On the Preliminary Scholastic Assessment Test (PSAT), which is used for choosing National Merit Scholars, a new method of scoring was recently introduced in the hope that more women might garner the prestigious award. The old formula, which assigned equal weight to the math and verbal sections of the test, was replaced by an index in which the verbal score, usually the higher one for female test takers, was doubled.

lion's share of educational resources and teachers' attention. Over the course of years, the uneven distribution of teacher time, energy, attention, and talent, takes its toll on girls.

Girls are the majority of our nation's schoolchildren. However, each time a girl opens a book and reads a womanless history, she learns she is worthless. Each time the teacher passes over a girl to elicit the ideas and opinions of boys, that girl is conditioned to be silent and to defer. As teachers use their expertise to question, praise, probe, clarify, and correct boys, they help those male students sharpen ideas, refine their thinking, gain their voice, and achieve more. When female students are offered the leftovers of teacher time and attention, morsels of amorphous feedback, they achieve less.

Sadker and Sadker point out the following:

- In high school, girls score lower on the SAT and ACT tests, which are crucial for college admis-

sion. The greatest gender gap is in the crucial areas of science and math.

- Boys are much more likely to be awarded state and national college scholarships.

- Women score lower on all sections of the Graduate Record Exam, which is necessary to enter many postgraduate programs. Women also trail on most tests needed to enter business school, law school, and medical school.

Christina Hoff Sommers (1994; 2000) believes these examples of bias are exaggerated. She has pointed out that 56% of college students are female. In 1971, women received 43% of bachelor's degrees, 40% of master's degrees, and 14% of doctorates. Today, women receive 53.9% of BAs, 53.6% of MAs, and 41% of doctoral degrees.

In addition, 15 years ago, roughly the same number of boys and girls were taking advanced placement

More widely publicized was the massive "recentering" of SAT scores that went into effect with the 1996 results. Though the declared aim was to create a better distribution of scores around the test's numerical midpoint, the practical effect was a windfall for students in almost every range.

But since neither "recentering" nor any other such device has succeeded in eliminating disparities in scores, opponents of tests have had to look elsewhere. The law school at the University of California at Berkeley, for instance, has introduced a selection system that will consider a "coefficient of social disadvantage" in ranking applicants.

Some schools go farther, hoping simply to do away with standardized tests altogether. There are, they insist, other, less problematic indicators of student merit. High-school grades are a starting point, but no less important are essays, interviews, and work portfolios that offer a window into personal traits no standardized test can reveal.

Defenders of the college admissions tests point out that the SAT is hardly the meaningless academic snapshot described by critics. Results from these tests have been shown to correspond with those on a whole range of other measures, including IQ tests. This led the National Academy of Science to conclude that standardized tests display no evidence whatsoever of cultural bias.

Nor do the tests fail to predict how minority students will ultimately perform in the classroom, defenders note. If, indeed, the purported bias in the tests were real, such students would earn better grades in college than what is suggested by their SAT scores; but that is not the case.

Foes of testing are correct when they point out that women end up doing better in college than their scores would indicate. But defenders of the test say the "under prediction" is very slight—a tenth of a grade point on the four-point scale—and only applies to less demanding schools. For more selective institutions, the SAT predicts the grades of both sexes equally accurately.

As for the claim that test scores depend heavily on income, defenders of the tests claim the facts again tell us otherwise. Though one can always point to exceptions, students who are not of the same race but whose families earn alike tend, on average, to perform very differently.

What about relying less on tests and more on other measuring rods like high-school grades? Unfortunately, as everyone knows, high schools across the country vary considerably, not only in their resources but in the demands they make of students. It was precisely to address this problem that a single nationwide test was introduced in the first place.

Source: "The War Against Testing," by D. W. Murray, September 1998, *Commentary*.

exams in U.S. high schools—a strong indicator of who is enrolling in the most intellectually challenging courses. By 1998, boys had fallen well behind. Boys today are less apt than girls to complete high school, to attend college, and to stay out of jail. They do less homework and come to school worse prepared. They read and write less well. Save for sports, they are less frequent participants in extracurricular activities of every sort. In those areas—like scores on math and science tests—where boys continue to hold an edge, the gap is fast narrowing. In those areas where girls are out in front, it seems to be widening.

Others, such as Carol Gilligan, have shifted the debate somewhat. She has pointed out that because girls possess a different moral sensibility—being more caring, empathetic, nurturing, sensitive, and community-minded than boys—they are devalued by society and discouraged from entering its most important institutions and honored vocations.

In recent years, even as schools have tried to put in place measures to end the last vestiges of discrimination against students, the debate over whether bias exists in the classroom continues.

The Gifted

The very term *gifted* is emotionally loaded. The word may evoke feelings that range from admiration to resentment and hostility. Throughout history, people have displayed a marked ambivalence toward the gifted. It was not unusual to view giftedness as either divinely or diabolically inspired. Genius was often seen as one aspect of insanity. Aristotle's observation "There was never a great genius without a tincture of madness" continues to be believed as common folklore.

People also tend to believe that intellectualism and practicality are incompatible. That belief is expressed in such sayings as "He (or she) is too smart for his (her) own good" or "It's not smart to be too smart." High intelligence is often assumed to be incompatible with happiness.

There is little agreement on what constitutes giftedness. The most common measure is performance on a standardized test. All those who score above a certain level are defined as gifted, though there are serious problems when this criterion alone is used. Arbitrary approaches to measuring giftedness tend to ignore the likelihood that active intervention could increase the number of candidates among females, minorities, and the disabled, groups that are often underrepresented among the gifted.

Ellen Winner (1996) has proposed that gifted children have three atypical traits. These include (1) *precociousness*—gifted children begin early to master some domain; (2) *nonconformity,* an insistence on doing things according to their own specific rules; and (3) *a rage to master,* or a desire to know everything there is to know about a subject.

Females tend to be underrepresented among the gifted because popular culture holds that high intelligence is incompatible with femininity; thus some girls quickly learn to deny, disguise, or repress their abilities. Minorities are hindered because commonly used assessment tools discriminate against ethnic groups whose members have had different cultural experiences or use English as a second language. The intellectual ability of disabled youngsters is often overlooked. Their physical handicaps may mask or divert attention from their mental potential, particularly when communication is impaired, as this is a key factor in assessment procedures.

Teachers often confuse intelligence with unrelated school behaviors. Children who are neat, clean, and well mannered, have good handwriting, or manifest other desirable but irrelevant classroom traits may often be thought to be very bright.

Teachers often associate giftedness with children who come from prominent families, have traveled widely, and have had extensive cultural advantages. Teachers are likely to discount high intelligence when it might be present in combination with poor grammar, truancy, aggressiveness, or learning disabilities.

The first attempt to deal with the gifted in public education took place in the St. Louis schools in 1868. The program involved a system of flexible promotions enabling high-achieving students not to remain in any grade for a fixed amount of time. By the early 1900s, special schools for the gifted began to appear.

There has never been a consistent, cohesive national policy or consensus on how to educate the gifted. Those special programs that have been instituted have reached only a small fraction of those who conceivably could benefit from them. A serious problem with the education of the gifted arises from

philosophical considerations. Many teachers are reluctant to single out the gifted for special treatment, as they feel that the children are already naturally privileged. Sometimes, attention given to gifted children is seen as antidemocratic.

No matter how inadequate it may seem, the effort to provide for the educational needs of learning-disabled children has far exceeded that expended for the gifted. Similarly, the time and money spent on research into educating the slower children far outstrip that set aside for research on materials, methodology for teaching, and so on for the gifted.

When schools do have enrichment programs, they are rarely monitored for effectiveness. Enrichment programs are often provided by teachers totally untrained in dealing with the gifted, for it is assumed that anyone qualified to teach is capable of teaching the gifted. Yet most basic teacher-certification programs do not require even one hour's exposure to information on the theory, identification, or methodology of teaching such children. Most administrators do not have the theoretical background or practical experience necessary to establish and promote successful programs for the gifted.

There is some evidence that the nation's population of gifted children—and possibly, prodigies—is growing. Researchers who test large numbers of children have detected a startling proportion in the 170- to 180-IQ range.

While psychologists agree that early exceptional ability should be nurtured to thrive, they do not necessarily think that the current movement to produce superbabies by force-feeding a diet of mathematics and vocabulary to infants is a good idea. Pediatricians have begun seeing children with backlash symptoms—headaches, stomachaches, hair-tearing, anxiety, depression—as a result of this pressure to perform.

History has shown that being an authentic child prodigy creates problems enough of its own. The fine line between nurturing genius and trying to force a bright but not brilliant child to be something he or she is not is clearly one that must be walked with care.

It appears that there are more than 2.5 million schoolchildren in the United States who can be described as gifted, or about 3% of the school population. Giftedness is essentially potential. Whether these children will achieve their potential intellectual growth will depend on many factors, not the least of which is the level of educational instruction they receive. We must question why we continue to show such ambivalence toward the gifted and why we are willing to tolerate incompetence and waste in regard to such a valuable resource (Baskin & Harris, 1980).

SUMMARY

Why should people get an education? The most common answer Americans give is to get a better job. Yet most jobs in our society do not require 12 or more years of training. Something else must be taking place during those years of schooling. Functionalists suggest that that something else consists of activities that are functional for the society as a whole.

One of the manifest functions of education is socialization. In nonindustrial societies, no real distinction is made between education and the socialization that occurs within the family. Another function of education is cultural transmission, in which major portions of society's knowledge are passed from one generation to the next. A third function of schools is to equip children with the academic skills needed to function as adults.

Learning how to think independently and creatively is probably one of the most valuable tools the educational institution can transmit. This is the function called innovation, and today it is a systematic enterprise generally carried out with government funding in institutions of higher education by highly trained specialists.

In addition to its manifest functions, schooling in America has developed a number of unintended consequences as well. One latent function of many public schools is to provide child care outside of the nuclear family. This has become increasingly important in recent years with the growing number of women in the labor force and the dramatic increase in single-parent families. A related service is to provide students with at least one nutritious meal a day.

Conflict theorists view society as an arena of conflict, in which certain groups dominate others, and social institutions become the instruments by which those in power can control the less powerful. Thus, the conflict theorist sees education as a means for maintaining the status quo by producing the kinds of people the system needs. This is accomplished through teaching the hidden curriculum—attitudes and values that prepare children to accept the requirements of adult life and to fit into the social, political, and economic statuses the society provides.

While the overall high-school graduation rate has been increasing, the dropout rate for minorities has remained high. Factors associated with dropping out include low educational and occupational attainment levels of parents, low family income, speaking a language other than English in the home, single-parent families, and poor academic achievement.

Violence in schools is a growing problem as a result of increased gang presence, more drugs, and more weapons. Many urban schools operate under a siege mentality, with doors chained shut and students afraid to be in the wrong place at the wrong time, inside of school or out.

In American schools, the standardized test is the most frequently used means of evaluating student aptitude, ability, and performance. Critics assert that these tests are academically invalid and culturally biased against minorities and the lower class.

 INTERNET RESOURCES

Go to http://web-enhanced.thomsonlearning.com which features Web links relevant to this chapter to expand and enhance the learning experience. This site will be monitored and updated periodically to ensure the links are correct and live.

At Virtual Society: The Wadsworth Sociology Resource Center, http://sociology.wadsworth.com, you will find a Career Center, Sociology in the News, Research Online, InfoTrac College Edition, Virtual Tours in Sociology, and a variety of book-specific student and instructor resources.

CHAPTER TWELVE STUDY GUIDE

KEY CONCEPTS

Match each concept or person with its definition, illustration, or explanation below.

a. De facto segregation
b. White flight
c. ESL
d. Hidden curriculum

e. Cultural transmission
f. The credentialized society
g. De jure segregation

h. Tracking
i. Crab/bucket syndrome
j. *Brown v. Board Education/Topeka*

_____ 1. The process in which major portions of a society's knowledge are passed from one generation to the next.

_____ 2. The social attitudes and values taught in school that prepare children to accept the requirements of adult life and to "fit into" the social, economic, and political statuses the society provides.

_____ 3. The stratification of students by ability, social class, and various other categories.

_____ 4. A form of racial separateness based on laws prohibiting interracial contact.

_____ 5. "In the field of education, the doctrine of separate but equal has no place."

_____ 6. A form of racial separateness resulting from residential housing patterns.

_____ 7. The exodus of large numbers of white Americans from the central cities to the suburbs.

_____ 8. English as a Second Language; the most common way schools respond to immigrant students.

_____ 9. The increasing trend in the United States for more and more jobs to require a degree regardless of whether possession of a degree increases job performance.

_____ 10. Used to describe the process whereby nonachieving students try to drag down toward failure those students who are attempting to climb out and escape their environment.

KEY THINKERS/RESEARCHERS

Match the thinkers/researchers with their main ideas or contributions.

a. R. Rosenthal and L. Jacobson
b. James Coleman
c. Jonathan Kozol
d. National Comm./Excellence in Ed.
e. Samuel Bowles and Herbert Gintis
f. 1983 National Commission on Education

g. Carol Gilligan
h. Christina Hoff Sommers
i. Herbert Spencer
j. Lester Frank Ward
k. Cedric Jennings

_____ 1. Their report, *A Nation at Risk,* instigated the current educational reform movement.

_____ 2. Conducted a famous study that demonstrated the powerful effect of the self-fulfilling prophecy in school.

_____ 3. They argue that educational success is more likely to be determined by possession of the appropriate personality traits than by merit or intelligence.

_____ 4. Author of a 1966 survey of 645,000 children that demonstrated substantial class and race differences in educational achievement and opportunity.

_____ 5. A report bitterly attacked the effectiveness of American education.

_____ **6.** Contends that five decades after the Supreme Court ordered integration, the United States continues to have two separate and unequal systems divided by class and race.

_____ **7.** Argues that girls posses a different moral sensibility than boys and are discouraged from entering its most important institutions and honored vocations.

_____ **8.** A young man who managed to beat the crab/bucket syndrome pervasive in his high school.

_____ **9.** Believes that the examples of a bias against girls are exaggerated and that boys actually fare less well in public schools than do girls.

_____ **10.** Pioneer in sociology who argued that the purpose of education was to prepare us for complete living, although little could be done to help those at the bottom of the social class ladder.

_____ **11.** First president of the American Sociological Association who believed that the purpose of education was to equalize society.

CENTRAL IDEA COMPLETIONS

Following the instructions, fill in the appropriate concepts and descriptions for each of the questions posed in the following section.

1. Using your college or university as an example, describe the manifest and latent functions of education. How might these functions vary across large and small as well as public and private institutions?

   a. Manifest: _____

   b. Latent: _____

   c. Institutional size: _____

   d. Public versus private: _____

2. What are the consequences of job postponement for the individual and the society?

   a. Individual: _____

   b. Society: _____

3. What is standardized testing? What are the consequences of standardized testing?

   a. Definition: _____

   b. Consequences: _____

4. What are the basic features of home schooling and why has its popularity been increasing in recent years?

   a. Home schooling: _____

   b. Popularity factors: _____

5. What are the major factors contributing to the dropout rate for high-school students?

   _____

   _____

   _____

6. To what extent do American minorities continue to face unequal access to education?

   _____

   _____

   _____

7. What is the major factor accounting for the differences in the number of college graduates in the United States and Japan? How does the number of persons attaining college degrees in the United States compare to the rest of the world?

a. U.S./Japan differences explained: _____

b. U.S. and world rate comparisons: _____

8. What do Bowles and Gintis mean by their assertion that the idea of educational success being determined by merit or intelligence is an illusion fostered in the schools?

_____

_____

9. What factors seem to account for violence in the schools?

_____

_____

10. How would a conflict theorist evaluate education in American public high schools?

_____

_____

11. Based on your reading of the text, how would you describe the fundamental issues involved in college admisstion tests? Overall, would you conclude that they are basically fair or basically unfair? How so?

_____

_____

12. In terms of violence within the school, approximately how many students in grades 9 through 12 report carrying a weapon at least once during the previous month? _____
Clearly a manifest outcome lies in the potential harm to students. What do you see as the latent dysfunctions for education of students carrying weapons to school?

_____

_____

13. How well do homeschoolers do?

_____

What factors affect the performance of these students? _____

_____

14. What evidence does Christina Hoff Sommers provide that challenges the notion that there is a bias in favor of boys and against girls in public schools? _____

_____

_____

15. Briefly compare and contrast Herbert Spencer and Lester Frank Ward on the purpose of education.

a. Spencer: _____

b. Ward: _____

16. Following your reading of the major goals of the 1994 Education America Act, how would you define *literacy?*

_____

_____

CRITICAL THOUGHT EXERCISES

1. How has the Internet transformed education? Log on to the Internet and conduct a search for sites relevant to the program(s) you are pursuing at your college or university. Select three sites and describe how they might be of special relevance to the courses you are taking this semester. Be specific about what aspects of the site make it useful for your studies. To what extent are your instructors integrating Internet-based learning experiences and materials into the courses you are presently taking? Based on your minisurvey of the Net, what advice could you give them to better integrate materials of the type you discovered into their college and university courses?

2. American schools have been characterized as "being under siege" regarding the levels of violence now common in many places. What factors contribute to violence in the schools? To what extent was violence commonplace in the high school you attended prior to attending this college or university? Go to the documents or educational research section of your library and find a program devoted to reducing high-school violence. What are its basic features? How would a sociologist evaluate its effectiveness?

3. Examine your college or university in terms of its socialization function. What activities are manifestly designed to socialize the student population? Provide examples of the latent socialization functions occurring at your institution. To what extent are these socialization activities functionally linked to the purposes and goals of your institution (refer to your institutional catalog for your institution's mission and goals statement).

4. Institutions of higher education are concerned about dropout rates, although they are likely to use the term "student persistence." Have any of your personal acquaintances "not persisted' or dropped out from your institution? To what extent are the reasons likely to be similar to or different from the reasons for students dropping out of high school? Interview an official in the Office of Student Services at your institution to learn about how student persistence is viewed and what programs, if any, are in place to reduce the dropout rates.

5. Innovation is a primary task of education as it involves enabling persons to learn how to think independently. Reflecting back on your high-school experiences, to what extent did you have the opportunity to think independently? How have things changed now that you are involved in higher education? What experiences have you had in innovative, independent, and creative thinking? What might your institution do to encourage greater opportunities for innovative thought? What is the role of student motivation in creative, independent thought?

ANSWERS

KEY CONCEPTS

1. e	3. h	5. j	7. b	9. f	10. i
2. d	4. g	6. a	8. c		

KEY THINKERS/RESEARCHERS

1. d.	3. e	5. f	7. g	9. h.	11. j.
2. a	4. b	6. c	8. k.	10. i.	

LEARNING OBJECTIVES

After studying this chapter, you should be able to do the following:

- Distinguish between authority and coercion.
- Understand the basic functions of the state.
- Know the basic features of capitalism.
- Distinguish between capitalism, socialism, and democratic socialism.
- Describe the basic features of political democracy.
- Contrast the functionalist and conflict theory views of the state.
- Describe the major features of the American political system.

The founders of the United States distrusted a strong unified government. The American Constitution, the oldest in the world, established a divided form of government. There was to be a president, two houses of Congress, and a federal high court. These actions represented a deliberate decision to create a weak political system. There were to be varying terms of office. The president was to be chosen every 4 years. Two senators from each state were to be chosen for 6-year terms, with one-third of the seats open every 2 years. The House of Representatives was to be filled every 2 years, with the number allotted to each state roughly proportional to its share of the national population (Lipset, 1996).

Thomas Jefferson thought that even this arrangement had to be reexamined frequently. He suggested that every two decades or so, a new generation should be required collectively to define its own values and redefine those of its forebears. Jefferson believed that a political system had to be taken out periodically, inspected, examined in the harsh daylight, and changed before it was bequeathed to a new generation. If it did not pass muster, then some other political system had to be found that could better guarantee life, liberty, and the pursuit of happiness. A new generation should not be burdened by the traditions of the past or the comfort of well-worn customs (Hart, 1993).

The founders of the United States realized that, in most societies, what laws are passed, or not passed, depend to a large extent on which categories of people have the power. The powerful in a society work hard to pass laws to their liking. They were trying to provide a prescription for how power was to be used and how it was to be passed from one generation to the next. It sounds like a fairly radical and idealistic view of how a country should be governed. However, the United States was born out of radical and idealistic conceptions of politics.

This chapter examines the political institution. **Politics** is *the process by which power is distributed and decisions are made.* This chapter will clarify what is unique about our two-party system and where the American political system fits into the whole spectrum of political institutions.

Politics, Power, and Authority

There are more than 800 candidates for the U.S. House of Representatives every other year. Americans also select senators, governors, and a host of other officials on a regular basis. Candidates ring our doorbells, shake our hands, stuff our mailboxes, and exhort us through our television sets. They make promises they often cannot keep. This is politics, American style. Small wonder that it has been said that politics, like baseball, is the great American pastime. Running for president of the United States is a political activity; so is enacting legislation; so is taxing property owners to subsidize the digging of sewers; so is going to war. The study of the political process, then, is the study of power.

Power

Max Weber (1958a) referred to **power** as *the ability to carry out one person's or group's will, even in the presence of resistance or opposition from others.* In this sense, power is the capability of making others comply with one's decisions, often exacting compliance through the threat or actual use of sanctions, penalties, or force.

In some relationships, the division of power is spelled out clearly and defined formally. Employers have specific powers over employees, army officers over enlisted personnel, ship captains over their crews, professors over their students. In other relationships, the question of power is defined less clearly and may even shift back and forth, depending on individual personalities and the particular situation: between wife and husband, among sisters and brothers, or among friends in a social clique.

Power is an element of many types of relationships, and is a complex phenomenon that covers a broad spectrum of interactions. At one pole is **authority**—*power that is regarded as legitimate by those over whom it is exercised, who also accept the authority's legitimacy in imposing sanctions or even in using force if necessary.* For example, in the United States few people are eager to pay income taxes, yet most do so regularly. Most taxpayers accept the authority of the government not only to demand payment but also to impose penalties for nonpayment.

At the other extreme is **coercion**—*power that is regarded as illegitimate by those over whom it is exerted.* Their compliance is based on fear of reprisals that are not recognized as falling within the range of accepted norms. Power based on authority is quite stable, and obedience to it is accepted as a social norm. Power based on coercion, in contrast, is unstable. People will obey only out of fear, and any opportunity to test this power will be taken. Power based on coercion will fail in the long run. The American Revolution, for example, was preceded by the erosion of the legitimacy of the existing system. The authority of the king of England was questioned, and his power, based increasingly on coercion rather than on acceptance as a social norm, inevitably crumbled.

Political Authority

An individual's authority often will apply only to certain people in certain situations. For example, a professor has the authority to require students to write term papers but no authority to demand the students' votes should he or she run for public office.

In the same sense, Weber pointed out that the most powerful states do not impose their will by physical force alone, but by ensuring that their authority is seen as legitimate. In such a state, people accept the idea that the allocation of power is as it should be and that those who hold power do so legitimately.

Weber (1957) identified three kinds of authority: legal-rational authority, traditional authority, and charismatic authority.

Legal-Rational Authority **Legal-rational authority** is *authority derived from the understanding that specific individuals have clearly defined rights and duties to uphold and implement rules and procedures impersonally.* Indeed, that is the key: Power is vested not in individuals but in particular positions or offices. There usually are rules and procedures designed to achieve a broad purpose. Rulers acquire political power through meeting requirements for office, and they hold power only as long as they obey the laws that legitimize their rule.

Traditional Authority **Traditional authority** is *rooted in the assumption that the customs of the past legitimate the present*—that things are as they always have been and basically should remain that way. Usually, both rulers and ruled recognize and support the tradition that legitimizes such political authority. Typically, traditional authority is hereditary, although this is not always the case. For example, throughout most of English history, the English crown was the property of various families. As long as tradition is followed, the authority is accepted. (See "Sociology at Work: The Importance of Presidential Concession Speeches.")

Charismatic Authority **Charismatic authority** derives from *a ruler's ability to inspire passion and devotion among followers.* Weber noted that a charismatic leader—who is most likely to emerge

The power of a charismatic leader derives from the ruler's force of personality and ability to inspire passion and devotion among followers. President John F. Kennedy was the closest the United States has come to having a charismatic leader.

during a period of crisis—will emerge when followers (1) perceive a leader as somehow supernatural, (2) blindly believe the leader's statements, (3) unconditionally comply with the leader's directives, and (4) give the leader unqualified emotional commitment. Others (Willner, 1984) have added that charismatic leaders also must perform seemingly extraordinary feats and have outstanding speaking ability.

Sitting Bull and Red Cloud, for example, were charismatic leaders of the Sioux Indians. Their people followed them because they led by example and inspired personal loyalty. However, individuals were free to disagree, to refuse to participate in planned undertakings, and even to leave and look for a group led by people with whom they were more likely to agree (Brown, 1974). This was not true in Russia under V. I. Lenin, in Germany under Adolf Hitler, or in Iran under the Ayatollah Ruholla Khomeini. These men all were charismatic rulers but also had the political authority necessary to enforce obedience or conformity to their demands.

Charismatic authorities and rulers emerge when people lose faith in their social institutions. Lenin led the Russian Revolution in the chaos left in the wake of World War I. Hitler rose to power in a Germany that had been defeated and humiliated in World War I and whose economy was shattered: Inflation was so bad that money was almost worth-

less. Khomeini rose to power in a country in which rapid modernization had undercut traditional Islamic norms and values, in which great poverty and great wealth existed side by side, and in which fear of the shah's secret police left the populace constantly anxious for its personal safety. The great challenge facing all charismatic rulers is to sustain their leadership after the crisis subsides and to create political institutions that will survive their death or retirement. Weber pointed out that if the program that the leader has implemented is to be sustained, the leader's charisma must be routinized in some form. For example, after Christ's death—and after it became apparent that his return to earth was not imminent—the apostles began to set up the rudiments of a religious organization with priestly offices.

Government and the State

Governments vary according to the relationship that exists between the rulers and the ruled. In some societies, political power is shared among most or all adults. This is true of the Mbuti pygmies and the !Kung San of Africa, for instance, for whom the group is its own authority and decisions are made by a consensus among adults. Among such societies, the concept of government is meaningless. However, in larger, more complex societies, government does exist.

In complex societies, the **state** is *the institutionalized way of organizing power within territorial limits.* Just as a true government is nonexistent in some societies, so too is the state limited to certain societies. Its presence indicates a high level of social and political development.

Modern thought regarding the nature of government and its forms is derived directly from the ideas of three Greek philosophers: Socrates, his student Plato, and Plato's student Aristotle. Socrates (469–399 B.C.) believed that evil and wrong actions arise from ignorance and the failure to investigate why people act as they do. In the *Republic,* written around 365 B.C., Plato discusses the form of government that would be most just. When Plato (ca. 428–348 B.C.) was writing, Athens had been through a period of political upheaval, so Plato was concerned with maintaining social order. Hence, he rejected democracy or rule by the majority because he believed that this form of government would lead to chaos. He also rejected *autocracy,* or rule by one person, because he thought that no single person could be wise or competent enough to make decisions for a whole society. Rather, he favored what he called **aristocracy** *(a form of oligarchy), or rule by a select few.*

Sociology at Work

The Importance of Presidential Concession Speeches

After every presidential election, we require the loser to engage in a public declaration of defeat. Why do we require this invasion of privacy? Why do we not allow the defeated candidate to suffer defeat away from the glare of the media? Could it be that this ritual serves an important social function ?

Dwelling on defeat contradicts a basic American commitment to success. After a presidential election, our gaze is on the triumphant winner; but instead of drawing a veil of silence over the crushed hopes of the losing candidate, we require one final ordeal, the concession speech.

The November 2000 presidential election was particularly unusual in that it took 36 days for a final winner to be declared. It was not considered completely over until Al Gore made the all-important speech. With a tone of respect, television news desks informed the viewing public that the defeated candidate was scheduled to concede:

> Just moments ago, I spoke with George W. Bush and congratulated him on becoming the 43rd president of the United States. I offered to meet with him as soon as possible so that we can start to heal the divisions of the campaign and the contest through which we just passed.
>
> I say to President-elect Bush that what remains of partisan rancor must now be put aside, and may God bless his stewardship of this country. Neither he nor I anticipated this long and difficult road. Certainly neither of us wanted it to happen. Yet it came, and now it has ended, resolved, as it must be resolved, through the honored institutions of our democracy.
>
> Now the U.S. Supreme Court has spoken. Let there be no doubt, while I strongly disagree with

the court's decision, I accept it. I accept the finality of this outcome, which will be ratified next Monday in the Electoral College. And tonight, for the sake of our unity of the people and the strength of our democracy, I offer my concession.

> I also accept my responsibility, which I will discharge unconditionally, to honor the new president elect and do everything possible to help him bring Americans together in fulfillment of the great vision that our Declaration of Independence defines and that our Constitution affirms and defends.
>
> I personally will be at his disposal, and I call on all Americans—I particularly urge all who stood with us to unite behind our next president. This is America. Just as we fight hard when the stakes are high, we close ranks and come together when the contest is done.
>
> And while there will be time enough to debate our continuing differences, now is the time to recognize that that which unites us is greater than that which divides us.
>
> While we yet hold and do not yield our opposing beliefs, there is a higher duty than the one we owe to political party. This is America and we put country before party. We will stand together behind our new president.

The words themselves are less important than the larger purpose they serve. They ritualize the passing of power and the legitimacy of the new authority. Throughout the world, the yielding of power is often a matter of life or death. The concession speech is not merely a report of an election result, it is a reaffirmation of the democratic process. There are often murderous consequences in societies where a candidate will not accept the results of an election. Haiti is a recent example of this situation, whereby

---

Plato called his proposed ruling class the guardians of society. The guardians, he argued, should be born from the most exemplary parents but then separated from them at birth. They should live in poverty for 30 years while being trained in mind and body, he said, and then they should fill positions of government in which they would execute their responsibilities wisely and without favoritism. It is important not to confound Plato's use of the term *aristocracy* with the modern usage. Plato explicitly rejected the ideal of inherited political power, which

he believed inevitably results in power falling to unqualified individuals—leading to an unjust society.

Aristotle (384–322 B.C.) tutored Alexander the Great and was a political scholar. Unlike Plato, he recognized that even in just societies, social-class interests produce class conflicts, which the state must control. Aristotle favored centering political power in the middle class (consisting of merchants, artisans, and farmers), but he insisted on defining the rights and duties of the state in a legal constitution (Laslett & Cummings, 1967).

Concession speeches ritualize the passing of power and the legitimacy of the new authority.

the United States had to send in troops to oust the military junta that had seized power illegally by throwing out the elected president and installing another man.

The presidential concession speech becomes a public duty prescribed by an unwritten law. It must be performed at the requisite time and invoke key themes, and it must be gracious. Nothing of what the candidate says or what the journalists write is surprising. Yet the function it serves is important.

Source: "Presidential Concession Speeches: The Rhetoric of Defeat," by P. E. Corcoran, 1994, *Political Communication*, *11*, pp. 109–131.

Functions of the State

Although a preindustrial society can exist without an organized government, no modern industrial society can thrive without those functions that the state performs: establishing laws and norms, providing social control, ensuring economic stability, setting goals, and protecting against outside threats.

Establishing Laws and Norms The state establishes laws that formally specify what the society expects and what it prohibits. The laws often repre-sent a codification of specific norms; for example, one should not steal from or commit violent crimes against others. Establishing laws also means exacting penalties for violating the laws.

Providing Social Control In addition to establishing laws, the state also has the power to enforce them. The police, courts, and various government agencies make sure that violators of those laws are punished. In the United States, the Internal Revenue Service seeks out tax evaders, the courts sentence

Remaking the World

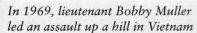

The Campaign to Ban Land Mines

In 1969, lieutenant Bobby Muller led an assault up a hill in Vietnam until a bullet severed his spinal cord. Miraculously, he survived. When the doctor came to see him he said, "we've got good news and bad news. We're pretty confident that you're going to live—but you're going be paralyzed. Muller remembers saying, "Don't worry about that. That's okay, I'm here. I'm here."

Despite his injury, he became head of the Vietnam Veterans of America Foundation. In that capacity, Muller traveled to Cambodia where he witnessed the devastating impact of land-mines on the local population. He launched himself into this cause and, along with cofounder Jody Williams, started the international campaign to ban land-mines. In 1997 they won the Nobel Peace Prize for their efforts.

Since 1975, land-mines have exploded under more than 1 million people and are currently thought to be killing 800 people a month. There seems little prospect of any end to the carnage. In 64 countries throughout the world, there are an estimated 110 million land-mines buried in the ground. Once laid, a mine may remain active for up to 50 years. Unless vigorous action is taken, mines placed today will still be killing and maiming people well into the middle of the next century.

There are basically two types of land-mines: anti-tank and anti-personnel. The most dangerous to children are the anti-personnel mines that explode even under the gentle pressure of a child's hand or foot. These come in a bewildering array of shapes and colors. Some look like stones, others like pineapples. They can become an interesting discovery for a curious child. One of the most infamous is the "butterfly" mine, designed to float to the ground from helicopters without exploding, but with a shape and color that also make it a deadly toy.

Land-mines are particularly lethal to children. Adults caught in the blast of an anti-personnel mine will often be maimed but survive with treatment. Children are less likely to survive. Their small bodies succumb more readily to the horrific injuries mines inflict. Children who manage to survive explosions are likely to be more seriously injured than adults, and often permanently disabled. Shrapnel may also cause blindness and disfigurement. All of this happens in countries that have difficulty offering the simplest medicines or pain-killers, let alone artificial limbs. In El Salvador, fewer than 20 percent of child victims receive any kind of remedial therapy; the rest have had to fend for themselves as best they can—often begging or stealing to survive.

Virtually all countries use land-mines. During the Persian Gulf war, the US and its allies laid about 1 million mines along the Iraq-Kuwait border and around the Iraqi city of Basra. Some of the largest numbers of mines lie in wait in Africa and Asia. The countries most devastated by land-mines are Afghanistan, Angola, and Cambodia. Afghanistan has an estimated 10–15 million mines in place. It is clear that many of these have been randomly scattered in inhabited areas precisely to cause civilian casualties and terrorize the population.

Mine removal is a lengthy and expensive business. Weapons that cost as little as $3 each to manufacture can cost up to $1,000 to remove. Land-mines can be blithely spread at rates of over 1,000 per minute, but it may take a skilled expert an entire day just to clear by hand a small area of mine-contaminated land.

In response to all of this devastation, in 1997 the United Nations produced a land-mine treaty banning the use, production, stockpiling, and transfer of antipersonnel land-mines. As of December 2000 the [treaty] has been ratified by 109 countries and signed by 139. Signing the treaty is only one step in the process. A country must also ratify it in order to be fully bound by the ban provisions. The United States so far has not signed the treaty.

As Bobby Muller notes: "My dream for the future is to make a real contribution in containing conflict. Conflict has been fundamentally transformed. Now it is the vulnerable and the innocent that are the targets of violence, instead of the military. This is absolutely unacceptable."

Sources: "The International Ban on Land Mines," by B. Muller, 2000, in K. K. Cuomo, *Speak Truth to Power* (pp. 88–91), New York: Crown Publishers; and interview with K. K. Cuomo, October 2000; Grac'a Machel, "Impact of Armed Conflict on Children," August 1996, New York: The United Nations.

criminals to prison, and the police attempt to maintain order.

Ensuring Economic Stability In the modern world, no individual can provide entirely for his or her own needs. Large workforces must be mobilized to build roads, dig canals, and erect dams. Money must be minted, and standards of weights and measures must be set and checked; merchants must be protected from thieves, and consumers from fraud. The state tries to ensure that a stable system of distribution and allocation of resources exists within the society.

Setting Goals The state sets goals and provides a direction for the society. If a society is to curtail its use of oil, for instance, the government must promote this as a goal. It must encourage conservation and the search for alternative energy sources and must discourage (perhaps through taxation or rationing) the use of oil. How is the government able to accomplish these tasks? How can it bring about individual and organizational compliance? Obviously, it would be best if the government could rely on persuasion alone, but this course seldom is enough. In the end, the government usually needs the power to compel compliance.

Protecting Against Outside Threats Historic data leave little doubt that the rise of the state was accompanied almost everywhere by the intensification of warfare (Otterbein, 1970, 1973). As early as the fourteenth century, Ibn Khaldun (1958), an Islamic scholar, noted this connection and even attributed the rise of the state to the needs of sedentary farmers to protect themselves from raids by fierce nomads. His views were echoed by Ludwig Gumplowicz (1899): "States have never arisen except through the subjugation of one stock by another, or by several in alliance." In any event, it is clear that one of the tasks of maintaining a society is to protect it from outside threats, especially those of a military nature. Hence, governments build and maintain armies. (For a discussion of how military actions can produce long-term threats to innocent civilians see "Remaking the World: The Campaign to Ban Land Mines.")

Although there is widespread agreement that the functions just described are tasks that the state should and usually does perform, not all social scientists agree that the state emerged because of the need for those functions.

The Economy and the State

Politics and economics are intricately linked, and the political form a state takes is tied to the economy.

In its simplest terms, the **economy** is *the social institution that determines how a society produces, distributes, and consumes goods and services.* Money, goods, and services do not flow of their own accord. People work at particular jobs, manufacture particular products, distribute the fruits of their labor, purchase basic necessities and luxury items, and decide to save or spend their money.

A very simple society may only produce and distribute food, water, and shelter. As a society becomes more complex and productive, the products produced and distributed become increasingly more elaborate. To be useful, all these goods and services must be distributed throughout the society. We depend on impersonal distribution systems to bring us such essential items as food, water, housing, clothing, health care, transportation, and communication, all of which we consume according to our ability to pay for them.

The political problem for any society is to decide how much to be involved in the production and distribution of goods and services. In most economies, markets play the major role in determining what is produced, how, and for whom. Which means, is there a demand for something? Who is willing to produce it? How much are people willing to pay for it? There are some economies that do not allow markets to function. In a **command economy** *the government makes all the decisions about production and consumption.*

Capitalism

In its classic form, **capitalism** is *an economic system based on private ownership of the means of production, and in which resource allocation depends largely on market forces.* The government plays only a minor role in the marketplace, which works out its own problems through the forces of supply and demand.

There are two basic premises behind capitalism. The first, as Weber noted, is production "for the pursuit of profit and ever renewed profit." Capitalism entitles people to pursue their own self-interests, and this activity is desirable and eventually benefits society through the "invisible hand" of capitalism. For example, pharmaceutical companies may have no other goal in mind than profit when they develop new drugs. The fact that their products eventually benefit society is an indirect benefit brought about by this invisible hand.

The second basic premise behind capitalism is that the free market will determine what is produced and at what price. If people can profit from the production of a product, it will be produced.

Adam Smith (1723–1790) is regarded as the father of modern capitalism. He set forth his ideas

One of the tasks a state must perform is to protect itself from outside threats, especially those of a military nature.

in his book *The Wealth of Nations* (1776), which is still used as a yardstick for analyzing economic systems in the Western world. According to Smith, capitalism has four features: private property, freedom of choice, freedom of competition, and freedom from government interference.

Private Property Smith believed that the ability to own private property acts as an incentive for people to be thrifty and industrious. These motivations, although selfish, will benefit society, because those who own property will respect the property rights of others.

Freedom of Choice Along with the right to own property comes the right to do with it what one pleases as long as it does not harm society. Consequently, people are free to sell, rent, trade, give away, or retain whatever they possess.

Freedom of Competition Smith believed society would benefit most from a free market featuring unregulated competition for profits. Supply and demand would be the main factors determining the course of the economy.

Freedom from Government Interference Smith believed government should promote competition and free trade and keep order in society but should not regulate business or commerce. The best thing the government can do for business, Smith said, is to leave it alone. *This view that government should

stay out of business is referred to as* **laissez-faire capitalism**. The French words *laissez-faire* mean "allow to act."

Laissez-faire capitalism and command economies are completely opposite ways for societies to deal with the distribution of goods and services. In the United States, the government plays a vital role in the economy. Therefore, the U.S. system cannot be seen as an example of pure capitalism. Rather, many have referred to our system as modified capitalism, also known as a mixed economy (Rachman, 1985).

A **mixed economy** *combines free-enterprise capitalism with governmental regulation of business, industry, and social welfare programs.* Although private property rights are protected, the forces of supply and demand are not allowed to operate with total freedom. Resources are distributed through a combination of market and governmental forces. Because there are few nationalized industries in this country (the Tennessee Valley Authority and Amtrak are two exceptions), the government uses its regulatory power to guard against private-industry abuse. Our government is also involved in such areas as antitrust violations, the environment, and minority employment. Ironically, this involvement may be even greater than it is in some of the more socialistic European countries.

Most countries have a mixed economy. Some countries, such as the United States and Taiwan, are closer to the free market end of the continuum, whereas China and Cuba are closer to the command economy end. The absolute extremes of a complete command economy or complete capitalism do not exist, whether it be in Cuba or the United States. Command economy countries let consumers choose some of the goods they buy and allow private agricultural markets to some extent. In the United States, in addition to regulating economic activity by setting minimum wage levels, requiring safety standards for the workplace, and enacting antitrust laws and farm price supports, the federal government has also assumed partial or total control of privately owned businesses when their collapse would significantly impact the people. Two examples include the government involvement in Amtrak and, for a time, the Chrysler Corporation.

The Marxist Response to Capitalism

Smith believed that ordinary people would thrive under capitalism; not only would their needs for goods and services be met, but they also would benefit by being part of the marketplace. In contrast, Karl Marx was convinced that capitalism produces a small group of well-to-do individuals, while the masses suffer under the tyranny of those who exploit them for profit.

Marx argued that capitalism causes people to be alienated from their labor and from themselves. Under capitalism, he said, the worker is not paid for part of the value of the goods produced. Instead, he said, this "surplus value" is taken by the capitalist as profit at the expense of the worker.

Workers also are alienated, according to Marx, by doing very small specific jobs, as on an assembly line, and thereby not feeling connected to the final product. The worker feels no relationship or pride in the product and merely works to obtain a paycheck and survive—a far cry from work being the joyous fulfillment of self that Marx believed it should be.

Marx believed that nineteenth-century capitalism contained several contradictions that were the seeds of its own destruction. The main problem with capitalism, he contended, is that profits will decline as production expands. This, in turn, will force the industrialist to exploit the laborers and pay them less in order to continue to make a profit. As the workers are paid less or are fired, Marx said, they are less able to buy the goods being produced. This causes profits to fall even further, leading to bankruptcies, greater unemployment, and even a full-scale depression. After an increasingly severe series of depressions, Marx argued, the workers will rise up and take control of the state. They then will create a socialist form of government in which private property is abolished and turned over to the state. The workers will then control the means of production, and the exploitation of workers will end.

The reality of capitalism has not matched Marxist expectations. As we have seen, earlier capitalist economies have become much more mixed economies than the capitalist model developed by Smith. This has prevented some of Marx's prophesied outcomes. The level of impoverishment that Marx predicted for the workers has not taken place, as labor unions have obtained higher wages and better working conditions for labor. Marx thought these changes could come about only through revolution. Labor-saving machinery has also led to higher profits without the predicted unemployment, and the production of goods to meet consumer demands has increased accordingly.

Marxists have offered a number of explanations for capitalism's continued success. Some have suggested that capitalism has survived because Western societies have been able to sell excess goods to developing countries, and that these sales have enabled the capitalists to maintain high prices and profits. However, Marxists see this as only a temporary solution to the inevitable decline of capitalism. Eventually the whole world will be industrialized, they contend, and the contradictions in capitalism will be revealed. They believe the movement toward socialism has not been avoided, but only postponed.

Command Economies

In a command economy, a government planning office decides what will be produced, how, and for whom and instructs workers and firms about what they are to produce. Such planning is an incredibly complicated task, and there are no complete command economies where all allocation is done by others. There is, however, a large dose of central direction and planning in communist countries, where the state owns factories and land and makes many of the decisions about what people should consume and how much they should work.

To appreciate the immensity of the task of central planning, try to imagine what it would be like in the city where you live. Think of the food, clothing, and housing allocations you would have to make. Then consider how you would know who should get what and how to produce it. Those decisions are currently being made by the market.

Total command economies, then, are not realistic ways of distributing goods and services. The distribution of goods and services is performed by both capitalistic and socialistic economic systems. In the next section, we will examine how these functions are performed in different ways by these two economic systems.

Socialism

Socialism is one type of command economy that is an alternative to and a reaction against capitalism. Whereas capitalism views profit as the ultimate goal of economic activity, socialism is based on the belief that economic activity should be guided by public needs rather than private profit. **Socialism** is *an economic system in which the government owns the sources of production—including factories, raw materials, and transportation and communication systems—and sets production and distribution goals.* Production and distribution are oriented toward output rather than profit. This orientation ensures that key industries run smoothly and that the public good is met. Individuals are heavily taxed to support a range of social-welfare programs that benefit every member of the society. Many socialist countries are described as having a "cradle-to-grave" welfare system.

Instead of relying on the marketplace to determine prices, under socialism prices for major goods and services are set by government agencies. Socialists believe that major economic, social, and political decisions should be made by elected representatives of the people in conjunction with the broader plans of the state. The aim is to influence the economic system so that wealth and income are distributed as equally as possible. The belief is that everyone should have such essentials as food, housing, medical care, and education before some people can have luxury items, such as cars and jewelry.

The Capitalist View of Socialism

Capitalists view the centrally planned economies of socialist societies as inefficient and concentrating power in the hands of one group whose authority is based on party position. Any disagreement with the policies of the group in power, such as strikes or organized efforts to change policies, is seen as disloyalty to the state. The workers thus are controlled economically and politically, having very little opportunity to improve their lifestyles or participate in political decisions.

Capitalists also argue that if essential goods and services are subsidized by the state and the consumers do not pay their full cost, nothing will prevent them from using more than they are entitled to and taking advantage of the system. If the producers of goods and services are immune from competition and have few incentives, what will encourage them to produce high-quality products?

The critics of socialism believe that the workers actually have very little freedom and that the centrally planned economy is ineffective, compared with a system based on market forces and individual incentives.

Types of States

Different types of states exist side by side and must deal with one another constantly in today's shrinking world. To comprehend their interrelationships, it is helpful to understand the structure of each main form of government—autocracy, totalitarianism, democracy, and socialism.

Autocracy

In an **autocracy,** *the ultimate authority and rule of the government rest with one person, who is the chief source of laws and the major agent of social control.* For example, the pharaohs of ancient Egypt were autocrats. More recently, the reigns of Ferdinand Marcos of the Philippines, Juan and Eva Peron of Argentina, and Sadam Hussein of Iraq have been autocratic.

In an autocracy, the loyalty and devotion of the people are required. To ensure that this requirement is met, dissent and criticism of the government and the person in power are prohibited. The media may be controlled by the government, and terror may be used to prevent or suppress dissent. For the most part, however, no great attempt is made to control the personal lives of the people. A strict boundary is set up between people's private lives and their pub-

lic behavior. Individuals have a wide range of freedom in pursuing such private matters as religion, family concerns, and many other traditional elements of life. At the same time, virtually all present-day autocracies have witnessed exploitation of the poor by the rich and powerful—a situation supported by the respective governments.

Totalitarianism

In a **totalitarian government,** *one group has virtually total control of the nation's social institutions.* Any other group is prevented from attaining power. Religious institutions, educational institutions, political institutions, and economic institutions all are managed directly or indirectly by the state. Typically, under totalitarian rule, several elements interact to concentrate political power.

1. *A single political party* controls the state apparatus. It is the only legal political party in the state. The party organization is itself controlled by one person or by a ruling clique.

2. *The use of terror* is implemented by an elaborate internal security system that intimidates the populace into conformity. It defines dissenters as enemies of the state and often chooses, arbitrarily, whole groups of people against whom it directs especially harsh oppression (for instance, the Jews in Nazi Germany or minority tribal groups in several recently created African states).

3. *The control of the media* (television, radio, newspapers, and journals) is in the hands of the state. Differing opinions are denied a forum. The media communicate only the official line of thinking to the people.

4. *Control over the military apparatus,* both the military personnel and the use of its weapons, is monopolized by those who control the political power of the totalitarian state.

5. *Control of the economy* is wielded by the government, which sets goals for the various industrial and economic sectors and determines both the prices and the supplies of goods.

6. *An elaborate ideology,* in which previous sociopolitical conditions are rejected, legitimizes the current state and provides more or less explicit instructions to citizens on how to conduct their daily lives. This ideology offers explanations for nearly every aspect of life, often in a simplistic and distorted way (Friedrich & Brzezinski, 1965).

Two distinct types of totalitarianism are found in the modern world. Though they share the same basic political features described previously, they differ widely in their economic systems. Under **totalitarian socialism,** *in addition to almost total regulation of* *all social institutions, the government controls and owns all major means of production and distribution: There is little private ownership or free enterprise. Totalitarian socialist forms of government are more commonly known as* **communism.**

Totalitarian capitalism, on the other hand, *denotes a system under which the government, while retaining control of social institutions, allows the means of production to be owned and managed by private groups and individuals.* However, production goals usually are dictated by the government, especially in heavy industry. Hitler's Germany is a good example of totalitarian capitalism. The mammoth Krupp industrial complex was owned and managed by the family of that name, but the government had the company gear its efforts toward producing munitions and heavy equipment to further the nation's political and militaristic aims. *Totalitarian capitalism is often referred to as* **fascism.**

One of the problems faced by totalitarian governments is that because their total control over their citizenry allows no organized independent opposition, they are never sure whether the populace's conformity to the laws is based on its acceptance of the government's legitimate authority or is motivated primarily by fear of coercion by a ruling power it considers illegitimate.

Democracy

Democracy has not always been regarded as the best form of government. People have often approved of the aims of democracy, yet have argued that democracy was impossible to attain. Others argued that it was logically unsound. Today, however, there is hardly a government anywhere in the world that does not claim to have some sort of democratic authority. In the United States, we regard our political system as democratic, and the same claim is made by leaders in communist countries. The word *democracy* seems to have so many different meanings today that we face the problem of distinguishing democracy from other political systems.

Democracy comes from the Greek words *demos,* meaning "people," and *kratia,* meaning "authority." By democracy, then, the Greeks were referring to a system in which rule was by the people rather than a few selected individuals. Because of the growth in population, industrialization, and specialization, it has become impossible for citizens to participate in politics today as they did in ancient Athens. Today, **democracy** refers to *a political system operating under the principles of constitutionalism, representative government, majority rule, civilian rule, and minority rights.*

Constitutionalism *means that government power is limited.* It is assumed that there is a higher law, which is superior to all other laws. The various agencies of the government can act only in specified legal ways. Individuals possess rights, such as freedom of speech, press, assembly, and religion, which the government cannot take away.

A basic feature of democracy is that it is rooted in **representative government,** which means that *the authority to govern is achieved through, and legitimized by, popular elections.* Every government officeholder has sought, in one way or another, the support of the **electorate** *(those citizens eligible to vote)* and has persuaded a majority of that group to grant its support *(through voting).* The elected official is entitled to hold office for a specified term and generally will be reelected as long as that body of voters is satisfied that the officeholder is adequately representing its interests.

Representative institutions can operate freely only if certain other conditions prevail. First, there must be what sociologist Edward Shils (1968) calls *civilian rule*—that is, every qualified citizen has the legal

One of the most important ways for people to participate in the political life of the country is to vote.

Global Sociology

When Violence and Politics Equal Democracy

The scene was straight out of a Wild West barroom brawl. Men were shouting and grabbing, pushing and kicking. Within minutes, dozens of people were leaping into the fray and four were rushed to the hospital, bleeding, unconscious, or otherwise injured.

This was no drunken melee among cardsharks and gunslingers. It was the culmination of a bitter dispute over constitutional reform in Taiwan's National Assembly.

Fistfighting in the two-chamber parliament has become an enduring feature of one of the newest and most open democracies in Asia. Violence is used whenever one party believes it is not being treated fairly. Politicians also use it to prove to the voters that they are fighting for their interests.

Fistfights in the political arena are an improvement over the previous era when martial law was in effect and there were more than 1,000 street demonstrations for democracy in a 10-year period.

Before we feel too smug about the decorum in the U.S. Congress, we must remember that democracy in the United States has not always been as staid as it now appears. A variety of physical confrontations plagued Congress in the first century of American democracy. There were at least two gun duels, four cane fights, one beating, an attack

with fire tongs, and various other violent melees, including the 1856 caning of Massachusetts Senator Charles Sumner by South Carolina lawmaker Preston Brooks. It took Sumner 3 years to recover.

Tensions ran high because there was a sense in the early Republic that the future of the nation hung in the balance of these disputes—a far cry from today when often there is a sense that nothing really matters. With cable television recording proceedings live, politicians today speak to the camera—often in nearly empty chambers—while 200 years ago they addressed each other, and insults were immediately heard by the offended party. Many of today's new governments are practicing democracy in precisely the same way we did in the early years.

In addition to the fistfights in Taiwan, lawmakers in the relatively young democracies of Russian and India have also resorted to violence to defend passionately held views. Today we may look at the filibuster as a normal way for a political party to block legislation. In many of these new democracies, the use of violence is just another method of accomplishing the same goal.

Source: "In Taiwan's Legislature Democracy Packs Punch," by I. A. R. Lakshmanan, August 2, 1997, *The Boston Globe,* pp. A1, A12.

---

right to run for and hold an office of government. Such rights do not belong to any one class (say, of highly trained scholars, as in ancient China), caste, sect, religious group, ethnic group, or race. These rights, with certain exceptions, belong to every citizen. Further, there must be public confidence that such organized agencies as the police and the military will not intervene in, or change the outcome of, elections.

In addition, *majority rule* must be maintained. Because of the complexity of a modern democracy, it is not possible for the people to rule directly. One of the most important ways for people to participate in the political life of the country is to vote. For this to happen people must be free to assemble, to express their views and seek to persuade others, to engage in political organizing, and to vote for whomever they wish.

Democracy also assumes that *minority rights* must be protected. The majority may not always act wisely, and it may be unjust. The minority abides by the laws as determined by the majority, but the minority must be free to try to change these laws.

Democratic societies contrast markedly with totalitarian societies. Ideally, democratic societies are open and culturally diverse, dissent is not viewed as disloyalty, there are two or more political parties, and terror and intimidation are not an overt part of the political scene.

The economic bases of democratic societies can vary considerably. Democracy can be found in a capitalistic country like the United States and in a more socialistic one like Sweden. However, it appears necessary for the country to have reached an advanced level of economic development before democracy can evolve. Such societies are most likely to have the sophisticated population and stability necessary for democracy (Lipset, 1960). (For a discussion of the growing pains inherent in fledgling democracies, see "Global Sociology: When Violence and Politics Equal Democracy.")

Democracy and Socialism

Critics of capitalism argue that true democracy is an impossible dream in a capitalist society. They claim

Chou Chia-chih, deputy of the opposition Democratic Progressive Party, punches rival Liu Men-Change at the National Assembly.

that although in theory all members of a capitalist society have the same political rights, in fact capitalist society is inherently stratified, and therefore wealth, social esteem, and even political power are unequally distributed (see Chapter 7). Because of this, some critics contend, true democracy can be achieved only under socialism—when the government owns and controls the major means of production and distribution (thus avoiding inequality of ownership among its members) and in which there is no social stratification (Schumpeter, 1950).

There is no obvious reason that socialist societies cannot be democratic. In fact, though, many societies whose economies are socialist tend to be communist states. One reason is that, historically, socialist societies often have been born in revolution. Political revolutions, by definition, mean a redistribution of power (see the discussion of political change later in this chapter). One group seizes power from another and then tries to prevent the old group from retaking it. Lenin argued that to consolidate power, the new group must use strong repressive measures against the old—in fact, build a dictatorship. A **dic-**

tatorship is *a totalitarian government in which all power rests ultimately in one person.* This person, or dictator, generally heads the only recognized political party, at least until all the economic resources of the old group are seized, its links to all political agencies are broken, and its claims to political legitimacy are wiped out. Lenin (1949) quoted Marx on this point:

> Between capitalist and communist society lies the period of the revolutionary transformation of the one into the other. Corresponding to this is also a political transition period in which the state can be nothing but the revolutionary dictatorship of the proletariat [the working class].

Lenin (1949) put it more graphically himself:

> The proletariat needs state power, a centralized organization of force, and organization of violence, both to crush the resistance of the exploiters and to lead the enormous mass of the population . . . in the work of organizing a socialist economy.

In China, Cuba, and more recently in African and Southeast Asian countries, socialist revolutions have resulted in dictatorships. Members of the previous ruling classes have been executed, jailed, "reeducated," or exiled, and their properties have been seized and redistributed. In none of these societies has the dictatorship proved to be temporary, nor has the state gradually withered away, as Marx and Friedrich Engels predicted it would after socialism was firmly established. Many Marxists contend that this will happen in the future, especially once capitalism has been defeated worldwide and socialist states no longer need to protect themselves against counterrevolutionary subversion and even direct military threats by capitalist nations. However, it is fair to observe that even the ancient Greeks knew that power corrupts and that those who have power are unlikely ever to give it up voluntarily. So we may expect that at least in the foreseeable future, socialist states that have emerged through revolution will—despite their disclaimers—remain totalitarian communist dictatorships.

Democratic Socialism

Democratic socialism is *a convergence of capitalist and socialist economic theory in which the state assumes ownership of strategic industries and services, but allows other enterprises to remain in private hands.* In Western Europe, democratic socialism has evolved as a political and economic system that attempts to preserve individual freedom in the context of social equality and a centrally planned economy.

With the parliamentary system of government present in many European countries, social democratic political parties have been able to win representation in the government. They have been able to enact their economic programs by being elected to office, as opposed to producing a workers' uprising against capitalism. The social democrats have also attempted to appeal to middle-class workers and highly trained technicians, as well as to industrial workers.

Under democratic socialism, the state assumes ownership of only strategic industries and services, such as airlines, railways, banks, television and radio stations, medical services, colleges, and important manufacturing enterprises. Certain enterprises can remain in private hands as long as government policies can ensure that they are responsive to the nation's common welfare. High tax rates prevent excessive profits and the concentration of wealth. In return, the population receives extensive welfare benefits, such as free medical care, free college education, or subsidized housing.

Democratic socialism flourishes to varying degrees in the Scandinavian countries, in Great Britain, and in Israel. These countries all have a strong private (that is, capitalist) sector in their economies, but they also have extensive government programs to ensure the people's well-being. Those programs pertain to such things as national health service, government ownership of key industries, and the systematic tying of workers' pay to increases in the rate of inflation. Many observers believe that the American political economy has been moving in this direction. The social democratic movement is an example of the convergence of the capitalist and socialist economic theories, a trend that has been evident for some time. Capitalist systems have seen an ever-greater introduction of state planning and government programs, and socialist systems have seen the introduction of market forces and the profit motive. The growing economic interdependence of the world's nations will help continue this trend toward convergence.

Functionalist and Conflict Theory Views of the State

Functionalists and conflict theorists hold very different ideas about the function of the state. As our discussion of social stratification in Chapter 7 revealed, functionalist theorists view social stratification—and the state that maintains it—as necessary devices that recruit workers to perform the tasks necessary to sustain society. Individual talents must be matched to jobs that need doing, and those with specialized talents must be given sufficiently satisfying rewards. Functionalists therefore maintain that the state emerged because society grew so large and complex that only a specialized, central institution (that is, the state) could manage society's increasingly complicated and intertwined institutions (Davis, 1949; Service, 1975).

Marxists and other conflict theorists take a different view. They argue that technological changes creating surplus production brought about the production of commodities for trade (as opposed to products for immediate use). Meanwhile, certain groups were able to seize control of the means of production and distribution of commodities, thereby establishing themselves as powerful ruling classes that dominated and exploited workers and serfs. Finally, the state emerged to coordinate the use of force, which allows the ruling classes to protect their institutionalized supremacy from the resentful and potentially rebellious lower classes. As Lenin (1949) explained, "The state is a special organization of force: it is an organization of violence for the suppression of some class."

There is evidence to support this view of the state's origins. The earliest legal codes of ancient states featured laws protecting the persons and properties of rulers, nobles, landholders, and wealthy merchants. The Code of Hammurabi of Babylon, dating to about 1750 B.C., prescribed the death penalty for burglars and for anybody who harbored a fugitive slave. The code regulated wages, prices, and fees to be charged for services. It provided that a commoner be fined six times as much for striking a noble or a landholder as for striking another commoner. It also condemned to death women who were proved by their husbands to be uneconomical in managing household resources (Durant, 1954).

Nevertheless, the functionalist view also has value. The state provides crucial organizational functions such as carrying out large-scale projects and undertaking long-range planning, without which complex society probably could not exist. Because it provides a sophisticated organizational structure, the state can—and does—fulfill many other important functions. In most modern societies, the state supports a public school system to provide a basic, uniform education for its members. The health and well-being of its citizens also have become the concern of the state. In our own country, as in many others, the government provides some level of medical and financial support for its young, old, and disabled, and it also sponsors scientific and medical research for the welfare of its people. Regulating industry and trade to some degree also has become a function of the modern state, and it has devised ways (different for each kind of state) of establishing, controlling, and even safeguarding the civil rights and liberties of its citizens. Certainly, one of the most important functions of any state is the protection of its people. Long gone are the days when cave dwellers, lords of the manor, or pioneers defended their territories and other members of their groups from attack or encroachment; the state now provides such protection through specialized agencies: armies, militias, and police forces.

When groups in a society develop sufficient dissatisfaction with their government and achieve the strength to influence the direction of that system, changes can take place. After such changes, the state may perform the same functions, but it may do so in a different way and under different leadership.

Political Change

Political change can occur when there is a shift in the distribution of power among groups in a society. It is one facet of the wider process of social change, the topic of the last part of this book. Political change can take place in a variety of ways, depending on the type of political structure the state has and the desire for change among the people. People may attempt to produce change through established channels within the government, or they may rise up against the political power structure with rebellion and revolution. Here we shall consider briefly three forms of political change: institutionalized change, rebellion, and revolution.

Institutionalized Political Change

In democracies, the institutional provision for the changing of leaders is implemented through elections. Usually, candidates representing different parties and interest groups must compete for a particular office at formally designated periods. There may also be laws that prevent a person from holding the same office for more than a given number of terms. If a plurality or a majority of the electorate is dissatisfied with a given officeholder, they can vote the incumbent out of office. Thus, the laws and traditions of a democracy ensure the orderly changeover of politicians and, usually, of parties in office.

In dictatorships and totalitarian societies, if a leader unexpectedly dies, is debilitated, or is deposed, a crisis of authority may occur. In dictatorships, illegal, violent means must often be used by an opposition to overthrow a leader or the government, because no democratic means exist whereby a person or group can be legally voted out of power. Thus, we should not be surprised that revolutions and assassinations are most likely to occur in developing nations that have dictatorships. Established totalitarian societies, such as the former Soviet Union and China, are more likely than dictatorships to offer normatively prescribed means by which a ruling committee decides who should fill a vacated position of leadership.

Rebellions

Rebellions are *attempts—typically through armed force—to achieve rapid political change that is not possible within existing institutions.* Rebellions typically do not call into question the legitimacy of power, but, rather, its uses. For example, consider Shays' Rebellion. Shortly after the American colonies won their independence from Britain, they were hit by an economic depression followed by raging inflation. Soon, in several states, paper money lost almost all its value. As the states began to pay off their war debts (which had been bought up by speculators), they were forced to increase the taxation of farmers, many of whom could not afford to pay those new taxes and consequently lost their land. Farmers began to band together to prevent courts from hearing debt cases, and state militias were called out to protect court hearings. Desperate farmers in

People may attempt to produce change through established channels within the government, or they may rise up against the political structure.

the Connecticut River Valley armed themselves under Daniel Shays, an ex-officer of the Continental Army (Blum et al., 1981). This armed band was defeated by the Massachusetts militia, but its members eventually were pardoned and the debt laws were loosened somewhat (Parkes, 1968). Shays' Rebellion did not intend to overthrow the courts or the legislature; rather, it was aimed at effecting changes in their operation. Hence, it was a typical rebellion.

Revolutions

Revolution is a powerful word. It evokes vivid images and strong emotions. It contains a mix of hope, excitement, and terror. Small wonder that many great works of art, literature and film have been inspired by revolution. In contrast to rebellions, **revolutions** are *attempts at rapidly and dramatically changing a society's previously existing structure.* Sociologists further distinguish between political and social revolutions.

Political Revolutions *Relatively rapid transformations of state government structures that are not accompanied by changes in social structure or stratification* are known as **political revolutions** (Skocpol, 1979). The American war of independence is a good example of a political revolution. The colonists were not seeking to change the structure of society or even necessarily to overthrow the ruling order. Their goal was to put a stop to the abuse of power by the British. After the war they created a new form of government, but they did not attempt to change the fact that landowners and wealthy merchants held the reins of political power—just as they had before

the shooting started. In the American Revolution, then, a lower class did not rise up against a ruling class. Rather, it was the American ruling class going to war in order to shake loose from inconvenient interference by the British ruling class. The initial result, therefore, was political, not social, change.

Social Revolutions In contrast, **social revolutions** are *rapid and basic transformations of a society's state and class structures.* They are accompanied and in part carried through by class-based revolts (Skocpol, 1979). Hence, they involve two simultaneous and interrelated processes: (1) the transformation of a society's system of social stratification, brought about by upheaval in the lower class(es), and (2) changes in the form of the state. Both processes must reinforce each other for a revolution to succeed. The French Revolution of the 1790s was a true social revolution. So were the Mexican Revolution of 1910, the Russian Revolution of 1917, the Chinese Revolution of 1949, and the Cuban Revolution of 1959—to name some of the most prominent social revolutions of the twentieth century. In all these revolutions, class struggle provided both the context and the driving force. The old ruling classes were stripped of political power and economic resources, wealth and property were redistributed, and state institutions were thoroughly reconstructed (Wolf, 1969).

Although the American Revolution did not arise from class struggle and did not result immediately in changes in the social structure, it did mark the beginning of a form of government that eventually modified the social stratification of eighteenth-century America.

The American Political System

The United States' political system is unique in a number of ways, growing out of a strong commitment to a democratic political process and the influence of a capitalist economy. It has many distinctive features that are of particular interest to sociologists. In this section, we will examine the role of the electorate and how influence is exerted on the political process.

The Two-Party System

Few democracies have only two main political parties. Besides the United States, there are Australia, New Zealand, and Austria. The other democracies all have more than two major parties, thus providing proportional representation for a wide spectrum of divergent political views and interests. In most European democracies, if a political party receives 12% of the vote in an election, it is allocated 12% of the seats in the national legislature. Such a system ensures that minority parties are represented.

The American two-party political system, however, operates on a winner-take-all basis. Therefore, groups with differing political interests must face not being represented if their candidates lose. Conversely, candidates must attempt to gain the support of a broad spectrum of political interest groups, because a candidate representing a narrow range of voters cannot win. This system forces accommodations between interest groups on the one hand and candidates and parties on the other.

Few, if any, individual interest groups (like the National Organization for Women, the National Rifle Association, or the Conservative Caucus) represent the views of a majority of an electorate, be it local, state, or national. Hence, it is necessary for interest groups to ally with political parties in which other interest groups are involved, hoping thereby to become part of a majority that can succeed in electing one or more candidates. Each interest group then hopes that the candidate(s) it has helped elect will represent its point of view. For most interest groups to achieve their ends more effectively, they must find a common ground with their allies in the party they have chosen to support. In doing so, they often have to compromise some strongly held principles. Hence, party platforms often tend to be composed of mild and noncontroversial issues, and party principles tend to adhere as closely as possible to the center of the American political spectrum.

When either party attempts to move away from the center to accommodate a very strong interest group with left- or right-wing views, the result generally is disaster at the polls. This happened to the Republican party in 1964, when the politically conservative Barry Goldwater forces gained control of its organizational structure and led it to a landslide defeat: In that year's presidential election, the Democrats captured 61.1% of the national vote. Eight years later, the Democratic party made the same mistake. It nominated George McGovern, a distinctly liberal candidate who, among other things, advocated a federally subsidized minimum income. The predictable landslide brought the Republicans and Richard Nixon 60.7% of the vote.

The candidates themselves have other problems. To gain support within their parties, they must somehow distinguish themselves from the other candidates. In other words, they must stake out identifiable positions. Yet to win state and national elections, they must appeal to a broad political spectrum. To do this, they must soften the positions that first won them party support. Candidates thus often find themselves justly accused of double-talking and vagueness as they try to finesse their way through this built-in dilemma. This is most true of presidential candidates. It is no accident that once they have their party's nomination, some candidates may express themselves differently and far more cautiously than they did before.

Voting Behavior

In totalitarian societies, strong pressure is put on people to vote. Usually there is no contest between the candidates, as there is no alternative to voting for the party slate. Dissent is not tolerated, and nearly everyone votes. In the United States, there is a constant progression of contests for political office, and in comparison with those of many other countries, the voter turnout is quite low. Since the 1920s, the turnout for presidential elections has ranged from about 50% to nearly 70% of registered voters. Why do people not bother to vote? More than 1 in 5 say they could not take time off from work or were too busy. Another reason increasingly cited for not voting is apathy about the political process (U.S. Bureau of the Census, 1998).

Considering the emphasis that Americans place on a democratic society, it is interesting that participation in national elections is declining steadily, a cause of grave concern to social scientists and political observers alike.

Voting rates vary with the characteristics of the people. For example, those 45 and older and with a college education and a white-collar job have high rates of voter participation. Hispanics, the young, and the unemployed have some of the lowest voter participation rates (see Table 13-1 for voter participation by selected characteristics). The age group

Table 13-1

Voter Participation by Selected Characteristics, 1998

CHARACTERISTIC	PERCENTAGE VOTING
Sex	
Male	41.4
Female	42.4
Race	
White	43.3
Black	39.6
Hispanic	20.0
Age	
18–20	13.5
21–24	Not available
25–34	28.0
35–44	40.7
45–64	53.6
65 and older	59.5
Education Attained	
Fewer than 8 years	24.6
High-school graduate	37.1
College graduate	57.2
Employment	
Unemployed	28.4
Employed	41.2

Source: *Statistical Abstract of the United States: 2000* (p. 290), U.S. Bureau of the Census, 2001, Washington, DC: U.S. Government Printing Office.

with the highest proportion of voters is 55- to 74-year-olds, with more than 7 in 10 casting ballots. The lowest voting rates belong to 18- to 24-year-olds, who had a ratio of slightly more than 1 in 3 voting.

The Democratic party has tended to be the means through which the less privileged and the unprivileged have voted for politicians whom they hoped would advance their interests. Since 1932, the Democratic party has tended to receive most of its votes from the lower class, the working class, blacks, those of Southern and Eastern European descent, Hispanics, Catholics, and Jews. Thus, almost all the legislation that has been passed to aid these groups has been promoted by the Democrats and opposed by the Republicans: Democrats have sponsored legislation supporting unions, Social Security, unemployment compensation, disability insurance, antipoverty legislation, Medicare and Medicaid, civil rights, and consumer protection.

The Republican party has tended to receive most of its votes from the upper-middle and lower-middle classes, Protestants, and farmers. Americans younger than 30 tend to vote the Democratic ticket; and those older than 49 tend to vote for the Republican

party, even though the Democratic party has been responsible for almost all the legislation to benefit older citizens. These voting patterns are, of course, generalizations and may change during any specific election.

Both parties tend to be most responsive to the needs of the best-organized groups with the largest sources of funds or blocks of votes. Thus, we would expect the Republicans to represent best the interests of large corporations and well-funded professional groups (such as the American Medical Association). The Democrats would logically be more closely aligned with the demands of unions.

Factors other than the social characteristics of voters and the traditional platforms of parties may affect the way people vote (Cummings & Wise, 1981). Indeed, the physical attributes, social characteristics, and personality of a candidate may prompt some people to vote against the candidate of the party they usually support. More important, the issues of the period may cause voters to vote against the party with which they usually identify. When people are frustrated by factors such as war, recession, inflation, and other international or national events, they often blame the incumbent president and the party he represents.

Since the 1960s there have been efforts to increase the number of minority members who register to vote and to improve their voting rate. The greater prominence of minority candidates has helped with this effort, as minority groups are more likely to vote if they feel that the elections are relevant to their lives. As minority group members increase their voting rates, they also become successful in electing members of their groups. Figure 13-1 shows the consistent rise in the number of African-American and Hispanic elected officials.

Women also have been successful in increasing their representation in state legislatures. Figure 13-2 shows that the number of women holding such offices doubled between 1975 and 1998.

Despite these advances, the members of Congress are still overwhelmingly white men older than 40. In 2000, the Senate had only 13 women members and one African American; the other 91 members were white men. The lack of women and African Americans in the House of Representatives was also striking (see Table 13-2 for a description of selected characteristics of members of Congress).

It would be wrong to conclude from the figures cited that Americans are politically inactive. There is more to political activity than voting. A study of American political behavior identified four modes of participation (Verba, 1972): (1) Some 21% of Americans eligible to vote do so more or less regularly in municipal, state, and national elections but

Figure 13-1 African-American and Hispanic Elected Officials, 1985–1995

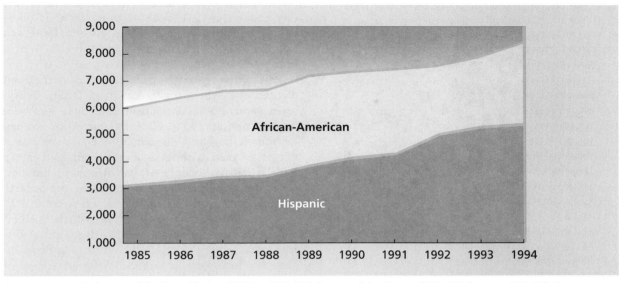

Source: *Statistical Abstract of the United States: 2000* (p. 288), U.S. Bureau of the Census, 2001, Washington, DC: U.S. Government Printing Office.

Figure 13-2 Women Holding Office Within State Legislatures, 1975–1999

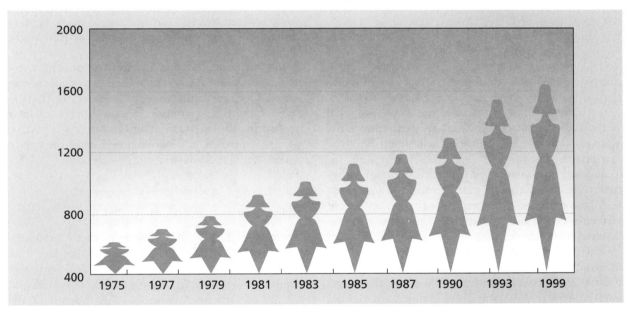

Source: *Statistical Abstract of the United States: 2000* (p. 287), U.S. Bureau of the Census, 2001, Washington, DC: U.S. Government Printing Office.

do not engage in other forms of political behavior. (2) Roughly 4% of the American electorate not only vote but also make the effort to communicate their concerns directly to government officials in an attempt to influence the officials' actions. (3) Some 15% of the electorate vote and also periodically take part in political campaigns. However, they do not involve themselves in ongoing local political affairs. (4) About 20% of eligible voters are relatively active in ongoing community politics. They vote but do not get involved in political campaigns.

African-Americans as a Political Force

African-Americans were the only race or ethnic group to defy the trend of declining voter participation in congressional elections, increasing their presence at

Table 13-2

Selected Characteristics of Members of Congress

	SENATORS	REPRESEN-TATIVES
Sex (2001)		
Male	87	363
Female	13	72
Race (1995)		
White	97	374
Black	1	40
Asian-Pacific Islander	2	4
Hispanic	0	17
Age (1999)		
Younger than 40	1	23
40–49	14	116
50–59	38	173
60–69	33	87
70 and over	13	35

Source: *Statistical Abstract of the United States: 2000* (p. 283), U.S. Bureau of the Census, 2001, Washington, DC: U.S. Government Printing Office.

the polls from 37% in 1994 to 40% in 1998 (U.S. Bureau of the Census, 2000). Part of the explanation for this increase may be that African-Americans always seem to receive attention before major political campaigns, as each candidate attempts to convince them that he or she takes their interests to heart. Yet African-Americans have often been skeptical of those preelection promises, sensing that once the election is won, their concerns will again be given low priority.

Indeed, the statistics on African-American progress in economic areas are still grim. Although progress has occurred in a variety of areas in recent years, the economic gap between whites and African-Americans remains large.

The picture is somewhat more promising for education, but when one considers such areas as infant health, adult mortality rates, and income, the movement of African-Americans into mainstream middle-class America is still a long way off.

Statistics show an improvement in income levels of African-American married-couple families, educational attainment and school enrollment, and home ownership among blacks during the previous decade. There have been setbacks also, demonstrated by high African-American unemployment, sharply increased divorce and separation rates, and a rise in family households maintained by black women.

A cursory look at income statistics makes blacks appear better off than they are. African-American married couples have closed the gap with white incomes, but the proportionate share of married

couples in the total black population is smaller than a decade ago. More African-Americans are in poverty today than a decade ago. In 1999, 7.7% of whites had incomes below the poverty level, but fully 23.6% of African-Americans did (U.S. Bureau of the Census, September 2000).

Part of the reason for this difference is the fragmentation of the African-American family. Forty-seven percent of all African-American families were married couples, 45% were maintained by women without a husband present, and 8% were maintained by men without a wife present. When we look at married-couple African-American families, we find that only about 7% are below the poverty level (U.S. Bureau of the Census, February 2000).

African-Americans also have fallen farther behind whites in the accumulation of wealth. Part of this was because they were less likely to own property and therefore were not able to take advantage of rapidly rising real estate values over the previous decade. Regardless of income level, African-Americans tend to have less wealth than whites. Even blacks who already own a home or eventually manage to purchase one must contend with the fact that most homes owned by blacks are in central cities, areas in which homes are less likely to appreciate than in the suburbs.

In the political arena, African-Americans are not just another interest group. They experience deep-seated economic differences that make them skeptical of political promises. Those differences have made African-Americans cautious about supporting white Democrats, no matter how liberal the candidates' voting records are.

Hispanics as a Political Force

There were 21.3 million Hispanics of voting-age in November 2000. This figure was 16% greater than in November 1996. However, only 35.7% were registered to vote compared with 67.7% of whites and 63.5% of African-Americans (U.S. Bureau of the Census, 2000).

Hispanic voters are crucial in many presidential elections. The main reason is the geographical distribution of the voters. Approximately 10 million Hispanic voters are registered in six major states—New York, New Jersey, Florida, Illinois, Texas, and California—which together can give a presidential candidate most of the electoral votes needed to win the election.

Hispanic voters have traditionally been overwhelmingly Democratic. Hispanics tend to support Democratic candidates because of the support those candidates give to social programs that help the poor. The fact that Hispanics, like African-Americans, have a harder time climbing the socioeconomic ladder keeps them in the Democratic camp.

With the Hispanic population expected to grow rapidly over the next few decades, Democrats and Republicans will be paying more attention to the needs of this group. It is also likely the Hispanics will be electing greater numbers of Hispanic candidates to political office.

A Growing Conservatism

Public opinion tends to change as the age composition of the population changes. Older people are typically more conservative than younger people. As the average age of the U.S. population has been increasing, we can expect that conservatism will increase. In addition, if the younger population is also growing more conservative in its own right, as many have claimed, the impact on conservatism doubles.

Data to help us decide whether these assumptions are correct exist in the General Social Survey (GSS) from the National Opinion Research Center (NORC). Twenty years of GSS results leave little doubt that a conservative mood has pervaded the country. All age groups participated in the shift toward conservatism. Overall, self-described liberals dropped from 30% of Americans in 1975 to 25% in 1985, and 19% in 1994, while self-described conservatives increased from 30% to 33%.

The leading edge of the post-World War II baby boom, those 51 to 56 years old in 2002, are already more conservative on capital punishment, the courts, welfare spending, gun control, pornography, and spending on the poor, the cities, and space exploration than they were in 1975 when they were in their 20s. There is only one issue on which they become more liberal as they age: spending for foreign aid. This same group, when asked to categorize themselves as conservative, moderate, or liberal, has grown decidedly more conservative. (See "Sociology at Work: Comparing the Political and Moral Values of the 1960s with Today.")

The Democratic party has undergone a good deal of soul-searching in recent years. Some have criticized the party for moving to the right in response to perceived shifts in public opinion. If the Democratic party is to continue as a viable political force, it must contend with the changing political climate. Labor unions, long a bastion of Democratic support, are in decline, while the new college-educated service workers are more likely to be Republican. The ranks of those who call themselves Republicans or independents are growing, while those who call themselves Democrats are declining. The next 5 years

should present us with a major turning point in American politics.

The Role of the Media

The media have contributed to a radical transformation of election campaigns in the United States. This transformation involves changes in how political candidates communicate with the voters, and in the information journalists provide about election campaigns.

On one side, we have candidates who are trying to present self-serving and strategically designed images of themselves and their campaigns. On the other side are television and print journalists who believe that they should be detached, objective observers, motivated primarily by a desire to inform the U.S. public accurately.

Although the norm is often violated, there is no doubt that the journalistic perspective is markedly different from that of the candidate. Campaign coverage is far more apt to be critical and unfavorable toward a candidate than favorable. Journalists strive to reveal a candidate's flaws and weaknesses and to uncover tasty tidbits of hushed-up information.

Journalists exercise considerable political power in four important ways. First, the most obvious way involves deciding how much coverage to give a campaign and the candidate involved. A candidate who is ignored by the media has a difficult time becoming known to the public and acquiring important political resources, such as money and volunteers. Such candidates have little chance of winning.

In the beginning stages of presidential nomination campaigns, for example, a candidate's goal is typically to do something that will generate news coverage and stimulate campaign contributions. These contributions can then be used for further campaigning, helping to convince the press and the public that the candidate is credible and newsworthy.

Second, the media decide which of many possible interpretations to give to campaign events. Since an election is a complex and ambiguous phenomenon, different conclusions can be drawn about its meaning. Here the journalists help the public to form specific impressions of the candidates.

For example, presidential candidates are always concerned with how the results of presidential primaries are interpreted. Candidates want to be seen as winners who are gaining momentum. At the same time, a candidate seen as the front-runner too early is open to being shot down later.

Third, the media exercise discretion in how favorably candidates are presented in the news. Although norms of objectivity and balance prevent

Sociology at Work

Comparing the Political and Moral Values of the 1960s with Today

We often hear about the era of the 1960s and the massive social changes that emerged from that period. It was a time of sweeping changes in civil rights legislation. The youth culture was challenging the moral authority of government in many areas. Social changes, whether they be in the area of family, sex, religion, or race, were taking place everywhere. Have we become more conservative as a society than we were during that period? In some areas, we clearly have moved away from the liberal views of the 1960s. Surprisingly, in other areas we are actually a much more tolerant society today than we were in the 1960s.

Examples of the more conservative mood include the distrust of the government. During the 1960s, only a distinct minority of Americans was distrustful of government. Today two-thirds of Americans do not trust government to do the right thing. The proportion of Americans who see big government as the biggest threat to the country's future has grown from 35% in the 1960s to 59% today. In addition, far fewer Americans today want government to do more to help minorities reduce income differences than in the 1960s.

On the other hand, there have been striking changes in the tolerance Americans express for different groups in society. Today, 87% of Americans are against laws that would prohibit marriages between African Americans and whites. In the 1960s, only 37% felt that way. As late as 1977, only 57%

of Americans thought homosexuals should have equal rights in terms of job opportunities. Today, 87% support equal job opportunities. Americans have even become more willing to vote for an atheist for president, increasing from 18 to 44%.

Table 13-3

American Values—The 1960s and Today

	THE 1960s	TODAY
Do not trust government in Washington to do the right thing	23%	66%
Would like government to do more to help minorities	50%	37%
Government should reduce differences in income	48%	30%
There should be no laws against blacks and whites being able to marry	36%	87%
Homosexuals should have equal rights in terms of job opportunities	56%	87%
Would vote for a well-qualified atheist for president	18%	44%

Source: "The 60s and the 90s: Americans' Political, Moral, and Religious Values Then and Now," by R. J. Blendon, J. M. Benson, M. Brodie, D. E. Altman, R. Moran, Claudia Deane, & N. Kjellson, Spring 1999, *Brookings Review*, pp. 14–17.

most campaign coverage from including biased assertions, a more subtle and pervasive slant or theme to campaign coverage is possible and can be significant.

Finally, newspaper editors and publishers may officially endorse a candidate. This support can be particularly important to the candidate if the newspaper is one of the major national publications.

Politicians need the mass media to get the coverage they need to win. At the same time, they can very easily fall victim to the intense scrutiny that is likely to result.

Although a great deal has been written about the power of the press, little is known about the upper echelons of the newspaper hierarchy—the top decision makers or the boards of directors of newspaper-owning corporations. Much of what is known about these individuals comes from official and unofficial

biographies of publishers, histories of particular newspapers, and journalistic accounts. Though these studies suggest that publishers and board members can influence the general tone of a paper as well as specific stories, there has been little systematic research on the characteristics of these people and how (or if) they are connected to other sectors of the U.S. power structure.

Special-Interest Groups

With the government spending so much money and regulating so many industries, special-interest groups constantly attempt to persuade the government to support them financially or through favorable regulatory practices. **Lobbying** refers to *attempts by special-interest groups to influence government policy.* Farmers lobby for agricultural subsidies, labor unions for higher minimum wages and laws favorable to union organizing and strike actions, corporate and big business interests for favorable legislation and less government control of their practices and power, the National Rifle Association to prevent the passage of legislation requiring the registration or licensing of firearms, consumer-protection groups for increased monitoring of corporate practices and product quality, the steel industry for legislation taxing or limiting imported steel, and so on.

Lobbyists Of all the pressures on Congress, none has received more publicity than the role of the Washington-based lobbyists and the groups they represent. The popular image of a lobbyist is an individual with unlimited funds trying to use devious methods to obtain favorable legislation. The role of today's lobbyist is far more complicated.

The federal government has tremendous power in many fields, and changes in federal policy can spell success or failure for special-interest groups. With the expansion of federal authority into new areas and the huge increase in federal spending, the corps of Washington lobbyists has grown markedly. The number of registered lobbyists swelled to 20,512 in 1999, according to the Senate Office of Public Records. In all there were more than 38 registered lobbyists and $2.7 million in lobbying money for every member of Congress.

Lobbyists usually are personable and extremely knowledgeable about every aspect of their interest group's concerns. They cultivate personal friendships with officials and representatives in all branches of the government, and they frequently have conversations with these government people, often in a semisocial atmosphere, such as over drinks or dinner.

The pressure brought by lobbyists usually has self-interest aims—that is, to win special privileges or financial benefits for the groups they represent. On some occasions the goal may be somewhat more objective, as when the lobbyist is trying to further an ideological goal or to put forth a group's particular interpretation of what is in the national interest.

There are certain liabilities associated with lobbyists. The key problem is that they may lead Congress to make decisions that benefit the pressure group but may not serve the interests of the public. A group's influence may be based less on the arguments for their position than on the size of the membership, the amount of their financial resources, or the number of lobbyists and their astuteness.

Lobbyists might focus their attention not only on key members of a committee but also on the committee's professional staff. Such staffs can be extremely influential, particularly when the legisla-

Lobbyists usually attempt to win special privileges or financial benefits for the groups they represent. The National Rifle Association (NRA) has been very effective in lobbying members of Congress to defeat gun-control legislation.

tion involves highly technical matters about which the member of Congress may not be knowledgeable. Lobbyists also exert their influence through testimony at congressional hearings. Those hearings may give the lobbyist a propaganda forum and also access to key lawmakers who could not have been contacted in any other way. The lobbyists may rehearse their statements before the hearing, ensure a large turnout from their constituency for the hearing, and may even give leading questions to friendly committee members so that certain points can be made at the hearing.

Lobbyists do perform some important and indispensable functions, such as helping inform Congress and the public about problems and issues that normally may not get much attention, thereby stimulating public debate and making known to Congress who would benefit and who would be hurt by specific legislation. Many lobbyists believe that their most important and useful role, both to the groups they represent and to the government, is the research and detailed information they supply. In fact, many members of the government find the data and suggestions they receive from lobbyists to be valuable in studying issues, making decisions, and even in voting on legislation. (While lobbying is often motivated by a desire to gain political influence, it usually does not involve outright payoffs and illegal

Controversies in Sociology

How Evil Are Bribery and Corruption?

Corruption has been common in societies since the days of ancient Greece and Rome. No type of society is immune, whether it be dictatorial or democratic; feudal, capitalistic, or socialistic. Corruption comes in many forms including bribery, extortion, fraud, and embezzlement. But it also takes the form of nepotism and cronyism. Corruption and bribery need not involve the exchange of money. Other gifts or advantages, such as membership in an exclusive club or promises of scholarships for children, have been used as bribes.

What harm does corruption inflict on a society? It erodes public confidence in the political institutions and leads to contempt for the rule of law. Money is not spent on what is needed, but on what can best line the pockets of corrupt officials, resulting in a devastating effect on investment, growth, and development. The poor in particular suffer from corrupt governments, because they are denied access to vital basic services. Corrupt governments hurt the education system, because corrupt public officials shift government expenditures to areas where bribes can more easily be collected. Larger, hard-to-manage projects, such as airports or highways, facilitate fraud. Education expenditures that are more visible are less attractive for officials seeking bribes.

Widespread corruption takes place when the legal system, the media, and civil society have no con-

Table 13-4

Ten Most Corrupt and Ten Least Corrupt Countries According to Corruption Perceptions Index, 2000

MOST CORRUPT COUNTRIES	LEAST CORRUPT COUNTRIES*
1. Nigeria	1. Finland
2. Yugoslavia	2. Denmark
3. Ukraine	3. New Zealand
4. Azerbaijan	4. Sweden
5. Indonesia	5. Canada
6. Angola	6. Iceland
7. Cameroon	7. Norway
8. Russia	8. Singapore
9. Kenya	9. Netherlands
10. Mozambique	10. United Kingdom

*The United States was 14th on the least corrupt list.

Source: "TI Press Release: Transparency International Releases Year 2000 Corruption Perception Index," September 13, 2000, Transparency International, Berlin.

trol over government agencies. When corruption permeates a country's political and economic institutions, it is no longer a matter of a few dishonest individuals. The country starts to be governed for the benefit of the few at the expense of the many.

Source: "Corruption, Culture and Markets," by S. M. Lipset & G. S. Lenz, 2000, in L. E. Harrison & S. P. Huntington, *Culture Matters* (pp. 112–124), New York: Basic Books.

activity. For a discussion of such actions see "Controversies in Sociology: How Evil is Bribery and Corruption?")

Political Action Committees Special-interest groups called **political action committees (PACs)** are *organized for the purpose of raising and spending money to elect and defeat candidates.* Most PACs represent business, labor, or ideological interests. PACs have been around since 1944, when the first one was formed to raise money for the reelection of President Franklin D. Roosevelt. In 1998 there were 3,798 PACs in operation; they are an extremely influential form of special-interest group. Between 1997 and 1998, PACs contributed $219.9 million to congressional and senatorial political campaigns (*Statistical Abstract of the United States: 1999, 2000*).

Several criticisms have been leveled at these special-interest groups. Among the most prominent is that they represent neither the majority of the American people nor all social classes. Most PACs represent groups of affluent and well-educated individuals or large organizations. Only about 10% of the population is in a position to exert this kind of pressure on the government. Disadvantaged groups—those that most need the ear of the government—have no access to this type of political action (Cummings & Wise, 1981).

PACs also tend to favor incumbents. Two-thirds of all PAC contributions in recent elections have gone to incumbents. Challengers therefore end up being much more dependent on small donations from individuals or from the Democratic and Republican National Committees. PACS ultimately may diminish the role of the individual voter.

SUMMARY

In most societies, what laws are passed or not passed depends largely on which categories of people have power. Politics is the process by which power is distributed and decisions are made. Power, as Weber defined it, is the ability of a person or group to carry out its will, even in the face of resistance or opposition. Power is exercised in a broad spectrum of ways. At one pole is authority, or power that is regarded as legitimate by those over whom it is exercised, who also accept the authority's legitimacy in imposing sanctions or even in using force if necessary. At the other extreme is coercion, or power that is regarded as illegitimate by those over whom it is exerted. Power based on authority is quite stable, and obedience to it is accepted as a social norm. Power based on coercion is unstable. It is based on fear; any opportunity to test it will be taken, and in the long run it will fail.

In societies where political power is shared among most or all adults, the concept of government is meaningless. In modern, complex societies, however, government is necessary, and the state is the institutionalized way of organizing power within territorial limits.

The state has a variety of functions. One is to establish laws that formally specify what is expected and what is prohibited in the society. Another function is to enforce those laws and make sure that violations are punished. In modern societies, the state must also try to ensure that a stable system of distribution and allocation of resources exists. The state also sets goals and provides a direction for society, usually through the power to compel compliance. Finally, the state must protect a society from outside threats, especially those of a military nature.

Politics and economics are intricately linked, and the political form a state takes is tied to the type of economy. The economy is the social institution that determines how a society produces, distributes, and consumes goods and services.

Democracy has many different meanings, and there is hardly a government anywhere that does not claim some form of democratic authority. Today, democracy refers to a political system operating under the principles of constitutionalism, representative government, civilian rule, majority rule, and minority rights.

Functionalists view social stratification—and the state that maintains it—as necessary for the recruitment of workers to perform the tasks required to sustain society. Marxists and other conflict theorists argue that the state emerged as a means of coordinating the use of force, by means of which the ruling classes could protect their institutionalized supremacy from the resentful and potentially rebellious lower classes. While there is historical evidence for the conflict view of the state's origins, the functionalist view points to the important societal functions fulfilled by the state.

Growing out of a strong commitment to a democratic political process and the influence of a capitalist economy, the political system in the United States is unique in a number of ways. Few democracies, for instance, have a winner-take-all two-party system. In this system, political candidates are forced to gain the support of a broad spectrum of interest groups to be elected. These interest groups must find common ground and join together in coalitions or face the prospect of not being represented at all if their candidate loses.

INTERNET RESOURCES

Go to http://web-enhanced.thomsonlearning.com which features Web links relevant to this chapter to expand and enhance the learning experience. This site will be monitored and updated periodically to ensure the links are correct and live.

At Virtual Society: The Wadsworth Sociology Resource Center, http://sociology.wadsworth.com, you will find a Career Center, Sociology in the News, Research Online, InfoTrac College Edition, Virtual Tours in Sociology, and a variety of book-specific student and instructor resources.

Chapter Thirteen Study Guide

KEY CONCEPTS

Match each concept with its definition, illustration, or explanation below.

a. Revolution
b. Aristocracy
c. Fascism
d. Totalitarian capitalism
e. Dictatorship
f. Laissez-faire capitalism
g. Legal-rational authority
h. Rebellion
i. Lobbying
j. Politics
k. Totalitarian socialism
l. Constitutionalism

m. Capitalism
n. Authority
o. Social revolution
p. Mixed economy
q. Democracy
r. The state
s. Coercion
t. Traditional authority
u. Democratic socialism
v. Power
w. Autocracy
x. Representative Government

y. Totalitarian government
z. Economy
aa. Political revolution
bb. Communism
cc. Political action committee
dd. Charismatic authority
ee. Code of Hammurabi
ff. Command economy
gg. Civilian rule
hh. Socialism
ii. Electorate
jj. FDR presidency

_____ 1. The process by which power is distributed and decisions are made.

_____ 2. The ability of a person or group to get their way, even in the face of resistance or opposition.

_____ 3. Power that is regarded as legitimate by those over whom it is exercised.

_____ 4. Power that is regarded as illegitimate by those over whom it is exerted.

_____ 5. A form of authority derived from the understanding that specific individuals have clearly defined rights and duties to uphold and implement rules and procedures impersonally.

_____ 6. A form of authority rooted in the assumption that the customs of the past legitimate the present.

_____ 7. A form of authority derived from a ruler's ability to inspire passion and devotion among followers.

_____ 8. The institutionalized way of organizing power within territorial limits.

_____ 9. A form of government in which ultimate authority rests with one person, who is the chief source of laws and the major agent of social control.

_____ 10. A form of government in which a select few rule.

_____ 11. The social institution that determines how a society produces, distributes, and consumes goods and services.

_____ 12. An economic system based on private ownership of the means of production, and re-source allocation through the market.

_____ 13. The view that government should stay out of business affairs.

_____ 14. A combination of free-enterprise capitalism and governmental regulation of business and provision of social welfare programs.

_____ 15. An economic system in which the government makes all the decisions about production and consumption.

_____ 16. An economic system in which the government owns the sources of production and sets production and distribution goals.

_____ 17. A situation in which one group has virtually total control of a nation's social institutions.

_____ **18.** A situation in which the government, in addition to almost total regulation of all social institutions, owns and controls all major means of production and distribution.

_____ **19.** Another name for totalitarian socialism.

_____ **20.** A situation in which the government, while retaining control of social institutions, allows private ownership of the means of production.

_____ **21.** Another name for totalitarian capitalism.

_____ **22.** A political system operating under the principles of constitutionalism, representative government, majority rule, civilian rule, and minority rights.

_____ **23.** A situation in which government power is limited by law.

_____ **24.** The situation in which the authority to govern is achieved through, and legitimized by, popular elections.

_____ **25.** Those citizens eligible to vote.

_____ **26.** A situation in which every qualified citizen has the legal right to run for and hold an office of government.

_____ **27.** A totalitarian government in which all power rests ultimately in one person.

_____ **28.** A convergence of capitalist and socialist economic theory in which the state assumes ownership of strategic industries and services, but allows other enterprises to remain in private hands.

_____ **29.** An attempt—typically through armed force—to achieve rapid political change not possible within existing institutions.

_____ **30.** An attempt to change a society's previously existing structure rapidly and dramatically.

_____ **31.** A relatively rapid transformation of state or government structures that is not accompanied by changes in social structure or stratification.

_____ **32.** A rapid and basic transformation of a society's state and class structures.

_____ **33.** Attempts by special-interest groups to influence government policy.

_____ **34.** A special-interest group concerned with very specific issues that usually represents corporate, trade, or labor interests.

_____ **35.** His 1944 reelection campaign saw the beginnings of political action committees in the United States.

_____ **36.** Showed the class bias of law through proscribing the death penalty for persons harboring fugitive slaves and for wives who poorly manage husband's households.

KEY THINKERS/RESEARCHERS

Match the thinkers with their main ideas or contributions.

a. **Adam Smith**   d. **Plato**   f. **Lenin**
b. **Karl Marx**   e. **Aristotle**   g. **Max Weber**
c. **Thomas Jefferson**

_____ **1.** Developed the sociological definitions of power and authority.

_____ **2.** Constructed a philosophical argument for aristocracy as the best form of government.

_____ **3.** Argued that power should be centered in the middle class and that the rights and duties of the state should be defined in a legal constitution.

_____ **4.** Regarded as the father of modern capitalism; discussed many of the basic premises of this system.

_____ **5.** A severe critic of capitalism who argued that it was based on alienation and exploitation.

_____ **6.** Argued the proletariat needed state power and a centralized government.

_____ **7.** Argued that a political system had to be taken out periodically, inspected, examined, and changed if necessary before passing it on to the next generation.

CENTRAL IDEA COMPLETIONS

Following the instructions, fill in the appropriate concepts and descriptions for each of the questions posed in the following section.

1. What are the three main reasons given by people when questioned about why they did not bother to vote?

_____

_____

_____

2. What is the relationship between the marital status among African-American couples and the income gap between African-Americans and white Americans?

_____

_____

3. What is the number of Hispanics of voting age as of November 2000? _____

How does the number of Hispanic American registered voters during this same time period compare with the numbers of white and African-American registered voters? _____

_____

4. What are the five functions of the state?

a. _____

b. _____

c. _____

d. _____

e. _____

5. According to Weber, there are two functions of capitalism. What are they?

a. _____

b. _____

6. What are the four features of capitalism outlined by Adam Smith?

a. _____

b. _____

c. _____

d. _____

7. Discuss the vision of the future of capitalism held by Karl Marx:

_____

_____

_____

8. What are the essential similarities and differences between socialism and communism?

   a. Similarities: _____

   _____

   b. Differences: _____

   _____

9. What are the basic points of the critiques of socialism by capitalists?

   a. _____

   b. _____

   c. _____

10. What are the six specific elements totalitarian governments use to concentrate political power?

    a. _____

    b. _____

    c. _____

    d. _____

    e. _____

    f. _____

11. What are the two simultaneous and interrelated processes of social revolutions?

    a. _____

    b. _____

12. What are the different forms of authority described by Max Weber? How are these forms related to the use of coercion for gaining compliant behavior?

    a. Forms of authority: _____

    _____

    b. Coercion: _____

    _____

13. Briefly list the issues about which the post WWII baby boomers have changed their political opinions since the time when they were in their 20s. Be sure to include the one issue in which they have become more liberal as they have aged.

    _____

    _____

    _____

14. Present a brief listing with a short explanation of the ways in which the media have contributed to a radical transformation of election campaigns in the United States:

    _____

    _____

15. Present examples supporting the statement that the founders of the United States purposively made decisions to create a weak political system:

    _____

    _____

    _____

CRITICAL THOUGHT EXERCISES

1. Go to your library and find two examples of very close elections. Next, look up two examples of presidential concession speeches that were offered after those close elections. Compare these speeches with the recent concession speech made by Vice-President Al Gore. Do you find any common features of content? If there are common features, what are they? What are the manifest and latent functions of concession speeches following presidential elections in the United States? What specific phrases were made by each of these contenders in their concession speeches that could be linked to the manifest and latent function you previously listed?

2. Beginning with FDR's first term and continuing into the election of George W. Bush, the media have played an important role in American presidential politics. Compare and contrast the role of the media as it has evolved through these 12 men. What media variables have been relatively constant and what ones are either new or specific to a particular president(s)? Following the discussion in your text, what changes regarding the role of the media in American presidential politics would you forecast for the next election?

3. Discuss the American two-party system by focusing on the voting behaviors and power distributions of African Americans and Hispanic Americans. To what extent have they been able to function as effective political forces and in what areas have their agendas been sidetracked by the dominant white majority?

4. Directing your attention to Table 13-3, what have been the major changes in American values when comparing the 1960s to today? What do these changes in values tell one about the culture and structure of American society in the twentieth-first century?

5. Examine the box titled "Remaking the World: The Campaign to Ban Land Mines." What were the roles of Bobby Muller and Jody Williams in this campaign? What does the deployment of land mines say about how opposing countries view the health and safety of noncombatant members of a population with which one is at war?

6. Data presented in Table 13-4 offer an intriguing picture of the perception of countries throughout the world. What common factors appear to be operating in the perceptions of corruption across the 10 most corrupt nations compared with those common factors that appear to be operating across the 10 least corrupt nations? How do you interpret the fact that the United States ranked only 14th on the least corrupt list? In your opinion, is that a reasonable ranking given the information presented in your text and what you have come to know by virtue of being a citizen of the United States?

ANSWERS

KEY CONCEPTS

1. j	7. dd	13. f	19. bb	25. ii	31. aa
2. v	8. r	14. p	20. d	26. gg	32. o
3. n	9. w	15. ff	21. c	27. e	33. i
4. s	10. b	16. hh	22. q	28. u	34. cc
5. g	11. z	17. y	23. l	29. h	35. jj
6. t	12. m	18. k	24. x	30. a	36. ee

KEY THINKERS/RESEARCHERS

1. g	3. e	4. a	5. b	6. f	7. c
2. d					

PART 5

Social Change and Social Issues

At this point in your voyage of sociological discovery, you have mapped out much of the structure of society, and you have discovered many important characteristics of the natives who inhabit the territory. What remains is to explore some of the most compelling social issues within contemporary society—problems and situations that demand attention and represent driving forces of social change. How we address these issues will undoubtedly affect the quality of life in our future societies. To inhibit damaging trends and enhance productive ones, we must understand not only the problems themselves, but also the process of social change. Only then can we hope to reshape society in ways that effectively promote human growth and development.

Chapter 14 deals with the dynamics of human populations. How do we measure such things as population growth, stability, and decline? What social factors influence each trend? Do we face a problem of overpopulation on this planet? In this chapter you will see, once again, that there are competing definitions and explanations for a given social problem. The perspective we adopt, obviously, affects the solutions we generate.

One important trend in population dynamics is urbanization: the increasing tendency of humans to live in cities and for urban culture and urban concerns to dominate national agendas. In Chapter 15, you will examine this phenomenon, as well as some of the most significant urban problems in the United States today: central-city decay, suburban sprawl, and homelessness.

Staying healthy is an important human issue. Who gets sick, however, is powerfully influenced by social factors such as gender, race, social class, and age. In addition, access to health care is distributed unequally. These are some of the issues you will investigate in Chapter 16. Additionally, you will explore two of the most compelling health issues in the world today: the spread of AIDS, and the tenuous health of mothers and children in developing countries.

You have seen throughout this book that humans are social animals. How do humans behave in large groups? How and why does group size make a difference? In Chapter 17, you will explore several sociological theories of collective behavior, and you will examine a variety of collective behaviors: crowds, fads, rumors, public opinion, mass hysteria, and social movements. In the latter case, you will be looking at how people consciously and collectively organize to make social change. Chapter 18, the final chapter of this book, will introduce you to major theories and significant sources of social change in contemporary society.

When you have finished Part 5 of this book, you will have completed your maiden voyage of sociological exploration. The knowledge and insight that you have gained to this point should be immensely useful in navigating through your society and daily social interactions. As an educated citizen of your society and the world, however, you are challenged to continue to expand your knowledge and insight in order to act more intelligently and effectively, both as an individual and in concert with others. Bon voyage!

CHAPTER FOURTEEN

Population and Demography

LEARNING OBJECTIVES

After studying this chapter, you should be able to do the following:

- Describe the phenomenon of exponential growth.
- Define the three major components of population change.
- Contrast the Malthusian and Marxist theories of population.
- Summarize the demographic transition model and explain why there might be a second demographic transition.
- Discuss the determinants of fertility and family size.
- Discuss the problems of overpopulation and possible solutions.

The Chinese words *wan, xi, shao* mean later, longer, fewer: later marriage, longer periods of time between pregnancies, and fewer children.

In 1971, that slogan launched a birth-control campaign—unprecedented in scale in all of history—in the People's Republic of China. Posters proclaiming the message appeared virtually everywhere, even in the smallest village outposts.

At first the goal was not specified, but a three-child family was considered acceptable. Then, two. Now the emphasis is on the last word: *shao,* fewer—as few as possible, as fast as possible. *Shao* has been reinterpreted three times from the original "fewer" to "one is best" and finally to "one is enough."

When Mao Ze Dong was in power he said there could never be too many Chinese. He assumed that the sheer number of people would be China's greatest defense in the imminent third world war. As a result, the population grew from 540 million in 1950 to more than 850 million by 1970. In the 1970s, the average Chinese woman of childbearing age gave birth to six children (Hesketh & Zhu, 1997).

As the political situation in China changed, the government decided to take drastic action to ensure future survival, by setting a population goal of no more than one child per couple in most urban areas. To accomplish this goal, China instituted a nationwide campaign to persuade couples to follow the government's guidelines.

To promote acceptance of the one-child limit, the Chinese leaders devised a reward-punishment system that makes daily life easier and richer for those who comply and burdensome for those who do not. A nationwide campaign to promote the one-child family included such slogans as "With two children you can afford a 14-inch TV, with one child you can afford a 21-inch TV" (Hesketh & Zhu, 1997).

Chinese couples must obtain permission to be married as well as to have a child. Couples are also encouraged to obtain the Glorious One Child Certificate, which shows that they have signed an agreement to limit their family to a single child in exchange for extensive benefits. The certificate entitles the couple to a monthly child-rearing allowance until the child reaches age 14.

In some areas, particularly cities, the one-child policy is often promoted through incentives, such as extra salary or larger houses for couples who pledge to have just one child. The government generally pays for birth control and abortions (and a woman who has an abortion receives a vacation with pay). Failure to abide by the policy may result in job loss or demotion.

Only certain families may have more than one child: rural families whose first child is a daughter, families where both parents are themselves only children, and families in which the first child is handicapped. Others who have a second child pay stiff fines. In Shanghai, the fine is three times the combined annual salary of the parents (Rosenthal, 1998).

If a couple has a second child without permission or paying a fine, the second child cannot be registered and, therefore, does not legally exist. The child cannot attend school and later will have difficulty obtaining permission to marry, to relocate, and for other life choices requiring the government's permission (Cato Institute, 2000).

Penalties are particularly harsh for those who have a third child. Ten percent of their pay or work points are deducted from the fourth month of pregnancy until the third child is 14 years old. The same penalties are imposed on couples who have a child out of wedlock.

An additional punishment for the three-or-more-child couple is denial of any job promotion and loss of all work bonuses for at least 3 years. This is to make sure that couples who exceed the state limit for children do so at the price of personal sacrifice. They cannot prosper by doing extra work. In 1996, only 6.6% of Chinese births were third or higher-order children (Population Reference Bureau, 1997).

There is no question that these policies are working. The total fertility rate—the expected number of live births for a woman in her childbearing career—fell from 6.7 in the mid-1960s in rural China to 1.81 in 1996. This is just below the number required for the population to remain stable (Population Reference Bureau, 1997).

Population Dynamics

Chinese leaders have given the one-child policy top priority, ahead of competing social goals, because population growth overshadows all other issues in that country. As of 2000, China's 1.26 billion people accounted for 21% of the world's inhabitants. The dramatic impact that the one-child policy can have on population growth can be seen if we compare population projections based on a three-child family and a one-child family. If for the next 100 years, couples were to have an average of three children, China's population would reach 4.2 billion by the year 2080. If, however, the planned one-child family were to prevail over that time, China could bring its population down to a socially and economically manageable 370 million by 2080; a population number

Most of the world's population growth in the next few decades will occur in the developing countries.

still 130 million greater than that of the United States today in a land area virtually the same size.

China is responding to a basic demographic fact: Population problems become progressively more pressing because of what is referred to as exponential growth. The yearly increase in population is determined by a continuously expanding base. Each successive addition of 1 million people to the population requires less time than the previous addition required, even if the birthrate does not increase. The best way to demonstrate the effects of exponential growth is to use a simple example. Let us assume that you have a job that requires you to work eight hours a day, every day, for 30 days. At the end of that time, your job is over. Your boss offers you a choice of two different methods of payment. The first choice is to be paid $100 a day for a total of $3,000. The second choice is somewhat different. The employer will pay you 1 cent for the first day, 2 cents for the second day, 4 cents for the third, 8 cents on the fourth, and so on. Each day you will receive double what you received the day before. Which form of payment would you choose? The second form of payment would yield a significantly higher total payment—so high, in fact, that no employer could realistically pay the amount. Through this process of successive doubling, you would be paid $5.12 for your labor on the 10th day. Only on the 15th day would you receive more than the flat $100 a day you could have received from the first day under the alternative payment plan: The amount that day would be $163.84. However, from that day on, the daily pay increase is quite dramatic. On the 20th day you would receive $5,242.88, and on the 25th day your daily pay would be $167,772.16. Finally, on the 30th day, you would be paid $5,368,709.12, bringing your total pay for the month to more than $10 million.

This example demonstrates how the continual doubling in the world's population produces enormous problems. The annual growth rate in the world's population has declined from a peak of 2.04% in the late 1960s to 1.5% in 1997. This difference between global birth and death rates means that the world's population now doubles every 51 years instead of every 35 years. Although this is an improvement for the world as a whole, many portions of the world have not seen any improvement in their growth rates; in fact, the reverse may be true. In some countries in Africa, the national fertility rates have actually increased in the past decade, and Africa is now the area of the world with the most rapid population growth. In Ethiopia, for example, the average woman now has 6.7 children. When this fact is combined with the declining infant mortality rate, the country's population could balloon from 64.1 million in 2000 to 256.4 million

Table 14-1

Annual Percentage Increase and Years Needed to Double Current Population

	Yearly Percentage Increase	Years Needed to Double the Population
World		
More developed countries	0.1	809
Less developed countries	1.7	42
Country		
Oman	3.9	18
Palestinian Territory	3.7	19
Chad	3.3	21
Zaire	3.2	22
Saudi Arabia	3.0	23
Nicaragua	3.0	23
Jordan	2.9	24
Somalia	2.9	24
Yemen	2.8	25
Pakistan	2.8	25
Syria	2.8	25
Cambodia	2.6	27
Libya	2.5	28
Afghanistan	2.5	28
Ethiopia	2.4	29
El Salvador	2.4	29
Philippines	2.3	31
Mozambique	2.2	32
Kenya	2.1	33
Malawi	1.9	36
Israel	1.5	45
Brazil	1.5	45
Argentina	1.1	62
Thailand	1.0	70
Singapore	0.8	84
United States	0.6	116
Ireland	0.6	116
Canada	0.4	178
France	0.3	204
Japan	0.2	462
United Kingdom	0.1	546
Italy	0.1	—
Sweden	− 0.1	—
Russia	− 0.6	—
Bulgaria	− 0.6	—

Source: Population Reference Bureau, 2000 World Population Data Sheet.

in 2046. Enormous growth has been projected for Oman, Nigar, and Uganda. (See Table 14-1 for projected yearly population growth percentages for selected countries.)

Population growth rates differ greatly by region. The rate is highest in countries in the center of sub-Saharan Africa, where the rate is 3% a year, and lowest in Eastern Europe at −0.5%. Most of these

Figure 14-1 Past and Projected World Population, A.D. 1–2150

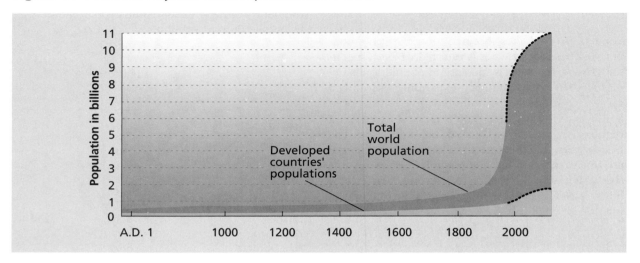

former Soviet republics are experiencing population loss—that is, more deaths than births. If sub-Saharan Africa continued its current rate of growth, it would take it 23 years to double in population. By contrast, at current growth rates, Western Europe's population would need 612 years to double in size (Population Reference Bureau, 2000 World Population Data Sheet).

As you may have deduced, most of the world's growth in population in the next few decades will take place in developing countries. The population of the richer countries will increase by 200 million by the year 2050, while the developing areas will have added about 6 billion (see Figure 14-1).

Symbolizing the shift in population from the developed to the less-developed world is São Paulo, Brazil. In 1950, this city was smaller than Manchester, England, but by the year 2000 Sao Paulo's population had reached about 24.0 million, making it one of the largest cities in the world. London, which was the second-largest city in the world in 1950, was not in the top 20 largest cities in 2000, due to current growth trends.

Our discussion of the population shifts in various countries falls into what is called demography. **Demography** is *the study of the size and composition of human populations, as well as the causes and consequences of changes in these factors.* Demography is influenced by three major factors: fertility, mortality, and migration.

Fertility

Fecundity is *the physiological ability to have children.* Most women between the ages of 15 and 45 are capable of bearing children. During this time, a woman could potentially have up to 30 children; however, the realistic maximum number of children

a woman can have is about 15. This number is a far cry from real life, though, where health, culture, and other factors limit childbearing. Even in countries with high birthrates, the average woman rarely has more than 8 children (McFalls, 1991).

Whereas fecundity refers to the biological potential to bear children, **fertility** refers to *the actual number of births in a given population.* One common way of measuring fertility is by using the **crude birthrate,** *the number of annual live births per 1,000 people in a given population.* The crude birthrate for the United States fell from 24.1 in 1950 to 14.6 in 1997 (Population Reference Bureau, *Population Today, January 1998*).

Another indicator of reproductive behavior is the **fertility rate,** *the annual number of births per 1,000 women of childbearing age, usually defined as 15 to 44.* In 1999, the fertility rate for the United States was 65.3 (Population Reference Bureau, *Population Today, November/December, 2000*).

As you will see later in this chapter, the fertility rate is linked to industrialization. Fertility declines with modernization, but not immediately. This lag is a source of tremendous population pressure in developing nations that have benefited from the introduction of modern medical technology, which immediately lowers mortality rates.

Mortality

Mortality is *the frequency of deaths in a population.* The most commonly used measure of this is the **crude death rate,** *the annual number of deaths per 1,000 people in a given population.* In 1999, 2,351,000 Americans died, producing a crude death rate of 8.6 per 1,000 (Population Reference Bureau, *Population Today, November/December 2000*). Demographers also look at **age-specific death rates,**

The fertility rate is linked to industrialization and tends to decline with modernization.

which *measure the annual number of deaths per 1,000 people in a population at specific ages.* For example, one measure used is the **infant mortality rate,** which *measures the number of children who die within the first year of life per 1,000 live births.* Of 1000 babies born this year in sub-Saharan Africa, 94 will die within 1 year. In the world's more developed countries, it will take 60 years for these 94 deaths to occur. The difference reflects a continuing gap in mortality levels between the world's more and less developed countries (Population Reference Bureau, *2000 World Population Data Sheet*).

In the United States, the infant mortality rate dropped from 47 in 1940 to 7 in 1999 (Population Reference Bureau, *Population Today, November/December 2000*). This rate does not apply to all infants, however. In 1940, whites had an infant mortality rate of 43.2 (below the national average at that time), and African Americans had an infant mortality rate of 73.8. Today, the rates are 6.0 for whites and 13.8 for blacks (U.S. Census Bureau, 2000). Those figures suggest that good infant medical care is not equally available to all Americans. Differing cultural patterns of childrearing may also affect infant mortality.

Although the infant mortality rate in the United States is considerably lower than the rates of developing countries, it is higher than those in some Asian or European countries. The lowest infant mortality rates in the world are found in Singapore (3.2) and Japan (3.5). Mortality is reflected in people's **life expectancy,** *the average number of years a person born in a particular year can expect to live* (see Table 14-2 for life expectancies in selected countries). The average life expectancy at birth in the United States is

77 years. The world's longest life expectancy is in Japan, at 81 years. The shortest life expectancies are in Rwanda, Swaziland, and Zambia (Population Reference Bureau, *2000 World Population Data Sheet*).

Life expectancy is usually determined more by infant mortality than by adult mortality. Once an individual survives infancy, life expectancy improves dramatically. In the United States, for example, only when individuals reach their 60s do their chances of dying approximate those of their infancy (U.S. Bureau of the Census, 1997).

Table 14-2

Life Expectancy for Children Born in 1999

	YEARS		
World	66		
COUNTRY	YEARS		YEARS
Japan	81	China	71
Australia	79	Saudi Arabia	70
Canada	79	Morocco	69
Sweden	79	Peru	68
France	78	Bolivia	65
Spain	78	India	61
Greece	78	Nigeria	52
Italy	78	Sudan	51
Netherlands	78	Angola	47
United Kingdom	77	Afghanistan	46
Germany	77	Uganda	42
United States	77	Rwanda	39
Chile	75	Zambia	37
Mexico	72		

Source: Population Reference Bureau, *2000 World Population Data Sheet.*

Table 14-3

Ten Leading Causes of Death in the United States, 1900 and 1997

1900		1997	
CAUSE OF DEATH	PERCENTAGE OF ALL DEATHS	CAUSE OF DEATH	PERCENTAGE OF ALL DEATHS
1. Pneumonia	12	1. Heart disease	31.4
2. Tuberculosis	11	2. Cancer	23.3
3. Diarrhea and enteritis	8	3. Stroke	6.9
4. Heart disease	8	4. Lung disease	4.7
5. Kidney disease	5	5. Accidents	4.1
6. Accidents	4	6. Pneumonia and influenza	3.7
7. Stroke	4	7. Diabetes	2.7
8. Diseases of early infancy	4	8. Suicide	1.3
9. Cancer	4	9. Kidney disease	1.1
10. Diphtheria	2	10. Liver disease	1.1

Source: "Deaths: Final Data for 1997," National Center for Health Statistics, June 30, 1999, *Monthly Vital Statistics* 47(19).

A fact that is often overlooked is that the rapid increase in population growth in Third World countries is caused by sharp improvements in life expectancy, not by rising birthrates. Disease took a dramatic toll on life expectancy in the United States in the not-too-distant past. For example, Abraham Lincoln's mother died when she was 35 and he was 9. Prior to her death she had three children: Abraham Lincoln's brother died in infancy and his sister died in her early 20s. Of the four sons born to Abraham and Mary Todd Lincoln, only one survived to maturity (see Table 14-3).

In developing countries, the proportion of infant and child deaths is quite high, resulting in a significantly lower life expectancy than that in developed countries. In Bangladesh, infant deaths account for more than one-third of all deaths. In the United States, that figure is about 1%. The high proportion of deaths in developing countries can be attributed to impure drinking water and unsanitary conditions. In addition, the diets of pregnant women and nursing mothers often lack proper nutrients, and babies and children are not fed a healthy diet. Flu, diarrhea, and pneumonia are common, as are typhoid, cholera, malaria, and tuberculosis. Many children are not immunized against common childhood disease (such as polio, measles, diphtheria, and whooping cough), and the parents' income is often so low that when the children do fall ill, they cannot provide medical care.

Life expectancies in developing nations vary greatly. They range from an average regional low of about 53 years in Africa to 69 years in Latin America. A major contributor to the overall death rate is infant mortality. A rapid decrease in infant mortality in a developing nation will result in a significant rise in the overall rate of population growth.

Migration

Migration is *the movement of populations from one geographical area to another.* We call it **emigration** *when a population leaves an area* and **immigration** *when a population enters an area.* All migrations, therefore, are both emigrations and immigrations. Of the three components of population change—fertility, mortality, and migration—migration historically exerts the least impact on population growth or decline.

Most countries do not encourage immigration. When they do permit immigration, it is often viewed as a way to provide needed skilled labor, or to provide unskilled labor for jobs the resident population no longer wishes to do. Exceptions to this trend are the traditional receiver countries, such as the United States, Australia, and Canada. These countries owe much of their growth to immigrant populations.

Where migration is a significant factor, it is necessary to take into account the age and sex of the immigrants and emigrants, as well as the number of migrants. Those characteristics tell us the number of potential workers among the migrants, the number of women of childbearing age, the number of school-age children, the number of elderly, and other factors that will affect society.

Sometimes it is important to distinguish between those movements of populations that cross national boundary lines from those that are entirely within a country. To make this distinction, sociologists use the term **internal migration** for *movement within a nation's boundary lines*—in contrast with immigration, in which boundary lines are crossed.

Since 1970, population growth in the United States has been greatest in the Sunbelt states, reflecting continued migration patterns toward the South

In only three decades, Korean immigration has changed the face of commercial areas in New York City and Los Angeles.

and West. California, Texas, and Florida are growing significantly faster than the United States as a whole because they attract many Northeastern and Midwestern residents. There is some indication, however, that internal migration patterns may be starting to change. Typically, Northern and Midwestern migrants moved to the three major Sunbelt states and then distributed themselves to the surrounding states. There is some indication that future migration patterns in the United States could resemble an enormous cyclone, with a long westward flow to California from the Northeast and Midwest, and a series of shorter eastward flows through the South beginning in California and continuing to Florida and back up the Atlantic coast (Sanders & Long, 1987).

Theories of Population

The study of population is a relatively new scholarly undertaking; it was not until the eighteenth century that populations as such were examined carefully. The first person to do so, and perhaps the most influential, was Thomas Malthus.

Malthus's Theory of Population Growth

Malthus (1776–1834) was a British clergyman, philosopher, and economist who believed that population growth is linked to certain natural laws. The core of the population problem, according to Malthus, is that populations will always grow faster than the available food supply. With a fixed amount of land, farm animals, fish, and other food resources, agricultural production can be increased only by cultivating new acres, catching more fish, and so on—an additive process that Malthus believed would increase the food supply in an arithmetic progression (1, 2, 3, 4, 5, and so on). Population growth, by contrast, increases at a geometric rate (1, 2, 4, 8, 16, and so on) as couples have 3, 4, 5, and more children. (A stable population requires 2 individuals to produce no more than 2.1 children: 2 to reproduce themselves and 0.1 to make up for those people who remain childless.) Thus, if left unchecked, human populations are destined to outgrow their food supplies and suffer poverty and a never-ending "struggle for existence" (a phrase coined by Malthus that later became a cornerstone of Darwinian and evolutionary thought).

Malthus recognized the presence of certain forces that limit population growth, grouping these into two categories: preventive checks and positive checks. **Preventive checks** are *practices that would limit reproduction.* Preventive checks include celibacy, the delay of marriage, and such practices as contraception within marriage, extramarital sexual relations, and prostitution (if the latter two are linked with abortion and contraception). **Positive checks** are *events that limit reproduction either by causing the deaths of individuals before they reach reproductive age or by causing the deaths of large numbers of people, thereby lowering the overall population.* Positive checks include famines, wars, and epidemics. Malthus's thinking assuredly was influenced by the plague that wiped out so much of Europe's population during the fourteenth and fifteenth centuries.

Malthus refuted the theories of the utopian socialists, who advocated a reorganization of society to eliminate poverty and other social evils. Regardless of planning, Malthus argued, misery and suffering are inevitable for most people. On the one hand, there is the constant threat that population will outstrip the available food supplies; on the other hand, there are the unpleasant and often devastating checks on this growth, which result in death, destruction, and suffering.

According to Thomas Malthus, population will always grow faster than the available food supply.

History proved Malthus wrong, at least for developed countries. Technological breakthroughs in the nineteenth century enabled Europe to avoid many of Malthus's predictions. The newly invented steam engine used energy more efficiently, and labor production was increased through the factory system. An expanded trade system provided raw materials for growing industries and food for workers. Fertility declined and emigration eased Europe's population pressures. By the end of the nineteenth century, Malthus and his concerns had been all but forgotten.

Marx's Theory of Population Growth

Karl Marx and other socialists rejected Malthus's view that population pressures and their attendant miseries are inevitable. Marxists argue that the sheer number of people in a population is not the problem. Rather, they contend, it is industrialism (and in particular, capitalism) that creates the social and economic problems associated with population growth. Industrialists need large populations to keep the labor force adequate, available, flexible, and inexpensive. In addition, the capitalistic system requires constantly expanding markets, which can be assured only by an ever-increasing population. As the population grows, large numbers of unemployed and underemployed people compete for the few available jobs, which they are willing to take at lower and lower wages. Therefore, according to Marxists, the norms and values of a society that encourages population growth are rooted in its economic and political systems. Only by reorganizing the political economy of industrial society in the direction of socialism, they contend, is there any hope of eliminating poverty and the miseries of overcrowding and scarce resources for the masses.

Demographic Transition Theory

Sweden has been keeping records of birth and deaths longer than any other country. Throughout the centuries, Swedish birth and death rates fluctuated widely. There were periods of rapid population growth, followed by periods of slow growth, and even population declines during famines. In the late 1800s, Sweden's death rate began a sustained

Figure 14-2 The Demographic Transition Theory

The demographic transition theory states that societies pass through four stages of population change. Stage 1 is marked by high birthrates and high death rates. In Stage 2, populations rapidly increase as death rates fall, but birthrates stay high. In Stage 3, birthrates begin to fall. Finally, in Stage 4, both fertility and mortality rates are relatively low.

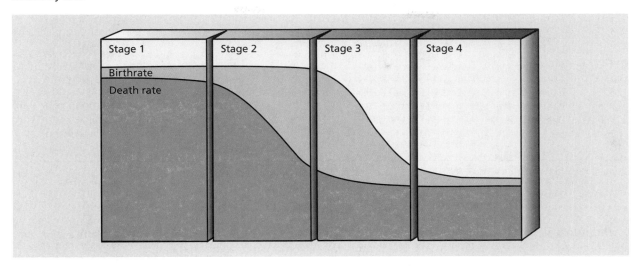

decline, while the birthrate remained high. Eventually, Sweden's birthrate also declined, so that today its births and deaths are virtually in balance.

The shift in Sweden's population can be explained by a theory of population dynamics developed by Warren Thompson. According to the **demographic transition theory**, *societies pass through four stages of population change from high fertility and high mortality to relatively low fertility and low mortality.* During Stage 1, high fertility rates are counterbalanced by a high death rate due to disease, starvation, and natural disaster. The population tends to be very young, and there is little or no population growth. During Stage 2, populations rapidly increase as a result of a continued high fertility that is linked to the increased food supply, development of modern medicine, and public health care. Slowly, however, the traditional institutions and religious beliefs that support a high birthrate are undermined and replaced by values stressing individualism and upward mobility. Family planning is introduced, and the birthrate begins to fall. This is Stage 3, during which population growth begins to decline. Finally, in Stage 4 both fertility and mortality are relatively low, and population growth once again is stabilized (see Figure 14-2).

As long as many developing nations remain in Stage 2 of demographic transition (high fertility but falling mortality), they will continue to be burdened by overpopulation, which slows economic development and creates widespread severe hunger.

Overpopulation undermines economic growth by disproportionately raising the **dependency ratio,** *the number of people of nonworking age in a society for every 100 people of working age.* Because populations at Stage 2 have a high proportion of children, as compared with adults, they have fewer able-bodied workers than they need. For example, 49% of Uganda's population is below the age of 15, compared with 21% in the United States (Population Reference Bureau, 2000 *World Population Data Sheet*). The economic development of countries with high dependency ratios is slowed further by the channeling of capital away from industrialization and technological growth and toward mechanisms for feeding the expanding populations.

Applications to Industrial Society The first wave of declines in the world's death rate came in countries experiencing real economic progress. Those declines gradually gained momentum as the Industrial Revolution proceeded. Advances in agriculture, transportation, and commerce made it possible for people to have a better diet, and advances in manufacturing made adequate clothing and housing more widely available. A rise in people's real income facilitated improved public sanitation, medical science, and public education.

Although the preceding explanation applies well to Western society, it does not explain the population trends in the underdeveloped areas of today's world. Since 1920, those areas have experienced a much faster drop in death rates than Western societies without a comparable rate of increase in economic development. The rapid rate of decline in the death rate in these countries has been due primarily

Table 14-4

Countries With the Largest Populations, 2002 and 2050

2002			2050		
RANK	COUNTRY	POPULATION	RANK	COUNTRY	POPULATION
1	China	1,284,303,705	1	India	1,533,000,000
2	India	1,045,845,226	2	China	1,517,000,000
3	United States	280,562,489	3	Pakistan	357,000,000
4	Indonesia	232,073,071	4	United States	348,000,000
5	Brazil	176,029,560	5	Nigeria	339,000,000
6	Russia	147,663,429	6	Indonesia	318,000,000
7	Pakistan	144,978,573	7	Brazil	243,000,000
8	Bangladesh	133,376,684	8	Bangladesh	218,000,000
9	Japan	129,934,911	9	Ethiopia	213,000,000
10	Nigeria	126,974,628	10	Iran	170,000,000

Sources: "World Population Prospects: 1996 Revision," 1996, U.S. Census Bureau, International Data Base; United Nations Secretariat.

to the application of medical discoveries made in and financed by the industrial nations. For example, the most important death threat being eliminated is infectious disease. Those diseases have been controlled through the introduction of vaccines, antibiotics, and other medical advances developed in the industrial nations. Those who are used to paying high prices for private medical care will find it hard to believe that preventive public health measures in underdeveloped countries can save millions of lives at costs ranging from a few cents to a few dollars per person.

Because the mortality rates in underdeveloped countries have been significantly reduced, the birthrate, which has not fallen as fast or as consistently, has become an increasingly serious problem. Often, this problem persists and worsens despite monumental government efforts to disseminate birth control information and contraceptive devices. In India, for example, despite the government's commitment to controlling population size through birth control and sterilization, the population is expected to increase from 717 million in 1982 to 1.3 billion in 2016. With the current rate of growth, India could surpass China as the most populated country somewhere between 2030 and 2035 (Population Reference Bureau, 1997). (See Table 14-4 for the countries with the largest populations in 2002 and projections for 2050.)

These failures have shown that the birthrate can be brought down only when attention is paid to the complex interrelationships of biological, social, economic, political, and cultural factors. For example, a study in Pakistan revealed that even among women who already had six children, if all six were daughters, there was a 46% chance that the mother would want more children. If, however, the six children

were boys, there was only a 4% chance that she would want additional children (Weeks, 1994).

A Second Demographic Transition

The original demographic transition theory ends with the final stage involving an equal distribution of births and deaths and a stable population. Some people (van de Kaa, 1987) have suggested that Europe has gone beyond the original theory and entered a second demographic transition. The start of this second demographic transition is arbitrarily set at 1965; the principal feature of this transition is the decline in fertility to a level well below replacement.

If fertility stabilizes below replacement, as seems likely in most of Europe, and barring major changes in immigration, the populations of those countries will decline. This second demographic transition is already the case in Estonia, Latvia, Hungary, Bulgaria, the Ukraine, and Italy, as well as most of Eastern Europe (Population Reference Bureau, *2000 World Population Data Sheet*).

The United States is not expected to experience a population decline in the near future. United Nations population projections to the year 2025 indicate that the United States will continue to grow modestly, even with continued low fertility, because of immigration levels.

The reasons for the second demographic transition center around a strong desire for individual advancement and improvement. In European societies, as in American society, this advancement is dependent on education and a commitment to develop and use one's talents. This holds for both men and women. Marrying and having children present a number of tradeoffs, especially for women. A child may interrupt the parents' career plans, as well as add to financial costs. In European societies, children no longer are expected or required to support the parents in old age or to help with the family finances. Therefore, the emotional satisfaction of parenthood can usually be satisfied by having one or perhaps two children. Multiplied on a large scale, this trend produces a birthrate below replacement level.

Pronatalist Policies In most of this chapter, we are presenting information about population growth rates that appear to be out of control. Yet scores of developed nations face an equally ominous threat: dwindling population because of low fertility. In some 50 countries, the average number of children born to each woman has fallen below 2.1, the number required to maintain a stable population. Nearly all of these countries are in the developed world, where couples have been discouraged from having large families by improved education and health care, widespread female employment, and rising costs of raising and educating children.

The implications are particularly dire for Europe, where the average fertility rate has fallen to 1.4 children per woman. Even if the trend reversed itself and the fertility rate returned to 2.1, the continent would have lost a quarter of its current population before it stabilized around the middle of the century.

With fewer children being born, the ratio of older people to younger people already is growing. These countries face the prospect of shrinking workforces and growing retiree populations, along with slower economic growth and domestic consumption.

Some European governments are quite concerned. Italy, France, and Germany have introduced generous child subsidies, in the form of tax credits for every child born, extended maternal leave with full pay, guaranteed employment upon resumption of work, and free child care.

Although it is most pronounced in Europe, the birth dearth affects a few countries in other parts of the world as well. Japan's population may fall from 126 million today to 55 million over the next century if its 1.4 fertility rate remains unchanged.

Apart from encouraging childbirth, the only other way a government can stop population loss is to open its doors to immigrants. While immigration has always played a major role in the United States, most European countries are more homogeneous and resistant to immigration. Japan has been particularly inhospitable to immigrants (*CQ Researcher*, July 1998).

Pronatalist policies hardly ever lead to spectacular long-term effects on the birthrate. They may, however, contribute toward slowing down the fertility decline and improving the living conditions of the parents and their children.

Current Population Trends: A Ticking Bomb?

Every minute, 249 babies are born in the world. That means about 358,988 new human beings a day—131.4 million a year who need to be fed, clothed, sheltered, educated, and employed. (See Table 14-5 for the world population clock.) The 6 billionth person arrived in the year 2000. Another billion people will be added every 11 to 13 years until the middle of the twenty-first century. It took nearly all of human history for the world's population to reach the first billion.

Population growth has continued throughout the past three decades in spite of the decline in fertility rates that began in many developing countries in the late 1970s and, in some countries, in spite of the toll taken by the HIV/AIDS pandemic. While the rate of increase is slowing, in absolute terms world population growth continues to be substantial. According to U.S. Census Bureau projections, world population will increase to a level of nearly 8 billion persons by the end of the next quarter century and will reach 9.3 billion persons—a number more than half again as large as today's total—by 2050. Ninety-nine percent of global natural increase—the difference

Table 14-5

World Population Clock, 2000

Time Unit	Births	Deaths	Natural Increase
Year	131,389,622	54,409,824	76,979,798
Month	10,949,135	4,534,152	6,414,983
Day	358,988	148,661	210,327
Hour	14,958	6,194	8,764
Minute	249	103	146
Second	4.2	1.7	2.4

Source: U.S. Bureau of the Census, International Data Base.

between numbers of births and deaths—occurs in the developing world.

The U.S. Census Bureau's projections indicate that early in the next century, crude death rates will exceed crude birthrates for the world's more developed countries. As the growth rate in the world's more affluent nations becomes negative, *all* of the net annual gain in global population will, in effect, come from the world's developing countries (U.S. Census Bureau, 2000).

"Short of nuclear war itself, population growth is the gravest issue that the world faces over the decade ahead. . . . Both can and will have catastrophic consequences. If we do not act, the problem will be solved by famine, riot, insurrection, war." This view was put forth by Robert S. McNamara, a former secretary of defense and president of the World Bank. Like many other specialists in this field, he believes that overpopulation is threatening the basic fabric of world order.

The problem is underscored by the increasing speed with which the world's population is multiplying. During the first 2 million to 5 million years of human existence, the world population never exceeded 10 million people. The death rate was about as high as the birthrate, so there was no population growth. Population growth began around 8000 B.C., when humans began to farm and raise animals. In A.D. 1650, there were an estimated 510 million people in the entire world. One hundred years later, there were 710 million, an increase of some 39%. By 1900, there were 1.6 billion. Only 100 years later the world population had spiraled to 6.08 billion, with 131.4 million people added each year (Population Reference Bureau, *2000 World Population Data Sheet*). By the year 2025, the global population will be greater than 8 billion. Of that total, approximately 7 billion will be residents of the poorest and least-developed nations. As you can see in Figure 14-3, the world population has been rising at an ever-increasing rate.

Right now, the world population is doubling about every 51 years. If it continues to expand this way, it will quadruple within 102 years and increase eightfold to an incredible 40 billion within 153 years—a situation in which widespread poverty and famine seem almost assured. In recent years, however, there has been a small but significant slowing in the rate of world population growth: Between 1965 and 1970, the annual rate of growth was 2.1%, but by 1999, it had dropped to 1.4% (Population Reference Bureau, 2000). If this slowing trend continues, the world growth rate will be far smaller than previously predicted. This more hopeful pattern is contingent on the average family size being limited to two children.

We have already seen a significant worldwide trend toward smaller families. The developing

Figure 14-3 The Population Explosion

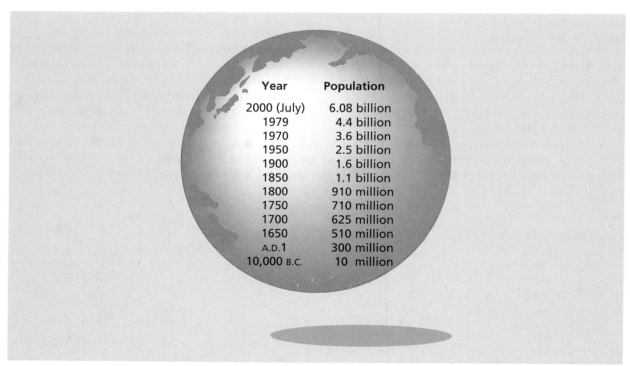

Year	Population
2000 (July)	6.08 billion
1979	4.4 billion
1970	3.6 billion
1950	2.5 billion
1900	1.6 billion
1850	1.1 billion
1800	910 million
1750	710 million
1700	625 million
1650	510 million
A.D. 1	300 million
10,000 B.C.	10 million

Sources: Data through 1970, *The World Almanac,* New York: Newspaper Enterprises, 1980, p. 734; data for 1979–2000, U.S. Bureau of the Census, *International Data Base.*

countries that are cutting their population growth most quickly are Costa Rica, Colombia, India, Mauritius, and Tunisia. Others include Thailand, Malaysia, Peru, and the Dominican Republic. Young women in those nations have, on the average, about two fewer children than their mothers and grandmothers had. Researchers credit increased education, government incentives, the rapid spread of contraceptive devices, mass communications (legitimizing new norms for families in previously isolated, tradition-dominated societies), and the diminishing necessity to have large numbers of children to ensure adequate labor for family agriculture as major reasons for this development (The World Bank, 1994).

A demographic trend seen worldwide is the continued migration of rural populations into cities. This trend will have serious sociological consequences. On the one hand, urbanization makes it easier to organize the delivery of goods and services and to develop specialized occupations to meet the needs of newly emerging economic institutions. On the other hand, urban crowding stresses many individuals and services and severely strains local environmental resources, an issue that we shall discuss shortly.

Determinants of Fertility

How many children a family has will affect that family's lifestyle in both developed and undeveloped countries. Many factors determine the typical family size in a country. In this section we will examine some of those factors to gain an insight into the variety of economic, social, and psychological issues that have to be addressed when trying to limit population growth. (See Table 14-6 for a summary of some of the determinants of fertility.)

Average Age of Marriage

Early marriage provides more years for conception to take place. It also decreases the parents' years of schooling and limits their employment opportunities. In South Asia and sub-Saharan Africa, about half of all women between 15 and 19 are or have been married. In Afghanistan, the mean age for women at marriage is 17.8, and thus the fertility rates in those areas are extremely high. By contrast, the average age of marriage in Sri Lanka is 24.4. Thus, in Afghanistan the average woman has 6.9 children, while in Sri Lanka she has 2.3 children (United Nations, *Women's Indicators and Statistics Database, 1997*).

To limit fertility some countries have tried to establish a minimum age for marriage. In 1950, China legislated the minimum age for marriage as 18 for

Table 14-6

Determinants of Fertility

Number of Children
1. Women's average age at first marriage
2. Breast-feeding
3. Infant mortality

Demand for Children
1. Gender preferences
2. Value of children
   a. Children as insurance against divorce
   b. Children as securers of women's position in family
   c. Children's value for economic gain
   d. Children's value for old-age support
3. Cost of children

Fertility Control
1. Use of contraception
2. Factors influencing fertility decisions
   a. Income level
   b. Education of women
   c. Urban or rural residence

women and 20 for men. It also recognized the effect that social controls can have on the marriage age and therefore increased institutional and community pressures for later marriage. In 1980, China again raised the legal minimum marriage ages, to 20 for women and 22 for men; it is one of the few countries that has been successful in raising the average age of marriage. When the Chinese youth were asked what would be the ideal age for their own marriage, the responses averaged 24.3 years for women and 25.4 for men (*Population Today*, 1995).

Breast-Feeding

Breast-feeding delays the resumption of menstruation and therefore offers limited protection against conception. It also avoids some of the considerable health risks connected with bottle-feeding, particularly when the powdered milk may be improperly prepared, adequate sterilization is not possible, and families cannot afford an adequate supply of milk powder. These risks have produced an outcry against certain large companies that produce powdered milk and that have been trying to encourage bottle feeding in Third World countries.

In Bangladesh, Pakistan, Nepal, and most of sub-Saharan Africa, where very few women use contraception, breast-feeding is one of the few controls on fertility rates.

In the United States, breast-feeding was relatively unpopular until the late 1960s. As the advantages of breast-feeding became known, however, better-educated women began to use it more. Today, college-educated women are the most likely to start breast-feeding and continue it for the longest periods.

Infant and Child Mortality

High infant mortality promotes high fertility. Parents who expect some of their children to die may give birth to more babies than they really want as a way of ensuring a certain sized family. This sets in motion a pattern of many children born close together, weakening both the mother and babies and producing more infant mortality.

In the short term, the prevention of 10 infant deaths may produce only 1 to 5 fewer births. Initially, lower infant and child mortality will lead to somewhat larger families and faster rates of population growth. However, the long-term effects are most important. With improved chances of survival, parents devote greater attention to their children and are willing to spend more on their children's health and education. Eventually, lower mortality rates help parents achieve the desired family size with fewer births and also lead them to want a smaller family.

Education and child health go hand in hand. In Indonesia, the infant mortality rate drops from 131 per 1,000 births for mothers with no schooling to 51 for those with the most education. The mother's education also has an effect on a child's health and nutrition. Children whose mothers complete secondary education are much less likely to be underweight for their age than those whose mothers have less education (Riley, 1997).

Gender Preferences

The effect of gender preference on fertility is actually more complicated than might appear at first glance. In most countries, there is a strong preference for male children. Logically, this does not make much sense. In most underdeveloped countries, daughters typically help their mothers with household chores. One would think that this would increase their worth to the family. Clearly, there must be countervailing factors to reduce the desire for daughters. It has been suggested that it may be tied in with the expense of a dowry or the early loss of the daughter's help through marriage. Spain is one of the few countries where daughters are preferred over sons (Gallup Poll, 1997).

The preference for sons, however, may not be all that important in countries with high fertility rates. In Bangladesh, for example, the preference for sons is extremely strong, but because couples desire large families and contraception is relatively rare, it appears to have little impact on fertility (Ahmed, 1981).

In countries with declining fertility rates, the preference for sons may cause the fertility rate to level off above replacement level. Couples may continue to have children until they have the desired son.

Girls have a better survival rate than boys, and women normally live longer than men. Even with the natural compensation of more male than female births, this still means that most nations have more women than men.

In a number of developing countries, there are far fewer women than men. This is not what would normally be expected and is often due to willful acts or discriminatory customs. One explanation would be lower survival rates for girls—either because of female infanticide or because girls receive less health care than boys. It also is related to the estimated half a million women who die each year of pregnancy and birth complications. Unsafe abortions kill another 100,000 each year. Other factors include widow burnings, in which the wife is burned when the husband dies, and dowry deaths, family disputes over the dowry that result in the wife being killed. Unfortunately, as appalling as these events sound, they take place regularly in some countries. Table 14-7 shows the nine countries where there are 95 or fewer women for every 100 men. This means that there are at least 10% fewer women than would normally be expected given women's tendency to live longer than men.

In China there are 113.8 boys born for every 100 girls. Worldwide data show that about 105.5 boys are born for every 100 girls, implying that 12% of all Chinese baby girls (about 1.7 million) are missing annually. This does not mean that they are killed. Some are not reported to the authorities, others are adopted informally by friends or relatives (Riley, 1996).

The major reason for the missing girls seems to be technology. As early as 1979, China began manufacturing ultrasound scanners. The scanners are meant to help doctors check whether the fetuses are developing normally. Ultrasound scanners also can be used to determine whether the fetus is male or female. By 1990, there were 100,000 scanners available throughout China. Women go to clinics, have an ultrasound exam, and if the fetus appears to be a girl, an abortion is performed in the same clinic. It is

Table 14-7

Countries with 10% Fewer Women Than Expected

COUNTRY	WOMEN PER 100 MEN
Pakistan	92
Papua New Guinea	93
India	94
Hong Kong	94
Bangladesh	94
Albania	94
China	94
Afghanistan	95
Nepal	95

Source: The Progress of Nations: 1993, UNICEF, 1993, New York: United Nations Statistical Office and Population Division.

estimated that between 500,000 and 750,000 unborn girls are aborted in China every year as a result of couples having access to the ultrasound scanner that reveals the sex of a fetus (Cox News Service, January 29, 1999). Combination ultrasound/abortion clinics are common in China, South Korea, India, and Afghanistan (Kristof & Wudunn, 1994).

We would be wrong to assume that a strong preference for a son is just limited to less developed countries. A recent Gallup Poll showed despite the significant advances women have made over the past century, and the modern emphasis on equality between the sexes, Americans today still harbor a preference for having male rather than female children, just as they did over a half century ago. Forty-two percent of Americans said they would prefer a boy if they could have only one child, while 27% said they would prefer a girl. One-quarter of Americans said it would not matter to them either way. In 1941, 38% said they would prefer to have a boy, compared with 24% who said they would choose a girl and 23% who said they did not have a preference.

Although one might expect older Americans to be more "patriarchal" than society as a whole, young adults are actually more likely than older people to prefer boys. Fifty-five percent of people between the ages of 18 and 29 say they would prefer a boy, compared to just 31% of those aged 65 and older (Gallup Poll, December 2000).

Benefits and Costs of Children

In underdeveloped countries, the benefits for an individual family of having children have generally been greater than the costs. However, those costs and benefits change with the second, third, fourth, and fifth child.

For a rural sample in the Philippines, three-quarters of the costs involved in rearing a third child come from buying goods and services; the other quarter comes from costs in time (or lost wages). Receipts from child earnings, work at home, and old-age support offset 46% of the total. The remaining 54%, the net cost of a child, is equivalent to about 6% of a husband's annual earnings. By contrast, a study of urban areas of the United States showed that almost half the costs of a third child are time costs. Receipts from the child offset only 4% of all costs. (In many countries, families depend on their children for needed income—see "Remaking the World: A World of Child Labor.")

Only economic costs and benefits are taken into account in these calculations. To investigate social and psychological costs, other researchers have examined how individuals perceive children. Economic contributions from children are clearly more important in the Philippines, where fertility is higher than in the United States; concern with the restrictions children impose on parents is clearly greatest in the United States.

In many countries, however, couples demonstrate a progression in the values they emphasize as their families grow. The first child is important to cement the marriage and bring the spouses closer together, as well as to have someone to carry on the family name if it is a boy. Thinking of the first child, couples also stress the desire to have someone to love and care for and the child's bringing play and fun into their lives.

In considering a second child, parents emphasize more the desire for a companion for the first child. They also place weight on the desire to have a child of the opposite sex from the first. Similar values are prominent in relation to third, fourth, and fifth children; emphasis is also given to the pleasure derived from watching children grow.

Beyond the fifth child, economic considerations predominate. Parents speak of the sixth child or later

The support given to family planning programs differs dramatically from country to country. The government of India is active in providing birth control information and devices, while countries in which the Roman Catholic faith is predominant follow anti-birth-control policies.

Remaking the World
A World of Child Labor

One day in April of 1995, 12-year-old Craig Kielburger picked up the daily newspaper and as he did every day was about to turn to the comics section. This morning however, an article about a boy his own age caught his eye. It was the story of Iqbal Masih, a Pakistani boy who, at the age of 4, was sold into slavery by his parents. For the next 6 years, the article read, Iqbal was shackled to a carpet loom, tying thousands upon thousands of tiny knots, 12 hours a day, 6 days a week. For this he was paid 3 cents a day. Amazingly, Iqbal's will was never broken. Eventually he escaped and began to reveal the horror of child labor. Iqbal began to gain international attention, and Pakistani carpet manufacturers began to lose orders. Shortly thereafter while riding a bicycle with his friends, Iqbal was ambushed and killed.

Craig had never heard about child labor and wasn't even certain where Pakistan was on the world map, but he could not help but notice the difference between his life and that of Iqbal. At that point he felt he had to do something, but what can a 12-year-old do?

Craig gathered a group of friends together, most of them 12 years old like himself, and founded the organization called *Free the Children*—not only to free children from abuse and exploitation, but to free children from the idea that they are not old enough or smart enough or capable enough to help change the world.

Craig learned that the International Labor Organization (ILO) estimates that 250 million children between the ages of 5 and 14 in developing countries work; 120 million work full time. Africa, the poorest region, has the highest incidence of child workers—some 40%—while the corresponding figure for both Asia and Latin America is about

Craig Kielburger.

20%. But Asia, being by far the most populous region, has the largest number of child workers. Some 60% of child workers are in Asia, 32% in Africa and 7% in Latin America.

Craig wanted to see the child labor working conditions firsthand. He decided he needed to go to South Asia, one of the centers of child labor. Quite a stretch for a boy whose parents did not allow him to take the subway by himself. He teamed up with a 25-year-old human-rights worker named Alam Rahman and convinced his reluctant parents to let him travel halfway around the world. For seven weeks Craig journeyed through the slums, sweatshops, and back alleys of Bangladesh, Thailand, India, Nepal, and Pakistan where many children live

children in terms of their helping around the house, contributing to the support of the household, and providing security in old age. For first to third children, the time taken away from work or other pursuits is the main drawback; for fourth and later children, the direct financial burden is more prominent than the time costs (The World Bank, 1984).

Contraception

Apart from the factors already mentioned, fertility rates eventually are tied to the increasing use of contraception. Contraception is partly a function of a couple's wish to avoid or delay having children and partly related to costs. People have regulated family size for centuries through abortion, abstinence, and even infanticide. However, the costs of preventing a birth, whether economic, social, or psychological, may be greater than the risk of having another child.

Use of contraception varies widely: 18% or fewer for married women in almost all of sub-Saharan Africa, but between 70 and 80% for women in Europe, Asia, and the United States (Population Reference Bureau, 2000).

in servitude performing menial and often danger-ous jobs.

Child prostitution is also common, particularly in Asia. Thailand, for example, has earned an inter-national reputation for this offense, with thousands of girls from China and Southeast Asia regularly be-ing kidnapped and sold to brothels in Bangkok and other Thai locales. This practice also takes place throughout the major urban centers of India, Paki-stan, Africa, and Latin America.

After the trip, Craig was transformed from a typ-ical middle-class kid into a committed advocate for children's rights.

Since 1995, *Free the Children* has grown into an international children's organization active in more than 20 countries. Over 100,000 children have par-ticipated in *Free the Children* campaigns and activi-ties. Craig continues to travel the world on behalf of *Free the Children*.

In 1999, President Clinton signed a proposal ap-proved by the U.S. Senate to "take immediate and effective measures" to eliminate the worst cases of child abuse. The United States is also the largest contributor to the ILO's program to eliminate child labor, having increased its contribution from $3 mil-lion to $30 million a year.

The American and Western revulsion over child labor is, in fact, by no means a universal concern. In some countries, child labor is defended as necessary for economic viability, often for survival itself.

Many developing countries resent the intrusion of the wealthy and industrial West, under the guise of "human rights," into their national workplace. Shabbin Jamal, for example, an adviser to Pakistan's Ministry of Labor, has deplored the Western world's "double standard" in failing to recognize what he sees as an economic need. Indeed, the family and

The detrimental aspects of child labor have received far less attention than they deserve.

national need for child labor is regarded as a neces-sity in most developing nations. Thus, many believe that for reformers to be effective, their first goal should be to end abusive and hazardous conditions, not necessarily the labor itself.

Sources: *Free the Children*, by C. Kielburger & K. Major, 1998, Toronto: McClelland & Stewart; *The World of Child Labor*, by J. J. Tierney, Jr., Vol. 15, The World & I, 08-01-2000, p. 54; International Conference on Child Labor, Oslo, October 27–30, 1997, Geneva: International Labour Office, 1998.

Contraception is most effective when such pro-grams are publicly subsidized. Not only do such programs address the economic costs of spreading contraception, but they also help communicate the idea that birth control is possible. These programs also offer information about the private and social benefits of smaller families, which helps reduce the desired family size.

The support given to family-planning programs differs dramatically from country to country. At one end of the family-planning spectrum are the gov-ernments of India and China, which provide birth-

control information and devices and actively support abortion. At the other end are Muslim countries such as Bangladesh or Roman Catholic ones such as Ireland, whose populations follow the anti-birth-control and anti-abortion teachings of the church.

Abortion is the single most widely used form of birth control, and it is common even when it is ille-gal. Worldwide, the demand for abortions is rising, and it has been estimated that one in three pregnan-cies ends in abortion. Abortion is legal in the world's three most populous countries (China, India, and the United States) as well as in Japan and all of

Figure 14-4 Abortions per 100 Pregnancies in Selected Countries

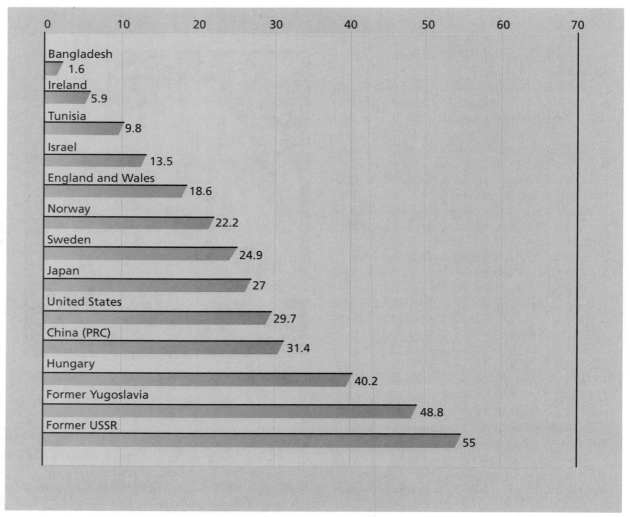

Sources: "Induced Abortion: A World Review, 1990," by S. Henshaw, *Family Planning Perspectives,* 22(2), pp. 76–89; "Population Change in the Former Soviet Republics," by C. Haub, December 1994, *Population Bulletin* 49(4).

Europe except Belgium and Ireland (Weeks, 1994). In Russia, where contraceptives are hard to find, there were 216 abortions for every 100 births in 1992 (Haub, 1994). (See Figure 14-4 for the number of abortions per 100 pregnancies throughout the world.)

Income Level

It is a well-established fact that people with higher incomes want fewer children. Alternative uses of time such as earning money, developing or using skills, and pursuing leisure activities become more attractive, particularly to women. The children's economic contributions become less important to the family welfare, as the family no longer needs to think of children as a form of social security in old age. It is not the higher income itself but, rather, the life change it brings about that lowers fertility.

This relationship between income and fertility holds true only for those with an income above a certain minimum level. If people are extremely poor, increases in income will actually increase fertility. In the poorest countries in Africa and Asia, families are often below this threshold. Above the threshold, though, the greatest fertility reduction with rising income takes place among low-income groups.

Education of Women

In nearly all societies, the amount of education a woman receives affects the number of children she has. Fertility levels are usually the lowest among the most highly educated women within a country. In Guatemala, for example, women with no formal schooling have an average of 6.9 children, while those with a secondary or higher education have only 2.7 children on average. In Zimbabwe, the

In underdeveloped countries, the benefits of having children have generally been greater than the costs for the individual family.

trend is similar with women with no schooling having 6.9 children and those with secondary or higher education having 3.8 (Riley, 1997).

Studies also show that women's level of education has a greater impact on fertility than does that of men. There are a number of reasons for this. In most instances, children have a greater impact on women, in terms of time and energy, than they do on men. The more educated a woman is, the more opportunities she may encounter that conflict with having children. Education also appears to delay marriage, which in itself lowers fertility. In 10 out of 14 developing countries, women with 7 or more years of education marry at least 3.5 years later than do women with no education. Educated women are also more likely to know about and adopt birth-control methods. In Mexico, 72% of women with 9 or more years of education are likely to use contraception, whereas only 31% of those with 5 or fewer years of education are likely to do so. A study of 37 countries showed that one additional year of schooling lowers desired family size by more than 0.1 children, so that women with 10 or more years of education want, on average, 1.3 fewer children than women with no formal education (Lutz, 1994).

Urban or Rural Residence

Rural lifestyles that involve farming and herding tend to perpetuate high birthrates because children are seen as contributing workers. Urban fertility tends to be lower in developing countries, although it still is at least twice as high as fertility in developed countries (Goliber, 1997). Urban dwellers usually have access to better education and health services, a wider range of jobs, and more avenues for self-improvement and social mobility than do their rural counterparts. They are exposed to new consumer goods and are encouraged to delay or limit childbearing to increase their incomes. They also face higher costs in raising children. As a result, urban fertility rates are usually one to two children lower than rural fertility rates.

The urban woman marries, on the average, at least 1.5 years later than the rural woman does. She is more likely to accept the view that fertility should be controlled, and the means for doing so are more likely to be at her disposal.

Problems of Overpopulation

The very word *overpopulation* is a problem for people in less-developed countries. If the world is overpopulated, then who are the unneeded? If people in the richer countries view those in poorer ones as surplus, then a bias is built into the term.

At first glance, the connection between population growth and environmental problems seems clear: More humans consume more resources and generate more waste. Twelve billion people could do a great deal more environmental damage than the 6 billion in the world of 2000. A closer look at the situation reveals that it is more complex than that. Some people have a far greater environmental impact than others. Consider the following facts:

■ People in industrialized nations, who account for 22% of the world's population, consume 70% of its energy, and 85% of its wood. They produce two-thirds of all greenhouse gasses and 90% of all ozone-depleting chloro-fluorocarbons.

There are inherent limits to the land, water, fertilizer, and energy available for food production. Overfarming and the indiscriminate destruction of forests have produced deserts. Careful use of the land can save many people from starvation.

- A child born in the United States will, in his or her lifetime, have 35 times the impact on the earth's environment on average than a child born in India and more than 250 times the impact of a child born in one of the countries of sub-Saharan Africa.

- The United States, with 5% of the world's population, produces 72% of the hazardous waste (*Population Today*, April 1996).

Some people have suggested that instead of being concerned about overpopulation, we should focus on the world's carrying capacity instead (Carty, 1994). How will the world's carrying-capacity problems be resolved in the future? Neo-Malthusians paint a gloomy picture of what lies ahead, contending that as we head toward the end of this century, the population inevitably will outpace the supply of food.

In the 1970s, Paul Ehrlich and Anne Ehrlich, outspoken critics of population growth and unlimited consumption by the wealthy countries, predicted global shortages in the future. "It seems certain," they noted, "that energy shortages will be with us for the rest of the century, and that before 1985 mankind will enter a genuine age of scarcity." Crucial materials would be nearly depleted during the 1980s,

they predicted. "Starvation among people will be accompanied by starvation of industries for the materials they require" (Ehrlich & Ehrlich, 1974).

Also in the 1970s, The Club of Rome, a group of scientists, businesspeople, and academics, used elaborate computer models to predict the world of the future. The Club concluded that if the then current trends continued, the limits of growth on this planet would be reached within the next hundred years. The result, they predicted, would be a sudden and uncontrollable decline in population and production capacity:

We have tried in every doubtful case to make the most optimistic estimate of unknown qualities, and we have also ignored discontinuous events such as wars or epidemics, which might act to bring an end to growth even sooner than our model would indicate. In other words, the model is biased to allow growth to continue longer than it probably can continue in the real world. We can thus say with some confidence that under the assumption of no major change in the present system, population and industrial growth will certainly stop within the next century, at the latest. (Meadows et al., 1972)

Controversies in Sociology

Have We Exaggerated the Extent of the Population Problem?

Throughout this chapter we have talked about the problems caused by the continuing growth in the world's population. At the same time that many people are alarmed by the growth rates, others, such as economist Stephen Moore, believe that population growth will level off and that we need not worry about it as much as was previously thought.

A recent *New York Times* story wails that if the world's population isn't curtailed soon, the globe will start to look as poor and crowded as Calcutta. Ted Turner says mankind is breeding like "a plague of locusts" and urges couples all over the world to limit themselves to one child. Zero Population Growth laments that the population of the U.S. is about twice the size it should be in order to protect the environment.

The mystery is why anyone takes these modern-day Chicken Littles seriously anymore. After all, every objective fact and environmental trend is running in precisely the opposite direction of what the widely acclaimed doomsayers of the 1960s—from Lester Brown to Paul Ehrlich to the Club of Rome—once predicted. Birth rates around the world are lower, not higher, today than at anytime in at least a century. Global per capita food production is 40 percent higher today than as recently as 1950. . . . In sum, the population bomb propagandists have all the intellectual credibility of the Flat Earth Society.

Yes, it is true that in just this past century the number of human beings on the planet has just about quadrupled. But . . . the simple and benign explanation is improved health and more wealth. Consider the trends in life expectancy, arguably the single best measure of human well-being. From about the time of the Roman Empire through about 1800 average human life expectancy was less than 30 years. In the U.S. today, life expectancy is 77. Even in poor countries, like India and China, life expectancy has risen to above 60. We have doubled the number of years of life in just the past 200 years.

Meanwhile, infant mortality rates in the U.S., and across the globe, have fallen by about tenfold in just the last century. A century ago, if a woman had three children, the likelihood was that at least one of them would have died at birth or before the fifth birthday. Nowadays the probability of childhood death is less than 1 in 100. . . .

[W]e are nowhere near running out of room on the planet. If every one of the 6 billion of us resided in Texas, there would be room enough for every family of four to have a house and 1/8th of an acre of land—and the rest of the globe would be vacant. True, if population growth continues, soon some of these people would have to spill over the border into Oklahoma.

The dreaded population bomb that emerged as a worldwide obsession in the 1960s and 1970s has been all but defused. The birthrate in developing countries has plummeted from just more than six children per couple in 1950 to just more than three per couple today. The major explanation for smaller family sizes, even in China, has been economic growth, not condom distributions or coercive birth control measures. . . .

We used to worry about our capacity to feed the planet, but . . . fewer than half as many people die of famine each year now than did a century ago—despite almost a quadrupling of the population. Virtually every natural resource has fallen steadily in price. . . According to the most recent EPA statistics, pollution of the air and water is not increasing, it is decreasing—even though there are more people.

The population controllers . . . regard mother nature as pure and fragile and man's footprint on the Earth as the despoiler of this natural state. They worship the created, not the Creator. And they are in many cases hostile to economic development and human progress. They celebrate the planting of a new tree as magnificent progress, but abhor the planting of another fetus in a woman's womb as anti-progress.

But the good news for those of us with two, three, or God forbid, four children or more is this: The Malthusians are wrong. There is no ethical, environmental or economic case for small families.

Source: "Defusing the Population Bomb," by S. Moore, October 13, 1999, *The Washington Times.*

Others believe that we have the technological means to provide all the world's people with food. They speak of a "Green Revolution," in which new breeds of grain and improved fertilizers will raise harvest yields and eliminate the threat of a food shortage. The only thing holding back the revolution is poor international cooperation in planning the production and distribution of food. (See "Controversies in Sociology: Have We Exaggerated the Extent of the Population Problem?")

Sources of Optimism

Critics point to a number of logical fallacies in the doomsday predictions and argue that the dire pronouncements ignore the role of the marketplace in helping to produce adjustments that bring population, resources, and the environment back in balance. In their optimistic view, population growth is a stimulus, not a deterrent, to economic advance. If imbalances exist, they are because markets are not allowed to operate freely to permit innovators, investors, and entrepreneurs to provide solutions (Simon & Kahn, 1984).

Others have questioned the basic assumptions of the doomsday model, namely exponential growth in population and production and absolute limits on natural resources and technological capabilities. The following are some of these counterarguments.

1. *Wider application of existing technology.* Greater efficiency and wider application of technology on a worldwide basis could continue to supply the world's needs far into the future. For example, between 1961 and 1994, global production of food doubled, and greater efficiency in land cultivation will ensure the ability to feed the world's population for many years. Prices for food have continually decreased since the end of the eighteenth century.

2. *Discovery of new resources.* The supply of natural resources, according to some critics, is not really as fixed as the doomsday predictors claim. These critics maintain that new resources will come into play that were not previously anticipated. Technology will stay ahead of resource use, and the doomsday scenario will never be enacted. For example, raw materials and energy resources are generally more abundant and less expensive today than they were 20 years ago. Companies have become more adept at discovering new resources and exploiting old ones. Exploring for oil used to be a hit or miss proposition, resulting in many dry holes. Today, oil companies can use seismic waves to help them create precise computer images of the earth. In effect, the more advanced the technology, the more reserves become known and recoverable. In addition, the more we learn about materials, the more efficiently we use them. Refrigerators sold in 1993 were 23% more efficient than those sold in 1990 and 65% more efficient than those sold in 1980.

3. *Exponential increase in knowledge.* Just as the doomsday predictors claim there will be exponential growth in population and production, others say there also will be exponential growth in knowledge that will enable societies to solve the problems associated with growth. New technological information and discoveries, furthermore, will alleviate new problems as they arise. For example, computer game consoles today have more computing power than the 1976 Cray supercomputer, which the United States tried to keep away from the Soviet Union for security reasons.

Population issues will continue to affect the developed and developing areas of the world. The likeliest path to helping populations is through economic development. Poor people who are unable to acquire food and fuel do whatever they have to in order to survive. Sensitive ecological systems such as rain forests often fall victim to these needs. Raising living standards produces lower fertility and lowers environmental deterioration from inefficient resource depletion. There is a strong trend toward an improving situation. In the next decade, we will see whether this trend is developing fast enough to prevent further outbreaks of famine and misery for millions of people.

SUMMARY

The size of a population tends to become a progressively greater problem due to exponential growth, in which a continuously expanding base rapidly doubles. The annual growth rate in world population has recently declined so that the world's population now doubles every 51 years instead of every 35 years. In most developing countries, however, the rate of growth is much faster. Much of the world's growth in population in the next few decades will take place in those countries.

Demography is the study of the size and composition of human populations as well as the causes and consequences of changes in those factors. Demography is influenced by three major factors: fertility, mortality, and migration.

Life expectancy is usually determined more by infant than adult mortality. Rapid population growth in the Third World is largely due to an increase in life expectancy rather than a rise in the birthrate.

Migration is the movement of populations from one geographical area to another. Historically, it has had less impact on population change in an area than either fertility or mortality.

The core of the population problem, according to Thomas Malthus, is that populations will always grow faster than the available food supply, because resources increase arithmetically while population increases geometrically.

According to demographic transition theory, societies pass through four stages of population change as they move from high fertility and mortality to relatively low fertility and mortality.

Demographic transition theory accurately describes the population changes that have occurred in Western society with the advance of industrialism. It does not, however, explain population trends in the underdeveloped world today. These latter countries have experienced a much faster drop in death rates than Western societies did, without a comparable rate of increase in economic development. Thus, the birthrate, which has not fallen as fast or as consistently as it did in Western countries, has become an increasingly serious problem.

Many factors determine the typical family size in any country. Those societies that have succeeded in raising the average age of marriage have decreased fertility. Breast-feeding delays the resumption of menstruation and therefore offers limited protection against conception. In many underdeveloped countries, breast-feeding is one of the few controls on fertility.

Lifestyle changes associated with higher incomes promote a desire for fewer children. If people are extremely poor, however, increases in income will increase fertility. The number of children per woman declines substantially as women's level of education increases, a much stronger effect than increasing levels of men's education. Urban fertility in developing countries tends to be lower than rural fertility. Urban dwellers have more opportunities and more encouragement for upward social mobility. They also face higher costs in raising children.

Some people believe that the population crisis will be avoided through wider application of existing technologies, the discovery of new resources, or an exponential growth in knowledge.

 ## INTERNET RESOURCES

Go to http://web-enhanced.thomsonlearning.com which features Web links relevant to this chapter to expand and enhance the learning experience. This site will be monitored and updated periodically to ensure the links are correct and live.

At Virtual Society: The Wadsworth Sociology Resource Center, http://sociology.wadsworth.com, you will find a Career Center, Sociology in the News, Research Online, InfoTrac College Edition, Virtual Tours in Sociology, and a variety of book-specific student and instructor resources.

CHAPTER FOURTEEN STUDY GUIDE

KEY CONCEPTS

Match each concept with its definition, illustration, or explanation below.

a. Emigration
b. Internal migration
c. Fecundity
d. Infant mortality rate
e. Capitalism
f. Mortality
g. Immigration
h. Life expectancy

i. Age-specific death rate
j. Fertility rate
k. Positive checks
l. Crude birthrate
m. Pronatalist policies
n. Demography
o. Dependency ratio
p. Breast-feeding

q. Migration
r. Fertility
s. Crude death rate
t. India
u. Second demographic transition
v. Preventive checks
w. Demographic transition theory
x. Japan

_____ 1. The study of the size and composition of human populations as well as the causes and consequences of changes in these factors.

_____ 2. The physiological ability to have children.

_____ 3. The actual number of births in a given population.

_____ 4. The number of annual births per 1,000 people in a given population.

_____ 5. The number of annual births per 1,000 women of childbearing age in a given population.

_____ 6. The frequency of deaths in a population.

_____ 7. The annual number of deaths per 1,000 people in a given population.

_____ 8. The number of deaths per 1,000 people at specific ages.

_____ 9. The number of children who die within the first year of life per 1,000 live births.

_____ 10. The average number of years that a person born in a given year can expect to live.

_____ 11. The movement of populations from one geographical area to another.

_____ 12. The phenomenon that occurs when part or all of a population leaves an area.

_____ 13. The phenomenon that occurs when a population enters a geographical area.

_____ 14. Movement of populations within a nation's boundaries.

_____ 15. In Malthus's theory, practices that limit reproduction.

_____ 16. In Malthus's theory, events that limit the population by causing death.

_____ 17. The concept that societies, as they industrialize, pass through four stages of population change, from high fertility and mortality to low fertility and mortality.

_____ 18. The number of people of nonworking age in a society for every 100 people of working age.

_____ 19. The world's longest life expectancy at birth is in this country.

_____ 20. The cause of the social and economic problems associated with population growth, according to Marxist theorists.

_____ 21. The decline of fertility to a level well below replacement.

_____ 22. Where very few women use contraceptives, one of the few controls on fertility.

_____ 23. Governmental policies in some countries designed to encourage people to have children.

_____ 24. Will be the world's most populated country somewhere between 2030 and 2035.

KEY THINKERS/RESEARCHERS

Match the thinkers/researchers with their main ideas or contributions.

a. **Club of Rome**   c. **Karl Marx**   e. **Warren Thompson**
b. **Paul Ehrlich**   d. **Thomas Malthus**

_____ 1. A pioneer in the study of population, he believed that population growth is linked to certain natural laws.

_____ 2. Argued that industrial capitalism was the cause of overpopulation.

_____ 3. Developed the demographic transition model.

_____ 4. A group of scientists, academics, and businesspeople who have predicted worldwide economic and ecological collapse.

_____ 5. One of the first to warn against the catastrophic consequences of unchecked population growth.

CENTRAL IDEA COMPLETIONS

Following the instructions, fill in the appropriate concepts and descriptions for each of the questions posed in the following section.

1. Briefly discuss the factors leading to population growth in the Third World:

_____
_____
_____

2. Explain the difference between fecundity and fertility:

   a. Fecundity: _____
   b. Fertility: _____

3. Explain the causes infant mortality. Include consideration of how infant mortality is related to population growth in the Third World:

_____
_____
_____

4. What type of subsidies are being instituted by Italy, France, and Germany to enhance their population growth?

_____
_____
_____

5. Using Pakistan as an example, what is the social effect of gender on population growth?

_____
_____
_____

6. What are the two effects of migration of rural populations into urban centers?

   a. _____
   b. _____

7. Using Afghanistan and Sri Lanka as cases in point, how does the average age of marriage relate to population increases?

_____

8. In what ways does the educational level of women have a greater impact on fertility than the educational level of men?

_____

9. Your text notes that "the connection between population growth and environmental problems is more complex than it would appear at first glance." List and explain three facts that illustrate this complexity:

a. _____

b. _____

c. _____

10. Your text discusses the phenomenon of exponential growth. Precisely what is it and why is it a significant concern for population research and theory?

_____

_____

11. What evidence does your text provide that a preference for male children exists not only in many Third World countries, but in the United States as well?

_____

_____

12. With fewer children being born in certain countries, the ratio of older people to younger people is growing. What are three consequences of this changing ratio?

_____

_____

CRITICAL THOUGHT EXERCISES

1. Compare and contrast the twin problems of increasing population growth rates and the threat of dwindling populations because of low fertility. Based on your reading of the text, name three nations that are likely to be found in each category by the year 2030. Which situation would most easily respond to social intervention: population increase or population decrease? Elaborate on your choice by considering the manifest and latent consequences following social intervention with either category.

2. Take and defend the position, "We have exaggerated the extent of the population problem." Be sure to include specific points in your argument based on the readings in your text as well as two outside sources that support this position.

3. You have been called before a U.S. Senate select committee investigating infant mortality in the United States. The senators believe *pockets of poverty* to be the cause. Your task is to educate the senators about the actual causes of infant mortality.

Your presentation will be titled, "Social and Behavioral Factors Contributing to Low Infant Birthrates in the United States." You should have enough information from Chapter 14 to develop most of your presentation. At the conclusion of the "facts" portion of your presentation, present a series of possible solutions to the problem. Be certain to explore the ethical dimensions of any intervention programs you propose.

4. Your text provided considerable detail about, the tragedy of child labor. Given the evidence, what should a leading industrialized nation like the United States do to combat worldwide child labor? Could, for example, a national campaign similar to that tackling cigarette manufacturing and smoking be developed for child labor? Develop a paper in which you discuss this question, as well as the related issues of the functions and dysfunctions of child labor throughout the world—not only for the developing nations, but for more developed nations as well.

ANSWERS

KEY CONCEPTS

1. n	5. j	9. d	13. g	17. w	21. u
2. c	6. f	10. h	14. b	18. o	22. p
3. r	7. s	11. q	15. v	19. x	23. m
4. l	8. i	12. a	16. k	20. e	24. t

KEY THINKERS/RESEARCHERS

1. d	2. c	3. e	4. a	5. b

Urban Society

LEARNING OBJECTIVES

After studying this chapter, you should be able to do the following:

- Describe the history of urbanization.
- Contrast preindustrial and industrial cities.
- Know the different terms that are used to describe urban environments.
- Describe the various theories of urban development.
- Understand the issues surrounding homelessness in American cities.
- Describe trends in urban growth in the United States.

Stand in your window and scan the sights,
On Broadway with its bright white lights.
Its dashing cabs and cabarets,
Its painted women and fast cafes.
That's when you really see New York.
Vulgar of manner, overfed,
Overdressed and underbred.
Heartless and Godless, Hell's delight,
Rude by day and lewd by night.

          —Anonymous, 1916

. . . They tell me you are wicked and I believe
   them, for I have seen your painted women
   under the gas lamps luring farm boys.
And they tell me you are crooked and I answer:
   Yes, it is true I have seen the gunman kill
   and go free to kill again.
And they tell me you are brutal and my reply
   is: On the faces of women and children I
   have seen the marks of wanton hunger.
And having answered so I turn once more to
   those who sneer at this my city, and I give
   them back the sneer and say to them:
Come and show me another city with lifted
   head singing so proud to be alive and
   coarse and strong and cunning. . . .

          —Carl Sandburg,
          Chicago Poems, 1916

The two poems that you have just read display the range of responses to the city. We are fascinated by cities, as well as terrified and revolted by them. Most of us live in urban environments, and we have found ways of adjusting to city life. In this chapter, we will attempt to gain a better understanding of our urban society and the impact it has on people.

The Development of Cities

According to archaeologists, people have been on earth for a couple of million years. During the vast majority of these years, human beings lived without cities. Although we accept cities as a fundamental part of human life, cities are a relatively recent addition to the story of human evolution, appearing only within the past 7,000 to 9,000 years.

The city's dominance in social, economic, and cultural affairs is even more recent. Nonetheless, what we label as "civilization" emerged only during the time span that coincides with the city, encompassing the whole history of human triumphs and tragedies. The very terms *civilization* and *civilized* come from the Latin *civis,* which means "a person living in a city."

The cities of the past still were very unusual in an overwhelmingly rural world of small villages. In 1800, 97% of the world lived in rural areas of fewer than 5,000 people. By 1900, 86% of the world still lived in rural areas (Palen, 1992).

England was the first country to undergo urban transformation. One hundred years ago it was the only predominantly urban country. Not until 1920 was the United Sates that urbanized. Today we are on the threshold of living in a world that will for the first time be more urban than rural. The most rapid change is occurring in the developing world. Within the next decade, more than half of the world's population, an estimated 3.3 billion, will be living in urban areas. As recently as 1975, just over one-third of the world's people lived in urban areas.

Not all parts of the world are urbanizing at the same pace, though. In the more industrialized areas—North America, Europe, and the republics of the former Soviet Union—urban growth has stopped or slowed considerably. For example, in 1970, 73.5% of the U.S. population lived in urban areas, while in 1990, the figure had only increased to 75.2% (*Statistical Abstract of the United States,* 1997).

The greatest urban growth is now in the non-industrial world, such as Africa, Latin America, the Middle East, and Asia. The United Nations projects that world population will increase from 6.1 billion in 2000 to 7.8 billion in 2025. Ninety percent of this growth will occur in urban areas of less developed countries. By 2020, a majority of the population of less developed countries will live in urban areas (United Nations, 2000).

The population of the less developed countries will become increasingly concentrated in large cities of 1 million or more people. Already there are hundreds of cities of more than 1 million people in the less developed countries. Most of us have never heard the names of more than a few of those cities. In 2000, there were an estimated 292 "million-plus" cities in less developed countries (Brockerhoff, 2000).

In some respects, the growth is similar to what took place a century ago in Europe and North America. Many of these cities sustained growth, which was as fast as that now underway in the developing world. What is different, however, is the number of countries undergoing rapid urbanization and the sheer number of people involved. As a measure of comparison, consider that in 1950 only two cities in the world had populations that exceeded 8 million, New York and London. By 2000, there were 19 such cities. By 2015, there will be 34 such megacities. Only six of these megacities will be in the developed world, the same as in 1985 (United Nations, 2000).

The rapid transformation from a basically rural to a heavily urbanized world and the urban lifestyles that accompany this shift are having a dramatic impact on the world's peoples. In this section, we will examine the historical development of cities and urbanization trends.

The Earliest Cities

Two requirements had to be met for cities to emerge. The first was that there had to be a surplus of food and other necessities. Farmers had to produce more food than their immediate families needed to survive. This surplus made it possible for some people to live in places where they could not produce their own food and had to depend on others to supply their needs. Those settlements could become relatively large, densely populated, and permanent.

The second requirement was that there had to be some form of social organization that went beyond the family. Even though there might be a surplus of food, there was no guarantee that it would be distributed to those in need of it. Consequently, a form of social organization adapted to those kinds of living environments had to emerge.

The world's first fully developed cities arose in the Middle Eastern area, mostly in what is now Iraq, which was the site of the Sumerian civilization. The land is watered by the giant Tigris and Euphrates rivers, and it yielded an abundant food surplus for the people who farmed there. In addition, this area (called Mesopotamia) lay at the crossroads of the

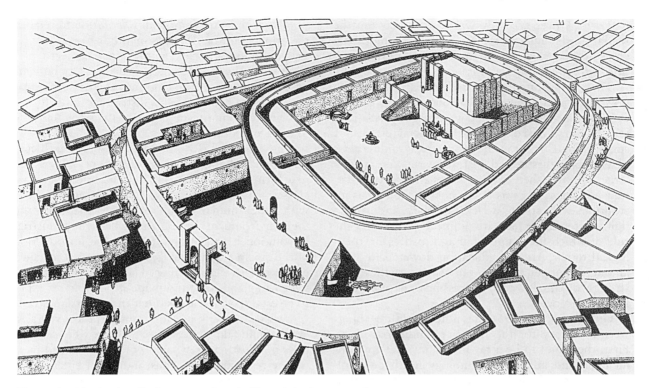

The early cities had walls that were closed off at night for protection.

trade networks that already, 6,000 years ago, tied together East and West. Not only material goods, but also the knowledge of technological and social innovations, traveled along these routes.

Sumerian cities were clustered around temple compounds that were raised high up on brick-sheathed mounds called ziggurats. The cities and their surrounding farmlands were believed to belong to the city god, who lived inside the temple and ruled through a class of priests who organized trade caravans and controlled all aspects of the economy. In fact, these priests invented the world's first system of writing as well as numerical notation late in the fourth millennium B.C. to keep track of their commercial transactions. Because warfare both among cities and against marauders from the deserts was chronic, many of these early cities were walled and fortified, and they maintained standing armies. In time, the generals who were elected to lead these armies were kept permanently in place, and their positions evolved into hereditary kingships (Frankfort, 1956).

The early Sumerian cities had populations that ranged between 7,000 and 20,000. However, one Sumerian city, Uruk, extended over 1,100 acres and contained as many as 50,000 people (Gist & Fava, 1974). By today's standards, the populations of those early cities seem rather small. They do, however, present a marked contrast with the small nomadic and seminomadic bands of individuals that existed prior to the emergence of these cities.

Within the next 1,500 years, cities arose all across the ancient world. Memphis was built around 3200 B.C. as the capital of Egypt, and between 2500 and 2000 B.C. major cities built in what is now Pakistan. The two largest, Harappa and Mohenjo-Daro, were the most advanced cities of their day. They were carefully planned in a grid pattern with central grain warehouses and elaborate water systems, including wells and underground drainage. The houses of the wealthy were large and multi-storied. Built of fired brick, they were in neighborhoods separated from the humble dried-mud dwellings of the common laborers. Like the Sumerian cities, these cities were supported by a surplus-producing agricultural peasantry and were organized around central temple complexes.

By 2400 B.C., cities were established in Europe; by 1850 B.C., in China. No fully developed cities were erected in the Americas until some 1,500 years later during the so-called Late Preclassic times (300 B.C. to A.D. 300). In Africa, cities of prosperous traders appeared around A.D. 1000 in what are now Ghana and Zimbabwe.

Preindustrial Cities

Preindustrial cities—*cities established prior to the Industrial Revolution—often were walled for protection and densely packed with residents whose occupations, religion, and social class were clearly evident from symbols of dress, heraldic imagery, and* *manners.* Power typically was shared between the feudal lords and religious leaders. Preindustrial cities housed only 5 to 10% of a country's population. Most had populations of fewer than 10,000.

These cities often served as the seats of political power and as commercial, religious, and educational centers. Their populations were usually stratified into a broad-based pyramid of social classes: A small ruling elite sat at the top; a small middle class of entrepreneurs rested just beneath; and a very large, impoverished class of manual laborers (artisans and peasants) was at the bottom. Religious institutions were strong, well established, and usually tightly interconnected with political institutions, the rule of which they supported and justified in theological terms. Art and education flowered (at least among the upper classes), but these activities were strongly oriented toward expressing or exploring religious ideologies.

Gideon Sjoberg (1956) has noted that three things were necessary for the rise of preindustrial cities. First, there had to be a favorable physical environment. Second, advanced technology in either agricultural or nonagricultural areas had to have developed to provide a means of shaping the physical environment—if only to produce the enormous food surplus necessary to feed city dwellers. Finally, a well-developed system of social structures had to emerge so that the more complex needs of society could be met: an economic system, a system of social control, and a political system.

Industrial Cities

Industrial cities are *cities established during or after the Industrial Revolution and are characterized by large populations that work primarily in industrial and service-related jobs.*

We use the term *Industrial Revolution* to refer to the application of scientific methods to production and distribution, wherein machines came to perform work that had formerly been done by humans or farm animals. Food, clothing, and other necessities could be produced and distributed quickly and efficiently, freeing some people—the social elites—to engage in other activities.

The Industrial Revolution of the nineteenth century forever changed the face of the world. It created new forms of work, new institutions, and new social classes, and multiplied many times over the speed with which humans could exploit the resources of their environment. In England, where the Industrial Revolution began in about 1750, the introduction of the steam engine was a major stimulus for such changes. This engine required large amounts of coal, which England had, and made it possible for cities to be established in areas other than ports and trade centers through its use for transportation vehicles. Work could take place wherever there were coal deposits, industries grew, and workers streamed in to fill the resulting jobs. Thus, industrial cities arose, cities with populations that were much larger than those of preindustrial cities.

Nineteenth-century urban industrialization produced industrial slums, which were seen as some of the worst results of capitalism. Friedrich Engels, a close associate of Karl Marx, described the horrors of one of these areas.

> The view from this bridge—mercifully concealed from smaller mortals by a parapet as high as a man—is quite characteristic of the entire district. At the bottom the Irk flows, or rather stagnates. It is a narrow, coal-black stinking river full of filth and garbage, which it deposits on the lower-lying bank. In dry weather, an extended series of the

The Industrial Revolution created new forms of work. In this photo, taken in 1911, a young woman is taking home piecework.

Table 15-1

A Comparison of the Preindustrial City and the Industrial City

	PREINDUSTRIAL CITY	INDUSTRIAL CITY
Physical Characteristics	A small, walled, fortified, densely populated settlement, containing only a small part of the population in the society	A large, expansive settlement with no clear physical boundaries, containing a large proportion of the population in the society
Transportation	Narrow streets, made for travel by foot or horseback	Wide streets, designed for motorized vehicles
Functions	Seat of political power; commercial, religious, and educational center	Manufacturing and business center of an industrial society
Political Structure	Governed by a small, ruling elite, determined by heredity	Governed by a larger elite made up of business and financial leaders and some professionals
Social Structure	A rigid class structure	Less rigidly stratified but still containing clear class distinctions
Religious Institutions	Strong, well established, tightly connected with political and economic institutions	Weaker, with fewer formal ties to other social institutions
Communication	Primarily oral, with little emphasis on record keeping beyond mercantile data; all records handwritten	Primarily written, with extensive record keeping; use of mechanical print media
Education	Religious and secular education for upper-class men	Secular education for all classes but with differences related to social class

most revolting blackish green pools of slime remain standing on this bank, out of whose depths bubbles of miasmatic gases constantly rise and give forth a stench that is unbearable even on the bridge forty or fifty feet above the level of the water. (Engels, 1845)

Modern industrial cities are large and expansive, often with no clear physical boundary separating them from surrounding towns and suburbs. Like the preindustrial cities before them, industrial cities are divided into neighborhoods that reflect differences in social class and ethnicity. (See Table 15-1 for a comparison of the preindustrial and industrial city.)

The industrial cities of today have become centers for banking and manufacturing. Their streets are designed for autos and trucks as well as for pedestrians, and they feature mass transportation systems. They are stratified, but class lines often become blurred. The elite is large and consists of business and financial leaders as well as some professionals and scientists. There is a large middle class consisting of white-collar salaried workers and professionals such as sales personnel, technicians, teachers, and social workers.

Formal political bureaucracies with elected officeholders at the top govern the industrial city. Religious institutions no longer are tightly intertwined with the political system, and the arts and education

are secular with a strong technological orientation. Mass media disseminate news and pattern the consumption of material goods as well as most aesthetic experiences. Subcultures proliferate, and ethnic diversity often is great. As the industrial city grows, it spreads out, creating a phenomenon known as urbanization.

Urbanization

The vast majority of the people in the United States live in urban areas. We should not confuse urban areas with cities. In fact, several terms often are used inappropriately when cities or urban areas are discussed. Thus, when we discuss cities, we are referring to something that has a legal definition. A **city** is *a unit that typically has been incorporated according to the laws of the state within which it is located* (see Table 15-2 for a list of the world's largest megacities, 1995 and 2015). Legal and political boundaries may be quite arbitrary, however, and a city-type, or urban, environment may exist in an area that is not known officially as a city. For example, New England contains many places known as towns—such as Framingham, Massachusetts, with a population of nearly 65,000—that for all practical purposes should be known as cities. Yet they do not adopt the city form of government with a

Table 15-2

World's Largest Megacities, 1995 and 2015

1995		2015	
CITY	**POPULATION (MILLIONS)**	**CITY**	**POPULATION (MILLIONS)**
Tokyo-Yokohama, Japan	28.4	Bombay, India	28.2
Mexico City, Mexico	23.9	Tokyo-Yokohama, Japan	26.4
Sao Paulo, Brazil	21.5	Lagos, Nigeria	23.2
Seoul, South Korea	19.1	Dhaka, Bangladesh	23.0
New York City, United States	14.6	Sao Paulo, Brazil	20.4
Osaka-Kobe-Kyoto, Japan	14.1	Karachi, Pakistan	19.8
Bombay, India	13.5	Mexico City, Mexico	19.2
Calcutta, India	12.9	Delhi, India	17.8
Rio de Janeiro, Brazil	12.8	New York City, United States	17.4
Buenos Aires, Argentina	12.2	Jakarta, Indonesia	17.3

Source: *Statistical Abstract of the United States: 1997*, U.S. Bureau of the Census, 1997, Washington, DC: U.S. Government Printing Office; United Nations, *World Urbanization Prospects: The 1999 Revision*, 2000.

mayor, a city council, and so on because of an attachment to the town council form of government.

Classification of Urban Environments

The legal boundaries of a city seldom encompass all the people and businesses that have an impact on that city. The U.S. Bureau of the Census has realized that for many purposes, it is necessary to consider the entire population in and around the city that may be affected by the social and economic aspects of an urban environment. As a result, the bureau recognized the need for a complex set of terms to describe and classify urban environments. The terms the Bureau of the Census applies to urban data include urbanized area, urban population, metropolitan statistical area, primary metropolitan statistical area, and consolidated metropolitan statistical area (Federal Committee on Standard Metropolitan Statistical Areas, 1979).

An **urbanized area** contains *a central city and the continuously built-up, closely settled surrounding territory that together have a population of 50,000 or more.* The term, thus, refers to the actual urban population of an area regardless of political boundaries such as county or state lines. This term is often confused with **urbanization,** which refers to *a process whereby a population becomes concentrated in a specific area because of migration patterns.* More simply, *urbanized area* refers to a certain place, and *urbanization* refers to a set of events that are taking place. **Urban population** refers to *the inhabitants of an urbanized area and the inhabitants of incorporated or unincorporated areas with a population of 2,500 or more.*

As a result of the 1990 census, 33 new urbanized areas were recognized, bringing the total number of urbanized areas in the United States to 396. When an area receives the urbanized area designation, it generally becomes recognized as having a large enough urban population to warrant extra attention by government planners and business marketers (U.S. Department of Commerce News, August 16, 1991).

A **metropolitan area** is *a core area containing a large population nucleus and adjacent communities that are economically and socially integrated into that nucleus* (U.S. Census Bureau, May 1997). A metropolitan area emerges as an industrial city expands ever outward, incorporating towns and villages into its systems of highways, mass transportation, industry, and government.

A **metropolitan statistical area (MSA)** contains *one or more central cities, each with a population of at least 50,000, or a single urbanized area that has at least 50,000 people that is part of a larger metropolitan population of at least 100,000 (75,000 in New England).* Each MSA also contains at least one central county; more than half the population of an MSA resides in these central counties. Some outlying counties may be rural but have close economic and social ties to the central counties, cities, and urbanized areas in the MSA (see Table 15-3 for a list of the largest cities in the United States). A **primary metropolitan statistical area (PMSA)** is *a large, urbanized county or cluster of counties that is part of an MSA with 1 million people or more.*

Expanding metropolitan areas draw on surrounding areas for their labor pool and other resources to such an extent that all levels of planning—from the building of airports, sports complexes, and highway and railroad systems to the production of electric power and the zoning of land for industrial use—must increasingly be undertaken with their possible effects on entire regions kept in mind. In the United States, the most dramatic examples of this trend are the Los Angeles metropolitan area, the Dallas-Fort Worth area, and the 500-mile northeast corridor

Table 15-3

Largest Cities in the United States, 1999

CITY	SIZE (MILLIONS)	RANK 1980	RANK 1999
New York	7.4	1	1
Los Angeles	3.6	2	2
Chicago	2.8	3	3
Houston	1.8	5	4
Philadelphia	1.4	4	5
San Diego	1.2	8	6
Phoenix	1.2	9	7
San Antonio	1.1	11	8
Dallas	1.1	7	9
Detroit	.96	6	10
San Jose	.87	17	11
San Francisco	.75	13	12
Indianapolis	.74	12	13
Jacksonville	.70	22	14
Columbus	.70	10	15

Source: Population Estimates Program, Population Division, U.S. Bureau of the Census, October 2000.

from Washington, D.C., to Boston, Massachusetts ("Boswash")—an area that includes 60 million people and that is forming an enormous metropolis, sometimes called a megalopolis, with New York City as its hub. A **megalopolis** is *a metropolitan area with a population of 1 million or more that consists of two or more smaller metropolitan areas*. The federal government uses the phrase **consolidated metropolitan statistical area (CMSA)** to refer to *a megalopolis*.

Despite the much-heralded flow of population to nonmetropolitan areas, more than one-third of all Americans live in the country's twenty-three megalopolises or CMSAs. New York-Northern New Jersey-Long Island ranks first among the megalopolises/CMSAs in population size, and it will continue to be number one well into the future. However, the Houston-Galveston-Brazoria, Texas, area should experience the most rapid rate of population growth of any CMSA.

Basically, then, the United States can be seen as a three-tiered system of metropolitan areas: 255 freestanding MSAs; 73 larger PMSAs; and 18 very large CMSAs *(Office of Management and Budget, June 30, 1996).*

Before a city grows outward to form a metropolitan area or become part of a megalopolis, it goes through certain internal developments that establish the placement of various types of industrial, commercial, and residential areas.

The Structure of Cities

The community and the city have been two of the primary areas of study since the beginning of American

The legal boundaries of a city seldom encompass all the people and businesses that depend on the city or have an impact on it. The U.S. Bureau of the Census has realized that it is necessary to consider the entire population in and around the city.

sociology. In the 1920s, classical *human ecology* blossomed under the leadership of Robert E. Park and Ernest W. Burgess at the University of Chicago. The early human ecologists were attempting to systematically apply the basic theoretical scheme of plant and animal ecology to human communities.

Theories of human communities were developed that were analogous to theories explaining plant and animal development. For example, if you were to drive from the mountains to the desert, you would find that different soil, water, and climate conditions produce entirely different types of vegetation. By analogy, driving from a city's business district to its suburbs, you also will notice different types of communities based on a competition for specific types of land uses.

In fact, the human ecologists told us, human communities could be understood from a Darwinian perspective. Communities and cities have evolved and changed as a consequence of competition for prime space, invasion, succession, and segregation of new groups.

Park and Burgess and other members of the Chicago school of sociologists studied the internal structure of cities as revealed by what they called the ecological patterning (or spatial distribution) of urban groups. In investigating the ways in which cities are patterned by their social and economic systems and by the availability of land, these sociologists proposed a theory based on concentric circles of development.

Concentric Zone Model   The concentric zone model, sometimes irreverently called the bull's-eye model, is illustrated in Figure 15-1 (Park, Burgess, & McKenzie, 1925). The **concentric zone model** is *a theory of city development in which the central city is made up of (1) a business district, and radiating from this district is (2) a zone of transition with low-income, crowded and unstable, residential housing with high crime rates, prostitution, gambling, and other vices; (3) a working-class residential zone; (4) a middle-class residential zone; and (5) an upper-class residential zone in what we would now think of as the suburbs.* These zones reflect the fact that urban groups are in competition for limited space and that not all space is equally desirable in terms of its location and resources.

The concentric zone model initially was quite influential in that it did reflect the structure of certain cities, especially those like Chicago that developed quickly early in the Industrial Revolution, before the development of mass transportation and the automobile introduced the complicating factor of increased mobility. It did not, however, describe many other cities satisfactorily, and other models were needed.

Sector Model   In the 1930s Homer Hoyt (1943) developed a modified version of the concentric zone model that attempted to take into account the influence of urban transportation systems. He agreed with the notion that a business center lies at the heart of a city but abandoned the tight geometrical symmetry of the concentric zones. Hoyt suggested that the structure of the city could be better represented by a **sector model**, *in which urban groups establish themselves along major transportation arteries (railroad lines, waterways, and highways).* Then, as the city becomes more crowded and desirable land is even farther from its heart, each sector remains associated with an identifiable group but extends its boundaries toward the city's edge (see Figure 15-2).

Multiple Nuclei Model   A third ecological model, developed at roughly the same time as the sector model, stresses the impact of land costs, interest-rate schedules, and land-use patterns in determining the structure of cities. This multinuclei model (Harris & Ullman, 1945) emphasizes the fact that different industries have different land-use and financial requirements, which determine where they establish themselves (see Figure 15-3). Thus, *the **multiple nuclei model** holds that as similar industries are established near one another, the immediate*

Figure 15-1 Concentric Zone Model

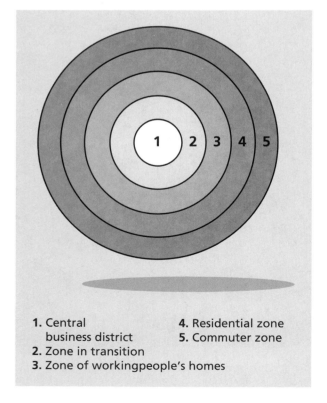

1. Central
   business district
2. Zone in transition
3. Zone of workingpeople's homes
4. Residential zone
5. Commuter zone

Figure 15-2 **Sector Model**

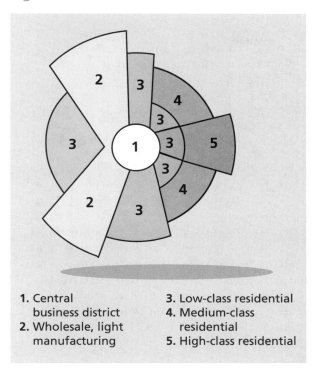

1. Central business district
2. Wholesale, light manufacturing
3. Low-class residential
4. Medium-class residential
5. High-class residential

Figure 15-3 **Multiple Nuclei Model**

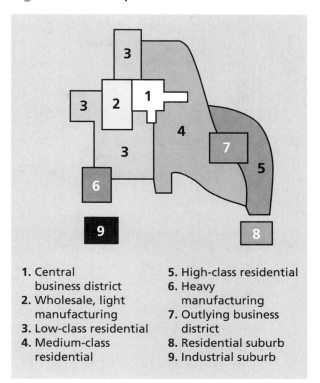

1. Central business district
2. Wholesale, light manufacturing
3. Low-class residential
4. Medium-class residential
5. High-class residential
6. Heavy manufacturing
7. Outlying business district
8. Residential suburb
9. Industrial suburb

neighborhood is shaped by the nature of its typical industry, becoming one of a number of separate nuclei that together constitute the city. For example, some industries, such as scrap metal yards, need to be near railroad lines. Others, such as plants manufacturing airplanes or automobiles, need a great deal of space. Still others, such as dressmaking factories, can be squeezed into several floors of central business district buildings. In this model, a city's growth is marked by an increase in the number and kinds of nuclei that compose it.

The limitation of the ecological approach to studying urban structure is that it downplays variables that often strongly influence urban residential and land-use patterns. For instance, the ethnic composition of a city may be a powerful influence on its structure: A city with but one or two resident ethnic groups will look very different from a city with many groups. Another important variable is the local culture—the history and traditions that attach certain meanings to specific parts of the city. For example, the north end of Boston has become the Italian section of the city. People of Italian descent who normally might move to the suburbs have remained in the neighborhood because of strong ties to the traditions associated with that area. Indeed, cultural factors are important contributors to the continuing trend of urbanization.

The early ecologists could not predict some of the trends that have taken place since World War II.

Since that time, our Eastern and Midwestern cities have declined in population while the sprawling Sunbelt cities of the South and West have gained population and business. In addition, cities everywhere are more decentralized because of the automobile. The central business districts of cities have become less important over time. As the cities spread out, urban areas become linked to one another in a manner more complex than the early urban ecologists could have imagined. (See "Sociology at Work: How to Ruin a City.")

Contemporary urban ecologists (Berry & Kasarda, 1977; Exline, Peters, & Larkin, 1982; Hawley, 1981) have developed more advanced theories that take into account some contemporary developments. Computers and modern statistical techniques are used now to analyze the variables that influence the growth and development of a city.

The Nature of Urban Life

Ever since sociologists began writing about communities, they have been concerned with differences between rural and urban societies and with changes that take place as society moves away from small, homogeneous settlements to modern-day, urban centers. These changes have been accompanied by a shift in the way people interact and cooperate.

Sociology at Work

How to Ruin a City

William H. Whyte is best known for his classic book The Organization Man. *Later, his interests moved into the area of urban environments and the impact that changes in that environment have on people and cities themselves. In this section, Whyte discusses the blurring of lines between ugly and pleasant city centers.*

Whyte believes some cities are improving their downtown areas, reinforcing the role of the street, and in general reasserting the dominance of the center. But a growing number are doing the opposite. They are changing the city structure, gearing it more to the car, taking the pedestrian and retailing off the street. They are doing almost everything, indeed, to eliminate the structured advantages of the center they inherited.

As a way of distinguishing which camp a city is in, Whyte prepared a checklist of eight yes or no questions. The answers will help determine whether a city has been moving decisively in one direction or in the other.

1. Was much of downtown successfully razed under urban renewal?
2. Is at least half of downtown devoted to parking?
3. Have municipal and county offices been relocated to a campus?
4. Have streets been demapped for superblock developments?
5. Have the developments included an enclosed shopping mall?
6. Have they been linked together with skyways?
7. Have they been linked together with underground concourses?
8. Is an automated people-mover system being planned?

Whyte believes that the higher the score, the more likely it is that the city has lost its ego, its sense and pride of place, its awareness of where it has come from and where it is going. It is a city with so little assurance that it is prey to what could be billed as bold new approaches and to architectural acrobatics of all kinds.

Small and medium-sized cities seem particularly vulnerable. Their downtowns, for one thing, are more subject to the dominance of the freeways than are very large cities.

The worst problem is parking. The blight of parking lies in what is not there—people, activity, function. The daytime storage of vehicles is not a highest and best use of space, but is treated as if it were.

In some American cities, so much of the center has been cleared to make way for parking that there is more parking than there is city. If they clear away any more of what is left, there would not be much reason to go there and park.

Whyte noted that in their zeal to woo the car, developers and municipalities grabbed off some of the best-located parcels of downtown. Supply has so conditioned demand that parking has become an end in itself, with people's bondage to it more psychological than physical. But suppose, just suppose, that Americans were to extend their walking radius by only a few hundred feet. The result could be the emancipation of downtown. Instead of being sequestered for the storage of vehicles, prime space would be released for positive activities. In our wastefulness, in sum, lies opportunity. There is really a great deal of space in the city if we have the wit to see it.

Source: *City: Rediscovering the Center* (pp. 310–316), by W. H. Whyte, 1988, New York: Doubleday.

The Chicago school of sociology, as it was called, produced a large number of studies dealing with human interaction in city communities. Those sociologists were interested in discovering how the sociological, psychological, and moral experiences of city life reflected the physical environment.

Gemeinschaft and Gesellschaft

The Chicago sociologists, in their studies of the city, used some of the concepts developed by Ferdinand Tönnies (1865–1936), a German sociologist.

In his book *Gemeinschaft und Gesellschaft*, Tönnies examined the changes in social relations attributable to the transition from rural society (organized around small communities) to urban society (organized around large impersonal structures).

Tönnies noted that in a **Gemeinschaft** *(community), relationships are intimate, cooperative, and personal.* For example, author Philip Roth, in his book *American Pastoral,* describes such a community:

About one another, we knew who had what kind of lunch in the bag in his locker and who ordered

A small town is likely to produce a *Gemeinschaft,* in which relationships are intimate, cooperative, and personal; a city is likely to produce a *Gesellschaft,* in which relationships are impersonal and people look out for their own interests.

what on his hot dog at Syd's; we knew one another's physical attributes—who walked pigeon-toed and who had breasts, who smelled of hair oil and who oversalivated when he spoke; we knew who among us was belligerent and who was friendly, who was smart and who was dumb; we knew whose mother had an accent and whose father had a mustache, whose mother worked and whose father was dead; somehow we even dimly grasped how every family's different set of circumstances set each family a distinctive difficult human problem. (p. 43)

In a Gemeinschaft the exchange of goods is based on reciprocity and barter, and people look out for the well-being of the group as a whole. Among the Amish, for example, there is such a strong community spirit that, should a barn burn down, members of the community quickly come together to rebuild it. In just a matter of days a new barn will be standing—the work of community members who feel a strong tie and responsibility to another community member who has encountered misfortune.

In a **Gesellschaft** *(society), relationships are impersonal and independent.* People look out for their own interests, goods are bought and sold, and formal contracts govern economic exchanges. Everyone is seen as an individual who may be in competition with others who happen to share a living space.

Tönnies saw *gesellschaft* as the product of mid-nineteenth-century social changes that grew out of industrialization, in which people no longer automatically wanted to help one another or to share freely what they had. There is little sense of identification with others in a *gesellschaft,* in which each individual strives for advantages and regards the accumulation of goods and possessions as more important than the qualities of personal ties. Modern urban society is, in Tönnies's terms, typically a Gesellschaft, whereas rural areas retain the more intimate qualities of Gemeinschaft.

In small, rural communities and preliterate societies, the family provided the context in which people lived, worked, were socialized, were cared for when ill or infirm, and practiced their religion. In contrast, modern urban society has produced many secondary groups in which these needs are met. It also offers far more options and choices than did the society of Tönnies's *Gemeinschaft:* educational options, career options, lifestyle options, choice of marriage partner, choice of whether to have children, and choice of where to live. In this sense, the person living in today's urban society is freer.

Mechanical and Organic Solidarity

Tönnies wrote about communities and cities from the standpoint of what we call an ideal type, in that no community or city actually could conform to the definitions he presented. They are basically concepts that help us understand the differences between the two. In the same sense, Émile Durkheim devised ideas about mechanical and organic solidarity.

According to Durkheim, every society has a **collective conscience**—*a system of fundamental beliefs and values.* These beliefs and values define for its members the characteristics of the good society, which is one that meets the needs for individuality, for security, for superiority over others, and for any of a host of other values that could become important to the people in that society. **Social solidarity** *emerges from the people's commitment and conformity to the society's collective conscience.*

A **mechanically integrated society** is one in which *a society's collective conscience is strong and there is a great commitment to that collective conscience.* In this type of society, members have common goals and values and a deep and personal involvement with the community. A modern-day example of such a society is that of the Tasaday, a food-gathering community in the Philippines. Theirs is a relatively small, simple society, with little division of labor, no separate social classes, and no permanent leadership or power structure.

In contrast, in an **organically integrated society,** *social solidarity depends on the cooperation of individuals in many different positions who perform* *specialized tasks.* The society can survive only if all the tasks are performed. With organic integration such as is found in the United States, social relationships are more formal and functionally determined than are the close, personal relationships of mechanically integrated societies.

Although we may take for granted the movement from *Gemeinschafts* to *Gesellschafts,* or mechanically integrated to organically integrated societies, it is only relatively recently in the course of human history that cities have become the dominant type of living arrangement.

Social Interaction in Urban Areas

The anonymity of social relations and the cultural heterogeneity of urban areas give the individual a far greater range of personal choices and opportunities than typically are found in rural communities. People are less likely to inherit their occupations and social positions. Rather, they can pick and choose and even improve their social positions through education, career choice, or marriage. Urbanism creates a complicated and multidimensional society, with people involved in many different types of jobs and roles.

Louis Wirth proposed what is now a widely accepted definition of city in his classic essay "Urbanism as a Way of Life" (1938). Wirth defined the city as a "relatively large, dense, and permanent settlement of socially heterogeneous individuals." For years urban studies tended to accept Wirth's view of the city as an alienating place where, because of population density, people hurry by one another without personal contact. However, in *The Urban Villagers* (1962), Herbert Gans helped refocus the way sociologists see urban life. Gans showed that urbanites can and do participate in strong and vital community cultures, and a number of subsequent studies have supported this view. For example, researchers in Britain found that people who live in cities actually have a greater number of social relationships than do rural folk (Kasarda & Janowitz, 1974). Other investigators have discovered that the high population density typical of city neighborhoods need not be a deterrent to the formation of friendships; under certain circumstances, city crowding may even enhance the likelihood that such relationships will occur. Gerald Suttles (1968) showed that in one of the oldest slum areas of Chicago, ethnic communities flourish with their own cultures— with norms and values that are well adapted to the poverty in which these people live.

Of course, increased population size can lead to increased superficiality and impersonality in social relations. People interact with one another because they have practical rather than social goals. For example, adults patronize neighborhood shops primarily to purchase specific items rather than to chat and share information. As a result, urbanites rarely know a significant number of their neighbors. As Georg Simmel (1955) noted, in rural society people's social relationships are rich because they interact with one another in terms of several role relationships at once (a neighbor may be a fellow farmer, the local baker, and a member of the town council). In urban areas, by contrast, people's relationships tend to be confined to one role set at a time (see Chapter 5).

With increasing numbers of people, it also becomes possible for segments or subgroups of the population to establish themselves—each with its own norms, values, and lifestyles—as separate from the rest of the community. Consequently, the city becomes culturally heterogeneous and increasingly complex. As people in an urban environment come into contact with so many different types of people, typically they also become more tolerant of diversity than do rural people.

Although urban areas may be described as alienating places in which lonely people live in crowded, interdependent, social isolation, there is another side to the coin, one that points to the existence of vital community life in the harshest urban landscapes. Further, urban areas still provide the most fertile soil for the arts in modern society. The close association of large numbers of people, wealth, communications media, and cultural heterogeneity provide an ideal context for aesthetic exploration, production, and consumption.

Urban Neighborhoods

People sometimes talk of city neighborhoods as if they were all single, united communities, such as Spanish Harlem or Little Italy in New York City, or Chinatown in San Francisco. Those communities do display a strong sense of identity, but to some extent this notion is rooted in a romantic wish for the good old days when most people still lived in small towns and villages that were in fact communities and that gave their residents a sense of belonging. Yet although the sense of community that does develop in urban neighborhoods is not exactly like that in small, closely-knit rural communities, it is very much present in many sections throughout a city. Urban dwellers have a mental map of what different parts of their city are like and who lives in them.

Gerald Suttles (1972) found that people living in the city draw arbitrary (in terms of physical location) but socially meaningful boundary lines between local neighborhoods, even though these lines do not always reflect ethnic group composition, socioeconomic status, or other demographic variables. In Suttle's view, urban neighborhoods attain such symbolic importance in the local culture because they provide a structure according to which city residents organize their expectations and their behavior. For example, in New York City the neighborhood of Harlem (once among the most fashionable places to live) begins east of Central Park on the north side of Ninety-sixth Street and is a place that has symbolic significance for all New Yorkers. Whites tend to think of it as a place where they are not welcome and that is inhabited by African-American and Spanish-speaking people. For many African Americans and Hispanics, however, Harlem represents the "real" New York City and is where most of their daily encounters take place.

Even those urban neighborhoods that are well known, that have boundaries clearly drawn by very distinctive landmarks, and that have local and even national meaning are not necessarily homogeneous communities. For example, Boston's Beacon Hill neighborhood is divided into four (or possibly five) subdistricts (Lynch, 1960), and New York's Greenwich Village consists of several communities defined in terms of ethnicity, lifestyle (artists), and subculture (especially homosexual).

On the whole, Jane Jacobs's observations in *The Death and Life of American Cities* (1961) generally seem to hold true. She has argued that the social control of public behavior and the patterning of social interactions in terms of what might be called community life are to be found on the level of local blocks rather than entire neighborhoods. Once city dwellers venture beyond their own block, they tend to lose their feelings of identification. In fact, one of the typical features of urban life is the degree to which people move through many neighborhoods in their daily comings and goings rushing here and there without much attention or attachment to their surroundings. Occasionally a city as a whole may have meaning to all or most of its residents and may, for this reason, assume some community-like qualities. Consider, for example, the community spirit expressed in spontaneous celebrations for homecoming World Series or Super Bowl winners.

Although urban blocks and neighborhoods may offer a rich context for community living, there are some inescapably unpleasant facts about urban America that make many people decide to live elsewhere. Cities and urban areas in general can be crowded, noisy, and polluted; they can be dangerous; and they may have poorer schools than those in the suburbs. Consequently, many families, especially those with children, choose the suburbs as an alternative to urban life. Other city dwellers, such as the elderly living on fixed incomes, may be forced to remain despite their wish to move.

Urban Decline

A grim circle of problems threatens to strangle urban areas. Since World War II there has been a migration of both white and African-American middle-class families out of the cities and into the suburbs. The number of African-American middle-class families moving to the suburbs increased sharply in the aftermath of the civil rights movement of the 1960s. This migration pattern of the middle class has led to a greater concentration of poor people in the central cities, which is reflected in the loss of revenues for many large cities. As the more affluent families leave urban areas, so do their tax dollars and the money they spend in local businesses. In fact, many businesses have followed the middle class to the suburbs, taking with them both their tax revenues and the jobs that are crucial to the survival of urban neighborhoods. This has meant a shrinking of the central cities' tax base, while at the same time creating conditions (such as the loss of jobs) that force people to rely on government assistance.

It is not easy to entice suburbanites to move back into cities, even though the U.S. Department of Housing and Urban Development has created several financial incentive programs designed to do so. Because most U.S. suburbs developed in the last few decades, their facilities and physical plants (schools and hospitals) are still relatively new, clean, and attractive. So are the shopping centers that continue to mushroom across the country and that offer their suburban patrons local outlets of prestigious downtown stores as well as supermarkets, discount warehouses, specialty shops, and even entertainment centers. In the central cities many buildings are old, and apartment dwellers must pay higher rents for accommodations that are inferior to those of their suburban counterparts. (See "Our Diverse Society: Disorderly Behavior and Community Decay.")

Gans (1977) has suggested that if we are to save the central cities, we can do so only by mobilizing resources at the national level and raising central-city residents to middle-class economic status. He believes this is more likely to succeed than trying to persuade suburbanites to move back into urban areas. Many people believe, however, that the vicious circle outlined really is a sinking spiral, and that some of our cities have already spun downward and out of control. They see no way of resurrecting them and forecast their gradual demise as the population spreads itself out across the country, particularly into the Sun Belt of the South and Southwest.

Our Diverse Society

Disorderly Behavior and Community Decay

What contributes to the decay of neighborhoods? Crime, violence, drugs, or minor offenses? At a time when people suggest we should do more to solve serious crime problems, George Kelling and Catherine Coles have suggested that the minor offenses may be more serious than we may think.

Starting in the 1960s, disorderly behavior began to be seen as a sign of cultural pluralism. In 1967, President Johnson's Commission on Crime pointed out that even though we have to do something about disorder, it is not the primary business of the police. The role of the police, according to the commission, was to arrest people for serious crimes and put them into the criminal-justice system. Of course, it is nice to think that the police are not going to be too intrusive. But the result was that city streets were abandoned by police, and therefore citizens as well.

What is disorder? In its broadest sense, disorder is incivility, boorish and threatening behavior that disturbs life, especially urban life. Urban life is characterized by the presence of many strangers, and in such circumstances citizens need minimum levels of order. Most citizens have little difficulty balancing civility, which implies self-imposed restraint and obligation with freedom. Yet a few are either unable or unwilling to accept any limitations on their own behavior. At the extreme are predatory criminals who murder, assault, rape, rob, and steal. Less extreme is disorderly behavior such as panhandling, drunkenness, and sleeping on the street. Most of the latter are either ignored or punishable by fines or community service.

How should we respond to a panhandler or a drunk person walking down the street? Should we be sympathetic, helpful, or fearful? Some would see these individuals as reminders of the shortcomings of society, who are making a statement about how society has failed them. Others may not think of these less severe offenses in political terms, but as minor inconveniences with little in the way of serious consequences.

George Kelling and Catherine Coles have suggested that we use the image of broken windows to explain how neighborhoods might decay into disorder and even crime if no one attends faithfully to their maintenance. If a window in a factory is broken and not fixed, passersby will conclude that no one cares and no one is in charge of taking care of the property. Over time, youths in the neighborhood throw rocks at the building and break some more windows. Eventually all the windows are broken. Now it seems that not only is no one in charge of the building, but no one is in charge of the street that it faces. The message is that only the foolhardy or criminal have any business on this unprotected street, and so more and more citizens abandon the street to those they assume prowl it. In this way small disorders lead to larger ones and eventually to crime. However, if a neighborhood appears orderly, people feel safer, and wrongdoers are less likely to commit crimes.

Source: *Fixing Broken Windows: Restoring Order and Reducing Crime in Our Communities,* by G. L. Kelling & C. M. Coles, 1996, New York: Martin Kessler Books (The Free Press).

A small countertrend has been noticed since the early 1990s. Many middle-class young adults have begun to find urban life attractive again. Most of these people are single or are married with no children. This trend has produced *an upgrading of previously marginal urban areas and the replacement of some poor residents with middle-class ones, a process known as* **gentrification.** Critics contend that gentrification depletes the housing supply for the poor. Others counter that it improves neighborhoods and increases a city's tax base. So far, however, this trend has been limited to a handful of cities. The trend may be short-lived, because those young adults may once again abandon the city once they

start having children and find city life unsuitable for child rearing. If the trend continues to grow, however, it clearly will have a major impact on the future of urban life and could serve as a convincing argument against doomsday predictions about the city.

Homelessness

Two decades ago every city had its "skid row" with "bums," "derelicts," and "vagrants." Aside from the occasional story about the executive who became an alcoholic and ended up sleeping in "flophouses," little interest or sympathy was expressed for the denizens of these marginal areas of the city.

Today, not only have the words that we use to describe these people changed, but so have our thinking and attitudes about them. The "bum" or "hobo" of the past has become today's "homeless person." The sense of personal responsibility for their fate attributed to them before has been replaced with a view that the homeless are the victims of a selfish, even ruthless society.

What has really changed? Has society become more heartless and created more victims, or have we become more compassionate and become more aware of the problem? Have the numbers of homeless gone up so much that we are forced to recognize the issue?

The movement to the suburbs of post–World War II America emphasized the suburban ideal of a single-family home. The city was where people worked during the day; once nightfall came, they left for the safety of the suburban community.

If the movement to the suburbs required abandoning the downtown streets at nightfall, there were many people who did not leave the central city. There are, of course, the working-class neighborhoods of the older cities where family life goes on in close proximity to the central business district, under somewhat less private and more crowded conditions than in the suburbs. There are also the marginal people for whom downtown provided alternatives not available elsewhere. Commercial and industrial areas, as well as fringe areas in decaying working-class districts, have tended to provide the housing vital to poor people not living in conventional families. Single-room-occupancy hotels, rooming houses, and even skid row flophouses have provided low-cost single accommodations for those who might not be able to come by them elsewhere.

Downtowns in many older cities traditionally have contained the cities' skid rows and red-light districts, which provided shelter and a degree of tolerance for deviant individuals and activities. Being close to transportation and requiring little initial outlay (often renting by the week), single-room housing traditionally has been utilized by the elderly poor, the seasonally employed, the addicted, and the mentally handicapped.

As the old skid rows decrease in size or disappear, the traditional skid-row population of single older men is being supplemented with large numbers of people of both sexes, many of whom are mentally ill or drug addicted. The deinstitutionalization of the mentally ill caused many of those who previously would have been committed to institutions to become homeless street people. They are not necessarily physically dangerous, but rather are disturbed or marginally competent individuals without supportive families.

Advocates for the homeless often claim that these people are on the streets because of a lost job, a low minimum wage, or a lack of affordable housing. Others point to problems that cannot be corrected by economic measures.

In recent years, the downtown sections of many American cities have undergone extensive renovation and revitalization. This gentrification movement has been both hailed as an urban renaissance and condemned for disrupting urban neighborhoods and displacing inner-city residents. As city land becomes more desirable, space declines in what is usually considered to be the nation's least desirable housing stock, namely single-room-occupancy hotels, rooming houses, and shelters. Although these places have long been seen as the very symbols of urban decay, they serve the vital needs of people with few resources or alternatives. Gentrification has placed these powerless people in direct competition for inner-city space. These trends offer a partial explanation for the growing ranks of the homeless on the streets of many cities.

Advocates for the homeless often assert that these people are on the streets because of a lost job, a low minimum wage, or a lack of affordable housing—all things outside the control of the homeless. Certainly there are many situations where these conditions are the cause, particularly among those homeless for a short spell. However, for the broader category of the homeless, we will find very few auto workers laid off from well-paid jobs. Most homeless people have histories of chronic unemployment, poverty, family disorganization, illiteracy, crime, mental illness, and welfare dependency—problems that cannot be corrected by quick or simple economic measures.

Controversies in Sociology

What Produces Homelessness?

Are the homeless like you and me? Is the lack of affordable housing the main reason for the problem? Many advocates for the homeless state that a little bit of bad luck could cause any one of us to suffer the same plight that befalls many of those we see in our urban centers. Christopher Jencks, the John D. MacArthur Professor of Sociology at Northwestern University, has done a thorough study of homelessness. He notes that mental illness is one of the main reasons that homelessness has increased over the past two decades.

As soon as Americans noticed more panhandlers and bag ladies on the streets, they began trying to explain the change. Since the most noticeable of these people behaved in quite bizarre ways, and since everyone knew that state mental hospitals had been sending their chronic patients "back to the community," many sidewalk sociologists initially assumed that the new homeless were mostly former hospital inmates.

Although deinstitutionalization mostly meant that patients were released from mental hospitals after a few weeks instead of remaining there for months, years, or even a lifetime, it also meant that some people who would once have been sent to a mental hospital were now sent to the psychiatric service of a general hospital or were treated as outpatients. It follows that considerably more than a quarter of today's homeless might have spent time

in a mental hospital if we still ran the system the way we ran it in the 1950s.

Before blaming homelessness on deinstitutionalization, however, we must explain one awkward fact: hospitalization rates for mental illness began to fall in the late 1950s, not in the late 1970s or early 1980s. Since deinstitutionalization caused very little homelessness from 1955 to 1975, how could it have suddenly begun to cause a lot of homelessness after that? The answer is that deinstitutionalization was not a single policy but a series of different policies, all of which sought to reduce the number of patients in state mental hospitals but each of which did so by moving these patients to a different place. The policies introduced before 1975 worked quite well. Those introduced after 1975 worked very badly.

By 1975 most state hospitals had discharged almost everyone they thought they could house elsewhere. Their 200,000-odd remaining inmates were of two kinds: long-term residents who were so disturbed nobody else would take them, and short-term patients who were admitted, medicated, observed for a couple of weeks, and discharged. Some of the short-term patients were readmitted fairly regularly, often because they stopped taking their medication, but they spent the bulk of their time outside hospitals.

Although the number of patients in state mental hospitals fell from 468 per 100,000 adults in 1950 to 119 in 1975, advocates of deinstitutionalization

The largest single category of homeless people is made up of the mentally ill. Many of today's homeless were dumped onto the streets as a result of the deinstitutionalization process. Others who have reached adulthood since then have never been institutionalized, but would have been during previous decades.

Homelessness was not the intended result of this process. The original idea was to free the patients from the wretched and abusive conditions in mental hospitals and to allow them, instead, to be treated in the community. However, funding for community treatment never materialized and many people ended up on the streets, uncared for and unable to care for themselves. (For more on the cause of homelessness see "Controversies in Sociology: What Produces Homelessness?")

The National Coalition for the Homeless believes that on a typical night there are about 700,000 homeless people sleeping on the streets or in shel-

ters. That number is a broad estimate, based on street interviews and counts at soup kitchens, shelters, and other services for the homeless. Other groups estimate the number of homeless people to be closer to 300,000.

Equally difficult to pinpoint are the homeless population's demographics, since most people do not remain homeless permanently and the population is therefore always shifting. According to the U.S. Conference of Mayors (USCM), a coalition of mayors from cities with populations of 30,000 or more, about half of all homeless people are single men, while 14% are single women; families with children make up about 36% of the homeless population. More than half (58%) of all homeless people are black, while 29% are white, 10% are Hispanic, 2% are Native American, and 1% are Asian (*Issues and Controversies on File,* January 21, 2000).

The homeless have a variety of problems. Homeless people are more likely than the general

were far from satisfied. Rather than simply continuing their campaign to alter physicians' clinical judgments about who should be hospitalized, reformers increasingly turned to the courts, challenging physicians' right to commit anyone at all. These challenges began to influence medical practice in some states during the early 1970s, but their main impact came in the late 1970s, when they precipitated a fourth round of deinstitutionalization.

Once America restricted involuntary commitment, many seriously disturbed patients began leaving state hospitals even when they had nowhere else to live. When their mental condition deteriorated, as it periodically did, these patients were also free to break off contact with the mental-health system. In many cases they also broke with the friends and relatives who had helped them deal with public agencies. The mentally ill are seldom adept at dealing with such agencies on their own, so once they lost touch with the people who had acted as their advocates, they often lost (or never got) the disability benefits to which they were theoretically entitled. In due course some ended up not only friendless but penniless and homeless.

If the courts had not limited involuntary commitment and if state hospitals had not started discharging patients with nowhere to go, the proportion of the adult population living in state hospitals would probably be about the same today as in 1975. Were that the case, state hospitals would

have sheltered 234,000 mental patients on an average night rather than 92,000. It follows that 142,000 people who would have been sleeping in a state hospital under the 1975 rules are sleeping somewhere else today. On any given night, some of these people were in the psychiatric wards of general hospitals, and a few were in private psychiatric hospitals, but many were in shelters or on the streets.

Almost everyone agrees that what happened to the mentally ill after 1975 was a disaster. Both liberals and conservatives blame this disaster on their opponents, and both are half right. It was the insidious combination of liberal policies aimed at increasing personal liberty with conservative polices aimed at reducing government spending that led to catastrophe.

Had politicians been committed to keeping the mentally ill off the streets, they could have used the money that hospitals once spent on these patients to provide SRO [single-room occupancy] rooms and out-patient services. Some states did try this. In most states, however, political leaders mouthed clichés and looked the other way.

Source: Excerpted from *The Homeless* (pp. 1–11), by C. Jencks, 1994, Cambridge, MA: Harvard University Press.

population to be mentally ill, suffer from chronic illness, and be addicted to drugs or alcohol. A 1999 survey by the U.S. government's Housing and Urban Development office found that:

- Nearly two-thirds of homeless people suffer from chronic or infectious diseases. Another 9% are suffering from AIDS.
- Thirty-nine percent are mentally ill.
- One-quarter were abused as children and 21% were homeless as children.

In recent years, many cities have imposed restrictions on the homeless as "compassion fatigue" has set in. People have grown disillusioned with programs for the homeless because they appear to have little effect. Instead, policy makers and the public have used more aggressive ways of dealing with the problem. By 1999, all 50 of the largest cities in the United States had approved regulations limiting ac-

tivities associated with the homeless, such as loitering and sleeping on sidewalks.

The deinstitutionalized, the ex-offender, the addicted, the poor, the sick, and the elderly all bring to the central city a lifestyle incompatible with that of the new urban middle class. Yet these people will not go away simply because their housing is eliminated. They remain on our streets and tax the strained resources of the remaining shelters. Unlike the suburb, the newly gentrified inner city cannot close its gates to marginal members of society. It therefore becomes imperative that new alternatives be provided.

Future Urban Growth in the United States

What will metropolitan areas look like in the future? Which cities will grow the most? Which cities will lose the most population? One way to answer such questions would be to extend into the future the trends of the 1970s, 1980s, and 1990s. Yet doing so would miss some important changes that have been taking place in the United States.

Two important trends have had an important impact on cities. The first trend is the sharp rise in immigration to the United States. Each year about one million people, predominantly Latin American and Asian in origin, arrive in the United States, most settling in urban areas.

The impact of immigration is apparent by looking at urban areas experiencing the greatest population gains between 1990 and 1999. The population gains in Los Angeles, Houston, San Diego, Miami, and Dallas came entirely from international migration and natural increase. Were it not for immigration, the population of these areas would have been far smaller or would have outright declined.

The second trend involves the aging of the baby-boomer generation, the 76 million people born between 1946 and 1961. The immigrants are settling in the cities while the baby boomers are aging in the suburbs.

These two trends complicate both urban/suburban relationships as well as race and ethnic dynamics. The city of Los Angeles provides an example of how the two trends interact. In Los Angeles, the out-migration of whites to the suburbs has coincided with waves of immigration of new ethnic minorities. Los Angeles County's elderly population is still mostly white, its working-aged population is only about one-third white, and its child population is predominantly Hispanic but includes children from other racial and ethnic groups. Reflecting their age, the growing racial and ethnic groups in Los Angeles are concerned about affordable housing, good

schools, and neighborhoods conducive to the raising children. The older white population is more concerned with health and social support services for an aging, dependent population.

Whatever else the new urban population profiles show, the old models of dealing with cities and suburbs will need to be revised to adapt to new demographic forces in America today (Frey, 2000).

Metropolitan trends vary widely according to region. Some metropolitan areas are expected to maintain high growth rates over the next two decades. Those areas are primarily in the South and West. However, constraints to this expansion are appearing. The rate of growth for Houston and other southern and western metropolitan areas should slow somewhat, and the same pattern will occur in most other Sunbelt areas.

Suburban Living

Suburbs are *incorporated or unincorporated spatial communities that lie outside the central city but within the metropolitan area.* According to this definition, 60 million people lived in the suburbs in 1960, 74 million in 1970, and more than 143 million in 2000. In fact, most residents of metropolitan areas live in suburbs rather than in the central cities (Palen, 1995). (See Figure 15-4.)

Originally, farmers mocked the small-scale agriculture of suburban gardeners. City newspaper editors derided the lack of cultural facilities in the suburbs. Suburbanites inhabited a territory that did not fit any traditional definition of city or country.

In many respects, at least until recently, suburbs have served as a dramatic contrast to city life.

Figure 15-4 **Percentage of Americans Living in Suburbs, 1910–2000**

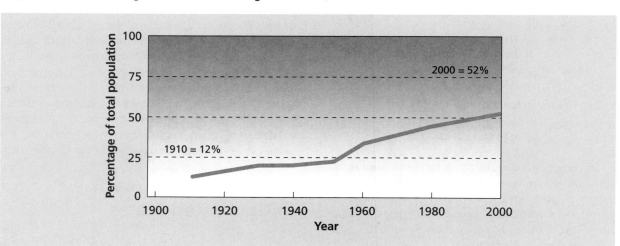

Source: U.S. Bureau of the Census.

Suburbs generally are cleaner than cities, less crowded, less noisy, and less crime ridden. Often, their school systems are newer and better. Many characteristics of urban life, however, have followed people to the suburbs.

Today we accept without question the sprawling landscape of single-family houses on small lots and assume that suburbs have always been with us. We give little thought to the suburb's origin. Suburbs, as we know them, developed relatively recently and largely without any planning. They were a direct response to changes that made commuting easier. Well-surfaced roads and hundreds of new bridges began to appear in the early nineteenth century. The steamship also changed matters dramatically in the New York City region. Individuals discovered a new lifestyle that included steaming down the Hudson River in the early morning and cruising slowly upstream in early evening.

Railroads were also instrumental in the development of the suburbs. Trains made it possible to travel in all kinds of weather, prompting thousands of upper-class Americans to use them. Originally, the railroad companies discouraged short-haul commuting, but eventually demand created the commuter train. Suburbs began to blossom all along the railroad routes that led to the city. Thousands of middle-income families settled in the suburbs, which offered the advantages of both rural and urban living.

The typical suburban house was part farmhouse, part urban residence, and it reflected a desire for open space, sanitation, and security. It usually had many large closets, a large cellar, an attic, a pantry, and a back hall—even guest rooms.

Suburbanites also started to copy the front lawn that was typical of estates and English country homes. The lawn mower began to appear in the 1880s, and magazines explained the use of the new machines. By the 1920s the idea of a smooth, green lawn became widely accepted. Lawn-mower manufacturers determined that grass height should be about 1.5 inches, and suburbanites were quick to follow the recommendation.

Suburban growth slowed during the depression years and the subsequent war years when gasoline rationing took place. Once the war ended, however, suburban growth resumed. By the mid-1950s, the automobile made it possible for suburbs to exist far beyond the range of railroads and trolleys. Popular television shows such as *Father Knows Best* and *Leave It to Beaver* responded to the public sentiment, which was turning against the city, and shifted their locales to the suburbs.

The way in which homes were purchased also influenced the growth of the suburbs. Before the 1920s, a home buyer borrowed 30 to 40% of the cost of the property. Mortgage interest payments of 5 or 6% were made semiannually. At the end of anywhere from 3 to 8 years, the principal was repaid in one lump sum or the mortgage was renegotiated. After the 1920s, savings and loan associations replaced the informal, individual lender. Following World War II, the government guaranteed mortgages to veterans. Ex-GIs needed only the smallest of down payments to purchase a home, and a building boom emerged.

For the children of those families, the suburbs represented the world, a world ruled by women more than men. Men went to work in the city and returned every evening to the suburban landscape. As commuting time increased, the fathers looked for jobs in the suburbs and began deserting the city altogether. When white-collar workers started to appear in the suburbs in large numbers, suburbs entered their present stage.

The growth of the nation's suburbs continued throughout the century. The share of the U.S. population that lived in the suburbs doubled from 1900 to 1950 and doubled again from 1950 to 2000.

People used to believe the stereotype that suburbs were predominantly white. This is no longer the case and ignores the heterogeneous growth patterns within the suburbs. During the 1980s and 1990s, some areas saw explosive growth in the number of African-American suburbanites. Today, there are more than 40 metropolitan areas with at least 50,000 African-American suburbanites. The largest suburban black population is found in Washington, D.C. A second large African-American suburb is located outside of Atlanta, where suburban African-Americans live in such neighborhoods as Brook Glen, Panola Mill, and Wyndham Park. Los Angeles also has large numbers of suburban African Americans.

One characteristic that still separates suburban from city neighborhoods is that suburbs tend to be homogeneous with regard to the stage of development of families in their natural life cycle. For example, in the suburbs, retired couples rarely live among young couples that are just starting to have children. Aside from the rather dulling sameness of many suburban tracts that results, the situation also creates problems in the planning of public works. For example, a young suburb might well invest money in school buildings, which, some 20 years later, are likely to stand empty when the children have left home.

A problem of the late twentieth century has been suburban sprawl. Suburban sprawl involves the expansion of new residential subdivisions and commercial areas, increased use of cars, and segregated use of land according to activities.

Levittown, Pennsylvania, became the symbol of the post–World War II suburban building boom. With little or no money down and low monthly payments attracting former GIs and their families, the U.S. urban landscape and society were changed.

The increased dependence on cars produced by sprawl has changed how we live. Many Americans now think of the second car as a necessity, not a luxury. The typical adult averages 72 minutes every day behind the wheel, and most of this time is spent alone. Some (Putnam, 2000) believe that our love affair with our cars has reduced our involvement with other people and the community.

Now in the 21st century, most Americans live in suburbs. But they have not exactly left the cities behind. Cities and suburbs alike face daunting challenges. Our suburbs—from wealthy gated communities, to gritty blue-collar bungalow and industrial communities, to new housing tract developments on the urban fringe—are inexorably linked to the fate of nearby cities. In fact, many of the places we call suburbs are really small cities. They, too, confront serious problems: fiscal challenges, poverty, environmental concerns, crime, housing shortages, traffic congestion, ethnic tensions, and struggling public schools (Villaraigosa, 2000).

Exurbs

Problems that were once associated with central cities—traffic congestion, overcrowded schools, and the loss of open space—have emerged in many suburbs. This situation has motivated some people to move to the more rural areas also knows as exurbs. **Exurbs** are *middle and upper-middle class communities that can be found in outlying semi-rural suburbia.* These areas are located in a newer, second ring beyond the old suburbs. These communities sometimes form around old villages or small towns. The inhabitants, as a rule, are affluent, well-educated professionals (Palen, 1995).

The exurb is taking shape largely within metropolitan boundaries, but it differs sharply from the traditional suburb. For one thing, development in the exurbs is much less dense, emerging near farms and rural land in the remotest fringes of metropolitan areas. For another, suburbs traditionally are dependent on the city for jobs and services. Not so in the exurb, which creates its own economic base in its shopping malls, office complexes, and decentralized manufacturing plants. While the earlier wave of fringe development may have been an expansion of the city, the new areas may be anticity (Townsend, 1987).

Why are so many people moving still farther out from the central city? Recent research suggests a variety of reasons, but on the whole, reasons that are not very surprising when we consider the suburban exodus of the early postwar era. Basically, exurbanites are similar to people a century ago who were seeking the better life in romantic suburbs and rustic settings. Today these people expect the services, schools, and shopping areas to follow them out of the city.

North Carolina could be the prototype for America's future. The state offers scenic beauty alongside economic vitality, thanks in part to a conscious policy of dispersed development. To avoid the problems of large urban areas, the state decentralized the university system into 16 campuses, improved roads even in the remotest areas, and encouraged scattered industrial development. As a result, North Carolina is a leading manufacturing and computer technology state, but has no city larger than Charlotte, with 396,000 people (Herbers, 1986).

In short, the exurb, like the suburb before it, is seemingly another step in the quest for the American dream.

Urbanism has become an American way of life. Although the shape and form of metropolitan areas continue to change, their influence continues to dominate the manner in which we interact with our environment.

SUMMARY

For vast majority of time that humans have lived on earth, they have lived without cities. Cities have appeared only within the past 7,000 to 9,000 years, coinciding with the rise of what we know as civilization. The industrialized areas of the world are already heavily urban, and the rate of urbanization is slowing. The pace of urbanization is accelerating rapidly, however, in the still largely rural nonindustrial world.

Two requirements had to be met for cities to emerge. First, there had to be a surplus of food and other necessities, so that some people could afford to live in settlements where they did not produce their own food but could depend on others to meet their needs. The second requirement was some form of social organization beyond the family that was capable of distributing the surplus to those who needed it. The world's first fully developed cities arose 6,000 years ago in the Middle East in what is now Iraq, which was the site of the Sumerian civilization.

Preindustrial cities—cities established before the Industrial Revolution—often were walled for protection and densely packed with residents whose occupations, religion, and social class were clearly evident from symbols of dress and manners. Industrial cities are cities established during or after the Industrial Revolution and are characterized by large populations that work primarily in industrial and service-related jobs. The Industrial Revolution of the nineteenth century created new forms of work, as well as new institutions and social classes, and multiplied many times over the speed with which humans could exploit the resources of their environment. Modern industrial cities are thus large and expansive, often with no clear physical boundary separating them from surrounding towns and suburbs. Modern industrial cities have become centers for banking and manufacturing.

The vast majority of people in the United States live in urban areas, though not in cities. A city is a unit that typically has been incorporated according to the laws of the state within which it is located. The legal boundaries of a city seldom encompass all the people and businesses that may be affected by the social and economic aspects of an urban environment.

Although some urban blocks and neighborhoods offer a rich community life, there are also many serious problems in the cities—crime, pollution, noise, poor educational and social services, and so on—that make many people want to live elsewhere. Since World War II there has been a migration of both black and white middle-class families out of the cities. Many businesses have followed those people to the suburbs, taking with them their tax revenues and the jobs that are crucial to the survival of urban neighborhoods.

INTERNET RESOURCES

Go to http://web-enhanced.thomsonlearning.com which features Web links relevant to this chapter to expand and enhance the learning experience. This site will be monitored and updated periodically to ensure the links are correct and live.

At Virtual Society: The Wadsworth Sociology Resource Center, http://sociology.wadsworth.com, you will find a Career Center, Sociology in the News, Research Online, InfoTrac College Edition, Virtual Tours in Sociology, and a variety of book-specific student and instructor resources.

CHAPTER FIFTEEN STUDY GUIDE

KEY CONCEPTS

Match each concept with its definition, illustration, or explanation below.

a. Gentrification
b. Urbanized area
c. *Gemeinschaft*
d. Skid row
e. Multiple nuclei model
f. Sector model
g. Exurbs
h. City
i. Collective conscience
j. Preindustrial cities
k. Megalopolis

l. Ecological patterning
m. Social solidarity
n. Industrial cities
o. Metropolitan statistical area (MSA)
p. Urban population
q. Concentric zone model
r. Urbanization
s. Mechanically integrated society

t. Consolidated metropolitan statistical area (CMSA)
u. *Ziggurats*
v. *Gesellschaft*
w. Metropolitan area
x. Human ecology
y. Primary metropolitan statistical area (PMSA)
z. Suburbs
aa. Organically integrated society

_____ 1. Cities established before the Industrial Revolution.

_____ 2. Cities established during or after the Industrial Revolution.

_____ 3. A unit that typically has been incorporated according to the laws of the state within which it is located.

_____ 4. An area that contains a central city and the continuously built-up, closely settled surrounding territory that together have a population of 50,000 or more.

_____ 5. The process whereby a population becomes concentrated in a specific area because of migration patterns.

_____ 6. Inhabitants of an urbanized area and places with a population of 2,500 or more.

_____ 7. An area that has a large population nucleus and adjacent communities that are economically and socially integrated into that nucleus.

_____ 8. Counties that have at least one central city or urbanized area with a population of 50,000 or more, as well as any outlying counties that have close economic and social ties to the central urbanized area.

_____ 9. A large, urbanized county or cluster of counties with a population of 1 million or more.

_____ 10. A metropolitan area with a population of a million or more that encompasses two or more smaller metropolitan areas.

_____ 11. The federal government's name for a megalopolis.

_____ 12. The theoretical attempt to explain human communities by the dynamics of plant and animal communities.

_____ 13. A model of urban development in which distinct, class-identified zones radiate from a central business district.

_____ 14. A model of urban development in which groups establish themselves along transportation arteries.

_____ 15. A model of urban development in which similar industries locate near one another and shape the characteristics of the immediate neighborhood.

_____ 16. A type of living situation in which relationships are intimate, cooperative, and personal.

_____ 17. A type of living situation in which relationships are impersonal and independent.

_____ **18.** According to Durkheim, a society's system of fundamental beliefs and values.

_____ **19.** According to Durkheim, a product of people's commitment and conformity to the collective conscience.

_____ **20.** A situation in which a society's collective conscience is strong and there is a great commitment to that collective conscience.

_____ **21.** A situation in which social solidarity depends on the cooperation of individuals in many different positions who perform specialized tasks.

_____ **22.** The process by which middle- and upper-class people upgrade marginal areas by displacing the poor.

_____ **23.** An area of the city that has traditionally provided shelter and a degree of tolerance for deviant individuals and activities.

_____ **24.** Those territories that are part of the metropolitan area but outside the central city.

_____ **25.** Middle- and upper-middle-class semirural communities located beyond the old suburbs.

_____ **26.** Temple compounds raised high on brick-sheathed mounds.

_____ **27.** The spatial distribution of cities by social and economic systems and availability of land.

KEY THINKERS/RESEARCHERS

Match the thinkers/researchers with their main ideas or contributions.

a. **Herbert Gans**
b. **C. D. Harris and E. L. Ullman**
c. **Georg Simmel**
d. **Gideon Sjoberg**
e. **Kasanda and Janowitz**
f. **Jane Jacobs**
g. **Louis Wirth**
h. **Homer Hoyt**
i. **Émile Durkheim**
j. **Christopher Jencks**
k. **Gerald Suttles**
l. **Robert Park and Ernest Burgess**
m. **Ferdinand Tönnies**
n. **U.S. Bureau of the Census**

_____ **1.** Analyst of the preindustrial city.

_____ **2.** Developed a set of terms to describe and classify urban environments.

_____ **3.** Pioneers in human ecology; developed the concentric zone model of urban development.

_____ **4.** Developed the sector model of urban development.

_____ **5.** Developed the multiple nuclei model of urban development.

_____ **6.** Developed the concepts of *Gemeinschaft* and *Gesellschaft* to explain the rural to urban transition.

_____ **7.** Developed the concept of the collective conscience and mechanical and organic integration as forms of identification with it.

_____ **8.** Proposed a widely accepted definition of the city as a "relatively large, dense, and permanent settlement of socially heterogeneous individuals."

_____ **9.** Helped refocus the way sociologists see urban life by showing that urban residents can and do participate in strong and vital community cultures.

_____ **10.** Showed that even in poor neighborhoods, people can have a vital culture with norms and values well adapted to the poverty in which the residents live; also that people draw mental maps of the city's neighborhoods.

_____ **11.** Argued that rural social relationships are rich because they encompass a number of role relationships at once, whereas urban relationships tend to be confined to one role set at a time.

_____ **12.** Argued that social control of public behavior and the patterning of the social interactions of community life take place on the level of blocks rather than entire neighborhoods.

_____ **13.** Found people who live in British cities have greater numbers of social relationships than those living in rural areas.

_____ **14.** Argued that the deinstitutionaliztion of mental hospitals in 1975 had the net effect of increasing the number of homeless people on the streets of many urban environments.

CENTRAL IDEA COMPLETIONS

Following the instructions, fill in the appropriate concepts and descriptions for each of the questions posed in the following section.

1. Today, where would one expect to find the greatest urban growth across the world?

_____

2. How did the excerpt from Philip Roth's novel illustrate the concept of *Gemeinschaft?*

_____

_____

3. What is the largest category of homeless persons in the United States?

_____

What is the major factor causing these people to be homeless?

_____

_____

4. What two important trends have an important impact on cities?

_____

_____

5. Profile the new suburb in the United States as it has evolved during the 1980s and 1990s:

_____

_____

6. What are *exurbs?* Profile the shape of exurbs:

_____

_____

7. What two requirements must be met for cities to emerge?

_____

_____

8. Your author notes that the suburban dream life is showing signs of strains. What are they?

_____

_____

9. What were three main outcomes of the deinstitutionalization of the mentally ill?

a. _____

b. _____

c. _____

10. Compare and contrast the basic features of suburbs and exurbs:

    a. Similarities: _____

    _____

    b. Dissimilarities: _____

11. What are the changing patterns of skid row, as discussed in your text?

    _____

    _____

12. What major factor may limit the process of gentrification in cities?

    _____

13. What did Gerald Suttles (1972) mean by proposing people living in cities draw arbitrary but socially meaningful boundary lines between local neighborhoods?

    _____

    _____

14. Define and present examples of Durkheim's concepts of mechanical and organic solidarity:

    a. Mechanical: _____

        example: _____

    b. Organic: _____

        example: _____

15. Define and present an example of a *Gesellschaft* society and a *Gemeinschaft* society:

    a. *Gessellschaft:* _____

        Example: _____

    b. *Gemeinschaft:* _____

        Example: _____

CRITICAL THOUGHT EXERCISES

1. You have been asked by a local community government organization to present a program that addresses the question, "What contributes to the decay and disorder of neighborhoods?" Present your argument based on the information presented in Chapter 15 of your text.

2. In many urban areas around the country, the homeless are viewed as a significant social problem. Drawing on the information presented in your text, particularly the box titled "Controversies in Sociology: What Produces Homelessness?" what can you say about the causes of homelessness? What steps can be taken by a city to reduce the size of its homeless population as well as the underlying factors producing the homeless?

3. This is a fairly extensive project and may necessitate that you work with two or more classmates. First, go to the city hall or police department of the community where your college or university is located and ask for a map of the community. Next, with your map in hand, take a driving tour with one of your classmates and explore the boundaries of your community. (Depending on your community's size, you may need to divide up the drive-mapping with other class members.) As you drive, and later when you are back at your institution, try to apply one of the three ecological models (concentric zone, sector theory, multiple nuclei) discussed in your text to the spatial patterns you observe. Develop a report discussing the extent to which any of the models fit your community. For the model that fits closest, what changes in the structure and assumptions of the model would one need to make in order to attain a better fit? Are there any special ecological features of your community that make it more difficult to "fit" into one of the three spatial models? To what extent is the size of your community a factor in its "goodness of fit" with the models? How would you classify your community: rural, town, city, suburb, exurb, metropolis? Are there "natural areas" within the community? Is your campus a "natural area"? Finally, develop your own "ecological model" that graphically accounts for the spatial and social patterns in your campus community.

4. This is a fairly extensive project and may necessitate that you work with two or more classmates. First, go to the city hall or police department of the community were your college or university is located and ask for a map of the community. Along with a classmate make a quick drive throughout the residential areas of the community. Mark off these areas on your map. Select any two map sections and conduct a walking-photo tour of the residential areas. Being certain to ask permission before photographing any residents who might be outside their homes, take a selection of pictures that summarize the social class and social organizational features of the areas on your walking tour. Maintain a detailed set of field notes about your observations and the conditions surrounding your photographs. When you return to your institution, write a report that, supported by your photographs, presents a profile of the social and ecological characteristics of the neighborhoods you visited. If your community was large enough to permit several teams to conduct similar examinations of other neighborhoods, develop a section of your report presenting comparisons and contrasts of the differing residential locals.

5. This project involves combining the information gained from your author's discussion of urban life with those research techniques covered Chapter 2. First, conduct a brief automobile tour of the community in which your college or university is located. Next, drawing on your urban and methodological knowledge, develop a Quality of Residential Life Survey. Plan your survey as though it were to be given to residents and businesses in your community. Based on your tour (and the fact that you are a resident), what community issues might be covered in your survey? As you develop your items consider the spatial and organizational issues that may be impacting on your community as you develop your survey. What are the current patterns of population growth or decline in your community? Do those patterns affect equally the residential, business, service, and manufacturing sectors of your community? What emerging trends may change your community 10 years from now?

6. Often we can discover how to improve something if we more fully understand how to make it worse. Drawing on the paper by William H. Whyte, develop a lecture for the chamber of commerce of the community where your college or university is located titled, "How to Ruin This Community."

ANSWERS

KEY CONCEPTS

1. j	6. p	11. t	16. c	21. aa	25. g
2. n	7. w	12. x	17. v	22. a	26. u
3. h	8. o	13. q	18. i	23. d	27. l
4. b	9. y	14. f	19. m	24. z	
5. r	10. k	15. e	20. s		

KEY THINKERS/RESEARCERS

1. d	4. h	7. i	9. a	11. c	13. e
2. n	5. b	8. g	10. k	12. f	14. j
3. l	6. m				

CHAPTER SIXTEEN

Health and Aging

LEARNING OBJECTIVES

After studying this chapter, you should be able to do the following:

- Know what sociologists mean by the sick role.
- Describe the basic characteristics of the U.S. health-care system.
- Understand the link between demographic factors and health.
- Describe the difference between the evolution of AIDS in the developed and developing world.
- Describe the three major models of illness prevention.
- Know the factors affecting the health of infants and children in developing countries.
- Describe the basic demographic features of the older population in the United States.

Tierney looks radiant in a new black-and-white striped dress, smiling and chatting about the gentle bulge in her normally flat stomach. She lies on a table in the obstetrics room of St. Francis Hospital. Her husband Greg sits in a chair beside her, watching as a technician begins the routine ultrasound test that is their last task before a weeklong vacation on Martha's Vineyard.

The technician, Maryann, strokes Tierney's belly with a sonogram wand, using sound waves to create a picture of the life inside. Greg studies the video screen, fascinated by the details emerging from what looks like a half-developed Polaroid: an arm here, a leg there, a tiny face in profile; then the internal organs: brain, liver, kidneys.

"I'm having a hard time seeing the heart. Maybe the baby's turned," Maryann says calmly. Then it appears, pumping in a confident rhythm. She stops the moving image, capturing a vivid cross-section, and Greg remembers his high-school biology.

"All mammals have four chambers in the heart," Greg thinks to himself. "There are only three chambers there."

"You know," Maryann says tactfully, "I'm not as good at this as some other people. Maybe somebody else should take a look." She tries to mask her alarm as she leaves the room.

The inescapable truth is staring at Tierney and Greg from the silent screen: There is no fourth chamber. There is a hole in the heart. And not just any hole, they will soon learn. It is a telltale sign of Down syndrome, a genetic stew of physical defects and mental retardation.

A hole in the heart—in their baby's and, suddenly, in their own.

Tierney and Greg Fairchild have just entered a world of technological wizardry and emotional uncertainty called "prenatal screening." It is a confusing place where even the name is misleading; abortion screening is more accurate.

Most disorders tested for today—including Down syndrome, muscular dystrophy, and cystic fibrosis—cannot be corrected. That means the most common question prompted by distressing prenatal test results is not, "How can we fix it?" It is: "Should this pregnancy continue?"

In the weeks ahead, Tierney and Greg will make a journey through uncharted terrain, filled with fears and tears. They will be tested and torn, changing their minds repeatedly as they confront a new reality amid the ache of lost dreams.

Researchers say they are close to deciphering the blueprint of human development—the genetic code that acts as the operating instructions for creating life. That achievement is expected to drive prenatal screening into the realm of science fiction. Then what? Does a woman carry to term a baby susceptible to mental illness? Cancer? Obesity? Infertility?

Already, hard science has far outpaced the emotional side of the equation. People can learn a great deal about their unborn children, but no one tells them how to handle that knowledge, or what the future might hold (Zuckoff, 1999).

Medicine and health-care issues are intertwined with our social and cultural life. In this chapter, we will examine these interactions as we explore health and illness.

The Experience of Illness

Illness not only involves the body, but it also affects the individual's social relationships, self-image, and behavior. Being defined as "sick" has consequences that are independent of any physiological effects. Talcott Parsons (1951) has suggested that to prevent the potentially disruptive consequences of illness on a group or society, there exists a sick role. The **sick role** is *a shared set of cultural norms that legitimates deviant behavior caused by the illness and channels the individual into the health-care system.* According to Parsons, the sick role has four components. First, the sick person is excused from normal social responsibilities, except to the extent that he or she is supposed to do whatever is necessary to get well. Second, the sick person is not held responsible for his or her condition and is not expected to recover by an act of will. Third, the sick person must recognize that being ill is undesirable and must want to recover. Finally, the sick person is obligated to seek medical care and cooperate with the advice of the designated experts, notably the physicians. In this

sense, sick people are not blamed for their illnesses, but they must work toward regaining their health.

The sick role concept is based on the perspective that all roads lead to medical care. It tends to create a doctor-centered picture, with the illness being viewed from outside the individual. Some (Schneider & Conrad, 1983; Strauss & Glaser, 1975) have suggested that the actual subjective experience of being sick should be examined more closely. These researchers suggest that we should focus more on individuals' perceptions of illness, their interactions with others, and the effects of the illness on the person's identity.

At the same time, society tends to define what should be considered a medical issue. Over the years, many issues have become medical issues that were not considered such before. **Medicalization** is *the process by which nonmedical problems become defined and treated as medical problems, usually in terms of illness or disorder.* Of late, such things as alcoholism, drug abuse, battering, gender confusion, obesity, anorexia and bulimia, and a host of reproductive issues from infertility to menopause have undergone medicalization (Conrad, 1992).

Health Care in the United States

The United States has the most advanced health-care resources in the world. We have 6,097 hospitals, 756,700 highly trained physicians, and 2,203,000 nurses; we are thus prepared to treat illness and injury with the most modern techniques available (*Statistical Abstract of the United States: 1999, 2000*). We can scan a brain for tumors, reconnect nerve tissues and reattach severed limbs through microsurgery, and eliminate diseases like poliomyelitis, which crippled a president. We can do all this and much more; yet many consider our health-care system wholly inadequate to meet the needs of all Americans. Critics maintain that the U.S. health-care system is one that pays off only when the patient can pay.

The American health-care system has been described as "acute, curative, [and] hospital based" (Knowles, 1977). This statement implies that our approach to medicine is organized around the cure or control of serious diseases and repairing physical injuries, rather than caring for the sick or preventing disease. The American medical-care system is highly technological, specialized, and increasingly centralized.

Medical-care workers include some of the highest-paid employees (physicians) in our nation and some of the lowest-paid. About three-quarters of all

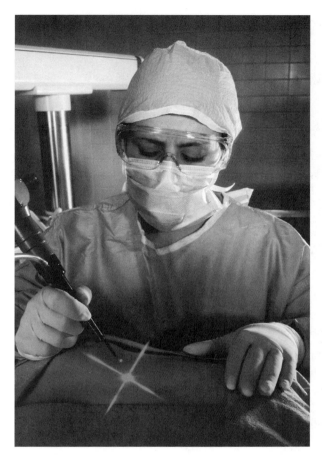

The United States has the most advanced health-care resources in the world, yet Americans are not the healthiest people.

the triangle, there are increasing numbers of much lower-paid workers with little or no authority in the health-care organization. This triangle is further layered with more than 300 licensed occupational categories of medical workers. There is practically no movement of workers from one category to another, because each requires its own specialized training and qualifications (Conrad & Kern, 1986).

Even with our large investment in health care, Americans are not the healthiest people in the world. The high death rate among African Americans causes the United States to rank 18th in worldwide infant mortality. U.S. life expectancy ranks in the lower half of the global scale (*Statistical Abstract of the United States,* 1997) (see Table 16-1).

Gender and Health

You probably are aware of the striking differences in life expectancy between men and women. In 1998, life expectancy at birth was 73.9 for males and 79.4 for females. The life expectancy of women in the United States has been continuously higher than that for men since records began being kept. Even though the life expectancy for both men and women has increased, the increase has been greater for women. Between 1940 and 1998, life expectancy at birth has increased by 14.2 years for women and 12.9 years for men (*Statistical Abstract of the United States: 2000,* 2001).

The death rates of males exceed those of females at all ages and in relation to nearly every one of the leading causes of death, such as heart disease, cancer, cerebrovascular diseases, accidents, and pneumonia. The overall death rate for women ages 25 to 44 is 59% lower than for men in this age group. For those over 65, it is 41% lower (see Figure 16.1).

Women suffer from illness and disability more frequently than men, but their health problems are usually not as life threatening as those encountered by men. Women, of course, do suffer from most of the same diseases as men. The difference is at what point in life they encounter those diseases. For example, coronary heart disease is a leading cause of death for

medical workers are women, although the majority of doctors are men. Many of the workers are members of minority groups, and most come from lower-middle-class backgrounds. The majority of the physicians are white and upper-middle-class.

The medical-care workforce can be pictured as a broad-based triangle, with a small number of highly paid physicians and administrators at the top. Those people control the administration of medical-care services. As one moves toward the bottom of

Table 16-1

Changes in U.S Mortality Rates, 1920–1998

MORTALITY DATA	1920	1950	1980	1998
Death rates (per 1,000 population)	13.0	9.6	8.8	8.67
Life expectancy (total)	54.1	68.2	73.7	76.7
Infant mortality (per 1,000 live births)	*	29.2	12.6	7.2
Maternal mortality (per 1,000 live births)	*	83.3	9.2	6.1

*Data not available.

Sources: U.S. Department of Health, Education, and Welfare, National Center for Health Statistics, Health, United States, 2000.

Figure 16-1 Death Rates for Selected Illnesses for People 25 to 44 by Gender

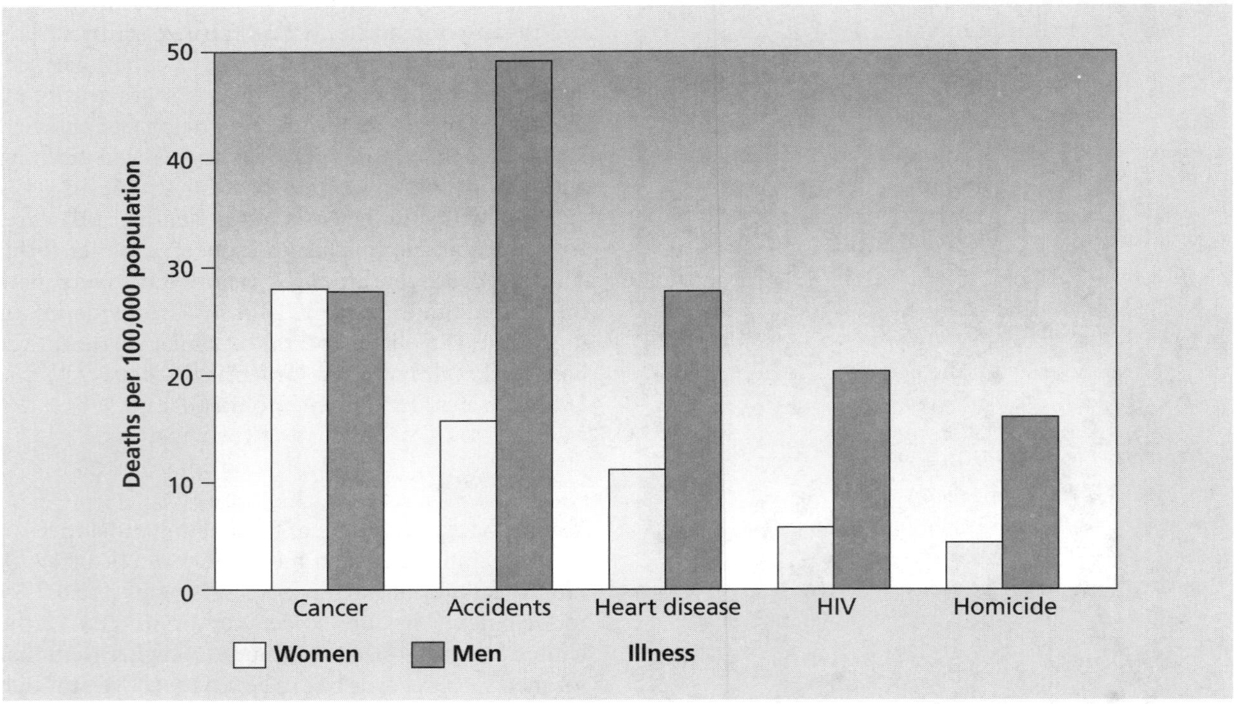

Source: "Deaths: Final Data for 1997," Centers for Disease Control and Prevention, June 30, 1999, *National Vital Statistics Reports,* 47(19).

women older than 66, but for men it is the leading killer after age 39.

Men appear to have lower life expectancies than women because of biological and sociological reasons. Males are at a biological disadvantage to females, as seen by the higher mortality rates from the prenatal and neonatal (newborn) stages onward. Although the percentages may vary from year to year, the chances of dying during the prenatal stage are approximately 12% greater among males than females, and 130% greater during the newborn stage. Neonatal disorders common in male rather than female babies include respiratory diseases, digestive diseases, certain circulatory disorders of the aorta and pulmonary artery, and bacterial infections. The male seems to be more vulnerable than the female even before being exposed to the different social roles and stress situations of later life.

There are a number of sociological factors that also play an important role in the different life expectancies of men and women. Men are more likely to place themselves in dangerous situations both at work and during leisure activities. Therefore it should be no surprise that accidents cause more than three times as many deaths among younger males than among females. Men also are concentrated in some of the most dangerous jobs, such as structural steel workers, loggers, bank guards, coal miners, and state police. The rates of alcohol

use, high-speed driving, and participation in violent sports are also much higher among men and contribute to the differences (Cockerham, 1989).

While men have shorter life expectancies, women appear to be sick more often. Women have high rates of acute illnesses, such as infectious and parasitic diseases, digestive problems, and respiratory conditions, as well as chronic illnesses, such as hypertension, arthritis, diabetes, and colitis. Some have suggested that women may not be sick more often, but may just be more sensitive to bodily discomforts and more willing to report them to a doctor.

According to the National Institute of Mental Health, men are just as vulnerable to psychiatric problems as women. Earlier studies found higher rates among women, partly because alcoholism, drug dependence, and antisocial personality disorders were not included. There still is a gender difference in mental disorders. Men, when emotionally disturbed, are likely to act out through drugs, liquor, and antisocial acts, while women display behaviors that show an internalization of their problems, such as depression or phobias (Riche, 1987).

Race and Health

There are significant differences in the health of the various racial groups in the United States. Asian-

Americans have the best health profile, followed by whites. African-Americans and Native Americans display the worst health data.

There are glaring disparities in childhood mortality of the white and black populations. The infant mortality rate for African-Americans is more than twice that for white infants. In some cities, such as Washington, D.C., the infant mortality rate is higher than in less-developed countries such as Cuba, Costa Rica, and Chile (*Statistical Abstract of the United States,* 1999, 2000).

Many of the same problems that face mothers in less-developed countries are factors in the high infant mortality rate among African-Americans. For example, black women giving birth are 2.5 times as likely as white mothers to be younger than 18, nearly one-third have fewer than 12 years of education, and 40% have not received prenatal care during the crucial first trimester of pregnancy. Consequently, low-birth-weight babies are more than twice as common among African-Americans than whites.

Life expectancies for whites and blacks also differ markedly. The African-American male has the lowest life expectancy of any racial category. In 1998, a black male baby had a life expectancy of 67.8 years. For the white male baby, the figure was 73.9 years. The data for white and black females are 79.4 and

75.0 years, respectively (see Figure 16-2) (*Statistical Abstract of the United States: 2000,* 2001).

The health situation for African-American men is particularly bad. Black men aged 15 to 29 die at a higher rate than any other group except those 85 and older. The biggest differences in causes of deaths of black men versus white men are in AIDS and homicide. Health statistics for African-Americans and whites also show higher ratios of deaths for blacks due to hypertension, heart disease, cancer, diabetes, and accidents (Centers for Disease Control and Prevention, 2000).

The health situation for Hispanic Americans is also worse than that for Anglo Americans. Studies show that Hispanic Americans have a higher infant mortality rate, a shorter life expectancy, and higher rates of death from influenza, pneumonia, diabetes, and tuberculosis. Hispanic death rates for heart disease and cancer are lower than those for whites.

Native Americans have shown an improvement in overall health since 1950. Native Americans have the lowest cancer rates in the United States, and their mortality rates from heart disease are lower than those of the general population as well. There are other areas in which Native Americans fare much worse than other groups, however. Their mortality rates from diabetes are 2.3 times that of the general

Figure 16-2 Life Expectancy by Gender and Race

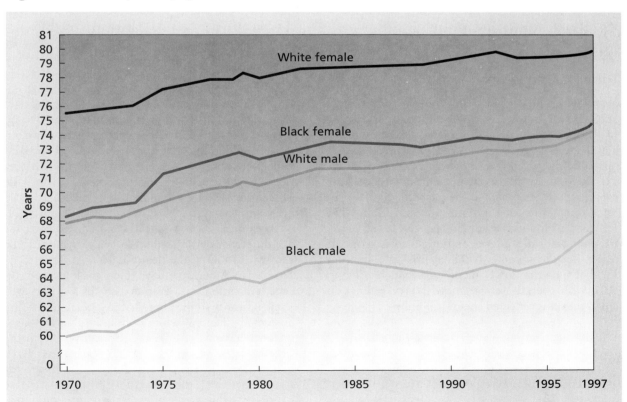

Source: "Deaths: Final Data for 1997," Centers for Disease Control and Prevention, June 30, 1999, *National Vital Statistics Reports,* 47(19).

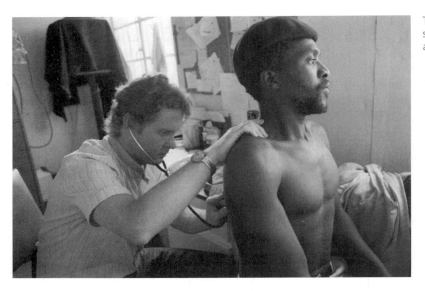

The African-American male has the shortest life expectancy of people in any racial group.

population. The complications from diabetes take a further toll by increasing the probability of kidney disease and blindness. Native Americans also suffer from high rates of venereal disease, hepatitis, tuberculosis, and alcoholism and alcohol-related diseases such as cirrhosis of the liver, gastrointestinal bleeding, and dietary deficiency.

The suicide rate for Native Americans is 20% higher than for the general population. Native-American suicide victims are generally younger than those of other groups, with their suicide rate peaking between ages 15 and 39. For the general population, suicide peaks after age 40.

Social Class and Health

It should surprise no one that poverty and the difficult life circumstances that accompany it, such as inadequate housing, malnutrition, stress, and violence, increase your chance of getting sick. Recognition of the link between poverty and disease has existed for more than a century. One difference, however, is that researchers in the past saw high disease rates among the poor as a moral weakness. In 1891, John Shaw Billings, the surgeon general of the United States at that time, after examining records from different hospitals, came to the surprising conclusion for that time that lower-income patients had higher death rates than those with more money. Billings decided that this reflected the moral failing of

> a distinct class of people who are structurally and almost necessarily idle, ignorant, intemperate and more or less vicious, who are failures or the descendants of failures. (Billings, 1891)

Part of the reason why the poor have higher death rates is because they have less access to high-quality

medical care, good nutrition, and are less likely to feel they have control over their life circumstances. They are also more likely to smoke, be overweight, and have low physical activity levels. Studies that have tried to account for these risk factors have shown that this is only part of the explanation, however. This suggests that the relationship between social class and health is more complex. (Centers for Disease Control, *Health, United States: 1998,* 2000)

Poverty contributes to disease and a shortened life span, both directly and indirectly. An estimated 25 million Americans do not have enough money to feed themselves adequately and as a result suffer from serious nutritional deficiencies that can lead to illness and death.

Poverty also produces living conditions that encourage illness. Pneumonia, influenza, alcoholism, drug addiction, tuberculosis, whooping cough, and even rat bites are much more common in poor minority populations than among middle-class ones. Inadequate housing, heating, and sanitation all contribute to those acute medical problems, as does the U.S. fee-for-service system that links medical care to the ability to pay.

An example of how social class can account for race differences with respect to health issues can be seen by examining the health data of Asian-Americans. Asian-Americans have the highest levels of income, education, and employment of any racial or ethnic minority in the United States, often exceeding those of the general white population. At the same time, the lowest age-adjusted mortality rates in the United States are among Asian-Americans. Even though heart disease is the leading cause of death for Asian-Americans, their mortality from this disease is less than that for whites and for other minorities. Deaths from homicide and suicide are particularly low for Asian-Americans. The infant mortality

Many health problems are the result of environmental factors related to poverty.

rates for Asian-Americans range between 5 and 6 per 1,000, which are lower than those for whites. Although infant mortality rates are only one indicator of health within a group, they are nevertheless an important measure of the quality of life experienced by that population.

Studies of life expectancy show that on every measure, social class influences longevity. At age 65, white men in the highest income families can expect to live 3.1 years longer than white men in the lowest income families. (See "Our Diverse Society: Why Isn't Life Expectancy in the United States Higher?")

Age and Health

As advances in medical science lengthen the life span of most Americans, the problem of medical care for the aged becomes more acute. In 1900, there were

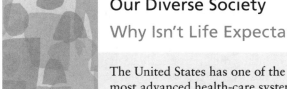

Our Diverse Society

Why Isn't Life Expectancy in the United States Higher?

The United States has one of the most advanced health-care systems in the world. There are thousands of highly qualified doctors, and medical research produces life-saving breakthroughs on a regular basis. Yet, according to the World Health Organization, U.S. life expectancy is 24th in the world. The Japanese have the longest life expectancy and people in Australia, France, Spain, and Italy also live longer than they do in the United States.

One of the ways to begin to understand why the U.S. life expectancy is not higher is to realize that the United States includes many groups who have vastly different life experiences. Some groups have life expectancies significantly above the average, others significantly below the average. Some of the reasons why the average life expectancy in the United States is not higher include the following:

■ In the United States, some groups, such as Native Americans, rural African-Americans,

and the inner-city poor, have extremely poor health, more characteristic of a poor developing country rather than a rich industrialized one.

■ The United States is one of the leading countries for cancers related to tobacco, especially lung cancer. Tobacco use also causes chronic lung disease.

■ A high coronary heart disease rate, which has dropped in recent years but remains high.

■ The HIV epidemic causes a higher proportion of death and disability to U.S. young and middle-aged than in most other advanced countries.

■ Fairly high levels of violence, especially homicides, when compared to other industrial countries.

Source: *World Health Organization Press Release,* June 4, 2000.

only 3.1 million Americans 65 or older, a group that constituted a mere 4% of the total population. Today, however, there are more than 35 million Americans aged 65 or older, a full 13% of the total population. This change in the age structure of the American population has had important consequences for health and health care (*Current Population Reports*, 2000).

At the turn of the twentieth century, more Americans were killed by pneumonia, influenza, tuberculosis, infections of the digestive tract, and other microorganism diseases than by any other cause. By comparison, only 8% of the population died of heart disease and 4% of cancer. Today, this situation is completely reversed. Heart disease, cancer, and stroke and its related disorders are now the three most common causes of death. Those diseases are tied to the bodily deterioration that is a natural part of the aging process.

The result of these changes in health patterns is increased hospitalization for those older than 65. Only 1% percent of all Americans are institutionalized in medical facilities. However, of those 65 and older, 5% are institutionalized in convalescent homes, homes for the aged, hospitals, and mental hospitals. The elderly are 30 times as likely to be in nursing homes as people under 65.

About 1.5 million residents are receiving care in 16,700 nursing homes. The residents are elderly, predominantly female, 75 years old and over, white, non-Hispanic, and widowed. The likelihood of living in a nursing home increases with age. Less than 2% of the 65-to-74-year-old population lives in nursing homes, compared with19% of those aged 85 to 89 (Center for Disease Control, Nursing Home Survey, 1997). (For a discussion of the health problems of another age group, see "Sociology at Work: Binge Drinking as a Health Problem.")

Women in Medicine

Women have always played a major role in U.S. health care, and by the late nineteenth century the United States was a leader in the training of female doctors. In the twentieth century, the number of women in medical schools dropped markedly when a review of medical education closed many medical schools that had served them. There were fewer female physicians in Boston in 1950 than there had been in 1890.

By the 1970s, medical schools had altered their discriminatory admissions policies and women began to be admitted once again. The number of female physicians in the United States increased 310% between 1970 and 1990, from 25,400 to 104,200. In 1996, 26.4% of all U.S. physicians were women, up from 13% in 1980. A good predictor of this trend

continuing is that in 1996, 42% of new entrants into medical school were women. The proportion of women on medical school faculties increased from 13% in 1967 to 24% in 1994 (Knox, 1996; *Statistical Abstract of the United States*, 1997).

Even with these advances, some problems still remain. The Association of American Medical Colleges questioned the 1996 medical school graduates and found that 29% of the women verses 8% of the men reported experiencing sexual harassment. With respect to specific types of harassment 18% of the women (2% of the men) reported offensive sexist remarks directed at them personally; 7% of the women (3% of the men) said they were "denied opportunities for training or rewards because of their gender"; and 10% of the women (1% of the men) reported "unwanted sexual advances by school personnel" (AAMC, *Medical School Graduation Questionnaire*, 1996).

Female physicians earn less than male doctors. According to the American Medical Association, salaries of female family practitioners in 1998 averaged 86% of that earned by men ($123,546 versus $143,061). The earnings ratio is higher for women who have been practicing medicine for 20 years or more, but these experienced doctors still earn less than their male peers (Medical Group Management Association, 1999).

Structural discrimination against women accounts for some of this earnings disprepancy, but there are other factors. The average woman in medical practice is younger and less experienced than the average man. She also earns less than a man partly because she works fewer hours, 55 a week compared with 61 for men. Doctors usually do not get paid by the hour, but those in private practice get paid according to how many patients they see; fewer hours translate into fewer patients and less income.

Women doctors may work fewer hours than men for the same child-care-related reasons other women do. Two-thirds of practicing female physicians have children. Although they may have broken some traditional barriers at work, they still remain the primary family caretakers, being in charge of three times as much child care as men and more than twice the other household duties.

Studies also show that female doctors spend more time with each patient. Their per-visit fees are not necessarily smaller, but they see fewer patients per hour than men. Male physicians saw 117 patients per week compared with 97 for women. This may explain why studies show that patients consider women physicians "more sensitive, more altruistic, and less egoistic" than men. Patients who see a female physician report a significantly higher total satisfaction level than those who see a male physician. These patients may be picking up on the fact that

Sociology at Work

Binge Drinking as a Health Problem

Every semester when we read about the deaths of college students from acute alcohol overdoses, we are reminded that despite progress in reducing under-age drinking in the United States, alcohol consumption by college students remains a persistent problem.

In a national random survey of college students (Weschler, 1999), 44% of students reported binge drinking during the previous two weeks. Binge drinking consisted of five or more drinks at a single occasion for men and four or more for women. Nearly half the students in this survey were under 21. Yet the amount of binge drinking did not differ between those under or over 21. This percentage of binge drinking is even greater than that for non-students of the same age (36%) or high-school seniors (28%). The survey also found that 78.9% of students living in fraternities or sororities were binge drinkers.

As a group, binge drinkers were significantly more likely than non-binge drinkers to report, after drinking, having suffered hangovers, done things they regretted, missed classes, forgotten where they were or what they did, lagged behind in schoolwork, argued with friends, been hurt or injured, had unplanned sexual relations, not used protection when having sex, damaged property, and been in trouble with the police.

Even the non-binge-drinking students were affected by the drinkers. At schools with high binge levels, the non-drinking students were more likely to experience assaults, property damage, interrupted sleep, unwanted sexual advances, serious quarrels, and having to take care of a drunk student than those at low-binge-level schools.

College-age binge drinking also leads to traffic fatalities. Car crashes are the leading cause of death in the United States for people under age 25. In 1996, 10,431 people between ages 15 and 24 died in motor vehicle crashes, and 4,461 (45%) of those were alcohol-related deaths. National research comparing the blood alcohol level of drivers in single-vehicle fatal crashes have found that each 0.02% increase in blood alcohol level nearly doubles the risk of a fatal crash. For drivers under age 21, the risk of a fatal crash increases even more rapidly than it does for older drivers. Such drivers have had less road experience and as a group more often take risks such as speeding or failing to wear seat belts.

How should college campuses respond to under-age drinking and driving after drinking? This question has been a contentious topic of debate on many campuses. Most of the students who engage in binge drinking do not feel they have a problem and the large majority consider themselves moderate drinkers instead of binge drinkers. They believe five drinks do not constitute a binge and instead believe out-of-control drinking is what defines binge drinking.

Some argue that tough campus alcohol restriction drives alcohol consumption off campus and into the surrounding communities, where it could produce even greater dangers. Some even suggest that the drinking age should be lowered so that teenagers can learn to drink safely before they leave home for college.

To address the problem, students must themselves become involved in the solution. If only city and college officials deal with the issue, the initiatives may appear paternalistic and engender resistance among the students. Student leaders need to be involved in educating their peers about the risks posed by alcohol, not only to the frequent binge drinkers, but to the college community in general. Emphasis should be placed on protecting the rights of those negatively affected by binge drinkers, very much like we emphasize the rights of innocent drunk driving victims.

Sources: "College-Age Drinking Problems," by R. W. Hingson, Public Health Reports 113, January/February 1998; and "1999 College Alcohol Studies," by H. Wechsler, Harvard School of Public Health.

---

when men were asked why they entered medicine, they responded, "prestige and salary," whereas women said, "helping people." The interesting side effect of this differing approach to practicing medicine is that women physicians are far less likely to be sued for malpractice than are male physicians.

The type of practice a woman is in also determines her earnings. Female physicians are twice as likely as male physicians to be employed by a hospital, health maintenance organization (HMO), or group practice. Men are more likely to be in independent practices. Independent practitioners generally earn more money and tend to work longer hours.

Women also tend to be concentrated in lower-paying specialties such as internal medicine, pediatrics, and family practice. Specialists such as radiologists, surgeons, and cardiologists typically earn more than primary-care practitioners.

Female physicians tend to be concentrated in lower-paying specialties such as pediatrics, internal medicine, and family practice.

As the number of women entering the field continues to increase, we should see changes in the education of physicians and the manner in which medical care is carried out.

Contemporary Health-Care Issues

The American health-care system is among the best in the world. The United States invests a large amount of social and economic resources into medical care. It has some of the world's finest physicians, hospitals, and medical schools. It is no longer plagued by infectious diseases and is in the forefront in developing medical and technological advances for the treatment of disease and illness.

At the same time, there are many issues that the American health-care system must address. In this section we will discuss a few of them.

Acquired Immunodeficiency Syndrome (AIDS)

The Centers for Disease Control (CDC) defines AIDS as a specific group of diseases or conditions that are indicative of severe immunosuppression related to infection with the human immunodeficiency virus (HIV). Through June 2000, the cumulative number of AIDS cases reported to the CDC was 753,907. Of those, 438,795 patients have died. Newly developed antiretroviral drugs have caused the deaths from AIDS to drop over the past 5 years (see Figure 16-3).

AIDS, a disease now known virtually everywhere in the world, was only identified in 1981, and the retrovirus that causes it was only discovered in 1983. Yet this disease could transform the global future in ways no one imagined even a few years ago. For many developing countries, the impact of this disease could be as great as that of a major war, unless a vaccine or cure can be developed soon.

AIDS is caused by the human immunodeficiency virus (HIV), which is a member of the retrovirus family. HIV gradually incapacitates the immune system by infecting at least two types of white blood cells. The depletion of white blood cells leaves the infected person vulnerable to a multitude of other infections and certain types of cancers. Those infections and cancers rarely occur, or produce only mild illness, in individuals with normally functioning immune systems. HIV also causes disease directly by damaging the central nervous system.

HIV is transmitted through sexual contact, piercing the skin with HIV-contaminated instruments, transfusion of contaminated blood products, and transplantation of contaminated tissue. An infected mother can transmit the virus to her child before, during, or shortly after giving birth. There is no evidence that HIV is transmitted by casual contact. According to the Centers for Disease Control, the majority of AIDS cases have occurred among homosexual or bisexual males (47%) or heterosexual intravenous drug users (25%) (see Figure 16-4) (Centers for Disease Control, 2000). The vast majority of those infected are young adults who do not realize they are infected. Within 6 to 10 years, half of those infected will have developed AIDS. Unless treated with the newly developed antiretroviral drugs, death often follows within a few years after the disease emerges.

AIDS is transmitted sexually and through intravenous drug use, but not as easily as is commonly assumed, perhaps because carriers are only intermittently infectious. Women are much more likely to contract the disease from infected men than men from infected women. It takes as long as 2 years before half of the spouses or regular sex partners of infected people become infected.

The Centers for Disease Control of the U.S. Public Health Service estimate that 800,000 to 900,000 people are infected with HIV in the United States, and most are expected to develop AIDS within 10 to 15 years. This amounts to 1 in 30 men aged 30 to

Figure 16-3 **AIDS Deaths in the United States, 1982–1999**

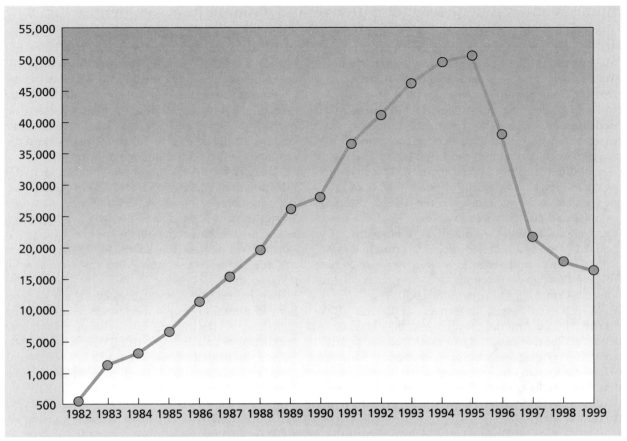

Source: *HIV/AIDS Surveillance Report*, U.S. Centers for Disease Control and Prevention, December 2000, *12*(1).

Figure 16-4 **AIDS Deaths in the United States by Exposure Category, June 1981–June 2000**

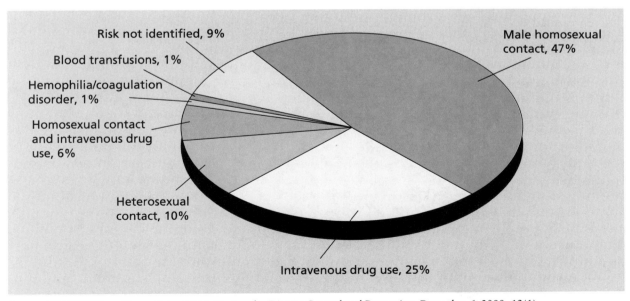

Source: *HIV/AIDS Surveillance Report*, U.S. Centers for Disease Control and Prevention, December 6, 2000, *12*(1).

50 carrying the infection. On a national scale, the spread of AIDS is concentrated in a number of metropolitan areas. San Francisco (52 cases per 100,000 population), New York City (68.1), and Miami (58.3) have infection rates that are 8 to 15 times higher than those of cities such as Cincinnati (4.1), Youngstown (2.2), or Scranton (2.6) (Centers for Disease Control and Prevention, 2000).

Initially, the majority of U.S. AIDS victims were male homosexuals. AIDS, however, is not a gay disease; it can be transmitted between two people of any sex by the exchange of infected blood or semen. Racial differences are appearing in the transmission of AIDS. Sixty-two percent of white men and 53% of Asian men were exposed to the disease through homosexual contact and fewer than 9 percent from intravenous drug use. About 40% of Hispanic men and 31% of black were infected through homosexual contact and about 19% were exposed to the disease through intravenous drug use (Centers for Disease Control and Prevention, 2000).

In the United States, the impact of HIV and AIDS in the African-American community has been devastating. Representing only an estimated 13% of the total U.S. population, African-Americans made up 47% of all AIDS cases reported in the United States in 1999. Blacks are 10 times more likely than whites to be diagnosed with AIDS, and 10 times more likely to die from it.

The Centers for Disease Control now believes that one in every 50 black American men is infected with HIV. AIDS is the leading cause of death for African-Americans between the ages of 25 and 44.

Hispanic groups are also disproportionately represented in the AIDS population. In 1999, Hispanics represented 13% of the U.S. population, but accounted for 19% of the total number of new U.S.

AIDS cases reported that year (Centers for Disease Control and Prevention, November 2000).

In 1999, the number of AIDS deaths dropped by one-third from 1995 among white men. The majority of AIDS deaths in that year were to black and Hispanic men. We will see if American society will continue to be as concerned about AIDS as the disease shifts from one that affects mainly white gay men to one whose victims are minority group drug users. Race, class, and gender issues have an impact on how an illness is perceived and treated, and discrimination factors are not insignificant.

There is a growing gap developing between HIV infection in the developed world and in the developing countries. In North America, Western Europe, Australia, and New Zealand, newly available anti-retroviral drugs are reducing the speed at which HIV-infected people develop AIDS. In Western Europe and the United States, new AIDS cases are slowing considerably.

For the rest of the world, the news is not as good. Some 5.3 million people worldwide were newly infected with HIV during 2000, bringing the total number of people living with HIV or AIDS to 36.1 million, up from 34.3 million in 1999. Since the epidemic was identified, 21.8 million people have died from AIDS.

Around the world, the epidemic varies. East Asia and the Pacific Rim countries are still keeping the epidemic at bay. Infection rates are lower, but the numbers are still large. Over half the world's population lives in the region, and even though the percentage of the population with HIV may be lower, the absolute number of people is very large. In India, infection rates are at about 1% of the adult population, or somewhere between 3 and 5 million people. Even with the lower number of 3 million, India is the

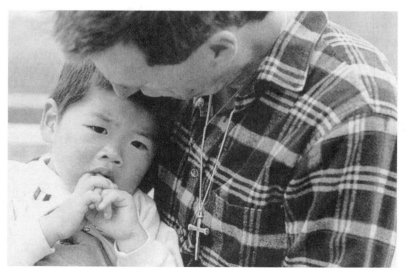

AIDS can be transmitted from parents to children. This child acquired the disease from his mother.

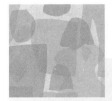

Global Sociology

HIV/AIDS: Worldwide Facts

New estimates show that worldwide infection with the human immunodeficiency virus (HIV) is far more common than previously thought. The World Health Organization estimates that more than 36 million people were living with the HIV infection at the end of 2000. Included in the 36-million figure are 2 million children under the age of 15. During 2000, AIDS caused the deaths of an estimated 3 million people, including 500,000 children under 15. An estimated 21.8 million people have died from AIDS since the epidemic began. Of all persons living with HIV, 90% live in sub-Saharan Africa, southeast Asia, or Latin America, and most of these do not know they are infected. Table 16-2 shows some examples.

Source: UNAIDS, *AIDS Epidemic Update,* December 2000.

Table 16-2

World Area	Number of People Infected	Percentage of Total Population Infected	Percentage Women
Sub-Saharan Africa	25.3 million	8.8%	55%
South and Southeast Asia	5.8 million	0.56	35
Latin America	1.4 million	0.5	25
East Asia and Pacific	640,000	0.07	13
Caribbean	390,000	2.3	35
North Africa and Middle East	400,000	0.2	40

country with the largest number of HIV-infected people in the world.

In South and Southeast Asia, there are approximately 5.8 million people living with HIV or AIDS. In Indonesia, Malaysia, the Philippines, and Singapore, the infection rate is well under 1%. In Thailand, a country that has a large group of sex workers and drug users, the infection rate is 2.3%. The picture is also very bleak in Cambodia, with at least 2% of the population infected. The AIDS epidemic arrived late in Asia, and today the region accounts for 20% of all infections worldwide.

In Latin America and the Caribbean, the epidemic is complex, driven by heterosexual and homosexual sex and injecting drug use. Some 1.4 million people were estimated to be living with HIV/AIDS. In the Caribbean, HIV infection rates continue to be higher than anywhere in the world outside Africa, although awareness about the epidemic is on the rise.

In the world's wealthier countries, prevention efforts have stalled. About 30,000 adults are believed to have become infected in Western Europe, and 45,000 in North America, the bulk of whom are thought to be injecting drug users. There are also signs that safer sex in gay communities is on the wane, leading to increased infections among homosexual men after many years during which the epidemic had remained stable or reduced.

Sub-Saharan Africa is the region of the world where the epidemic is moving fastest. The region is thought to have fully two-thirds of the total world number of people living with HIV. An estimated 3.8 million people became infected with HIV in sub-Saharan Africa during 2000, bringing the total number of people living with HIV or AIDS in the region to 25.3 million. At the same time, 2.4 million people died in Africa of AIDS in 2000, up from 2.2 million in 1999.

The region has reached the unprecedented level of 7.4% of all those between ages 15 and 49 being infected with HIV. Heterosexual transmission accounts for most infections throughout Africa (*AIDS Epidemic Update: December 2000,* The World Health Organization). (See "Global Sociology: HIV/AIDS: Worldwide Facts.")

One of the greatest causes for concern is that over the next few years, the epidemic is bound to get worse before it gets better. The region faces a triple challenge: providing care for the growing population of people infected with HIV, bringing down the number of new infections through more effective prevention, and coping with the impact that 17 million deaths has had on the continent.

AIDS is systematically cutting down life expectancy in the countries where the disease is most common. Life expectancy in Zimbabwe has fallen by 25 years due to AIDS, and it will be shortened by a third in South Africa. Life expectancy in Botswana rose from 43 years in 1955 to 61 years in 1990. Now, with larger numbers of adults infected with HIV, life expectancy is going back to late-1960 levels.

AIDS does not just affect the people with the disease. It also has a major impact on families. As life expectancies fall, the social safety net provided by extended families breaks down, health and education systems falter, and businesses and governments lose their most productive staff.

AIDS is shattering family structures. It is estimated that since the beginning of the epidemic, more than 8 million children under age 15 have lost their mothers to AIDS. Increasingly, elderly grandparents and older children are the sole caretakers of scores of younger children.

Even though it appears that in the United States medical science is finding ways to control the spread of the disease, for the rest of the world the worst is still to come. The vast majority of the people living with HIV live in the developing world where access to the new antiretroviral drugs is difficult or impossible. Therefore, unless there is some major cost-effective medical advance, many if not most of the millions currently infected will die within the next decade.

Health Insurance

Most people pay for their health services through some form of insurance. Poor people, however, cannot afford premiums or the out-of-pocket expenses required before insurance coverage begins. They receive coverage through the government-sponsored Medicare and Medicaid programs

Most of the money spent on medical care in the United States comes in the form of third-party payments as differentiated from direct or out-of-pocket payments. Third-party payments are those made through public or private insurance or charitable organizations. Essentially, insurance is a form of mass financing that ensures that medical care providers will be paid and people will be able to obtain the medical care they need. Insurance involves collecting small amounts of money from a large number of people. That money is put into a pool, and when any of the insured people get sick, that pool pays for the medical services.

The United States has private and public insurance programs. Public insurance programs include Medicare and Medicaid, which are funded with tax money collected by federal, state, and local governments. The country also has two types of private insurance organizations: nonprofit tax-exempt Blue Cross and Blue Shield plans, and for-profit commer-

cial insurance companies. Blue Cross and Blue Shield emerged out of the Depression of the 1930s as a mechanism to pay medical bills to hospitals and physicians. People made regular monthly payments to the plan, and if they became sick their hospital bills were paid directly by the insurance plan. Blue Cross and Blue Shield originally set the cost of insurance premiums by what was called community rating, giving everybody within a community the chance to purchase insurance at the same price. Commercial insurance companies appeared after World War II and based their prices on experience rating, which bases premiums on the statistical likelihood of the individual needing medical care. People more likely to need medical care are charged more than those less likely to need it. Eventually, in order to compete, Blue Cross and Blue Shield had to follow the path of the commercial insurers.

One unfortunate result of the use of experience ratings was that those who most needed insurance coverage, "the old and the sick," were least able to afford or obtain it. Congress created Medicare and Medicaid in 1965 to help with this problem. Medicare pays for medical care for people older than 65, while Medicaid pays for the care of those too poor to pay their own medical costs.

Even with these forms of government insurance, many people believe that the way health care is delivered in the United States produces problems. Critics note that the United States is the only leading industrial nation that does not have an organized, centrally planned health-care delivery system. Attempts by the Clinton administration to resolve this problem were met with resistance from the medical establishment and the general public.

In addition, the care that the poor receive is inferior to that received by the more affluent. Many doctors will not accept Medicare or Medicaid assignments, or ask patients to reimburse them for the difference between the insurance coverage and their bill. Moreover, most doctors will not practice in poor neighborhoods, with the result that the poor are relegated to overcrowded, demeaning clinics. Under conditions like these, it is not a surprise that the poor generally wait longer before seeking medical care than do more affluent patients, and many poor people seek medical advice only when they are seriously ill and intervention is already too late.

At various times, a national health insurance program has been suggested. The American Medical Association is a leading opponent of such a plan. Doctors prefer a fee-for-service system of remuneration. The fee-for-service approach has produced health-care delivery problems such as the uneven geographic distribution of doctors and the overabundance of specialists. For example, only 8.3% of the nation's physicians are general practitioners, even though many more are needed to treat the total population, especially the poor and elderly. Part of the reason is that salaries for male general practitioners, for example, were $143,061, while specialists such as radiologists were paid $284,696 (Medical Group Management Association, 1999).

Health maintenance organizations (HMOs) and other types of managed care have dramatically changed the face of health care in the United States in the past decade. HMOs try to control the cost of medical care by strictly regulating patients' access to doctors and treatments. This shift has meant lower insurance premiums for many individuals and employers. Many people, however, say those savings have come at too great a price. The HMOs have been accused of being more concerned with profits than patients.

Critics of HMOs and managed care include the public, doctors' groups such as the American Medical Association, and many lawmakers. The public and lawmakers have supported efforts to regulate HMOs. These patients' rights laws are designed to give patients more access to care and hold the HMOs accountable for their decisions.

Preventing Illness

Our cultural values cause us to approach health and medicine from a particular vantage point. We are conditioned to distrust nature and assume that aggressive medical procedures work better than other approaches. This situation has caused about one-quarter of all births in the United States to be by caesarean section. The rate of hysterectomy in the United States is twice that in England and three times that in France. Sixty percent of these hysterectomies are performed on women younger than 44. Our rate of coronary bypass operations is five times that of England. We tend to be a can-do society in which doctors emphasize the risk of doing nothing and minimize the risk of doing something. This approach is coupled with the fact that Americans want to be in perfect health. The result is that far more surgery is performed in the United States than in any other county. We tend to think that if something is removed, then our health will return (Payer, 1988).

Yet our experience with heart disease has shown us that significant health benefits can be obtained from such adjustments as changing diets or engaging in healthier practices. At the moment, the best way to deal with the AIDS crisis is through prevention techniques that limit the spread of the illness.

If the health-care system is to reorient its perspective from one that seeks cures to one that focuses on prevention, it must place greater emphasis on sociological issues. Illness and disease are socially as well as biophysiologically produced. During the past century, the medical system has devoted a great deal of effort to combating germs and viruses that cause specific illnesses. We are starting to see that there are limitations to this viewpoint. We must now investigate environments, lifestyles, and social structures for the causes of diseases with the same commitment we have shown to investigating germ theory. This is not to say that we should ignore established biomedical knowledge; rather, we should focus greater attention on the interaction of social environments and human physiology. (See "Remaking the World: The Lost Art of Healing.")

At first glance, most of the factors that come to mind when thinking about preventing disease are little more than healthful habits. For example, if people adopt better diets, with more whole grains and less red meat, sugar, and salt; and if people stop smoking, exercise regularly, and keep their weight down, they will surely prevent illness.

Being overweight and obesity are risk factors for a variety of chronic health conditions, including hypertension and diabetes mellitus. Despite public health efforts to encourage Americans to attain and maintain a healthy weight, it appears that much work remains to be done.

In 1997, two-thirds (62.3%) of men and just under one-half (46.6%) of women were overweight. Almost 1 in 5 adults (18.8% of men and 19.3% of women) were obese. The likelihood of being overweight increased steadily with age, peaking in the 45-to-64-year-old age group and declining somewhat thereafter.

Among women, African-American women have the highest percentage of overweight individuals (64.5%), followed by Hispanic women (56.8%), white women (43%), and Asian/Pacific Islander women (25.2%). In addition, black women are nearly twice as likely as white women and more than 5 times as likely as Asian/Pacific Islander women to be obese. There is little difference among white, Hispanic, or African American men, with more than 60% of each group being overweight. Only Asian/Pacific Islander men had significantly lower rates of being overweight (35.2%) than each of the other three groups.

Being overweight or obese is correlated with educational levels. The prevalence of being overweight ranged from approximately 60% among women who had not finished high school, down to about 29% for women who had earned post-graduate college degrees. In contrast, among men, the prevalence of being overweight remained high (and even increased) at all levels of education, dropping noticeably only among men who had attained college

Remaking the World

The Lost Art of Healing

Dr. Bernard Lown is the world re-nowned cardiologist who developed the defibrillator, the device doctors use to restart failing hearts, which has become a fixture of emergency rooms, and television shows and movies that depict emergency rooms. Lown is also the inventor of the Lown cardioverter, which is used to correct heart rhythm abnormalities. Lown also won the Nobel Peace Prize as the cofounder of the International Physicians for the Prevention of Nuclear War.

People from around the world seek out Lown to treat their heart conditions. Yet Lown believes much more than diagnostic skills and the ability to mobilize technology is needed in the healing process. Lown believes the doctor-patient relationship is vitally important and that we need to put a human face on medicine. This approach can cure as many ills as all the wonders of modern technology, and it can contain costs more readily than any health-care reform plan. We seem to recognize that the doctor-patient relationship is important. Yet so often doctors seem to think that talking with a patient is not important and that the technology is all that really matters in the healing process. Lown notes that it begins with medical school when the doctor is taught to be a scientist; as scientists, the medical students are encouraged to adopt the view that sci-ence can cure all that ails human beings. Thereby they start on the wrong track.

Therefore doctors do not think it is necessary to give time, because technology is the substitute for time and in the doctor's mind the most effective way to use time. Furthermore, it is much more financially rewarding to use technology than to talk to the patient. Below Dr. Lown tells of an early lesson he learned about the importance of doctor-patient relationships:

I believe we have lost sight of the fact that the most efficient way to provide medical care is to invest time in the patient, because the amount of technology used and specialist referred to is an inverse function of the time spent with the patient. The less time spent with the patient, the less the doctor knows about the patient and the more he has to refer the patient to somebody else for tests.

Words are the most powerful tool a doctor possesses, but words, like a two-edged sword, can maim as well as heal. I first witnessed the catastrophic power of words early in my medical career when I was training in cardiology. I was making the rounds with my mentor, Dr. Samuel Levine. We came to Mrs. S., a woman in her 40s who had been a patient of Dr. Levine for the last 30 years. A childhood bout with rheumatic fever had left her with a badly

---

degrees. Even then, well over 50% of the men were overweight (Centers for Disease Control and Prevention, National Health Interview Survey, 1997). (See Figure 16-5.)

We must also think of illness prevention as involving at least three levels: medical, behavioral, and structural (see Table 16-3). Medical prevention is directed at the individual's body, behavioral prevention is directed at changing people's behavior, and structural prevention is directed at changing the society or the environments within which people work and live.

We hear a great deal about trying to prevent disease on a behavioral level. While this is an important level of prevention, we have little knowledge about how to change people's unhealthful habits. Education is not sufficient. Most people are aware of the health risks of smoking or not wearing seat belts, yet roughly 30% of Americans smoke, and 80% do not use seat belts regularly. Sometimes individual habits are responses to complex social situations, such as coping mechanisms to stressful and alienating work environments. Behavioral approaches to prevention focus on the individual and place the burden of change on the individual.

The structural factors related to health care in the United States do not get as much attention as the medical and behavioral factors. We discussed the issues of gender, race, and class earlier in this chapter. Increasingly, people are realizing that health care and the prevention of illness take place on a number of different levels.

World Health Trends

The World Health Organization defines health as "a state of complete mental, physical, and social well-being." This concept may appear straightforward, but it does not easily lend itself to measurement. Consequently, to describe the state of health in the world, we must look at trends. When we look at human health from this perspective, we find that the

scarred and narrowed tricuspid valve. When this valve is constricted, blood backs up into the liver, the abdomen, as well as the arms and legs. Patients who have this problem, which is known as tricuspid stenosis, experience bloating and often have swollen bellies resembling an advanced stage of pregnancy. People live a long time with this condition.

Dr. Levine came into the woman's room with a group of physicians in training trailing him and listening to every word. This particular morning Levine seemed overwhelmed and harassed. This morning Levine was more hurried than usual and his examination quite perfunctory. He looked at Mrs. S. and barked out that this was a case of TS, medical jargon for tricuspid stenosis. After everyone left, the woman grew anxious and visibly agitated, murmuring, "This is the end."

I asked her why she was so upset.

"Dr. Levine said that I have TS."

"Yes, of course you have TS," I affirmed. She began to cry quietly, as though bereft of any hope. "What do you think TS means?" I inquired.

"It means terminal situation," she answered.

Reexamining her, I was dismayed to detect moist rales one-third of the way up her chest, denoting severe congestion. Only minutes earlier her lungs had been completely clear. She was promptly treated, yet none of the usual measures, such as oxygen, morphine, or diuretics, made much difference. I mustered the courage to call Levine and told him what had happened. He thought the story sounded outlandish, but promised to come by after finishing with his other patients. Before he arrived, the woman was overwhelmed with pulmonary edema and died. Patients with tricuspid stenosis waste away and die slowly, not with frothing congestion of the lungs. Such congestion is invariably the result of a failing left ventricle, but her left ventricle was not diseased. As she died, I stood transfixed, helpless, and aghast.

This made the deepest impression on me. I realized that words have such enormous power to maim. In order to heal, you have to listen. You cannot begin to imagine how profoundly complex and difficult the art of listening is. Nothing in medicine, physics, chemistry, or mathematics is as complicated as mastering the art of listening. Listening is not just done with your ears, it is done with your total presence. You have to embrace another human being. You have to care to begin with, because if you do not care, you cannot listen.

Sources: The Lost Art of Healing, by B. Lown, 1996, Boston: Houghton Mifflin; and an interview with the author, September 1996.

twentieth century has seen unprecedented gains in health and survival. On a worldwide basis, the average life expectancy for a newborn more than doubled, from 30 years in 1900 to 66 years in 1996. For a country like China, this has meant moving from conditions at the turn of the century, when scarcely 60% of newborns reached their fifth birthday, to the present, when more than 60% can expect to reach their seventieth birthday.

Health advances in some countries have reached the point where it appears that the population is approaching the upper limit of average life expectancy. In Japan, where life expectancy is nearly 80 years, a newborn has only a 4 in 1,000 chance of dying before its first birthday and less than a 1 in 1,000 risk of dying by age 40 (see Table 16-4).

Unfortunately, the same cannot be said for less-developed countries. More than 300 million people live in 24 countries where life expectancy is less than 50 years. In these countries, 1 out of 10 newborns die by age 1, and 3 million a year do not survive for one week. In some African villages, deaths among infants and young children occur 10 times more frequently than deaths among the aged.

Currently, 80% of the world's population does not have access to any health care. Malnutrition and parasitic and infectious diseases are the principal causes of death and disability in the poorer nations. The problems these diseases cause are largely preventable, and most of these conditions could be dramatically reduced at a relatively modest cost.

The Health of Infants and Children in Developing Countries

When we look at infant and child health from a global perspective, we find that death among children is overwhelmingly a problem of the developing countries in Africa, Asia, and Latin America. Those countries account for 98% of the world's deaths among children younger than 5. To make matters worse, UNICEF estimates that 95% of these deaths are preventable. Let us examine the causes for these deaths.

Figure 16-5 **Percentage of People Overweight by Gender and Education**

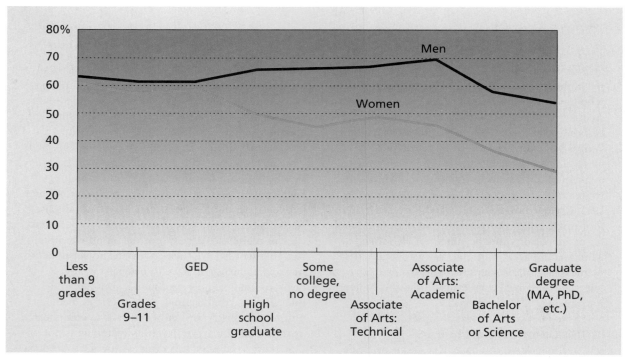

Source: Centers for Disease Control and Prevention, National Health Interview Survey, 1997.

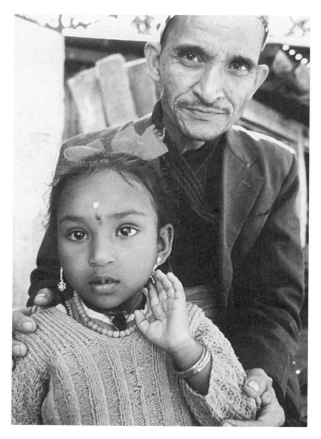

Many infectious diseases that commonly afflict people in less-developed countries are unknown in the United States.

Child Killers Although infectious diseases account for only 1% of deaths in developing countries, they are the major killers in the developing world accounting for a staggering 16.3 million deaths, about 41.5% of all deaths. Almost two-thirds of the deaths among children in the developing world each year are caused by just three diseases: pneumonia, diarrhea, and measles. All three could be treated by available and affordable means. Malaria, extremely rare in developed countries, accounts for 2 million deaths annually.

Pneumonia, with more than 3.5 million young victims a year, is the biggest single killer. In 80 to 90% of the cases, the problem is bacterial and can be controlled by a 5-day course of antibiotics treatment costing 25 cents. Diarrheal diseases claim 3 million children a year. Teaching parents to use the almost cost-free technique of oral rehydration therapy could prevent half of those deaths. Vaccination against the third killer, measles, has produced spectacular progress over the past 5 years. The poorest children are often not reached, however, so 800,000 children die of the disease each year. The measles vaccine costs less than 50 cents per child.

Vitamin A deficiency, which blinds a quarter of a million children each year and threatens up to 10 million with serious illness and disease, could be brought under control by small changes in diet or by six monthly vitamin A capsules. The cost would be only a few cents per child.

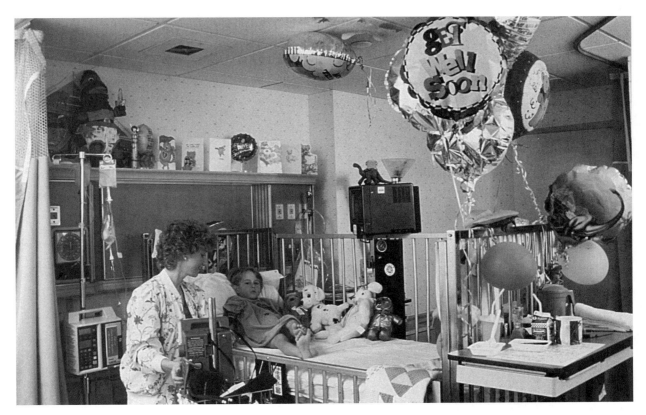

Unlike developing countries, where 95% of infant and child deaths are preventable, in the United States a sick child will usually receive proper medical care.

Table 16-3

Types of Illness Prevention

LEVEL OF PREVENTION	TYPE OF INTERVENTION	SITE OF INTERVENTION	EXAMPLES OF INTERVENTION
Medical	Biophysiological	Individual's body	Vaccinations; medical procedures
Behavioral	Social psychological	Individual's behavior or lifestyle	Change of habits or behaviors
Structural	Sociological	Social structure and social systems	Legislative controls; environmental changes

Source: *The Sociology of Health and Illness*, 2d ed., by P. Conrad & R. Kern, eds., 1986, New York: St. Martin's Press.

Iodine deficiency, the world's biggest cause of mental retardation, could also be eliminated worldwide at a total cost of approximately $100 million, about the cost of one stealth fighter plane (UNICEF, The State of the World's Children: 1993).

There are some success stories. Smallpox has been eradicated, and the World Health Organization believes polio is on the verge of being eradicated. Those are the exceptions, however.

Maternal Health Infant and child survival is critically dependent on the nurturing provided by the mother during pregnancy and childhood. When an expectant mother suffers from illness or malnutrition during pregnancy, she will be more likely to have a low-birth-weight baby. In developing countries,

Table 16-4

Preventable Deaths of Children Before Age 5 in Developing Countries

CAUSE	ANNUAL NUMBER OF CHILD DEATHS	PERCENTAGE PREVENTABLE AT LOW COST
Pneumonia	3,560,000	70%
Diarrhea	3,000,000	90
Measles	880,000	85
Malaria	800,000	70
Neonatal tetanus	560,000	90
Whooping cough	360,000	80
Tuberculosis	300,000	65

Source: *The State of the World's Children: 1993*, UNICEF, 1993, New York: Oxford University Press.

about 17% of the births involve low birth weight. In addition to maternal malnutrition, other causes of low-birth-weight babies are the mother's excessively heavy work, malaria infection, severe anemia related to an iron-deficient diet, and hookworm infestation.

Maternal Age   An infant is at high health risk if the mother is in her teens or older than 40, if she has given birth more than seven times, or if the interval between births is less than 2 years. The interval between births is the most important factor in infant and child mortality. Infants born to mothers at less than 2-year intervals between births are 80% more likely to die than children born at birth intervals of 2 to 3 years.

This information suggests that family planning programs that discourage early childbearing can substantially reduce infant and child mortality by preventing births to high-risk mothers. When childbearing is spread out, the woman has more time and energy to devote to her own health and the care of her other children.

Maternal Education   Children's chances of surviving improve as their mother's education increases. An increase of 3 years of education produces 20 to 30% declines in the mortality of children younger than 5. A mother's education affects her child's health in a number of ways. Better-educated mothers know more about good diet and hygiene. They are also more likely to use maternal and child health services—specifically prenatal care, delivery care, childhood immunization, and other therapies (Cleland & Ginneken, 1988).

Breast-Feeding   In developing countries, it is much safer for infants to be breast-fed during the first 6 months of life than to be bottle-fed. Bottle-fed babies are three to six times more likely to experience respiratory infections or diarrhea than breast-fed babies. Bottle-feeding has been introduced into developing countries by hospitals that are trying to follow Western practices as well as by commercial concerns that aggressively promote breast milk substitutes. The World Health Organization and UNICEF have tried to combat this trend, with limited success.

The Aging Population

For the first time in our history, there are more people aged 65 and over in the population than there are teenagers. When the first U.S. Census was taken in 1790, there were about 50,000 Americans over age 65, representing 2% of the population of 2.5 million. One hundred years later, the over-65 population had grown to 2.4 million, or just under 4% of the population. The life expectancy of a newborn white child was only 50 years then; it was less than 35 for African-Americans. In just three decades, the average life expectancy shot up to 60 years for whites and 50 years for African-Americans. The number of elderly people more than doubled, to 6.7 million, about 5% of the total population.

By 1960, the elderly population more than doubled again. There were 16.7 million people over age 65, about 9% of the population. By 1996, the number had doubled again to 33. 9 million, representing 12.8% of the population.

By 2030, there will be 70 million older people in the United States. The Census Bureau estimates that almost one in five Americans will be 65 or older by 2025. By 2045, there will be more elderly in the

For the first time in our history, there are more people aged 65 and older in the population than there are teenagers.

Figure 16-6 United States Population Age 65 and Over, 1900–2050

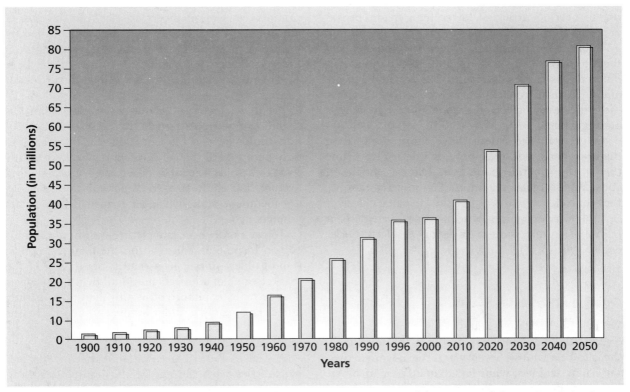

Sources: *Historical Statistics of the United States: Colonial Times to 1970*, U.S. Bureau of the Census; and Current Population Reports, pp. 25–1104, Table 2.

population than the total population of the United States in 1900 (Longino, 1994). Projections for the future usually are conservative and do not take into account declines in mortality that produce longer life spans. In fact, then, the over-65 population may be considerably higher (see Figure 16-6).

There are several basic reasons for the growth in the older population. First, a large number of people who were born when the birthrate was high are now reaching age 65. Second, many people who immigrated to the United States before World War II are also reaching age 65. Third, improvements in medical technology have created a dramatic increase in life expectancy.

Composition of the Older Population

People tend to think of the older population as a homogeneous group with common concerns and problems. This is not the case. We can divide the older population into three groups: the young-old, ages 65 to 74 (18.7 million); the middle-old, ages 75 to 84, (11.4 million); and the old-old, ages 85 and over (3.8 million). The older population as a whole is itself getting older. Whereas in 1960 about one-third

of the older population was over 75, by the year 2000, nearly 50% of the older people were over 75. In that same period of time, the old-old will have nearly tripled in numbers. In 1996, one in ten elderly people was 85 or older. By 2045, the oldest old will be one in five (Administration on Aging, *Profile of Older Americans: 1997*).

Aging and the Sex Ratio

The mortality rate for both men and women has been declining, but it has been declining much more rapidly for women. Women at any age are less likely to die than men. This has helped to produce a substantial sex ratio imbalance in the over-65 group, which continues to become more apparent as we move farther up the age scale. Among those over age 65, there are 65 men per 100 women. Among those over age 85, there are only 38 men per 100 women. Only 24% of men and women over 85 are married, and most live alone (Pol, May, & Hartranft, 1992).

Aging and Race

About 85% of the people over 65 in the United States are white, 7.9% are African American; 4.7%

Hispanic; 1.9% Asian or Pacific Islander, and 1% Native American. The African-American percentage of the over-65 group is lower than might be expected. That is, if African-Americans make up 13% of the total population, then we would expect them to make up 13% of the older population. This difference is caused by the higher recent birthrate among African-Americans, as well as the fact that life expectancy is lower among blacks than it is among whites.

In general, the elderly have more assets than the nonelderly, and their poverty rate is also below that of the general population. There is, however, considerable racial variation. The poverty rate for the white elderly is 9.4%, for the Hispanic elderly it is 24.4%, and for the African-American elderly it is 25.3% (Administration on Aging, *Profile of Older Americans: 1997*).

Aging and Marital Status

Women in the United States outlive men by nearly 6 years. The older a woman becomes, the more likely she is to be a widow. Seventy percent of women over age 75 are widows, while less than 20% of men that age are widowers. In addition to the longer life expectancy of women, this disparity also is caused by the fact that men tend to marry women younger than themselves. This situation also is explained by the fact that a widow will find few older men available to marry whereas a widower has many older women available to remarry. Consequently, it is more likely that those older people living alone or in institutions are women.

Aging and Wealth

Special-interest groups lobby Washington politicians claiming that millions of older adults could be thrown into poverty if there are cuts in Social Security. Meanwhile, cruise-ship companies, automobile manufacturers, and land developers present images of energetic, attractive, silver-haired couples with substantial amounts of money.

Which of these two images is correct? Both of these presentations are accurate depictions of some older Americans. The lobbyists are talking about the poorest 20% of households containing the elderly. These households have an average net worth of about $3,400. Luxury cruise lines are interested in the richest 20%, who are worth almost 90 times as much.

Some of the confusion over the economic condition of older Americans stems from the ways in which they use and save money. The median income of the average American household peaks when its primary working members are between the ages of 45 and 54. The incomes of households headed by people age 65 and older are two-fifths as high, on average. This would seem to indicate that older Americans are significantly less well off than younger Americans.

Yet lower incomes do not necessarily mean less spending power. Although the median income of households headed by people aged 65 and older is only about 40% that of households headed by 45- to 54-year-olds, the older group has a much smaller average household size, so its per capita discretionary income is actually higher than it is among the younger group.

To get a more accurate picture, we should not focus on income, but instead on the financial condition of older Americans—that is, net worth, or the market value of all assets minus all debts. This gives a more realistic picture of wealth than does income alone. For example, older adults can finance major purchases by cashing in on stocks, real estate, or other assets.

A Census Bureau survey found that median net worth rises with the age of the household, moving from $7,428 for households younger than 35, to $61,248 for those 45 to 54, to $106,408 for those 65 to 69. From this point on it starts to drop, but even those households headed by people over age 75 have a median net worth of $81,600 (Household Net Worth and Ownership, February, 2001). Looking at net worth shows us that the oldest households control more wealth than do households in other age groups. Moreover, the share of wealth controlled by working-age Americans is eroding, while the share controlled by the elderly is increasing.

Three factors have caused the elderly to control a substantial and increasing portion of the nation's wealth. First, the share of households headed by the elderly has been increasing, thereby increasing the aggregate wealth of older Americans. Second, the stock market growth has benefited the affluent elderly who control a large portion of individual stock holdings. Finally, the escalation in home values in many states has boosted the net worth of the elderly because most older Americans own their own homes.

These facts paint a picture that is different than what is normally thought of as being the case. As with so much in sociology, the conclusion one draws appears to depend on which factors one considers to be most important in understanding the issue (Longino & Crown, 1991). (See "Sociology of Work: Stereotypes About the Elderly.")

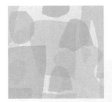

Sociology at Work

Stereotypes About the Elderly

The elderly are the most heterogeneous of all age groups because we tend to age differently biologically, sociologically, and psychologically. Yet people tend to have fairly specific stereotypes of the elderly and what it must be like to be old. Surprisingly, the elderly hold the same negative stereotypes about their age group that the general population does.

In one study, people over 65 were asked if any of the following was a serious problem for them: fear of crime, not having enough money to live on, poor health, loneliness, being needed, and keeping busy. Next, they were asked if they thought these issues were serious problems for other elderly people. Table 16-5 presents the results.

There were vast differences between how serious a problem these issues were for the elderly person being questioned and how much of a problem the person thought it was for other elderly people. For example, 57% of the elderly people questioned thought poor health was a serious problem for other elderly people, but only 15% said it was a serious problem for them. Fifty-five percent thought not having enough money was a serious problem for other elderly people, but only 12% said it was a serious problem for them.

We could understand that younger people would have distorted stereotypes about the elderly, but why would the elderly themselves also be so one-sided in their views? Part of the answer is that the ageism of the larger society, the belief that the elderly represent the least capable, least healthy, and least alert members of society, has seeped into the thinking of the elderly as well. In addition, there is an ironic twist here in that when elderly people believe that other older people are worse off than they are, it produces a higher level of life satisfaction. In a sense, holding the ageist stereotypes makes some elderly people feel better if they can say the stereotype does not apply to them. It is time for us all, no matter what our age, to rethink the distorted views we hold of the elderly.

Table 16-5

Percentage of People Over 65 Who Believe the Problem Is "Very Serious"

Problem	Serious for Others	Serious for Self	Difference
Fear of crime	69%	37%	32%
Not enough money	55	12	33
Loneliness	46	6	40
Poor health	57	15	42
Being needed	41	8	33
Keeping busy	26	4	22

Source: "Images of Aging in America," American Association of Retired People, 1994.

Global Aging

The numerical and proportional growth of older populations around the world is indicative of major achievements—decreased fertility rates, reductions in infant and maternal mortality, reductions in infectious and parasitic diseases, and improvements in nutrition and education—that have occurred, although unevenly, on a global scale.

The rapidly expanding numbers of older people represent a social phenomenon without historical precedent. The world's elderly population numbers half a billion people today and is expected to exceed 1 billion by the year 2020. In most countries, the elderly population is growing faster than the population as a whole. Almost half of the world's elderly live in China, India, the United States, and the countries of the former Soviet Union. China alone is home to more than 20% of the global total.

The oldest old (85 plus) are the fastest-growing segment of the population in many countries worldwide. Unlike the elderly as a whole, the oldest old today are more likely to reside in developed than developing countries, although this trend too is changing.

The percentage of the elderly population living alone varies widely among nations. In developed countries, percentages generally are high, ranging from 9% in Japan to 40% in Sweden. In developing countries, few elderly people live alone. In China 3% of the elderly live alone, in South Korea 2%, and in Pakistan 1%. Living alone in those countries is often the result of having a spouse, siblings, and even children who have died.

To date, population aging has been a major issue mainly in the industrialized nations of Europe, Asia, and North America. In at least 30 of these countries, 15% or more of the entire population is

Figure 16-7 **World Population 65 and Older, 1995 and 2025**

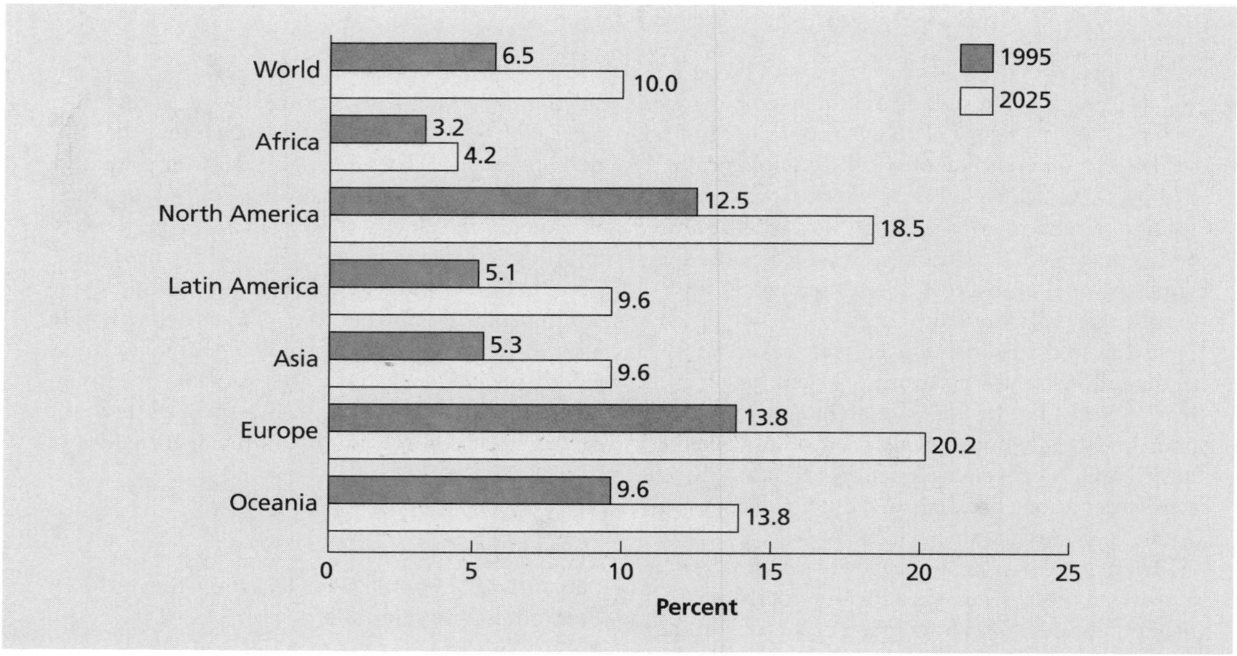

Source: World Population Prospects: The 1996 Revision (Annex II and II), 1996: New York: United Nations.

60 and older. Those nations have experienced intense public debate over elder-related issues such as social security costs and health care provisions (see Figure 16-7).

Future Trends

Relatively stagnant birthrates and big jumps in life expectancy have contributed to the growth in the older population. In addition, the enormous baby-boom generation, born between 1946 and 1961, which now represents about one-third of the population, is moving into middle age. The number of people between the ages of 30 and 44 has surged to 20%, and their total is 60 million (U.S. Bureau of Census).

As the baby-boom population ages, its large numbers will cause cycles of relative growth and decline at each stage of life. The aging of the baby-boom generation will push the median age, now at 34, to more than 38 by 2050.

The years from 1985 through 2020 cover the most economically productive period for the baby-boom generation. In 2010, the oldest members of the baby boom will be nearing 65, while the youngest will be around 50. As the baby-boom cohorts begin to reach age 65 starting in 2011, the number of elderly people will rise dramatically.

In about 2030, the final phase of the elderly explosion caused by the baby boom will begin. At that point, the population aged 85 and older will be the only older age group still growing. It will increase from 2.7 million now to 8.6 million in 2030, to more than 16 million by 2050 (U.S. Bureau of the Census). To put it another way, today 1 in 100 people is 85 years or older; in 2050, 1 in 20 people could be so old. People 85 and older could constitute close to one-quarter of the older population by then.

Concerns associated with problems of the older population are exacerbated by the large excess of women over men in the older ages. Among the aged, women outnumber men 3 to 2. That imbalance increases to more than 2 to 1 for people 85 and over.

Increased longevity and the aging of the baby-boom generation will mean that many people will find themselves caring for very old people after they themselves have reached retirement age. Assuming that generations are separated by about 25 years, people 85, 90, or 95 would have children who are anywhere from 60 to 70 years old. It is estimated that every third person 60 to 64 years old will have a living elderly parent by 2010.

The rising cost of caring for the elderly will become a bigger problem. Without serious health-care reform, the United States could spend 20% of its gross national product on health care by the year 2003. Currently, more than one-third of total

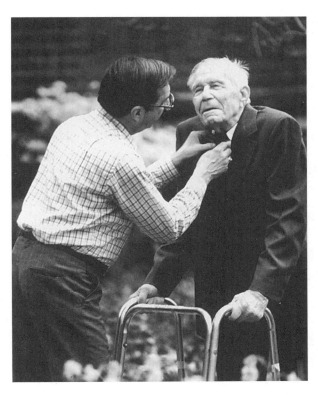

Increased longevity and the aging of the baby-boom generation will mean that many people will find themselves caring for very old persons after they themselves have reached retirement age.

health-care expenditures are spent on the 12% of the population aged 65 and older (Longino, 1994).

Now, there are more than twice as many children as there are elderly people. By 2030, the proportion of children will have shrunk and the proportion of elderly will have grown until these two groups are approximately equal, at just over one-fifth of the population each (U.S. Bureau of Census).

There are few certainties about the future, but the demographic outlook seems relatively clear, at least for people already born. Careful consideration of the impact of our aging population can be an important tool in planning for the future.

SUMMARY

Medicine and health-care issues are intertwined with social and cultural customs and reflect the society of which they are a part. Illness not only involves the body, but it also affects an individual's social relationships, self-image, and behavior. Talcott Parsons has suggested that to prevent the potentially disruptive consequences of illness for a group or society, there exists a sick role, which is a shared set of cultural norms that legitimates deviant behavior caused by illness and channels the individual into the health-care system.

The United States has the most advanced health-care resources in the world. Critics maintain that the system pays off, though, only when the patient can pay. The system has been described as acute, curative, and hospital based, in that the focus is on curing or controlling serious diseases rather than on maintaining health.

Male death rates exceed female death rates at all ages and for the leading causes of death such as heart disease, cancer, cerebrovascular diseases, accidents, and pneumonia. Women seem to suffer from illness and disability more than men, but their health problems are usually not as life threatening. Some have suggested that women may not be sick more often, but may in fact be more sensitive to bodily discomforts and more willing to report them to a doctor.

Male infants are biologically more vulnerable than female infants, in both the prenatal and neonatal stages. Sociologically, men are more likely to have dangerous jobs and more likely than women to place themselves in dangerous situations during both work and leisure. Men and women are equally vulnerable to psychiatric problems, but emotionally disturbed men are likely to act out through drugs, liquor, and antisocial acts, whereas women display behaviors such as depression or phobias that indicate an internalization of their problems.

Asian-Americans have the best health profile, followed by whites. African-Americans and Native Americans have the worst health profiles. On the average, life expectancy for African-American infants is about 6 years less than that of white infants, for both males and females.

Poverty contributes to disease and a shortened life span both directly and indirectly. An estimated 25 million Americans do not have enough money to feed themselves adequately, and therefore suffer from serious nutritional deficiencies that lead to illness and death. Diseases such as tuberculosis, influenza, pneumonia, whooping cough, and alcoholism are more common among the poor than the middle class. Inadequate housing, heating, and sanitation all contribute to those medical problems, as does the U.S. fee-for-service system that links medical care to the ability to pay. For all of these reasons, social class and life expectancy are highly correlated.

Most Americans pay for their health services through some form of health insurance or third-party payments. Individuals—or their employers—pay premiums into a pool, which is used to finance the medical care of those covered by the insurance. Public insurance programs include Medicare, for those older than 65, and Medicaid, for those who are below or near the poverty level. Private insurance is provided by nonprofit, tax-exempt organizations like Blue Cross and Blue Shield or for-profit, commercial insurance companies. The United States is the only leading industrial nation that does not have an organized, centrally planned health-care delivery system. In addition, the care that the poor receive is inferior to that received by the more affluent.

Recently, there has been a shift among health-care providers from the cure orientation to a prevention orientation. Prevention has three levels: Medical prevention is directed at the individual's body; behavioral prevention is directed at changing the habits and behavior of individuals; structural prevention involves changing the social environments where people work and live.

For the first time in our history, there are more people aged 65 and older in the population than there are teenagers. The older population can be divided into the young-old (age 65 to 74), the middle-old (age 75 to 84), and the old-old (age 85 and older). By 2000, 50% of the older population was older than 75. Because mortality rates are higher for men, there is a substantial sex ratio imbalance in the over-65 group that becomes more apparent as one goes up the age scale. Due to higher mortality rates among African Americans, the older population is disproportionately white.

INTERNET RESOURCES

Go to http://web-enhanced.thomsonlearning.com which features Web links relevant to this chapter to expand and enhance the learning experience. This site will be monitored and updated periodically to ensure the links are correct and live.

At Virtual Society: The Wadsworth Sociology Resource Center, http://sociology.wadsworth.com, you will find a Career Center, Sociology in the News, Research Online, InfoTrac College Edition, Virtual Tours in Sociology, and a variety of book-specific student and instructor resources.

CHAPTER SIXTEEN STUDY GUIDE

KEY CONCEPTS

Match each concept with its definition, illustration, or explanation below.

a. Fee-for-service system
b. Health
c. Sick role
d. Medicalization
e. Binge drinking

f. Third-party payments
g. Acquired immunodeficiency syndrome (AIDS)
h. Experience rankings
i. Structural discrimination
j. Female physicians

_____ 1. A shared set of cultural norms that legitimates deviant behavior caused by illness and channels the individual into the health care system.

_____ 2. A specific group of diseases or conditions that indicate severe immune suppression related to infection with the human immunodeficiency virus.

_____ 3. A system in which the costs of an individual's health care are paid for by some form of public or private insurance or charitable organization.

_____ 4. A system by which doctors are paid only for treating illness.

_____ 5. A state of complete mental, physical, and social well-being.

_____ 6. The process by which nonmedical problems become defined and treated as medical problems, usually in terms of illness or disorder.

_____ 7. Regarded by patients as "more sensitive, more altruistic, and less egoistic."

_____ 8. Demonstrated by the net income of female family physicians in 1998, which averaged 86% of that earned by men ($123,546 versus $143,061).

_____ 9. A process used by health companies to distribute insurance coverage across different groups.

_____ 10. A major national health problem across American colleges and universities

KEY THINKERS/RESEARCHERS

Match the thinkers/researchers with their main ideas or contributions.

a. Talcott Parsons
b. Centers for Disease Control (CDC)

c. World Health Organization (WHO)
d. Dr. Bernard Lown

_____ 1. American sociologist who developed the concept of the sick role.

_____ 2. A part of the U.S. Public Health Service, this agency is charged with monitoring communicable ailments.

_____ 3. Developed the defibrillator, won a Nobel Prize, and writes about the importance of the doctor-patient relationship.

_____ 4. An international organization that monitors health issues.

CENTRAL IDEA COMPLETIONS

Following the instructions, fill in the appropriate concepts and descriptions for each of the questions posed in the following section.

1. Describe the change in life expectancy in the United States between 1940 and 1998.

   a. Males: _____

   b. Females: _____

2. What biological factors account for the differences in life expectancy between men and women?

_____

_____

_____

_____

3. What sociological factors account for the differences in life expectancy between men and women?

_____

_____

_____

_____

4. What two factors result in U.S. life expectancy ranking in the lower half of the global scale of healthiest societies?

a. _____

_____

_____

b. _____

_____

_____

5. Your text notes that in the United States, medical science is finding ways to control AIDS. What does your author say about the AIDS situation from a global perspective?

_____

_____

_____

_____

6. Explain the difference between Medicare and Medicaid:

a. Medicare: _____

_____

_____

b. Medicaid: _____

_____

_____

7. Define and present an example of what your text refers to as the sick role:

_____

_____

_____

_____

8. Your author notes that the U.S. health-care system needs to reorient itself from a cure perspective to a prevention perspective. Discuss what your author had in mind with that statement, and provide examples as part of your discussion:

_____

_____

_____

_____

9. Provide specific examples supporting the statement that American culture causes us to approach health and medicine from a particular vantage point:

_____

_____

_____

_____

10. Define AIDS and indicate its transmission patterns and control prospects:

  a. Definition: _____

  _____

  b. Transmission: _____

  _____

  c. Control prospects: _____

  _____

11. Although the percentage of women in medical schools has increased, your text notes that some problems still remain. In a few sentences, discuss those problems:

_____

_____

_____

_____

_____

12. List and discuss the three major models of illness prevention:

  a. _____

  _____

  b. _____

  _____

  c. _____

13. What are some of the major differences in the causes of death between black and white American males?

  a. _____

  _____

  b. _____

  _____

c. _____

_____

d. _____

_____

14. What is the use of experience ratings, and how is this concept related to the situation of health insurance in the United States?

Experience ratings: _____

Effect on health insurance: _____

15. How is AIDS related to race in the United States?

_____

_____

_____

16. What is the triple challenge facing sub-Saharan Africa relative to its AIDS epidemic?

_____

_____

_____

17. What factors are correlated with obesity in the United States?

_____

_____

_____

_____

CRITICAL THOUGHT EXERCISES

1. Discuss the changes in patterns of death among Americans since the turn of the twentieth century to the present time. What have been the results of changes in health patterns? How have these changes been manifested across social class and racial groups in the United States?

2. Profile the American female physician. In what ways is she similar to and different from her male counterpart? What does research evidence tell us about the responses of patients to the sex of their physician? What are the major remaining gender-related problems facing American medicine?

3. Profile the basic demographic features of elderly Americans. What special concerns does this population face? Are the present demographic trends likely to continue in the next couple of decades? Why? Using your library, examine the U.S. Census statistics and projections regarding the geographic location of elderly Americans. What social structural outcomes might one find in geographic locations with higher concentrations of American elderly? How might such concentrations impact on the health-care delivery systems in those areas?

4. Congratulations! Due to your expertise on the elderly and nursing homes, you have been nominated to a presidential commission on aging in America. Using your text as a beginning, then supplementing your knowledge with material from your academic library, develop a "needs assessment" for nursing homes in the United States for the year 2020. Based on your research of current trends, what do you project will be the demand for nursing homes in 2020? To what extent will those nursing homes be similar to and different from today's nursing homes?

5. Your author notes that for the first time in American history there are more people age 65 and older in the population than there are teenagers. Based on the discussion in this chapter as well as in the earlier chapter on culture, what current and emerging changes in American culture would you project for the next decade? Using your library as well as the Internet for sources, present a brief overview of the music and fashion fads as they existed 50 years ago, when today's 65-year-old was 15. In what ways are the music and fashion trends of 50 years ago similar to and different from the music and fashion fads existing in America today? To

what extent, if any, might the increasing proportion of elderly in America influence music and fashion tastes during the next decade?

6. Present a detailed analysis of why, according to the World Health Organization, life expectancy in the United States is ranked 24th in the world. After you have presented the factors, develop a plan for improving this ranking.

7. Discuss binge drinking as a national health concern across colleges and universities in the United States. How common is binge drinking at your college or university? What major factors are associated with binge drinking? If you were in charge of developing a program to reduce binge drinking, what would you include in your plan? How might the type of institution (small college, state university, etc.) be a factor in the type of program you develop?

ANSWERS

KEY CONCEPTS

1. c	3. f	5. b	7. j	9. h	10. e
2. g	4. a	6. d	8. i		

KEY THINKERS/RESEARCHERS

1. a	2. b	3. d	4. c

Collective Behavior and Social Movements

LEARNING OBJECTIVES

After studying this chapter, you should be able to do the following:

- Understand the theories of collective behavior.
- Describe the attributes and types of crowds.
- Know what the various dispersed forms of collective behavior are.
- Describe the major types of social movements.
- Understand the life cycle of social movements.

In 1990, men in Lagos, Nigeria, could be seen walking in public holding onto their penises, either openly or discreetly with their hands in their pockets. This was in response to the widespread fear of genital snatchers. The fear was so strong that incidental body contact with a stranger could cause the "victim" to feel strange scrotum sensations and believe his penis had been taken. The victim would then confront the perpetrator accusing him of being a genital thief. As a crowd gathered, the victim would strip naked to convince bystanders that the penis was really missing. When the penis was still there, the victim would claim that it had been returned by the thief or that, although the penis was now back, it was shrunken and probably the wrong one. The accused was often threatened or beaten and, in some instances, killed (Bartholomew & Goode, 2000).

Jennifer Sorento was waiting patiently with about 40 other brides for her turn to use the mirror in the ladies room. She wore a filmy veil, a white lace dress, and white lace gloves, and when asked about her husband, Luke Hester, she giggled like a new bride anticipating her honeymoon. But for Jennifer and her new husband—just as for the 2,500 other couples married by the Reverend Sun Myung Moon, leader of the Unification Church, in a mass wedding ceremony in a Washington, D.C., stadium in November of 1997—there would be no honeymoon. For that matter, there would be no living together, at least for the foreseeable future. Jennifer and Luke were first introduced to each other only 3 days before. Theirs is a church-arranged marriage—the Unification movement's most blessed sacrament (Talbot, 1997).

During the winter of 1691–1692, several young girls in Salem, Massachusetts, began to exhibit strange and inexplicable behavior. Their symptoms included "blasphemous" screaming, convulsive seizures, and trancelike states. Unable to determine any physical cause for this behavior, physicians concluded that the girls were under the influence of Satan. Pressured to reveal the source of their affliction, the girls identified three women as witches. Within the next few weeks, other townspeople came forward and testified that they too had been harmed by some community members. For the better part of the year, trials and condemnations

509

A crowd becomes a social entity greater than the sum of its parts.

continued and Salem was involved in a serious witch-hunt. Of the 27 convictions for witchcraft, 19 people were hanged and 4 others died in jail. Eventually, Governor William Phipps put an end to the trials.

By this time in your reading of this book, you have probably noticed that sociology makes us aware of the fact that most social behavior is patterned and follows agreed-upon rules. We may interact with each other based on specific social statuses and roles, or participate in rituals of social solidarity. At other times, however, it appears that people act in ways that seem to escape the control of common expectations, and their behavior strikes us as dangerous, bizarre, and unpredictable. How can we begin to explain the violence and hysteria in Lagos, Nigeria, and of the Salem witch trials? Or the willingness of people to engage in a mass marriage wedding ceremony to people they hardly know? These actions fall under the general category of collective behavior.

Collective behavior refers to *relatively spontaneous social actions that occur when people respond to unstructured and ambiguous situations.* Collective behavior has the potential for causing the unpredictable, and even the improbable, to happen. Collective actions are capable of unleashing powerful social forces that catch us by surprise and change our lives, at times temporarily but at other times even permanently. It is the more dramatic forms of collective behavior that we tend to remember: riots, mass hysterias, lynchings, or panics. Fads, fashion, and rumor, however, also are forms of collective behavior.

Theories of Collective Behavior

There are two ways of approaching the topic of collective behavior. We could think of collective be-

havior such as riots, demonstrations, and religious revivals as the means for improving society. Those actions provide the needed push to overcome the inertia of established institutions in dealing with human problems. Social reform comes about when the system is pressured by large groups. Thus, it can be argued that the civil rights legislation of the 1960s occurred only because legislators were pushed by the protest demonstrations of that period.

On the other hand, we can argue that collective behavior is pathological and destructive to the fabric of society. Gustave Le Bon espoused this view when he identified the crowd as something of a reversion to the bestial tendency in human nature. People are swept away in the contagious excitement of the crowd. Civilized members of society, who would never engage in antisocial acts as individuals, engage in just such acts under the cloak of anonymity provided by the crowd.

In the next section, we will examine several theories that have been devised to account for crowd behavior. These include the contagion (or "mentalist") theory, the emergent norm theory, the convergence theory, and the value-added theory.

Contagion ("Mentalist") Theory

Gustave Le Bon (1841–1931) was a French sociologist whose major interest was the role played by collective behavior in shaping historical events such as the storming of the Bastille in 1789, a turning point in the launching of the French Revolution. In 1895, Le Bon published his classic *The Psychology of Crowds* (1960), in which he argued that once individuals experience the sense of anonymity in a crowd, they are transformed. Hence, they think, feel, and act quite differently from the way they would if alone. According to this **contagion theory,** *members of a crowd acquire a crowd mentality, lose their characteristic inhibitions, and become highly receptive to group sentiments.* Concerns for proper behavior or norms disappear, and individuals give up their personal moral responsibilities to the will of the crowd. When this happens, the crowd becomes a social entity greater than the sum of its individual parts.

Herbert Blumer (1946) explains the contagion that sweeps through a crowd as what he calls the "circular reaction." In his view, a crowd begins as a collectivity of people more or less waiting for something to happen. Sooner or later an exciting event stirs them, and people react to it without the caution and critical judgment they would ordinarily use if they were experiencing the event alone. Individuals become excited, the excitement spreads, the original event is invested with even greater emotional significance, and people give in to the "engulfing mood, impulse, or form of conduct." In this manner, a crowd can spiral out of control, as when a casual crowd of onlookers observing the arrest of a drunken driver is transformed into an acting crowd of rioters.

There are a number of problems with the contagion theory. Le Bon did not specify under what conditions contagion would sweep through a crowd. In addition, the theory does not account for events that could limit the spread of contagion or for the fact that contagion may affect only one portion of a crowd. Finally, research has not borne out Le Bon's basic premise that the average person can be transformed through crowd dynamics from a civilized being into an irrational and violent person.

Emergent Norm Theory

Rather than viewing the formation of crowd sentiments and behavior as inherently irrational, as Le Bon and Blumer did, Ralph H. Turner (1964) and other sociologists espouse the emergent norm theory of collective behavior. **Emergent norm theory** notes that *even though crowd members may have different motives for participating in collective behavior, they acquire common standards by observing and listening to one another.* In this respect, contagion does play a role in establishing the crowd's norms. A few leaders may help in the emergence of these norms by presenting the crowd with a particular interpretation of events. However, even without leaders, the crowd can still develop shared expectations about what behavior is appropriate.

The emergent norm theory provides the basis for analyzing the factors that push a crowd in one direction or another. If people bring with them into a crowd situation a set of expectations about the norms that are likely to be established, then the emergence of such norms will not be just a matter of collective processes of the moment (Lang & Lang, 1961). Thus, many hockey fans attending the Boston Bruins games in The Fleet Center expect to vent hostile feelings against opposing players; expect that members of the crowd may throw paper cups and other debris; and expect that management will encourage this fanaticism by playing "Charge!" music on the public address system, by flashing violence-oriented slogans on the scoreboard, and by selling alcoholic beverages. In other words, the fans expect to become frenzied. Predictably, fights often occur in the stands, sportswriters from out of town are subjected to abuse, and players from opposing teams are harassed. As one journalist has put it, "A sense of hostility . . . pervades the arena" (Fischler, 1980).

Convergence Theory

Whereas contagion theory assumes that a crowd mentality arises when people are gathered in a specific area and interact in ways that produce common

According to Herbert Blumer, a crowd is a collectivity of people more or less waiting for something to happen. Eventually something stirs them, and they react without the kind of caution and critical judgments they would normally use as individuals.

perceptions and common behavior, **convergence theory** views *collective behavior as the outcome of situations in which people with similar characteristics, attitudes, and needs are drawn together*. In contrast with contagion theory, convergence theory holds that it is not the crowd situation that produces unusual behavior but, rather, that certain kinds of people who are predisposed to certain kinds of actions have been brought together. Consequently, if violent or unusual collective behavior takes place during and after a pop music concert, it is because people who are predisposed to this type of behavior have been drawn to the event.

Convergence theory is helpful because it stresses the role of the individual and points out that no matter how powerful a group's influence may be, not everyone will respond to it. Therefore, it is unlikely that a group of conservative bankers who may be attending the above-mentioned music event will be part of any unusual collective action.

The problem with convergence theory is that it cannot explain why crowds often pass through a number of stages, from disorganized milling to organized action against specific targets. If the participants' characteristics do not change, what does produce the changes in the crowd behavior? Convergence theory also does not tell us which events will ignite a crowd with common characteristics into action and which will thwart collective behavior.

Value-Added Theory

Of all the attempts to understand collective behavior, the value-added theory of sociologist Neil Smelser (1962) is, in many ways, the most comprehensive. **Value-added theory** *attempts to explain whether collective behavior will occur and what direction it will take*. Smelser has suggested that when combined, the following six conditions shape the outcome of collective behavior.

1. *Structural conduciveness*. This refers to the conditions that may promote or encourage collective behavior. Structural conduciveness is tied to the arrangement of the existing social order. For example, in the example of the Los Angeles riots, the fact that the news media were quick to report on the progress of the trial and its outcome was important. The people in the African-American neighborhoods of South Central Los Angeles were also able to gauge each other's reactions to the verdict. All of this provided a fertile ground for collective action.

2. *Structural strain*. When a group's ideals conflict with its everyday realities, structural strain occurs. For the African-American community in Los Angeles, the disparity between their hopes and dreams and the reality of their lives produced structural strain.

3. *Growth and spread of a generalized belief*. People develop explanations for the structural strains under which they must struggle to exist. When these explanations are clearly expressed and widely shared, collective behavior may take the shape of well-organized social movements, such as the civil rights and labor movements. The less clearly these explanations are expressed or the more competing explanations that exist, the more likely that collective behavior will emerge in an unstructured form, such as a riot. In the Los Angeles African-American community, the widely shared beliefs included a strong resentment of the police and the hope of a guilty verdict for the four police officers accused of beating Rodney King. At the same time, many people suspected that an all-white jury would not produce a fair verdict for a black man. Once the outcome of the trial was known, and the fact that any other types of structured collective behavior were not likely to change the verdict, conditions leading to the riot began to appear.

4. *Precipitating factors*. In all cases of collective behavior there is an event, or a related set of events, that triggers a collective response. In the Los Angeles riots, the news of the not-guilty verdict caused people to unite and take action.

5. *Mobilization for action*. A group of people must be mobilized or organized into taking action. When there are no previously recognized leaders to take charge, a group is easily swayed by its more boisterous members. In Los Angeles, the not-guilty verdict mobilized people into unplanned, expressive acts that included setting scattered fires, looting, and committing random destruction. By the time community leaders attempted to intervene, the riot was out of control.

6. *Mechanism of social control*. At this point, the course that collective behavior follows depends on the various ways that those in power respond to the action in order to reestablish order. In the Los Angeles riot, Mayor Tom Bradley imposed a curfew, the National Guard was brought in, and President George Bush sent in federal troops. This show of force was eventually enough to quell the riot.

According to Smelser, the outcome of collective behavior depends on how each of the six determinants has built on the previous one. Each becomes a necessary condition and an important part of the next determinant.

Crowds: Concentrated Collectivities

A **crowd** is *a temporary concentration of people who focus on some thing or event, but who also are attuned to one another's behavior.* There is a magnetic quality to a crowd: It attracts passersby, who often interrupt whatever they are doing to join. Think, for example, of the crowds that gather "out of nowhere" at fires or accidents. Crowds also fascinate social scientists, because crowds always have within them the potential for unpredictable behavior and group action that erupts quickly and often seems to lack structure or direction—either from leaders or from institutionalized norms of behavior.

Attributes of Crowds

In his study *Crowds and Power* (1978), Elias Canetti attributed to crowds the following traits:

1. *Crowds are self-generating.* Crowds have no natural boundaries. When boundaries are imposed artificially—for example, by police barricades intended to isolate a street demonstration—there is an ever-present danger that the crowd will erupt and spill over the boundaries, thereby creating chaos. So, in effect, crowds always contain threats of chaos, serious disorder, and uncontrollable force.

2. *Crowds are characterized by equality.* Social distinctions lose their importance within crowds. Indeed, Canetti believes that people join crowds specifically to achieve the condition of equality with one another, a condition that carries with it a charged and exciting atmosphere.

3. *Crowds love density.* The circles of private space that usually surround each person in the normal course of events shrink to nothing in crowds. People pack together shoulder to shoulder, front to back, touching one another in ways normally reserved for intimates. Everyone included within the crowd must relinquish a bit of his or her personal identity to experience the crowd's fervor. With a "we're all in this together" attitude, the crowd discourages isolated factions and detached onlookers.

4. *Crowds need direction.* Many crowds are in motion. They may move physically as they do in a marching demonstration or psychologically as at a rock concert. The direction of movement is set by the crowd's goals, which become so important to crowd members that individual and social differences lessen or disappear. This constant need for direction contains the seeds of danger: Having achieved or abandoned one goal, the crowd may easily seize on another, perhaps destructive, one. The direction that a crowd will take depends on the type of crowd involved.

Types of Crowds

In his essay on collective behavior, Herbert Blumer (1946) classified crowds into four types: acting, expressive, conventional, and casual.

Acting Crowd An **acting crowd** is *a group of people whose passions and tempers have been aroused by some focal event, who come to share a purpose, and who feed off one another's arousal, often erupting into spontaneous acts of violence.*

In 2000, hundreds of Fat Tuesday revelers taunted and threw beer bottles at police in riot gear as Mardi Gras festivities in Seattle turned ugly. Under the glare of a helicopter spotlight, police gradually pushed crowds away from the epicenter of the disorder where as many as 500 had gathered to celebrate the final day of Mardi Gras. The disturbance began around midnight when people gathered at First Avenue and Yesler Street. A woman standing on a newspaper vending box fell and hit her head. Officers who tried to come to her aid were pelted with rocks and bottles (Associated Press, 2000).

Acting crowds can become violent and destructive, as 400 million worldwide television viewers discovered in the summer of 1985. Sixty thousand soccer fans had assembled in Brussels to watch the European Cup Finals between Italy and Great Britain. Verbal taunts quickly turned into rocks and bottles being thrown. Suddenly, British fans stormed the fence and surged toward the Italian fans, trampling hundreds of helpless spectators. Before the horror could be stopped, 38 people were dead and another 400 injured (*Time*, 1985).

A **threatened crowd** is *an acting crowd that is in a state of alarm, believing that some kind of danger is present.* Such a crowd is in a state of panic, as when a crowded nightclub catches fire and everybody tries to get out, jamming exits and trampling one another in their rush to escape. A threatened crowd created havoc when a busboy accidentally ignited an artificial palm at the Coconut Grove Night Club in Boston on November 28, 1942, spreading fire instantaneously throughout the club. The fire lasted only 20 minutes, but 488 people died. Most died needlessly when panic gripped the crowd. Fire investigators found that the club's main entrance—a revolving door—was jammed by hundreds of terrified patrons. With their escape route blocked, those people died of burns and smoke inhalation only feet away from possible safety (Veltfort & Lee,

Officials must attempt to minimize the widespread panic that could erupt during a major earthquake.

1943). In this as well as other threatened crowds, there is a lack of communication regarding escape routes.

Expressive Crowd An **expressive crowd** is *drawn together by the promise of personal gratification through active participation in activities and events.* For example, many rock concert audiences are not content simply to listen to the music and watch the show. In a very real sense, they want to be part of the show. Many dress in clothing calculated to draw attention to themselves, take drugs during the performance, body surf or slam dance in packed masses up against the stage, and delight in giving problems to security personnel.

Conventional Crowd A **conventional crowd** is *a gathering in which people's behavior conforms to some well-established set of cultural norms, and gratification results from a passive appreciation of an event.* Such crowds include the audiences attending lectures, the theater, and classical music concerts, where everybody is expected to follow traditional norms of etiquette.

Casual Crowd A casual crowd is the inevitable outgrowth of modern society, in which large num-

bers of people live, work, and travel closely together. A **casual crowd** is *any collection of people who happen, in the course of their private activities, to be in one place at the same time and focus attention on a common object or event.* On Fifth Avenue in New York City at noon, many casual crowds gather to watch an accident, a purse snatcher, the construction of a new building, or a theatrical performer. A casual crowd has the potential of becoming an acting crowd or an expressive crowd; the nature of a crowd can change if events change.

The Changeable Nature of Crowds

Although the typology presented is useful for distinguishing kinds of crowds, it is important to recognize that any crowd can shift from one type to another. For example, if a sidewalk musician starts playing a violin on Fifth Avenue, part of the aggregate walking by will quickly consolidate into a casual crowd of onlookers. Or an expressive crowd at a rock concert will become a threatened crowd if a fire breaks out.

Changing times may also affect the nature of crowds. For example, until the 1970s, British soccer matches generally attracted conventional crowds who occasionally turned into expressive crowds chanting team songs. Since the late 1970s, however, British soccer fans have become active crowds: fighting in the stands is epidemic, charging onto the field to assault players and officials has become common, and rioting has taken place. In 2000, Danish police and British soccer fans clashed in Copenhagen prior to a match between the Turkish and British soccer teams. Police used tear gas, dogs, and batons on about 100 British soccer fans. Minutes before that clash, British fans fought with Turkish fans near Copenhagen's town square. As a precaution against further rioting, authorities erected iron fencing outside the 39,000-seat stadium to separate British and Turkish fans. Another fence was put up inside around the playing field (CNN News, May 17, 2000).

Because they are relatively concentrated in place and time, crowds present rich materials for sociological study (even if much of the data must be tracked down after the dust has settled). However, when collective behavior is widely dispersed among large numbers of people whose connection with one another is minimal or even elusive, the sociologist must then deal with phenomena that are extremely difficult to study, including fads and fashions, rumors, public opinion, panics, and mass hysteria.

Dispersed Collective Behavior

In this age of mass media, with television and other systems of communication spreading information instantaneously throughout the entire population, collective behavior shared by large numbers of people who have no direct knowledge of one another has become commonplace. Sociologists use the term **mass** to describe *a collection of people who, although physically dispersed, participate in some event either physically or with a common concern or interest.*

A nationwide television audience watching a presidential address or a Super Bowl game is a mass. So are those individuals who rush out to buy the latest best-selling CD and the fashion-conscious whose hemlines, lapel widths, and clothes always reflect the "in" look. In other words, dispersed forms of collective behavior seem to be universal. (See "Society and the Internet: Social Movements on the Internet.")

Fads and Fashions

Fads and fashion are transitory social changes (Vago, 1980), patterns of behavior that are widely dispersed among a mass but that do not last long enough to become fixed or institutionalized. Yet it would be foolish to dismiss fads and fashions as unimportant just because they fade relatively quickly. In modern society, fortunes are won and lost trying to predict fashions and fads—in clothing, in entertainment preferences, in eating habits, in choices of investments.

Probably the easiest way to distinguish between fads and fashions is to look at their typical patterns of diffusion through society. Fads are social changes with a very short life span marked by a rapid spread and an abrupt drop in popularity. This was the fate of the Hula Hoop in the 1950s and the dance known as the "twist" in the 1960s. The roller-skating fad that emerged in 1979 rolled off into the pages of history sometime in the 1980s, as did the Rubik's cube, and Coleco, the company that made Cabbage-Patch dolls, which went bankrupt in 1988. The Tickle Me Elmo stuffed toys were quickly forgotten after the 1996 Christmas season. While you or your parents may remember such past fads as yo-yos and Mr. Potato Head, recent fads include the introduction of Furby, Teletubbies, Pokemon, and Dragonball Z.

Some fads may seem particularly absurd. During the Great Depression of the 1930s, when many Americans were having trouble putting food on the table, college students started engaging in the practice of swallowing goldfish. The fad was started by a Harvard freshman who swallowed a single, live fish as fellow students looked on in disgust. Three weeks later a student at Franklin and Marshall College swallowed three fish. The practice quickly escalated and new records were set daily, with 89 being swallowed in one sitting at Clark University. Eventually a pathologist at the U.S. Public Health Service cautioned that goldfish may contain tapeworms that could cause intestinal problems and anemia. The fad disappeared shortly thereafter (Levin, 1993).

A fad that is especially short-lived may be called a **craze**. The Mohawk hairstyle among both young males and females was a relatively short-lived craze, as was streaking, or running naked down a street or through a public gathering, in 1974. One streaker even ran on stage during the Academy Awards presentation.

At the peak of their popularity, fads and crazes may become competitive activities. For example, when streaking was a craze, individual streaking was followed by group streaking, streaking on horseback, and parachuting naked from a plane.

On other occasions, what appears to be a fad actually signals a trend and a change in social values. In 1922, newspapers reported the shocking news that smoking was common among female college students. The University of Wisconsin's dean of women said the smoking fad, most popular among

Society and the Internet

Social Movements on the Internet

If there has been one development in recent years that has changed the way in which groups organize, recruit, raise funds, and generally get the message out, it is the Internet. Collective action now can involve people who do not see each other and have little knowledge about where the organizers of a movement actually reside. The Internet can further collective behavior and social movements in the following ways:

1. *Public relations.* Many groups have recognized the public relations aspect of being able to tell their story. It is a way to communicate with other like-minded people, and it is a way to avoid filtering and distortion by the media. In the past, social movements were often helped or hindered by the interpretation that the media put on their activities. Now groups can put their messages out themselves.

2. *Recruitment.* A significant role for the Internet is the part it plays in helping people of like minds to find each other. By having some sort of activity on the Net, groups can be located and contacted by potential members.

3. *Member communication.* Many groups would not be able to exist without the Internet. The availability of cheap, fast, and effective communications with their membership allows them to exist when they might be too thinly spread to exist from a geographical standpoint. In addition, the ability to communicate quickly allows a group to appear large, well funded, and well organized, regardless of whether that is actually the case. The Electronic Freedom Foundation,

which waged a successful campaign against Congress's attempt to pass the Communications Decency Act in 1996, was made up of only eight people. The act, which sought to ban pornography and other offensive material on the Internet, was met with an abundance of blue ribbons on thousands of Web sites, all indicating support for free speech on the Net.

4. *Media relations.* One of the attractions of the Internet is that it provides a pathway to the mass media for groups that need one. While not all groups want media attention, many do, at least as a way to report on abuses of their rights.

5. *Fund-raising.* Most social movements in their early stages are chronically short of money. Fund-raising is an activity that is well suited to the Internet. Mailing lists can be developed and shared with other groups.

6. *Group communication.* Communication among groups with compatible goals is an important use of the Internet, and one that appears to be increasing as groups learn that by working together they can create the appearance of having more clout than they do separately.

7. *Political discussion.* Discussion groups are an important part of self-identification and provide a way for many groups to recruit members. In addition, they may serve to shore up morale by helping members of thinly scattered organizations realize that they are not fighting their battles alone.

Source: *Politics on the Net* (pp. 95–97), by W. Rash, Jr., 1997, New York: W. H. Freeman.

women of the "idle, blase, disappointed class," was already passing. She pointed out that an intelligent woman "cannot see herself rocking a baby or making a pie with a cigarette in her mouth, flicking ashes in the baby's face or dropping them in the pie crust." (*American Heritage,* February/March 1997).

Fashions relate to *the standards of dress or manners in a given society at a certain time.* They spread more slowly and last longer than fads. In his study of fashions in European clothing from the eighteenth to the present century, Alfred A. Kroeber (1963) showed that though minor decorative features come and go rapidly (that is, are faddish), basic silhouettes move through surprisingly predictable cycles that he

correlates with degrees of social and political stability. In times of great stress, fashions change erratically; but in peaceful times, they seem to oscillate slowly in cycles lasting about 100 years.

Georg Simmel (1957) believed that changes in fashion (such as dress or manners) are introduced or adopted by the upper classes, who seek in this way to keep themselves visibly distinct from the lower classes. Of course, those immediately below them observe these fashions and also adopt them in an attempt to identify themselves as upper class. This process repeats itself again and again, with the fashion slowly moving down the class ladder, rung by rung. When the upper classes see that their fashions

 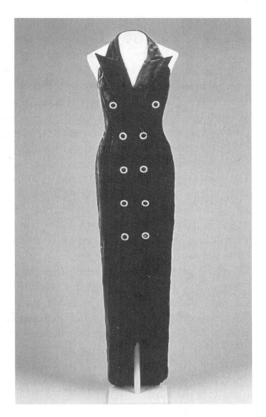

Fashions relate to standards of dress and manners during a particular time. Here we see how the dress of royalty has changed over time. At the left is a court dress worn between 1765 and 1775. At the right is a dinner dress worn by Diana, Princess of Wales, in 1991.

have become commonplace, they take up new ones, and the process starts all over again.

Blue jeans have shown that this pattern is no longer true today. Jeans started out as sturdy work pants worn by those engaged in physical labor. Young people then started to wear them for play and everyday activities. College students wore them to class. Eventually, fashion designers started to make fancier, higher-priced versions, known as designer jeans, worn by the middle and upper classes. In this way the introduction of blue jeans into the fashion scene represents movement in the opposite direction from what Simmel noted.

Of course, the power of the fashion business to shape consumer taste cannot be ignored. Fashion designers, manufacturers, wholesalers, and retailers earn money only when people tire of their old clothes and purchase new ones. Thus, they promote certain colors and widen and narrow lapels to create new looks, which consumers purchase.

Indeed, the study of fads and fashions provides sociologists with recurrent social events through which to study the processes of change. Because they so often involve concrete and quantifiable objects, such as consumer goods, fads and fashions

are much easier to study and count than are rumors, another common form of dispersed collective behavior.

Rumors

A **rumor** is *information that is shared informally and spreads quickly through a mass or a crowd.* It arises in situations that, for whatever reasons, create ambiguity with regard to their truth or their meaning. Rumors may be true, false, or partially true, but characteristically they are difficult or impossible to verify.

Rumors are generally passed from one person to another through face-to-face contact, but they can be started through television, radio, and the Internet as well. However, when the rumor source is the mass media, the rumor still needs people-to-people contact to enable it to escalate to the point of causing widespread concern (or even panic). Sociologists see rumors as one means through which collectivities try to bring definition and order to situations of uncertainty and confusion. In other words, rumors are "improvised news" (Shibutani, 1966).

In 1992, Snapple Beverage Corporation needed to counter potentially damaging and unsubstantiated rumors that proceeds from its operations were helping to finance Operation Rescue, the militant antiabortion group that attempted to close or impede the activity of abortion clinics. The company was successful in putting an end to the rumor.

Other rumors are not that easy to stop. Hard-to-believe rumors usually disappear first, but this is not always the case. For 103 years, Procter & Gamble, the maker of familiar household products such as Mr. Clean and Tide laundry detergent, used the symbol of the moon and 13 stars as a company logo on its products. Around 1979, a rumor began circulating that this symbol indicated a connection between the giant corporation and satanic religion. There was no evidence to substantiate this rumor, but unable to dispel it, the company finally decided to remove the logo from its products in 1985 (Koenig, 1985). In 1997 Procter & Gamble was still plagued by the rumor and filed the latest in a series of lawsuits, this time against Amway Corporation and some of its distributors for allegedly spreading rumors that Procter & Gamble is affiliated with the Church of Satan. Since the early 1980s, the company said it has filed 15 lawsuits to fight the rumors.

Public Opinion

The term *public* refers to a dispersed collectivity of individuals concerned with or engaged in a common problem, interest, focus, or activity. An opinion is a strongly held belief. Thus, **public opinion** refers to *the beliefs held by a dispersed collectivity of individuals about a common problem, interest, focus, or activity.* It is important to recognize that a public that forms around a common concern is not necessarily united in its opinions regarding this concern. For example, Americans concerned about abortion are sharply divided into pro and con camps.

Whenever a public forms, it is a potential source for, or opposition to, whatever its focus is. Hence, it is extremely important for politicians, market analysts, public relations experts, and others who depend on public support to know the range of public opinion on many different topics. Those individuals often are not willing to leave opinions to chance, however. They seek to mold or influence public opinion, usually through the mass media. Advertisements are attempts to mold public opinion, primarily in the area of consumption. They may create a need where there was none, as they did with fabric softeners, or they may try to convince consumers that one product is better than another when there

is actually no difference. *Advertisements of a political nature, seeking to mobilize public support behind one specific party, candidate, or point of view* technically are called **propaganda** (but usually by only those people in disagreement). For example, radio broadcasts from the former Soviet Union were habitually called "propaganda blasts" by the American press, but similar Voice of America programs were called "news" or "informational broadcasts" by the same American press.

Opinion leaders are *socially acknowledged experts whom the public turns to for advice.* The more conflicting sources of information there are on an issue of public concern, the more powerful the position of opinion leaders becomes. The leaders weigh various news sources and then provide an interpretation in what has been called the two-step flow of communication. Those opinion leaders can have a great influence on collective behavior, including voting (Lazarsfeld et al., 1968), patterns of consumption, and the acceptance of new ideas and inventions. Typically, each social stratum has its own opinion leaders (Katz, 1957). Jesse Jackson, for example, is an opinion leader in the African-American community. The mass media have turned news anchors like Dan Rather, Tom Brokaw, and Peter Jennings into accepted opinion leaders for a broad portion of the American public. Rush Limbaugh has emerged as one of the more influential opinion leaders, as the fortunes of political candidates are determined by his loyal listeners.

When rumor and public opinion grip the public imagination so strongly that facts no longer seem to matter, terrifying forces may be unleashed. Mass hysteria may reign, and panic may set in.

Mass Hysteria and Panic

At a summer program in Florida, 150 children would gather every day in a dining hall where they were served pre-packaged lunches. As lunch began one day, a girl complained that her sandwich didn't taste right: she felt nauseated, and came back from the restroom reporting that she had become sick. Others began to complain that their stomachs hurt too and that the sandwiches really did taste funny. Some children reported having headaches, tingling in their hands and feet, and abdominal cramping. The supervisor, obviously worried about all the complaints, announced to the horrified children that the food might be poisoned. They were told to stop eating immediately. Within 40 minutes, 63 children were sick. More than 25 of them had vomited. Ambulances were called and the children had to be divided up among three different hospitals.

But an hour later, it was all over. Every examination and test performed on the children was normal. Meal samples were analyzed, but no bacteria or pesticides were detected. Food processing and storage techniques had been faultless. And no one had become ill at any of the other 68 sites at which the very same food was served. It appears the children were all victims of a case of mass hysteria. (Feldman & Feldman,1998)

Mass hysteria occurs when *large numbers of people are overwhelmed with emotion and frenzied activity or become convinced that they have experienced something for which investigators can find no discernible evidence.* A **panic** is *an uncoordinated group flight from a perceived danger,* as in the public reaction to Orson Welles's 1938 radio broadcast of H. G. Wells's *War of the Worlds* and to the 508-point drop in stock prices that occurred in October 1987.

According to Irving Janis and his colleagues (1964), people generally do not panic unless four conditions are met. First, people must feel that they are trapped in a life-threatening situation. Second, they must perceive that the threat to their safety is so large that they can do little else but try to escape. Third, they must realize that their escape routes are limited or inaccessible. Fourth, there must be a breakdown in communication between the front and rear of the crowd. Driven into a frenzy by fear, people at the rear of the crowd make desperate attempts to reach the exit doors, and their actions often completely close off the possibility of escape.

The perception of danger that causes a panic may come from rational as well as irrational sources. A fire in a crowded theater, for example, can cause people to lose control and trample one another in their attempt to escape. This happened when fire broke out at the Beverly Hills Supper Club in Southgate, Kentucky, on May 28, 1977. When employees discovered an out-of-control fire, they warned the 2,500 patrons and tried to usher them out of the building. A panic resulted as people attempted to escape the overcrowded, smoke-filled building. People trampled each other trying to reach the exits, and 165 people died in the process.

Such extreme events are not very common, but they do occur often enough to present a challenge to social scientists, some of whom believe there is a rational core behind what at first glance appears to be wholly irrational behavior (Rosen, 1968). For example, sociologist Kai T. Erikson (1966) looked for the rational core behind the wave of witchcraft trials and hangings that raged through the Massachusetts Bay Colony beginning in 1692. Erikson joins most other scholars in viewing this troublesome episode in American history as an instance of mass hysteria (Brown, 1954). He accounts for it as one of a series of symptoms, suggesting that the colony was in the grip of a serious identity crisis and needed to create real and present evil figures who stood for what the colony was not—thus enabling the colony to define its identity in contrast and build a viable self-image.

Mass hysterias account for some of the more unpleasant episodes in history. Of all social phenomena, they are among the least understood—a serious gap in our knowledge of human behavior.

Social Movements

A **social movement** is *a form of collective behavior in which large numbers of people are organized or alerted to support and bring about, or to resist, social change.* By their very nature, social movements are an expression of dissatisfaction with the way things are or with changes that are about to take place.

Participation in a social movement is, for most people, only informal and indirect. Usually, large numbers of sympathizers identify with and support the movement and its program without joining any formal organizations associated with the movement. For people to join a social movement, they must think that their own values, needs, goals, or beliefs are being stifled or challenged by the social structure or specific individuals. The people feel that this situation is undesirable and that something must be done to set things right. Some catalyst, however, is needed to mobilize the discontent that people feel. Two major theories, relative deprivation theory and resource mobilization theory, attempt to explain how social movements emerge.

Relative Deprivation Theory

Relative deprivation is a term that was first used by Samuel A. Stouffer (1950). It refers to the situation in which deprivation or disadvantage is measured not by objective standards, but by comparison with the condition of others with whom one identifies or thinks of as a reference group.

Thus, **relative deprivation theory** assumes *social movements are the outgrowth of the feeling of relative deprivation among the large numbers of people who believe they lack certain things they are entitled to*—such as better living conditions, working conditions, political rights, or social dignity.

From the standpoint of relative deprivation theory, the actual degree of deprivation people suffer

Leadership is an important ingredient in the emergence of a social movement.

is not automatically related to whether people feel deprived and therefore join a social movement to correct the situation. Rather, deprivation is considered unjust when others with whom the people identify do not suffer the deprivation (Gurr, 1970).

Karl Marx expressed this view when he noted,

> A house may be large or small; as long as the surrounding houses are equally small it satisfies all social demands for a dwelling. But let a palace arise beside the little house, and it shrinks from a little house to a hut. . . . Our desires and pleasures spring from society; we measure them therefore, by society and not by the objects which serve for their satisfaction. Because they are of a social nature, they are of a relative nature. (Marx, 1968)

There is a flaw in the theory of relative deprivation, in that often the people who protest a situation or condition may not be deprived. Sometimes people protest because a situation violates their learned standards of justice. The white civil rights marchers in the 1960s were not personally the victims of antiblack discrimination. We could argue, however, that they were experiencing deprivation in the sense that the reality was not judged to be what it ought to be. They were experiencing a moral as opposed to a material or personal social deprivation (Rose, 1982).

Resource Mobilization Theory

The **resource mobilization theory** assumes that *social movements arise at certain times and not at others because some people know how to mobilize and channel the popular discontent.* Although discontent exists virtually everywhere, a social movement will not emerge until specific individuals actually mobilize resources available to a group, by persuading people to contribute time, money, information, or anything else that might be valuable to the movement. An organizational format must also be developed for allocating these resources. Leadership, therefore, becomes a crucial ingredient for the emergence of a social movement.

The leadership tries to formulate the resources and ideology in such an attractive fashion that many others will join the movement. It is not enough to make speeches and distribute fliers; the messages have to strike a respondent chord in others (Ferree & Hess, 1985).

One of the most successful people in terms of resource mobilization was Saul Alinsky, an activist who devoted his life to developing "people's organizations" at the neighborhood level to combat exploitation and poor living conditions.

Alinsky was not the sort of man people reacted to impartially. He was abrasive, forceful, witty, antagonistic, irreverent, and not above shocking or lying to people. Some saw him as a menace. Corporations hired detectives to follow him. What people were really responding to was not Alinsky the person, but the method of community organizing that came to be associated with him.

To accomplish his goals, he followed a number of guidelines. First, he believed that the professional organizer was the key catalyst for social change. Community organizing is difficult and requires crucial and correct decisions. He believed that democracy was important, but that the organizer was even more so.

Second, the goal was to use any tactics necessary to win. The end justified the means, Alinsky counseled. For example, Alinsky described the tactics used by one of his organizations in the following way:

> When [they have] a bunch of housing complaints they don't forward them to the building inspector. They drive forty or fifty members—the blackest ones they can find—to the nice suburb where the slumlord lives and they picket his home. Now we know the picket line isn't going to convert the slumlord. But we also know what happens when his white neighbors get after him and say, "We don't care what you do for a living—all we're telling you is to get those niggers out of here or you get out." That's the kind of jujitsu operation that forces the slumlord to surrender and gets repairs made in the slum. (Quoted in Fisher, 1984)

Throughout his life, Alinsky was involved in countless organizing efforts, and his methods were adopted by a vast array of groups. He was a clear example of a leader who knew how to mobilize and channel the feelings of deprivation and dissatisfaction that existed among the poor.

Types of Social Movements

In the politics chapter, we discussed rebellions and revolutions, which certainly qualify as social movements; but there are other kinds of social movements as well. Scholars differ as to how they classify social movements, but some general characteristics are well recognized. We shall discuss these characteristics according to William Bruce Cameron's (1966) four social-movement classifications: reactionary, conservative, revisionary, and revolutionary. In addition, we shall examine the concept of expressive social movements first developed by Herbert Blumer (1946).

For people to join a social movement, they must think that their own values, needs, goals, or beliefs are being stifled or challenged by the social structure or by specific individuals.

Although this classification is useful to sociologists in their studies of social movements, in practice it is sometimes difficult to place a social movement in only one category. This is because any social movement may possess a complex set of ideological positions in regard to the many different features of the society, its institutions, the class structure, and the different categories of people within that society. (See "Sociology at Work: Is Vegetarianism a Social Movement?")

Reactionary Social Movement **Reactionary social movements** *embrace the aims of the past and seek to return general society to yesterday's values.* Using slogans like the "good old days" and our "grand and glorious heritage," reactionaries abhor the changes that have transformed society and are committed to re-creating a set of valued social conditions that they believe existed at an earlier time. Reactionary groups such as the neo-Nazis and the Ku Klux Klan hold racial, ethnic, and religious values that are more characteristic of a previous historic period.

The Klan has sought to uphold white dominance and what it sees as traditional morality. To do this it has threatened, flogged, mutilated, and, on occasion, murdered. The main purpose of the Klansmen has been to defend and restore what they conceived as traditional cultural values. Their values legitimize prejudice and discrimination based on race, ethnicity,

Sociology at Work

Is Vegetarianism a Social Movement?

On many college campuses, as well as society in general, vegetarianism—the avoidance of food containing meat or animal products—has been growing.

Vegetarian diets vary widely in the strictness with which they restrict animal products. The strictest vegetarians, known as vegans, steer clear of all food that contains any ingredient derived from animals—including meat, eggs, milk, honey, and gelatin. Lacto-vegetarians consume dairy products but no meat or fish, while ovo-lacto-vegetarians do not eat beef or pork, but eat eggs and milk products.

The primary reason for the growing popularity of vegetarianism is publicity surrounding its health benefits. During the 1990s, a number of doctors and medical organizations vocally recommended limiting meat intake as a way of preventing certain degenerative diseases. They said that the high level of fat found in meat could contribute to heart disease, diabetes, and possibly cancer.

Many vegetarians are also motivated by concern for the welfare of animals. These so-called ethical vegetarians believe that no animal should die or suffer merely to provide people with the luxury of eating meat. Ethical vegetarians object to the modern methods of raising animals for food. They say animals typically live in dirty cramped pens, and that farmers treat the animals as machines rather than living beings. For this reason, these people swear off eggs, milk, cheese, and meat.

Many people disagree with some or all of the arguments for vegetarianism. They believe the ethical concerns over the use of animals for food are unfounded. Animals, they say, do not possess self-awareness in the same way that people do, and it is therefore foolish to think they have any thoughts on the issue. They also think it would be illogical and self-destructive for farmers to treat their animals badly, since unhealthy animals produce inferior meat, cheese, or eggs.

The idea that it is immoral for human beings to eat animals has existed for thousands of years. Until the nineteenth century, however, such beliefs were common only in Far Eastern nations such as India and China, where Hinduism and Buddhism are the dominant religions. According to Hindu beliefs, for example, slaughtering and eating animals violates the idea of *ahimsa,* or nonviolence, and certain animals, such as cows, are viewed as sacred. Buddhism, meanwhile, teaches that human beings share a close affinity with all living beings, including animals. Buddhism restricts the eating of meat less stringently than Hinduism.

Other views have also developed. Aristotle believed that since only people have the ability to think and reason, they have a natural right to use animals as food or tools. Christians also generally hold this view. The first book of the Bible states that man has "dominion over the fish of the sea, and over the birds of the air, and over the cattle."

Some groups supporting animal rights have challenged this interpretation of the Bible. People for the Ethical Treatment of Animals (PETA) claims that the Bible precludes Christians from eating meat. They interpret the statement that humans have "dominion" over animals to mean that they have "stewardship" over them. They note that Jesus was a member of the Essenes, a Jewish sect that disdained meat and animal sacrifice. Most Christian leaders have rejected PETA's claim. They see it as an example of a group manipulating the Bible to serve its own ends.

The number of people who consider themselves vegetarians has been rising. The Gallup Poll found that in 1999, 6% of Americans, about 16.5 million people, considered themselves vegetarians. As more doctors recommend meatless diets to their patients and as more people become concerned about the slaughter of animals, this trend should continue.

Source: "Vegetarianism," November 26, 1999, *Issues and Controversies on File,* pp. 481–488.

---

and religion, patterns that are now neither culturally legitimate nor legal (Chalmers, 1980).

Recently, this kind of reactionary fanaticism has been displayed by organized groups of teenagers who are part of a neo-Nazi political movement. The teens, known as skinheads, are white supremacists who commit a variety of hate crimes against blacks, immigrants, and Jews across the country.

Conservative Social Movement   **Conservative social movements** *seek to maintain society's current values by reacting to change or threats of change* *that they believe will undermine the status quo.* Many of the evangelical religious groups hold conservative views. For example, they often oppose the forces that promulgate equal rights for women. To preserve what they consider to be traditional values of the family and religion, these groups have threatened to boycott advertisers that sponsor television programs containing sex and violence and have mounted successful campaigns to defeat political candidates who oppose their views.

Conservative movements are most likely to arise when traditional-minded people perceive a threat of change that might alter the status quo. Reacting to what might happen if another movement achieves its goals, members of conservative movements mobilize an "anti" movement crusade. Thus, antigun control groups have waged political war against any group seeking to restrict access to guns, whether they be handguns or assault weapons. Although reactive in nature, conservative movements are far different from true reactionary movements, which attempt to restore values that have already changed.

Revisionary Social Movement **Revisionary social movements** *seek partial or slight changes within the existing order but do not threaten the order itself.* The women's movement, for example, seeks to change the institutions and practices that have imposed prejudice and discrimination on women. The civil rights movements, the antinuclear movement, and the ecology movement are all examples of revisionary social movements.

Revolutionary Social Movement **Revolutionary social movements** *seek to overthrow all or nearly all of the existing social order and replace it with an order they consider more suitable.* For example, the black guerrilla movement in Zimbabwe (formerly Rhodesia) was a revolutionary movement. Through the use of arms and political agitation, the guerrillas

Although both revolutionary and revisionary social movements seek change in society, they differ in the degree of change they seek.

were successful in forcing the white minority to turn over political power to the black majority and in creating a new form of government that guaranteed 80 of the 100 seats in the country's legislative body would be held by blacks.

Although revolutionary and revisionary social movements both seek change in society, they differ in the degree of change they seek. The American Revolution, for example, which sought to overthrow British colonial rule and led to the formation of our own government, differed significantly from the women's movement, which seeks change within the existing judicial and legislative structures.

Expressive Social Movement Though other types of movements tend to focus on changing the social structure in some way, **expressive social movements** *stress personal feelings of satisfaction or well-being and typically arise to fill some void or to distract people from some great dissatisfaction in their lives.* The Unification Church and the Hare Krishnas are religious movements of this type.

The Life Cycle of Social Movements

Social movements, by their nature, do not last forever. They rise, consolidate, and eventually succeed, fail, or change. Armand L. Mauss (1975) suggested that social movements typically pass through a series of five **life cycle stages:** *(1) incipiency, (2) coalescence, (3) institutionalization, (4) fragmentation, and (5) demise.* However, these stages are by no means common to all social movements.

Incipiency The first stage of a social movement, called **incipiency,** *begins when large numbers of people become frustrated about a problem and do not perceive any solution to it through existing institutions.* Incipiency occurred in the nineteenth century when American workers, desperate over their worsening working conditions, formed the U.S. labor movement. This stage is a time of disorder, when people feel the need for something to give their lives direction and meaning or to channel their behavior toward achieving necessary change. Disruption and violence may mark a social movement's incipiency. In 1886 and 1887, as the labor movement grew, workers battled private Pinkerton agents and state militia and called nationwide strikes. Although physically beaten, the workers continued to organize.

Incipiency is also a time when leaders emerge. Various individuals offer competing solutions to the perceived societal problem, and some people are more persuasive than others. According to Max Weber (see Chapter 13), many of the more successful leaders have charismatic qualities derived from exceptional personal characteristics. Samuel Gompers, who launched the American Federation of

Expressive social movements typically arise to fill some void or to distract people from some dissatisfaction with their lives.

Labor (AFL) in 1881, was such a leader, as was the Reverend Martin Luther King, Jr.

Coalescence During *the second stage,* known as **coalescence,** *groups form around leaders to promote policies and to promulgate programs.* Some groups join forces; others are defeated in the competition for new members. Gradually, a dominant group or coalition of groups emerges that establishes itself in a position of leadership. Its goals become the goals of many, its actions command wide participation, and its policies gain influence. The labor movement coalesced in 1905, when William D. Haywood organized the Industrial Workers of the World (IWW), which led its increasingly dissatisfied members in a number of violent strikes. Labor coalescence continued in 1935, when such militant industrial union leaders as John L. Lewis of the United Mine Workers and David Dubinsky of the International Ladies Garment Workers founded the Committee for Industrial Organization (CIO), which rapidly organized the steel, automobile, and other basic industries. Thus, through coalescence, the labor movement gradually created several large, increasingly powerful organizations.

Institutionalization During *the third stage,* known as **institutionalization,** *social movements reach the peak of their strength and influence and become firmly established.* Their leadership no longer depends on the elusive quality of charisma to motivate followers. Rather, it has become firmly established in formal, rational organizations (see Chapter 5) that have the power to effect lasting changes in the social order. At this point, the organizations themselves become part of the normal pattern of everyday life.

When the institutionalization of the U.S. labor movement became formalized with the legalization of unions in the 1930s, union leaders no longer used the revolutionary rhetoric that was necessary when unions were neither legitimate nor legal. Instead, they talked in pragmatic terms, worked within the political power structure, and sought reforms within the structure of the existing democratic, capitalistic system.

Not all social movements become institutionalized. In fact, social movements fail and disappear more often than they reach this stage. Institutionalization depends to a great degree on how the members feel about the movement—whether it reflects their goals and has been successful in achieving them—and on the extent to which the movement is accepted or rejected by the larger society.

Ironically, the acceptance of a social movement may also mark its end. Many members drop out or lose interest once a movement's goals have been reached. It can be argued that a certain amount of opposition from those in power reminds the members that they still must work to accomplish their goals. Movement leaders often hope for a confrontation that will clarify the identity of the opposition and show the members against what and whom they must fight. Movements that evoke an apathetic or disinterested response from the institutions controlling the power structure have few resources with which to unite their membership.

Fragmentation **Fragmentation** is *the fourth stage of a social movement, when the movement gradually begins to fall apart.* Organizational structures no longer seem necessary because the changes they sought to bring about have been institutionalized or the changes they sought to block have been prevented. Disputes over doctrine may drive out dissident members, as when the United Auto Workers (UAW) and the Teamsters left the AFL-CIO. Also, demographic changes may transform a once-strong social movement into a far less powerful force. Economic changes have been largely responsible for the fragmentation of the American labor movement. Unions now represent the smallest share of the labor force since World War II, even though the workforce continues to expand. Their lost power is due, in part, to a sharp decrease in the percentage of more easily unionized blue-collar workers in the labor force and a dramatic increase in the percentage of white-collar employees, who are largely resistant to unionization.

Demise **Demise,** *the last stage, refers to the end of a social movement.* The organizations that the movement created and the institutions they introduced may well survive—indeed, their goals may become official state policy—but they are no longer set apart from the mainstream of society. Transformed from social movements into institutions, they leave behind well-entrenched organizations that guarantee their members the goals they sought. This pattern of social-movement demise has occurred in parts of the American labor movement. The United Auto Workers, for example, is no longer a social movement fighting for the rights of its members from the outskirts of the power structure. Rather, it is now an institutionalized part of society. All unions have not followed this course. Labor is still very much a social movement, for it is trying to organize such previously unorganized groups as farm workers, nonunionized clerical professional workers, and all workers in the traditionally nonunion South. The American Federation of State, County, and Municipal Employees is a recent example of the labor movement's continued organizational efforts.

have similar characteristics, they are predisposed to similar kinds of actions.

Value-added theory presents a series of six conditions that shape its generation and outcome. In all cases of collective behavior there is an event, or series of events, that triggers a collective response. Then the group must be mobilized, or organized into taking action. Finally, the course of the collective action will depend on the mechanisms of social control employed by those in power. Each of these conditions is a necessary condition for, and a partial determinant of, the next succeeding condition.

A crowd is a temporary concentration of people who focus on some thing or event but who also are attuned to one another's behavior. Crowds have the potential for unpredictable behavior and group action that erupts quickly and often seems to lack structure or direction.

Because the mass media today spread information quickly among millions of people, collective behavior is often shared by large numbers of people who have no direct knowledge of or contact with one another.

A social movement is a form of collective behavior in which large numbers of people are organized or alerted to support and bring about or to resist social change. For people to join a social movement, they must feel that their own values, needs, goals, or beliefs are being stifled or challenged by the social structure or specific individuals and that things must be set right. Some catalyst, however, is needed to mobilize the discontent people feel.

In theory, social movements can be classified according to type; in practice, a given movement may possess a complex ideology that places parts of it in several different classifications.

SUMMARY

Collective behavior refers to relatively spontaneous social actions that occur when people respond to unstructured and ambiguous situations. A number of theories have been devised to explain collective behavior. Contagion theory argues that individuals are transformed by the experience of anonymity in a crowd. They acquire a crowd mentality, lose their inhibitions and sense of personal moral responsibility, and become highly susceptible to group sentiments.

Convergence theory views collective behavior as the outcome of situations in which people with similar characteristics are drawn together. Because they

INTERNET RESOURCES

Go to http://web-enhanced.thomsonlearning.com which features Web links relevant to this chapter to expand and enhance the learning experience. This site will be monitored and updated periodically to ensure the links are correct and live.

At Virtual Society: The Wadsworth Sociology Resource Center, http://sociology.wadsworth.com, you will find a Career Center, Sociology in the News, Research Online, InfoTrac College Edition, Virtual Tours in Sociology, and a variety of book-specific student and instructor resources.

CHAPTER SEVENTEEN STUDY GUIDE

KEY CONCEPTS

Match each concept with its definition, illustration, or explanation below.

a. Expressive social movement
b. Fragmentation
c. Structural conduciveness
d. Conventional craze
e. Craze
f. Value-added theory
g. Reactionary social movement
h. Crowd
i. Public opinion
j. Revolutionary social movements
k. Demise
l. Contagion theory

m. Acting crowd
n. Fad
o. Institutionalization
p. Social movement
q. Resource mobilization theory
r. Casual crowd
s. Revisionary social movement
t. Convergence theory
u. Threatened crowd
v. Fashion
w. Coalescence
x. Mass hysteria
y. Emergent norm theory

z. Expressive crowd
aa. Structural strain
bb. Mass
cc. Incipiency
dd. Collective behavior
ee. Propaganda
ff. Conservative social movement
gg. Relative deprivation theory
hh. Opinion leaders
ii. Rumor
jj. Panic
kk. Procter & Gamble

_____ 1. Relatively spontaneous social actions that occur when people respond to unstructured and ambiguous situations.

_____ 2. The theory that crowd behavior is caused by a kind of irrational group feeling that spreads among individuals, causing them to lose their inhibitions and become more receptive to group sentiments.

_____ 3. A theory that crowd members develop common standards by observing and listening to one another.

_____ 4. A theory that collective behavior is the result of people with similar characteristics being drawn together.

_____ 5. A theory that collective behavior occurs as a result of six necessary social conditions or processes that build on one another.

_____ 6. A condition in the existing social order that may promote collective behavior.

_____ 7. A condition in which a group's ideals conflict with its everyday realities.

_____ 8. A temporary concentration of people who focus on some thing or event but who also are attuned to one another's behavior.

_____ 9. A group of people whose passions and tempers have been aroused by some focal event, who come to share a purpose, and who feed off one another's arousal, often erupting into spontaneous acts of violence.

_____ 10. A crowd that is in a state of alarm, believing some kind of danger is present.

_____ 11. A group of people drawn together by the promise of personal gratification through active participation in activities and events.

_____ 12. A gathering in which people's behavior conforms to some well-established set of cultural norms, and gratification results from passive appreciation of an event.

_____ 13. Any collection of people who just happen, in the course of their private activities, to be in one place at the same time and focus their attention on a common object or event.

_____ **14.** A collection of people who, although physically dispersed, participate in some event either physically or with a common concern or interest.

_____ **15.** A social change with a very short life span marked by a rapid spread and an abrupt drop in popularity.

_____ **16.** A fad that is especially short-lived.

_____ **17.** The standards of dress or manners in a given society at a certain time.

_____ **18.** Information that is shared informally and spreads quickly through a mass or crowd.

_____ **19.** Beliefs held by a dispersed collectivity of individuals about a common problem, interest, focus, or activity.

_____ **20.** Advertisements of a political nature, seeking to mobilize public support behind one specific party, candidate, or point of view.

_____ **21.** Socially acknowledged experts to whom the public turns for advice.

_____ **22.** A condition in which large numbers of people are overwhelmed with emotion and frenzied activity or become convinced that they have experienced something for which investigators can find no discernible evidence.

_____ **23.** An uncoordinated group flight from a perceived danger.

_____ **24.** A form of collective behavior in which large numbers of people are organized or alerted to support and bring about, or to resist, social change.

_____ **25.** States that social movements occur when large numbers of people experience the feeling that they lack the living or working conditions, political rights, or social dignity to which they are entitled.

_____ **26.** Argues that social movements arise at certain times because skilled leaders know how to mobilize and channel popular discontent.

_____ **27.** Movements that embrace the aims of the past and seek to return the general society to yesterday's values.

_____ **28.** Movements that seek to maintain society's current values by reacting to change or threats of change they believe will undermine the status quo.

_____ **29.** Movements that seek partial or slight changes within the existing order but do not threaten the order itself.

_____ **30.** Movements that seek to overthrow all or nearly all of the existing social order and replace it with an order they consider more suitable.

_____ **31.** Movements that stress personal feelings of satisfaction or well-being and typically arise to fill some void or to distract people from some great dissatisfaction in their lives.

_____ **32.** The beginning stage of a social movement.

_____ **33.** The stage of a social movement when groups form around leaders, begin to promote policies, and promulgate programs.

_____ **34.** The stage of a social movement when it becomes firmly established through formal organizations.

_____ **35.** A company plagued by rumors that it is affiliated with the Church of Satan.

_____ **36.** The state of a social movement when the movement begins to fall apart.

_____ **37.** Refers to the end of a social movement.

KEY THINKERS/RESEARCHERS

Match the thinkers/researchers with their main ideas or contributions.

a. Saul Alinsky
b. Elias Canetti
c. Alfred A. Kroeber
d. Irving Janis
e. Ralph Turner

f. Neil Smelser
g. Armand L. Mauss
h. Kai Erikson
i. Herbert Blumer
j. Gustave LeBon

k. Georg Simmel
l. Samuel Stouffer
m. William Bruce Cameron

_____ 1. French sociologist who was a pioneer in the study of collective behavior and the theorist who developed the contagion explanation of crowd behaviors.

_____ 2. One of the first sociologists to develop the emergent norm theory of collective behavior.

_____ 3. Developed the value-added theory of collective behavior.

_____ 4. Described the importance of traits of crowds.

_____ 5. Showed that fashion moves through predictable cycles, which are correlated with degrees of political and social stability.

_____ 6. Argued that changes in fashion are adopted by the upper class as a way of keeping themselves distinct from the lower classes.

_____ 7. Described the conditions under which people collectively panic.

_____ 8. Explained the seventeenth-century Salem witchcraft trials as an episode of mass hysteria created by the Massachusetts Bay Colony's identity crisis.

_____ 9. First described the concept of relative deprivation.

_____ 10. A well-known activist leader who was particularly good at mobilizing community resources into social movements.

_____ 11. Classified social movements into four fundamental types.

_____ 12. Suggested that social movements typically pass though a series of stages that are the equivalent of a biological life cycle.

_____ 13. A major figure in the study of social movements. Known for discussions of contagion theory, classifications of crowd types, and the first discussions of expressive social movements.

CENTRAL IDEA COMPLETIONS

Following the instructions, fill in the appropriate concepts and descriptions for each of the questions posed in the following section.

1. What does your text mean when it says, "there is a flaw in the theory of relative deprivation"?

_____

_____

2. Differentiate between fads and fashions, providing specific examples of each:

   a. Fad: _____

   _____

   b. Fashion: _____

   _____

3. What are the seven ways Wayne Rash Jr. maintains that the Internet can further collective behavior?

a. _____

b. _____

c. _____

d. _____

e. _____

f. _____

g. _____

4. What are the basic ideas of Le Bon's contagion theory? Why do you think it has such wide appeal to the nonsociological audiences? What are the problems with this theory?

a. Contagion theory: _____

_____

b. Theory's appeal: _____

_____

c. Theory's limitation: _____

_____

5. Present the key elements of relative deprivation theory along with an example for which this theory might provide an explanation:

a. Theory: _____

_____

b. Example: _____

6. Explain the idea of the life cycle of a social movement:

_____

_____

_____

7. List and briefly identify the four major theories of collective behavior:

a. _____

_____

b. _____

_____

c. _____

_____

d. _____

_____

8. State and provide an example of each of the four attributes of crowds:

a. _____

Example: _____

b. _____

Example: _____

   c. _____

     Example: _____

   d. _____

     Example: _____

9. Define and illustrate by example each of the following types of crowds: acting, threatening, expressive, conventional, and casual:

   a. Acting: _____

   b. Threatening: _____

   c. Expressive: _____

   d. Conventional: _____

   e. Casual: _____

10. In what ways is a craze different from a fad? Provide an example of a recent craze that has affected the student body at your college or university:

   _____

   _____

11. Your book begins this chapter by describing how genital snatchers affected men in Lagos, Nigeria. How was this phenomenon related to collective behavior? _____

   _____

   _____

12. What is the last stage of a social movement? What are the possible outcomes at this last stage?

   _____

   _____

   _____

CRITICAL THOUGHT EXERCISES

1. Your text raises the question, "Is Vegetarianism a Social Movement?" Develop an argument, based specifically on the theoretical material presented in this chapter, in which you answer this question. What do you believe will be the status of vegetarianism (relative to social movements) by the year 2025?

2. Select a contemporary social movement, for example, Green Peace. Using your academic library as well as information from the World Wide Web, follow the material presented in your text and describe the history and development of Green Peace. To what extent does it fit the text's criteria of a social movement?

3. Select an issue in your college or university community paper. Following the discussions of Alinsky in your text, develop a plan of action to "sway" the community's constituents, or the student body, toward one outcome. Who would you need to con-

tact first in order to bring them over to your side? What will be the role of propaganda in your campaign? Are there any normative limits at your institution regarding what can and cannot be said during a community issues debate?

4. You text discusses social movements on the Internet. Develop a paper in which you examine (1) the nature of these movements, (2) the extent to which they are similar to and distinctive not only from other forms of social movements, but also among themselves on the Net. Finally, (3) select one social movement on the Internet for a descriptive analysis. How does it conform to the discussion by Tischler? To what extent, if any, does it represent a threat to any segments of American society? If you were the "director of the Internet," what would you recommend as a Net-management response to this social movement?

ANSWERS

KEY CONCEPTS

1. dd	8. h	14. bb	20. ee	26. q	32. cc
2. l	9. m	15. n	21. hh	27. g	33. w
3. y	10. u	16. e	22. x	28. ff	34. o
4. t	11. z	17. v	23. jj	29. s	35. kk
5. f	12. d	18. ii	24. p	30. j	36. b
6. c	13. r	19. i	25. gg	31. a	37. k
7. aa					

KEY THINKERS/RESEARCHERS

1. j	4. b	6. k	8. h	10. a	12. g
2. e	5. c	7. d	9. l	11. m	13. i
3. f					

LEARNING OBJECTIVES

After studying this chapter, you should be able to do the following:

- Describe the source of social change in society.
- Contrast differing ideologies and show how they influence social change.
- Explain the processes of cultural diffusion and forced acculturation.
- Understand the various theories of social change.
- Describe the process of modernization.
- Understand the impact of technological innovation.
- Summarize the changes that will occur in the U.S. labor force by the year 2005.

Social change is a common fact of life. While many people embrace it, equal numbers fear it. As common as social change may be, rebelling against it is also common. In 1811, the Luddites, English textile weavers, who had been displaced by the invention of the automatic loom, smashed those machines in a revolt against the Industrial Revolution. Many written works that we hold in high esteem were really expressions of distress over the way science was going to change the world for the worse. Henry David Thoreau's book *Walden* is essentially about one man's rebellion against the complicated world coming about because of social change. Mary Shelley's novel *Frankenstein,* published more than 150 years ago, is not really just about a monster created in the laboratory, but about a world where scientists can use technology to create havoc. Stanley Kubrick's classic film *2001* warns us of the same danger when we see HAL, the robot computer, going out of control. Ted Kaczynski, the person known as the Unibomber, who sent letter bombs to those he thought were responsible for some of the technological changes in society, was also rebelling in a vicious way against a changing world. All around us we have academics, writers, and journalists who refuse to use computers to write on and insist on using their manual typewriters or even legal pads to produce their works. All these people are saying that social change, while inevitable, should not be embraced wholeheartedly.

Twenty-five centuries ago, the Greek philosopher Heraclitus lectured on the inevitability of change. "You cannot step into the same river twice," he said, "for . . . waters are continually flowing on." Indeed, he observed, "everything gives way and nothing stays fixed" (quoted in Wheelright, 1959). Social and technological change is an important topic for sociological study. Sociologists must be aware of the impressive number of technological changes and the social changes growing out of them. In this chapter, we will try to examine how, why, and in what specific direction societies are likely to change.

Society and Social Change

What, exactly, is social change? The best way to analyze how sociologists define social change is through example. The invention of the steam locomotive was not in itself a social change, but the acceptance of the invention and the spread of railroad transportation were. Martin Luther's indictments of the Catholic Church, which he nailed to the door of Wittenberg Cathedral in 1517, were not in themselves social change, but they helped give rise to one of the major social changes of all time, the Protestant Reformation. Adam Smith's great work *An Inquiry Into the Nature and Causes of the Wealth of Nations* (first published in 1776) was not in itself social change, but it helped initiate a social change that altered the world—the Industrial Revolution. Thus, individual discoveries, actions, or works do not themselves constitute social change, but they may lead to alterations in shared values or patterns of social behavior or even to the reorganization of social relationships and institutions. When this happens, sociologists speak of social change. Hans Gerth and C. Wright Mills (1953) defined social change as "whatever happens in the course of time to the roles, the institutions, or the orders comprising a social structure, their emergence, growth and decline." To put it simply, using terms we defined in Chapter 5, **social change** consists of *any modification in the social organization of a society in any of its social institutions or social roles.*

The invention of the airplane was not in itself social change, but the acceptance of the invention and the spread of air travel were.

Some social changes are violent and dramatic, like the French Revolution of 1789 or the 1917 Russian Revolution. However, not all cases of violent social or collective behavior are instances of social change. Thus, the U.S. race riots of the late 1960s and early 1970s were not in themselves examples of social change. For that matter, not all social change need be violent. For example, the transformation of the family into its modern forms over the past 200 years represents an enormous social change that has profoundly affected both the general nature of society and each person's childhood and adult experiences. Similarly, the rise of computer technology and developments in telecommunications over the past few decades have resulted in the emergence of what sociologist Edward Shils (1971) described as "mass society," in which vast numbers of individuals share collectively in the community and in a common language.

It would be a mistake to think of social change only in terms of the social structure. Morris Ginsberg (1958) observed that "the term *social change* must also include changes in attitudes and beliefs, in so far as they sustain institutions and change with them." In addition, individual motivation always plays a real but immeasurable role in social change.

Sources of Social Change

What causes social change? Sociologists have cited several factors. Those factors, which we shall consider here, are categorized as internal and external sources of change.

Internal Sources of Social Change

Internal sources of social change include *those factors that originate within a specific society and that singly or in combination produce significant alterations in its social organization and structure.* The most important internal sources of social change are technological innovation, ideology, cultural conflicts, and institutionalized structural inequality.

Technological Innovation   Technological change in industrial society is advancing at a dizzying pace, carrying social organizations and institutions along with it. Computer- and communications-based electronic information technology has already transformed American life as we know it in the home, family, workplace, and school.

The Internet has rapidly changed our lives. By 2001, 104 million adults (56% of the adult population) and 30 million children (45% of the under-18 population) in the United States were using the

Internet. On a typical day, 58 million Americans were logging on, 9 million more than the previous year. Women, minorities, and people earning $30,000 to $50,000 were among the population segments that were growing the most rapidly.

Even so, there are significant disparities in online access by income and age. Eighty-two percent of those living in households earning more than $75,000 have Internet access, compared with 38% of those in households earning less than $30,000. Meanwhile, 75% of those from 18 to 29 years old have Internet access, compared with 15% of those 65 and over (Stellin, 2001).

Homes have been transformed into technological workstations in which people both work and live. By receiving and sending information through computer terminals and fax machines, workers have less need to establish a base of work operations in a separate office.

Already, signs of these trends are plentiful. We can use our home computers to transfer funds and pay bills without ever handling money. The computer has even changed crime patterns. Clever thieves have diverted funds using computer codes in banks. In fact, so much about society is changing so quickly as a result of computer technology that scholars are beginning to wonder whether humans have the psychological resilience to adapt to the social changes that must follow.

Technology increases the complexity in our lives. When taken individually, the new devices that surround us seem reasonable and helpful: microwave ovens, voice-mail systems, and the Internet just help us get more done, save time, and communicate with friends and colleagues more easily. But when taken together, and when used not just by us but also by everybody else in our society, these devices sometimes make our lives more complex.

These effects of new technology are not obvious at first. For example, e-mail and cell phones may widen the circle of people each of us can contact: today we can reach out to a far larger number of people than we could even 20 years ago, and a far larger number of people can easily reach us. As a result, we may find ourselves facing an expanding range of obligations and responsibilities, because it is easier for other people to make demands on our time. Voice-mail messages, e-mail letters, and faxes pile up. Scheduling calendars and Palm Pilots are crammed with appointments and meetings. Since we can all now do more, we feel we *must* do more or be left behind by our colleagues, neighbors, and competitors. Thus, the technologies that save us time and labor individually—that empower each of us—increase our obligations and demands (Homer-Dixon, 2000). (See "Society and the Internet: Do We Really Need The Information Highway?")

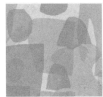

Predicting the Future of Computers—1967

One of the most exciting prospects for the future is that all of us, however lacking in engineering skill, will someday be able to operate a computer as easily as we now operate our automobiles, because the computer itself will show us how. It is only a matter of time . . . before all of us will have our homes hooked up to what the scientists are calling "information public utilities" and will have brain power piped in just as we now have electric power. What the world will then be like staggers the imagination.

Source: "Computers: Their Scope Today," by E. Havemann, October 1967, *Playboy.*

Ideology The term **ideology** most often refers to *a set of interrelated religious or secular beliefs, values, and norms that justify the pursuit of a given set of goals through a given set of means.* Throughout history, ideologies have played a major role in shaping the direction of social change.

Conservative (or traditional) **ideologies** try to *preserve things as they are.* Indeed, conservative ideologies may slow down social changes that technological advances are promoting.

Liberal ideologies seek *limited reforms that do not involve fundamental changes in the social structure of society.* Affirmative-action programs, for example, are intended to redress historical patterns of discrimination that have kept women and minority groups from competing on an equal footing with white males for jobs. Although far-reaching, these liberal ideological programs do not attempt to change the economic system that more radical critics believe is at the heart of job discrimination.

Radical (or revolutionary) ideologies reject liberal reforms as mere tinkerings that simply make the structural inequities of the system more bearable and therefore more likely to be maintained. Like the socialist political movement described in Chapter 13, **radical ideologies** seek *major structural changes in society.* Interestingly, radicals sometimes share the objectives of conservatives in their opposition to liberal reforms that would lessen the severity of a problem, thereby making major structural changes less likely. For example, conservative as well as many radical groups bitterly attacked President Franklin D. Roosevelt's New Deal policies in which federal funds were used to create jobs and bring the country out of the Great Depression. Conservatives attacked

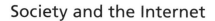

Society and the Internet

Do We Really Need the Information Highway?

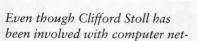

Even though Clifford Stoll has been involved with computer networks since their inception and is known for tracking and catching a German spy ring on the Internet, he remains skeptical about the power of the Internet to change our lives. "While the Internet beckons brightly, seductively flashing as an icon of knowledge-as-power" wrote Stoll, "this non-place lures us to surrender our time on earth."

Just when did the information age begin? Did it begin with the early personal computers in 1980? Was it back in 1964 when Marshall McLuhan first used the term? Were we living in the information age in October of 1929, when the stock market crash radioed around the world in minutes? Did 1848 mark the beginning of the information age when Morse sent the first telegram?

I'll bet we've always lived in the information age. But only recently have technocrats arrogantly proclaimed themselves the high priests of a new order. The Internet delivers a mountain of information, but it sure doesn't make anyone powerful.

Information isn't power. Who's got the most information in your neighborhood? Librarians, and they're famous for having no power at all. Who has the most power in your community? Politicians, of course. And they're notorious for being ill informed.

Information isn't money either. I've never met a panhadler on the street corner, hand outstretched, begging for information. Nor have I seen a corporate executive pounding his desk, complaining that the company's profits are going down because they don't produce enough information. Come to think of it, I haven't met anyone yearning for more junk mail. Many problems confront society, but too little information isn't one of them.

The Information Highway is often promoted as the savior of this country's education system, an invention destined to breathe new life into our library systems, create a meeting place for people with common interests, and enhance diversity and culture. This electronic creation, however, has fallen short of these expectations. Despite the myths that the Internet is an inexpensive, egalitarian method of communication used by millions, computers and on-line accessories remain pricey, and entire segments of the population have yet to log on.

Cyberspace communication tends to focus primarily on accruing information as quickly as possible, thereby shifting our cognitive processes away from more contemplative thought. Information is not knowledge. Instead of encouraging in-depth discussions, the Internet has sparked bursts of intelligence with little coherent direction; instead of strengthening our creative capacities, the recent explosion of networks has nourished drones; instead of fostering literacy, on-line communication gives birth to quick cryptic messages.

The Internet began as a technical community with convivial neighbors who would help each other. Its friendly anarchy promised to revolutionize social interaction and transcend political boundaries. With time, the Internet has evolved into a mass of possibly useful information—a created environment governed by no official monitors. It is as addictive and isolating as it is efficient and ultimately democratic, or so the myth goes. As Henry David Thoreau notes, "Our inventions are wont to be pretty toys, which distract our attention from serious things. They are but improved means to an unimproved end." Just because we have created this ubiquitous global network does not mean that we are any closer to solving our most pressing dilemmas.

Sources: Based on *High Tech Heretic* (pp. 141–142), by C. Stoll, 1999, New York: Doubleday; *Silicon Snake Oil*, by C. Stoll, 1995, New York: Doubleday; and an interview with the author, March 2000.

---

the New Deal as "creeping socialism," and radicals saw it as a desperate (and successful) attempt to save the faltering capitalist system and stave off a socialist revolution.

Reactions to Institutionalized Inequality When groups perceive themselves to be the victims of unjust and unequal societal patterns or laws, they are likely to demand social, economic, political, and cultural reforms. Such pressure for social change exists in U.S. society in a variety of forms. African Americans, Hispanics, and other minorities, for example, have often been the victims of institutionalized inequality. As these groups have asserted their rights, society has been forced to change. For example, federal, state, and local laws have been passed to make it illegal to discriminate against minorities in voting; in access to schools, the labor force, housing; and in other sectors of American life. The labor and civil rights movements, for example, arose because of institutionalized inequalities in American society.

External Sources of Social Change

As we noted in Chapter 3, diffusion, the process of transmitting traits from one culture to another, occurs when groups with different cultures come into contact and exchange items and ideas with one another. Diffusion is thus an example of an **external source of social change,** *changes within a society produced by events emanating from outside that society.*

It does not take the diffusion of many cultural traits to produce profound social changes, as the anthropologist Lauriston Sharp (1952) demonstrated with regard to the introduction of steel axes to the Yir Yoront, a Stone Age tribe inhabiting southeastern Australia. Before European missionaries introduced steel axes to the Yir Yoront, the natives made axes by chipping and grinding stone—a long, laborious process. Axes were very valuable, had religious importance, and also were the status symbol of tribal leaders. Women and young men had to ask permission from a leader to use an ax, which reinforced the patriarchal authority structure. However, anybody could earn a steel ax from the missionaries simply by impressing them as being "deserving." With women and young men thus having direct access to superior tools, the symbols representing status relations between male and female as well as young and old were devalued, and the norms governing those traditional relationships were upset. In

addition, introducing into the tribe valuable tools that did not have religious sanctions governing their use led to a drastic rise in the incidence of theft. In fact, the entire moral order of the Yir Yoront was undermined because their myths explained the origins of all important things in the world, but did not account for the arrival of steel axes. This, as Sharp observed, caused conditions fertile for the introduction of a new religion, a happy circumstance for the missionaries.

Diffusion occurs wherever and whenever different cultures come into contact with one another, though contact is not essential for traits to diffuse from one culture to another. For example, Native American groups below the Arctic smoked tobacco long before the arrival of the Europeans. However, in Alaska the Inuit (Eskimos) knew nothing of its pleasures. European settlers brought tobacco back to Europe, where it immediately became popular and diffused eastward across Central Europe and Eurasia, up into Siberia, and eventually across the Bering Strait to the Inuit.

Today, of course, when so many of the world's peoples increasingly are in contact with one another through all forms of mass communication, cultural traits spread easily from one society to another. The direction of diffusion, however, rarely is random or balanced among societies. In general, traits diffuse from more powerful to weaker peoples, from the more technologically advanced to the less so. *A social*

Cultural diffusion inevitably results when people from one group or society come into contact with another.

change that is imposed by might or conquest on weaker people is called **forced acculturation.**

Why do these internal and external social changes occur? Different theories offer some important insights into the process of social change.

Theories of Social Change

The complexity of social change makes it impossible for a single theory to explain all its ramifications. Because each theory views social change from an entirely different perspective, contradictions are common. For example, functionalist and conflict theories are diametrically opposed, but this does not make one theory right and the other wrong. Rather, the theories are complementary views that must be analyzed together in order for one to understand the total theoretical framework of social change.

Evolutionary Theory

By the middle of the nineteenth century, the concept of **evolution**—*the continuous change from a simpler condition to a more complex state*—was the dominant concern of European scholars in a variety of disciplines. The most influential evolutionary theorist was Charles Darwin, who in his 1859 volume, *On the Origin of Species,* described what he believed to be the biological evolutionary process that moved populations of organisms toward increasing levels of biological complexity.

Darwin's evolutionary theory influenced the work of sociologist Herbert Spencer, who used terms like "survival of the fittest" and "struggle for existence" to explain the superiority of Western cultures over non-Western ones. In Spencer's view, Western cultures had reached higher levels of cultural achievement because they were better adapted to compete for scarce resources and to meet other difficult challenges of life.

Late nineteenth-century and early twentieth-century philosophers continued to be influenced by what has come to be known as social-evolutionary thought. Although using different names, the theories they developed proposed similar stages through which societies progress. Two of the more influential social-evolutionary theorists, Émile Durkheim and Ferdinand Tönnies, were discussed in Chapter 15.

Durkheim argued that evolutionary changes affect the way society is organized, particularly with regard to work. Small, primitive societies whose members share a set of common social characteristics, norms, and values come together in a bond that Durkheim called mechanical solidarity. These people tend to be of the same ethnicity and religion and share similar economic roles. As society grows larger, it develops a more complex division of labor. People play different economic roles, a more complex class structure develops, and members of the society increasingly do not share the same beliefs, values, and norms. However, they still must depend on one another's efforts for all to survive. Durkheim called the new advanced form of cohesion *organic solidarity.*

Tönnies's views of social evolution paralleled those of Durkheim. In his view, societies shift from the intimate, cooperative relationships of small societies (characterized by *Gemeinschaft*) to the specialized impersonal relationships typical of large societies (*Gesellschaft*). Tönnies did not believe that these changes always brought progress (a feeling shared by Durkheim). Rather, he saw social fragmentation, individual isolation, and a general weakening of societal bonds as the direct results of the movement toward individualization and the struggle for power that characterize urban society.

Much of early evolutionary theory has been harshly criticized by contemporary sociologists, who contend that it uses the norms and values of one culture as absolute standards for all cultures. In response to these problems, modern evolutionists propose sequences of evolutionary stages that are much more flexible in allowing for historical variation among societies. Anthropologist Julian H. Steward (1955) proposed that social evolution is "multilineal," by which he meant that the evolution of each society or cultural tradition must be studied independently and must not be forced into broad, arbitrary, "universal" stages. Marshall D. Sahlins and Elman R. Service (1960) distinguished between "general" evolution (the trend toward increasing differentiation) and "specific" evolution (social changes in each specific society that may move either in the direction of greater simplicity or greater complexity).

Sociologists today realize that change occurs in many different ways and does not necessarily follow a specific course. Nor does change necessarily mean progress (Lenski & Lenski, 1982). All evolutionary theories suffer to a greater or lesser degree from an inability to give a convincing answer to the question, why do societies change? One approach that attempts to deal with this question is conflict theory.

Conflict Theory

According to conflict theory, conflicts rooted in the class struggle between unequal groups lead to social change. This change, in turn, creates conditions that lead to new conflicts.

Modern conflict theory is rooted in the writings of Karl Marx, whose theory of society and social conflict was introduced in Chapter 1. In *Das Kapital,* first published in 1867, Marx argued that social-class conflict is the most basic and influential source of all social change. The classes are in conflict because of the unequal allocation of goods and services. Those with money may purchase them; those without cannot. To Marx, it is a division between the exploiting and exploited classes.

Europe's transition from a feudal to a capitalistic society gave Marx the source of his model for social change. "Without conflict no progress: this is the law which civilization has followed to the present day" (Marx, 1959).

Several modern conflict theorists have modified Marx's theories of class conflict in light of recent historical events. Ralf Dahrendorf (1959) sees as too simplistic the view that all social change is the outgrowth of class conflict. He believes that conflict and dissension are present in nearly every part of society. For example, nonsocial-class conflict may involve religious groups, political groups, or even nations. Dahrendorf does accept, however, the basic principle of conflict theory that social conflict and social change are built-in structural features of society.

Conflict theory accounts for some of the major sources of social change within a society: the changing means of production and class conflict. Marx believed that social change within capitalist society would occur through a violent revolution of the workers against the capitalists. However, Marx did not foresee that those who controlled the means of production would tolerate the legalization of unions, collective bargaining, strikes, and integration of the less privileged into legal, reform-oriented parties of the Left. Marx also did not foresee that those who controlled the means of production would accept government regulation of corporations, welfare legislation, civil rights legislation, and other laws aimed at protecting employees and consumers.

Functionalist Theory

Functionalists view society as a **homeostatic system,** that is, *an assemblage of interrelated parts that seeks to achieve and maintain a settled or stable state* (Davis, 1949). A system that maintains a stable state is said to be in equilibrium. Because society is inherently an open system subject to influence from its natural and social environments, complete equilibrium never can be achieved. Rather, functionalists describe society as normally being in a condition of dynamic or near equilibrium, constantly making small adjustments in response to shifts or changes in its internal elements or parts (Homans, 1950).

Probably the best-known spokesperson for functionalist theory in America was Talcott Parsons (1951, 1954, 1966), who saw society as a "homeostatic action system" that seeks to integrate its elements and whose patterns of actions are maintained

by its culture. According to Parsons, it is the role of society to fulfill six basic needs: (1) member replacement, (2) member socialization, (3) production of goods and services, (4) preservation of internal order, (5) provision and maintenance of a sense of purpose, and (6) protection from external attack. These needs are in a constant state of equilibrium with one another, and when one changes, the others must accommodate. For example, when industrialization shifted the burden of socializing young people from the family to the school, schools enlarged their educational function to include the education of the whole child. Parent-teacher associations were established, and guidance counselors were hired to coordinate the function of the school with those of the family and other institutions. Thus, when the family became more specialized, the schools stepped in to fill the vacuum. In this case, as in all others, argued Parsons (1951, 1971), change promotes adaptation, equilibrium, and eventual social stability.

As functionalist theory developed, it began to trace the cause of social change to people's dissatisfactions with social conditions that personally affect them. Consider the area of medicine. Technological advances in medical science have made the practice of general medicine all but impossible and encouraged the development of medical subspecialties. Patients who were forced to see a different specialist for each of their health-care needs quickly became dissatisfied, even though the technical ability of each subspecialist was greater than that of the general practitioner. Responding to this dissatisfaction and to patients' conviction that their all-around health care was suffering as a result of the system of medical subspecializations, the medical profession created the "new" specialty of family medicine. Thus, the needs met by the old family doctor are once again being addressed by the new family medicine specialist.

Another example of the functionalist view of social change is the American civil rights movement of the 1960s, which gained strength outside the normal channels of political action. Hundreds of thousands of individuals joined in street protests, and many thousands were arrested. Occurring against the backdrop of race riots in inner-city neighborhoods, the civil rights movement threatened the society with widespread rebellion and chaos. In response, political and social leaders launched a series of important adjustments that functioned to reform the society's institutional structure. Those structural changes included the passage of the Civil Rights Act of 1964, which was intended to eliminate discrimination in public accommodations and in the labor force. Affirmative-action programs were established to integrate the labor force and provide equal access to higher education. The Voting Rights Act of 1965 was passed to attack discrimination in voting, in particular in the registration procedures in the South. In 1968, Congress passed the first laws to attack discrimination in housing. Institutions that did not comply with these new federal rulings faced the possible withdrawal of all federal funds. Hence, though American society was pushed toward disequilibrium in the 1960s and early 1970s, greater equilibrium was reestablished through selected institutional adjustments that diminished organized expressions of discontent.

Functionalist theory successfully explains moderate degrees of social change, such as the adjustments that diffused the civil rights movement in America. The concepts of equilibrium and homeostasis are not very helpful, however, in explaining major structural changes (Bertalanffy, 1968). This criticism was summed up by Mohammed Gnessous in 1967:

> [A]n equilibrium theory like that of Parsons can neither explain the occurrence of radical changes in society nor account for the phenomena that accompany them; it says nothing about what happens when a social system is in disequilibrium . . . it is tied to the image of a society whose historical development holds no surprises.

William F. Ogburn's (1964) concept of *cultural lag,* discussed in Chapter 3, attempts to deal with those criticisms and explain social change in functionalist terms. Although all elements of a society are interrelated, Ogburn asserted, some elements may change rapidly and others lag behind. According to Ogburn, technological change typically is faster than is change in the nonmaterial culture—that is, the beliefs, norms, and values that regulate people's day-to-day lives in friendship and kinship groups and in religion. Therefore, he argued, technological change often results in cultural lag. New patterns of behavior may emerge, even though they conflict with traditional values. When the birth control pill was developed, for example (a product of our material culture), orthodox religious norms forbade its use. Catholic women who wanted to limit their family size thus were on the horns of a dilemma. If they took the pill, they would violate the dictum of the church. If they did not, they would face additional pregnancies and the concomitant economic and family stress. Thus, even though Ogburn adopts a functionalist approach to social change, his theories incorporate the idea that stresses and strains, or "lack of fit," among the parts of the social order are inevitable.

Popular during the 1940s and 1950s, the functionalist theory of social change has been criticized widely in recent times. Aside from the criticisms that we have mentioned, critics argue further that functionalism is a conservative theory that overestimates the amount of consensus in society and underestimates the effects of social conflict.

Cyclical (Rise and Fall) Theory

Inherent in cyclical theories of social change is the assumption that the rise and fall of civilizations is inevitable and the notion that social change may not be for the good. Shocked by the devastation of World War I, people began to see social progress as the decline of society rather than as its enhancement. Those feelings were crystallized in the works of Oswald Spengler, Arnold Toynbee, and Pitirim A. Sorokin.

In his controversial work *The Decline of the West* (1932), German historian Oswald Spengler theorized that every society moves through four stages of development: childhood, youth, mature adulthood, and old age. Spengler felt that Western society had reached the "golden age" of maturity during the Enlightenment of the eighteenth century, and since then had begun the inevitable crumbling and decline that go along with old age. Nothing, he believed, could stop this process. Just as the great civilizations of Babylon, Egypt, Greece, and Rome had declined and died, so too would the West.

British historian Arnold Toynbee (1946) theorized that the rise and fall of civilizations were explicable through the interrelated concepts of societal challenge and response. Every society, he observed, faces both natural and social challenges from its environments. Are its natural resources plentiful or limited? Are its boundaries easy or difficult to defend? Are important trade routes readily accessible or difficult to reach? Are its neighbors warlike or peaceful? When a society is able to fashion adequate responses to those challenges, it survives and grows. When it cannot, it falls into a spiral of decline. According to Toynbee, as each challenge is met, new challenges arise, placing the society in a constant give-and-take interaction with its environments.

Pitirim A. Sorokin (1889–1968) theorized that cultures are divided into two groups: **ideational cultures,** which *emphasize spiritual values;* and **sensate cultures,** which *are based on what is immediately apparent through the senses.* In an ideational culture, progress is achieved through self-control and adherence to a strong moral code. In a sensate culture, people are dedicated to self-expression and the gratification of their immediate physical needs.

Sorokin believed societies are constantly moving between the two extremes of sensate and ideational cultures. The main reason for this back-and-forth movement is that neither sensate nor ideational culture provides the basis for a perfect society. As one culture begins to deteriorate, its weaknesses and excesses become apparent, and there is a movement in the opposite direction. Occasionally, however, a

Oswald Spengler believed that the disappearance of the Egyptian civilization was part of the inevitable crumbling and decline that is part of the life cycle of any society.

culture may reach an intermediate place between these extremes. Sorokin called this the **idealistic point,** at which *sensate and ideational values coexist in a harmonious mix.*

Although cyclical theories offer an interesting perspective on social change, they assume that social change cannot be truly controlled by those who experience it. There is a supposedly inevitable cycle that all societies follow. The actions of people are all part of an elaborate, predetermined cyclical progression of events that has a life of its own. This may well be true of the ways in which many individuals experience social changes. However, such changes are clearly rooted in concrete decisions made by many individuals. The study of modernization allows us to examine the dominant form of social change in the world today and reveals both social and intrapersonal dynamics at work.

Modernization: Global Social Change

Modernization refers to *a complex set of changes that take place as a traditional society becomes an industrial one.* Modernization as we know it is a phenomenon that first began with the Industrial Revolution some two centuries ago. Whereas the modernization of Western society evolved steadily over that time, the modernization of the Third World developing nations has proceeded at a much more rapid pace.

Modernization: An Overview

As modernization progresses from the first stages onward, many changes occur. In the first stages of modernization, farmers move beyond subsistence farming to produce surplus food, which they sell in the market for money instead of bartering for goods and services. In addition, a few limited cash crops and natural resources are exploited, bringing a steady flow of money into the economy. Simple tools and traditional crafts are replaced by industrialized technology and applied scientific knowledge. Whenever possible, human physical power is replaced by machines.

Work becomes increasingly specialized. New jobs—often requiring special training—are created, and people work for wages rather than living from the products of their labor. The economic system is freed from the traditional restraints and obligations rooted in kinship relations, and money becomes the medium of exchange. Educational institutions become differentiated from family life, and the population becomes increasingly literate. Cities rise as industrial and commercial centers and attract migrants from rural areas. Thanks to modern medicine, the death rate falls, but the birthrate stays the same (at least in the early stages of modernization), creating overcrowding.

Modernization reduces the role of the family in the socialization of young children. Nuclear families are cut off from extended kinship networks, and many traditional constraints on behavior, such as notions of family pride and religious beliefs, lose their potency. Frequently, social equality between men and women increases as new social statuses and roles allow for changes in institutionalized behavior. Wealth is unequally distributed between the upper and lower classes (Dalton, 1971; Moore, 1965; Smelser, 1971). As you may have noted, many of these developments are separate facets of the overall pattern of increasing differentiation, which is a key trait of modernizing societies. (For a discussion of how rational systems serve to deny human reason, see "Sociology at Work: The McDonaldization of Society.")

Modernization in the Third World

Whereas modernization was indigenous to most of Europe, it was forced on Third World nations by conquering armies, missionaries, plantation managers, colonial administrators, colonist groups, and industrial enterprises. Colonial administrators did not hesitate to destroy existing political structures whenever they seemed to endanger their rule. Missionaries used the threat of military force, bribery, and even good deeds (such as the construction of hospitals and schools) to draw people away from their traditions. They stamped out practices they did

Sociology at Work

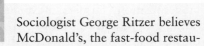

The McDonaldization of Society

Sociologist George Ritzer believes McDonald's, the fast-food restaurant chain, is undoubtedly one of the most influential creations of twentieth-century America. This is true not just as a fast-food restaurant, but also as a model for many other institutions, and for its principles that have been widely adopted throughout American society and much of the rest of the world.

George Ritzer uses the term *McDonaldization* to describe the process by which the principles of the fast-food restaurant are coming to dominate more and more sectors of American society, as well as the rest of the world. McDonaldization is not only affecting the restaurant business, but also education, work, travel, leisure-time activities, dieting, politics, religion, the family, and virtually every other sector of society. Ritzer believes McDonaldization has shown every sign of being an irresistible process as it sweeps through institutions and parts of the world that one would have thought impervious to it.

McDonald's has achieved its exalted position in the world because virtually all Americans, and many people from other countries, have passed through its golden arches. Furthermore, Ritzer has noted, we have been bombarded by commercials extolling McDonald's virtues. Thus, in a survey of schoolchildren, 96% were able to identify Ronald McDonald, making him second only to Santa Claus in name recognition.

Ritzer has noted that the McDonald's model has succeeded because it offers the consumer efficiency and predictability, and because it seems to offer the diner a good value. It also has flourished because it

has been able to exert control, through nonhuman technologies, over both employees and customers, getting them to behave the way the organization wishes them to. Thus, there are good, solid reasons why the process of McDonaldization continues unabated.

Ritzer has noted that the McDonaldization process as it permeates many areas of modern society also has a downside. We can think of efficiency, predictability, quantification, and control through nonhuman technology as the basic components of a rational system. However, rational systems often spawn irrationalities—that is, rational systems serve to deny human reason.

The fast-food restaurant, Ritzer has pointed out, is often a dehumanizing setting in which to eat and work. People lining up for a burger or waiting in the drive-through line often feel as if they are dining on an assembly line. Assembly lines are hardly human settings in which to eat, and they have been shown to be inhuman settings in which to work.

Ritzer believes we should question the headlong rush toward McDonaldization in our society and throughout the world. True, there are great gains to be made from McDonaldization. But there are great costs and enormous risks also, as Ritzer has pointed out. Ultimately, we must ask whether, in creating rationalized systems, we are not also creating an even greater number of irrationalities. At the minimum, Ritzer would like us to be aware of the costs associated with McDonaldization.

Source: *The McDonaldization of Society,* by G. Ritzer, 1992, Newbury Park, CA: Pine Forge Press.

---

not approve of (such as polygamy) and arbitrarily imposed European customs. Occasionally these missionary activities had comic consequences, as when women, unaccustomed to covering their breasts, simply cut holes in the fronts of their missionary-issued T-shirts. Other results were less humorous: People in tropical climates suffered skin infections after missionaries persuaded them to dress in Western clothes but neglected to introduce soap.

Until recently, modernization and Westernization were thought of as more or less the same thing. A developing country that wanted to adopt Western technology had to accept its cultural elements at the same time. However, as Third World nations have gained some measure of economic control over their resources, many have asserted political independence and insisted that modernization be guided by

their own traditional values. One need only think of the oil-rich Islamic countries to realize the different directions that the modernization of developing countries is now taking.

The goals and methods of modernization in the Third World vary widely from region to region and even from one nation to another. Nevertheless, because of the extreme abruptness and pervasiveness of the social changes created by modernization, certain common problems confront many developing nations.

Modernization and the Individual

Modernization has given people in developed countries improved health, increased longevity, more leisure time, and more affluence. Poverty, malnutrition, and disease, which were problems of Western nations as recently as 1890, have been eliminated or reduced dramatically for the bulk of the population. Life expectancy at birth increased from 47.6 years in 1900 to 77 years today, and the workweek has been reduced from about 62 hours in 1890 to about 37 hours today. Indeed, modernization has given many in Western society the luxury of turning their attention to the problems of affluence, including anxiety, obesity, degenerative diseases, divorce, high taxes, inflation, and pollution.

The positive psychological effects of modernization were demonstrated by Alex Inkeles and David H. Smith (1974), who interviewed factory workers in Argentina, Bangladesh, Chile, India, Israel, and Nigeria. These researchers found that attending school and going to work in a factory had been a valued, liberating experience for many of these workers. They had improved their standards of living, overcome their fears of new things and of foreign people, become more flexible about trying new ways of doing things, and adopted a more positive and action-oriented attitude toward their own lives.

Despite these benefits, modernization is not without its costs. Max Weber, who valued modernization as a means for making society more rational and efficient, nevertheless was painfully aware of its emotional costs, of its damaging impact on the spirit of the individual:

> Already now, throughout private enterprise in wholesale manufacture, as well as in all other economic enterprises run on modern lines, . . . rational calculation . . . is manifest at every stage. By it, the performance of each individual worker is mathematically measured, each man becomes a little cog in the machine and, aware of this, his one preoccupation is whether he can become a bigger cog. (Weber, 1956)

Anthropologists and others have documented the severe psychological dislocation suffered by many peoples around the world as a result of modernization. The collapse of traditional cultures under the pressures of modernization has left individuals emotionally adrift in a world they do not understand and cannot control. Probably the most horrifying account is Colin Turnbull's (1972) study of the *Ik*, a hunting and food-gathering people of Uganda who were relocated and forced to become farmers. Within 5 years their society, including its basic family unit, had disintegrated. Unable to feed themselves or their families in their traditional way, individuals starved, became demoralized, and lost their ability to empathize with one another.

Thus, it is clear that modernization has a profound psychological effect on people's lives. The *degree* of personal stress and dislocation that individuals experience as their society modernizes depends on many things, including the historical traditions of the culture, the conditions under which modernization is introduced, and the degree to which the masses are allowed to share in the material benefits of the change.

Social Change in the United States

There is virtually no area of life in the United States that has not changed in some respect since the relatively simple days of the 1950s. In addition, the pace of change will quicken even more as the turn of the century has passed. The following are a few of the major forces that are shaping future life in the United States.

Technological Change

In the past two decades, the personal computer has transformed the workplace and our lives; videocassette recorders, which are now present in most households, and compact discs have added dramatically to home entertainment activities; and biotechnology has helped us discover genetically engineered vaccines and a host of other benefits to society. What could possibly top what has already happened and the changes that have been produced?

The answer is that the technological innovations about to take place will be part of a new era that one author (Bylinsky, 1988) has called *the age of insight*, in which advances will help us understand how things work and how to make them work better. The immediate future will produce not just more and more data but also some startling discoveries. The computer is being transformed from a number cruncher into a machine for insight and discovery.

The age of insight will help us understand the workings of the human body in a way never before attainable. If we can decipher the body's own healing substances and the underlying causes of disease, researchers can develop new drugs and novel methods of treatment. Researchers will increasingly tap the body itself as a new source of medications that genetic engineers can copy and improve on. New insights into human diseases should essentially make it possible to prevent such autoimmune conditions as rheumatoid arthritis, multiple sclerosis, and insulin-dependent diabetes, in which the body mistakenly attacks its own tissue.

Many changes already have taken place in the area of telecommunications. We are entering a world linked by vast computerized networks that process voice, data, and video with equal ease. Desktop computers now have the power of what were once known as supercomputers.

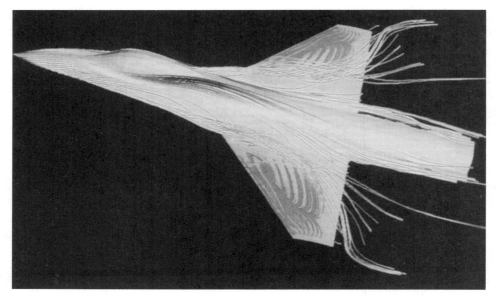

Many believe we are on the brink of the "age of insight," a period in which advances will help us understand how things work and how to make them work better. This is a computer image of the airflow around a vertical-takeoff-and-landing jet aircraft. This sophisticated technology makes it possible to design high-speed aircraft.

Even the individual scientist can now conduct research that was impossible just a short time ago. With a personal computer, compact computer disks that hold enormous amounts of information, and the Internet, huge databases can be tapped, making hard-to-find information available in minutes.

Computers themselves have already advanced to the point where they enable scientists to "see" objects on a smaller scale than microscopes can. This should make a vital contribution to chemistry, chemical engineering, molecular biology, and other fields. Computers also will be able to describe complex events, such as the chemistry involved in photosynthesis, in greater detail than is possible with today's instruments.

Visual computing in effect reproduces the world around us mathematically within a computer. The objects can then be both seen and manipulated in all sorts of ways. A researcher can then describe the nature and behavior of an object or phenomenon with equations and present it visually. The objects simulated can include anything from the wing of an airplane to the interior of the sun. This kind of computing will create new areas of scientific inquiry and the development of new consumer products.

When sociologists examine technological change, they often refer to **technological determinism,** *the view that technological change has an important effect on a society and impacts its culture, social structure, and even its history.* For example, the printing press, the automobile, the jet engine, and nuclear weapons have all had a phenomenal social impact.

We must be careful to realize that technology *influences* rather than determines social change. Technological innovation always occurs in the context of other forces—political, economic, or historical—which themselves help shape the technology and its uses. Technological innovation takes hold only when

there is some need for it and social acceptance for it. Technology itself is neutral; people decide whether and how to use it (Rybczynski, 1983).

The Workforce of the Future

Employment opportunities are affected by population trends, and changes in the size and composition of the U.S. population between 1998 and 2008 will influence the demand for goods and services. For example, the population aged 85 and over grew nearly 30% between 1990 and 1998 (U.S. Bureau of the Census, 1999).

Shifts in the age distribution of the population will mean that teenagers will represent a smaller share of the total population and those middle-aged and older will represent a larger share. Those workers age 45 to 64 will grow faster than the labor force of any other age group as the baby-boom generation (born between 1946 and 1964) continues to age. The number of workers 25 to 34 years of age is projected to decline by 2.7 million, reflecting the decrease in births in the late 1960s and early 1970s. (*Monthly Labor Review,* November 1999).

The number of older workers, aged 55 and older, is expected to grow about 15 times as fast as the older workers pool grew between 1979 and 1992. The baby-boom generation is expected to stay in the labor force longer than the previous generation of workers, because they are more highly educated and identify more strongly with their jobs (U.S. Department of Labor, 1994).

The proportion of minorities and immigrants in the workforce will increase. Many more workers will be of African-American, Hispanic, or Asian heritage, reflecting higher birthrates among African-Americans, Hispanics, and substantial numbers of immigrants. Non-Hispanic whites have historically been the largest sector of the workforce, but their share has been dropping and is expected to fall from 78% in 1992 to 71% in 2008.

Even though black workers will increase by 20% in the next 10 years, by 2008 the Hispanic labor force will be larger than the black labor force.

Women will continue to join the labor force in growing numbers. The percentage increase of women in the labor force between 1998 and 2008 will be larger than the percentage increase in the total labor force and will include nearly all age groups. Women's share of the labor force will increase from 46% in 1998 to 48% in 2008 (*Monthly Labor Review,* November 1999).

The workforce of the next decade will be influenced by a new generation of sophisticated information and communication technologies currently being introduced into a wide variety of work situations.

Nowhere is the effect of the computer revolution and reengineering of the workplace more pronounced than in the manufacturing sector. In the 1950s, 33% of all U.S. workers were employed in manufacturing. By 2008, only 12% of the workforce will be engaged in blue-collar manufacturing work. Contrary to popular opinion, the loss of manufacturing jobs is not just due to cheap foreign labor, but more so to automation and technologically based efficiency. At the same time as the number of manufacturing jobs has been declining, manufacturing productivity has been soaring.

Changes have been dramatic in the wholesale and retail sectors also. In many retail outlets, electronic bar codes and scanners have greatly increased the efficiency of cashiers and thereby significantly reduced the number of employees needed (Rifkin, 1995).

Service-producing industries will account for virtually all of the job growth between 1998 and 2008. Certain occupational groups are projected to grow faster than average between now and the year 2008: computer engineers and scientists, home health-care workers, medical assistants, and corrections officers. Other groups—agricultural workers, semiskilled labor, and private household workers—are expected to decline (see Figure 18-1 for the fastest growing occupations) (*Monthly Labor Review,* November 1999).

Occupations that require the highest levels of education and skill make up an increasing proportion of jobs. There will still be many jobs available for those with only a high-school diploma. However, the opportunities for those who do not finish high

Figure 18-1 **Projections for Fastest-Growing Occupations, 1998–2008**

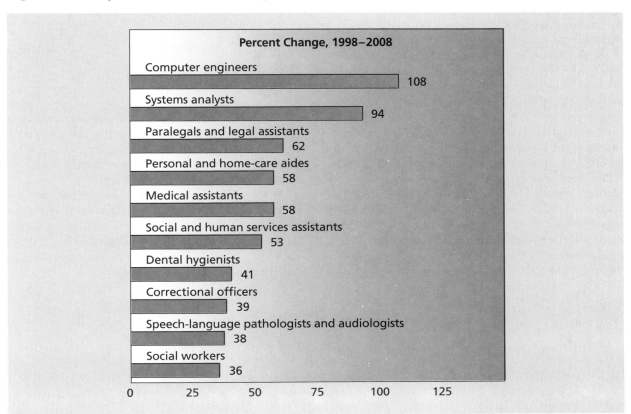

Sources: U.S. Bureau of Labor Statistics, *Monthly Labor Review,* November 1999; *Statistical Abstract of the United States: 2000* (p. 419), U.S. Bureau of the Census, 2001, Washington, DC: U.S. Government Printing Office.

school will be increasingly limited, and people who cannot read may not be considered for most jobs. Those who have not completed high school are likely to have low-paying jobs with little advancement potential.

SUMMARY

Social change consists of any modification in the social organization of a society in any of its social institutions or social roles. It can take place gradually or relatively quickly and it can be accomplished through violent or peaceful means. Changes in technology often result in social changes.

Social change can occur as a result of internal or external forces. Technological innovation is an important source of internal change that has transformed the way we live and work.

Ideology, a set of interrelated religious or secular beliefs, values, and norms that justify the pursuit of a given set of goals through a given set of means, is another important internal source of change.

External sources of social change originate outside of a society. Diffusion, the process of transmitting the traits of one culture to another, is an example of this.

The complexity of social change makes it impossible for a single theory to explain all its ramifications. Thus, a variety of complementary views must be analyzed together in order for one to understand the total theoretical framework of social change.

Evolution—the continuous change from a simpler condition to a more complex state—was the dominant concern of European scholars from a variety of disciplines in the second half of the nineteenth century.

Early evolutionary theory has been severely criticized on the basis that it uses the norms and values of one culture as absolute standards for all cultures. Thus, contemporary theorists have proposed evolutionary stages that are more flexible and allow for more variation among societies. All evolutionary theories have trouble, though, answering the question, why do societies change?

According to conflict theory, conflicts rooted in the class struggle between unequal groups lead to social change. This, in turn, creates conditions that lead to new conflicts. Functionalists view society as a homeostatic system—that is, an assemblage of interrelated parts that seeks to achieve and maintain a settled or stable state. A system that maintains a stable state is said to be in equilibrium, but—because society is an open system subject to many external and internal influences—it is more accurate to talk about a state of dynamic or near equilibrium. Functionalist theory successfully explains moderate degrees of social change, but is not helpful in explaining major structural changes.

Cyclical theories are built around the assumption that the rise and fall of civilizations is inevitable. How the society responds to those challenges determines whether it survives and flourishes or spirals into decline. The problem with cyclical theories is that they erroneously assume that social change is an inevitable process that cannot really be controlled by those who experience it.

There is virtually no area of life in the United States that has not changed in some respect since the 1950s. Moreover, the pace of change is likely to quicken as the new century begins. Technological innovation will continue, but will change qualitatively to produce not just more new and different things, but more insight into how things work and how to make them work better. This change will occur, among other places, in the realms of computers, human biology, and telecommunications.

INTERNET RESOURCES

Go to http://web-enhanced.thomsonlearning.com which features Web links relevant to this chapter to expand and enhance the learning experience. This site will be monitored and updated periodically to ensure the links are correct and live.

At Virtual Society: The Wadsworth Sociology Resource Center, http://sociology.wadsworth.com, you will find a Career Center, Sociology in the News, Research Online, InfoTrac College Edition, Virtual Tours in Sociology, and a variety of book-specific student and instructor resources.

CHAPTER EIGHTEEN STUDY GUIDE

KEY CONCEPTS

Match each concept with its definition, illustration, or explanation below.

a. Forced acculturation
b. Liberal ideologies
c. Sensate cultures
d. Evolution
e. Technological determinism
f. Conservative ideologies

g. Homeostatic system
h. External sources of social change
i. Radical ideologies
j. Modernization
k. Internal sources of social change
l. Ideational cultures

m. Ideology
n. Social change
o. Idealistic point
p. Luddites
q. McDonaldization
r. The age of insight

_____ **1.** Any modification in the social organization of a society in any of its social institutions or social roles.

_____ **2.** Those factors that originate within a specific society and that singly or in combination produce significant alterations in its social organization and structure.

_____ **3.** A set of interrelated religious or secular beliefs, values, and norms that justify the pursuit of a given set of goals through a given set of means.

_____ **4.** Ideologies that try to preserve things as they are.

_____ **5.** Ideologies that seek limited reforms that do not involve fundamental changes in the social structure of society.

_____ **6.** Ideologies that seek major structural changes in society.

_____ **7.** Changes within a society produced by events outside of it.

_____ **8.** A social change that is imposed by might or conquest on weaker peoples.

_____ **9.** The continuous change from a simpler condition to a more complex state.

_____ **10.** An assemblage of interrelated parts that seeks to achieve and maintain a settled or stable state.

_____ **11.** Cultures that emphasize spiritual values.

_____ **12.** Cultures that are based on what is immediately apparent through the senses.

_____ **13.** The point at which sensate and ideational values coexist in a harmonious mix.

_____ **14.** The idea that the rationalized business principles by which McDonald's is run are coming to dominate modern society.

_____ **15.** A period of technological innovations that will help us to understand how things work and how to make them work better.

_____ **16.** A complex set of changes that take place as a traditional society becomes an industrial one.

_____ **17.** The view that technology alone is largely responsible for social change.

_____ **18.** English textile weavers who smashed the machines that had replaced humans.

KEY THINKERS/RESEARCHERS

Match the thinkers/researchers with their main ideas or contributions.

a. Max Weber
b. Julian H. Steward
c. Ralf Dahrendorf
d. Sahlins and Service
e. Herbert Spencer
f. George Ritzer
g. Edward Shils

h. Pitirim Sorokin
i. Colin Turnbull
j. William Ogburn
k. Charles Darwin
l. Karl Marx
m. Lauriston Sharp
n. Arnold Toynbee

o. Alex Inkeles and David Smith
p. Ferdinand Tönnies
q. Talcott Parsons
r. Oswald Spengler
s. Clifford Stoll
t. Hans Gerth and C. Wright Mills
u. Émile Durkheim

_____ 1. Developed the definition of social change.

_____ 2. Developed the concept of mass society.

_____ 3. Described the disruptive effects of the diffusion of steel axes to the Yir Yoront tribe.

_____ 4. The most influential evolutionary theorist, he systematized the concept in his book *On the Origin of Species.*

_____ 5. Brought Darwin's idea of natural selection into sociology and suggested that Western cultures had reached higher levels of cultural development because they were better adapted.

_____ 6. A pioneering evolutionary sociologist, he described the transition from mechanical to organic solidarity.

_____ 7. Described social evolution as a shift from *Gemeinschaft* to *Gesellschaft.*

_____ 8. Proposed the concept of multilineal evolution as a way of avoiding ethnocentrism in the concept.

_____ 9. Distinguished between the general process of differentiation and the specific evolution of each society.

_____ 10. A pioneering conflict theorist, he argued that social change is rooted in class conflict.

_____ 11. Argued that conflict and dissent are present in nearly every part of society.

_____ 12. The best-known functionalist theorist in America, he saw society as a "homeostatic action system."

_____ 13. Developed the concept of cultural lag in an effort to provide a functionalist explanation for disequilibrium.

_____ 14. Theorized that every society moves through a life cycle.

_____ 15. Argued that the rise and fall of civilizations could be explained through the concepts of challenge and response.

_____ 16. Distinguished between ideational and sensate cultures.

_____ 17. Proposed the concept of the "McDonaldization" of society.

_____ 18. Demonstrated the positive psychological effects of modernization through a cross-cultural study.

_____ 19. Felt that the process of modernization through increasing rationality could have a damaging effect on the spirit of individuals.

_____ 20. Argued the Internet fosters sparked bursts of intelligence with little coherent direction.

_____ 21. Studied the cultural and social disintegration of the Ik of Uganda as they experienced forced acculturation.

CENTRAL IDEA COMPLETIONS

Following the instructions, fill in the appropriate concepts and descriptions for each of the questions posed in the following section.

1. It is often argued that cheap foreign labor and overseas production are the major factors causing the decline in the blue-collar portion of the American labor force. Based on what you have learned in this chapter, what are likely to be the *correct* factors affecting the decline in the American blue-collar labor force?

   _____

   _____

   _____

2. Define and provide an example of how each of the following ideologies influences social change:

   a. Radical ideology: _____

      Example: _____

   b. Conservative ideology: _____

      Example: _____

   c. Liberal ideology: _____

      Example: _____

2. What is social change?

   _____

3. What examples of social change have taken place on your campus during the past 5 years?

   _____

   _____

4. Define and present examples of ideational cultures and sensate cultures:

   a. Ideational: _____

      _____

   b. Sensate: _____

      _____

5. Define the concept of a homoeostatic system:

   _____

   _____

6. Define the McDonaldization of society, and present an example of this phenomenon that you have observed (different from text examples) over the past 5 years:

   _____

   _____

   _____

7. Compare and contrast evolutionary theory with conflict theory as explanations for social change:

   a. Evolutionary theory: _____

      _____

b. Conflict theory: _____

_____

8. Present a brief explanation of how the devices of modern technology can actually make our lives more complicated:

_____

_____

9. What is forced acculturation?

_____

_____

How does it differ from cultural diffusion? _____

_____

10. Profile the typical American user of the Internet:

_____

_____

11. What might be a latent function of a modern technological device such as the Palm Pilot?

_____

_____

12. What does Bylinsky argue will be the outcome(s) of the age of insight?

_____

_____

13. Profile the characteristics of the workforce of the future:

_____

_____

14. According to your text, what types of qualities often have been called information? What common definitions of information does your text suggest are incorrect?

_____

_____

CRITICAL THOUGHT EXERCISES

1. Discuss how and why some individuals—for example, Henry David Thoreau, Mary Shelley, Stanley Kubrick, and Ted Kaczynski—each in his or her own way had severe reservations about the ways in which science and technology were changing the world as they knew it. Include your own assessment of the pace of social change stimulated by science.

2. Present an overview of the major points presented by Clifford Stoll in "Society and the Internet: Do We Really Need the Information Highway?" Do you agree with Stoll's discussion of the properties of information? What changes do you envision for the Internet during the next 5 years?

3. After completing this chapter, what changes in the U.S. labor force would you envision by the year 2008? Where do you see yourself relative to the labor force at that time?

4. What do you see as the connection between ideology and social change? To what extent could the Internet be used to transfer ideology with the aid of technology? Log on to the World Wide Web and locate one site you believe to be a positive example of ideology transfer through technology and one site you regard as a negative example of the transfer of ideology through technology. Be sure to explain your selections and conclusions.

ANSWERS

KEY CONCEPTS

1. n	4. f	7. h	10. g	13. o	16. j
2. k	5. b	8. a	11. l	14. q	17. e
3. m	6. i	9. d	12. c	15. r	18. p

KEY THINKERS/RESEARCHERS

1. t	5. e	9. d	13. j	16. h	19. a
2. g	6. u	10. l	14. r	17. f	20. s
3. m	7. p	11. c	15. n	18. o	21. i
4. k	8. b	12. q			

Glossary

Abstract ideals Aspects of a religion that focus on correct ways of thinking and behaving, rather than on a belief in supernatural forces, spirits, or beings.

Achieved statuses Statuses obtained as a result of individual efforts.

Acting crowd A group of people whose passions and tempers have been aroused by some focal event, who come to share a purpose, who feed off one another's arousal, and who often erupt into spontaneous acts of violence.

Adaptation The process by which human beings adjust to the changes in their environment.

Adult socialization The process by which adults learn new statuses and roles. Adult socialization continues throughout the adult years.

Age-specific death rates The annual number of deaths per 1,000 people at specific ages.

Alienation The process by which people lose control over the social institutions they themselves invented.

Analysis The process through which scientific data are organized so that comparisons can be made and conclusions drawn.

Anglo conformity A form of assimilation that involves the renunciation of the ancestral culture in favor of Anglo-American behavior and values.

Animism The belief in animate, personalized spirits or ghosts of ancestors that take an interest in, and actively work to influence, human affairs.

Annihilation The deliberate practice of trying to exterminate a racial or ethnic group; also known as genocide.

Anomie The feeling of some individuals that their culture no longer provides adequate guidelines for behavior; a condition of "normlessness," in which values and norms have little impact.

Aristocracy The rule by a select few; a form of oligarchy.

Ascribed statuses Statuses conferred on an individual at birth or on other occasions by circumstances beyond the individual's control.

Assimilation The process whereby groups with different cultures come to have a common culture.

Associations Purposefully created special-interest groups that have clearly defined goals and official ways of doing things.

Authority Power regarded as legitimate by those over whom it is exercised, who also accept the authority's legitimacy in imposing sanctions or even in using force, if necessary.

Autocracy A political system in which the ultimate authority rests with a single person.

Bilateral system A descent system that traces kinship through both female and male family members.

Bilocal residence Marital residence rules allowing a newly married couple to live with either the husband's or wife's family of origin.

Blind investigator A researcher who does not know whether a specific subject belongs to the group of actual cases being investigated or to a comparison group. This is done to eliminate researcher bias.

Bourgeoisie The term used by Karl Marx to describe the owners of the means of production and distribution in capitalist societies.

Bureaucracy A formal, rationally organized social structure with clearly defined patterns of activity in which, ideally, every series of actions is fundamentally related to the organization's purpose.

Capitalism An economic system based on private ownership of the means of production, in which resource allocation depends largely on market forces.

Caste system A rigid form of social stratification, based on ascribed characteristics, that determines its members' prestige, occupation, residence, and social relationships.

Casual crowd A crowd made up of a collection of people who, in the course of their private activities, happen to be in the same place at the same time.

Charismatic authority The power that derives from a ruler's force of personality. It is the ability to inspire passion and devotion among followers.

City A unit typically incorporated according to the laws of the state within which it is located.

Class system A system of social stratification that contains several social classes and in which greater social mobility is permitted than in a caste or estate system.

Closed society A society in which the various aspects of people's lives are determined at birth and remain fixed.

Coalescence The second stage in the life cycle of a social movement, when groups begin to form around leaders, promote policies, and promulgate programs.

Coercion A form of conflict in which one of the parties in a conflict is much stronger than the others and imposes its will because of that strength.

Cognitive culture The thinking component of culture, consisting of shared beliefs and knowledge of what the world is like—what is real and what is not, what is important and what is trivial. One of the two categories of nonmaterial culture.

Cohabitation Unmarried couples living together out of wedlock.

Collective behavior Relatively spontaneous social actions that occur when people respond to unstructured and ambiguous situations.

Collective conscience A society's system of fundamental beliefs and values.

Command economy An economy in which the government makes all the decisions about production and consumption.

Communism The name commonly given to totalitarian socialist forms of government.

Companionate marriage Marriage based on romantic love.

Competition A form of conflict in which individuals or groups confine their conflict within agreed-upon rules.

Concentric zone model A theory of city development in which the central city is made up of a business district and, radiating from this district, zones of low-income, working-class, middle-class, and upper-class residential units.

Conditioning The molding of behavior through repeated experiences that link a desired reaction with a particular object or event.

Conflict The opposite of cooperation. People in conflict struggle against one another for some commonly prized object or value.

Conflict approach The view that the elite use their power to enact and enforce laws that support their own economic interests and go against the interests of the lower classes.

Conflict theory This label applies to any of a number of theories that assume society is in a constant state of social conflict, with only temporarily stable periods, and that social phenomena are the result of this conflict.

Consensus approach An approach to law that assumes laws are merely a formal version of the norms and values of the people.

Conservative ideologies Ideologies that try to preserve things as they are.

Conservative social movement A social movement that seeks to maintain society's current values.

Consolidated metropolitan statistical area (CMSA) A metropolitan area with a population of 1 million or more that consists of two or more smaller metropolitan areas.

Constitutionalism Government power that is limited by law.

Contagion theory The theory that states members of a crowd acquire a crowd mentality, lose their characteristic inhibitions, and become highly receptive to group sentiments.

Context The conditions under which an action takes place, including the physical setting or place, the social environment, and the other activities surrounding the action.

Conventional crowd A crowd in which people's behavior conforms to some well-established set of cultural norms and in which people's gratification results from passive appreciation of an event.

Convergence theory Views collective behavior as the outcome of situations in which people with similar characteristics, attitudes, and needs are drawn together.

Cooperative interaction A form of social interaction in which people act together to promote common interests or achieve shared goals.

Craze A fad that is especially short-lived.

Crime Behavior that violates a society's legal code.

Criminal justice system Personnel and procedures for arrest, trial, and punishment to deal with violations of the law.

Cross-sectional study An examination of a population at a given point in time.

Crowd A temporary concentration of people who focus on some thing or event but who also are attuned to one another's behavior.

Crude birthrate The annual number of births per 1,000 population.

Crude death rate The annual number of deaths per 1,000 population.

Cult A religious movement that often introduces totally new religious ideas and principles and involves an intense sense of mission. Cults usually have charismatic leaders who expect total commitment from the cult members.

Cultural lag A situation that develops when new patterns of behavior conflict with traditional values. Cultural lag may occur when technological change (material culture) is more rapid than are changes in norms and values (nonmaterial culture).

Cultural relativism The position that social scientists doing cross-cultural research should view and analyze behaviors and customs within the cultural context in which they occur.

Cultural traits Items of a culture such as tools, materials used, beliefs, values, and typical ways of doing things.

Cultural transmission The transmission of major portions of a society's knowledge, norms, values, and perspectives from one generation to the next. Cultural transmission is an intended function of education.

Cultural universals Forms or patterns for resolving the common, basic, human problems that are found in all cultures. Cultural universals include the division of labor, the incest taboo, marriage, the family, rites of passage, and ideology.

Culture All that human beings learn to do, to use, to produce, to know, and to believe as they grow to maturity and live out their lives in the social groups to which they belong.

Culture shock The reaction people may have when encountering cultural traditions different from their own.

De facto segregation Segregation of community or neighborhood schools that results from residential patterns in which minority groups often live in areas of a city where there are few whites or none at all.

De jure segregation Segregation that is an outgrowth of local laws that prohibit one racial group from attending school with another.

Demise The last stage in the life cycle of a social movement, when the movement comes to an end.

Democracy A political system operating under the principles of constitutionalism, representative government, majority rule, civilian rule, and minority rights.

Democratic socialism A political system that exhibits the dominant features of a democracy, but in which the control of the economy is vested in the government to a greater extent than under capitalism.

Demographic transition theory A theory that explains population dynamics in terms of four distinct stages from high fertility and high mortality to relatively low fertility and mortality.

Demography The study of the dynamics of human populations.

Denomination A religious group that tends to draw its membership from a particular "socially acceptable" class, or ethnic group, or at least to have its leadership positions dominated by members of such a group.

Dependency ratio The number of people of non-working age in a society for every 100 people of working age.

Dependent variable A variable that changes in response to changes in the independent variable.

Deviant behavior Behavior that fails to conform to the rules or norms of the group in which it occurs.

Dictatorship A totalitarian government in which all power rests ultimately in one person, who generally heads the only recognized political party.

Diffusion One of the two mechanisms responsible for cultural evolution. Diffusion is the movement of cultural traits from one culture to another.

Discrimination Differential treatment, usually unequal and injurious, accorded to individuals who are assumed to belong to a particular category or group.

Diversion Steering youthful offenders away from the juvenile justice system to nonofficial social agencies.

Double-blind investigator A researcher who does not know either the kind of subject being investigated or the hypothesis being tested.

Dramaturgy The study of the roles people play to create a particular impression on others.

Dyad A small group that contains only two members.

Ecclesia A church that shares the same ethical system as the secular society and that has come to represent and promote the interests of the society at large.

Economy An institution whose primary function is to determine the manner in which society produces, distributes, and consumes goods and services.

Ecumenism The trend among many religions to draw together and project a sense of unity and common direction.

Ego In Freudian theory, one of the three separately functioning parts of the self. The ego tries to mediate in the conflict between the id and the superego and to find socially acceptable ways for the id's drives to be expressed. This part of the self constantly evaluates social realities and looks for ways to adjust to them.

Electorate Those citizens eligible to vote.

Emergent norm theory A theory that notes that even though crowd members may have different motives for participating in collective behavior, they acquire common standards by observing and listening to one another.

Emigration The movement of a population from an area.

Empirical question A question that can be answered by observation and analysis of the world as it is known.

Empiricism The view that generalizations are valid only if they rely on evidence that can be observed directly or verified through our senses.

Endogamy Societal norms that limit the social categories from within which one can choose a marriage partner.

Environmental determinism The belief that the environment dictates cultural patterns.

Estate A segment of a society that has legally established rights and duties.

Estate system A closed system of stratification in which social position is defined by law and membership is based primarily on inheritance. A very limited possibility of upward mobility exists.

Ethnic group A group that has a distinct cultural tradition with which its members identify and which may or may not be recognized by others.

Ethnocentrism The tendency to judge other cultures in terms of one's own customs and values.

Ethnomethodology The study of the sets of rules or guidelines people use in their everyday living practices. This approach provides information about a society's unwritten rules for social behavior.

Ethology The scientific study of animal behavior.

Evolution The continuous change from a simpler condition to a more complex state.

Exchange interaction An interaction involving one person doing something for another with the express purpose of receiving a reward or return.

Exogamy Societal norms that require an individual to marry someone outside his or her culturally defined group.

Experiment An investigation in which the variables being studied are controlled and the researcher obtains the results through precise observation and measurement.

Expressive crowd A crowd that is drawn together by the promise of personal gratification for its members through active participation in activities and events.

Expressive leadership A form of leadership in which a leader works to keep relations among group members harmonious and morale high.

Expressive social movement A social movement that stresses personal feelings of satisfaction or well-being and that typically arises to fill some void or to distract people from some great dissatisfaction in their lives.

Expulsion The process of forcing a group to leave the territory in which it resides.

Extended families Families that include, in addition to nuclear family members, other relatives such as the parents' parents, the parents' siblings, and in-laws.

External means of social control The ways in which others respond to a person's behavior that channel his or her behavior along culturally approved lines.

External source of social change Changes within a society produced by events external to that society.

Exurbs The fast-growing area located in a newer, second ring beyond the old suburbs.

Fad A transitory social change that has a very short life span marked by a rapid spread and an abrupt drop from popularity.

Family of orientation The nuclear family in which one is born and raised. Also known as family of origin.

Family of procreation The family that is created by marriage.

Fascism A political-economic system characterized by totalitarian capitalism.

Fashion A transitory change in the standards of dress or manners in a given society.

Fecundity The physiological ability to have children.

Felonies Offenses punishable by a year or more in a state prison.

Fertility The actual number of births in the population.

Fertility rate The annual number of births per 1,000 women of childbearing age in a population.

Folkways Norms that permit a rather wide degree of individual interpretation as long as certain limits are not overstepped. Folkways change with time and vary from culture to culture.

Forced acculturation The situation that occurs when social change is imposed on weaker peoples by might or conquest.

Forced migration The expulsion of a group of people through direct action.

Formal negative sanctions Actions that express institutionalized disapproval of a person's behavior, such as expulsion, dismissal, or imprisonment. They are usually applied within the context of a society's formal organizations, including schools, corporations, and the legal system.

Formal positive sanctions Actions that express social approval of a person's behavior, such as public gatherings, rituals, or ceremonies.

Formal sanctions Sanctions that are applied in a public ritual, usually under the direct or indirect leadership of social authorities (for example, the award of a prize or the announcement of an expulsion).

Fragmentation The fourth stage in the life cycle of a social movement, when the movement gradually begins to fall apart.

Functionalism (structural functionalism) One of the major sociological perspectives, which assumes that society is a system of highly interrelated parts that operate (function) together rather harmoniously.

Funnel effect The situation in our criminal justice system whereby many crimes are committed but few criminals seem to be punished.

Game stage According to George Herbert Mead, the stage in the development of the self when the child learns that there are rules that specify the proper and correct relationship among the players.

Gemeinschaft A community in which relationships are intimate, cooperative, and personal.

Gender The social, psychological, and cultural attributes of masculinity and femininity that are based on biological distinctions.

Gender identity The view of ourselves resulting from our sex.

Gender-role socialization The lifelong process whereby people learn the values, attitudes, and behavior considered appropriate to each sex by their culture.

Generalized adaptability One of the two forms of adaptation. Generalized adaptation involves developing more complicated yet more flexible ways of doing things.

Generalized others The viewpoints, attitudes, and expectations of society as a whole or of a general community of people that we are aware of and who are important to us.

Genes The set of inherited units of biological material with which each individual is born.

Gentrification A trend that involves young, middle-class people moving to marginal urban areas, upgrading the neighborhood and displacing some of the poor residents, who become priced out of the available housing.

Gesellschaft A society in which relationships are impersonal and independent.

Ghetto A term originally used to refer to the segregated quarter of a city where the Jews in Europe were often forced to live. Today it is used to refer to any kind of segregated living environment.

Group A collection of specific, identifiable people.

Hidden curriculum The social attitudes and values learned in school that prepare children to accept the requirements of adult life and to fit into the social, political, and economic statuses of adult life.

Homeostatic system An assemblage of interrelated parts that seeks to achieve and maintain a settled or stable state.

Homogamy The tendency to choose a spouse with a similar racial, religious, ethnic, educational, age, and socioeconomic background.

Horizontal mobility Movement that involves a change in status with no corresponding change in social class.

Hypothesis A testable statement about the relationship between two or more empirical variables.

I The portion of the self that wishes to have free expression, to be active, and to be spontaneous.

Id In Freudian theory, one of the three separately functioning parts of the self. The id consists of the unconscious drives or instincts that Freud believed every human being inherits.

Idealistic point A term developed by Pitirim A. Sorokin to refer to the situation in which sensate and ideational values coexist in a harmonious mix.

Ideal norms Expectations of what people should do under perfect conditions. The norm that marriage will last "until death do us part" is an ideal norm in American society.

Ideal type A simplified, exaggerated model of reality used to illustrate a concept.

Ideational culture A term developed by Pitirim A. Sorokin to describe a culture in which spiritual concerns have the greatest value.

Ideologies Strongly held beliefs and values to which group members are firmly committed and that cement the social structure.

Ideology A set of interrelated religious or secular beliefs, values, and norms justifying the pursuit of a given set of goals through a given set of means.

Immigration The movement of a population into an area.

Incest The term used to describe sexual relations within families. Most cultures have strict taboos against incest, which is often associated with strong feelings of horror and revulsion.

Incest taboo A societal prohibition that forbids sexual intercourse among closely related individuals.

Incipiency The first stage in the life cycle of a social movement, when large numbers of people perceive a problem without an existing solution.

Independent variable A variable that changes for reasons that have nothing to do with the dependent variable.

Industrial cities Cities established during or after the Industrial Revolution, characterized by large populations that work primarily in industrial or service-related jobs.

Infant mortality rate The number of children who die within the first year of life per 1,000 live births.

Informal negative sanctions Spontaneous displays of disapproval of a person's behavior. Impolite treatment is directed toward the violator of a group norm.

Informal positive sanctions Spontaneous actions such as smiles, pats on the back, handshakes, congratulations, and hugs, through which individuals express their approval of another's behavior.

Informal sanctions Responses by others to an individual's behavior that arise spontaneously with little or no formal leadership.

Innovation Any new practice or tool that becomes widely accepted in a society.

Innovators Individuals who accept the culturally validated goal of success but find deviant ways of reaching it.

Instincts Biologically inherited patterns of complex behavior.

Institutionalization The third stage in the life cycle of a social movement, when the movement reaches its peak of strength and influence and becomes firmly established.

Institutionalized prejudice and discrimination Complex societal arrangements that restrict the life chances and choices of a specifically defined group.

Instrumental leadership A form of leadership in which a leader actively proposes tasks and plans to guide the group toward achieving its goals.

Interactionist perspective An orientation that focuses on how individuals make sense of or interpret the social world in which they participate.

Intergenerational mobility Changes in the social level of a family through two or more generations.

Internal means of social control A morality control in which a group's moral code becomes internalized and becomes part of each individual's personal code of conduct that operates even in the absence of reactions by others.

Internal migration The movement of a population within a nation's boundary lines.

Internal sources of social change Those factors that originate within a specific society and that singly or in combination produce significant alterations in its social organization and structure.

Interview A conversation between an investigator and a subject for the purpose of gathering information.

Intragenerational mobility Social changes during the lifetime of one individual.

Juvenile crime The breaking of criminal laws by individuals under the age of 18.

Labeling theory A theory of deviance that assumes that the social process by which an individual comes to be labeled a deviant contributes to causing more of the deviant behavior.

Laissez-faire capitalism The view of capitalism that believes government should stay out of business.

Latent function One of two types of social functions identified by Robert Merton, referring to the unintended or not readily recognized consequences of a social process.

Laws Formal rules adopted by a society's political authority.

Leader Someone who occupies a central role or position of dominance and influence in a group.

Legal code The formal body of rules adopted by a society's political authority.

Legal-rational authority Authority that derives from the fact that specific individuals have clearly defined rights and duties to uphold and implement rules and procedures.

Liberal ideologies Ideologies that seek limited reforms that do not involve fundamental changes in the structure of society.

Libido One of the two basic instincts of the id. According to Freud, the libido controls the erotic or sexual drive.

Life expectancy The average number of years that a person born in a particular year can expect to live.

Lobbying Attempts by special-interest groups to influence government policy.

Longitudinal research A research approach in which a population is studied at several intervals over a relatively long period of time.

Looking-glass self A theory developed by Charles Horton Cooley to explain how individuals develop a sense of self through interaction with others. The theory has three stages: (1) We imagine how our actions appear to others, (2) we imagine how other people judge these actions, and (3) we make some sort of self-judgment based on the presumed judgments of others.

Magic Interaction with the supernatural. Magic does not involve the worship of a god or gods, but rather an attempt to coerce spirits or control supernatural forces.

Majority rule The right of people to assemble to express their views and seek to persuade others, to engage in political organizing, and to vote for whomever they wish.

Mana A Melanesian/Polynesian concept of the supernatural that refers to a diffuse, nonpersonalized force that acts through anything that lives or moves.

Manifest function One of two types of social functions identified by Robert Merton, referring to an intended and recognized consequence of a social process.

Marital residence rules Rules that govern where a newly married couple settles down and lives.

Marriage The socially recognized, legitimized, and supported union of individuals of opposite sexes.

Mass A collection of people who, although physically dispersed, participate in some event either physically or with a common concern or interest.

Mass hysteria A condition in which large numbers of people are overwhelmed with emotion and frenzied activity or become convinced that they have experienced something for which investigators can find no discernible evidence.

Mass media Methods of communication, including television, radio, magazines, films, and newspapers, that have become some of society's most important agents of socialization.

Master status One of the multiple statuses a person occupies that dominates the others in patterning that person's life.

Material culture All the things human beings make and use, from small hand-held tools to skyscrapers.

Matriarchal family A family in which most family affairs are dominated by women.

Matrilineal system A descent system that traces kinship through the females of the family.

Matrilocal residence Marital residence rules that require a newly married couple to settle down near or within the wife's mother's household.

Me The portion of the self that is made up of those things learned through the socialization process from the family, peers, school, and so on.

Mechanically integrated society A type of society in which members have common goals and values and a deep and personal involvement with the community.

Mechanisms of social control Processes used by all societies and social groups to influence or mold members' behavior to conform to group values and norms.

Medicalization The process by which nonmedical problems become defined and treated as medical problems, usually in terms of illness or disorder.

Megalopolis *See* Consolidated metropolitan statistical area (CMSA).

Metropolitan area An area that has a large population nucleus, together with the adjacent communities that are economically and socially integrated into that nucleus.

Metropolitan statistical area (MSA) An area that has either one or more central cities, each with a population of at least 50,000, or a single urbanized area that has at least 50,000 people and that is part of an MSA with a total population of 100,000.

Middle-range theories Theories concerned with explaining specific issues or aspects of society instead of trying to explain how all of society operates.

Migration The movement of populations from one geographical area to another.

Minority A group of people who, because of physical or cultural characteristics, are singled out from others in the society in which they live for different and unequal treatment and who therefore regard themselves as objects of collective discrimination.

Minority rights The right of the minority to try to change the laws of the majority.

Mixed economy An economy that combines free-enterprise capitalism with government regulation of business, industry, and social-welfare programs.

Modernization The complex set of changes that take place as a traditional society becomes an industrial society.

Monogamous marriage The form of marriage in which each person is allowed only one spouse at a time.

Monotheism The belief in the existence of only one god.

Moral code The symbolic system, made up of a culture's norms and values, in terms of which behavior takes on the quality of being good or bad, right or wrong.

Moral order A society's shared view of right and wrong.

Mores Strongly held norms that usually have a moral connotation and are based on the central values of the culture.

Mortality The frequency of deaths in a population.

Multiple marriage A form of marriage in which an individual may have more than one spouse (polygamy).

Multiple nuclei model A theory of city development that emphasizes that different industries have different land-use and financial requirements, which determine where they establish themselves. As similar industries are established close to one another, the immediate neighborhood is strongly shaped by the nature of its typical industry, becoming one of a number of separate nuclei that together constitute the city.

Negative sanctions Responses by others that discourage the individual from continuing or repeating the behavior.

Neolocal residence Marital residence standards that allow a newly married couple to live virtually anywhere, even thousands of miles from their families of origin.

Nonmaterial culture The totality of knowledge, beliefs, values, and rules for appropriate behavior that specifies how a people should interact and how they may solve their problems.

Normal behavior Behavior that conforms to the rules or norms of the group in which it occurs.

Norms Specific rules of behavior that are agreed upon and shared within a culture to prescribe limits of acceptable behavior.

Nuclear family The most basic family form, made up of parents and their children, biological or adopted.

Nurture The entire socialization experience.

Oligarchy Rule by a few individuals who occupy the highest positions in an organization.

Open-ended interview *See* Semistructured interview.

Open society A society that provides equal opportunity to everyone to compete for the role and status desired, regardless of race, religion, gender, or family history.

Operational definition A definition of an abstract concept in terms of the observable features that describe the things being investigated.

Opinion leaders Socially acknowledged experts to whom the public turns for advice.

Organically integrated society A type of society in which social solidarity depends on the cooperation of individuals in many different positions who perform specialized tasks.

Panic An uncoordinated group flight from a perceived danger.

Pantheon The hierarchy of deities in a religious belief system.

Paradigm A basic model for explaining events that provides a framework for the questions that generate and guide research.

Participant observation A research technique in which the investigator enters into a group's activities while, at the same time, studying the group's behavior.

Patriarchal family A family in which most family affairs are dominated by men.

Patriarchal ideology The belief that men are superior to women and should control all important aspects of society.

Patrilineal system A family system that traces kinship through the males of the family.

Patrilocal residence Marital residence rules that require a newly married couple to settle down near or within the husband's father's household.

Peers Individuals who are social equals.

Personality The patterns of behavior and ways of thinking and feeling that are distinctive for each individual.

Play stage According to George Herbert Mead, the stage in the development of the self when the child has acquired language and begins not only to imitate behavior, but also to formulate role expectation.

Pluralism The development and coexistence of separate racial and ethnic group identities in a society in which no single subgroup dominates.

Political action committees (PACs) Special-interest groups concerned with very specific issues who usually represent corporate, trade, or labor interests.

Political revolutions Relatively rapid transformations of state or government structures that are not accompanied by changes in social structure or stratification.

Politics The process by which power is distributed and decisions are made.

Polyandrous family A polygamous family unit in which the central figure is female and the multiple spouses are male.

Polygamous families Nuclear families linked together by multiple marriage bonds, with one central individual married to several spouses.

Polygynous family A polygamous family unit in which the central person is male and the multiple spouses are female.

Polytheism The belief in a number of gods.

Positive checks Events, described by Thomas Robert Malthus, that limit reproduction either by causing the deaths of individuals before they reach reproductive age or by causing the deaths of large numbers of people, thereby lowering the overall population (for example, famines, wars, and epidemics).

Positive sanctions Responses by others that encourage the individual to continue acting in a certain way.

Power The ability of an individual or group to attain goals, control events, and maintain influence over others—even in the face of opposition.

Prayer A religious ritual that enables individuals to communicate with supernatural beings or forces.

Preindustrial cities Cities established before the Industrial Revolution. Such cities were usually walled for protection, and power was typically shared between feudal lords and religious leaders.

Prejudice An irrationally based negative, or occasionally positive, attitude toward certain groups and their members.

Preparatory stage According to George Herbert Mead, the stage in the development of the self characterized by the child's imitating the behavior of others, which prepares the child for learning social-role expectations.

Prestige The approval and respect an individual or group receives from other members of society.

Preventive checks Practices, described by Thomas Robert Malthus, that limit reproduction (for example, contraception, prostitution, and other "vices").

Primary deviance A term used in labeling theory to refer to the original behavior that leads to the individual being labeled as deviant.

Primary group A group characterized by intimate face-to-face association and cooperation. Primary groups involve interaction among members who have an emotional investment in one another and who interact as total individuals rather than through specialized roles.

Primary metropolitan statistical area (PMSA) A large urbanized county or cluster of counties that is part of a metropolitan statistical area with 1 million people or more.

Primary socialization The process by which children master the basic information and skills required of members of society.

Profane All empirically observable things that are knowable through ordinary everyday experiences.

Proletariat The label used by Karl Marx to describe the mass of people in society who have no other resources to sell than their labor.

Propaganda Advertisements of a political nature seeking to mobilize public support behind one specific party, candidate, or point of view.

Property crime An unlawful act that is committed with the intent of gaining property but that does not involve the use or threat of force against an individual.

Psychoanalytic theory A body of thought developed by Sigmund Freud that rests on two basic hypotheses: (1) Every human act has a psychological cause or basis, and (2) every person has an unconscious mind.

Public opinion The beliefs held by a dispersed collectivity of individuals about a common concern, interest, focus, or activity.

Race A category of people who are defined as similar because of a number of physical characteristics.

Radical ideologies Ideologies that seek major structural changes in society.

Random sample A sample selected purely on the basis of chance.

Reactionary social movement A social movement that embraces the aims of the past and seeks to return the general society to yesterday's values.

Real norms Norms that allow for differences in individual behavior. Real norms specify how people actually behave, not how they should behave under ideal circumstances.

Rebellions Attempts to achieve rapid political change that is not possible within existing institutions.

Rebels Individuals who reject both the goals of what to them is an unfair social order and the institutionalized means of achieving them. They propose alternative societal goals and institutions.

Recidivism Repeated criminal behavior after punishment.

Reference group A group or social category that an individual uses to help define beliefs, attitudes, and values, and to guide behavior.

Reformulation The process in which traits passed from one culture to another are modified to fit better in their new context.

Relative deprivation theory A theory that assumes social movements are the outgrowth of the feeling of relative deprivation among large numbers of people who believe they lack certain things they believe they are entitled to.

Reliability The ability to repeat the findings of a research study.

Religion A system of beliefs, practices, and philosophical values shared by a group of people that defines the sacred, helps explain life, and offers salvation from the problems of human existence.

Religious taboo A sacred prohibition against touching, mentioning, or looking at certain objects, acts, or people.

Replication Repetition of the same research procedure or experiment for the purpose of determining whether earlier results can be duplicated.

Representative government Government in which the authority to govern is achieved through, and legitimized by, popular elections.

Representative sample A sample that has the same distribution of characteristics as the larger population from which it is drawn.

Researcher bias The tendency for researchers to select data that support their hypothesis and to ignore data that appear to contradict it.

Research process A sequence of steps in the design and implementation of a research study, including defining the problem, reviewing previous research, determining the research design, defining the sample and collecting data, analyzing and interpreting the data, and preparing the final research report.

Resocialization An important aspect of adult socialization that involves being exposed to ideas or values that conflict with what was learned in childhood.

Resource mobilization theory A theory that assumes social movements arise at certain times and not at others because some people know how to mobilize and channel popular discontent.

Retreatists Individuals—such as drug addicts, alcoholics, drifters, and panhandlers—who have pulled back from society altogether and who do not pursue culturally legitimate goals.

Revisionary social movement A social movement that seeks partial or slight changes within the existing order but does not threaten the order itself.

Revitalization movements Powerful religious movements that stress a return to the religious values of the past. Those movements spring up when a society is under great stress or attack.

Revolutionary social movement A social movement that seeks to overthrow all or nearly all of the existing social order and replace it with an order it considers to be more suitable.

Revolutions Relatively rapid transformations that produce change in a society's power structure.

Rites of passage Standardized rituals that mark the transition from one stage of life to another.

Ritualists Individuals who de-emphasize or reject the importance of success once they realize they will never achieve it and instead concentrate on following and enforcing rules more precisely than ever was intended.

Rituals Patterns of behavior or practices related to the sacred.

Role conflict The situation in which an individual who is occupying more than one status at the same time is unable to enact the roles of one status without violating those of another.

Roles Culturally defined rules for proper behavior associated with every status.

Role sets The roles attached to a single status.

Role strain The stress that results from conflicting demands within a single role.

Rumor Information that is shared informally and spreads quickly through a mass or a crowd.

Sacred Things that are awe inspiring and knowable only through extraordinary experience. Sacred traits or objects symbolize important values.

Sample The particular subset of a larger population that has been selected for study.

Sampling A research technique in which a manageable number of subjects (a sample) is selected for study from a larger population.

Sampling error The failure to select a representative sample.

Sanctions Rewards and penalties used to regulate an individual's behavior. All external means of control use sanctions.

Sapir-Whorf hypothesis A hypothesis that argues that the language a person uses determines his or her perception of reality.

Science A body of systematically arranged knowledge that shows the operation of general laws. The term also refers to the logical, systematic methods by which that knowledge is obtained.

Scientific method The approach to research that involves observation, experimentation, generalization, and verification.

Secondary analysis The process of making use of research data that has been collected by others.

Secondary deviance A term used in labeling theory to refer to the deviant behavior that emerges as a result of a person having been labeled as deviant.

Secondary group A group that is characterized by an impersonal, formal organization with specific goals. Secondary groups are larger and much less intimate than are primary groups, and the relationships among members are patterned mostly by statuses and roles rather than by personality characteristics.

Sect A small religious group that adheres strictly to religious doctrine involving unconventional beliefs or forms of worship.

Sector model A modified version of the concentric zone model, in which urban groups establish themselves along major transportation arteries around the central business district.

Secularization The process by which religious institutions are confined to ever-narrowing spheres of social influence, while people turn to secular sources for moral guidance in their everyday lives.

Segregation A form of subjugation that refers to the act, process, or state of being set apart.

Selectivity A process that defines some aspects of the world as important and others as unimportant. Selectivity is reflected in the vocabulary and grammar of language.

Self The personal identity of each individual that is separate from his or her social identity.

Semistructured (open-ended) interview An interview in which the investigator asks a list of questions but is free to vary them or make up new ones that become important during the interview.

Sensate culture A term developed by Pitirim A. Sorokin to describe a culture in which people are dedicated to self-expression and the gratification of their immediate physical needs.

Sex The physical and biological differences between men and women.

Sick role A role that legitimizes any deviant behavior caused by an illness and that channels the sick individual into the health-care system.

Significant others Those people who are most important in our development, such as parents, friends, and teachers.

Signs Objects or things that can represent other things because they share some important quality with them. A clenched fist, for example, can be a sign of anger because fists are used in physical arguments.

Small group A relative term that refers to the many kinds of social groups that actually meet together and contain few enough members so that all members know one another.

Social action Anything people are conscious of doing because of other people.

Social aggregate People who happen to be in the same place but share little else.

Social attachments The emotional bonds that infants form with others that are necessary for normal development. Social attachments are a basic need of human beings and all primates.

Social change Any modification in the social organization of a society in any of its social institutions or social roles.

Social class A category of people within a stratification system who share similar economic positions, similar lifestyles, and similar attitudes and behavior.

Social Darwinism The application of Charles Darwin's notion of "survival of the fittest" to society. Darwin believed those species of animals best adapted to the environment survived and prospered, while those poorly adapted died out.

Social evaluation The process of making qualitative judgments on the basis of individual characteristics or behaviors.

Social function A social process that contributes to the ongoing operation or maintenance of society.

Social group A number of people who have a common identity, some feeling of unity, and certain common goals and shared norms.

Social identity The statuses that define an individual. Social identity is determined by how others see us.

Social inequality The uneven distribution of privileges, material rewards, opportunities, power, prestige, and influence among individuals or groups.

Social institutions The ordered social relationships that grow out of the values, norms, statuses, and roles that organize those activities that fulfill society's fundamental needs.

Social interaction The interplay between the actions of one individual and those of one or more other people.

Socialism An economic system under which the government owns and controls the major means of production and distribution. Centralized planning is used to set production and distribution goals.

Socialization The long and complicated processes of social interactions through which a child learns the intellectual, physical, and social skills needed to function as a member of society.

Social mobility The movement of an individual or a group from one social status to another.

Social movement A form of collective behavior in which large numbers of people are organized or alerted to support and bring about, or to resist, social change.

Social organization The web of actual interactions among individuals and groups in society that defines their mutual rights and responsibilities and differs from society to society.

Social revolutions Rapid and basic transformations of a society's state and class structures that are accompanied, and in part carried through, by class-based revolts.

Social sciences All those disciplines that apply scientific methods to the study of human behavior. The social sciences include sociology, cultural anthropology, psychology, economics, history, and political science.

Social solidarity People's commitment and conformity to a society's collective conscience.

Social stratification The division of society into levels, steps, or positions, which is perpetuated by the major institutions of society such as the economy, the family, religion, and education.

Social structure The stable, patterned relationships that exist among social institutions within a society.

Sociobiology An approach that tries to use biological principles to explain the behavior of social beings.

Sociological imagination The relationship between individual experiences and forces in the larger society that shape our actions.

Sociology The scientific study of human society and social interactions.

Specialization One of the two forms of adaptation. Specialization is developing ways of doing things that work extremely well in a particular environment or set of circumstances.

State The institutionalized way of organizing power within territorial limits.

Statement of association A proposition that changes in one thing are related to changes in another, but that one does not necessarily cause the other.

Statement of causality A proposition that one thing brings about, influences, or changes something else.

Statuses The culturally and socially defined positions occupied by individuals throughout their lifetimes.

Status inconsistency Situations in which people rank differently (higher or lower) on certain stratification characteristics than on others.

Status offenses Behavior that is criminal only because the person involved is a minor.

Stratified random sample A technique to ensure that all significant variables are represented in a sample in proportion to their numbers in the larger population.

Structural conduciveness One of sociologist Neil Smelser's six conditions that shape the outcome of collective behavior, structural conduciveness refers to the conditions within society that may promote or encourage collective behavior.

Structural strain One of Smelser's six conditions that shape the outcome of collective behavior, structural strain refers to the tension that develops when a group's ideals conflict with its everyday realities.

Structured interview An interview with a predetermined set of questions that are followed precisely with each subject.

Subculture The distinctive lifestyles, values, norms, and beliefs of certain segments of the population within a society.

Subgroups Splinter groups within a larger group.

Subjugation The subordination of one group and the assumption of a position of authority, power, and domination by another.

Suburbs Those territories that are part of a metropolitan statistical area but are outside the central city.

Superego In Freudian theory, one of the three separately functioning parts of the self. The superego consists of society's norms and values, learned in the course of a person's socialization, that often conflict with the impulses of the id. The superego is the internal censor.

Supernaturalism A belief system that postulates the existence of impersonal forces that can influence human events.

Survey A research method in which a population or a sample is studied in order to reveal specific facts about it.

Symbolic interactionism A theoretical approach that stresses the meanings people place on their own and one another's behavior.

Symbols Objects that represent other things. Unlike signs, symbols need not share any of the qualities of whatever they represent.

Taboo A sacred prohibition against touching, mentioning, or looking at certain objects, acts, or people.

Techniques of neutralization A process that makes it possible to justify illegal or deviant behavior.

Technological determinism The view that technological change has an important effect on a society and has an impact on its culture, social structure, and even history.

Theism A belief in divine beings—gods and goddesses—who shape human affairs.

Theory of differential association A theory of juvenile delinquency based on the position that criminal behavior is learned in the context of intimate groups. People become criminals as a result of associating with others who engage in criminal activities.

Threatened crowd A crowd that is in a state of alarm, believing itself to be in danger.

Total institutions Environments, such as prisons or mental hospitals, in which the participants are physically and socially isolated from the outside world.

Totalitarian capitalism A political-economic system under which the government retains control of the social institutions but allows the means of production and distribution to be owned and managed by private groups and individuals.

Totalitarian government A government in which one group has virtually total control of the nation's social institutions.

Totalitarian socialism A political-economic system under which, in addition to almost total regulation of all social institutions, the government controls and owns all major means of production and distribution.

Totem An ordinary object, such as a plant or animal, that has become a sacred symbol to a particular group, which not only reveres the totem but identifies with it.

Tracking The stratification of students by ability, social class, and various other categories.

Tracks The academic and social levels typically assigned to and followed by the children of different social classes.

Traditional authority Power that is rooted in the assumption that the customs of the past legitimize the present.

Traditional ideology An ideology that tries to preserve things as they are.

Triad A small group containing three members.

Unconscious In Freudian theory, the part of the mind where thoughts and feelings exist that a person is unaware of holding.

Universal church A church that includes all the members of a society within one united moral community.

Urbanization A process whereby a population becomes concentrated in a specific area because of migration patterns.

Urbanized area An area that contains a central city and the continuously built-up and closely settled surrounding territory that together have a population of 50,000 or more.

Urban population The inhabitants of an urbanized area and the inhabitants of incorporated or unincorporated areas with a population of 2,500 or more.

Validity The ability of a research study to test what it was designed to test.

Value-added theory A theory that attempts to explain whether collective behavior will occur and what direction it will take.

Values A culture's general orientation toward life—its notion of what is good and bad, what is desirable and undesirable.

Variable Anything that can change (vary).

Vertical mobility Movement up or down in the social hierarchy that results in a change in social class.

Victimless crimes Acts that violate those laws meant to enforce the moral code.

Violent crime Crime that involves the use of force or the threat of force against the individual.

White-collar crime Crime committed by individuals who, while occupying positions of social responsibility or high prestige, break the law in the course of their work for illegal, personal, or organizational gain.

REFERENCES

Ahlberg, Dennis A., & De Vita, Carol J. (1992, August). "New Realities of the American Family." *Population Bulletin.*

American Association of Retired People. (1994). *Images of Aging in America.* Washington, DC: American Association of Retired People.

Ansolabehere, Stephen, & Iyengar, Shanto. (1995). *Going Negative: How Political Advertisements Shrink and Polarize the Electorate.* New York: The Free Press.

Antonovsky, Aaron. (1972). "Social Class, Life Expectancy and Overall Mortality." In E. Gatly Jaco (Ed.), *Patients, Physicians and Illness* (2nd ed., pp. 5–30). New York: The Free Press.

Aristotle. (1908). *The Politics & Economics of Aristotle.* In Edward English Walford & John Gillies. (Trans.). London: G. Bell & Sons.

Armstrong, Karen. (1994). *A History of God.* New York: Ballantine Books.

Asch, Solomon. (1955). "Opinions and Social Press." *Scientific American, 193,* 31–35.

Axtell, Roger E. (1998). *Gesture: The Do's and Taboos of Body Language Around the World.* New York: John Wiley and Sons.

Baker, Paul J., Anderson, Louis E., & Dorn, Dean S. (1993). *Social Problems: A Critical Thinking Approach* (2nd ed.). Belmont, CA: Wadsworth.

Baldus, David C., Woodworth, George, & Pulaski, Charles A., Jr. (1990). *Equal Justice and the Death Penalty: A Legal and Empirical Study.* Boston: Northeastern University Press.

Bales, R. F. 1958. "Task Roles and Social Roles in Problem-Solving Groups." In E. E. Maccoby, T. M. Newcomb, & E. L. Hartley (Eds.), *Readings in Social Psychology* (3rd ed.). New York: Holt, Rinehart and Winston.

Bandura, Albert. (1969). *Principles of Behavior Modification.* New York: Holt, Rinehart and Winston.

Barnett, Rosalind C., & Baruch, Grace K. (1987). "Social Roles, Gender, and Psychological Distress." In Rosalind C. Barnett, Lois Biener, & Grace K. Baruch (Eds.), *Gender and Stress.* New York: The Free Press.

Barnett, R. C., & Marshall, N. L. (1991). "The Relationship Between Women's Work and Family Roles and Their Subjective Well-Being and Psychological Distress." In M. Frankenhaeuser, V. Lundberg, & M. Chesney (Eds.), *Women, Work, and Health: Stress and Opportunities.* New York: Plenum.

Baron, Salo W. (1976). "European Jewry Before and After Hitler." In Yisrael Gutman & Livia Roch-kirchen (Eds.), *The Catastrophe of European Jewry.* Jerusalem: Yad Veshem.

Bartholet, Elizabeth. (1993). *Family Bonds: The Politics of Adoptive Parenting.* Boston: Houghton Mifflin.

———. (1993, Fall). "Blood Knots: Adoption, Reproduction, and the Politics of the Family." *The American Prospect, 15.*

———. (1991, May). "Where Do Black Children Belong? The Politics of Race Matching in Adoption." *University of Pennsylvania Law Review.*

Bartholomew, Robert E., & Goode, Erich. (2000, May/June). "Highlights From the Past Millennium." *Skeptical Inquirer Magazine, 24*(3).

Baskin, Barbara H., & Harris, Karen. (1980). *Books for the Gifted Child.* New York: R. R. Bowker.

Bateson, Mary Catherine. (1994). *Peripheral Visions.* New York: HarperCollins.

Becker, Howard. (1963). *Outsiders: Studies in the Sociology of Deviance.* New York: The Free Press.

Beckwith, Burnham P. (1986, July/August). "Religion: A Growing or Dying Institution?" *The Futurist, 20*(4), 24–25.

Bedau, Hugo Adam (Ed.). (1997). *The Death Penalty in America: Current Controversies.* New York: Oxford University Press.

Benedict, Jeffrey. (1997). *Public Heroes, Private Felons: Athletes and Crimes Against Women.* Boston: Northeastern University Press.

Benedict, Jeffrey, & Klein, Alan. (1997). "Arrest and Conviction Rates for Athletes Accused of Sexual Assault." *Sociology of Sport Journal, 14,* 86–94.

Benedict, Ruth. (1961/1934). *Patterns of Culture.* Boston: Houghton Mifflin.

———. (1938). "Continuities and Discontinuities in Cultural Conditioning." *Psychiatry, 1,* 161–67.

Berger, Peter. (1967). *The Sacred Canopy.* New York: Doubleday.

———. (1963). *Invitation to Sociology: A Humanistic Perspective.* New York: Doubleday.

Berger, Suzanne E. (1996). *Horizontal Woman.* Boston: Houghton Mifflin.

Bernard, L. L. (1924). *Instinct.* New York: Holt, Rinehart and Winston.

Berry, Brewton, & Tischler, Henry L. (1978). *Race and Ethnic Relations* (4th ed.). Boston: Houghton Mifflin.

Berry, Brian J., & Kasarda, John D. (1977). *Contemporary Urban Ecology.* New York: Macmillan.

Bertalanffy, Ludwig von. (1968). *General System Theory.* New York: George Braziller.

Bianchi, Suzanne. (1991, March). "Family Disruption and Economic Hardship." *U.S. Bureau of the Census,* Series P-70, No. 23.

Bierstadt, Robert. (1974). *The Social Order* (4th ed.). New York: McGraw-Hill.

Billings, John S. (1891, February). "Public Health and Municipal Government." *Annals of the American Academy of Political and Social Science* (Supplement).

Black, Dan, Oates, Gary, Sanders, Seth, & Taylor, Lowell. (2000, May). "Demographics of the Gay and Lesbian Population of the United States: Evidence From Available Systematic Data Sources." *Demography, 37,* 139–54.

Blank, Jonah. (1992). *Arrow of the Blue-Skinned God.* Boston: Houghton Mifflin.

Blankenhorn, David. (1995). *Fatherless America.* New York: Basic Books.

Blau, Peter M. (1964). *Exchange and Power in Social Life.* New York: John Wiley.

Blendon, Robert J., Benson, John M., Brodie, Mollyann, Altman, Drew E., Moran, Richard, Deane, Claudia, & Kjellson, Nina. (1999, Spring). "The 60s and the 90s: Americans' Political, Moral, and Religious Values Then and Now." *Brookings Review,* 14–17.

Blum, John M., Morgan, Edmund S., Rose, Willie Lee, Schlesinger, Arthur M., Jr., Stamp, Kenneth M., & Van Woodard, C. (1981). *The National Experience: A History of the United States* (5th ed.). New York: Harcourt Brace Jovanovich.

Blumer, Herbert. (1946). "Collective Behavior." In Alfred McClung Lee (Ed.), *Principles of Sociology.* New York: Barnes & Noble.

Blumstein, Phillip, & Schwartz, Pepper. (1983). *American Couples.* New York: Morrow.

Borjas, George J. (1999). *Heaven's Door.* Princeton, NJ: Princeton University Press.

Bornstein, Kate. (1994). *Gender Outlaw.* New York: Routledge.

Bowles, Samuel, & Gintis, Herbert. (1976). *Schooling in Capitalist America: Educational Reform and the Contradictions of Economic Life.* New York: Basic Books.

Braus, Patricia. (1994, November). "How Women Will Change Medicine." *American Demographics,* 40–47.

Brockerhoff, Martin P. (2000). *An Urbanizing World.* Washington, DC: Population Reference Bureau.

Brotz, Howard. (1966). *Negro Social and Political Thought 1850–1920.* New York: Basic Books.

Brown, Dee. (1971). *Bury My Heart at Wounded Knee.* New York: Holt, Rinehart and Winston.

Brown, Lester R. (1974). *In the Human Interest.* New York: Norton.

Brown, Roger W. (1954). "Mass Phenomena." In Gardner Lindzey (Ed.), *Handbook of Social Psychology.* Cambridge, MA: Addison-Wesley.

Bullough, Vern L. (1973). *The Subordinate Sex.* Chicago: University of Chicago Press.

Bumpass, Larry L. (1990). "What's Happening to the Family? Interactions Between Demographic and Institutional Change." *Demography, 27*(4), 485E.

Bumpass, Larry L., Sweet, James A., & Cherlin, Andrew. (1991, November). "The Role of Cohabitation in Declining Rates of Marriage." *Journal of Marriage and the Family, 53*(4), 913–27.

Buss, David M. (1994). *The Evolution of Desire.* New York: Basic Books.

Bylinsky, Gene. (1988). "Technology in the Year 2000." *Fortune,* July 18, 92–98.

Califano, Joseph A., Jr. (1994). *Radical Surgery.* New York: Random House.

Caplow, Theodore, Hicks, Louis, & Wattenberg, Ben J. (2001). *The First Measured Century: An Illustrated Guide to Trends in America. 1900–2000.* Washington, DC: American Enterprise Institute.

Cameron, William Bruce. (1966). *Modern Social Movements: A Sociological Outline.* New York: Random House.

Canetti, Elias. (1978/1960). *Crowds and Power.* New York: Seabury Press.

Carlson, Darren K. (2001, February 14). "Over Half of Americans Believe in Love at First Sight." Princeton, NJ: Gallup News Service.

Carr, P., Ash, A., Friedman, R., et al. (2000). "Faculty Perceptions of Gender Discrimination and Sexual Harassment in Academic Medicine." *Annals of Internal Medicine, 132,* 889–96.

Carty, Win. (1994, October). "Population Lingo Can Push 'Hot Buttons.'" *Population Today,* 3.

Casper, Lynne M., & Cohen, Philip. (2000, May). "How Does POSSLQ Measure Up? Historical Estimates of Cohabitation." *Demography, 37*(2), 237–45.

Cassidy, Tina. (1999, September 28). "Job Complaint Recalls 'Racial Charade.'" *The Boston Globe,* B1, B5.

Cato Daily Dispatch. (2000, December 12). "China's One-Child Policy."

Chaiken, Jan M., & Chaiken, Marcia R. (1982). *Varieties of Criminal Behavior.* Santa Monica, CA: RAND Corporation.

Chalmers, David. (1980, August). "The Rise and Fall of the Invisible Empire of the Ku Klux Klan." *Contemporary Review, 237,* 57–64.

Chambliss, William J. (1973). "Elites and the Creation of Criminal Law." In William J. Chambliss (Ed.), *Sociological Readings in the Conflict Perspective.* Reading, MA: Addison-Wesley.

Chandler, David L. (1992, November 2). "Polling: The Methods Behind the Madness." *The Boston Globe*, 35, 37.

Cherlin, Andrew J. (1992). *Marriage, Divorce, and Remarriage*. Cambridge, MA: Harvard University Press.

Cherlin, Andrew J., & Furstenberg, Frank F., Jr. (1994, January 1). "Stepfamilies in the United States: A Reconsideration." *Annual Review of Sociology, 20*.

Chomsky, N. (1975). *Language and Mind*. New York: Harcourt Brace Jovanovich.

Clark, Robert D., & Hatfield, Elizabeth. (1989). "Gender Differences in Receptivity to Sexual Offers." *Journal of Psychology and Human Sexuality, 2*, 39–55.

Cleland, John G., & Van Ginneken, Jerome K. (1988). "Maternal Education and Child Survival in Developing Countries: The Search for Pathways of Influence." *Social Science and Medicine, 27*(12), 1, 357–68.

Cohen, Yehudi A. (1981, January/February). "Shrinking Households." *Society, 18*(2), 51.

Colapinto. John. (2000). *As Nature Made Him*. New York: HarperCollins.

Coleman, James S. (1977). *Parents, Teachers, and Children*. San Francisco: San Francisco Institute for Contemporary Studies.

———. (1966). *Equality of Educational Opportunity*. Washington, DC: U.S. Government Printing Office.

Collins, Randall. (1979). *The Credential Society: An Historical Sociology of Education and Stratification*. New York: Academic Press.

———. (1975). *Conflict Sociology: Toward an Explanatory Science*. New York: Academic Press.

Comte, Auguste. (1968/1851). *System of Positive Policy* (Vol. 1) (John Henry Bridges, Trans.). New York: Burt Franklin.

Conrad, Peter. (1992). "Medicalization and Social Control." *Annual Review of Sociology, 1992, 18*, 209–32.

Conrad, Peter, & Kern, Rochelle (Eds.). (1986). *The Sociology of Health and Illness* (2nd ed.). New York: St. Martin's Press.

Cooley, C. H. (1909). *Social Organization*. New York: Scribner.

Corcoran, Paul E. (1994). "Presidential Concession Speeches: The Rhetoric of Defeat." *Political Communication, 11*, 109–31.

Cose, Ellis. (1997). "Census and the Complex Issue of Race." *Commentary, 34*(6), 9–13.

Coser, L. A. (1977). *Masters of Sociological Thought* (2nd ed.). New York: Harcourt Brace Jovanovich.

———. (1967). *Continuities in the Study of Social Conflict*. New York: The Free Press.

———. (1956). *The Functions of Social Conflict*. Glencoe, IL: The Free Press.

CQ Researcher. (1998, July 17). " Population and Environment." *8*(26).

Crano, William D., & Aronoff, Joel. (1978, August). "A Cross-Cultural Study of Expressive and Instrumental Role Complementarity in the Family." *American Sociological Review, 43*, 463–71.

Crichton, Judy. (1998). *America 1900*. New York: Henry Holt and Company.

Crosby, F. J. (1991). *Juggling: The Unexpected Advantages of Balancing Career and Home for Women and Their Families*. New York: The Free Press.

Crossen, Cynthia. (1994). *Tainted Truth: The Manipulation of Fact in America*. New York: Simon & Schuster.

Cummings, Milton C., & Wise, David. (1981). *Democracy Under Pressure: An Introduction to the American Political System* (4th ed.). New York: Harcourt Brace Jovanovich.

Curtiss, Susan. (1977). *Genie: A Psycholinguistic Study of a Modern-Day Wild Child*. New York: Academic Press.

Cuzzort, R. P., & King, E. W. (1980). *Twentieth Century Social Thought* (3rd ed.). New York: Holt, Rinehart and Winston.

Dahrendorf, R. (1959). *Class and Conflict in Industrial Society*. Stanford, CA: Stanford University Press.

———. (1958, September). "Out of Utopia: Toward a Reorientation of Sociological Analysis." *American Journal of Sociology, 64*, 158–64.

Daly, M., & Wilson, M. (1988). *Homicide*. Hawthorne, NY: Aldine de Gruyter.

D'Andrade, Roy G. (1966). "Sex Differences and Cultural Institutions." In Eleanor Emmons Maccoby (Ed.), *The Development of Sex Differences*. Stanford, CA: Stanford University Press.

Darwin, Charles. (1964/1859). *On the Origin of Species*. Cambridge, MA: Harvard University Press.

Davis, F. James. (1979). *Understanding Minority-Dominant Relations*. Arlington Heights, IL: AHM Publishing.

Davis, Kingsley. (1949). *Human Society*. New York: Macmillan.

———. (1940). "Extreme Social Isolation of a Child." *American Journal of Sociology, 45*, 554–65.

Davis, Kingsley, & Moore, W. E. (1945). "Some Principles of Stratification." *American Sociological Review, 10*, 242–49.

Deacon, Terrence W. (1997). *The Symbolic Species*. New York: W. W. Norton.

Dershowitz, Alan M. (1996). *Reasonable Doubts.* New York: Simon & Schuster.

Diamond, Milton, & Sigmundson, H. Keith. (1997, March). "Sex Reassignment at Birth: A Long Term Review and Clinical Implications." *Archives of Pediatric and Adolescent Medicine, 151,* 298–304.

Domhoff, G. William. (1983). *Who Rules America Now?* Englewood Cliffs, NJ: Prentice-Hall.

Droege, Kristen. (1995, Winter). "Child Care: An Educational Perspective." *MIJCF, Jobs and Capital, 4,* 1–8.

Duncan, Greg J., & Hoffman, Saul D. (1988, November). "What Are the Economic Consequences of Divorce?" *Demography, 25*(4), 641.

Durant, Will. (1954). "Our Oriental Heritage." *The Story of Civilization* (Vol. 1). New York: Simon & Schuster.

Durkheim, Émile. (1961/1915). *The Elementary Forms of Religious Life.* New York: Collier Books.

———. (1960/1893). *The Division of Labor in Society* (G. Simpson, Trans.). New York: The Free Press.

———. (1958/1895). *The Rules of Sociological Method.* Glencoe, IL: The Free Press.

Eberstadt, Nicholas. (1992, January 20). "America's Infant Mortality Problem: Parents." *The Wall Street Journal,* A14.

Edmondson, Brad. (1996, October)." How to Spot a Bogus Poll." *American Demographics, 18*(10), 10, 12–15.

Ehrlich, Isaac. (1975). "The Deterrent Effect of Capital Punishment: A Question of Life and Death." *American Economic Review,* 397–417.

———. (1974). *The End of Affluence.* New York: Simon & Schuster.

Ekman, Paul, Friesen, William V., & Bear, John. (1984, May). "The International Language of Gestures." *Psychology Today,* 64–69.

El-Badry, Samia. (1994, January). "Understanding Islam in America." *American Demographics,* 10–11.

Elkind, David. (1981). *The Hurried Child.* Reading, MA: Addison-Wesley.

Ellis, Joseph J. (1997). *American Sphinx: The Character of Thomas Jefferson.* New York: Knopf.

Ember, Carol R., & Ember, Melvin. (1981). *Anthropology* (3rd ed.). Englewood Cliffs, NJ: Prentice-Hall.

Embree, Edwin R. (1967/1939). *Indians of the Americas.* Boston: Houghton Mifflin.

Engels, Friedrich. (1973/1845). *The Condition of the Working Class in England in 1844.* Moscow: Progress.

———. (1942/1884). *The Origin of the Family, Private Property and the State.* New York: International Publishing.

Erikson, Erik H. (1968). *Identity, Youth and Crisis.* New York: Norton.

———. (1964). *Childhood and Society.* New York: Norton.

Erikson, Kai T. (1966). *Wayward Puritans: A Study in the Sociology of Deviance.* New York: John Wiley.

Erkel, R. Todd. (1994, July/August). "The Mighty Wedge of Class," *Family Therapy Networker.*

Fenster, M. (1993). "Genre and Form: The Development of the Country Music Video." In Simon Firth, Andrew Goodwin, & Lawrence Grossberg (Eds.), *Sound and Vision: The Music Video Reader.* New York: Routledge.

Ferree, Myra Marx, & Hess, Beth B. (1985). *Controversy and Coalition: The New Feminist Movement.* Boston: G. K. Hall.

Festinger, Leon, Rieken, Henry W., & Schacter, Stanley. (1956). *When Prophesy Fails.* New York: Harper Torchbooks.

Fine, Mark A. (1994, May). "An Examination and Evaluation of Recent Changes in Divorce Laws in Five Western Countries: The Crucial Role of Values." *Journal of Marriage and the Family, 56,* 249–63.

Fisher, Helen. (1994, October 16). "'Wilson,' They Said, 'You're All Wet!.'" *New York Times Book Review,* 15–17.

———. (1992). *Anatomy of Love.* New York: Ballantine Books.

Fisher, Robert. (1984). *Let the People Decide: Neighborhood Organizing in America.* Boston: G. K. Hall.

Ford, Clellan S. (1970). "Some Primitive Societies." In Georgene H. Seward & Robert C. Williamson (Eds.), *Sex Roles in Changing Society.* New York: Random House.

Fortes, M., Steel, R. W., & Ady, P. (1947). "Ashanti Survey, 1945–46: An Experiment in Social Research." *Geographical Journal, 110,* 149–79.

Fouts, Roger. (1997). *Next of Kin: What Chimpanzees Have Taught Me About Who We Are.* New York: William Morrow.

Frankfort, H. (1956). *The Birth of Civilization in the Near East.* Garden City, NY: Doubleday/Anchor.

Fredrickson, George M. (1971). *The Black Image in the White Mind.* New York: Harper & Row.

Freedman, Samuel G. (1990). *Small Victories.* New York: Harper & Row.

Freeman, James M. (1974, January). "Trial by Fire." *Natural History,* 54–63.

Freud, Sigmund. (1930). "Civilization and Its Discontents." *Standard Edition of the Complete Psychological Works of Sigmund Freud* (Vol. 29). London: Hogarth Press.

———. (1928). *The Future of an Illusion.* New York: Horace Liveright and the Institute of Psycho-Analysis.

———. 1923. "The Ego and the Id." *Standard Edition of the Complete Psychological Works of Sigmund Freud* (Vol. 19). London: Hogarth Press.

———. (1920). "Beyond the Pleasure Principle." *Standard Edition of the Complete Psychological Works of Sigmund Freud* (Vol. 14). London: Hogarth Press.

———. (1918). *Totem and Taboo.* New York: Moffat, Yard.

Frey, Darcy. (1994). *The Last Shot.* Boston: Houghton Mifflin.

Frey, William H. (2000, Summer). "The New Urban Demographics: Race Space & Boomer Aging" *The Brookings Review, 18*(3), 18–21.

Fried, Morton. (1967). *The Evolution of Political Society.* New York: Random House.

Friedrich, Carl J., & Brezinski, Zbigniew. (1965). *Totalitarian Dictatorship and Autocracy* (Vol. 2). Cambridge, MA: Harvard University Press.

Galinsky, Ellen, Howe, Caroline, Koutos, Susan, & Shinn, Marybeth. (1994). *The Study of Children in Family Child Care and Relative Care: Highlights and Findings.* New York: Families and Work Institute.

Gallup, George, Jr., & Castelli, Jim. (1987). *The American Catholic People: Their Beliefs, Practices, and Values.* Garden City, NY: Doubleday.

The Gallup Poll. (2001, February 14). "Over Half of Americans Believe in Love at First Sight."

———. (2000, August 24–27 and March 17–19). "Religion."

———. (1997). "Public Attitudes Toward Public Schools."

———. (1997, October). "Many Women Cite Spousal Abuse: Job Performance Affected."

———. (1997, March). "Religious Belief Widespread But Many Skip Church."

Gans, Herbert J. (1979, May). "Deception and Disclosure in the Field." *The Nation, 17,* 507–12.

———. (1977, February 12). "Why Exurbanites Won't Reurbanize Themselves." *New York Times,* 21.

———. (1962). *The Urban Villagers.* New York: The Free Press.

Gardner, Howard. (1978). *Developmental Psychology.* Boston: Little, Brown.

———. (1972). "Studies of the Routine Grounds of Everyday Activities." In David Snow (Ed.), *Studies in Social Interaction.* New York: The Free Press.

———. (1967). *Studies in Ethnomethodology.* Englewood Cliffs, NJ: Prentice-Hall.

Gargan, Edward A. (1988, November 2). "Beijing Admits Easing of Birth Limits." *The New York Times,* A3.

Geertz, C. (1973). *The Interpretation of Cultures.* New York: Basic Books.

Gehrke, Robert. (2001)."Utah Case Tests American Indian Law." *The Associated Press.* January 8.

Gelles, Richard J. (1996). *The Book of David.* New York: Basic Books.

Gelles, Richard J., & Conte, J. R. (1990, November). "Domestic Violence and Sexual Abuse of Children: A Review of Research in the Eighties." *Journal of Marriage and the Family, 52,* 1045–58.

Gelles, Richard J., & Straus, Murray A. (1988). *Intimate Violence.* New York: Simon & Schuster.

Gerth, Hans, & Mills, C. Wright. (1953). *Character and Social Structure.* New York: Harcourt Brace.

Gill, Richard T. (1991, Fall). "Day Care or Parental Care?" *The Public Interest.*

Gilligan, Carol. (1982). *In a Different Voice.* Cambridge, MA: Harvard University Press.

Ginsberg, Morris. (1958). "Social Change." *British Journal of Sociology, 9*(3), 205–29.

Gist, Noel P., & Fava, Sylvia Fleis. (1974). *Urban Society* (6th ed.). New York: Crowell.

Glazer, Nathan, & Moynihan, Daniel P. (Eds.). (1975). *Ethnicity: Theory and Experience.* Cambridge, MA: Harvard University Press.

Gnessous, Mohammed. (1967). "A General Critique of Equilibrium Theory." In Wilburt E. Moore & Robert M. Cooke (Eds.), *Readings on Social Change.* Englewood Cliffs, NJ: Prentice-Hall.

Goffman, E. (1971). *Relations in Public.* New York: Basic Books.

———. (1963). *Behavior in Public Places.* New York: The Free Press.

———. (1961). *Asylums: Essays on the Social Situation of Mental Patients and Other Inmates.* Chicago: Aldine.

———. (1959). *The Presentation of Self in Everyday Life.* Garden City, NY: Doubleday.

Goldman, Ari L. (1991). *The Search for God at Harvard.* New York: Times Books.

Good, Kenneth. (1991). *Into the Heart.* New York: Simon & Schuster.

Goode, W. J. (1963). *World Revolution and Family Patterns.* New York: The Free Press.

———. (1960, August 25). "A Theory of Role Strain." *American Sociological Review,* 902–14.

Gordon, Milton M. (1975/1961). "Assimilation in America: Theory and Reality." In Norman R. Yetman & C. Hoy Steele (Eds.), *Majority and Minority: The Dynamics of Racial and Ethnic Relations.* Boston: Allyn & Bacon.

———. (1964). *Assimilation in American Life.* New York: Oxford University Press.

Gough, Kathleen. (1952). "Changing Kinship Usages in the Setting of Political and Economic Change Among the Nayars of Malabor." *Journal of Royal Anthropological Institute of Great Britain and Ireland, 82*, 71–87.

Gould, H. (1971). "Caste and Class: A Comparative View." *Module, 11*, 1–24.

Gould, Stephen Jay. (1976, May). "The View of Life: Biological Potential Versus Biological Determinism." *Natural History Magazine, 85*, 34–41.

Gouldner, Alvin W. (1970). *The Coming Crisis of Western Sociology.* New York: Avon.

Greenberg, J. (1980). "Ape Talk: More Than Pigeon English?" *Science News, 117*(19), 298–300.

Gumplowicz, Ludwig. (1899). *The Outlines of Sociology.* Philadelphia: American Academy of Political and Social Sciences.

Gurr, Ted Robert. (1970). *Why Men Rebel.* Princeton, NJ: Princeton University Press.

Hall, Edward T. (1974). *Handbook for Proxemic Analysis.* Washington, DC: Society for the Anthropology of Visual Communication.

———. (1969). *The Hidden Dimension.* Garden City, NY: Doubleday.

Hall, Edward T., & Hall, Mildred Reed. (1990). *Hidden Differences.* New York: Anchor Books.

Hampden-Turner, Charles, & Trompenaars, Fons. (2000). *Building Cross-Cultural Competence.* New Haven, CT: Yale University Press.

Harlow, Harry F. (1959, June). "Love in Infant Monkeys." *Scientific American*, 68–74.

Harlow, Harry F., & Harlow, M. (1962). "The Heterosexual Affectional System in Monkeys." *American Psychologist, 17*, 1–9.

Harris, C. D., & Ullman, E. L. (1945). "The Nature of Cities." *Annals of the American Academy of Political and Social Science, 242*, 12.

Harris, Marvin. (1975). *Culture, People, and Nature: An Introduction to General Anthropology* (2nd ed.). New York: Crowell.

———. (1966). "The Cultural Ecology of India's Sacred Cattle." *Current Anthropology, 7*, 51–63.

The Harris Poll. (2000). "Religious Affiliation Year 2000."

———. (1998, May 15–19) "Prestige of Occupations."

Hart, Gary. (1993). *The Good Fight.* New York: Random House.

Harvard Law Review. (1993, June). "Notes." *Harvard Law Review, 106*, 1905–25.

Hassan, S. (1988). *Combatting Cult Mind Control.* Rochester, VT: Park Street Press.

"Hate Speech on the Internet. (1999)." *Issues and Controversies on File. June 4.*

Hatfield, E. (1988). "Passionate and Companionate Love." In R. J. Steinberg & M. L. Barnes (Eds.), *The Psychology of Love.* New Haven, CT: Yale University Press.

Haub, Carl. (1994, December). "Population Change in the Former Soviet Republics." *Population Bulletin, 49*(4).

Havemann, Ernest. (1967) "Computers: Their Scope Today." *Playboy Magazine.* October.

Hawes, Alex. (1995). "Machiavellian Monkeys and Shakespearean Apes: The Question of Primate Language." *Zoogoer, 24*(6), 1–10.

Hawkins, Sanford A., & Hastie, Robert. (1990). "Hindsight: Biased Judgments of Past Events After the Outcomes Are Known." *Psychological Bulletin, 107*, 311–327.

Hawley, Amos. (1981). *Urban Society* (2nd ed.). New York: John Wiley.

Heer, David M. (1980). "Intermarriage." *Harvard Encyclopedia of American Ethnic Groups.* Cambridge, MA: Harvard University Press, 513–21.

Henig, Robin Marantz. (1997). *The People's Health: A Memoir of Public Health and Its Evolution at Harvard.* Washington, DC: Joseph Henry Press.

Henshaw, S. (1991). "Induced Abortion: A World Review, 1990." *Family Planning Perspectives, 22*(2), 76–89.

Herbers, John. (1986). *The New Heartland.* New York: Times Books.

———. (1981, May 31). "Census Finds More Blacks Living in Suburbs of Nation's Large Cities." *New York Times*, 1, 48.

Herring, Susan. (1994, June). "Making the Net *Work*: Is There a Z39.50 in Gender Communication?" Paper presented at the American Library Association annual convention, Miami.

Hesketh, Therese, & Zhu, Wei Xing. (1997). "The One Child Policy: The Good, the Bad, and the Ugly." *British Medical Journal, 314*(7095).

Hingson, Ralph W. (1998, January/February). "College-Age Drinking Problems." *Public Health Reports, 113*.

Hirschi, Travis. (1969). *Causes of Delinquency.* Berkeley: University of California Press.

Hirschi, Travis, & Gottfredson, Michael. (1993). "Commentary: Testing the General Theory of Crime." *Journal of Research in Crime and Delinquency, 30*, 47–54.

Hochschild, Arlie Russell. (1997). *Time Bind: When Work Becomes Home and Home Becomes Work.* New York: Henry Holt.

———. (1997, April 20). "There's No Place Like Work." *The New York Times Magazine*, 51–55, 81, 84.

Hoebel, E. Adamson. (1960). *The Cheyennes: Indians of the Great Plains.* New York: Holt, Rinehart and Winston.

Hoecker-Drysdale, Susan. (1992). *Harriet Martineau: First Woman Sociologist.* Oxford, England: Berg.

Holden, Constance. (1997, March). "Changing Sex Is Hard to Do." *Science, 275*(5307), 1745.

Homans, G. C. (1950). *The Human Group.* New York: Harcourt Brace.

"Homelessness." (2000, *January 21). Issues and Controversies on File.*

Homer-Dixon. Thomas. (2000). *The Ingenuity Gap.* New York: Knopf.

Howe, Irving. (1976). *World of Our Fathers.* New York: Simon & Schuster.

Howes, C., & Hamilton, C. (1991). "Child Care for Young Children." In B. Spodek (Ed.), *Handbook of Research in Early Childhood Education.* New York: Macmillan.

Hoyt, H. (1943). "The Structure of American Cities in the Post-War Era." *American Journal of Sociology, 48,* 475–92.

Immerwahr, John & Johnson, Jean. (1996). "Incomplete Assignment. America's Views on Standards: An Assessment by Public Agenda." *The Progress of Education Reform, 8.*

"Income Gap." (1999, December 17.). *Issues and Controversies on File, 4*(3), 489–96.

International Conference on Child Labor. (1998). Oslo. October 27–30, 1997. Geneva: International Labour Office.

Irvine, Martha. (1999, April). "Facts About Violence Among Youth and Violence in Schools." Centers for Disease Control and Prevention.

Itard, J. (1932). *The Wild Boy of Aveyron* (G. Humphrey & M. Humphrey, Trans.). New York: Appleton-Century-Crofts.

Jacob, H. (1988). *Silent Revolution: The Transformation of Divorce Law in the United States.* Chicago: University of Chicago Press.

Jacobs, J. (1961). *The Death and Life of American Cities.* New York: Vintage.

Jamison, Kay Redfield. (1995). *An Unquiet Mind.* New York: Knopf.

Janis, I., Chapman, Dwight W., Gillin, John P., & Spiegel, John P. (1964). "The Problem of Panic." In Duane P. Schultz (Ed.), *Panic Behavior.* New York: Random House.

Jencks, Christopher. (1994). *The Homeless.* Cambridge, MA: Harvard University Press.

Johnson, George. (1995, June 6). "Chimp Talk Debate: Is It Really Language?" *The New York Times,* C1.

Josephy, Alvin M., Jr. (1994). *500 Nations: An Illustrated History of North American Indians.* New York: Knopf.

Kasarda, John D., & Janowitz, M. (1974). "Community Attachment in Mass Society." *American Journal of Sociology, 48,* 328–39.

Katz, Elihu. (1957). "The Two-Step Flow of Communication: An Up-to-Date Report on an Hypothesis." *Public Opinion Quarterly, 21,* 61–78.

Kelling, George L., & Coles, Catherine M. (1996). *Fixing Broken Windows: Restoring Order and Reducing Crime in our Communities.* New York: Martin Kessler Books (The Free Press).

Kennedy, John F. (1961). "Introduction." In William Brandon (Ed.), *The American Heritage Book of Indians.* New York: Dell.

Kennedy, Randall. (1997). *Race, Crime and the Law.* New York: Pantheon Books.

———. (1994, Spring). "Orphans of Separatism: The Painful Politics of Transracial Adoption." *The American Prospect, 17,* 38–45.

Khaldun, Ibn. (1958). *The Mugaddimah.* Bolligen Series 43. Princeton, NJ: Princeton University Press.

Kielburger, Craig, & Major, Kevin. (1998). *Free the Children.* Toronto: McClelland & Stewart.

Kiely, John L. (1995, December 15). "Poverty and Infant Mortality: United States, 1988." *Morbidity and Mortality Weekly Report, 44*(49), 922–27.

Kilker, Ernest Evans. (1993). "Black and White in America: The Culture and Politics of Racial Classification." *International Journal of Politics,Culture and Society, 7*(2), 229–57.

Knox, Richard A. (1996, October 24). "Women Doctors' Research 'Ceiling.'" *The Boston Globe,* A3.

Koenig, Frederick. (1985). *Rumor in the Marketplace: The Social Psychology of Commercial Heresy.* Dover, MA: Auburn House.

Kohlberg, Lawrence. (1969). "Stage and Sequence: The Cognitive-Developmental Approach to Socialization." In David A. Goslin (Ed.), *Handbook of Socialization Theory and Research.* Chicago: Rand McNally.

———. (1967). "Moral and Religious Education in the Public Schools: A Developmental View." In T. Sizer (Ed.), *Religion and Public Education.* Boston: Houghton Mifflin.

Kohn, Hans. (1956). *Nationalism and Liberty: The Swiss Example.* London: Allen and Unwin.

Kohn, Melvin L., & Schooler, Carmi. (1983). *Work and Personality: An Inquiry Into the Impact of Social Stratification.* New York: Ablex Press.

Kosmin, Barry A., & Lachman, Seymour P. (1993). *One Nation Under God.* New York: Harmony.

Kotkin, Joel, & Kishimoto, Yoriko. (1988). *The Third Century: America's Resurgence in the Asian Era.* New York: Crown.

Kozol, Jonathan. (1992). *Savage Inequalities: Children in America's Schools.* New York: Harper Perennial.

———. (1989). *Death at an Early Age.* New York: Plume Books.

Kraft, Susan. (1994, July). "Hide-and-Seek With Illegal Aliens." *American Demographics,* 10–11.

Krause, Michael. (1966). *Immigration: The American Mosaic.* New York: Van Nostrand-Reinhold.

Kristof, Nicholas, & Wudunn, Sheryl. (1994). *China Wakes.* New York: Random House.

Kroeber, Alfred A. (1963/1923). *Anthropology: Culture Patterns and Processes.* New York: Harcourt, Brace & World.

Kronholz, June. (1998, January 9). "Bilingual Schooling Faces Big Test in California as Voters Consider Measure to Abolish Program." *The Wall Street Journal,* A18.

Kutner, Lawrence. (1988, June 30). "For Children and Stepparents War Isn't Inevitable." *New York Times,* C8.

Lakshmanan, Indira A. R. (1997, August 2). "In Taiwan's Legislature Democracy Packs Punch." *The Boston Globe,* A1, A12.

———. (1997, August 26). "Love No Match for Custom: India's Young Favor Arranged Unions." *The Boston Globe,* A1, A10.

Landes, Richard. (1998). *While God Tarried: Disappointed Millennialism and the Making of the Modern West.* New York: Houghton Mifflin.

———. (1997, April 6). "Countdown 2000." *The Boston Globe.*

Lang, Kurt, & Lang, Gladys. (1961). *Collective Dynamics.* New York: Crowell.

Lantz, Herman R. (1982, Spring). "Romantic Love in the Pre-Modern Period: A Sociological Commentary." *Journal of Social History,* 349–70.

Larson, Jan. (1992, July). "Understanding Stepfamilies." *American Demographics,* 36–40.

Lasch, Christopher. (1977). *Haven in a Heartless World: The Family Besieged.* New York: Basic Books.

Laslett, P. (1965). *The World We Have Lost: England Before the Industrial Age.* New York: Scribner's.

Laslett, P., & Cummings, Phillip W. (1967). "History of Political Philosophy." In Paul Edwards (Ed.), *The Encyclopedia of Philosophy* (Vols. 5, 6). New York: Macmillan.

Lasswell, Thomas. (1965). *Class and Stratum.* Boston: Houghton Mifflin.

Lattimore, Pamela K., & Nahabedian, Cynthia A. (2000). "The Nature of Homicide: Trends and Changes." *Sourcebook of Criminal Justice Statistics: 1999.* Washington, DC: U.S. Department of Justice.

———. (1997). *The Nature of Homicide: Trends and Changes.* U.S. Department of Justice. Washington, DC: U.S. Government Printing Office.

Lazarsfeld, Paul F., Berelson, Bernard, & Gaudet, Hazel. (1968). *The People's Choice* (3rd ed.). New York: Columbia University Press.

Le Bon, Gustave. (1960/1895). *The Crowd: A Study of the Popular Mind.* New York: Viking.

Lee, Alfred McClung. (1978). *Sociology for Whom?* New York: Oxford University Press.

Lee, Richard Borshay. (1980). *The !KungSan.* Berkeley and Los Angeles: University of California Press.

Lemert, Edwin. (1972). *Human Deviance, Social Problems and Social Control* (2nd ed.). Englewood Cliffs, NJ: Prentice-Hall.

Lenin, Vladimir I. (1949/1917). *The State and Revolution.* Moscow: Progress.

Lenski, Gerhard. (1966). *Power and Privilege: A Theory of Social Stratification.* New York: McGraw-Hill.

Lenski, Gerhard, & Lenski, Jean. (1982). *Human Societies* (4th ed.). New York: McGraw-Hill.

Leonard, Ira M., & Parmet, R. D. (1972). *American Nativism: 1830–1860.* New York: Van Nostrand-Reinhold.

Leslie, Gerald R. (1979). *The Family in Social Context* (4th ed.). New York: Oxford University Press.

Levin, Jack. (1993). *Sociological Snapshots.* Newbury Park, CA: Pine Forge Press.

Levine, Robert. (1993, February). "Is Love a Luxury?" *American Demographics,* 27, 29.

Levinson, Daniel J. (with Judy Levinson). (1996). *The Seasons of a Woman's Life.* New York: Knopf.

Levinson, Daniel J. (with Charlotte N. Darrow, Edward B. Klein, Maria H. Levinson, & Braxton McKee). (1978). *The Seasons of a Man's Life.* New York: Ballantine.

Lewin, Kurt. (1948). *Resolving Social Conflicts.* New York: Harper.

Lewis, David Levering. (2000). *W. E. B. DuBois: The Fight for Equality and the American Century, 1919–1963.* New York: Henry Holt.

———. (1993). *W. E. B. DuBois: Biography of a Race/1868–1919.* New York: Henry Holt.

Lindesmith, Alfred R., & Strauss, Anselm L. (1956). *Social Psychology.* New York: Holt, Rinehart and Winston.

Lines, Patricia M. (2000, Summer). *The Public Interest* (140), 74–85.

Linton, R. (1936). *The Study of Man.* New York: Appleton-Century-Crofts.

Lipset, Seymour Martin. (1996). *American Exceptionalism: A Double-Edged Sword.* New York: W. W. Norton.

———. (1960). *Political Man.* Garden City, NY: Doubleday.

Lipset, Seymour Martin, & Lenz, Gabriel Salman. (2000)." Corruption, Culture and Markets." In Lawrence E. Harrison & Samuel P. Huntington (Eds.), *Culture Matters* (pp. 112–24). New York: Basic Books.

Lipsey, R. G., & Steiner, P. D. (1975). *Economics.* New York: Harper & Row.

Liska, Allen E. (1991). *Perspectives on Deviance* (3rd ed.). Englewood Cliffs, NJ: Prentice-Hall.

Lombroso-Ferrero, Gina. (1972). *Criminal Man* (Reprint ed.). Montclair, NJ: Patterson Smith.

Longino, Charles F., Jr. (1994, August). "Myths of an Aging America," *American Demographics,* 36–43.

Longino, Charles F., Jr., & Crown, William H. (1991, August). "Older Americans: Rich or Poor?" *American Demographics,* 48–53.

Lown, Bernard. (1996). *The Lost Art of Healing.* Boston: Houghton Mifflin.

Lynch, K. (1960). *The Image of the City.* Cambridge, MA: MIT Press.

Maccoby, Eleanor Emmons, Buchanan, C. M., Mnookin, R. H., & Dornbusch, S. M. (1993). "Post-divorce Roles of Mothers and Fathers in the Lives of Children." *Journal of Family Psychology, 7,* 24–38.

Maccoby, Eleanor Emmons, & Jacklin, Carol Nagy. (1975). *The Psychology of Sex Differences.* Stanford, CA: Stanford University Press.

Machel, Grac'a. (1996, August). *Impact of Armed Conflict on Children.* New York: The United Nations.

MacMurray, V. D., & Cunningham, P. H. (1973). "Mormons and Gentiles." In Donald E. Gelfand & Russell D. Lee (Eds.), *Ethnic Conflicts and Power: A Cross-National Perspective.* New York: John Wiley.

Mackenzie. Doris Layton. (2000). "Sentencing and Corrections in the 21st Century: Setting the Stage for the Future." U.S. Department of Justice.

Maddux, Cleborne D. (1997). "The World Wide Web and School Culture: Are They Incompatible?" *Computers and the Schools, 13*(1/2), 7–10.

Madsen, William. (1973). *The Mexican-Americans of South Texas* (2nd ed.). New York: Holt, Rinehart and Winston.

Maguire, Kathleen, & Pastore, Ann L. (Eds.). (2000). *Sourcebook of Criminal Justice Statistics—1999.* U.S. Department of Justice. Bureau of Justice Statistics. Washington. DC: U.S. Government Printing Office.

———. (1997). *Sourcebook of Criminal Justice Statistics—1996.* U.S. Department of Justice, Bureau of Justice Statistics. Washington, DC: U.S. Government Printing Office.

———. (1995). *Sourcebook of Criminal Justice Statistics—1994.* U.S. Department of Justice, Bureau of Justice Statistics. Washington, DC: U.S. Government Printing Office.

Males, Mike. (1993, September 20). "Public Enemy Number One." *In These Times.*

Malinowski, Bronislaw. (1954). *Magic, Science and Religion.* New York: The Free Press.

———. (1922). *Argonauts of the Western Pacific.* New York: Dutton.

Martin, Philip, & Midgley, Elizabeth. (1994, September). "Immigration to the United States: Journey to an Uncertain Destination." *Population Bulletin, 49*(2).

Marx, Karl. (1968). "Wage, Labour and Capital." In Karl Marx & Friedrich Engels, *Selected Works in One Volume.* New York: International Publishers.

———. (1967/1867). *Capital: A Critique of Political Economy.* Friedrich Engels (Ed.). New York: New World.

———. (1967/1867). *Das Kapital* (Vols. 1–3). Friedrich Engels (Ed.). New York: International Publishing.

———. (1959/1847). *Class and Class Conflict in Industrial Society.* Stanford, CA: Stanford University Press.

Marx, Karl, & Engels, Friedrich. (1961/1848). "The Communist Manifesto." In Arthur P. Mendel (Ed.), *Essential Works of Marxism.* New York: Bantam Books.

Mauss, Armand I. (1975). *Social Problems of Social Movements.* Philadelphia: Lippincott.

Mayberry, Maralee. (1991). "Conflict and Social Determinism: The Reprivatization of Education." Paper presented at the American Educational Research Association Meeting, Chicago.

McNeill, William H. (1976). *Plagues and People.* New York: Anchor/Doubleday.

Mead, George H. (1934). *Mind, Self, and Society.* C. W. Morris (Ed.). Chicago: University of Chicago Press.

Mead, Margaret. (1943). "Our Educational Emphases in Primitive Perspectives." *American Journal of Sociology, 48,* 633–39.

———. (1935). *Sex and Temperament in Three Primitive Societies.* New York: Morrow.

Meadows, Donelle H., Meadows, Dennis L., Randers, Jorgan, & Behrens, William, III. (1972). *The Limits of Growth: A Report of the Club of Rome's Project on the Predicament of Mankind.* New York: Universe Books.

Mednick, Sarnoff A., Moffitt, Terrie E., & Stacks, Susan A. (Eds.). (1987). *The Causes of Crime: New Biological Approaches.* Cambridge, England: Cambridge University Press.

Mednick, Sarnoff A. (1977). "A Biosocial Theory of the Learning of Law-Abiding Behavior." In Sarnoff A. Mednick & Karl O. Christiansen (Eds.), *Biosocial Bases of Criminal Behavior.* New York: Gardner Press.

Merton, Robert K. (1969/1949). *Social Theory and Social Structure.* New York: The Free Press.

———. (1938). "Social Structure and Social Action." *American Sociological Review, 3,* 672–82.

Michels, R. (1966/1911). *Political Parties* (Eden Paul & Adar Paul, Trans.). New York: The Free Press.

Migration News. (2000, February). "North America." 7(2).

Mills, C. Wright. (1959). *The Sociological Imagination*. New York: Oxford University Press.

Mishel, Lawrence, & Bernstein, Gared. (1995). *The State of Working America, 1994–95*. Washington, DC: Economic Policy Institute Service, M. E. Sharpe.

Monaghan, Peter. (1993, November 11). "Free After 6 Months: Sociologist Who Refused to Testify Is Released." *Chronicle of Higher Education*.

Money, John, & Tucker, Paul. (1975). *Sexual Signatures: On Being a Man or Woman*. Boston: Little, Brown.

Montagu, Ashley (Ed.). (1973). *Man and Aggression* (2nd ed.). London: Oxford University Press.

———. (1964a). *The Concept of Race*. New York: Collier Books.

———. (1964b). *Man's Most Dangerous Myth: The Fallacy of Race*. New York: Meridian.

Moore, Stephen. (1999). "Defusing the Population Bomb." *The Washington Times*. October 13.

Moore, Wilbert E. (1965). *The Impact of Industry*. Englewood Cliffs, NJ: Prentice-Hall.

Morris, Desmond. (1970). *The Human Zoo*. New York: McGraw-Hill.

Moskos, Charles C., & Butler, John Sibley. (1996). *All That We Can Be: Black Leadership and Racial Integration in the Army*. New York: Basic Books.

Murdock, George P. (1949). *Social Structure*. New York: Macmillan.

———. (1937). "Comparative Data on the Division of Labor by Sex." *Social Forces, 15*, 551–53.

Murray, Charles. (1994, December). "What to Do About Welfare." *Commentary, 98*(6), 26–34.

Murray, David W. (1998). "The War Against Testing" *Commentary*. September 1998.

Muson, H. (1979, February). "Moral Thinking—Can It Be Taught?" *Psychology Today*, 26–29.

Nakao, Keoko, & Treas, Judith. (1993). *General Social Surveys, 1972–1991: Cumulative Codebook* (pp. 827–35). Chicago: National Opinion Research Center.

———. (1990). "Occupational Prestige in the United States Revisited: Twenty-five Years of Stability and Change." Paper presented at the annual meeting of the American Sociological Association, Washington, DC.

National Association of Black Social Workers. (1994). "Preserving African-American Families." *Position Paper*.

———. (1988, Spring). *Newsletter*, 1–2.

———. (1972, April). *Position Paper*.

National Center for Health Statistics. (1999). "Health, United States, 1998, With Socioeconomic Status and Health Chartbook ."

———. (1999, June 30) "Deaths: Final Data for 1997. "*Monthly Vital Statistics, 47*(19).

———. (1997). "Report of Final Mortality Statistics, 1995." *Monthly Vital Statistics Report, 45*(11), supplement 2.

———. (1997). *Vital Statistics of the United States*.

———. (1996). "Advance Report of Mortality Statistics." *Monthly Vital Statistics Report, 45*(3) (supplement), 63.

———. (1974, June 27). "Mortality Statistics, 1950." *Monthly Vital Statistics Report, 22*(13).

National Commission on Excellence in Education. (1983). *A Nation at Risk*. Washington, DC: U.S. Government Printing Office.

National Institutes of Health. (1997, April 3). "Results of NICHD Study of Early Child Care Reported at Society for Research in Child Development Meeting." *NIH News Release*.

National Opinion Research Center. (1993). *General Social Surveys, 1972–1993: Cumulative Codebook* (pp. 927–45). Chicago: National Opinion Research Center.

Neugarten, Bernice, L. (1979). "Timing, Age, and the Life Cycle." *American Journal of Psychiatry, 136*, 887–94.

Niebuhr, Gustav. (2000, October 31). "Marriage Issue Splits Jews, Poll Finds." *The New York Times*.

"No-Fault Divorce?" (1999, May 7). *Issues and Controversies on File*, 187–89.

Nock, Albert Jay. (1996). *Jefferson*. New York: The John Day Company.

Novit-Evans, Bette, & Welch, Ashton Wesley. (1983). "Racial and Ethnic Definition as Reflections of Public Policy." *Journal of American Studies, 17*(3), 417–35.

Ogburn, William F. (1964). *On Culture and Social Change*. Chicago: University of Chicago Press.

O'Hare, William P. (1996, September). "A New Look at Poverty in America." *Population Bulletin, 5*(2).

O'Hare, William P., & Frey, William H. (1992, September). "Booming, Suburban, and Black America." *American Demographics*, 30–38.

Oliver, Melvin L., & Shapiro, Thomas M. (1997). *Black Wealth/White Wealth*. New York: Routledge.

Organization for Economic Co-operation and Development. (1997). INES Project, International Indicators Project. *The Condition of Education 1997* (Indicator 23).

Ortner, Sherry. (1974). "Is Female to Male as Nature Is to Culture?" In Michelle Zimbalist Rosaldo & Louise Lampheres (Eds.), *Woman, Culture and Society*. Stanford, CA: Stanford University Press.

Ostling, Richard N. (1999, Spring). "America's Ever Changing Religious Landscape." *The Brookings Review*, 10–13.

Otterbein, Keith. (1973). "The Anthropology of War." In John J. Honigmann (Ed.), *Handbook of*

Social and Cultural Anthropology. Chicago: Rand McNally.

———. (1970). *The Evolution of War.* New Haven, CT: Human Relations Area Files.

Palen, John J. (1995). *The Suburbs.* New York: McGraw-Hill.

———. (1992). *The Urban World* (2nd ed.). New York: McGraw-Hill.

Papert, Seymour. (1993). *The Children's Machine.* New York: Basic Books.

Park, R., Burgess, E., & McKenzie, R. (Eds.). (1925). *The City.* Chicago: University of Chicago Press.

Parker, R. N. (1995). *Alcohol and Homicide: A Deadly Combination of Two American Traditions.* Albany, NY: State University Press.

Parkes, Henry Bamford. (1968). *The United States of America: A History* (3rd ed.). New York: Knopf.

Parrillo, Vincent N. (1997). *Strangers to These Shores* (5th ed.). Boston: Allyn & Bacon.

Parsons, Talcott. (1971). *The System of Modern Societies.* Englewood Cliffs, NJ: Prentice-Hall.

———. (1966). *Societies: Evolutionary and Comparative Perspectives.* Englewood Cliffs, NJ: Prentice-Hall.

———. (1954). *Essays in Sociological Theory* (Rev. ed.). New York: The Free Press.

———. (1951). *The Social System.* New York: The Free Press.

Parsons, Talcott, & Bales, Robert F. (1955). *Family Socialization and Interaction Process.* New York: The Free Press.

Patterson, Orlando. (1997). *The Ordeal of Integration.* Washington: DC: Civitas Counterpoint.

Pavlov, I. P. (1927). *Conditioned Reflexes* (G. V. Anrep, Trans.). New York: Oxford University Press.

Payer, Lynn. (1988). *Medicine and Culture.* New York: Henry Holt.

Perry, T. (1993–1994). "The Transracial Adoption Controversy: An Analysis of Discourse and Subordination." *New York University Review of Law and Social Change, 21,* 30–34.

Petersen, William. (1975). "On the Subnations of Western Europe." In Nathan Glazer & Daniel P. Moynihan (Eds.), *Ethnicity: Theory and Experience.* Cambridge, MA: Harvard University Press.

Peterson, Richard R. (1996, June). "A Re-Evaluation of the Economic Consequences of Divorce." *American Sociological Review, 61,* 528–36.

———. "Reply to Weitzman." (1996, June). *American Sociological Review, 61,* 539–40.

Pettigrew, Thomas F., & Green, Robert C. (1975). "School Desegregation in Large Cities: A Critique of the Coleman White Flight Thesis." *Harvard Educational Review, 46*(1), 1–53.

Piaget, J., & Inhelder, B. (1969). *The Psychology of the Child.* New York: Basic Books.

Pinker, Steven. (1995). *The Language Instinct.* New York: Harper Perennial.

Pollan, M. (1997, December 14). "Town-Building Is No Mickey Mouse Operation." *The New York Times Magazine,* 56–63, 76, 78, 80–81, 88.

Population Reference Bureau. (2000). *2000 World Population Data Sheet.*

———. (1997). *1997 World Population Data Sheet.*

———. (1995, April). "Women, Children, and AIDS." *Population Today, 23*(4).

Prejean, Helen. (1993). *Dead Man Walking.* New York: Random House.

Provence, Sally. (1972). "Psychoanalysis and the Treatment of Psychological Disorders of Infancy." In S. Wolman (Ed.), *A Handbook of Child Psychoanalysis: Research, Theory, and Practice.* New York: Van Nostrand.

Putka, Gary. (1984, April 13). "As Jewish Population Falls in U.S., Leaders Seek to Reverse Trend." *The Wall Street Journal,* 1, 10.

Putnam, Robert D. (2000) *Bowling Alone.* New York: Simon & Schuster.

———. (1996, Winter). "The Strange Disappearance of Civic America." *The American Prospect, 24.*

Quinney, Richard. (1974). *Critique of Legal Order.* Boston: Little, Brown.

Radelet, Michael L., Bedau, Hugo Adam, & Putnam, Constance E. (1992). *In Spite of Innocence.* Boston: Northeastern University Press.

Raines, Howell. (1988, June 17). "British Government Devising Plan to Curb Violence by Soccer Fans." *The New York Times,* A1.

Rainwater, Lee, & Smeeding, Timothy M. (1995). "Doing Poorly: The Real Income of Children in Comparative Perspective." Working Paper No. 127, Luxembourg Income Study. Maxwell School of Citizenship and Public Affairs. Syracuse, NY: Syracuse University.

Ramo, Joshua Cooper. (1996, December 16). "Finding God on the Web." *Time,* 60–64, 66–67.

Rash, Wayne, Jr. (1997). *Politics on the Net* (pp. 95–97). New York: W. H. Freeman.

Ravitch, Diane. (2000). *Left Back : A Century of Failed School Reforms.* New York: Simon & Schuster.

Rawls, Phillip. (2000, November 15). "Baptists Leaders Affirm New Creed." *The Associated Press.*

Reid, Sue Titus. (1991). *Crime and Criminology* (6th ed.). Fort Worth, TX: Harcourt Brace Jovanovich.

Reiman, Jeffrey H. (1990). *The Rich Get Richer and the Poor Get Prison: Ideology, Class, and Criminal Justice* (3rd. ed.). New York: John Wiley and Sons.

Reinharz, Shulamit. (1993, September). *A Contextualized Chronology of Women's Sociological Work*

(2nd ed.). Waltham, MA: Brandeis University Women's Studies Program, Working Papers Series.

"Religious Freedom Abroad" (2000, January 21). *Issues and Controversies on File,* 17–24.

Rhymer, Russ. (1993). *Genie: Abused Child's Flight from Silence.* New York: Basic Books.

Riche, Martha Farnsworth. (1987, November). "Behind the Boom in Mental Health Care." *American Demographics,* 34–37, 60–61.

Ricks, Thomas E. (1997). *Making the Corps.* New York: Scribner.

Rifkin, Jeremy. (1995). *The End of Work: The Decline of the Global Labor Force and the Dawn of the Post-Market Era.* New York: Putnam.

Riley, Glenda. (1991). *Divorce: An American Tradition.* New York: Oxford University Press.

Riley, Nancy E. (1997, May). "Gender, Power, and Population Change." *Population Bulletin,* 52(1).

———. (1996, February). "China's 'Missing Girls': Prospects and Policy." *Population Today,* 4–5.

Ritzer, George. (1992). *The McDonaldization of Society.* Newbury Park, CA: Pine Forge Press.

Robinson, Jacob. (1976). "The Holocaust." In Yisrael Gutman & Livia Rothkirchen (Eds.), *The Catastrophe of European Jewry.* Jerusalem: Yad Veshem.

Roos, Patricia. (1995, August). "Occupational Feminization, Occupational Decline?" Paper presented at the annual meeting of the American Sociological Association.

Rosaldo, Michelle Zimbalist. (1974). "Woman, Culture and Society: A Theoretical Overview." In Michelle Zimbalist Rosaldo & Louise Lamphere (Eds.), *Woman, Culture and Society.* Stanford, CA: Stanford University Press.

Rose, Jerry D. (1982). *Outbreaks: The Sociology of Collective Behavior.* New York: The Free Press.

Rosen, George. (1968). *Madness in Society.* New York: Harper Torchbooks.

Rosenthal, Elizabeth. (1998, November 1) "For One Child Policy: China Rethinks Iron Hand." *The New York Times,* 1, 20.

Rosenthal, R., & Jacobson, L. (1966). "Teachers' Expectancies: Determinants of Pupils' I.Q. Gain." *Psychological Reports, 18,* 115–18.

Rosenzweig, Jane. (1999, July/August). "Can TV Improve Us?" *The American Prospect* (45).

Rossides, Daniel W. (1990). *Social Stratification.* Englewood Cliffs, NJ: Prentice-Hall.

Roth, Philip. (2000). *The Human Stain.* New York: Houghton Mifflin.

———. (1998). *American Pastoral.* New York: Vintage Books.

Rubin, Zick. (1973). *Liking and Loving.* New York: Holt, Rinehart and Winston.

———. (1970). "Measurement of Romantic Love." *Journal of Personality and Social Psychology, 16*(2), 265–73.

Rumberger, Russell W. (1987, Summer). "High School Drop-outs: A Review of Issues and Evidence." *Review of Educational Research,* 57(2), 101–21.

Sadker, Myra, & Sadker, David. (1994). *Failing at Fairness.* New York: Charles Scribner's Sons.

Sahlins, Marshall D., & Service, Elman R. (Eds.). (1960). *Evolution and Culture.* Ann Arbor: University of Michigan Press.

Sailer, Steve. (1997, July 14). "Is Love Colorblind?" *National Review.*

"Same-Sex Partnerships." (2000, February 18). Issues and Controversies on File, 49–56.

Samovar, Larry A., Porter, Richard, & Jain, Nemi C. (1981). *Understanding Intercultural Communication.* Belmont, CA: Wadsworth.

Sandburg, Carl. (1916). *Chicago Poems.* New York: Henry Holt.

Sanders, Alvin J., & Long, Larry. (1987, January). "New Sunbelt Migration Patterns." *American Demographics,* 38–41.

Sandor, Gabrielle. (1994, June). "The 'Other' Americans." *American Demographics,* 36–42.

Sapir, Edward. (1961). *Culture, Language and Personality.* Berkeley, CA, and Los Angeles: University of California Press.

Scarce, Rik. (1994, July). "(No) Trial (But) Tribulations," *Journal of Contemporary Ethnography,* 23(2), 123–49.

Scheck, Barry, Neufeld, Peter, & Dwyer, Jim. (2000). *Actual Innocence.* New York: Random House.

Schneider, Joseph W., & Conrad, Peter. (1983). *Having Epilepsy: The Experience and Control of Illness.* Philadelphia: Temple University Press.

Schumpeter, Joseph A. (1950). *Capitalism, Socialism and Democracy* (3rd ed.). New York: Harper Torchbooks.

Schur, Edwin M., & Bedau, Hugo A. (1974). *Victimless Crimes: Two Sides of a Controversy.* Englewood Cliffs, NJ: Prentice-Hall.

Science Service. (1998). Westinghouse Science Talent Search Science Service Database. Westinghouse Foundation.

Scott, Lawrence. (1990). "Rational Decision Making About Marriage and Divorce." *Virginia Law Review, 76,* 9–94.

Service, Elman R. (1975). *Origins of the State and Civilization.* New York: Norton.

Sharp, Lauriston. (1952). "Steel Axes for Stone-Age Australians." *Human Organization, 11,* 17–22.

Shattuck, R. (1980). *The Forbidden Experiment.* New York: Farrar, Straus & Giroux.

Shaw, Clifford R., & McKay, Henry D. (1942). *Juvenile Delinquency and Urban Areas.* Chicago: University of Chicago Press.

———. (1931). "Social Factors in Juvenile Delinquency." In *National Committee on Law Obser-*

vance and Law Enforcement, Report on the Causes of Crime (Vol. 2). Washington, DC: U.S. Government Printing Office.

Sheldon, W. H., Hartl, E. M., & McDermott, E. (1949). *The Varieties of Delinquent Youth.* New York: Harper.

Sheldon, W. H., & Stevens, S. S. (1942). *The Variety of Temperament.* New York: Harper.

Sheldon, W. H., & Tucker, W. B. (1940). *The Varieties of Human Physique.* New York: Harper.

Sheler, Jeffrey L. (1997, December 15). "Dark Prophecies." *U.S. News and World Report,* 62–63, 64, 68–71.

Shelton, Deborah L. (1999, March 2). "Scientists Study Why Women Get Certain Diseases More Readily Than Men Do." *Los Angeles Times.*

Shepardson, Mary. (1963). *Navajo Ways in Government.* Manasha, WI: American Anthropological Association.

Shibutani, Tamotsu. (1966). *Improvised News: A Sociological Study of Rumor.* Indianapolis: Bobbs-Merrill.

Shils, Edward. (1971/1960). "Mass Society and Its Culture." In Bernard Rosenberg & David Manning (Eds.), *Mass Culture Revisited.* New York: Van Nostrand.

———. (1968). *Political Development in the New States.* The Hague: Mouton.

Shorto, Russell. (1997, December 7). "Muslims in the United States." *The New York Times Magazine,* 60–61.

Simmel, Georg. (1957). "Fashion." *American Journal of Sociology, 62,* 541–88.

———. (1950). *The Sociology of Georg Simmel.* Kurt Wolff (Ed.). New York: The Free Press.

Simon, Rita J., & Alstein, Howard. (1987). *Transracial Adoptees and Their Families: A Study of Identity and Commitment.* New York: Praeger.

Simon, Julian L., & Kahn, Herman (Eds.). (1984). *The Resourceful Earth: A Response to Global 2000.* New York: Basil Blackwell.

Simpson, George E., & Yinger, Milton. (1972). *Racial and Cultural Minorities: An Analysis of Prejudice and Discrimination* (4th ed.). New York: Harper & Row.

Sjoberg, Gideon. (1956). *Preindustrial City: Past and Present.* New York: The Free Press.

Skocpol, Theda. (1979). *States and Social Revolutions.* New York: Cambridge University Press.

Slater, Philip. (1966). *Microcosm: Structural, Psychological, and Religious Evolution in Groups.* New York: John Wiley.

Smelser, Neil J. (1971). "Mechanisms of Change and Adjustment to Change." In George Dalton (Ed.), *Economic Development and Social Change.* Garden City, NY: Natural History Press.

———. (1962). *Theory of Collective Behavior.* New York: The Free Press.

Smith, Adam. (1969/1776). *An Inquiry Into the Nature and Causes of the Wealth of Nations* (Vols. 1, 2). Chicago: University of Chicago Press.

Smith, Ted J., & Scarborough, Melanie. (1992, May 19). "A Startling Number of American Children in Danger of Starving: A Case History of Advocacy Research." Paper presented at the meetings of the American Association for Public Opinion Research.

Smith, Tom. (1984, June). "America's Religious Mosaic." *American Demographics,* 19–23.

Sniffen, Michael J. (1999). "Most Female Crime Is Simple Assault." *Associated Press.* December 6.

Solecki, Ralph. (1971). *Shanidar: The First Flower People.* New York: Knopf.

Sommers, Christina Hoff. (2000). *The War Against Boys: How Misguided Feminism Is Harming Our Young Men.* New York: Simon & Schuster.

———. (1994). *Who Stole Feminism?* New York: Simon & Schuster.

Sosin, Michael. (1986). *Homelessness in Chicago.* Chicago: School of Social Service Administration, University of Chicago.

Spengler, Oswald. (1932). *The Decline of the West.* New York: Knopf.

Spicer, Edward H. (1962). *Cycles of Conquest.* Tucson: University of Arizona Press.

Spitz, Rene A. (1945). "Hospitalism: An Inquiry Into the Genesis of Psychiatric Conditions in Early Childhood." In Anna Freud et al. (Eds.), *The Psychoanalytic Study of the Child.* New York: International University Press.

Sprecher, S., & Metts, S. (1989). "Development of the 'Romantic Beliefs Scale' and Examination of the Effects of Gender and Gender-Role Orientation." *Journal of Personal and Social Relationships, 6,* 387–411.

Squire, Peverill. (1988, Spring). "Why the 1936 Literary Digest Poll Failed." *Public Opinion Quarterly, 52,* 125–33.

Stark, Rodney, & Bainbridge, William Sims. (1985). *The Future of Religion: Secularization, Revival, and Cult Formation.* Berkeley, CA: University of California Press.

Stark, Rodney, & Glock, Charles Y. (1968). *American Piety: The Nature of Religious Commitments.* Berkeley and Los Angeles: University of California Press.

Starr, Paul. (1992). "Social Categories and Claims in the Liberal State." *Social Research, 59* (2), 263–96.

———. (1982). *The Social Transformation of American Medicine.* New York: Basic Books.

Stellin, Susan. (2001, February 19). "Number of New Internet Users Is Growing." *The New York Times,* C3.

Steward, Julian H. (1955). *The Theory of Culture Change: The Methodology of Multilineal Evolution.* Urbana, IL: University of Illinois Press.

Stinchcombe, Arthur L. (1969). "Some Empirical Consequences of the Davis-Moore Theory of Stratification." In Jack L. Roach, Llewellyn Gross, & Orville R. Gursslin (Eds.), *Social Stratification in the United States*. Englewood Cliffs, NJ: Prentice-Hall.

Stoll, Clifford. (1999) *High Tech Heretic*. New York: Doubleday.

———. (1995). *Silicon Snake Oil*. New York: Doubleday.

Stouffer, Samuel A. (Ed.). (1950). *The American Soldier*. Princeton, NJ: Princeton University Press.

Stout, David. (1999, September 10). "U.S. Report Details World's Religious Persecution." *The New York Times*, A14.

Strauss, Anselm, & Glaser, Barney. (1975). *Chronic Illness and the Quality of Life*. St. Louis: Mosby.

Stulgoe, John R. (1984, February/March). "The Suburbs." *American Heritage, 35* (2), 21–36.

Sullivan, Andrew. (1995). *Virtually Normal: An Argument About Homosexuality*. New York: Knopf.

Sulloway, F. J. (1996). *Born to Rebel: Birth Order, Family Dynamics and Creative Lives*. New York: Pantheon Books.

Susskind, Ron. (1998). *A Hope in the Unseen*. New York: Broadway Books.

Sutherland, Edwin H. (1961). *White Collar Crime*. New York: Holt, Rinehart and Winston.

———. (1940). "White Collar Criminality." *American Sociological Review, 40*, 1–12.

———. (1924). *Criminology*. New York: Lippincott.

Sutherland, Edwin H., & Cressey, D. R. (1978). *Principles of Criminology* (10th ed.). Chicago: Lippincott.

Suttles, G. (1972). *The Social Construction of Communities*. Chicago: University of Chicago Press.

———. (1968). *The Social Order of the Slum*. Chicago: University of Chicago Press.

Suzuki, David, & Knudtson, Peter. (1989). *Genetics: The Clash Between the New Genetics and Human Values*. Cambridge, MA: Harvard University Press.

Szymanski, Albert T., & Goertzel, Ted George. (1979). *Sociology: Class, Consciousness, and Contradictions*. New York: Van Nostrand.

Talbot, Margaret. (1997, December 22). "Washington Scene: Married in a Mob." *The New Republic*.

Tannen, Deborah. (1998). *The Argument Culture: Moving From Debate to Dialogue*. New York: Random House.

———. (1994). *Talking From 9 to 5*. New York: William Morrow.

———. (1990). *You Just Don't Understand: Women and Men in Conversation*. New York: William Morrow.

Taylor, John Wesley. (1987). *Self-Concept in Home Schooling Children*. Doctoral dissertation, Andrews University.

Taylor, O. L. (1990). *Cross-Cultural Communication: An Essential Dimension of Effective Education* (Rev. ed.). Washington, DC: The Mid-Atlantic Equity Center.

Teigen, K. H. (1986). "Old Truths or Fresh Insight? A Study of Students' Evaluation of Proverbs." *British Journal of Social Psychology, 25*, 43–50.

Terrace, Herbert S., Petitto, L. A., Sanders, R. J., & Bever, T. G. (1979). "Can an Ape Create a Sentence?" *Science, 206*, 891–902.

Thernstrom, Stephen, & Thernstrom, Abigail. (1997). *America in Black and White*. New York: Simon & Schuster.

Thomas, Karen. (1994, April 6). "Learning at Home: Education Outside School Gains Respect." *USA Today*, 5D.

Thomas, W. I. (1928). *The Child in America*. New York: Knopf.

Tierney, John J., Jr. (2000, August). "The World of Child Labor." *The World & I, 15*, 54.

Tiger, Lionel, & Fox, Robin. (1971). *The Imperial Animal*. New York: Holt, Rinehart and Winston.

Tilly, Charles. (1985). "Does Modernization Breed Revolution?" In Jack A. Goldstone (Ed.), *Revolutions: Theoretical, Comparative, and Historical Studies*. New York: Harcourt Brace Jovanovich.

———. (1978). *From Mobilization to Revolution*. Reading, MA: Addison Wesley.

Times Mirror Center for the People and the Press. (1991). *The Pulse of Europe: A Survey of Political and Social Values and Attitudes* (sec. VIII). Washington, DC: Times Mirror Center.

Tischler, Henry L. (1994, March). "The Message Behind the Image: A Comparison of Rock Music Videos With Country Music Videos." Paper presented at the Eastern Sociological Society Meetings.

Tönnies, Ferdinand. (1963). *Community and Society*. New York: Harper & Row. (Originally published in German as *Gemeinschaft und Gesellschaft* in 1887.)

Torrey, E. Fuller. (1988, March). "Homelessness and Mental Illness." *USA Today Magazine*, 26–30.

Townsend, Bickley. (1987, March). "Back to the Future." *American Demographics*, 10.

Toynbee, Arnold. (1946). *A Study of History*. New York and London: Oxford University Press.

Trachtenberg, Joshua. (1961). *The Devil and the Jews*. New York: Meridian Books.

Transparency International. (2000, September 13). "TI Press Release: Transparency International Releases Year 2000 Corruption Perception Index." Berlin.

Turnbull, Colin. (1972). *The Mountain People*. New York: Simon & Schuster.

Turner, Ralph H. (1964). "Collective Behavior." In R. E. L. Faris (Ed.), *Handbook of Modern Sociology*. Chicago: Rand McNally.

Tylor, E. (1958/1871). *Primitive Culture: Researches Into the Development of Mythology, Philosophy, Religion, Art and Custom* (Vol. 1). London: John Murray.

Ulc, Otto. (1975/1969). "Communist National Minority Policy: The Case of the Gypsies in Czechoslovakia." In Norman R. Yetman & C. Hoy Steele (Eds.), *Majority and Minority: The Dynamics of Racial and Ethnic Relations*. Boston: Allyn & Bacon.

UNAIDS. (2000). *AIDS Epidemic Update*. December . New York: United Nations

———. (1997, December). *Report on the Global HIV/AIDS Epidemic*. New York: United Nations.

UNICEF. (1997). *The State of the World's Children: 1997*. New York: United Nations Statistical Office and Population Division.

———. (1993). *The State of the World's Children: 1993*. New York: Oxford University Press.

———. (1993). *The Progress of Nations: 1993*. New York: United Nations Statistical Office and Population Division.

The United Nations. (2000). *World Urbanization Prospects: The 1999 Revision*. New York: United Nations.

———. (1997). *Women's Indicators and Statistics Database*. New York: United Nations.

———. (1996). "World Population Prospects: The 1996 Revision (Annex II and II)." New York: United Nations.

U.S. Bureau of the Census. (2001). *Statistical Abstract of the United States: 2000*. 120th ed. Washington. DC.

———. (2000, January 13). "Projections of the Resident Population by Race, Hispanic Origin, and Nativity: Middle Series, 2001 to 2005."

———. (2000). *Statistical Abstract of the United States: 1999*. 119th ed. Washington, DC.

———. (1999). "Educational Attainment in the United States" March. P205–28.

———. (1976). *Historical Statistics of the United States: Colonial Times to 1970*. Washington, DC: U.S. Government Printing Office.

———. (1998, March). "Income in 1998 by Educational Attainment for People 18 Years Old and Over by Age, Sex, Race, and Hispanic Origin."

———. (1998, March). "Marital Status and Living Arrangements." Current Population Survey.

———. (1998, March). *World Population Profile: 1998*.

———. (1997, September 29). "Country of Origin and Year of Entry Into the U.S. of the Foreign Born: March 1997."

———. (1997). *Current Population Reports*, Series P-25, no. 1130. Washington, DC: U.S. Government Printing Office.

———. (1997). *Current Population Reports*, Series P-25, no. 1018. Washington, DC: U.S. Government Printing Office.

———. (1997, March). *Current Population Survey*. Washington, DC: U.S. Government Printing Office.

———. (1997). *Statistical Abstract of the United States: 1997* (117th ed.). Washington, DC: U.S. Government Printing Office.

———. (1997). *World Population Profile: 1996*. Washington, DC: U.S. Government Printing Office.

———. (1996). International Data Base; United Nations Secretariat. "World Population Prospects: 1996 Revision."

———. (1996). "Marital Status and Living Arrangements 1994." *Current Population Reports*, Series P204–84.

———. (1996). *Statistical Abstract of the United States: 1996* (116th ed.). Washington, DC: U.S. Government Printing Office.

———. (1994). *Educational Attainment in the United States: March 1993 and 1992*. Washington, DC: U.S. Government Printing Office.

———. (1993). "Population Profile of the United States: 1993." *Current Population Reports*, Series P-23, No. 185. Washington, DC: U.S. Government Printing Office.

———. (1991). "Marital Status and Living Arrangements: March 1990." *Current Population Reports*, Series P-20, No. 450 (p. 17). Washington, DC: U.S. Government Printing Office.

———. (1990). "Detailed Ancestry Groups for States." *Current Population Survey* (pp. 1–2). Washington, DC: U.S. Government Printing Office.

U.S. Bureau of the Census, Population Division. (1999, May). "Is Childlessness Among American Women on the Rise?" Working Paper No. 37.

———. (1997, November). *Population Estimates Program*. Washington, DC: U.S. Government Printing Office.

U.S. Centers for Disease Control and Prevention. (2000). Fact Book for the Year 2000: Working to Prevent and Control Injury in the United States.

———. (2000). Health. United States: 1998.

———. (2000, December). *HIV/AIDS Surveillance Report*. 12(1).

———. (1999, June 30). National Vital Statistics Reports. "Deaths: Final Data for 1997." 47(19).

———. (1997, January). *HIV/AIDS Surveillance Report*. Washington, DC.

———. (1997). Nursing Home Survey.

———. (1997). National Health Interview Survey. Washington, DC. U.S. Government Printing Office.

———. (1995, February). *HIV/AIDS Surveillance Report, 6*(1). Washington, DC: U.S. Government Printing Office.

U.S. Department of Commerce. (1991, September 26). "1990 Median Household Income Dips, Census Bureau Reports in Annual Survey." *U.S. Department of Commerce News.*

U.S. Department of Health, Education, and Welfare. National Center for Health Statistics. (2000). "Health. United States."

U.S. Department of Labor. Bureau of Labor Statistics. (2000, May). "Highlights of Women's Earnings in 1999." Report 943.

U.S. Department of Justice. Bureau of Justice Statistics. (2000, December). "Capital Punishment, 1999."

———. (2000, October). "Urban, Suburban, and Rural Victimization, 1993–98."

———. (2000, August). "Prisoners in 1999."

———. (1999, December) "Capital Punishment, 1998."

———. (1998, January) *Crime in the United States, 1997.*

———. (1998, January). "Prison and Jail Inmates at Mid-Year 1997." *Bureau of Justice Statistics Bulletin.*

———. (1997). *Criminal Victimizations in the United States, 1996.* Washington, DC: U.S. Government Printing Office.

———. (1997, December). "Capital Punishment, 1996." *Bureau of Justice Statistics Bulletin.*

———. (1996, May). "Households Touched by Crime, 1995." *Bureau of Justice Statistics Bulletin.*

———. (1994, March). "Women in Prison." *Bureau of Justice Statistics Bulletin.*

U.S. Department of Justice, Office of Justice Programs. (1996). *The Nature of Homicide: Trends and Changes.* Washington, DC: U.S. Government Printing Office.

U.S. Office of Management and Budget. (2001). *Budget of the United States Government Fiscal Year 2000.* Washington, DC: U.S. Government Printing Office.

Usdansky, Margaret L. (1993, April 12). "Gay Couples by the Numbers." *USA Today,* 8A.

Vago, Steven. (1988). *Law and Society* (2nd ed.). Englewood Cliffs, NJ: Prentice Hall.

———. (1980). *Social Change.* New York: Holt, Rinehart and Winston.

Van de Kaa, Dirk J. (1987, March). "Europe's Second Demographic Transition." *Population Bulletin.*

Vanfossen, Beth E. (1979). *The Structure of Social Inequality.* Boston: Little, Brown.

Van Lawick-Goodall, Jane. (1971). *In the Shadow of Man.* Boston: Houghton Mifflin.

"Vegetarianism." (1999). Issues and Controversies on File. November 26. 481-88.

Veltfort, Helene, & Lee, George E. (1943, April). "The Coconut Grove Fire: A Study in Scapegoating." *Journal of Abnormal and Social Psychology, Clinical Supplement, 38,* 138–54.

Verba, Sidney. (1972). *Small Groups and Political Behavior: A Study of Leadership.* Princeton, NJ: Princeton University Press.

Villaraigosa, Antonio R. (2000, Summer). "America's Urban Agenda: A View From California" *Brookings Review, 18*(3). 46–49.

Walker, T. B., & Elrod, L. D. (1993). "Family Law in the Fifty States: An Overview." *Family Law Quarterly, 26,* 319–421.

The Wall Street Journal. (1994, May 24). "Medical Schools Put Women in Curricula," B1.

Wallerstein, Judith S., Lewis, Julia M., & Blakeslee, Sandra. (2000). *The Unexpected Legacy of Divorce.* New York: Hyperion.

Wallerstein, Immanuel. (1991). *Geopolitics and Geoculture: Essays on the Changing World.* Cambridge, England: Cambridge University Press.

——— (1979). *The Capitalist World Economy.* New York: Cambridge University Press.

——— (1974). *The Modern World System: Capitalist Agriculture and the Origins of European World Economy in the Sixteenth Century.* New York: Academic Press.

Watkins, Paul. (1993). *Stand Before Your God: A Boarding School Memoir.* New York: Random House.

Watson, James L. (Ed.). (1997). *Golden Arches East: McDonald's In East Asia.* Berkeley: University of California Press.

Watson, J. B. (1925). *Behavior.* New York: Norton.

Weber, Max. (1958/1921). *The City.* New York: Collier.

———. (1957). *The Theory of Social and Economic Organization.* New York: The Free Press.

———. (1956). "Some Consequences of Bureaucratization." In J. P. Mayer (Trans.), *Max Weber and German Politics* (2nd ed.). New York: The Free Press.

———. (1930/1920). *The Protestant Ethic and the Spirit of Capitalism* (Talcott Parsons, Trans.). New York: Scribner.

———. (1968/1922). *Economy and Society* (Ephraim Fischoff, Trans.). New York: Bedminster Press.

Wechsler, Henry. (2000). "1999 College Alcohol Studies." Harvard School of Public Health.

Weeks, John R. (1994). *Population* (5th ed.). Belmont, CA: Wadsworth.

Weiss, Robert S. (1979). *Going It Alone: The Family Life and Social Situation of the Single Parent.* New York: Basic Books.

Weitzman, Lenore J. (1985). *The Divorce Revolution: The Unexpected Social and Economic Consequences for Women and Children in America.* New York: The Free Press.

Westoff, Charles F., & Jones, Elise F. (1977, September/October). "The Secularization of Catholic Birth Control Practice." *Family Planning Perspectives, 9,* 96–101.

Wheelwright, Philip. (1959). *Heraclitus.* Princeton, NJ: Princeton University Press.

Whitebook, M., Phillips, D., & Howe, C. (1993). *National Child Care Staffing Study Revisited: Four Years in the Life of Center-Based Care.* Oakland, CA: Child Care Employee Project.

Whorf, B. (1956). *Language, Thought, and Reality.* Cambridge, MA: MIT Press.

Whyte, William Foote. (1943). *Street Corner Society.* Chicago: University of Chicago Press.

Whyte, W. H. (1988). *City: Rediscovering the Center.* New York: Doubleday.

———. (1956). *The Organization Man.* New York: Simon & Schuster.

Williams, Lena. (2000). *It's the Little Things.* New York; Harcourt.

———. (1997, December 14). "It's The Little Things." *The New York Times.*

Wilson, Barbara Foley, & Clarke, S. C. (1992, June). "Remarriages: A Demographic Profile." *Journal of Family Issues, 13,* 123–41.

Wilson, Doris James, & Ditton, Paula M. (1999, January). "Truth in Sentencing in State Prisons." U.S. Department of Justice. Office of Justice Programs. Bureau of Justice Statistics. Special Report. NCJ 1700–32.

Wilson, Edmund. (1969). *The Dead Sea Scrolls 1947–1969* (Rev. ed.). London: W. H. Allen.

Wilson, Edward O. (1994). *Naturalist.* Washington, DC: Island Press.

———. (1979). *Sociobiology* (2nd ed.). Cambridge, MA: Belknap.

———. (1978). *On Human Nature.* Cambridge, MA: Harvard University Press.

———. (1975). *Sociobiology: The New Synthesis.* Cambridge, MA: Harvard University Press.

Wilson, James Q. (1996, March). "Against Homosexual Marriage." *Commentary.*

Wilson, James Q., & Herrnstein, Richard J. (1985). *Crime and Human Nature.* New York: Simon & Schuster.

Wilson, Woodrow. (1915, May 10). "Americanism." Speech given at Convention Hall, Philadelphia, PA.

Winner, Ellen. (1996). *Gifted Children: Myths and Realities.* New York: Basic Books.

Wirth, Louis. (1944, March). "Race and Public Policy." *Scientific Monthly, 58*(4), 303.

———. (1938). "Urbanism as a Way of Life." *American Journal of Sociology, 64,* 1–24.

Wolf, Eric. (1969). *Peasant Wars of the Twentieth Century.* New York: Harper & Row.

World Bank. (1984). "Measuring the Value of Children." In *World Development Report 1984.* New York: Oxford University Press.

The World Health Organization. (2000, December). *AIDS Epidemic Update.*

———. (2000, June 4). "WHO Issues New Healthy Life Expectancy Rankings, Japan Number One in New 'Healthy Life' System."

———. (1996). *The World Health Report 1995: Bridging the Gaps.*

Yinger, J. Milton. (1970). *The Scientific Study of Religion.* New York: Macmillan.

Zeitlin, I. M. (1981). *Social Condition of Humanity.* New York: Oxford University Press.

Zimring, Franklin E., & Hawkins, Gordon. (1997). *Crime Is Not the Problem: Lethal Violence in America.* New York: Oxford University Press.

Zuckoff, Mitchell. (1999, December 12). "Choosing Naia: A Family's Journey." *The Boston Globe.*

LITERARY CREDITS

Photo Credits

Chapter 1
p. 2 © Henry Horentstien / Photonica; p. 5 © George Belleros / Stock, Boston; p. 9 © T. Howard / Woodfin Camp & Associates; p. 10 © Bettmann / Corbis; p. 13 © Bettmann / Corbis p. 14 © Bettmann / Corbis; p. 17 Brown Brothers; p. 18 Culver Pictures

Chapter 2
p. 34 © Steve Chenn / Corbis; p. 35 © James Carroll / Stock, Boston; p. 36 © Jeff Greenberg / PhotoEdit; p. 37 © Antonio Mari / Gamma Liaison; p. 40 © David Weintraub / Photo Researchers

Chapter 3
p. 56 Boston Filmworks; p. 59 © Tony Freeman / PhotoEdit; p. 60 © M. & E. Bernheim / Woodfin Camp & Associates; p. 62 (left) © Minosu-Scorpio / Sygma / Corbis; p. 62 (right) © George Holton / Photo Researchers; p. 63 Foto du Monde / Picture Cube; p. 64 Neil Adams / George Washington University; p.68 © Wartenberg / Picture Press / Corbis; p. 70 © Jonathan Nourok / PhotoEdit; p. 71 © J. R. Heller / Stock, Boston; p. 73 Roger and Deborah Fouts; p. 77 © Bruce Rosemblum / Picture Cube; p. 78 Kenneth Good

Chapter 4
p. 86 Boston Filmworks; p. 88 © Matusou / Monkmeyer; p. 90 The Granger Collection, New York; p. 93 © J.W. Myers / Stock Market / Corbis; p. 94 Boston Filmworks; p. 95 Department of Special Collections, University of Chicago Library; p. 96 Boston Filmworks; p. 99 © Robert Frerck / Tony Stone Images; p. 101 © Picture Press / Corbis; p.103 © Owen Franken / Corbis; p. 105 © Raymond Gehman / Corbis; p. 110 © Charles Gupton / Stock, Boston

Chapter 5
p. 118 Boston Filmworks; p. 120 © P. Damien / Tony Stone Images; p. 123 © Bob Daemmrich / Image Works; p. 127 © Frank Sitman / Stock, Boston; p. 128 AP / Wide World; p. 131 © Bard Martin / Corbis; p. 139 © Bob Daemmrich / Image Works; p. 141 © Mark Richards / PhotoEdit; p. 143 (left) © Michael Newman / PhotoEdit; p. 143 (right) © Spencer Grant / Monkmeyer

Chapter 6
p. 154 © Superstock; p. 156 © Carey / Image Works; p. 157 (left) Karen Preuss; p. 157 (right) Boston Filmworks; p. 158 © Robert Capa / Magnum; p. 169 © Jonathan Kaplan / Panos Pictures; p. 172 Boston Filmworks; p. 173 © Bettmann / Corbis; p. 178 Finger Matrix; p. 182 AP / Wide World

Chapter 7
p. 196 © Linonel Delevingne / Stock, Boston; p. 198 © Bettmann / Corbis; p. 199 © Andrew Holbrooke / Black Star; p. 202 © AFP / Corbis; p. 204 © Neal Preston / Corbis; p. 209 Boston Filmworks; p. 212 © Bob Daemmrich / Image Works; p. 214 © John Lei / Stock, Boston; p. 215 © Mary Ellen Mark / Falkland Road; p. 221 © Michael Dyer / Stock, Boston; p. 222 © Gloria Karlson / Picture Cube

Chapter 8
p. 230 © Patrick Giardino / Corbis; p. 232 © David Young-Wolff / PhotoEdit; p. 236 © Raghu Rai / Magnum; p. 242 Harvard University, Office of New and Public Affairs; p. 245 (left) National Antropological Archives, Smithsonian Institution, Washington, D.C.; p. 245 (right) National Antropological Archives, Smithsonian Institution, Washington, D.C.; p. 246 Brown Brothers; p. 248 © Bettmann / Corbis; p. 253 © Hazel Hankin / Stock, Boston; p. 258 Brown Brothers; p. 259 © Lionel Delevingne / Stock, Boston

Chapter 9
p. 268 © Superstock; p. 270 © Dana White / PhotoEdit; p. 271 The Hutchinson Library; p. 274 © Jim Whitmer / Stock, Boston; p. 276 © The Cartoon Bank; p. 278 Boston Filmworks; p. 280 Boston Filmworks; p. 281 © Mike Adaskaveg / Boston Herald

Chapter 10
p. 294 ©Larry Burrows / LIFE Magazine / Time Inc.; p. 318 Jan Doyle; p. 298 Boston Filmworks; p. 301 © Raghu Rai / Magnum; p. 297 © Trinette Reed / Corbis; p. 302 © Laura Dwaight / Corbis; p. 304 © Bettmann / Corbis; p. 310 Eliot Holtzman

Chapter 11
p. 330 © Eric Van Den Brulle / Photonica; p. 332 Dennis Barnes; p. 333 © Lindsey Hebberd / Corbis; p. 336 Robert Harding Picture Library; p. 338 Betty Press / Monkmeyer; p. 340 Robert Freid Photography; p. 343 © Joseph Nettis / Tony Stone Images; p. 346 © G. Giansanti / Tony Stone Images; p. 348 Boston Filmworks; p. 349 © M. Oppersdorff / Photo Researchers; p. 351 Ali K. Nomachi / Pacific Press Service

Chapter 12
p. 358 © Ed Honowitz / Tony Stone Images; p. 361 (left) Brown Brothers; p. 361 (right) A. States; p. 362 © O'Brien Productions / Corbis; p. 363 © Devorah L. Martin / Unicorn Stock Photography; p. 364 © Jean-Claude Lejeune / Stock, Boston; p. 367 ©Bonnie Kamin; p. 369 Brown Brothers; p. 370 Thomas Victor; p. 372 © M. Yamashita / Woodfin Camp & Associates; p. 375 © Bob Daemmrich / Stock, Boston; p. 378 © Robin Sachs / PhotoEdit

Chapter 13
p. 388 © Craig Aurness / Corbis; p. 391 AP / Wide World Photos; p. 393 © Reuters NewMedia Inc./ Corbis; p. 396 © Bettmann / Corbis; p. 399 © Steve Liss / TIME Magazine; p. 401 Reuters / Yuean Manwang / Bettmann / Corbis; p. 404 Bettmann / Corbis; p. 411 © Bob Daemmrich / Image Works

Chapter 14
p. 420 © Sebastio Salgado / Contact Images; p. 422 © A. Reininger / Woodfin Camp & Associates; p. 425 Brown Brothers; p. 427 Danny Clinch Photography; p. 428 AP / Wide World Photos; p. 435 © Baldev / Sygma / Corbis; p. 436 AP / Wide World Photos; p. 437 Boston Filmworks; p. 439 Brown / Offshoot Stock; p. 440 © Lionel Delevingne / Stock, Boston

Chapter 15
p. 448 © Doug Wilson / Corbis; p. 451 S. Lloyd, The Art of the Ancient Near East, Oxford University Press, 1961, p. 264; p. 452 George Eastman House Collection; p. 455 © John Elk / Stock, Boston; p. 459 © B. Aspen / Envision; p. 463 © Gloria Karlson / Picture Cube; p. 468 Levitt Homes

Chapter 16
p. 476 © Nick Vedros / Tony Stone Images; p. 479 Corbis; p. 482 © David Turnley / Corbis; p. 483 © Owen Franken / Stock, Boston; p. 486 © K. Preuss / Image Works; p. 488 C. Nacke; p. 494 © MacDuff Everton / Image Works; p. 495 © Tom McCarthy / PhotoEdit; p. 496 © Walter Hodges / Corbis; p. 501 © Burk Uzzle / Zuma Press

Chapter 17
p. 508 © Reuters NewMedia Inc./ Corbis; p. 510 © Bettmann / Corbis; p. 511 © L. Mason / Sygma / Corbis; p. 514 © Catherine Karnow / Woodfin Camp & Associates; p. 517 (left) The Elizabeth Day McCormick Collction, Museum of Fine Arts, Boston; p. 517 (right) Pine Street Inn, Museum of Fine Arts, Boston; p. 520 New York Daily News Picture Collection; p. 521 © Michael Siluk / Picture Cube; p. 523 Alek Keplicz / AP / Wide World Photos; p. 524 © B. Glinn / Magnum

Chapter 18
p. 532 © Jill Fehrenbacher; p. 534 Brown Brothers; p. 537 © R. Lord / Image Works; p. 541 © Roger Wood / Corbis; p. 544 NASA, Ames Research Center

AUTHOR INDEX

SUBJECT INDEX

A

Abortion
    as form of birth control, 437–438
    in selected countries, 438
Abstract ideals religion, 337
Academic skills, 362–364
ACLU (American Civil Liberties Union), 238
ACT (American College Test), 378
Acting crowd, 513
Adams, John, 247
Adaptation
    cultural, 69–70
    Merton's typology of individual modes of, 163
Adolescence
    binge drinking during, 485
    development during, 96
    peer group support during, 100–101
    socialization during, 280
    See also Children
Adoption
    high-tech fertility treatment vs., 309
    transracial, 260–261
Adult socialization, 107–108, 110
AFL-CIO, 524
African-Americans
    aging and, 498
    Black middle class and, 254
    health/life expectancy of, 481, 482
    holding political office, 407
    immigration of, 252–253
    interpretations of nonverbal behavior by, 125
    interracial marriage and, 235
    perceptions of police by, 178
    as political force, 407–408
    poverty levels/impact on, 214, 221
    preventing illness among, 491
    Social Darwinism justification for repression of, 12
    suburban living and, 467
    transracial adoption and, 260–261
    unequal access to education and, 371–373
    See also Minority groups
Age
    distribution of arrests (1998) by, 173
    fertility and marriage, 433
    health and, 483–484
    infant health risk and maternal, 496
    marriage and medium, 302
    voting patterns by, 406
"Age of insight," 543, 544
Age-specific death rates, 424–425
Aging
    cultural differences regarding, 110
    future trends regarding, 500–501
    global, 499–500
    marital status and, 498
    race and, 497–498
    sex ratio and, 497
    stereotypes about elderly and, 499
    wealth and, 498

Aging population, 496–500
AIDS (acquired immunodeficiency syndrome), 486–490
"AIDS in the Priesthood" series (*Kansas City Star*), 40–41
Alienation, 341
Alinsky, Saul, 520–521
All Religions Are True (Gandhi), 331
Altruistic suicide, 14–15
Alturistic behavior, 89
AMA (American Medical Association), 204
"American Education: Making It Work" (1988), 362
American exceptionalism, 64–65
American Family Association, 238
American Jewish community, 350
American Pastoral (Roth), 458–459
American Protestant Union, 251
American Revolution, 404
American society
    aspects of religion in, 344–345
    childhood socialization in, 99–107, 278, 280, 297
    comparing Japanese and, 109
    diversity of, 260–262
    homicides in, 176–177
    See also Society; United States
American Sociological Association, 7, 42
American Sociological Association's Code of Ethics, 48
American system of politics
    African-American as force in, 407–408
    growing conservatism in, 409
    Hispanics as force in, 408–409
    minorities/women holding office in, 407
    role of media in, 409–410
    special-interest groups and, 410–412
    two-party system of, 405
    voting behavior in, 405–407
American values (1960s/today), 410
Americanization Movement, 252
Amnesty International, U.S.A., 184
Anglo conformity, 243
Animal Liberation Front, 48
Animals
    culture/language and, 71–74
    ethics of research on, 73
    ethology study of, 272
    socialization research on, 91
Animism, 336
Anna case study, 90
Annihilation, 246–247
Anomie theory, 163, 341
Anti-Semitism
    Jewish immigration due to European, 256
    of Nazi Germany, 245–247
Appellate courts, 179
Apple Computer, 199
Arapesh tribe (New Guinea), 275
Argument culture, 129
Aristocracy (oligarchy type), 391–392

Arranged marriage, 300
Arrest age distribution (1998), 173
Ascribed statuses, 131
Asian-Americans
    immigration of, 256–258
    interracial marriage and, 235
    See also Minority groups
ASL (American Sign Language), 72
Assimilation, 240–241, 243
Association of American Medical Colleges, 484
Association groups, 139–141
Attachment disorder, 91–92
Authority
    peer group vs. family, 100–101
    political, 390–391
Autocracy, 398
Average/mean, 44

B

Baby-boom aging generation, 500–501
Ban land mines campaign, 394
Behavior
    African/white Americans interpretations of, 125
    altruistic, 89
    conditioning instinctive, 88–89
    control over group member's, 137–138
    ethnomethodology of, 123, 126
    gender differences in violent, 105, 107
    illness risk factors due to, 491–492
    nonmaterial culture governing, 61–62
    personality as patterns of, 88
    reference groups and, 138–139
    See also Collective behavior; Deviant behavior
Behavioral theories, 162
Beliefs
    Apocalyptic, 344
    in magic, 334–335
    religious, 333
    social movement and, 521
    widespread American religious, 344–345
    See also Values
The Bell Curve (Hernstein and Murray), 242
Benjamin N. Cardozo School of Law, 180
Beverly Hills Supper Club fire (1977), 519
Bilateral family system, 298
Bilingual education, 373
Bilocal residence, 301
Binge drinking, 485
Biology
    culture and, 58–59, 88–92
    deprivation/development research, 90–92
    deviant behavior and, 160–161
    nature vs. nurture debate and, 88–90
    sociobiology and, 89–90
Birth order, 132
Birth-control policies
    abortion as element of, 437–438
    Chinese campaign of, 421–422

PRACTICE TESTS

CHAPTER ONE The Sociological Perspective

SOCIOLOGY AS A POINT OF VIEW

1. T F Although sometimes appearing similar, sociological knowledge is quite different from common sense.

2. T F The social science most closely related to sociology is cultural anthropology.

3. T F Science is a body of systematically collected knowledge.

4. T F The sociological imagination is a process through which sociologists apply their own, carefully reasoned opinions, to solve complex social issues.

5. T F The social sciences focus on phenomenon which are far less complex than those studied by the physical sciences.

6. T F Sociology asks you to broaden your personal perspective on the world.

---

7. Which of the following is generally a major concern of the field of psychology?
  a. human society
  b. helping people solve problems
  c. the operations of government
  d. individual motivation and behavior
  e. all of the above

8. The most common unit of focus of sociology is
  a. groups.
  b. rare events
  c. individuals.
  d. unusual people

9. "What are the basic characteristics of the perpetrators of domestic assault?" This is a question most likely asked by a(n)
  a. journalist.
  b. economist.
  c. talk show host.
  d. sociologist.

10. Who developed the concept of the sociological imagination?
  a. Auguste Comte
  b. C. Wright Mills
  c. Karl Marx
  d. Émile Durkheim
  e. W.E.B. DuBois

11. The systematically arranged knowledge demonstrating the operation of general laws is
  a. general laws production.
  b. solutions to problems.
  c. science.
  d. gremlins.

12. Which of the following is not a component of the scientific method?
  a. experimentation
  b. verification
  c. observation
  d. generalization
  e. personalization

13. Proponents of applied sociology believe the work of sociology
  a. can and should be used to solve "real world" problems.
  b. can create new markets for the employment of sociologists.
  c. can develop sets of abstract theories explaining behavior.
  d. a and c only
  e. all of the above

14. Empiricism is the feature of science that stresses
  a. observable entities.
  b. abstract concepts.
  c. philosophical doctrines.
  d. moral principles.

15. Sociology is a science because it
   a. uses systematic methods.
   b. is a social science.
   c. considers findings tentative until verified.
   d. a and c only
   e. all of the above

16. What is the main difference between sociology and social work?
   a. Sociology uses theory and social work does not.
   b. Social work overlaps with psychology while sociology does not.
   c. Social workers help people solve their problems while sociologists try to understand why the problems exist.
   d. There really is no difference between sociology and social work.

17. Both sociologists and psychologists are interested in alcoholism; however, in investigating alcoholism, a sociologist would be most interested in
   a. particular persons and how much liquor they consumed.
   b. the patterns of alcohol consumption among individuals and social groups.
   c. the history of alcohol consumption in America since independence.
   d. the impact of alcohol on the destruction of an individual.

THE DEVELOPMENT OF SOCIOLOGY

18. T F Auguste Comte invented the term "sociology."

19. T F The development of symbolic interactionism is associated with G. H. Mead.

20. T F Max Weber and Karl Marx were in close agreement in their explanations of the role of religion in society.

21. T F C. Wright Mills conducted a famous sociological study of suicide.

22. T F Max Weber maintained that a system of beliefs he called "the Protestant Ethic" enabled the development of capitalism where the "Ethic" existed.

---

23. According to Durkheim, social integration refers to how
   a. individuals are bonded to social groups.
   b. educational institutions are racially integrated.
   c. the component parts of the personality are aligned.
   d. the social order is without conflict.

24. Karl Marx's primary focus was on
   a. the reasons behind human social interaction.
   b. the relationship of people to the means of production.
   c. human social groups.
   d. human social life.
   e. helping persons solve personal problems.

25. Which of the following is true of the thought of Auguste Comte?
   a. He was an advocate of empirical social philosophy.
   b. He believed all societies move through fixed stages of development.
   c. He was less concerned with defining sociology's subject matter than with showing how it would improve society.
   d. He separated the analysis of society into social statics and social dynamics.
   e. These are all true of the thought of Auguste Comte.

26. Which of the following is not true of the work of Harriet Martineau?
   a. She believed that scholars should use their research to bring about social reform.
   b. She translated into English Comte's major work.
   c. Her major research involved observing day-to-day life in the United States.
   d. She compared social stratification systems in Europe with those in America.
   e. These are all true of the work of Harriet Martineau.

27. The sociologist who pioneered the notion that society was like a living organism and who is associated with the idea of "Social Darwinism" is _____.
   a. Herbert Spencer
   b. Karl Marx
   c. George Herbert Mead
   d. Harold Garfinkel

28. According to _____, the working class sometimes participates in its own oppression by believing in the economic, political, and religious ideologies that justify ruling-class privileges.
   a. Harold Garfinkel
   b. Karl Marx
   c. Talcott Parsons
   d. Herbert Spencer

29. Karl Marx believed that the history of human societies could be viewed mainly as a history of _____.
   a. functional interdependence
   b. class struggles
   c. great leaders
   d. increasing solidarity

30. A suicide resulting from a sense of moral and social disorientation and aimlessness would be called an _____ suicide by Durkheim.
   a. egoistic
   b. anomic
   c. altruistic
   d. atomic

31. People who kill themselves out of a sense of duty to the group or self-sacrifice would be termed _____ suicides by Durkheim.
   a. egoistic
   b. altruistic
   c. anomic
   d. atomic

32. _____ was the ideology that Max Weber said justified economic success through rational, disciplined hard work.
   a. The Iron Law of Wages
   b. Social Darwinism
   c. Communism
   d. The Protestant Ethic

THE DEVELOPMENT OF SOCIOLOGY IN THE UNITED STATES

33. **T F** American sociologist Robert Merton is an important proponent of conflict theory.

34. **T F** During the early years of sociology in the United States, most of the field's development took place at the University of Chicago.

35. **T F** Manifest functions are the unintended or not readily recognized consequences of a social process.

36. **T F** The work of Talcott Parsons is most closely related to that of Marx because of Parsons' interest in conflict and exploitation.

37. **T F** Anomic suicide results from a sense of feeling disconnected from society's values.

---

38. Which of the following is **not** true of the work of W.E.B. DuBois?
   a. Because of institutionalized racism, he was never able to obtain a college degree.
   b. He advocated militant resistance to white racism.
   c. He believed that doctrines and theories had a powerful effect on social conditions.
   d. He felt that sociological studies of African Americans would have a positive effect on white public opinion.

39. High crime rates cause problems for many people. However, they also help to create jobs for large numbers of people in fields such as law enforcement, criminal law, and corrections. This latter phenomenon is an example of
   a. latent function.
   b. social conflict.
   c. social interdependence.
   d. manifest function.

40. Which of the following is true of the work of Talcott Parsons?
   a. His viewpoint elaborated on Durkheim's functionalist perspective.
   b. He portrayed society as a stable system of well-ordered, interrelated parts.
   c. He presented English translations of European thinkers such as Weber and Durkheim.

d. a and c only

e. a, b, and c

41. Egoistic suicide comes from

a. over involvement with others.

b. a general uncertainty from norm confusion.

c. overall feelings of depression resulting from economic setbacks.

d. low group solidarity and under involvement with others.

THEORETICAL PERSPECTIVES

42. T F Durkheim was one of the early proponents of the modern perspective known as functionalism.

43. T F Functionalists view society as a system of highly interrelated parts that generally operate together harmoniously.

44. T F The interactionist perspective is most closely associated with Herbert Spencer.

45. T F The Protestant Ethic was Marx's terms for the religion of the owners.

---

46. The most important features of human interaction involve people's efforts to create and manage impressions, according to

a. Erving Goffman.

c. Talcott Parsons.

b. Robert K. Merton.

d. C. Wright Mills.

47. A paradigm is an

a. unintended consequence of a scientific discovery.

b. amateur social scientist.

c. explanatory framework to guide scientific research.

d. unsuccessful attempt at suicide, according to Durkheim.

48. If, when answering the phone, you decided to test callers' taken-for-granted notions of social interaction by saying "I'm listening" instead of "Hello," you would be conducting an exercise in

a. dramaturgy.

c. symbolic interactionism.

b. ethnomethodology.

d. functionalism.

49. Conflict theory

a. attempts to understand and eliminate the sources of social conflict.

b. sees conflict in society as normal.

c. suggests that most social conflicts can be solved by effective communication.

d. sees society operating on the basis of "survival of the fittest."

50. _____ are concerned with explaining specific issues or aspects of society.

a. Grand theories

c. Inclusive theories

b. Middle-range theories

d. Micro-level theories

ESSAY QUESTIONS

51. Choose a social problem and develop at least five questions that a sociologist might ask about that problem.

52. Compare and contrast the subject matter of sociology with the other social sciences, then briefly present an overview of the career options available for a sociology major.

53. Discuss three important criteria which a researcher could use to determine whether the results of a research study were accurate.

54. Discuss the major findings from Durkheim's classic study of suicide.

55. Using common proverbs and their opposites as examples, briefly discuss how sociological thinking moves beyond common sense in accounting for human behavior.

CHAPTER TWO Doing Sociology: Research Methods

THE RESEARCH PROCESS

1. **T F** Independent and dependent variables are found in statements of causality but not in statements of association.

2. **T F** A testable statement about the relationship between two or more empirical variables is known as a hypothesis.

3. **T F** Most research problems of interest to sociologists are most easily and accurately studied by means of controlled experiments.

4. **T F** An empirical question is an issue posed in such a way that it can be studied through observation.

5. **T F** The mean is the number that occurs most often in a data set.

6. **T F** The phenomenon studied by the researcher is called the independent variable.

7. **T F** A self-fulfilling prophecy is produced when a researcher who is strongly inclined toward a particular point of view communicates that attitude to the research subjects such that their responses wind up consistent with the initial point of view.

8. **T F** A structured interview is a form of research conversation in which the questionnaire is followed rigidly.

9. **T F** Validity is the extent to which a study tests what it was intended to test.

10. **T F** The failure of *Literary Digest* to accurately predict the 1936 Landon-Roosevelt presidential election was due to having too small a sample.

11. **T F** Secondary Analysis is useful for collecting historical and longitudinal data.

12. **T F** A factor affecting the accuracy of the *Kansas City Star* Catholic priest study was the low response rate and large percentage of non-respondents in the study.

13. **T F** The major factor affecting the response rate in the *Kansas City Star* study was the voluntary nature of priest participation in the survey.

---

14. Which of the following questions would one ask to determine the quality of any opinion survey or poll?
    a. Did you ask the right people?
    b. What is the margin of error?
    c. What were the questions which were asked?
    d. all of the above
    e. none of the above

15. Which of the following is a goal of science?
    a. to propose and test theories that help us understand things and events
    b. to tell us what the proper course of action is
    c. to describe in detail particular things or events
    d. a and c only
    e. a, b, and c

16. A subset of the population that exhibits, in correct proportion, the significant characteristics of the population as a whole is known as a
    a. dependent variable.
    b. representative sample.
    c. nonrandom sample.
    d. cross-sectional study.

17. The figure that falls in the exact middle of a ranked series of scores is the
    a. median.
    b. mode.
    c. mean.
    d. average.

18. A longitudinal study is research
    a. aimed at predicting the future.
    b. investigating a population over a period of time.

c. examining a population at a given point in time.

d. conducted without the participants' knowledge.

19. Which of the following steps in the research process must come **last?**
   a. developing hypotheses
   b. defining the problem
   c. analyzing the data and drawing conclusions
   d. reviewing previous research
   e. determining the research design

20. Which of the following steps in the research process must come **first?**
   a. developing hypotheses
   b. defining the problem
   c. determining the research design
   d. reviewing previous research
   e. analyzing the data and drawing conclusions

21. "Men who live in cities are more likely to marry young than are men who live in the country." In this hypothesis, the dependent variable is
   a. a place of residence (city or country).
   b. marital status (single or married).
   c. age at marriage.
   d. sex.

22. Which of the following would be the best research method for gaining a deep and detailed understanding of everyday life within a small group?
   a. a survey with a structured interview schedule
   b. participant observation
   c. a controlled experiment
   d. Any of these methods would be equally well-suited for such a study.

23. Which of the following is a measure of central tendency that is commonly referred to as the "average"?
   a. median
   b. mode
   c. mean
   d. meridian

24. In a series of research conversations, an investigator varies the questions asked and even makes up new ones on occasion. This is an example of
   a. researcher bias.
   b. a semistructured interview
   c. a controlled experiment.
   d. random sampling.

25. Which of the following is a variable?
   a. the proportion of Americans opposed to abortion
   b. a baseball player's batting average
   c. the number of students enrolled at your college or university
   d. a and c only
   e. a, b, and c

26. Which of the following research methods, despite its advantages, is the most subjective?
   a. surveys
   b. controlled experiments
   c. participant observation

27. The result of failing to achieve a representative sample is known as
   a. researcher bias.
   b. sampling error.
   c. subjectivity.
   d. random sampling.

28. An investigation of Americans' evaluation of the president's effectiveness in foreign policy six months after taking office would be
   a. a cross-sectional study.
   b. an exit poll.
   c. an operational definition.
   d. a longitudinal study.

29. Once a researcher has defined the problem to be studied, the next step is
   a. gathering the data.
   b. selecting the sample.
   c. determining the research design.
   d. reviewing the previous research on the subject.

30. Investigators who are kept unaware of both the kinds of subjects they are studying and the hypothesis being tested are called
a. blind investigators.
b. double-blind investigators.
c. subjective researchers.
d. participant nonobservers.

31. If the results of a study cannot be replicated in subsequent studies, then we say that the findings are not
a. justified.
b. valid.
c. appropriate.
d. reliable.

32. Which of the following is a statement of causality?
a. Rural areas have fewer services than urban areas.
b. This sociology course is difficult.
c. Poverty produces low self-esteem.
d. Mean income in New York is higher than in Florida.

33. Researchers ride in unmarked police cars to collect data on their drug dealers. They are using the
_____ method of research.
a. longitudinal survey
b. laboratory experiment
c. participant observation
d. semistructured interview

34. Given the following array of numbers, 18, 12, 3, 45, 6, 21, 4, 38, 4, the mean is (rounded):
a. 17
b. 4
c. 25
d. 14

OBJECTIVITY IN SOCIOLOGICAL RESEARCH

35. T F Objectivity is a condition in which it is assumed that researchers have no biases.

36. T F Tables often omit headings for rows and columns.

37. A researcher's work would be influenced by
a. the cultural, social, economic, and political environment.
b. the scientific tradition within which the scientist is educated.
c. the scientist's own temperament, inclinations, interests, concerns, and experiences.
d. b and c only
e. a, b, and c

38. Most sociologists agree that eliminating all subjectivity from sociological research is
a. possible and desirable.
b. neither possible nor necessarily desirable.
c. part of the research process.
d. what produces the most valuable results.

39. A good way to decide if the information in a table is reliable is to check its
a. source.
b. title.
c. headings.
d. head notes.

40. To make tables really useful, it is necessary to do more than just understand the specific meaning of each row and column. It is also valuable to
a. compare the data.
b. draw conclusions.
c. pose new questions.
d. b and c only
e. a, b, and c

ETHICAL ISSUES IN SOCIOLOGICAL RESEARCH

41. T F The American Sociological Association's Code of Ethics states that since sociologists are not free from arrest, confidentiality can be broken if there is a risk of being sent to jail for contempt of court.

42. T F Social scientists regard deception of research participants as a necessary evil in most social research.

43. T F One advantage of social research is that it frequently benefits the research subjects directly and immediately.

44. T F Most subjects of sociological research belong to groups with little or no power.

45. T F It is important for sociologist to examine the role they play in contemporary social processes and how these social processes affect their research findings.

---

46. Which of the following is an ethical issue that sometimes arises in the course of sociological research?
   a. the potential disclosure of confidential or personally harmful information
   b. the extent to which the subjects should be deceived
   c. the degree to which subjects risk pain or harm
   d. b and c only
   e. a, b, and c

47. The discussion of Richard R. Peterson's attempt to replicate the Lenore Weitzman study about changes in divorce laws illustrated
   a. the dangers of personal involvement in the research process.
   b. the requirement of science for openness in the replication of research studies.
   c. the responsibility of scientists to correct errors which might lead to faulty conclusions.
   d. the process by which scientists continually check and monitor each other's studies.
   e. all of the above

48. Richard Scarce, a doctoral student in sociology, was sentenced to jail for
   a. presenting fraudulent research findings.
   b. destroying computer files.
   c. releasing laboratory animals as a protest against animal research.
   d. failing to report all of the findings from his study.
   e. refusing to give authorities information he felt would violate his agreement of confidentiality with his subjects.

ESSAY QUESTIONS

49. Drawing on your Sociology at Work reading, discuss the techniques you would use to determine whether an opinion poll is "bogus."

50. Develop a short critical essay in which you make the case that there is *beauty* in research.

51. Discuss three important criteria which a person could use to determine whether the results of a research study were accurate.

52. Generate a hypothesis about a current social issue. Identify the dependent and independent variables and indicate whether the hypothesis is a statement of causality or association.

53. Develop a well-reasoned essay within which you describe the basic differences between "Truth in the Courtroom" and "Truth in the Social Sciences."

CHAPTER THREE Culture

THE CONCEPT OF CULTURE

1. T F Human beings, like most other species, pass a wide variety of behavioral patterns from generation to generation through their genes.

2. T F Social scientists have identified some societies that apparently do not have a culture.

3. T F Sociologists are in agreement over the fact that animals in the wild do not use tools.

4. **T F** Unlike material aspects of culture, nonmaterial aspects of culture take on the same meanings from one country to another.

5. Cultural relativism means that
   a. the concepts of right or wrong do not exist with regard to cultural practices.
   b. some cultures are clearly superior in relation to others.
   c. cultures must be studied on their own terms before being compared or judged.
   d. all cultures must be understood as consisting of many subcultures.

6. Most nonhuman animal species rely upon _____ to meet their needs.
   a. norms
   b. instinct
   c. culture
   d. mores

7. Anthropologist Kenneth Good was disgusted by some of the cultural practices of the Yanomamo. This is an example of
   a. cultural lag.
   b. cultural relativism.
   c. culture shock.
   d. material culture.

8. The process of making judgments about other cultures based on the customs and values of one's own culture is called
   a. ethnocentrism.
   b. cultural relativism.
   c. cultural innovation.
   d. reformulation.

9. Most definitions of culture emphasize that it is
   a. shared.
   b. instinctual.
   c. transmitted from one generation to the next.
   d. a and c only
   e. a, b, and c

10. Time would be a Western example of
    a. material aspect of culture.
    b. technological measurement.
    c. a universal shared by all cultures.
    d. a nonmaterial aspect of culture.

11. Washoe was famous as
    a. the location of ancient cultures.
    b. a chimp that learned sign language.
    c. the gesture in India for "I understand."
    d. the first chimpanzee to learn to use tools.

12. This scientist was famous for studying the behavior of chimpanzees in their natural habitat.
    a. Jane Goodall
    b. Ralph Linton
    c. Margaret Mead
    d. Sapir-Whorf

13. Pinker's theory about culture and language runs counter to much social science thinking,
    a. because he bases most of his research on animal experimental studies.
    b. because he argues, that with proper training, many animals can actually acquire a language of human levels of complexity.
    c. because he believes that essentially there is a language instinct.
    d. because he believes research on language acquisition is largely a "futile waste of time."

COMPONENTS OF CULTURE

14. **T F** Mores are expectations of what people should do under perfect conditions.

15. **T F** The Sapir-Whorf hypothesis suggests that other animals are incapable of really learning human language.

16. **T F** Mores are strongly held rules of behavior that have a moral connotation.

17. **T F** Culture is transmitted by means of language.

18. **T F** Even though all spoken languages are symbolic systems, it is not always possible to translate a word precisely from one language into another language.

19. The concept that professors should show up for every scheduled class meeting and make the best use of the allotted time is an example of a
a. norm.
b. value.
c. belief.
d. custom.

20. A society's beliefs, knowledge, and values are known as its
a. cognitive culture.
b. material culture.
c. nonmaterial culture.
d. normative culture.

21. It is considered polite to send regrets when you are invited to a party but cannot attend. Failure to do so, however, is not considered a moral lapse. This type of norm is called a
a. custom.
b. folkway.
c. taboo.
d. subculture.

22. Which of the following is an example of material culture?
a. the U.S. Capitol building
b. the right to bear arms
c. the findings of the latest Gallup Poll
d. b and c only
e. a, b, and c

23. In American culture, such things as freedom, individualism, and equal opportunity are deemed to be highly desirable. In sociological terms these concepts are
a. values.
b. beliefs.
c. folkways.
d. norms.

24. People sometimes disobey the DON'T WALK sign at street crossings because they see that there are no cars coming. This behavior reflects
a. cultural universals.
b. real norms.
c. ideal norms.
d. instincts.

25. Jamaal lives in a society where there is no word for "privacy." According to the Sapir-Whorf hypothesis,
a. this society is primitive.
b. privacy is not important in this society.
c. this society is not capable of understanding the concept of privacy.
d. an American entering this society could have privacy even though no one else could.

26. The translation of one language into another often presents problems, and direct translations are often impossible because
a. translators do not have the abilities necessary.
b. words have a variety of meanings.
c. many words and ideas are culture bound.
d. b and c above
e. all of the above

THE SYMBOLIC NATURE OF CULTURE

27. Morse code, in which letters and numbers are represented in patterns of dots and dashes, is a system of
a. symbols.
b. ideologies.
c. signs.
d. beliefs.

28. The picture of a woman on the door to the women's restroom is an example of using a(n) _____ to communicate.
a. belief
b. symbol
c. cultural universal
d. ideology

29. Gestures, like a nod of the head, are
a. universal and always have the same meaning.
b. understood in the same way by all cultures.
c. defined uniquely by each culture.
d. instinctively understood by humans as well as animals.

30. T F As a basic human emotional response, the practice of kissing in public is rather universal through-out the world.

CULTURE AND ADAPTATION

31. T F Specialization involves developing ways of doing things that work extremely well in a particular environment.

32. T F There are some forms or patterns found in all human cultures, although they may differ in their details.

33. T F Industrial technology has the advantage over less-developed technologies in that it is adaptable to a wider range of environments.

34. T F Generalized adaptability is the movement of culture traits from one culture to another.

35. T F Environmental determinism is the belief that the natural setting dictates cultural patterns.

36. T F Cultural lag is inevitable in rapidly changing societies.

---

37. Attempts to teach language to apes, such as Penny Patterson's work with the gorilla Koko, have
    a. been unsuccessful because of the very limited intelligence of apes.
    b. proven that apes have some intelligence but no symbolic ability.
    c. succeeded in teaching them fairly large vocabularies but not grammar and syntax.
    d. shown that apes can learn symbols as well as grammar and syntax.

38. The Japanese are very fond of the American game of baseball. However, they have modified many aspects of the game to fit their own culture. This is an example of
    a. innovation.
    b. invention.
    c. diffusion.
    d. reformulation.

39. Innovation and diffusion are mechanisms of
    a. cultural lag.
    b. culture shock.
    c. cultural evolution.
    d. cognitive culture.

40. The process by which humans adjust to changes in their environment is known as
    a. specialization.
    b. adaptation.
    c. innovation.
    d. environmental determinism.

41. T F Noam Chomsky believes teaching apes language is a worthwhile activity which in time should prove humans are not alone in the world regarding their use of language.

42. T F Compared to American customers, most diners in other countries spend a good deal of time in restaurants.

43. T F A latent function of not serving alcohol has resulted in McDonald's being viewed as a safe, socially acceptable place for women from traditional cultures to meet.

SUBCULTURES

44. T F Subcultures are partial, limited fragments of the larger culture of society.

45. T F Members of a religious cult participate in a subculture.

---

46. Rap artists, soapbox derby enthusiasts, ethnic clubs, and stamp collectors are all examples of
    a. cognitive cultures.
    b. secondary cultures.
    c. micro cultures.
    d. subcultures.

47. Which of the following is a type of subculture?
    a. political
    b. social class
    c. geographical
    d. occupational
    e. These are all types of subcultures.

UNIVERSALS OF CULTURE

48. Culturally recognized stages of life through which individuals move are called
- a. cultural universals.
- b. cultural normative proscriptions.
- c. cultural ideologies.
- d. formalized ritualism.
- e. rites of passage.

49. T F The incest taboo prohibits sexual relations between family members.

50. T F Time is a cultural universal because all cultures understand it in the same way—as a steady progression of past, present, and future events.

51. T F Selectivity is the process by which human societies apportion basic tasks.

---

52. Which of the following is a cultural universal?
- a. incest taboo
- b. division of labor
- c. rites of passage
- d. a and b only
- e. a, b, and c

53. People do not just drift from one stage in life to another. Rather they experience
- a. a division of labor.
- b. cultural evolution.
- c. cultural lag.
- d. rites of passage.

54. The most common rites of passage among human cultures are those relating to
- a. puberty, marriage, and death.
- b. educational achievements.
- c. military service.
- d. work achievements.

55. The ways in which humans divide their tasks is referred to as
- a. the division of labor.
- b. the level of technological development.
- c. the degree of cooperation.
- d. all of the above
- e. none of the above

CULTURE AND INDIVIDUAL CHOICE

56. T F Very little human behavior is instinctual or biologically programmed.

57. T F Human individual choices are bounded and influenced by culture.

ESSAY QUESTIONS

58. Discuss how McDonald's has both adapted to and modified their menus to the cultures in those societies throughout the world were they have established their restaurants.

59. Present an essay in which you argue that research with animals is *unethical*.

60. Present a pro and con discussion of the question: "Is there a language instinct?"

61. List and discuss each of the subcultures presented in your text. What do they have in common and what is unique about each? How many of the subcultures are you a member?

62. Discuss how examples of material culture, nonmaterial culture, and cognitive culture can be seen in rock videos.

63. American football seems to be derived from the English game of rugby. Over time, however, the game has gone through many changes, including the development of the passing game and offensive and defensive specialists. Explain the development of football in America in terms of the mechanisms of cultural change.

64. Discuss the differing norms concerning kissing in public as these norms vary in Mexico, the United States, Latin America, and France. What do these norms suggest?

65. Some linguists believe animal language experiments are motivated by the ideological conviction that humans should be knocked off their self-appointed thrones and that animals are capable of higher-level thinking. Based upon the evidence presented in this chapter, what is your position on this issue?

CHAPTER FOUR Socialization and Development

BECOMING A PERSON: BIOLOGY AND CULTURE

1. **T F** Most of our bodily processes are the result of the interaction of genes and our environment.

2. **T F** Sociobiologists such as Edward O. Wilson believe that human behavior can be understood as continuing attempts to ensure the transmission of one's genes to a new generation.

3. **T F** Although socialization may play a role, some people are clearly born with talents for business, tendencies toward crime, and so forth.

4. **T F** Conditioning can be used to mold the behavior of animals, but not that of human beings.

---

5. Harlow's study of rhesus monkeys demonstrated the
    a. severe impact of early social deprivation.
    b. importance of genetic inheritance in determining personality.
    c. reversibility of negative impacts of deprivation during infancy.
    d. ability of other primates to learn human emotions.

6. A critic of sociobiology, he has argued that culture is a far more important influence on human behavior than genetics.
    a. Edward O. Wilson
    b. Charles Horton Cooley
    c. Daniel Levinson
    d. Stephen Jay Gould

7. Isolated children develop slowly because they
    a. do not receive proper nutrition.
    b. lack social interaction and stimulation.
    c. are mentally and physically impaired to begin with.
    d. are physically abused during their isolation.

8. Biologically inherited patterns of complex behavior are known as
    a. genes.
    b. instincts.
    c. the personality.
    d. the unconscious.

9. The process of social interaction that teaches a child the intellectual, physical, and social skills needed to function as a member of society is called
    a. identification.
    b. social adjustment.
    c. socialization.
    d. social conditioning.

10. The patterns of behavior and ways of thinking and feeling that make us distinctive individuals are known as our individual
    a. social identity.
    b. self.
    c. consciousness.
    d. personality.

11. Experiments by Ivan Pavlov and John Watson showed that so-called instinctual behavior could be molded through
    a. administration of selected drugs.
    b. conditioning.
    c. imitation.
    d. resocialization.

12. Meaningful interaction with other human beings, which is often rare in the lives of institutionalized children, is referred to as
    a. significant interaction.
    b. affiliation.
    c. social attachment.
    d. nurture.

13. In his explanation for why elders in early hunting and gathering societies remained behind to die during times of ecological stress, biologist Stephen Jay Gould proposed that
    a. they did so due to a biological predisposition toward altruism.
    b. what happened was accidental and therefore neither biologically not culturally driven.

c. their behavior was a form of early instinctual development which eventually disappeared from the human species.

d. they had been socially conditioned from earliest childhood to the possibility and appropriateness of altruistic sacrifices.

THE CONCEPT OF SELF

14. **T F** According to Piaget, the operational stage is when the infant relies on touch and the manipulation of objects for information about the world.

15. **T F** The shared view of right and wrong that exists in a society is called the moral order.

16. **T F** According to Lawrence Kohlberg, concepts such as good and bad or right and wrong, once established, have the same meaning throughout our lives.

17. **T F** The ability to reason abstractly and anticipate possible consequences of behavior is characteristic of the formal logical stage of cognitive development, according to Jean Piaget.

18. According to Kohlberg, the most advanced state of moral reasoning involves making judgments on the basis of a
    a. personal set of values.
    b. desire to avoid punishment.
    c. desire to achieve a reward.
    d. sense of the values of one's peers.

19. The child's discovery, around the age of two years, that the objects of the world can be symbolized and talked about as well as experienced directly marks the beginning of what Piaget called the _____ stage of cognitive development.
    a. sensorimotor
    b. preoperational
    c. operational
    d. formal logic

20. Each individual's social identity consists of
    a. all the behaviors they have learned to imitate.
    b. their changing yet enduring view of themselves.
    c. the sum total of all the statuses they occupy.
    d. all the ways that other people view them.

21. Which of the following is a dimension of the concept of self, according to most researchers?
    a. ability to organize one's knowledge and beliefs
    b. awareness of the existence, appearance, and boundaries of one's own body
    c. knowledge of one's needs and skills
    d. ability to step back and look at one's being as others do
    e. These are all dimensions of the concept of self, according to most researchers.

22. The view of our self resulting from our sex is known as our
    a. sex identity.
    b. id.
    c. sex role.
    d. gender identity.

THEORIES OF DEVELOPMENT

23. **T F** John Watson argued against nurture and in favor of nature when he spoke of the chances of an infant actually becoming a doctor, lawyer, artist, merchant, beggar, or thief.

24. **T F** Overall, your text argues that gender identity and sex roles are far more a matter of nurture than nature.

25. **T F** In Freudian theory, the ego is the repository of the thoughts and feelings that are out of our awareness.

26. **T F** According to Mead, the "me" is a portion of the self that is made up of everything learned through the socialization process.

27. **T F** According to Levinson's theory of adult socialization, divorce and career changes are typical during the Early Adult period.

28. **T F** Overall, the fields of academic sociology and anthropology have not been quick to accept the positions advanced by theorists in sociobiology.

---

29. The view that the human mind and human behavior are shaped by unconscious impulses and resistance to or repression of them is most closely associated with
   a. Edward O. Wilson.
   b. Sigmund Freud.
   c. Ivan Pavlov.
   d. Lawrence Kohlberg.

30. People who are important to an individual's development were termed _____ by Mead.
   a. peers
   b. significant others
   c. primary socializers
   d. generalized others

31. In Freud's view of the self, the "internal censor" or conscience that embodies society's norms and values as learned primarily from the parents is called the
   a. ego.
   b. id.
   c. superego.
   d. "I."

32. Which parts of the self are shaped through socialization?
   a. the "I" only
   b. the "me" only
   c. both the "I" and the "me"
   d. neither the "I" nor the "me"

33. Which of the following is **not** one of the components of the *looking-glass self,* according to Cooley?
   a. our imagination of what we must really be like
   b. our imagination of how our actions appear to others
   c. our imagination of how others judge our actions
   d. a self-judgment in reaction to the imagined judgments of others

34. This theorist proposed a series of pro and con internal arguments taking place through six stages during which an individual moved to higher levels of moral development.
   a. Jean Piaget
   b. George Herbert Mead
   c. Sigmund Freud
   d. Erik Erikson
   e. Lawrence Kohlberg

35. He argued that the self develops through the interaction with significant others, from whom we learn increasingly sophisticated role expectations.
   a. Daniel Levinson
   b. Erik Erikson
   c. Sigmund Freud
   d. George Herbert Mead

36. Jason and Alison are teammates on a soccer team. According to Mead, for them to be able to play together successfully, they must be at the stage where they
   a. have developed advanced physical coordination.
   b. can understand and evaluate abstract strategies of play.
   c. know the rules and understand the expectations of the various positions.
   d. can demonstrate an ethical understanding of the concept of good sportsmanship.

37. Which of the following is a problem with "stage theories" of human development?
   a. They fail to take account of individual differences.
   b. They often describe only the experiences of people born during a certain time span.
   c. They do not take into account the process of social change.
   d. b and c only
   e. a, b, and c

38. Jennifer is five years old. She often pretends she is the mommy—enthusiastically correcting the "bad" behavior of her dolls and stuffed animals by imitating the way Jennifer's mother admonishes her. According to Mead, Jennifer is at the _____ stage of development.
   a. play
   b. game
   c. preparatory
   d. preoperational

39. According to Daniel Levinson's theory of socialization, adulthood is primarily characterized by
   a. the same stages that occurred early on.
   b. new and unpredictable tasks that must be worked on.

c. new but predictable tasks that must be worked through.

d. a slowing down of development since most skills have already been learned

40. The *generalized other* refers to
a. members of one's extended family.
b. the general attitudes we have about other people.
c. the sum total of all our experiences and interactions in society.
d. attitudes and expectations of groups that are important to us.

41. The most valuable contribution of Erikson's theory of socialization is that it
a. breaks development down into eight simple stages.
b. shows that socialization is completed in early childhood.
c. shows that socialization is a lifelong process.
d. demonstrates that the most important stages of socialization occur after age thirty.

42. The developmental point where the individual begins to reappraise his or her life and explore new possibilities.
a. Age 30 Transitional   c. Entering the Adult World
b. Early Adult Period   d. Settling Down

43. A major transitional period representing the turning point between young and middle adulthood.
a. Age 30 Transitional   d. Mid-life Transition
b. Early Adult Period   e. Beginning of Middle Adulthood
c. Settling Down

EARLY SOCIALIZATION IN AMERICAN SOCIETY

44. T F Young people who feel ignored by their parents seem to be more vulnerable to peer pressure.

45. T F Many children going to day-care centers do not encounter enduring stable relationships with child-care workers.

46. T F Individuals who are social equals are referred to as generalized others.

47. T F Research shows that it is nearly impossible to provide high-quality care in an organized day-care facility.

---

48. During adolescence, the socializing agent that has the strongest influence is the
a. family.   c. school.
b. peer group.   d. mass media.

49. Jenny comes from a white-collar household while Cheryl's parents hold blue-collar jobs. Cheryl's parents are more likely than Jenny's to encourage
a. obedience to authority.   c. intellectual curiosity.
b. flexibility in behavior and thought.   d. development of social skills.

50. The most important agent of primary socialization is the
a. peer group.
b. family.
c. school.
d. mass media.

51. The primary function of schools as a socializing agent is that they provide children with
a. academic instruction.
b. a model of adult relationships that is bureaucratic and impersonal.
c. relationships that are full of love and affection.
d. day care while their busy parents work.

52. By the time we reach age eighteen, most of us will have spent more time _____ than doing anything else.
a. talking with peers   c. going to school
b. reading   d. watching television

ADULT SOCIALIZATION

53. T F Even though aging is a biological process, becoming old is a social and cultural one.

54. T F Succeeding in a career is mainly a matter of fine-tuning the knowledge and skills acquired earlier in life.

55. T F Marriage is a public statement that both partners are committed to each other and to stability and responsibility.

56. T F Attachment disorder results in children being unable to form relationships.

57. Which of the following is a total institution?
   a. prisons
   b. mental hospitals
   c. military bases
   d. b and c only
   e. a, b, and c

58. Basic training in a military boot camp, where recruits must be taught to follow orders unquestioningly and to kill on command, involves a process of learning known as
   a. secondary socialization.
   b. primary socialization.
   c. resocialization.
   d. antisocialization.

59. Adult socialization differs from primary socialization in that adults
   a. are generally immune to the process of socialization.
   b. usually are more hostile to socialization attempts.
   c. often have more control over how they wish to be socialized.
   d. are more likely to be influenced by their peers.

60. Which of the following is a major concern of the elderly in the United States?
   a. having to acquire, late in life, a social identity that is not valued
   b. uncertainty about where they will live and who will care for them when they are sick
   c. having to stop productive work simply because they have reached a certain age
   d. b and c only
   e. a, b, and c

61. In psychological terms, the aspect of adult socialization that gives many people a "second chance" is
   a. marriage.
   b. retirement.
   c. parenthood.
   d. changing careers.

ESSAY QUESTIONS

62. What are the negative effects on children's development when they are deprived of meaningful human contact? Describe these effects and the variety of conditions human infants need to grow and develop normally.

63. Relate Piaget's stages of cognitive development to Mead's stages of development of the self, showing how each theorist's stages complement and can be integrated into those of the other.

64. After reading your text, what is your position on whether television should be used to teach values to viewers?

65. Describe the major socialization agents of adolescent children in the United States, and some of the positive and negative influences of each agent.

66. What are the key points of the *nature-nurture debate?*

67. Discuss Levinson's model of the periods in the adult life cycle.

68. Present an essay in which you discuss the pros and cons of day care in terms of its potential or harm to children.

CHAPTER FIVE Social Interaction and Social Groups

UNDERSTANDING SOCIAL INTERACTION

1. T F Human behavior is not random. It is patterned and, for the most part, quite predictable.

2. T F Studies have demonstrated that males prefer to sit across a table from a stranger, while females prefer to sit side-by-side with a stranger.

---

3. To an observer, two people engaging in a fistfight on the street would mean something entirely different from the same two people fighting in a boxing ring. This illustrates how the meaning of social interaction is dependent upon
   a. personalities.
   b. status.
   c. competition.
   d. context.

4. A group of autoworkers meets to discuss various labor problems. During the meeting two of the workers remain silent. In sociological terms these two would be described as
   a. having no influence on the interaction.
   b. conveying a message through their silence.
   c. inactive members of the group.
   d. being uncooperative.

5. Max Weber's method for analyzing social action, *Verstehen,* entails
   a. constructive criticism.
   b. face-to-face questioning.
   c. statistical analysis.
   d. sympathetic understanding.

6. In a large lecture class, it is generally expected that students will raise their hands and be called upon before speaking. This illustrates the operation of
   a. bureaucracy.
   b. roles.
   c. norms.
   d. statuses.

7. The process of two or more people taking each other's actions into account is called
   a. *Verstehen.*
   b. social interaction.
   c. social organization.
   d. social action.

8. Which of the following is true of Amish community life today?
   a. Life revolves around primary face-to-face ties.
   b. The conventional marks of social status are missing.
   c. A cardinal value of the Amish is separation from the world.
   d. Traditional patterns of authority supersede the formal arrangements of modern society.
   e. These are all true of Amish community life today.

TYPES OF SOCIAL INTERACTION

9. T F When people do something for each other with the express purpose of receiving something in return, they are engaged in exchange.

10. T F Communication by means of gestures is pretty much the same the world over.

11. T F Coercion is a form of conflict in which one of the parties is much stronger than the other and can impose its will on the weaker party.

12. T F Competition is a form of conflict within agreed-upon rules.

---

13. When unionized workers agree to a "no-strike clause" in their labor agreement in exchange for certain improvements in wages or benefits, they are engaging in what kind of cooperation with their employers?
   a. contractual
   b. spontaneous
   c. directed
   d. traditional

14. We tend to maintain the most eye contact when we are speaking to someone of _____ status than us.
a. higher
b. lower
c. approximately the same

15. In many small towns, it has long been customary for neighbors to bring gifts of food to the home of someone who is ill. This effort involves _____ cooperation.
a. spontaneous
b. traditional
c. directed
d. contractual

16. Staring, smiling, nodding one's head, and using one's hands while talking are all examples of
a. nonverbal communication.
b. cooperation.
c. instinctual behavior.
d. exchange.

17. The main difference between cooperation and exchange is that
a. cooperation is based on a shared goal.
b. relationships of exchange are voluntary.
c. cooperation has no material rewards.
d. exchange provides benefits for individuals only.

18. Ahmad agrees to tutor James in calculus so that James can graduate. James in turn agrees to teach Ahmad French. This is a relationship of
a. cooperation.
b. obligation.
c. exchange.
d. exploitation.

19. Your instructor divides the class into groups for the purpose of completing a class assignment. This is an example of
a. exchange.
b. spontaneous cooperation.
c. competition.
d. directed cooperation.

ELEMENTS OF SOCIAL INTERACTION

20. T F Socially defined positions, such as teacher, student, athlete, and daughter are called roles.

21. T F For many students, keeping up good grades, holding down a job, and spending enough time with significant others involves role conflict.

22. T F A master status refers to one of the multiple statuses a person occupies that seems to dominate the others in patterning his or her life.

23. T F The status of *mayor* is basically defined by any person who occupies that position.

24. College professors are usually expected to engage in at least three types of activities. They are expected to teach their classes effectively. They are expected to conduct research and publish the results in the form of professional articles and books. And they are expected to devote time to institutional governance and civic activities. Taken together, these activities constitute a
a. status set.
b. role set.
c. master status.
d. secondary group.

25. A college professor finds herself taking time away from class preparation in order to keep producing published research, or shirking committee responsibilities in order to grade papers for class. This professor is experiencing
a. role strain.
b. status anxiety.
c. role conflict.
d. status inconsistency.

26. Which of these multiple statuses would be **most likely** to function as a *master status?*
a. president of the United States
b. father
c. working mother
d. husband
e. none of the above as they are all equal statuses

27. T F Because they share a particular characteristic common only to them, left-handed people constitute that which sociologists would classify as a *social group.*

28. Which of the following is an ascribed status?
a. male
b. student
c. employee
d. shortstop

29. The relationship between roles and statuses is that
   a. a status may include a number of roles.
   b. a role may include many statuses.
   c. not all statuses have roles attached.
   d. statuses are dynamic while roles are not.

30. Marissa, a police officer, begins to suspect that her teenage son is involved in selling illegal drugs. She is torn between responding as a parent and responding as a police officer. Marissa is experiencing
   a. role conflict.
   b. role strain.
   c. status conflict.
   d. role playing.

THE NATURE OF GROUPS

31. T F Social groups and social aggregates both cease to exist when members are apart from one another.

32. T F Primary groups are usually more willing than secondary groups to tolerate members who deviate from group expectations.

33. Which of the following would most likely be a secondary group?
   a. a family
   b. a large lecture class
   c. a juvenile gang
   d. a high school clique

34. What is the sociological name for strangers waiting in line to buy concert tickets?
   a. group
   b. clique
   c. social aggregate
   d. social category

35. Which of the following is **not** a fundamental characteristic of social groups?
   a. members have a common identity
   b. members enjoy each other's company
   c. there is some feeling of unity
   d. there are common goals and shared norms

36. A group of friends get together once a month to play cards and talk. In sociological terms, they would constitute a
   a. regular aggregate.
   b. secondary group.
   c. primary group.
   d. status set.

37. The most important characteristic of primary groups that is missing in secondary groups is
   a. intimacy.
   b. small size.
   c. interaction.
   d. shared expectations.

FUNCTIONS OF GROUPS

38. T F A reference group is a purposefully created special-interest group that has clearly defined goals and official ways of doing things.

39. T F Instrumental leadership is more important to the success of a group than expressive leadership.

40. Asch's study of group pressure and conformity, in which subjects were asked to judge the length of lines, found that
   a. people were surprisingly eager to disagree with the judgments of the group.
   b. the majority was eager to expel those who disagreed with their judgments.
   c. about one-third of the people were willing to give incorrect answers in order to conform to the group.
   d. about one-quarter of the people were unable or unwilling to determine the length of the lines at all.

41. The process of orienting one's behavior and attitudes toward those of a group one plans to join is called _____ socialization.
   a. primary
   b. secondary
   c. anticipatory
   d. adult

42. Which of the following is one of the things groups must do in order to function effectively?
   a. choose leaders
   b. define boundaries
   c. set goals

    d. assign tasks
    e. These are all things that groups must do in order to function effectively.

43. Corinne receives a graduate degree in business, lands an excellent job on Wall Street, and finds herself making more money at age twenty-four than she ever thought possible. At first she is delighted. But soon she discovers that her salary is not especially high nor her new life-style especially luxurious compared with those of her new business associates. Before long, she is feeling dissatisfied and underpaid. This illustrates the operation of
    a. ideal types.
    b. reference groups.
    c. primary groups.
    d. anticipatory socialization.

44. Which of the following functions of a group is most likely to undermine the existence of the group if it is not carried out appropriately?
    a. controlling members' behavior
    b. defining boundaries
    c. assigning tasks
    d. setting goals

SMALL GROUPS

45. T F The smallest possible group is a dyad.

46. T F New members are more threatening to small groups than to large groups.

47. Three students stay after class to go over their notes. This group would be called a
    a. trio.
    b. dyad.
    c. triage.
    d. triad.

48. What happens when a group grows beyond five to seven members?
    a. It splits into subgroups.
    b. It ceases to exist as a group.
    c. It adopts a formal means of controlling communication.
    d. a and c only
    e. a or b or c

49. Alliances and group pressure initially become possible at the level of the
    a. triad.
    b. dyad.
    c. association.
    d. secondary group.

LARGE GROUPS: ASSOCIATIONS

50. T F All bureaucracies have networks of people who help one another by bending rules and taking procedural shortcuts.

51. T F Associations tend to have goals that are not clearly defined.

52. "Covering" or "looking the other way" when a co-worker makes a mistake is an example of
    a. bending a formal rule.
    b. intolerable behavior in organizations.
    c. alienation at work.
    d. ritualism at work.

BUREAUCRACY

53. T F An ideal type refers to the way sociologists wish society were organized.

54. T F Bureaucracies encourage workers to be generalists, capable of doing several jobs equally well, rather than specialists.

55. Michels' Iron Law of Oligarchy holds that bureaucracies
    a. eventually lead to political dictatorship.
    b. are inevitable in any democratic society.
    c. are outdated forms of social organization.
    d. always become dominated by self-serving elites.

56. A fluorescent light in the sociology department burns out, and the department secretary asks the head of maintenance to have it replaced. Even though he as an ample supply of replacements and could change it himself in five minutes, he insists that the secretary fill out an official work order and wait two weeks or so until an electrician can do the job. This episode provides an illustration of bureaucratic
   a. inertia.   c. alienation.
   b. ritualism.   d. conformity.

57. Which of the following is **not** a major characteristic of bureaucracy, according to Weber?
   a. heavy reliance on written rules and regulations
   b. a clear-cut division of labor
   c. a tendency toward favoritism and nepotism
   d. a clear distinction between public and private spheres
   e. These are all major characteristics of bureaucracy, according to Weber.

58. All societies must meet certain fundamental needs, but different societies may adopt different ways of doing so. The various vehicles for accomplishing basic social tasks are known as
   a. social institutions.   c. social aggregates.
   b. secondary groups.   d. formal structures.

59. The relatively stable pattern of social relationships among individuals and groups is called
   a. formal organization.   c. an institution.
   b. social organization.   d. an association.

ESSAY QUESTIONS

60. Present the case that American society has developed the characteristic of too much and too frequent arguing.

61. Discuss the ways in which the context of the college classroom structures at least three different types of social interaction.

62. Distinguish between primary and secondary groups, and weigh the relative advantage of each. That is, describe at least two circumstances or contexts each in which primary groups are more necessary, advantageous, or effective than secondary groups, or vice versa.

63. Responding to an ad in the newspaper, a group of strangers gets together to form a computer users club. Discuss, from a sociological point of view, the six tasks these people must successfully accomplish to create a viable group.

64. Drawing on the readings in this chapter, create an essay in which you describe the many ways Blacks and Whites in American society offend each other without realizing it.

CHAPTER SIX Deviant Behavior and Social Control

DEFINING NORMAL AND DEVIANT BEHAVIOR

1. T F Behavior can be classified as normal or deviant only with reference to the group in which it occurs.

2. T F From the work of Durkheim, we may conclude that a society without any deviant behavior is both desirable and possible.

3. T F After decades of careful investigation, sociologists have discovered absolute characteristics of deviance which are present across time and place.

4. To Émile Durkheim, the reason deviant behavior is found in all societies is that
   a. human beings are naturally selfish and aggressive.
   b. it performs useful social functions.

c. human beings have not yet developed effective methods of behavior control.
d. harmful actions are inevitable in mechanically integrated societies.

5. Which of the following is a dysfunction of deviance?
a. It makes social life difficult and unpredictable.
b. It diverts valuable resources that could be used elsewhere.
c. It causes confusion about the norms and values of society.
d. It is a threat to the social order.
e. all of the above

6. Deviance is defined as behavior that
a. occurs rarely.
b. violates group norms.
c. is psychologically abnormal.
d. is illegal.

7. Which of the following is **not** true of deviance?
a. It helps to maintain group boundaries.
b. It helps to reinforce appropriate behavior.
c. It is rare in highly structured societies.
d. It provides a societal safety valve.

MECHANISMS OF SOCIAL CONTROL

8. T F Ostracism and ridicule are formal negative sanctions.

9. T F In order for a group's moral code to work properly, it must be internalized by its members.

10. Which of the following is an internal means of social control?
a. ridicule
b. imprisonment
c. guilt
d. exclusion from the group

11. A criminal trial at which a defendant is found guilty and sentenced to a prison term is an example of a(n) _____ sanction.
a. informal positive
b. informal negative
c. formal positive
d. formal negative

12. In 1999, which of the following crimes was *least* likely to have been reported to the police?
a. murder
b. motor vehicle theft
c. burglary
d. robbery
e. theft of less than $50

13. T F The United States violent crime rate in 1999 reached the *highest level* since the Bureau of Justice Statistics started measuring it in 1973.

14. Approximately what percent of the victims of rape and sexual assault know their offender as an acquaintance, friend, relative or intimate?
a. 10%
b. 25%
c. 50%
d. 70%
e. 90%

15. Which category of persons is disproportionally affected by violent and property crime?
a. rural residents
b. urban residents
c. suburban residents
d. a and b only
e. all of the above

THEORIES OF CRIME AND DEVIANCE

16. T F Sutherland and Cressey's differential association theory maintains that deviance is learned, mostly within primary groups.

17. T F Sykes and Matza are well known for developing the control theory of deviance.

18. T F Neutralization theory allows a person to break the law and get away with it by neutralizing any ensuing penalties.

19. **T F** A person becomes deviant because of an excess of definitions favorable to violating the law over definitions unfavorable to violating the law.

20. **T F** Shaw and McKay found that it was the new ethnic groups moving into areas which was the cause of increasing crime rates in urban areas.

21. "I'm not really a bad person. It's just that I was drunk and I didn't know what I was doing." This is an example of which technique of neutralization?
    a. denial of responsibility
    b. denial of injury
    c. appeal to a higher principle
    d. denial of the victim

22. Which of the following theories is less concerned with the causes of norm violations than with the way others react to the deviance?
    a. psychoanalytic theory
    b. cultural transmission theory
    c. control theory
    d. anomie theory
    e. labeling theory

23. A youth who wants to achieve the sort of affluent lifestyle he sees on television but, unable to find a well-paying job, turns to drug dealing to pay for it would be classified as a(n) _____ in Merton's typology.
    a. retreatist
    b. innovator
    c. rebel
    d. conformist

24. Wilson and Herrnstein's individual choice theory of crime considers it to be the result of
    a. poverty.
    b. lack of equal educational opportunities for all individuals.
    c. an individual calculation that immediate benefits outweigh long-term risks.
    d. individual definitions of crime and deviance by law enforcement officers.

25. According to control theory, a youngster who stays out of trouble with the law is most likely to be one who
    a. had few unpleasant experiences during childhood.
    b. has no delinquent peers to put pressure on him or her.
    c. has strong relationships with parents, teachers, and peers.
    d. has confidence in his or her future occupational success.

26. Durkheim saw anomie as a condition of
    a. weak law enforcement.
    b. normlessness.
    c. overemphasis on the welfare of the group.
    d. dependency.

27. Lombroso and Sheldon, each in his own way, attempted to explain deviant behavior on the basis of
    a. psychological orientation.
    b. anatomical characteristics.
    c. early childhood experiences.
    d. differential association.

28. Hirschi and Gottfredson propose a theory of crime looking at one form of control:
    a. self-control.
    b. parental-supervised control.
    c. peer-group control.
    d. state-sponsored control.

29. Which theory attributes behavior, deviant or otherwise, to the rewards and punishments that come about as consequences of people's actions?
    a. behavioral
    b. individual choice
    c. psychoanalytic
    d. biological

30. Fourteen-year-old Janet is arrested for shoplifting. Even though the charges are later dropped, Janet's teachers designate her a troublemaker and someone not to be trusted. Subsequently, Janet begins to skip school frequently and to get into fights when she is there. Lemert and others would call these latter behaviors
    a. primary deviance.
    b. secondary deviance.
    c. status offenses.
    d. a case of recidivism.

31. The research of Shaw and McKay, which linked crime to certain types of urban neighborhoods, provided the foundation for _____ theories of deviance.
   a. labeling
   b. control
   c. genetic transmission
   d. cultural transmission

32. According to labeling theorists, the likelihood that you will receive a deviant label for your misbehavior depends upon
   a. the seriousness of what you did.
   b. your social identity.
   c. the context of your act.
   d. a and b only
   e. a, b, and c

33. Psychoanalytic approaches to deviance are criticized because they
   a. are too specific in their focus.
   b. cannot easily be tested.
   c. ignore the role of society.
   d. ignore the individual's past experiences.

THE IMPORTANCE OF LAW

34. "Truth-in-sentencing" laws are
   a. an attempt to inform the public about the exact punishment offenders receive.
   b. result in offenders being required to serve substantial portions of their sentences prior to becoming eligible for parole.
   c. are connected with congressional programs designed to build more prisons.
   d. all of the above
   e. none of the above

35. T F At this point, approximately one third of the states have established truth-in-sentencing laws.

36. T F A society's moral code refers to the formal rules, or laws, adopted by its political authority.

37. T F In every case, breaking a law involves violating a society's moral code as well.

---

38. The conflict approach sees law as a product of
   a. universal truths.
   b. a natural process of evolution.
   c. agreed-upon social values.
   d. powerful interest groups.

39. T F Older type blood tests, mistaken eyewitness identification, and unmonitored laboratory practices have contributed to the wrongful conviction of those who are innocent.

40. The idea that laws represent the codification of shared moral beliefs is called the _____ approach.
   a. consensus
   b. labeling theory
   c. conflict
   d. cultural transmission theory

CRIME IN THE UNITED STATES

41. T F The most frequent reason given by victims for not reporting crime to the authorities is the belief that the crime was not important enough.

42. T F Theft of a person's parked automobile would be classified as a property crime.

43. T F *The National Crime Victimization Survey,* which is considered by many to be superior to the *Uniform Crime Reports,* collects information on crimes from local police departments.

44. T F Crime can be defined as behavior that violates society's moral code.

45. T F Research on *serial murderers* has shown that the majority suffer from insanity.

46. T F One major difference between *serial and mass murderers* is the number of victims at one point in time.

---

47. Relatively minor crimes that are usually punishable by a fine or less than a year's confinement are called
   a. civil offenses.
   b. larcenies.
   c. misdemeanors.
   d. felonies.

48. Which of the following is **not** a criticism of the *Uniform Crime Report* figures?
   a. Statistics may not be comparable from one year to the next.
   b. Police departments across the country differ in the way crime reports are handled.
   c. Federal offenses are not included in the reports.
   d. Less than half of the police departments in the country report their crime statistics to the FBI.

49. Which of the following would **not** be classified as a violent crime?
   a. rape   c. burglary
   b. homicide   d. a and b only

KINDS OF CRIME IN THE UNITED STATES

50. T F A major difference between adult and juvenile crime is that juveniles are much more likely to commit offenses in groups.

51. T F Females are much more likely to be victims of serious crimes than are males.

52. T F Ninety percent of all crime in the United States is against property, not against a person.

53. T F Juveniles commit a relatively small proportion of the serious crimes that occur in the United States.

54. T F The United States has the highest homicide rate in the industrialized world.

---

55. Steering offenders away from the justice system and into social agencies is known as
   a. diversion.   c. labeling.
   b. recidivism.   d. rehabilitation.

56. Status offenses are acts that
   a. are criminal only because the person involved is a minor.
   b. attempt to use lawbreaking to increase one's social position or reputation.
   c. are committed by individuals occupying positions of social responsibility or high prestige.
   d. use trickery rather than violence to obtain the desired ends.

57. Overall, victims of violent crimes are disproportionately
   a. low-income.   d. a and b only
   b. African American.   e. a, b, and c
   c. elderly.

58. Violation of laws meant to enforce the moral code, such as public drunkenness, prostitution, gambling, and possession of illegal drugs, are called _____ crimes.
   a. moral   c. victimless
   b. organized   d. status

59. The least frequent violent crime is
   a. homicide.   c. robbery.
   b. rape.   d. assault.

CRIMINAL JUSTICE IN THE UNITED STATES

60. T F The United States has a higher rate of imprisonment than any other country in the world.

61. T F Deterrence refers to the practice of steering offenders away from the justice system to social agencies.

62. T F Research shows that the deterrent effect of the well-publicized execution of a criminal is short-lived.

63. T F Women are more likely to commit property as opposed to violent crimes.

---

64. Which of the following best describes the social origins of most police officers in the United States?
   a. Most come from middle-class families.   d. b and c only
   b. Most have a high school education or less. e. a, b, and c
   c. Most are white males.

65. Which of the following is **not** a component of the U.S. court system?
   a. federal district courts
   b. state appellate courts
   c. federal circuit courts
   d. the U.S. Supreme Court
   e. These are all components of the U.S. court system.

66. Which of the following is a goal of imprisonment?
   a. punishment of criminal behavior
   b. deterrence of criminal behavior
   c. rehabilitation of criminals
   d. separation of criminals from society
   e. These are all goals of imprisonment.

67. In a program called *Scared Straight,* hardened criminals were brought into junior and senior high schools to give first-person accounts of the negative aspects of being imprisoned. This is an example of
   a. deterrence.
   b. recidivism.
   c. diversion.
   d. the funnel effect.

68. The _____ effect is the characteristic of the criminal justice system by which few crimes are reported, few criminals are caught, few of those caught are convicted, and few of those convicted are imprisoned.
   a. diversion
   b. funnel
   c. deterrence
   d. recidivism

69. The most urgent problem with the nation's prisons at the present time is
   a. too many women in them.
   b. too many juveniles in them.
   c. they are overcrowded.
   d. they are designed mostly to handle misdemeanor offenders, not felony offenders.

ESSAY QUESTIONS

70. Discuss the *Innocence Project.* Include consideration of the scientific test which makes the project possible.

71. Since the 1960s, America has seen a succession of youth cultures, each of which has been self-consciously deviant from mainstream attitudes and values. Evaluate the functional and dysfunctional aspects of deviant youth culture.

72. Contrast two psychological theories of deviance with two sociological theories of deviance. Describe the differences and evaluate the benefits of one theory over another.

73. Using three different types of crime as your examples, compare and contrast the *conflict* with the *consensus* theories of law.

74. Describe Merton's five modes of individual adaptation to the discrepancy between cultural goals and institutionalized means.

75. What are the major components of law?

76. Present the arguments *in favor* of and *against* capital punishment.

CHAPTER SEVEN   Social Class in the United States

STUDYING SOCIAL STRATIFICATION

1. T F Many Americans believe that we live in a classless society.

2. Which of the following is a basis for class distinctions in the United States?
   a. race
   b. family name

c. career choice

d. education

e. These are all bases for class distinctions in the United States.

SOCIAL CLASS IN THE UNITED STATES

3. T F In recent years, it is reasonable to say that "the rich got richer and the poor got poorer" in the United States.

4. T F Most sociologists agree about how many classes there are.

---

5. In 1998, the richest fifth of the U.S. population received about _____ of all income.

a. 12%

b. 22%

c. 44%

d. 66%

6. Members of the lower class tend to blame _____ for their lack of success.

a. the rich

b. "the system"

c. bad luck

d. themselves

7. The upper class in the United States is approximately _____ of the population.

a. 1 to 3%

b. 7 to 9%

c. 13 to 15%

d. 21 to 23%

8. Of these four classes in American society, which one is the smallest?

a. upper-middle

b. upper

c. working

d. lower

9. Managers and other people who work in professional and technical fields are members of the _____ class.

a. upper

b. upper-middle

c. lower-middle

d. working

10. The wealthiest 5% of the U.S. population owns _____ of all the nation's wealth.

a. 5%

b. 25%

c. 55%

d. 75%

11. Hector is a master electrician. He holds a high school diploma, owns a modest home, but struggles to give his family the things they need and want. He considers himself politically conservative and fairly religious. According to the typology in the text, Hector belongs to the _____ class.

a. upper-middle

b. lower-middle

c. working

d. lower

12. When working-class people experience upward social mobility, they also experience

a. feelings of self-doubt.

b. a new class culture not very different from their old one.

c. lack of support from their working-class family and friends.

d. a and c only

e. a, b, and c

POVERTY

13. T F The poverty index was developed to determine which families and individuals were economically needy.

14. T F Over 40 percent of adults who are officially poor work.

15. T F Poverty refers to a condition in which people cannot afford the necessities of life.

16. T F Unlike African-American children, white American children have experienced a declining poverty rate over the past 20 years.

17. T F Almost half of all female-headed families with children under 18 years old are officially poor.

18. The poverty index is distorted in that it
   a. excludes non-cash income.
   b. underestimates non-food expenses.
   c. is based on household data.
   d. a and b only
   e. a, b, and c

19. In 1999, about _____ of all Americans lived below the official poverty line.
   a. 5%
   b. 8%
   c. 12%
   d. 20%
   e. 25%

20. The median net worth of African-Americans is about _____ of white median net worth.
   a. 10%
   b. 25%
   c. 33%
   d. 65%
   e. 50%

21. Poverty is especially concentrated among families headed by women who
   a. have never married.
   b. are divorced.
   c. are widowed.
   d. have been married more than twice.

22. Statistics on poverty in the United States show that
   a. most poor people do not work even though they could.
   b. most poor people who can work are working.
   c. unwed mothers are the largest single group of poor people.
   d. African Americans and Hispanics make up the bulk of the poor.

23. Which of the following is a myth about the poor in America?
   a. Most poor people are black and most black people are poor.
   b. Most poor people live in inner-city ghettos.
   c. Most people are poor because they are too lazy to work.
   d. Most poor people live in female-headed households.
   e. These are all myths about the poor.

24. The "feminization of poverty" refers to
   a. a trend in which females represent an increasing proportion of the poor.
   b. the fact that more feminists are poor these days.
   c. a phenomenon in which nearly all of the social workers who work with the poor are female.
   d. the fact that females seem to cope with poverty better than males.

GOVERNMENT ASSISTANCE PROGRAMS

25. T F Non means-tested assistance is a form of assistance in which individuals and families receive services instead of money.

26. T F Most government benefits go to the middle class.

27. T F Welfare programs provide a comfortable lifestyle for most of the poor.

28. By far the largest amount of the federal government assistance budget goes to
   a. Social Security retirement.
   b. Aid to Families with Dependent Children.
   c. Medicaid.
   d. Food stamps.

29. Means-tested cash assistance programs aimed at the poor account for _____ of the federal budget.
   a. a small and declining portion
   b. a small but ever-increasing portion
   c. about one-third
   d. more than half

30. Means-tested assistance is
   a. based on need.
   b. available regardless of need.
   c. based on political affiliation.
   d. based on ownership of the means of production.

THE CHANGING FACE OF POVERTY

31. T F Economic rewards are distributed more unequally in the United States than anywhere else in the Western industrialized world.

32. The United States has been most successful at holding down poverty among
   a. children.
   b. working-age adults.
   c. the elderly.
   d. women.

33. Of the following nations, the highest rate of poverty among children is found in
   a. the United States.
   b. Canada.
   c. the United Kingdom.
   d. Germany.

34. Which of the following is **not** true of people aged 65 to 69? They have the highest
   a. rate of poverty.
   b. rate of home ownership.
   c. median household net worth.
   d. rate of ownership of government securities, municipal and corporate bonds.
   e. These are all true of people aged 65 to 69.

35. T F The actual number of persons 65 and older living *below* the poverty line in 1999 had actually *declined* substantially since the 1970s.

36. Relative to other Americans, the elderly are
   a. worse off economically.
   b. about the same in terms of economic levels.
   c. better off economically.
   e. none of the above

CONSEQUENCES OF SOCIAL STRATIFICATION

37. T F Poor people are more likely to be arrested than members of higher social classes.

38. T F People with higher levels of education tend to have fewer children than those with less education.

39. T F Philip Roth suggested that the inability to read was a major factor which affected one's chances of future upward mobility.

---

40. Which of the following is true of the lower class in relation to the other social classes in the United States? People in the lower classes
   a. have shorter life expectancies.
   b. are more likely to spend time in a mental hospital.
   c. are more likely to be convicted of a crime.
   d. received longer prison sentences.
   e. These statements are all true of the lower class in relation to other social classes in the United States.

ESSAY QUESTIONS

41. Describe the occupation, educational background, and living circumstances of five families or individuals, one from each of the social classes discussed in your text.

42. Discuss five myths about the poor in the United States, showing why each is a myth.

43. Describe at least three non-economic costs of poverty to the poor.

44. Drawing on information from the 2000 Census cited in your text, describe the changing profile of the elderly relative to their economic standing and relationship to poverty.

45. What are the consequences resulting from the increasing income gap between the wealthiest Americans and the poorest?

46. What factors account for the high rates of poverty among children in America?

CHAPTER EIGHT Racial and Ethnic Minorities

THE CONCEPT OF RACE

1. T F Racial classifications are not simple or cut-and-dried; a person considered "white" in one society might be categorized as "black" in another.

2. T F Differences in traits such as skin color, hair texture, and nose type have proved useful in making biological classifications of human beings.

3. T F Johann Blumenbach was one of the early physiologists who argued that racial categories did not reflect the actual divisions among human groups.

4. T F Your text forecasts a decline in ethnic-identity movements in America during the coming years.

5. T F In recent years we have seen a rise in hate sites on the World Wide Web.

6. The term *race* refers to a category of people who are defined as similar because they
   a. have a unique and distinctive genetic makeup.
   b. share a number of physical characteristics.
   c. exhibit similar behaviors.
   d. express comparable attitudes.

7. Legal definitions of race have generally been constructed in order to
   a. determine who would be identified as black.
   b. determine who was not white.
   c. reinforce existing biological differences.
   d. a and c only
   e. a, b, and c

8. The U. S. Census relies on a _____ definition of race.
   a. social   c. genetic
   b. legal   d. reputational

9. The text's example of Paul and Philip Malone was used to illustrate
   a. the failure of multiculturalism.
   b. a humorous look at the problems of birth records.
   c. the difficulties in accurately assigning race.
   d. that racial appearance and racial definition generally match closely.

THE CONCEPT OF ETHNIC GROUP

10. T F To qualify as an ethnic group, members must be unified politically.

11. T F Ethnic groups often form subcultures with a high degree of internal loyalty.

12. A group that has a distinct cultural tradition with which its own members identify and which may or may not be recognized by others is known as a(n)
   a. subculture.   c. minority.
   b. race.   d. ethnic group.

THE CONCEPT OF MINORITIES

13. T F While the status of *minority group* presents some handicaps, it usually does not exclude members of such a group from full social participation.

14. Can women be regarded as a minority group in American society?
   a. No, because they are a numerical majority.
   b. No, because they are not a racial or ethnic group.

c. Yes, because they are singled out for discriminatory treatment.

d. Yes, because they are a numerical minority of the U.S. population.

15. Which of the following would qualify as a minority group in American society?
a. intellectuals
b. the aged
c. homosexuals
d. adolescents
e. These would all qualify as minorities in American society.

PROBLEMS IN RACE AND ETHNIC RELATIONS

16. T F Discrimination is best understood as an especially bad form of prejudice.

17. T F Prejudice is a rational but negative attitude.

18. In addition to sites that are specifically anti-black, anti-Jewish, or anti-immigrant, there are hate sites expressing views with strong neofascist overtones. These are known as
a. Third Position sites.
b. Third World sites.
c. Moral Center sites.
d. none of the above
e. all of the above

19. T F Whenever prejudice exists, discrimination will occur.

20. T F Prejudicial feelings are more likely to develop between groups that are competing against each other for scarce resources.

21. T F Because they are concerned with Internet hate group negative stereotyping of racial, ethnic, and religious groups, the American Civil Liberties Union advocates a ban on such sites.

22. T F David Goldman argues that hate-group sites have not been as effective as believed as can be seen in their inability to use the Internet to increase their membership.

---

23. Many white shopkeepers in U.S. southern towns during the 1950s and 1960s depended upon African-American customers for a large part of their business but considered them social inferiors. These merchants are examples of
a. unprejudiced nondiscriminators.
b. unprejudiced discriminators.
c. prejudiced nondiscriminators.
d. prejudiced discriminators.

24. J. T. is a member of a minority group. He is unable to obtain a well-paying, secure job not because of outright racism, but rather because, like many others of his minority group, he attended a less-than-adequate school and lacks "connections." J. T. is a victim of
a. subtle and unrecognized personal prejudice.
b. unfortunate accidental discrimination.
c. institutionalized prejudice and discrimination.
d. bad luck that has nothing to do with his being a minority.

25. An individual who feels very uncomfortable when friends tell a racist joke and yet does not speak out for fear of being ridiculed would be classified by Merton as a(n)
a. unprejudiced nondiscriminator.
b. unprejudiced discriminator.
c. prejudiced nondiscriminator.
d. prejudiced discriminator.

26. Which of the following is a function of prejudice?
a. It makes it easier for a group to denigrate its competitors.
b. It facilitates a feeling of in-group solidarity.
c. It allows people to project their own negative qualities onto another group.
d. a and b only
e. a, b, and c

PATTERNS OF RACIAL AND ETHNIC RELATIONS

27. T F Genocide is synonymous with annihilation.

28. T F Canada has two official languages, English and French. This is an example of pluralism.

29. T F Segregation is a form of subjugation.

30. During the nineteenth century, Native Americans were pushed off of land desired by white settlers and onto small and distant reservations. This exemplifies the process of
a. forced migration.
b. segregation.
c. subjugation.
d. assimilation.

31. Most of the nineteenth-century immigrants to America had distinctive subcultures with their own unique language, style of dress, norms, and values. The children of these immigrants, however, rapidly learned to speak English and to adopt mainstream American cultural styles. This is an example of
a. segregation.
b. pluralism.
c. assimilation.
d. subjugation.

32. _____ is most often imposed upon a minority group by the dominant majority. But sometimes it is at least partially voluntary, as in the development of ethnic neighborhoods such as Chinatowns and Little Italys.
a. Segregation
b. Anglo conformity
c. Pluralism
d. Assimilation

33. The policy of the Nazis toward the Jews during the Holocaust of the 1930s and 1940s was one of
a. pluralism.
b. Anglo conformity.
c. annihilation.
d. subjugation.

34. In order to advance his career in radio, Jaime Fernandez learns to speak English without a trace of a Spanish accent and changes his name to Jim Fox. This is an example of
a. subjugation.
b. Anglo conformity.
c. Chicano conformity.
d. annihilation.

RACIAL AND ETHNIC IMMIGRATION TO THE UNITED STATES

35. T F The United States is unlike most other nations in that its ethnic minorities have been fully assimilated.

36. T F As a result of the civil rights movement, African-Americans have reached social and economic equality with other Americans.

37. T F Even during its more restrictive periods, the United States has had one of the more open immigration policies in the world.

38. T F People in the first wave of Cuban immigration have been relatively successful in America because they arrived with more marketable skills and money than immigrants typically possess.

39. T F The *new migration* consists of people from southern and eastern Europe who came here between 1880 and 1920.

40. T F Chicanos are people from Central America.

41. T F More than half of all Native Americans live on or near reservations administered by the U.S. government.

42. T F With an "Executive Order," President Roosevelt empowered the military to remove U.S. citizens of Japanese descent to internment camps.

43. T F Median family income of U.S. Cubans is nearly a third higher than that of other Hispanics.

44. T F Most Asian Americans are concentrated in small towns and rural areas.

45. *Old migration* to the United States consisted of people from
a. the ancient civilizations of the Mediterranean.
b. northern Europe who came prior to 1880.
c. eastern Europe who came after 1880.
d. age 50 upward.

46. Currently, the number of immigrants from _____ is increasing faster than any other population of immigrants to the United States.
a. Latin America
b. Asia
c. Africa
d. the West Indies

47. The largest segment of the U.S. Hispanic population is composed of people of _____ descent.
a. Mexican
b. Puerto Rican
c. Cuban
d. Central and South American

48. In which decade did the largest number of people immigrate to the United States?
a. 1881–1890
b. 1901–1910
c. 1921–1930
d. 1981–1990

49. Almost one-third of all illegal immigrants to the United States come from
a. Mexico.
b. Central America.
c. Southeast Asia.
d. the Caribbean area.

50. Between 1882 and 1943 no _____ immigrants were legally allowed into the United States.
a. Eastern European
b. Chinese
c. African
d. Latin American

51. African-Americans make up approximately _____ of the total population of the United States.
a. 2%
b. 6%
c. 13%
d. 18%
e. 24%

52. Which of the following groups has the highest rate of business ownership?
a. Asian-Americans
b. Jews
c. Hispanics
d. African-Americans

53. Relative to white family income, the median income of African-American families has recently
a. been declining.
b. been rising.
c. remained unchanged.

54. Compared to Europeans, Americans are _____ toward African-Americans.
a. more prejudicial
b. less prejudicial
c. equally prejudicial
d. slightly more prejudicial

55. According to Census 2000, one-third of the foreign-born population was from
a. Florida, Illinois, and California.
b. Russia and Eastern Europe.
c. China and Southeast Asia.
d. England, Wales, and Ireland.
e. Mexico or a Central American country.

PROSPECTS FOR THE FUTURE

56. T F When examining the history of ethnic immigration to the United States, it is most appropriate to view this country as a giant melting pot.

57. T F Most ethnic groups that immigrated to the United States managed to escape prejudice and discrimination.

58. U.S. Census data indicate that since 1980 the number of black and white interracial married couples has
a. increased substantially.
b. remained about the same.
c. decreased slightly.
d. decreased substantially.

ESSAY QUESTIONS

59. Choose a specific group, category, or subculture and explain how it does or does not qualify as a race, ethnic group, or minority.

60. Choose an ethnic or racial group mentioned in the text and show how that group has been a victim of prejudice, discrimination, and institutionalized prejudice and discrimination.

61. List the six major patterns of racial and ethnic relations and give examples of each.

62. Discuss Orlando Patterson's argument that the debate over race and intelligence is "not worthwhile."

63. What effect did the change in the 2000 census, where people could check off more than one race, have on the final count?

64. Discuss the issue of hate groups on the Internet.

CHAPTER NINE Gender Stratification

ARE THE SEXES SEPARATE AND UNEQUAL?

1. T F Women's status appears to be highest in those societies that firmly differentiate between a private domestic sphere and a public sphere of power and authority.

2. T F The example of Kate Bornstein illustrated the logic of viewing gender in terms of two alternatives: male and female.

3. T F Ethology is the scientific study of the genetic sources of human behavior.

4. T F The biblical story of creation has been used as a theological justification for patriarchal ideology.

5. T F Pioneering sociologist Auguste Comte was one of the first to view women as equal to men.

6. A patriarchal ideology is
   a. the study of powerful males in past societies.
   b. the belief that there are differences in the social behavior of men and women.
   c. an attempt to find a genetic basis for human behavior.
   d. the belief that men are superior to women and should control all aspects of society.

7. T F Thomas Jefferson believed in the equal creation of all men, yet felt that women should be denied the right to vote.

8. T F Congress has made it illegal to exclude women as subjects in medical research.

9. T F Dr. John Money's theory of psychosexual neutrality in children was challenged by a graduate student who was eventually proven correct.

10. Gender is best understood as a(n) _____ status.
   a. achieved
   b. ascribed
   c. ideal
   d. peripheral

11. In stressful situations
   a. males and females react with similar intensity.
   b. males and females react with similar behaviors.
   c. men react more slowly.
   d. women react more slowly.

12. Supporters of the view that basic differences between men and women are biologically determined have sought evidence from studies of
   a. other animal species.
   b. physiological differences between the sexes.
   c. cultural practices that have caused genetic mutations.
   d. a and b only
   e. a, b, and c

13. Critics of sociobiology assert that
  a. cultural factors have to be accounted for among nonhuman primates.
  b. sex differences have no biological basis.
  c. gender differences do not exist among nonhuman primates.
  d. it is not valid to generalize from animal to human behavior.

14. Which of the following occurs more frequently among women than among men?
  a. osteoporosis
  b. Alzheimer's disease
  c. diabetes mellitus
  d. a and c only
  e. a, b, and c

15. **T F** Differences in hormones in men's and women's brains might explain why aging women suffer cognitive decline (e.g., Alzheimer's disease) at much higher incidence than men.

16. **T F** Research suggests that "befriending" behavior among females may be caused by their hormones.

WHAT PRODUCES GENDER INEQUALITY?

17. Functionalists argue that the family functions best when
  a. the father assumes the instrumental role and the mother assumes the expressive role.
  b. each parent performs relatively similar roles.
  c. the father assumes both the instrumental and the expressive role.
  d. the mother assumes the instrumental role and the father assumes the expressive role.

18. Conflict theorists argue that gender inequality is _____ based.
  a. biologically
  b. functionally
  c. economically
  d. psychologically

19. Critics of the functionalist view of gender argue that
  a. instrumental roles are not necessary in a family.
  b. expressive roles are far more important to a group than was previously thought.
  c. cross-cultural studies show that gender stratification is not inevitable.
  d. gender roles in modern industrial society are equal.

20. **T F** Anthropologist Michelle Rosaldo argues that the status of females will be lowest in societies which have a "firm differentiation between domestic and public spheres."

21. **T F** In some cultures, tradition requires men to be in control of women.

GENDER-ROLE SOCIALIZATION

22. **T F** According to Deborah Tannen, women use language primarily to create connections with others, whereas men use language mainly to convey information.

23. **T F** Maines and Hardesty argue that men tend to operate in a linear temporal word, while women's temporal word is contingent.

24. **T F** According to John Money, it is possible to raise a child a gender different from that which he or she was born.

25. Research indicates that a core gender identity seems to be established
  a. at birth.
  b. within the first year of life.
  c. by the second or third year.
  d. at puberty.

26. Carol Gilligan argues that women's ethical systems are characterized by a(n)
  a. inability to make firm decisions.
  b. emphasis on the interconnectedness of actions and relationships.
  c. concern with not hurting others.
  d. b and c only
  e. a, b, and c

27. The lifelong process whereby people learn the values, attitudes, motivations, and behavior considered appropriate for each sex in their culture is known as
   a. gender-role socialization.
   b. cultural socialization.
   c. gender identification.
   d. sex-role determination.

28. Social psychologist Erik Erikson argued that in Western society it is more difficult for girls than for boys to
   a. achieve a positive identity.
   b. learn to moderate their innate aggression.
   c. learn to be nurturing.
   d. develop behaviors to attract a suitable mate.

GENDER INEQUALITY AND WORK

29. T F When occupational segregation is controlled for, the difference between men's and women's wages disappears.

30. T F Women are more likely than men to work part-time.

---

31. As of 1998, approximately _____ of American women were in the paid labor force.
   a. 28%
   b. 48%
   c. 58%
   d. 78%

32. Which of the following is **not** one of the ways women experience discrimination in the business world?
   a. During the hiring process women are given jobs with lower occupational prestige than men with equivalent qualifications.
   b. Women receive less pay than men for equivalent work.
   c. Women are fired more often than men.
   d. Women find it more difficult than men to advance up the career ladder.

33. On the average, women with college degrees earn
   a. more than men with college degrees.
   b. about the same as men with college degrees.
   c. only slightly more than men with high school diplomas.
   d. about the same as women with high school diplomas.

34. In terms of college, which gender earns the greatest share of undergraduate degrees?
   a. females
   b. males
   c. there is no difference as they are exactly equal

35. The gap between men's and women's pay has narrowed rapidly in recent years primarily because
   a. men's wages have fallen significantly.
   b. occupational segregation has largely been eliminated.
   c. most women have received pay increases greater than those of most men.
   d. women workers no longer dominate the lowest-paid jobs.

ESSAY QUESTIONS

36. Discuss at least two arguments in support of the sociological notion that gender is an achieved status.

37. Discuss at least three problems posed by traditional gender-role socialization for women, men or both.

38. Discuss the question, "Can you have a gender if you do not have a body?" Use the text information about the issues of gender in cyberspace. Consider the differences in communication style of males and females covered in your text and by Tannen.

39. Describe the major patterns of job discrimination in the United States which are related to gender.

40. Discuss the differences in gender-related behaviors among the Arapesh, Tchambuli, and Mundugumor.

41. Compare and contrast the functionalist and the conflict perspectives on gender inequality.

42. Discuss honor killings and the efforts being made to eliminate them.

43. Discuss the case of Frankie and indicate what it tells us about gender identity.

CHAPTER TEN Marriage and Alternative Family Lifestyles

THE NATURE OF FAMILY LIFE

1. T F Murdock's study of 250 societies found that the family does not exist in some human cultures.

2. T F As in most European societies, the large, extended family was the norm in the United States in the eighteenth and nineteenth centuries.

3. T F According to the U.S. Bureau of the Census, a family and a household are basically the same.

4. T F The nuclear family is found only in industrial societies.

5. T F Polyandry is a very rare form of family structure, existing in only a few societies.

6. T F Most of the world's societies, including the United States, have bilateral systems of descent.

7. T F The patriarchal family refers to a situation in which most family affairs are dominated by men.

8. T F No society permits random sexual behavior.

---

9. A(n) _____ family structure consists of a married couple and their children.
   a. patriarchal
   b. nuclear
   c. extended
   d. matrilineal

10. Which of the following is one of the functions of the family?
   a. socializing children
   b. providing social status
   c. patterning reproduction
   d. organizing production and consumption
   e. These are all functions of the family.

11. A society that traces descent through the mother's side of the family would be characterized as a _____ system.
   a. matriarchal
   b. matrilineal
   c. polygynous
   d. matrilocal

12. _____ is a norm that forbids sexual intercourse between closely related individuals.
   a. The incest taboo
   b. Exogamy
   c. The endogamy prohibition
   d. Monogamy

13. The family is best suited to socialize young children because
   a. there are always at least two adults available for the process.
   b. family members are most knowledgeable of the child's needs and abilities.
   c. family members have more time than anyone else to interact with the children.
   d. parents are the experts in socializing children.

14. Dwayne and Katrina are a married couple. They have two children and live with Katrina's brother's family. This is an example of a(n) _____ family.
   a. extended
   b. polygamous
   c. nuclear
   d. blended

DEFINING MARRIAGE

15. T F Most of the world's cultures allow individuals, mostly males, to have more than one spouse.

16. T F Romantic love is closely associated with marriage in most societies.

17. T F Almost all societies allow for divorce.

18. T F Throughout history, most people have freely chosen their own marriage partners.

19. T F Unlike many European nations, none of the American states has ever had legislation restricting marriage between people of different religions.

---

20. Marriage is
   a. found in all societies.
   b. a legitimized union between two individuals.
   c. intended to be a stable and enduring relationship.
   d. b and c only
   e. a, b, and c

21. Homogamy refers to
   a. stable, marriage-like relationships between members of the same sex.
   b. marriage rules that limit each individual to one spouse at a time.
   c. the tendency to choose a spouse with a similar social and cultural background.
   d. customs that prohibit the marriage of same-sex partners.

22. In the United States today, the most common form of residence for newly married couples is
   a. patrilocal.
   b. matrilocal.
   c. bilocal.
   d. neolocal.

23. When young people know that their parents expect them to marry someone of their own religion, ethnic group, social class, and so forth, they are confronting rules of
   a. endogamy.
   b. monogamy.
   c. exogamy.
   d. polygamy.

24. Which of the following is **not** a feature of marriage in all societies?
   a. a public, usually formal aspect
   b. sexual intercourse as an explicit element of the relationship
   c. romantic love as an important characteristic of the relationship
   d. the intention that it should be a stable and enduring relationship

25. The most common type of interracial marriage in the U.S. involves a marriage between a
   a. white man and an African-American woman.
   b. white man and a nonwhite woman who is not African-American.
   c. white woman and an African-American man.
   d. white woman and a nonwhite man who is not African-American.

26. A person's _____ refers to the family in which he or she was raised.
   a. nuclear family
   b. family of orientation
   c. family of procreation
   d. patriarchal family

27. Which of the following is a feature of romantic love?
   a. love at first sight
   b. love wins out over all
   c. idealization of the loved one
   d. the notion of a one and only
   e. These are all features of romantic love.

28. Social class homogamy in the United States is maintained by
   a. the environment of the typical high school.
   b. the differential experience of college.
   c. explicit prohibitions against dating across class lines.
   d. a and b only
   e. a, b, and c

THE TRANSFORMATION OF THE FAMILY

29. T F The United States has the highest divorce rate in the world.

30. In 1999, about what amount of all births were to unmarried women?
   a. about 10%
   b. about one-quarter
   c. about one-third
   d. about one-half
   e. close to 75%

31. The divorce rate in 1998 was about _____ .
    a. 63.6%
    b. 40.2%
    c. 25.1%
    d. 10.4%
    e. 3.6%

32. T F The vast majority of states make no-fault divorces possible with virtually no waiting periods.

33. T F Critics of no-fault divorce believe it has led to a skyrocketing divorce rate.

34. T F In divorce cases in the United States today, legal custody is given to the father about as often as it is given to the mother.

35. T F A large majority of divorced persons remarry.

36. T F Joint custody laws require children to spend equal time living with each of the divorced parents.

37. T F Companionate marriage is marriage based upon the ideal of romantic love.

38. T F The modern period has witnessed the transfer of functions from the family to outside institutions.

39. T F There are more households in the United States consisting of childless couples than of couples with children younger than 18.

40. T F Children who are victims of abuse are more likely to be abusive as adults than children who have not already experienced family violence.

41. T F The presence of children lowers the probability of remarriage for women, but not for men.

---

42. Industrial society requires a family structure that allows for
    a. geographic mobility
    b. social mobility.
    c. bilateral inheritance.
    d. b and c only
    e. a, b, and c

43. T F Research suggests that Catholics are no less likely to divorce than non-Catholics.

44. T F Regarding the connection between education and divorce, people who have successfully completed an undergraduate college degree are the least divorce prone.

45. No-fault divorce laws
    a. automatically provide the wife with alimony and child support.
    b. blame both husband and wife equally for the failure of the marriage.
    c. are based on the principles of equity, equality, and economic need.
    d. compel the parent with the higher income to pay alimony.

46. T F Divorce rates were already on the rise in the United States even before no-fault divorce laws.

47. Which of the following occupations did **not** have a work force more than 50 percent female in 1999?
    a. cashiers
    b. lawyers
    c. social workers
    d. food preparation workers

ALTERNATIVE LIFESTYLES

48. T F Lesbians and gay men, like heterosexuals, are typically interested in forming stable "couple" relationships.

49. T F Cohabitation refers to a situation in which a newly married couple settles down near or within the husband's father's household.

50. T F The number of Americans living alone more than doubled between 1970 and 1998.

51. T F The United States has the highest rate of births to unmarried women in the Western world.

---

52. In 1998, approximately _____ of all U.S. children lived with only one parent.
    a. 7%
    b. 17%
    c. 27%
    d. 47%

53. The number of people living alone
   a. reflects a current trend to reject marriage.
   b. doubled between 1970 and 1998.
   c. has declined dramatically as many individuals can no longer afford to live alone.
   d. a and c only
   e. a, b, and c

THE FUTURE: BRIGHT OR DISMAL?

54. T F Recent evidence seems to point to the conclusion that the family as an institution is in decline.

---

55. The stereotypical traditional family with a working dad, a homemaker mom, and two or more children accounts for about _____ of all American families today.
   a. 5%   d 25%
   b. 10%   e. 55%
   c. 15%

56. About _____ of American children live in homes without a father present.
   a. 10%   d. 40%
   b. 20%   e. 50%
   c. 30%

ESSAY QUESTIONS

57. Describe the six functions of the family and indicate how each has changed in contemporary American society.

58. How has the tradition of arranged marriages undergone modification in modern India today?

59. Describe at least three alternative lifestyles in U.S. society and explain why they are growing.

60. What factors lead Arlie Hochschild and others to argue that the office "has become a substitute for the family"?

CHAPTER ELEVEN Religion

THE NATURE OF RELIGION

1. T F All religions include a belief in the existence of beings or forces that are beyond the ability of human beings to experience.

2. T F According to Durkheim, the profane consists of objects that people are prohibited from touching, looking at, or even mentioning.

3. T F Drug use plays a central role in some religions.

4. T F Research shows that the concept "God" has had a fixed, unchanging meaning through human history.

---

5. Émile Durkheim observed that all religions, regardless of their particular doctrines, divide the universe into two mutually exclusive categories.
   a. the natural and the unnatural.   c. the routine and the remarkable.
   b. the conventional and the unconventional.   d. the sacred and the profane.

6. Standardized behaviors or practices such as receiving holy communion, the singing of hymns, praying while bowing toward Mecca, and the Bar Mitzvah ceremony are examples of
   a. totems.   c. rituals.
   b. magic.   d. shamanism.

7. Which of the following is an element of religion?
   a. prayer
   b. emotion
   c. belief
   d. a and c only
   e. a, b, and c

8. Religion is
   a. a system of beliefs, practices, and values.
   b. shared by a group of people.
   c. absent in some primitive societies.
   d. a and b only
   e. a, b, and c

9. All religions
   a. worship a god or gods.
   b. utilize magic in their rituals.
   c. promote social equality.
   d. demand some public, shared participation.

MAGIC

10. T F Magic has historically been an important part of the Christian religion.

11. Magic
   a. tends to flourish under conditions of uncertainty and fear.
   b. is usually a means to an end, rather than an end in itself.
   c. has lost respectability as more scientific attitudes have proliferated.
   d. a and c only
   e. a, b, and c

12. T F We find magic where the element of danger is conspicuous.

MAJOR TYPES OF RELIGIONS

13. T F Only three world religions are known to be monotheistic: Judaism and its two offshoots, Christianity and Islam.

14. T F A mana is an ordinary object that has become a sacred symbol for a group or clan.

15. T F The earliest archaeological evidence of religious practice has been found in northern Europe.

16. T F Taboos exist in a wide variety of religions.

17. Which of the following is an example of a religion of abstract ideals?
   a. Christianity
   b. Judaism
   c. totemism
   d. Buddhism

18. The largest of the world's religions, in terms of the size of its membership, is
   a. Christianity.
   b. Islam.
   c. Hinduism.
   d. Confucianism.

19. Most theistic societies practice
   a. monotheism.
   b. supernaturalism.
   c. polytheism.
   d. ecumenism.

20. In animistic religions, the shamans are able to cure illness because they
   a. use powerful medicines.
   b. manipulate the populace to believe in their power.
   c. have the status of gods.
   d. have a special relationship with the spirits that cause illness.

21. Members of a hunting and gathering band believe that a certain sacred spear has the power to draw the prey hunted by the band to its vicinity. We could say that the spear possesses
   a. mana.
   b. animism.
   c. abstract ideals.
   d. profane power.

A SOCIOLOGICAL APPROACH TO RELIGION

22. T F Karl Marx referred to religion as "the opiate of the masses."

23. T F According to Durkheim, when people worship supernatural entities, they are really worshiping their own society.

24. T F Alienation refers to the process by which people lose control over the social institutions they themselves invented.

25. Sigmund Freud said religion was useful in
   a. helping individuals to control dangerous impulses.
   b. bringing about group cohesion.
   c. eliminating anxiety from people's lives.
   d. counterbalancing legal restrictions on individuals.

26. According to Karl Marx,
   a. religion makes humans, humans do not make religion.
   b. religious doctrines justify ruling-class authority, thus preventing opposition and revolt.
   c. religious doctrines can be used to expose ruling-class hypocrisy, thus stimulating revolt.
   d. religion fails to provide comfort even to the oppressed.

27. For Durkheim, the most important function of religion was to
   a. ensure that people behaved morally.
   b. bring about social cohesion.
   c. suppress social revolt.
   d. prevent suicide.

28. According to Max Weber, which of the following belief systems fostered a world view that promoted the development of capitalism?
   a. Judaism
   b. Catholicism
   c. Lutheranism
   d. Calvinism

29. Which of the following is **not** a function of religion?
   a. Emotional integration and the reduction of personal anxiety
   b. Helping a society adjust to its natural environment
   c. Legitimizing arrangements in the secular society
   d. Establishing a world view that helps to explain the purpose of life
   e. Creation of organizational structures providing employment

30. Marvin Harris has shown that the Hindu taboo against eating beef ensures
   a. a large supply of cow dung that is a source of fertilizer and fuel.
   b. that people eat healthier foods.
   c. a large supply of beef for the future, when it will be needed to feed a growing population.
   d. that cows do not have to be fed.

31. The Ghost Dance of the Plains Indians is an example of a
   a. millenarian movement.
   b. revitalization movement.
   c. religious sect.
   d. universal church.

ORGANIZATION OF RELIGIOUS LIFE

32. T F Millenarian movements typically prophesize the end of the world, the destruction of all evil people and their works, and the saving of the just.

33. The animistic beliefs and rituals of a Native American tribe, in which all members of the tribe participate, are an example of a(n)
   a. ecclesia.
   b. universal church.
   c. sect.
   d. denomination.

34. The Church of England, or Anglican Church, is the official church of that country, and its titular head is the king or queen of England. This would make the Anglican Church a(n)
   a. ecclesia.
   b. denomination.
   c. universal church.
   d. sect.

35. Sects differ from denominations in that they
   a. are less tolerant of other religious groups.   d. b and c only
   b. participate less in secular society.   e. a, b, and c
   c. have beliefs that are less conventional.

ASPECTS OF AMERICAN RELIGION

36. T F Although most Americans say that religion is losing its influence in society, 90 percent have a religious preference.

37. T F Ecumenism, the condition in which religious influence is lessened, is a significant trend in modern American society.

38. T F Unlike in Europe, ecumenism has flourished in the United States because the boundaries between denominations here are less rigid and more fluid.

MAJOR RELIGIONS IN THE UNITED STATES

39. T F Because the many Protestant denominations in the United States have so many similar beliefs, they really differ in name only.

40. T F The Catholic Church officially condemns artificial means of birth control, and most American Catholics support this ban.

41. T F For many Jews today, identification with the state of Israel has come to be a secular replacement for religiosity.

---

42. Which major U.S. denomination has the highest percentage that are central-city residents?
   a. Catholics   c. Jews
   b. Baptists   d. Methodists

43. Which of the following Protestant denominations is currently increasing in membership?
   a. United Methodists   d. a and c only
   b. Assemblies of God   e. a, b, and c
   c. United Presbyterians

44. Approximately what percent of Americans say they have a religious preference?
   a. 30%   d. 75%
   b. 50%   e. 90%
   c. 66%

45. Which of the following is **not** a typical characteristic of cults?
   a. members experience manipulation
   b. members are made to feel that they are part of an elite group
   c. morality is reduced to a matter of opinion, without absolute guidelines
   d. group goals take precedence over individual goals

46. The largest single religious denomination in the United States is
   a. Roman Catholic.   c. Baptist.
   b. Methodist.   d. Jewish.

47. "There is no god but Allah, and Mohammed is his Prophet," is the basic tenet of the
   _____ faith.
   a. Muslim   c. Eastern Orthodox
   b. Jewish   d. Jehovah's Witness

48. T F In terms of moderate and conservative views on social issues, Catholics share greater similarity with moderate Protestants than do Baptists.

49. Which of the following represent the third largest faith in America?
   a. Catholics   d. Muslims
   b Baptists   e. Presbyterians
   c. Jews

50. Worldwide, which of the following religions has the lowest membership?
  a. Judaism
  b. Hinduism
  c. Sikhism
  d. Anglicanism
  e. Buddhism

51. T F Generally, the greater the Orthodoxy among American Jews, the greater the disapproval of intermarriage.

ESSAY QUESTIONS

52. List and define the basic elements of religion. Which element do you see as most important?

53. Choose a religion with which you are familiar and show how it fulfills the four major functions of religion.

54. Compare and contrast the four major religions in the United States.

55. Explain how religions are rushing online and making substantial use of the Internet.

56. Briefly explain how the concept of God has changed through history.

CHAPTER TWELVE Education

EDUCATION: A FUNCTIONALIST VIEW

1. T F The functionalist perspective stresses the role of schools in perpetuating class differences from generation to generation.

2. T F The American educational system helps to slow the entry of young adults into the labor market.

3. T F Margaret Mead observed that many preliterate societies make no distinction between education and socialization.

4. T F Women and people older than 25 are the fastest-growing groups of college students.

5. T F Lester Frank Ward believed the main purpose of education was to equalize society.

---

6. Which of the following is a latent function of education?
  a. providing child care
  b. teaching basic academic skills
  c. transmitting cultural knowledge
  d. generating innovation

7. Proponents of bilingual education argue that it
  a. maintains cultural diversity in American society.
  b. forces reluctant Americans to learn a foreign language.
  c. eases the transition of non-English-speaking children into the all-English mainstream.
  d. a and c only
  e. a, b, and c

8. What, according to Lester Frank Ward, is the main source of inequality in society?
  a. the intellectual abilities of those at the top and bottom of society
  b. the differences in the way rich and poor families socialize their children
  c. the unequal distribution of knowledge
  d. the poor academic skills of teachers in the inner cities
  e. all of the above

9. T F Today, recent surveys indicate that Americans overwhelmingly support reforming the public education system.

10. Which of the following is **not** a function of education, according to the functionalist view?
  a. transmitting cultural knowledge
  b. screening and tracking students

c. teaching academic skills

d. socializing children

e. These are all functions of education, according to the functionalist view.

11. What was the basic message of the report titled *A Nation at Risk?*

a. It attacked the Japanese educational system.

b. It praised the American educational system.

c. It encouraged schools to add more electives to the curriculum.

d. It attacked the effectiveness of the American educational system.

12. Which of the following is true of progress toward educational goals in the United States in recent years?

a. Over half of American high school students now take calculus.

b. High school graduation rates rose steadily in the 1980s and 1990s.

c. Over 80 percent of American school districts reported an increase in incidents of violence over the past five years.

d. b and c only

e. a, b, and c

13. The single most important element in the phenomenon of continuing innovation in American society is the

a. work done by garage and basement hobbyists.

b. continuous effort to recruit foreign geniuses.

c. performance of high-caliber academic and research universities.

d. increased attention to standardized testing in science education.

14. Recent survey research indicates that approximately _____ of American adults cannot locate an intersection on a street map or locate two pieces of information in a sports article.

a. 2%   d. 21%

b. 5%   e. 58%

c. 15%

15. T F Research indicates that today the percentage of women attending college is lower than during the 1960s.

16. T F The areas in school where girls excel over boys seems to be widening.

17. As of 1998, just over _____ of high school graduates in the United States were enrolled in college.

a. 10%   d. 60%

b. 20%   e. 80%

c. 40%

THE CONFLICT THEORY VIEW

18. T F To the conflict theorist, the function of school is to produce the kind of people the system needs.

19. T F According to conflict theorists, what is important about obtaining a degree from Harvard or Yale or some other elite college is that it is a guarantee that a person has received quality training.

20. T F According to conflict theorists, the hidden curriculum of schooling subtly promotes creativity and imagination.

---

21. Rosenthal and Jacobson found that student performance is substantially affected by the

a. location of their school.   c. level of their innate intelligence.

b. occupational status of their parents.   d. expectations of their teachers.

22. Bowles and Gintis argue that school success is *least* affected by which of the following?

a. intelligence and effort   c. conformity to school norms

b. possession of appropriate personality traits   d. teachers' expectations of performance

23. Jonathan Kozol argues that, while there is a deep-seated reverence for fair play in the United States, people actually want the game to be unfair and rigged in favor of the advantaged in the area of
a. health care.
b. inheritance of wealth.
c. education.
d. a and c only
e. a, b, and c

24. Which of the following is an aspect of "the credentialized society"?
a. Credentials are necessary for the competent performance of just about every job.
b. A degree or certificate has become necessary to obtain a large number of jobs.
c. Colleges and universities have become gatekeepers, allowing only certain people to obtain credentials.
d. b and c only
e. a, b, and c

25. The stratification of students by ability, social class, and various other categories is referred to as
a. discrimination.
b. tracking.
c. status allocation
d. functional differentiation.

26. According to the conflict theory view, in order to succeed in school the typical American student must learn
a. the official academic curriculum.
b. to design his or her own curriculum.
c. the hidden social curriculum.
d. a and c only
e. a, b, and c

27. T F Boys today are less apt than girls to complete high school, to attend college, and to stay out of jail.

28. Arif has returned to school for certification in computer programming despite the fact that he has ten years of work experience in this field. The schooling will do very little for his performance, but it will get him a raise. This is an example of
a. functional illiteracy.
b. the hidden curriculum.
c. de facto segregation.
d. the credentialized society.

ISSUES IN AMERICAN EDUCATION

29. T F Giftedness is a fairly well-defined concept in educational research and practice.

30. T F Cross-district busing of schoolchildren was a direct outgrowth of the *Coleman Report* of 1966.

31. T F The United States adopted the concept of mass public education long after it had been accepted in Europe.

32. T F When children attend neighborhood schools, they often encounter de facto segregation.

33. T F The *Coleman Report* of 1966 found that minority students perform better when they go to predominantly minority schools.

34. T F Studies of gifted children show that beginning rigorous formal instruction in the preschool years enhances later academic and occupational success.

35. T F Standardized tests are reasonably accurate in measuring intelligence and abilities, especially among younger children.

36. T F The amount of money spent on education per pupil has a significant impact on students' academic success or failure.

37. T F In general, the younger that foreign-born children are when they enter school, the better they will perform.

38. Urban school segregation has been especially difficult to eliminate due to the phenomenon called
a. de jure segregation.
b. white flight from the central cities.
c. massive immigration of minorities.
d. the hidden curriculum.

39. What is the relationship between education and lifetime earnings?
   a. Each higher level of education attained brings higher lifetime earnings.
   b. Level of education attained has little effect on lifetime earnings.
   c. While obtaining a high school diploma increases lifetime earnings, going to college results in little additional earnings.
   d. While obtaining a four-year college degree increases lifetime earnings, post-graduate degrees add little in additional lifetime earnings.

40. Carol Gilligan argues that girls are devalued by society because
   a. they possess a different moral sensibility.
   b. they take jobs and positions males will not take.
   c. more males graduate with bachelor's degrees than females.
   d. young girls are not as serious about school grades as are young boys.
   e. all of the above

41. Since 1980, average combined Scholastic Aptitude Test (SAT) scores have
   a. fallen.   c. stabilized.
   b. risen.

42. Which of the following is **not** a social consequence of dropping out of high school?
   a. increased crime   c. increased intergenerational mobility
   b. decreased tax revenues   d. reduced political participation

43. Which of the following statements about the gifted is true?
   a. People tend to display a marked ambivalence toward them.
   b. Females, minorities, and the disabled are under represented among those identified as gifted.
   c. Teachers tend to associate middle-class cultural traits with giftedness.
   d. There is evidence that the nation's population of gifted children is growing.
   e. These statements about the gifted are all true.

44. High school dropout rates are higher for
   a. members of cultural minorities.   d. a and c only
   b. females.   e. a, b, and c
   c. Native Americans.

45. As of 1999, approximately what percent of the American adult population have completed high school?
   a. 88%   d. 50%
   b. 75%   e. 44%
   c. 66%

46. T F Nearly 20% of all students in grades 9–12 reported they had carried a weapon at least once during the previous month.

47. The *Coleman Report* of 1966 concluded that
   a. the quality of school experiences for black and white students had become approximately equal.
   b. the only way to improve the quality of school experiences for blacks was to spend more money on their schools.
   c. the most important influences on academic success are beyond the control of the schools.
   d. academic success is most powerfully influenced by individual merit rather than social factors.

48. Dropping out of high school affects not only those who leave school, but also society in general because dropouts
   a. pay less in taxes, because of their lower earnings.
   b. are less likely to vote.
   c. have poorer health.
   d. increase the demand for social services.
   e. all of the above

49. In its famous *Brown v. Board of Education* decision in 1954, the United Status Supreme Court banned
   a. de jure segregation.   c. busing.
   b. de facto segregation.   d. standardized testing.

50. Which of the following is **not** a factor associated with dropping out of high school?
   a. speaking a language other than English
   b. low family income
   c. low educational attainment of parents
   d. poor academic achievement
   e. being female and pregnant

51. About _____ of the nation's white students attend central city schools.
   a. 3%
   b. 13%
   c. 23%
   d. 33%
   e. 53%

52. Bilingual education has been used to teach academic subjects to immigrant children in their native languages
   a. in order to cover subjects which could not be taught in English.
   b. while slowly and simultaneously adding English instruction.
   c. because it is no longer necessary for immigrant children to learn English.
   d. has been used as a last-resort teaching technique.

53. Over the last 75 years the percentage of Americans completing high school has
   a. declined slightly.
   b. risen dramatically.
   c. risen slightly.
   d. remained about the same.

ESSAY QUESTIONS

54. Compare and contrast the functionalist and conflict theory views of education as socialization.

55. Discuss the issue of gender bias in the classroom. Based upon your reading of this chapter, what is your position on this issue?

56. Many current issues in American education involve, in one way or another, access to education. Describe at least four categories of students who currently find access to education difficult and explain why they are experiencing this problem.

57. Developed a well-reasoned essay in which you argue that the Internet has had a transforming effect on modern education. Be sure to support your position, especially on the idea of a transformation, with specific information gained from your text.

58. Discuss the issues involved in public school violence, from students carrying weapons to gang-related problems, across many schools.

59. Discuss the pro and con issues connected with the phenomenon of home schooling.

CHAPTER THIRTEEN Political and Economic Systems

POLITICS, POWER, AND AUTHORITY

1. T F Power that is regarded as illegitimate by those over whom it is exercised is called coercion.

2. T F Power based on fear is the most stable and enduring form of power.

3. T F Legitimacy refers to the condition in which people accept the idea that the allocation of power is as it should be.

4. T F Thomas Jefferson believed that, once established, a political system should be changed only in unusually dire circumstances.

5. T F In most societies, what laws are passed, or not passed, depends to a large extent on which categories of people have power.

6. Regardless of his personal abilities or popularity, Charles, Prince of Wales, will become King of England when his mother, Elizabeth II, abdicates the throne or dies. This is an example of _____ authority.
   a. rational-legal
   b. traditional
   c. appointive
   d. charismatic

7. After inspirational civil rights leader Martin Luther King Jr. was assassinated, the civil rights movement in the United States faced the difficult problem of
   a. institutionalized political change.
   b. routinization of charisma.
   c. reinventing the institution.
   d. redefining goals.

8. Although Ronald Reagan was considered by many to have a magnetic personality and to be a symbol of the conservative movement in America, as president he was nevertheless limited by the Constitution and the system of checks and balances established there. Thus, in Weber's terms, he is best thought of as a _____ authority.
   a. legal-rational
   b. traditional
   c. representative
   d. charismatic

9. _____ is the ability to carry out one's will, even in the face of opposition.
   a. Power
   b. Force
   c. Politics
   d. Authority

GOVERNMENT AND THE STATE

10. Plato's ideal society would be one ruled by
   a. the people as a whole.
   b. a military elite.
   c. a carefully bred aristocracy.
   d. religious leaders and philosophers.

11. Which of the following is **not** one of the basic functions of the state?
   a. establishing laws and norms
   b. protecting against outside threats
   c. ensuring economic stability
   d. socializing the young

12. With regard to politics, Aristotle argued that
   a. power should be centered in the middle class.
   b. all citizens should be equal.
   c. the rights and duties of the state should be defined in a legal constitution.
   d. a and c only
   e. a, b, and c

13. The institutionalized way of organizing power within territorial limits is known as
   a. politics.
   b. the state.
   c. authority.
   d. the market.

THE ECONOMY AND THE STATE

14. **T F** In theory, "the invisible hand of the market" means that a multiplicity of self-interested acts by individuals produces a social benefit.

15. **T F** In socialist societies, consumer goods tend to be expensive, while necessities are kept affordable.

16. According to Adam Smith, everyone in a society benefits most from
   a. competition among producers.
   b. centralized government planning.
   c. local government control of economic processes.
   d. democratic decision making in the workplace.

17. Which of the following is characteristic of mixed economies?
   a. all economic decisions are made by central planners
   b. private property is virtually abolished
   c. the government intervenes to prevent industry abuses
   d. a and c only
   e. a, b, and c

18. Marx predicted that, as production expands in capitalist economies,
    a. profits will decline and wages will fall, leading to revolution.
    b. wages will increase and profits will decline, leading to bankruptcy.
    c. profits and wages will both increase, leading to inflation.
    d. profits and wages will be become irrelevant, as a decent standard of living for all is attained.

19. Which of the following is one of the basic premises behind capitalism?
    a. Producers attempt to serve the best interests of society as a whole.
    b. Production is in the pursuit of profit.
    c. Free markets decide what is produced, and for what price.
    d. b and c only
    e. a, b, and c

20. Rosa is paid $10 per hour for her work on an assembly line. During that time she produces goods that will be sold by her employer for $45. The $35 difference between Rosa's wages and the value of what she produces would be termed _____ by Karl Marx.
    a. return on investment   c. surplus value
    b. overhead   d. alienation

21. A _____ is a mechanism for determining the supply, demand, and price of goods and services through consumer choice.
    a. command economy   c. legal-rational authority
    b. market   d. representative government

22. The U.S. economic system is best described as
    a. laissez-faire capitalism.   c. democratic socialism.
    b. a mixed economy.   d. a command economy.

23. Which of the following is a basic principle of socialism?
    a. Everyone should have the essentials before some can have luxury items.
    b. Major decisions should be made by elected representatives in conjunction with the state.
    c. The goal is to ensure that wealth and income are distributed as evenly as possible.
    d. a and c only
    e. a, b, and c

TYPES OF STATES

24. **T F** Communism is a form of totalitarian socialism.

25. **T F** Democracy has always been regarded as the best form of government.

26. **T F** There are both capitalist and socialist states that are totalitarian in nature.

27. **T F** Representative government is a form of government in which a select few rule.

28. **T F** According to Edward Shils, civilian rule means that no member of the military may hold public office.

29. **T F** Democracy can exist only in capitalist societies.

---

30. Under democratic socialism,
    a. private ownership of means of production is abolished.
    b. taxes are kept low.
    c. the state assumes ownership of strategic industries.
    d. little effort is made to expand social welfare programs or redistribute income.

31. _____ refers to the principle of limited government.
    a. The invisible hand   c. Autocracy
    b. Constitutionalism   d. Democracy

32. In which of the following state forms of government does one group exercise virtually complete control over a nation's institutions?
    a. autocracy   c. democracy
    b. totalitarianism   d. dictatorship

33. Totalitarian capitalism is often referred to as
    a. fascism.
    b. laissez-faire capitalism.
    c. a mixed economy.
    d. a command economy.

34. Which of the following is **not** a characteristic of democratic political systems?
    a. majority rule
    b. direct government by the people
    c. civilian rule
    d. constitutionalism

35. Socialists argue that
    a. true democracy is impossible in capitalist society.
    b. with appropriate modifications capitalism can be democratic.
    c. democracy is just an illusion in any form of society.
    d. socialism and democracy are in theory incompatible.

FUNCTIONALIST AND CONFLICT THEORY VIEWS OF THE STATE

36. **T F** Functionalists maintain that the state emerged to manage and stabilize an increasingly complex society.

37. **T F** Historical evidence indicates that the earliest legal codes were enacted to protect the property of the wealthy.

38. Conflict theorists see which of the following as the key to the origin of the state?
    a. the need to coordinate increasingly large and complex societies
    b. the desire of populations to control their own destiny
    c. the nature of human nature, in which some will always dominate others
    d. the need for a way to protect elite control over surplus production

POLITICAL CHANGE

39. When the Chinese Communists under Mao Zedong took power in China in 1949, they sought to institute an entirely new way of life for their people, transforming politics, economics, and culture. This was an example of a
    a. rebellion.
    b. political disorder.
    c. political revolution.
    d. social revolution.

40. One thing that democracies have that dictatorships tend to lack is a mechanism for
    a. dealing with dissent.
    b. implementing laws and official policies.
    c. protection against foreign powers.
    d. institutionalized political change.

41. The American Revolution is an example of a
    a. rebellion that produced social change.
    b. rebellion that produced political change.
    c. revolution that produced social change.
    d. revolution that produced political change.

42. Rebellions differ from revolutions in that they
    a. do not make use of force.
    b. question the uses of power, but not its legitimacy.
    c. are rarely successful.
    d. attack social, but not political issues.

THE AMERICAN POLITICAL SYSTEM

43. **T F** Most democracies have a two-party political system like the United States.

44. **T F** Lobbyists are people paid by special-interest groups to attempt to influence government policy.

45. **T F** The number of women elected to state legislatures has doubled in the past 15 years.

46. **T F** With its strong tradition of popular democracy, the United States has one of the highest levels of voter turnout in the world.

47. Which of the following statements about two-party political systems is true?
   a. They encourage compromise and relatively moderate party platforms designed to appeal to a broad spectrum of citizens.
   b. They are superior to multiple-party systems for ensuring that minority positions are heard.
   c. They are based on proportional representation, so that a party that receives 49 percent of the popular vote gets 49 percent of the legislative seats.
   d. a and b only
   e. a, b, and c

48. Which of the following groups has the highest rate of voting?
   a. eighteen- to twenty-year-olds
   b. the unemployed
   c. college graduates
   d. those with high school diplomas but no college education

49. Which of the following statements concerning the political influence of Hispanics is correct?
   a. They have virtually no influence because of their very low numbers in the electorate.
   b. They have rather little influence because most of them don't speak English.
   c. They have disproportionate influence because of their concentration in states with many electoral votes.
   d. They have disproportionate influence because of their large numbers and their relatively high rate of voter registration.

50. Which of the following is a criticism of Political Action Committees (PACs)?
   a. they tend to favor incumbents
   b. they do not represent the majority of Americans
   c. they tend to diminish the role of the individual voter
   d. a and c only
   e. a, b, and c

51. Which of the following is a way in which journalists exercise political power?
   a. They decide which of many possible interpretations to give to events.
   b. They exercise discretion over how favorably politicians are presented in the news.
   c. They determine how much coverage a politician will receive in the media.
   d. a and b only
   e. a, b, and c

52. Data on members of Congress indicate that they are disproportionately
   a. white.
   b. male.
   c. under 50 years of age.
   d. a and b only
   e. a, b, and c

53. What is the purpose of PACs?
   a. raise and spend money to elect and defeat candidates
   b. protect presidential and congressional candidates
   c. engage in political corruption of election officials
   d. all of the above
   e. none of the above

54. Which of the following voting-age groups has the lowest rate of voter registration?
   a. whites
   b. Hispanics
   c. African-Americans
   d. a and b only
   e. a, b, and c

55. Although they have become more conservative as they have grown older, the post-World War II baby boomers have become more liberal about
   a. capital punishment.
   b. spending on the poor.
   c. pornography.
   d. space exploration.
   e. spending for foreign aid.

56. T F Between 1975 and 1998, the number of women holding state political offices doubled.

ESSAY QUESTIONS

57. Describe the form of power and authority likely to be found in each of four different types of state.

58. Discuss the relationship of democracy to capitalism and socialism.

59. Discuss four specific ways in which the media influence the results of presidential elections in the United States.

60. What are the reasons why people do not bother to vote?

61. Discuss how the structure of the family is associated with the degree to which African-American families have closed the income gap with white incomes.

62. What are the essential similarities and differences between the political-moral values of the '60s with those of today?

CHAPTER FOURTEEN   Population and Demography

POPULATION DYNAMICS

1. T F The United States has achieved the lowest infant mortality rate in the world.

2. T F The crude death rate is the annual number of deaths per 1,000 people in a given population.

3. T F Immigration is the phenomenon that occurs when people enter a geographical area.

4. T F Contemporary Chinese fertility-reduction policies have not succeeded in lowering the birth rate very much.

5. T F Life expectancy is largely determined by the rate of infant mortality.

6. T F Up through the 1970s, China, the world's most populous society, actually had a policy of encouraging population growth.

7. T F The live birthrate for childbearing women in China has fallen from the 1960s rates.

8. The Chinese are attempting to lower their national birthrate to an average of
    a. one child for every three married couples.   c. two children per married couple.
    b. one child per married couple.   d. one child per adult in the family.

9. The technique used in China to lower its national birthrate is
    a. encouragement to obtain the Glorious One Child Certificate.
    b. preferential housing allotments for small families.
    c. larger retirement pensions for parents with fewer children.
    d. preference in school admission for single children.
    e. all of the above

10. Which of the following exerts the least impact on a country's population growth or decline?
    a. migration   c. fecundity
    b. fertility   d. mortality

11. A three-percent yearly increase in population means that a country's population will double every _____ years.
    a. 3   c. 23
    b. 18   d. 30

12. The movement of a population within a nation's boundary lines is known as
    a. internal migration.   c. emigration.
    b. immigration.   d. migration.

13. Which of the following is an example of the phenomenon of exponential growth?
a. 2, 4, 6, 8, 10, 12
b. 1, 3, 5, 7, 9, 11
c. 1, 2, 4, 7, 11, 16
d. 2, 4, 8, 16, 32, 64

14. Fecundity refers to the
a. physiological ability to bear children.
b. life expectancy of a woman after the birth of her first child.
c. actual number of births in a given population.
d. total number of children a woman in a given society can be expected to bear.

15. The rapid rates of population growth in the Third World in recent decades are largely the result of
a. sharp rises in birthrates.
b. sharp improvements in life expectancy.
c. immigration.
d. internal migration.

16. The infant mortality rate refers to
a. babies who are born dead.
b. miscarriages before the ninth month of pregnancy.
c. babies who die within the first year of life.
d. a and c only
e. a, b, and c

17. Which of the following is **not** one of the incentives being used by the Chinese authorities to limit the birthrate in that country?
a. monthly bonuses for urban couples who comply
b. preferential treatment in work and living arrangements for those who comply
c. fines for those who fail to comply
d. long prison sentences for those who fail to comply

18. Which area of the world is currently experiencing the *lowest* population growth?
a. Africa
b. North America
c. South America
d. Southeast Asia
e. Eastern Europe

19. Life expectancy is
a. the average life span of a person.
b. the average number of years one can expect to live.
c. the average number of years a person born in a particular year can expect to live.
d. all of the above

20. According to data presented in Table 14.2, at least _____ countries have higher life expectancies than the United States for children born in 1999.
a. 25
b. 15
c. 9
d. 2
e. no country has higher life expectancy for children than the United States for children born in 1999

21. Which of the following was not among the leading causes of death in the United States in 1998?
a. stroke
b. cancer
c. heart disease
d. accidents
e. These were all among the leading causes of death in the United States in 1998.

22. Migration is
a. the importation of people into a new area.
b. population exchanges.
c. the historical trends in where people are located.
d. the movement of populations from one geographic area to another.

THEORIES OF POPULATION

23. T F The second demographic transition has occurred primarily in North America.

24. T F Population is now declining in several European countries as a result of very low birthrates.

25. T F Utopian socialists were people who, during the Industrial Revolution, advocated a reorganization of society in order to eliminate poverty and other social evils.

26. T F In the developed countries, history has proved the correctness of Malthus's theory of population.

27. T F The second demographic transition may be resulting in fertility stabilizing below replacement levels.

28. T F The worldwide demographic trend of modern times seems to be continued migration of rural populations into cities.

29. T F Early marriage is unrelated to worldwide population growth.

30. T F Breast-feeding is a traditional form of contraception in developing nations.

31. T F Gender preference in children is related to fertility in developing nations.

32. T F In the developing world, education and child health go hand in hand.

---

33. In countries where a second demographic transition is occurring, governments have begun to institute
   a. pronatalist policies.   c. preventive checks.
   b. antinatalist policies.   d. immigration restrictions.

34. The size of the U.S. population is expected to _____ through the year 2025.
   a. grow rapidly   c. remain stable
   b. grow modestly   d. decline somewhat

35. During the first stage of the demographic transition,
   a. fertility rates are low, mortality rates are low, and population size is stable.
   b. fertility rates are low, mortality rates are high, and population size is declining.
   c. fertility rates are high, mortality rates are low, and population size is increasing rapidly.
   d. fertility rates are high, mortality rates are high, and population size is stable.

36. Karl Marx believed that the source of the population problem was
   a. the sheer number of people in the world.
   b. insufficient birthrates to reproduce the working class.
   c. the rise of totalitarian socialist regimes.
   d. industrial capitalist exploitation.

37. According to Thomas Malthus, the core of the population problem is that
   a. people are not educated.
   b. governments are not involved in population control.
   c. forced migration is a necessity.
   d. populations will always grow faster than the available food supply.

38. The demographic transition refers to the
   a. simultaneous decline in birth and death rates as a country industrializes.
   b. decline in birth rates followed later by decline in death rates as a country industrializes.
   c. decline in death rates followed later by a decline in birth rates as a country industrializes.
   d. rise in death rates followed later by a decline in birth rates as a country industrializes.

39. Population increases most rapidly in stage _____ of the demographic transition.
   a. 1   c. 3
   b. 2   d. 4

40. According to Malthus, preventive checks on population growth
   a. are practices that limit reproduction.
   b. include celibacy and the delay of marriage.
   c. involve the deaths of individuals before they reach reproductive age.
   d. a and b only
   e. a, b, and c

CURRENT POPULATION TRENDS: A TICKING BOMB?

41. T F World population is currently just over 3 billion people.

42. T F Most of the world's population growth in the next few decades will take place in the poorer countries.

43. In the middle of which century did world population first reach one billion?
  a. fourteenth
  b. seventeenth
  c. nineteenth
  d. twentieth

44. Which of the following is a significant worldwide demographic trend?
  a. migration of rural populations into cities
  b. smaller family size
  c. slowing in population growth
  d. a and b only
  e. a, b, and c

DETERMINANTS OF FERTILITY

45. T F One of the strongest factors in reducing fertility is the education of women.

46. With regard to the gender of offspring, in most countries there is
  a. a strong preference for males.
  b. a strong preference for females.
  c. no particular gender preference.
  d. a preference for the first child to be female.

47. The greatest consideration for having a second or third child in the United States is
  a. cementing the marriage.
  b. developing greater consumer clout.
  c. creating companions for other children in the family.
  d. carrying on the family name.

48. Which of the following is **not** associated with lower fertility?
  a. living in an urban area
  b. delaying marriage
  c. increasing the incomes of the very poor
  d. breast-feeding an existing child

49. Which of the following is true of the current world population problem?
  a. Less than a quarter of the world's population lives in nations whose standards of living have improved dramatically in the last century.
  b. Average per-capita wealth in poor countries is about one-fifteenth that of the rich nations.
  c. Infant mortality rates are 5 to 20 times higher in poor nations than in rich ones.
  d. b and c only
  e. a, b, and c

PROBLEMS OF OVERPOPULATION

50. T F The dependency ratio is the number of children per family.

51. T F The Green Revolution refers to attempts to dramatically remove pollution from the natural environment.

52. T F Infant death is a continuing problem with more than 90% occurring in developing countries.

53. The Club of Rome predicted
  a. a Green Revolution.
  b. the demographic transition.
  c. worldwide economic and ecological collapse.
  d. technological breakthroughs to solve pollution problems.

54. Why has the Green Revolution failed to materialize?
  a. There are inherent limits to agricultural inputs, such as land, water, and energy.
  b. It has failed to make use of the latest scientific breakthroughs in agricultural research.

  c. Increased agricultural production in the present has diminished resources for future production.
  d. a and c only
  e. a, b, and c

55. Critics of the doomsday model state that
  a. only the poorest countries will experience a decline in population and productivity.
  b. technological advances will create new sources of oil and coal.
  c. population growth will spur human ingenuity and technology will stay ahead of resource use.
  d. imbalances exist because markets operate freely; more government planning and coordination can solve the problem.

ESSAY QUESTIONS

56. Discuss the stages in the demographic transition theory. What is meant by the second demographic transition?

57. Discuss at least five factors associated with lower fertility rates.

58. Compare and contrast Malthus's and Marx's theories of population growth in relation to the problem of overpopulation.

59. Discuss the problems of child labor in developing countries.

60. Discuss how the education of women is associated with family size.

61. Discuss the relationship between population growth and environmental problems.

62. Discuss the argument presented by Stephen Moore that we need not worry about growth in the world population.

CHAPTER FIFTEEN Urban Society

THE DEVELOPMENT OF CITIES

  1. **T F** More than two-thirds of the world's population currently lives in urban areas.

  2. **T F** Cities have existed in one form or another for as long as humans have lived on this planet.

  3. **T F** Gerald Suttles is an important analyst of the preindustrial city.

  4. **T F** For all intents and purposes, civilization began with the rise of cities.

  5. **T F** The most rapid urban growth is currently occurring in the industrialized countries.

  6. **T F** In industrial cities, religious institutions became tightly connected with political and economic institutions.

  7. **T F** Early cities were often connected with religious worship.

  8. In preindustrial cities, the largest social class consisted of
  a. the ruling elite.   c. manual laborers.
  b. a middle class of shopkeepers.   d. slaves.

  9. Which of the following is a requirement that had to be met before cities could appear on the social landscape?
  a. the development of a factory system of production
  b. some form of social organization beyond the family
  c. the capacity to produce a surplus of food and other necessities
  d. b and c only
  e. a, b, and c

10. Ziggurats were
   a. zoned areas in early cities reserved for the poor and elderly.
   b. temple compounds elevated on brick-sheathed mounds, in early cities.
   c. the water transportation systems of early Roman cities.
   d. the inner belt of the working class surrounding preindustrial cities.

11. The world's first fully developed cities arose in
   a. the Middle East, mostly in what is now Iraq.
   b. East Africa, mostly in what is now Kenya.
   c. northern China.
   d. western Europe.

12. Modern-day industrial cities
   a. have clear physical boundaries.
   b. are business and manufacturing centers.
   c. are more stratified than older industrial cities.
   d. have a small middle class.

13. By the year 1900, about _____ of the world's population lived in urban areas.
   a. 14%
   b. 24%
   c. 54%
   d. 74%

14. The United Nations projects that world population will increase to _____ by 2025.
   a. 1.1 billion
   b. 3.4 billion
   c. 5.2 billion
   d. 6.1 billion
   e. 7.8 billion

URBANIZATION

15. T F The concentric zone model is a model of urban development in which distinct, class-identified zones radiate out from a central business district.

16. T F The terms urban area and city can basically be used interchangeably.

---

17. Sometimes metropolitan areas grow into one another to form a larger unit, as in southern California and the strip from Washington, D.C., through the Boston area. These megalopolises are known to the Bureau of the Census as
   a. urbanized areas.
   b. metropolitan statistical areas.
   c. primary metropolitan statistical areas.
   d. consolidated metropolitan statistical areas.

18. _____ suggested a model of urban development in which groups establish themselves in linear sectors that correspond to major transportation arteries.
   a. Ferdinand Tonnies
   b. Louis Wirth
   c. Homer Hoyt
   d. Herbert Gans

19. The largest megalopolis in the United States is centered on
   a. Chicago.
   b. Los Angeles.
   c. Houston.
   d. New York.

20. _____ developed the multiple-nuclei model of urban development, in which similar industries locate near one another and shape the characteristics of the immediate neighborhood.
   a. Park and Burgess
   b. Gerald Suttles
   c. Harris and Ullman
   d. Gideon Sjoberg

21. Early American sociologists of the Chicago School attempted to explain urban development on the basis of
   a. energy flow models borrowed from physics.
   b. the dynamics of plant and animal communities, as discussed by ecologists.
   c. psychological theories concerning the motivation of migrants to the city.
   d. the symbolic meanings attached to urban living by the various subcultures.

22. Which of the following contains a central city?
   a. metropolitan statistical area
   b. urbanized area
   c. suburban area
   d. a and b only
   e. a, b, and c

23. In the concentric zone model, the area farthest from the center is called the _____ zone.
   a. transitional
   b. working-class residential
   c. industrial
   d. commuter

THE NATURE OF URBAN LIFE

24. T F According to William H. Whyte, the parking of automobiles has become an end in itself in many U.S. city centers.

25. T F Because no reliable data have been collected on the homeless, it is impossible to say exactly how many there are.

26. T F In societies with little division of labor and simple group organization, we would expect to find mechanical solidarity, according to Durkheim.

27. T F In American cities today the number of skid rows is on the increase.

28. T F Large metropolitan areas are built around the sort of social relations that Tonnies called *Gemeinschaft.*

29. T F Urban life usually provides more personal freedoms to individuals than are found in rural communities.

30. T F Jane Jacobs argued that social control of public behavior and community life in the city take place on the level of the block, not the neighborhood.

31. T F Simmel argued that urban social relationships lack richness because they tend to be confined to one role set at a time.

32. Which category of persons makes up the largest number of the urban homeless?
   a. The mentally ill
   b. Laid-off factory workers
   c. Displaced minorities
   d. Young rural migrants unable to find employment

33. T F As a city becomes more complex and diverse, people tend to grow increasingly intolerant of diversity.

34. Which of the following social problems characteristize the homeless in urban areas?
   a. 9% are infected with AIDS
   b. 39% are mentally ill
   c. close to a quarter were abused as children
   d. many suffer from chronic alcoholism and drug addiction
   e. all of the above

35. T F A significant portion of the urban homeless are the result of the deinstitutionalization of mental hospitals beginning in 1975.

---

36. Which of the following is **not** among the factors that differentiate the homeless from the poor in general?
   a. extreme poverty
   b. chronic alcohol abuse
   c. fewer years of schooling
   d. less family support

37. _____ refers to the trend in which middle-class young adults are finding urban living more attractive and are moving back into marginal central city areas, thereby displacing the poor.
   a. Urbanization
   b. Gentrification
   c. Mechanical integration
   d. Exurbanization

38. Which of the following, according to William H. Whyte, is an indication that the automobile has triumphed over people in the city center?
   a. At least half of downtown is devoted to parking.
   b. New development includes an enclosed shopping mall.
   c. Office buildings and shopping areas are linked together with skyways and/or people movers.
   d. a and c only
   e. a, b, and c

39. In his book *Urban Villagers,* Herbert Gans showed that
   a. city dwellers often participate in strong community cultures.
   b. city life tends to be alienating and lacking in close personal contacts between people.
   c. the cultural diversity of cities is just as likely to be found in small towns and villages.
   d. urban community life is possible only in well-to-do neighborhoods.

40. According to Tonnies, in a *Gesellschaft*
   a. relationships are intimate.
   b. exchange is based on barter.
   c. formal contracts govern economic exchanges.
   d. people tend to look out for one another.

41. Which of the following is a component of Louis Wirth's definition of a city?
   a. a permanent settlement
   b. relatively large
   c. composed of socially heterogeneous individuals
   d. a and b only
   e. a, b, and c

42. According to Durkheim, _____ is a product of people's commitment and conformity to the collective conscience.
   a. a mechanically integrated society
   b. an organically integrated society
   c. social solidarity
   d. the city

43. Herbert Gans argues that in order to save the central city we need to
   a. lure the middle class back to live there.
   b. lower taxes as much as possible.
   c. lure the upper class back to live and shop there.
   d. raise the economic status of those already living in central cities.

44. Persons who settle in American cities typically are _____, while those settling in suburbs typically are _____.
   a. immigrants, baby boomers
   b. Hispanics, Asians
   c. wealthy, poor
   d. older Americans, younger Americans

FUTURE URBAN GROWTH IN THE UNITED STATES

45. T F Exurbs are those territories that are part of the metropolitan area but outside the central city.

46. T F Suburban growth doubled between 1900 and 1950, but has since come to a standstill.

47. Most residents of U.S. metropolitan areas live in
   a. small towns.
   b. suburbs.
   c. central cities.
   d. exurbs.

48. Which of the following states has been predicted to be the prototype for America's future urban decentralization?
   a. New York
   b. Illinois
   c. New Mexico
   d. North Carolina

49. The percentage of Americans who live in metropolitan areas is expected to _____ in the future.
   a. decline
   b. remain the same
   c. increase

50. T F Exurbanites of today expect the services, schools, and shopping areas to follow them out of the city.

51. Which of the following is a characteristic of an exurb?
   a. It is less densely populated than a traditional suburb.
   b. It remains economically dependent upon the central city for jobs and services.
   c. Its residents are white, relatively wealthy, highly educated, and in professional occupations.
   d. a and c only
   e. a, b, and c

ESSAY QUESTIONS

52. Compare and contrast the preindustrial and industrial city and describe three models for the structure of industrial cities.

53. Discuss three theoretical perspectives on the differences between urban and rural life.

54. Discuss three social problems faced by metropolitan areas in the United States today.

55. Discuss the concepts of mechanical and organic solidarity along with *Gemeinschaft* and *Gesellschaft* relationships.

56. Explain the factors leading to the development and growth of suburbs in the United States.

57. Describe the factors which contribute to disorderly behavior and community decay in American cities.

58. How would you answer the question, "What produces homlessness?"

CHAPTER SIXTEEN Health and Aging

THE EXPERIENCE OF ILLNESS

1. T F Three social factors—race, income, and poverty-nonpoverty residence—are major factors in the quality of a person's health.

2. T F In the United States, although sick people may not be blamed for their illness, they do not need to work toward regaining their health.

3. T F The sick role is a shared set of cultural norms that legitimates deviant behavior caused by the illness.

---

4. Which of the following is (are) a component of the sick role, according to Parsons?
   a. The sick person is not held responsible for his or her condition.
   b. The sick person must cooperate with the advice of designated experts.
   c. The sick person is excused from normal responsibilities.
   d. The sick person must want to get better.
   e. These are all components of the sick role, according to Parsons.

5. Parsons' theory of the sick role is
   a. doctor-centered.   c. society-centered.
   b. patient-centered.   d. mystic-centered.

6. Critics of Parsons' theory of the sick role suggest researchers should focus more on the:
   a. sick individual's interactions with others.   d. a and c only
   b. effect of the illness on the person's identity. e. a, b, and c
   c. sick individual's perception of their illness.

HEALTH CARE IN THE UNITED STATES

7. T F Due to medical advances, hypertension is no longer a major problem for African Americans.

8. T F Studies of life expectancy show that on every measure social class influences longevity.

9. T F Males suffer from illness and disability more frequently than females.

10. T F The United States has the most advanced health-care resources in the world.

11. T F Women are more likely than men to have psychiatric disorders.

12. T F The infant mortality rate among African Americans is currently double the rate among whites.

13. **T F** Because men tend to need more care taking, more than two-thirds of the nursing home popula-tion is male.

14. **T F** Due to the quality of its medical care, U.S. life expectancy ranks near the top of the global scale.

15. The American health-care system has been described as
    a. curative.
    b. acute.
    c. hospital-based.
    d. a and b only
    e. a, b, and c

16. Which of the following is **not** true of the American medical-care workforce?
    a. About three-quarters of all medical workers are women.
    b. The majority of physicians are white, upper-middle-class males.
    c. There is a great deal of mobility among licensed medical occupations.
    d. Medical-care workers include some of the lowest-paid employees in the country.

17. As of 1997, female life expectancy in the United States stood at about _____ years.
    a. 56
    b. 68
    c. 79
    d. 84

18. Males have lower life expectancies than females
    a. due to biological factors.
    b. due to higher male prenatal and neonatal mortality rates.
    c. because men have a higher rate of fatal accidents.
    d. b and c only
    e. a, b, and c

19. As of 1998, the difference between white and African-American male life expectancy was about _____ years.
    a. 2
    b. 6
    c. 10
    d. 15

20. Currently, about _____ of Americans are age 65 and older.
    a. 2%
    b. 7%
    c. 13%
    d. 21%

21. Which of the following groups has the best health profile?
    a. whites
    b. Native Americans
    c. African Americans
    d. Asian Americans

22. John Running Deer is a Native American. He is more likely than someone in the general U.S. population to suffer from
    a. dietary deficiency.
    b. hepatitis.
    c. alcoholism
    d. diabetes.
    e. venereal disease.

23. **T F** Studies of life expectancy show that on every measure, social class influences longevity.

24. Which of the following is a reason that female physicians earn less than male physicians?
    a. The average female physician works fewer hours.
    b. The average female physician is younger and less experienced.
    c. Female physicians are less likely to be in private practice.
    d. Female physicians are concentrated in lower-paying specialties.
    e. These are all reasons that female physicians earn less than male doctors.

25. What are the recent findings from the National Institute of Mental Health on the issue of gender and psychiatric problems?
    a. Men are just as vulnerable to psychiatric problems as women.
    b. Men are less vulnerable to psychiatric problems compared to women.
    c. There are no gender differences related to psychiatric problems.
    d. Older men are more vulnerable to psychiatric problems than older women.

CONTEMPORARY HEALTH-CARE ISSUES

26. **T F** The American Medical Association has been a leading advocate of national health insurance.

27. **T F** Recently, the incidence of AIDS has been shifting from white, gay men to minority drug users.

28. **T F** More surgery is performed in the United States than in any other country.

29. **T F** In the vast majority of AIDS cases in Africa the virus was transmitted heterosexually.

30. **T F** Medicare is a program legislated by Congress to pay the medical bills of people over age 65.

31. **T F** Any baby born to a mother infected with HIV will also be infected.

32. **T F** Recent evidence indicates that HIV can sometimes be transmitted through casual contact.

33. **T F** About half of the nation's physicians are in general practice.

---

34. The virus that causes AIDS is known as
    a. hepatitis-C.
    b. human immunodeficiency virus (HIV).
    c. AIDS-generating virus (AGV).
    d. influenza-B2.

35. From a health standpoint, the major problem with the fee-for-service system of health care is that
    a. doctors' fees are too high.
    b. doctors have a vested interest in pathology rather than health.
    c. doctors waste too much time trying to collect their fees.
    d. people are required to pay the doctor's fee before receiving any health services.

36. Blue Cross and Blue Shield were originally developed to ensure that
    a. everyone had access to affordable health care.
    b. only competent health-care professionals provided health services.
    c. physicians and hospitals got paid.
    d. socialized medicine would one day be possible.

37. Behavioral illness prevention
    a. is directed at changing people's habits or lifestyle.
    b. attempts to manage the threat from mentally ill people who act out dangerous behaviors.
    c. places the burden of change on the individual.
    d. a and c only
    e. a, b, and c

38. Estimates are that AIDS is _____ in state and federal prisons than in the general population.
    a. significantly lower
    b. slightly lower
    c. 4 times higher
    d. 14 times higher

39. Third-party payments are
    a. payments made by insurance or health-care organizations to health-care providers.
    b. a system of national health insurance not supported by either the Democrats or the Republicans.
    c. health-care payments made exclusively by the federal government.
    d. payments made by health-care consumers to their insurance companies.

40. According to the World Health Organization, 60 percent of the world's AIDS cases are currently found in
    a. the United States.
    b. Asia.
    c. Africa.
    d. Europe.

41. An example of intervention at the structural level of illness prevention would be
    a. funding of mass transit as a method of reducing air pollution from autos.
    b. developing low-cost or free comprehensive prenatal care programs.
    c. education campaigns encouraging people not to smoke or chew tobacco.
    d. a and b only
    e. a, b, and c

42. Compared with the situation in the United States, for the rest of the world, the AIDS epidemic is expected to
    a. become worse before it improves.
    b. slowly follow the pattern of the United States.
    c. dramatically improve due to a pattern of treatment developed in the United States.
    d. all of the above
    e. none of the above

43. The human immunodeficiency virus causes AIDS by
    a. directly attacking the major organs of the body.
    b. seeding the growth of particularly virulent forms of cancer.
    c. incapacitating the body's immune system by destroying white blood cells.
    d. altering the genetic code in white blood cells so that they attack and eventually destroy the body of origin.

WORLD HEALTH TRENDS

44. How has the control of the growth of AIDS throughout the world compared with trends in North America, Western Europe, Australia, and New Zealand?
    a. The rest of the world is seeing a faster trend in daily new infections.
    b. The rest of the world is actually seeing a decrease in daily new infections.
    c. The rest of the world now has daily new infections only from women as they have controlled the spread for men.
    d. b and c above
    e. a and b above

45. UNICEF estimates that _____ of the world's deaths among children under 5 years of age are preventable.
    a. 25%   d. 75%
    b. 33%   e. 95%
    c. 50%

46. Currently _____ of the world's population does not have any access to health care.
    a. 20%   c. 60%
    b. 40%   d. 80%

47. The World Health Organization defines health
    a. as the absence of any negative conditions in the body.
    b. in cultural terms, as relative to local standards and perceptions.
    c. as a state of complete mental, physical, and social well-being.
    d. relative to the existing standard of living in a country.

48. Infants are at a high health risk if their mothers
    a. are in their teens or over the age of 40.
    b. have had more than seven births.
    c. have had the baby less than two years after the previous birth.
    d. a and b only
    e. a, b, and c

49. At the end of the nineteenth century, the common explanation for the differences in quality of health of higher and lower income patients was:
    a. differences in housing affected exposure to illness.
    b. patients with more money enjoyed better diets and, therefore, better health.
    c. poverty contributes to disease both directly and indirectly.
    d. lower income patients suffered from moral failings and were ignorant, intemperate, and more or less vicious descendants of failures.

THE AGING POPULATION

50. **T F** Approximately 25% of Americans over sixty-five are institutionalized.

51. **T F** The likelihood of living in a nursing home increases with age.

52. **T F** The three most common causes of death in America today are influenza, heart disease, and cancer.

53. **T F** Almost one in five Americans will be age 65 or older by 2045.

---

54. Which of the following are basic reasons for the growth in the older population?
    a. A large number of persons who were born when the birthrate was high are now reaching age 65.
    b. Many immigrants to the U.S. before World War II are reaching age 65.
    c. Improvements in medical technology are increasing life expectancy.
    d. all of the above
    e. none of the above

55. The African-American percentage of the over 65 group is lower than their percentage of the general population because:
    a. the life expectancy of African-Americans is lower than it is among Whites.
    b. more African-Americans live in rural areas.
    c. The recent birthrate among African-Americans is higher than for Whites.
    d. a and c above
    e. b and c above

56. The lower median income of older Americans does not always mean that they are less well off compared to younger Americans because:
    a. as their household sizes shrink, their disposable income often increases.
    b. older Americans consume much less food than younger Americans.
    c. older Americans spend much less money on automobiles than younger Americans.
    d. a and c above
    e. b and c above

57. Census Bureau survey data indicate what regarding median income and age?
    a. Households ages 55–64 have the greatest incomes.
    b. Households ages 45–54 have the greatest incomes.
    c. Households ages 35–44 have the greatest incomes
    d. Households ages younger than 35 have the greatest incomes.

58. Which of the following was stated as a "very serious" problem in the text data from the American Association of Retired People?
    a. fear of crime
    b. not enough money
    c. loneliness
    d. keeping busy
    e. all of the above

ESSAY QUESTIONS

59. What does it mean to say that the U.S. health-care system is "acute, curative [and] hospital-based"? Explain this statement by describing the basic dimensions of the system as they now exist, including the orientation of the health-care system, how care is delivered, by whom it is delivered, and how it is paid for.

60. Explain at least four ways in which sociological variables affect people's health in the United States.

61. Describe the three basic models of preventing illness and give an example of each.

62. Discuss the problem of stereotyping the elderly.

63. Describe the larger social problems connected with the AIDS epidemic affecting Africa.

CHAPTER SEVENTEEN Behavior and Social Movements

THEORIES OF COLLECTIVE BEHAVIOR

1. T F Neil Smelser was the sociologist who developed the value-added theory of collective behavior.

2. T F Collective behavior has the potential for causing the unpredictable, and even the improbable, to happen.

---

3. "Birds of a feather flock together" might be seen as a common-sense statement of
_____ theory.
   a. convergence
   b. contagion
   c. value-added
   d. emergent norm

4. During World War II, women were very active in the labor force. However, after the war they were pushed out of good jobs by returning veterans. Thus many women who had experienced a taste of their employment capacity found themselves confined to traditional roles once again. In terms of value-added theory, this most likely caused
   a. structural strain
   b. generalized belief.
   c. mobilization for action.
   d. structural conduciveness.

5. Which theory of collective behavior assumes irrational behavior as an important part of its explanation?
   a. value-added theory
   b. convergence theory
   c. emergent norm theory
   d. contagion theory

6. In the value-added theory of collective behavior, which of the following conditions occurs last?
   a. mobilization for action
   b. growth and spread of a generalized belief
   c. precipitating factors
   d. structural strain

7. Gustave Le Bon's thesis in *The Psychology of Crowds* was that
   a. crowd behavior is a constructive force in society.
   b. people become more concerned with proper behavior when in crowds.
   c. individuals are transformed by the anonymity they feel in crowds.
   d. crowd behavior is simply the sum of the individual personalities and motivations that make up the crowd.

8. The objective conditions within society that may promote or discourage collective behavior are known as
   a. structural strain.
   b. collective factors.
   c. structural conduciveness.
   d. precipitating factors.

9. According to emergent norm theory,
   a. crowd formation is an irrational process.
   b. leadership is not essential in a crowd.
   c. contagion is a factor in crowd behavior.
   d. b and c only
   e. a, b, and c

10. Which of the following is a criticism of the convergence theory of collective behavior?
   a. It does not explain how crowds pass through stages.
   b. It does not explain which precipitating events will lead a crowd to action.
   c. It does not explain the individual's motive for accepting the group's influence.
   d. a and b only
   e. a, b, and c

CROWDS: CONCENTRATED COLLECTIVITIES

11. T F The fans attending a Super Bowl game make up an expressive crowd.

12. T F Convergence theory is especially effective in explaining why some crowds act and others don't.

13. T F Research has shown that the average person is quite likely to become irrational in an acting crowd.

14. T F A mob is an example of an acting crowd.

15. Which of the following is **not** a common feature of crowds, according to Elias Canetti?
    a. They are self-generating.
    b. They tend to develop very strong social distinctions among their members.
    c. They thrive on density.
    d. They tend to lose sight of their purpose and require redirecting.

16. A sizable number of passers-by stop to gawk at an auto accident. In sociological terms, this is a(n) _____ crowd.
    a. expressive   c. conventional
    b. casual   d. acting

17. The most frightening of Blumer's crowd types is the _____ crowd, which can easily become violent.
    a. acting   c. expressive
    b. casual   d. conventional

18. Students gathering on the lawn of a campus building between classes would be an example of a(n) _____ crowd.
    a. expressive   c. casual
    b. conventional   d. acting

19. Herbert Blumer is a major figure in the study of collective behavior and social movements. Which of the following aspects of the subject did he discuss?
    a. contagion theory   d. a and c only
    b. expressive social movements   e. a, b, and c
    c. classification of crowds

DISPERSED COLLECTIVE BEHAVIOR

20. T F A craze is a fad that is especially short-lived.

21. T F Opinion leaders are socially acknowledged experts to whom the public turns for advice.

22. T F When large numbers of people in a particular part of the country claim to have seen Elvis, we have an example of a fad.

23. T F An understanding of the dynamics of rumors has sometimes helped to prevent riots.

24. T F Alfred Kroeber first showed that fashion moves through predictable cycles correlated with degrees of political and social stability.

25. T F Public opinion seeks to mobilize public support behind one specific party, candidate, or point of view.

26. T F All of the people watching the Olympics on television constitute a mass.

27. T F Fads and fashions are transitory and have little social impact.

28. Kai Erikson's study of the Massachusetts witchcraft trials of 1692 concludes that they were the result of
    a. a fad.   c. mass hysteria.
    b. contagion.   d. a reactionary social movement.

29. Sociologist Georg Simmel argued that changes in clothing fashions occur because
    a. people have an insatiable desire for novelty.
    b. changes in the physical environment make new types of clothing necessary
    c. the young feel a constant need to be different from their elders, who also try to look young.
    d. the upper classes attempt to distinguish themselves from the lower classes, who try to mimic them.

30. Someone yells "Fire!" in a crowded movie theater and people immediately begin a feverish and chaotic run for the exits. This is an example of
    a. mass hysteria.   c. a rumor.
    b. a panic.   d. mobilization.

31. The difference between crowds and masses of people is that masses
    a. are inherently unstable.
    b. work to resist social change.
    c. do not require close proximity.
    d. consist of people with lower levels of education.

32. Propaganda is information presented to the public to
    a. prevent the spread of rumors.
    b. evaluate public opinion.
    c. clarify political issues.
    d. deliberately influence opinion.

33. Which American company has been fighting rumors of a connection with satanic religion?
    a. Ford Motor Corporation
    b. Amway Corporation
    c. Microsoft Corporation
    d. Snapple Beverage Corporation
    e. Procter and Gamble Corporation

SOCIAL MOVEMENTS

34. T F When a social movement opens a lobbying office in Washington, D.C., it has reached the institutionalization phase.

35. T F The idea behind relative deprivation theory is that people feel distressed in comparison with significant reference groups.

36. T F One flaw in the theory of relative deprivation is that often the people who protest a situation or condition may not be deprived.

37. Social movements like the civil rights movement, which accept most of society's values but seek partial change in the existing social order, are called _____ movements.
    a. expressive
    b. revisionary
    c. reactionary
    d. revolutionary

38. A resource mobilization theorist would place the most emphasis on which of the following?
    a. organizing talent
    b. social injustice
    c. increasing discontent
    d. public support

39. The first stage in the life cycle of social movements, in which the need for change is felt but no means for achieving it is readily available, is called
    a. coalescence.
    b. fragmentation.
    c. incipiency.
    d. institutionalization.

40. Anti-gun-control groups, who protect existing opportunities to buy and carry guns, are an example of a(n) _____ movement.
    a. reactionary
    b. conservative
    c. expressive
    d. revisionary

41. Resource mobilization theory assumes that
    a. protest can arise spontaneously.
    b. discontent exists only at certain times.
    c. in order for a movement to arise, people must know how to channel discontent.
    d. only the upper classes possess the required resources to mobilize a social movement.

42. Groups that seek a return to a remembered past, like the neo-Nazis, skinheads, and the Ku Klux Klan, would best be characterized as _____ social movements.
    a. revolutionary
    b. expressive
    c. revisionary
    d. reactionary
    e. conservative

43. _____ social movements seek to fill a void and distract people from problems.
    a. Reactionary
    b. Expressive
    c. Revisionary
    d. Revolutionary

44. The charisma of a leader is especially important during a social movement's
   a. institutionalization.
   b. coalescence.
   c. incipiency.
   d. fragmentation.

45. Which of the following is the best example of relative deprivation, and thus a possible precursor to social movement activity?
   a. A social group experiences a long-term decline in its standard of living.
   b. A social group is unable to increase its size or attract allies.
   c. A social group is unable to share in the rising standard of living of the surrounding society.
   d. A social group is cut off from contact with members of its extended families.

46. When social movements achieve their goals, they often undergo
   a. institutionalization.
   b. fragmentation.
   c. demise.
   d. disappearance.

47. A social movement that seeks to overthrow all or nearly all of the existing social order and to replace it with an order considered to be more suitable is known as a(n) _____ social movement.
   a. reactionary
   b. expressive
   c. revisionary
   d. revolutionary

48. The period in the life cycle of a social movement when groups begin to form around leaders, promote policies, and promulgate programs is known as
   a. demise.
   b. coalescence.
   c. incipiency.
   d. institutionalization.

49. In Saul Alinsky's style of activism, the most important factor is
   a. politeness.
   b. publicity.
   c. the organizer.
   d. voting.

ESSAY QUESTIONS

50. Discuss how each of the main theories of collective behavior would explain each of the types of crowds, noting strengths and weaknesses in the explanation.

51. Define and give an example of each of the following: fad, craze, fashion, rumor, propaganda, mass hysteria, and panic. Briefly discuss why and under what circumstances each occurs.

52. Discuss how each of the two major theories of social movements would explain each of the typical stages in the life cycle of a social movement.

53. List and discuss Cameron's four social-movement classifications.

54. What are the main points one would need to present to argue in the affirmative that Vegetarianism was a social movement? What are the main negative points one would need to present an argument that it is not a social movement?

CHAPTER EIGHTEEN Social Change

SOCIETY AND SOCIAL CHANGE

1. T F The invention of the printing press and the computer, by themselves, are examples of social change.

2. T F Back in 1967, a writer for *Playboy Magazine* predicted that in the future we would be able to operate a computer as easily as an automobile of his day.

3. T F All cases of violent social or collective behavior result in social change.

4. T F The concept of social change refers only to those events or processes that modernize a society.

5. T F Social change can occur as a result of internal or external forces.

6. T F Sociologically, social change is not always equated with social progress.

7. T F Max Weber was one of the first sociologists to discuss the damaging effects of modernization on the individual.

---

8. Any alteration in society's social organization or any of its social institutions or social roles is referred to as
a. productivity.
b. social change.
c. evolution.
d. modernization.

9. Social change
a. includes changes in attitudes and beliefs.
b. modifies the social structure.
c. is affected by individual motivation and behavior.
d. a and c only
e. a, b, and c

10. What do Henry David Thoreau, Mary Shelley, Stanley Kubrick, and Ted Kaczynski all have in common?
a. They all broke the law in one way or another.
b. All ran off from organized society to live their lives in isolation.
c. In their own way, all made a statement about social and technological change.
d. Actually, they shared nothing in common.

SOURCES OF SOCIAL CHANGE

11. T F To illustrate the benefits of cultural diffusion: When the Yir Yiront tribe was given steel axes they were able to increase their productivity while retaining their traditional cultural practices.

12. T F Diffusion occurs whenever and wherever different cultures come into contact with one another.

13. T F Ideology, whether as religious or secular beliefs, can be a source of social change.

---

14. When Native Americans were moved onto reservations, they were often compelled to wear European-style clothing and to speak English. This is an example of
a. forced acculturation.
b. evolutionary change.
c. an internal source of change.
d. innovation.

15. T F The creation of the computer-based home office is a relatively recent outcome of social change.

16. Which of the following is **not** an internal source of social change?
a. technological innovation
b. diffusion
c. ideology
d. institutionalized structural inequality
e. These are all internal sources of social change.

17. In social scientific terms, the introduction of "fast food" into Third World countries would be an example of
a. progress.
b. cultural lag.
c. diffusion.
d. innovation.

18. Affirmative action programs, designed to compensate for past discrimination, are an example of a _____ ideology.
a. radical
b. liberal
c. conservative
d. reactionary

19. Which of the following is an external source of social change?
a. technological innovation
b. diffusion
c. institutionalized structural inequality
d. ideology

20. The sustainable agriculture movement, which argues that food production in U.S. society must be fundamentally restructured, is driven by a _____ ideology.
a. radical
b. liberal
c. conservative
d. reactionary

THEORIES OF SOCIAL CHANGE

21. T F Pioneering sociologists Durkheim and Tonnies both felt that social change always brought progress.

22. T F According to Arnold Toynbee, the rise and fall of civilizations can be explained through the concepts of challenge and response.

23. T F Sociologist Ralf Dahrendorf developed the concept of mass society.

24. T F Evolutionary theories present a model of how societies change, but they have trouble explaining why they change.

25. T F According to Sorokin's model, sensate cultures are those that emphasize spiritual values.

26. T F According to Marx, the most basic source of all social change is class conflict.

27. T F Cyclical theories are built around the idea of linear stability in the life span of societies.

---

28. In developing his notions of social evolution, Herbert Spencer was strongly influenced by the ideas of
a. Karl Marx.
b. Sigmund Freud.
c. Émile Durkheim.
d. Charles Darwin.

29. Which of the following theories views society as a homeostatic system?
a. evolutionary theory
b. functionalist theory
c. conflict theory
d. cyclical theory

30. Conflict theory locates the source of social change in
a. the phenomenon of progress.
b. society's need to adapt to changing conditions.
c. the struggle between groups over resources and power.
d. oscillation between two opposite sets of dominant cultural values.

31. Durkheim's idea of mechanical solidarity would most closely resemble Tonnies's idea of
a. *Gemeinschaft.*
b. *Gesellschaft.*
c. cultural lag.
d. sensate culture.

32. The continuous change from a simpler condition to a more complex state is referred to as
a. productivity.
b. social change.
c. evolution.
d. modernization.

33. Which of the following is one of the basic needs of society, according to Talcott Parsons?
a. member replacement
b. production of goods and services
c. preservation of internal order
d. provision and maintenance of a sense of purpose
e. These are all basic needs of society, according to Parsons.

34. What does sociologist George Ritzer mean when he speaks of the "McDonaldization" of American society?
a. It appears that McDonald's will eventually drive just about every other competing fast-food chain out of business.
b. An increasingly steady diet of fast food will have a serious negative impact on the physical and mental health of Americans.
c. The principles of the fast-food restaurant—efficiency, quantification, control through nonhuman technology—are coming to dominate more and more aspects of American life.
d. As McDonald's increasingly pours its profits into diversification, it threatens to become the dominant economic factor in every major industry in America.

35. Which of the following most accurately states the functionalist interpretation of the civil rights movement that transformed America in the 1950s and 1960s?
a. It represented a systemic adjustment toward a new equilibrium.
b. It was an inevitable uprising by an oppressed and exploited population.
c. It was a dysfunctional episode that resulted in harmful state interference in the normal operations of the social system.
d. Functionalist theory is unable to explain social movements.

36. Recent predictions that U.S. dominance in world affairs is on the decline have received a great deal of popular attention. These predictions would best fit with the _____ theory of social change.
a. evolutionary
b. conflict
c. functionalist
d. cyclical

37. Early evolutionary theories, like that of Herbert Spencer, defined the process of evolution as
a. biologically driven.
b. movement toward Western standards.
c. movement toward greater social equality.
d. economically driven.

38. In order to avoid the ethnocentric bias in the notion that social evolution can proceed only in the fashion that it has in the Western industrialized countries, social scientists such as Julian Steward have proposed the notion of _____ evolution.
a. specific
b. homeostatic
c. multicultural
d. multilineal

39. Technological change often seems to outrun our shared social norms. For instance, fax machines, computer networks, and electronic mail make it possible for people to communicate in entirely new ways. Yet we are still trying to work out an etiquette of appropriate and inappropriate ways to communicate through this technology. This is an example of
a. ideational vs. technical culture.
b. multilineal evolution.
c. cultural lag.
d. forced acculturation.

MODERNIZATION: GLOBAL SOCIAL CHANGE

40. T F In the course of modernization, work tends to become more specialized.

41. T F Modernization in the Third World has proceeded at a slower pace than modernization in Western societies.

42. T F Whereas modernization was indigenous to most of Europe, it was forced on Third World nations.

43. Which of the following is a change that accompanies modernization?
a. The labor force becomes more specialized.
b. The role of the family is enhanced.
c. Machines are increasingly substituted for human physical power.
d. a and c only
e. a, b, and c

44. The degree of personal stress and dislocation that individuals experience as their society modernizes depends upon, among other things, the
a. conditions under which modernization is introduced.
b. historical traditions of the culture.
c. degree to which the masses share in the material benefits of the change.
d. a and b only
e. a, b, and c

SOCIAL CHANGE IN THE UNITED STATES

45. T F For the most part, occupations that require the most education will grow the most rapidly between now and the year 2005.

46. T F By the year 2005, people with only a high school education will probably not be able to get a job.

47. T F In addition to being helpful, the new technologies such as microwave ovens, voice-mail systems, and the Internet help reduce the complexity of our lives.

48. T F Service-producing industries will account for nearly all of the projected job growth in the next decade.

49. T F The Age of Insight is an era in which technological advances will help us understand how things work and how to make them work better.

50. T F Contrary to popular opinion, more factory jobs have been displaced by automation than by cheap foreign labor or imports.

---

51. Which of the following statements about the relationship between technology and social change is not true?
  a. Technological innovation takes hold only when there is some need for and social acceptance of it.
  b. Technology itself is neutral; people decide whether and how to use it.
  c. Technology influences rather than determines social change.
  d. Technological innovation always occurs in the context of other forces—political, economic, or historical—which in turn help to shape the technology and its uses.
  e. These statements about the relationship between technology and social change are all true.

52. Why did the labor force grow rapidly over the past 15 years?
  a. The baby-boom generation came of age.
  b. Large numbers of women entered the labor force.
  c. Immigration reached an all-time high.
  d. a and b only
  e. a, b, and c

53. How will the labor force of the year 2008 be different from the current labor force?
  a. It will have more minorities and women.
  b. It will have fewer young workers.
  c. It will have fewer blue-collar workers.
  d. a and b only
  e. a, b, and c

54. Which of the following is **not** an occupation projected to increase between now and the year 2008?
  a. travel agents
  b. human service assistants
  c. home health aides
  d. corrections officers
  e. These occupations are all projected to increase between now and the year 2008.

55. T F In addition to computer engineers, social workers too are projected to be a fast-growing occupation between 1998 and 2008.

ESSAY QUESTIONS

56. Discuss two major examples of social change in a society, one caused primarily by internal factors, and one caused primarily by external factors. Briefly explain how the factors led to the social change.

57. Compare and contrast two theories of social change on the phenomenon of modernization.

58. What are the essential criticisms made about the "Information Highway" in this chapter's Society and the Internet feature, "Do We Really Need the Information Highway?" by Clifford Stoll?

59. Describe at least two examples of social change in the direction of McDonaldization. Discuss these examples in light of modernization, technological change, and implications for the workforce of the future.

60. Discuss the process of modernization as it impacts on the individual.

61. Present a profile of the workforce of the future.

PRACTICE TEST ANSWERS

CHAPTER ONE

1. t	11. c	21. f	31. b	41. d
2. t	12. e	22. t	32. d	42. t
3. t	13. e	23. a	33. f	43. t
4. f	14. a	24. b	34. t	44. f
5. f	15. d	25. e	35. f	45. f
6. t	16. c	26. e	36. f	46. a
7. d	17. b	27. a	37. t	47. c
8. a	18. t	28. b	38. a	48. b
9. d	19. t	29. b	39. a	49. b
10. b	20. f	30. b	40. e	50. b

51. Determine the available data for your social problem then raise the following types of questions:
    a. What patterns can be observed in the occurrence of this problem?
    b. How can these patterns be connected with other problems?
    c. What are the background characteristics of people involved in the problem?
    d. What different interpretations of this problem exist in the research literature?
    e. What are the social characteristics of people in groups holding differing opinions and how do these social characteristics explain these differing views?

52. **Psychology:** Concentrates on individual attitudes and motivations as factors affecting human behavior.
    History: Looks at past events in terms of causes and patterns to forecast future outcomes.
    Cultural Anthropology: Examines interactions and relationships across societies characterized by simpler levels of technology.
    Social Work: Concerned with securing programs and legislation designed to reduce social problems as well as help individuals work through their problems.
    Political Science: Focuses on political parties, theories and processes of government, and the phenomenon of elections.
    Sociology: Primary focus is on the group rather than the individual. Sociologists attempt to understand the forces, often unseen, which operate throughout society. These forces mold individuals, shape their behaviors, and determine social events. Sociologists are employed in research firms, criminal justice facilities, private consulting firms, governmental agencies, and most prominently, university teaching. Often persons can increase their career options through combining their backgrounds in sociology with the other social sciences.

53. Poverty can be analyzed form a **Conflict Perspective** in which one might focus on how dominant groups use coercion, constraint, and even force to control members of society through laws and social arrangements designed to maintain their power and wealth. The **Interactionist Perspective** would look at how persons make sense out of and interpret the social world in which the live. This perspective might also examine how the poor interact with each other as well as members of society, for example, social workers, who are charged with assisting them with daily living. The **Functionalist Perspective** would look at the interrelated structure or parts of society which affect and are affected by poverty; for example, government, the family, law enforcement, and social work agencies.

54. Durkheim found that Protestants had higher rates of suicide than Catholics (in countries where they were the majority); single persons higher rates than married; military persons higher than civilians, and people without children higher than those with children. In addition, Durkheim proposed several forms of suicide: egoistic, altruistic, and anomic. He proposed that the extent to which persons were connected to meaningful social relationships and norms was important to suicide rates.

55. "Birds of a feather flock together" and "opposites attract" are common explanations for why people get married. A sociologist would go beyond such common sense explanations and examine how religion, education, and social class similarities might affect the mate selection process. The role of group structures in making possible or reducing the chances for dating and courtship might also be examined. In all, the information provided by sociologist would show patterns of decision making more complex than simple common sense explanations would lead one to believe.

CHAPTER TWO

1. f	11. t	21. c	31. d	41. f
2. t	12. t	22. b	32. c	42. f
3. f	13. t	23. c	33. c	43. f
4. t	14. d	24. b	34. a	44. t
5. f	15. d	25. e	35. f	45. t
6. f	16. b	26. c	36. f	46. e
7. t	17. a	27. b	37. e	47. e
8. t	18. b	28. a	38. b	48. e
9. t	19. c	29. d	39. a	
10. f	20. b	30. b	40. e	

49. You would want to know what type of people were questioned in the poll; i.e., to what extent was it a representative sample? Did the order and wording of poll's questions influence its results? What was the margin of error in the poll?

50. In your essay, you would want to take the statements by William Lipscomb, Albert Einstein, David Hilbert, Poincare, and J.W.N. Sullivan. Interpret each of these quotes in your own words and apply them to the idea that there can be beauty in research.

51. *Reliability:* Findings must be able to be replicated. *Validity:* The study actually tests what it was intended to test. *Representative Sample:* One that shows, in equivalent proportion, the significant variables that characterize the population as a whole.

52. *Hypothesis:* Testable statement about the relationship between empirical variables. *Dependent variable:* The phenomenon being studied. *Independent variables:* Change for reasons having nothing to so with the dependent variable. *Causality:* Something brings about, influences, or changes something else. *Association:* Changes in one variable related to changes in another but not causality

53. *Truth* in the courtroom is based upon a series of rules and procedures, for example, the exclusionary rule, which governs the procedures of presenting information (evidence). Evidence, whether empirically correct or not, would be thrown out if it did not conform to these procedures. While sociologists have procedures for collecting and evaluating data, they also have procedures testing the accuracy of the data (facts) and eliminating or revising them should the data themselves be incorrect. In the courtroom, the procedures are more important that the accuracy of the data (facts).

CHAPTER THREE

1. f	10. d	19. a	28. b	37. c
2. f	11. b	20. a	29. c	38. d
3. f	12. a	21. b	30. f	39. c
4. f	13. c	22. a	31. t	40. b
5. c	14. f	23. a	32. t	41. f
6. b	15. f	24. b	33. t	42. t
7. c	16. t	25. b	34. f	43. t
8. a	17. t	26. d	35. t	44. f
9. d	18. t	27. a	36. t	45. t

46. d	49. t	52. e	54. a	56. t
47. e	50. f	53. d	55. a	57. t
48. e	51. f			

58. Discuss how McDonald's has both adapted to and modified the cultures in those societies throughout the world where they have established their restaurants; for example in offering *ayran* in Turkey, teriyaki burgers in Japan, *Maharaja Macs* in India, and *McAloo Tikki* burgers for Hindus who are vegetarians. Also discuss the use of local personnel as staff; the use of multiple menus for differing dietary norms among customers, and the cultural habits revolving around eating which change the American concept of "fast food."

59. In your essay, consider the issue of animal suffering, the problems of attachment between scientists and the animal research subjects, the effects of separation of animal children from other animal family members, and the possibilities that researchers have a moral obligation to consider the emotional needs of their animal research subjects.

60. Some scientists argue that the human brain is "hard wired" to learn language automatically from birth. Citing the variety of language structures, the complexity of meanings, and the fact that symbols are always relative to the cultural context in which they exist, sociologists generally reject the "hard wire" notion as too simplistic, as it omits the significant influences of culture and environment.

61. The following subcultures were presented in your text: *Ethnic*—ways immigrant groups have maintained their group identities while at the same time adjusting to the demands of the wider society; *Occupational*—certain occupations seem to involve people in a distinctive lifestyle even beyond their work; *Religious*—certain religious groups, though continuing to participate in the wider society, nevertheless practice lifestyles that set them apart; *Political*—small, marginal political groups may so involve their members that their entire way of life is an expression of their political convictions; *Geographic*—large societies often show regional variations in culture; *Social Class*—the ways in which it is possible to discern cultural differences among the classes, e.g., in linguistic styles, values, and norms; *Deviant*—groups that are marginal to society in one way or another and whose lifestyles clash with that of the wider society in important ways.

62. *Material culture:* musical instruments, home and car stereos, fashion, and so on. *Nonmaterial culture:* ideal and real norms regarding concerts, listening to rock music. *Cognitive culture:* rock values and beliefs. Language: distinctive subcultural language of rock music.

63. *Diffusion* of rugby from England to America; *reformulation* of rules into American football; continuing *adaptation* or rules, equipment, and playing styles, including extreme *specialization* of players and positions (including officials).

64. In the United States kissing between men and women in public is common. At the other extreme, in Japan, kissing is considered an intimate sexual act and not permissible in public. In Mexico, a kissing sound is used to summon a waiter while in France, a man kisses a woman's hand without actually touching it. Throughout Latin America, you kiss the air—that is, lips make the sound of kissing, but do not actually press against the cheek.

65. Some linguists believe that animals are capable of higher-level thinking and that the evidence continues to mount that they are capable of using language. Others, for example Noam Chomsky, suggest trying to teach linguistic skills to apes is like trying to teach people to flap their wings and fly. Scientists, agreeing with Chomsky, argue that animal language experiments are exercises in wishful thinking because the animals have not learned anything more than ways to get the hairless animals to give them a treat. Perhaps the argument is best resolved by noting that while animals may use language and tools to some limited degrees, humans have reined, and use to a far greater degree, their use of both material and nonmaterial culture relative to other animals.

CHAPTER FOUR

1. t	3. f	5. a	7. b	9. c
2. t	4. f	6. d	8. b	10. d

11. b	22. d	32. b	42. a	52. d
12. c	23. f	33. a	43. d	53. t
13. d	24. t	34. e	44. t	54. f
14. f	25. f	35. d	45. t	55. t
15. t	26. t	36. c	46. f	56. t
16. f	27. f	37. e	47. f	57. e
17. t	28. t	38. a	48. b	58. c
18. a	29. b	39. c	49. a	59. c
19. b	30. b	40. d	50. b	60. e
20. c	31. c	41. a	51. b	61. c
21. e				

62. Negative effects: slowed or impaired physical, cognitive, emotional, and social development. Needs: social attachment and affiliation.

63. Piaget's *sensorimotor* (learning about cause and effect) and Mead's *preparatory* (imitation is key to learning about social cause and effect). Piaget's *preoperational* (learning symbols for objects) and Mead's *play* (using language to formulate norms in simple absolute terms). Piaget's *operational* (beginning to use abstract thought) and *formal logic* (fully capable of abstract, logical thought) and Mean's *game* (learning complex norms and roles).

64. First of all, the picture of an American shown on television is unrepresentative of the society as whole. In a typical night on Fox, one would see exclusively photogenic young people with disintegrating nuclear families and liberal attitudes about sex. In addition, all of the violence, sex, and pathology lead one to conclude that antisocial behavior is the norm rather than the exception. On the other hand, television has been used in positive ways relative to teaching values, for example, in the designated driver campaign that has had a major effect on reducing alcohol-related fatalities on the highways. Finally it is difficult to decide which side of controversial issues, for example, abstinence versus condoms, having abortions, or the desirability of interfaith marriages, television should advocate. One needs to be aware the embedding messages about moral values or social behavior can have potent effects—for good or for ill.

65. *Peer groups:* support for moving away from family and achieving one's own identity; information and skills regarding cultural adaptations; inappropriate/dangerous subcultural norms. *School:* knowledge, skills, social support; model of impersonal bureaucracy. *Mass media:* information, common culture; exposure to mindless violence and impersonal sex. *Family:* social support and encouragement; stifling personal development.

66. The *nature side* argues that there are sets of reflexes and social dispositions that are biologically based and that affect the development of the individual. The *nurture side* argues that there are no specific genes linked to specific behaviors, which in humans develop from an interplay of language, culture, and environment. Many scientists take a middle position, arguing that the debate is really a false issue, with genetic endowment offering potential which is then developed in the environment.

67. Levinson proposes six stages of development in which there is a close relationship between individual development and one's position in society at a particular time. The stages are *early adult period* (18–22); *getting into the adult world* (22–28), *age 30 transitional period* (28–32); *settling down* (33–40); *age 40 transitional period* (38–42); *beginning of middle adulthood* (mid-40s)

68. In this essay you would want to discuss the three types of child care (own home, family day care, and day care centers). Next, you would want to discuss sociological findings that not all parenting is uniformly good while not all day care is mediocre. Studies show that children who spend time in high quality day care are just as likely to have secure attachment relationships as children who stay at home and quality day care can lead to better mother-child interaction. Children in day care learn cooperation skills better and sooner than children who are cared for at home and also have more advanced language skills and tend to be less timid and fearful. On the other hand, problems arise in day care centers where there is a high turnover of children and day care workers who typically earn low wages. Parents often experience considerable role conflict and role strain as they attempt to meet the challenges of home and work; therefore, the issues of day care and child development are likely to persist in the future.

CHAPTER FIVE

1. t	13. a	25. a	37. a	49. a
2. f	14. a	26. a	38. f	50. t
3. d	15. b	27. f	39. f	51. f
4. b	16. a	28. a	40. c	52. a
5. d	17. a	29. a	41. c	53. f
6. c	18. c	30. a	42. e	54. f
7. b	19. d	31. f	43. b	55. d
8. e	20. f	32. t	44. a	56. b
9. t	21. t	33. b	45. t	57. c
10. f	22. t	34. c	46. t	58. a
11. t	23. f	35. b	47. d	59. b
12. t	24. b	36. c	48. d	

60. Deborah Tannen believes we have become an "argument culture." The argument culture urges us to approach the world and the people in it in an adversarial frame of mind. It rests on the assumption that opposition is the best way to get anything done. Most issues in American society, drugs for example, are framed as a war or a conflict. Conflict and competition are weighed as more important than cooperation and agreement in the argument culture. Finally when you are arguing with someone, you are not trying to understand what the other person is saying, but are trying to get your own response ready in order to win the argument.

61. Choose from among nonverbal behavior, exchange, spontaneous cooperation, directed cooperation, contractual cooperation conflict, coercion, or competition. Drawing upon your own classroom experiences indicate how any three of these forms represent differing types of social interaction.

62. Explain how the group will (1) define boundaries, (2) choose leaders, (3) make decisions, (4) set goals, (5) assign tasks, (6) control members' behaviors.

63. Primary groups involve interaction among members who have an emotional investment in one another and in a situation, who know one another intimately and interact as total individuals rather and through specialized roles. Primary group relationships are especially important in family and family-like relationships. Secondary groups are characterized by much less intimacy among its members. They usually have specific goals, are formally organized and are impersonal. Secondary groups are larger than primary groups and are necessary in situations where members behaviors and roles are more important than personality characteristics; for example in a large corporation such as General Motors.

64. Black Americans dislike it when white Americans call them by the wrong name or use the informality of first name greetings. White Americans are put off by the excessive macho image of black Americans. Black Americans dislike it when whites try to "talk" like black Americans. White Americans are tired of being regarded as "hopelessly uncool" or hearing about everything innocuous turned into "the race thing." Black Americans dislike "the look" often given to them by white Americans which often says that "you do not look like you fit the stereotype of a black American." Black Americans also get tired of being asked to give opinions that represent the entire African-American community. Clearly these everyday slights are areas that we all need to work toward eliminating.

CHAPTER SIX

1. t	8. f	15. b	22. e	29. b
2. f	9. t	16. t	23. b	30. b
3. f	10. c	17. f	24. c	31. d
4. b	11. d	18. t	25. c	32. e
5. e	12. e	19. t	26. b	33. b
6. b	13. f	20. f	27. b	34. d
7. c	14. b	21. a	28. a	35. t

36. f	**43.** f	**50.** t	**57.** d	**64.** d
37. f	**44.** f	**51.** f	**58.** c	**65.** b
38. d	**45.** f	**52.** t	**59.** a	**66.** e
39. t	**46.** t	**53.** f	**60.** t	**67.** a
40. a	**47.** c	**54.** t	**61.** f	**68.** b
41. t	**48.** c	**55.** a	**62.** t	**69.** c
42. t	**49.** c	**56.** a	**63.** t	

70. This is a project founded in 1921 by law professors Barry Scheck and Peter Neufeld. They are currently involved in over 200 cases and have a backlog of more than a thousand. Essentially they are providing free legal assistance to inmates who are challenging their convictions. Older type blood tests, mistaken eyewitness identification, and unmonitored laboratory practices have contributed to the wrongful conviction of the innocent. Today, forensic DNA evidence is often the deciding factor in exonerating the wrongfully convicted. It can provide scientific and irrefutable proof of innocence; however there is often resistance on the part of the police and prosecutors who are reluctant to reopen cases.

71. *Functions:* creates separate identity and group solidarity among youth; clarifies, by way of contrast, what it means to be "grown up"; provides social safety valve during extended adolescence. *Dysfunctions:* undermines social order and consensus about basic values; creates unpredictability; potentially wastes resources.

72. *Psychological theories:* psychoanalytic, individual choice. *Sociological:* anomie, strain, cultural transmission, techniques of neutralization, or labeling theory. Emphasize the implications of the different ways each theory approaches and explains deviance.

73. The *Consensus Approach:* Assumes that laws are merely a formal version of the norms and values of the people. Stealing would be an example of a behavior about which there is consensus that it should be a crime. The *Conflict Approach* to law assumes that the elite use their power to enact and enforce laws that support their own economic interests and go against the interests of the lower classes. Vagrancy laws would fall into conflict approach as they control the poor and clearly reflect the interests of the people who require cheap labor.

74. Merton proposes that individuals adapt to the gap between society's goals and means through *conformity* (accepting goals and means); *innovations* (accepting the goals, rejecting the means); *retreatism* (rejecting the means and the goals); *ritualism* (rejecting the goals, accepting the means) and *rebellion* (rejecting the goals and substituting new goals and doing the same for the means).

75. The components of law consist of the formal rules adopted by a society's political authority. Another component of law is the mechanisms for enforcing the law and for punishing law violators.

76. *Pro-death penalty:* acts as a deterrent; prevents future murders; represents a statement from society about ultimate deviance and crime. *Anti-death penalty:* does not deter; is arbitrary in terms of class, race and gender; is not cost effective; errors are made and cannot be corrected.

CHAPTER SEVEN

1. t	**11.** b	**21.** a	**31.** t
2. e	**12.** d	**22.** b	**32.** c
3. t	**13.** f	**23.** a	**33.** a
4. f	**14.** t	**24.** a	**34.** a
5. c	**15.** t	**25.** f	**35.** t
6. d	**16.** f	**26.** t	**36.** c
7. a	**17.** t	**27.** f	**37.** t
8. a	**18.** e	**28.** a	**38.** t
9. b	**19.** c	**29.** a	**39.** f
10. c	**20.** a	**30.** a	**40.** e

41. *The Upper Class:* Have great wealth going back generations. Recognize one another and are recognized by reputation. Historically Protestant, especially Episcopalian or Presbyterian. Includes chief executives as well as founders of new technology companies. Comprise about 1 to 3% of the population. *The*

Upper-Middle Class: Just below top in organizational hierarchy. Religiously are likely to be Presbyterian, Episcopalian, Congregationalist, Jewish, Unitarian. Live in comfortable homes and are active in civic groups. The largest group is two income couples, both of which are college educated and employed as corporate executives, governmental officials, business owners, or professionals. *The Lower-Middle Class:* Similar to upper middle but without achieving same lifestyle due to economic or economic shortcomings. Likely to be Baptists, Methodists, and Lutherans or sometimes Catholic or Greek Orthodox. Not wealthy but have some savings. Make up 25 to 30% of U.S. population and are politically and economically conservative. *The Working Class:* Made up of skilled and semiskilled laborers, factory employees, and other blue-collar workers. More than half belong to labor unions. Feel powerless, have little time for civic organizations but are much involved with their extended families. Many have not finished high school. Their religious backgrounds are similar to the lower middle class. *The Lower Class:* people at the bottom of the economic ladder. Have little education and few occupational skills. Are often unemployed or underemployed. They have little knowledge of world events, are not involved in their communities and usually do not identify with other poor people. They often have many social problems for example, alcoholism, broken homes, and criminal activities.

42. *Ten myths* about the poor were covered in the text. (1) People are poor because they are too lazy to work. (2) Most poor people are African-American and most African-Americans are poor. (3) Female-headed households are the source of most of the poor. (4) Most people in poverty live in inner-city depressed areas. (5) Federal budget benefits for the poor have been increasing in recent years. (6) Benefits for the poor make up the largest part of the governmental assistance budget. (7) The size of welfare families is growing as mothers have more babies to gain more benefits. (8) Once people get on welfare they stay there indefinitely. (9) The United States has the most generous child welfare programs in the world. (10) The elderly are more likely to be poor than other groups in the U.S. population.

43. (1) Higher rates of disease. (2) Shorter life expectancies (3) Higher infant mortality rates. (4) Greater likelihood of being arrested, charged, and convicted of crime. (5) More likely to be sentenced to prison and given longer prison terms. (6) Dependent on overworked court-appointed lawyers or public defenders. (7) More likely to suffer from certain mental illnesses.

44. Not only are the elderly as a group not poor, but also they are actually better off than the average American. They are more likely than other age groups to possess money-market accounts, certificates of deposits, U.S. government securities, and municipal and corporate bonds. The median household net worth of those 65 to 69 is the highest of any age group. Seventy-seven percent of those 65 to 74 own their own homes and most of those homes are paid in full.

45. During the nineteenth and twentieth centuries, the richest 1% of the U.S. adults held between 20 and 30% of all private wealth in the country. The income gap between the rich and the poor is growing and has reached the highest levels since the 1970s. Some argue that one should not tamper with the economic system in the United States, maintaining that the truly motivated workers can lift themselves into higher brackets. Others take note the fact that it is getting more difficult for people without special skills to land jobs that pay relatively high wages. One consequence of this income gap is the tendency for people to segregate themselves along socioeconomic lines. The Earned Income Tax Credit, which provides tax rebates for low-paid workers in nearly 20 million households, and attempts to increase the minimum wage includes two governmental approaches to dealing with the income gap between the rich and the poor.

46. There has been an enormous growth in the number of children born to unwed mothers in their teens. These mothers are often poor to begin with and children decrease the income earning possibilities. Second, the United States has more poor immigrants than any other country. A third factor is that children, who do not vote, do not have the political power to get government to address their problems. The large gap between the rich and the poor in the United States also contributes to the problem of children born into and living in poverty.

CHAPTER EIGHT

1. t	3. t	5. t	7. b	9. c
2. f	4. f	6. b	8. a	10. f

11. t	21. f	31. c	41. t	51. c
12. d	22. t	32. a	42. t	52. a
13. f	23. c	33. c	43. t	53. a
14. c	24. c	34. b	44. f	54. b
15. e	25. b	35. f	45. b	55. e
16. f	26. e	36. f	46. a	56. f
17. f	27. t	37. t	47. a	57. f
18. a	28. t	38. t	48. d	58. a
19. f	29. t	39. t	49. a	
20. t	30. a	40. f	50. b	

59. *Race:* a category of people who are defined as similar because of a number of physical characteristics. *African-Americans* qualify as a racial group. *Ethnic group:* a group with a distinct cultural tradition with which its own members identify and which may or may not be recognized by others. *Jewish Americans* qualify as an ethnic group. *Minority:* a group of people who, because of physical or cultural characteristics, are singled out for differential and unequal treatment, and who therefore regard themselves as objects of collective discrimination. *Women* qualify as a minority group.

60. See the section "Racial and Ethnic Immigration to the United States" for information on specific groups, and the section "Problems in Race and Ethnic Relations" for definitions of prejudice, discrimination, and institutionalized prejudice and discrimination.

61. (1) assimilation; (2) pluralism; (3) subjugation; (4) segregation; (5) expulsion; (6) annihilation. Give examples from the text or your own experience/knowledge.

62. Significant differences exist in measured average IQ of whites in Tennessee and rural Georgia compared with the northeast, yet we do not insult, neglect, or make a national issue about them. The same holds for the intelligence of the elderly. The principle of infrangibility should cause us to make a commitment to all members of society rather than singling out African Americans.

63. In the 2000 census, 6.8 million people (2.4% of the population) identified themselves as belonging to two or more races. The most common combination was "white" and "some other race" (32%). This was followed by "white" and Native American"(16%), "white and Asian" (13%) and "white and African-American" (11%). "Other" category provides these groups of people with an option.

64. Recent years have witnessed a rise of hate groups on the Internet. In addition to sites that are specifically anti-black, anti-Jewish, or anti-immigrant, there are hate sites known as "Third Position" sites. They can be divided into two groups: those affiliated with the European Liberation Front (often skinheads) and those affiliated with the British group: the International Third Position. There is controversy between the ACLU and other groups who believe these sites are protected speech and others who believe they should be restricted or even banned outright. Some researchers believe the entire issue is overblown and that groups that survived and thrived in the shadows do not do well on the Internet where they are subjected to public scrutiny. Finally, Goldman and others maintain that the hate groups have been very unsuccessful in gathering recruits via the Internet.

CHAPTER NINE

1. f	8. t	15. t	22. t	29. f
2. f	9. t	16. t	23. t	30. t
3. f	10. b	17. a	24. t	31. c
4. t	11. d	18. c	25. c	32. c
5. f	12. d	19. c	26. d	33. c
6. d	13. d	20. t	27. a	34. a
7. t	14. e	21. t	28. a	35. a

36. Cross-cultural variation in gender (e.g., Margaret Mead's research); (2) cases of erroneous gender assignment (e.g., Frankie); (3) sex and gender changers (e.g., Kate Bornstien); (4) differential socialization of males and females in childhood and adolescence.

37. (1) Problems for adolescent girls in developing a sense of identity; (2) lack of encouragement of assertiveness, autonomy, and achievement orientation in girls and women; (3) contingent temporal world that hinders women's career aspirations; (4) socialization of girls and women to choose careers that are less well-paying than male-oriented careers; (5) communication difficulties between men and women (see Tannen).

38. Although they technically do not "have a body," communications styles of males and females often give information about the person at the other end of the keyboard. In addition, women who do attempt to participate on an equal basis are often intimidated by attacks and avoid participating as a result. Finally, the information and "feelings" conveyed by men and women in cyberspace differ, thus often indicating the presence of one's original gender.

39. More women than men are employed part-time, which reduces their earnings and permits job discrimination due to lower-status positions more often being part-time. Women and men are concentrated in different occupational groups, with the ones having more women generally being the lower-paying occupations. Women are also discriminated against in the hiring process and the awarding of promotions process, and the unequal pay process, where they receive less pay than men for equivalent work.

40. The *Arapesh* were characterized as gentle and home loving, with a belief that men and women were of similar temperament. The *Mundugamor* assumed a natural hostility between members of the same sex and only slightly less hostility between the sexes. Both sexes were expected to be tough, aggressive, and competitive. The *Tchambli* believed that the sexes were temperamentally different, but their gender roles were reversed relative to the Western pattern.

41. The functionalist viewpoint argues that it was useful to have men and women fulfill different roles in *preindustrial* societies. It was developed during the 1950s, when American families were very traditional. The conflict perspective argues that by subordinating women, men gain greater control in the areas of economic, social, and political power. Conflict theorists believe that the root cause of gender inequality is economic inequality.

42. In Jordan, women who are raped or are thought to have participated in illicit sexual activity are seen as having compromised their family's honor. Fathers, brothers, and sons see it as their duty to avenge the offense by murdering the victim. Since the author started writing articles about these honor killings, former Queen Noor took up the cause to protect women, and King Abdullah asked the Prime Minister to amend all laws that discriminate against women.

43. Frankie was misclassified as a male at birth and was socialized as a male. At age 5, "he" was brought to the hospital for examination and reclassified as a female whose clitoris had been mistaken for a small penis. The nurses tried to treat her as a female, but she showed a decided preference for the company of little boys in the children's ward and a disdain for little girls and their "sissy" activities. The nurses were uncomfortable treating Frankie as a girl. The current view is that people are not gender neutral at birth and that they are predisposed to be male or female.

CHAPTER TEN

1. f	13. b	24. c	35. t	46. t
2. f	14. a	25. b	36. f	47. b
3. f	15. t	26. b	37. t	48. t
4. f	16. f	27. e	38. t	49. f
5. t	17. t	28. d	39. t	50. t
6. f	18. f	29. t	40. t	51. f
7. t	19. t	30. c	41. t	52. c
8. t	20. e	31. e	42. e	53. b
9. b	21. c	32. t	43. t.	54. f
10. e	22. d	33. t	44. t	55. a
11. b	23. a	34. f	45. c	56. d
12. a				

57. *Regulating sexual behavior:* almost universally, incest rules prohibit sex between parents and their children and between brothers and sisters. There are exceptions: the royal families of ancient Egypt, the Inca nation, and Hawaii did allow sex and marriage between brothers and sisters. *Patterning reproduction:* by permitting or forbidding certain forms of marriage (multiple wives or multiple husbands, for example) a society can encourage or discourage reproduction. *Organizing production and consumption:* In almost all societies, the family consumes food and other necessities as a social unit. Therefore, a society's economic system and family structures are often closely correlated. *Socializing children:* The society must ensure that its children are encouraged to accept the lifestyle it favors, to master the skills it values, and to perform the work it requires. The family provides this almost universally, because its members know the child from birth and are aware of its special abilities and needs. *Providing care and protection:* the family meets all the physical and emotional needs in life: food and shelter, people who care for us emotionally, who help us with the problems that arise in daily life, and who back us up when we come into conflict with others. *Providing social status:* Simply by being born into a family, each individual inherits both material goods and a socially recognized position defined by ascribed statuses, which includes social class or caste membership and ethnic identity.

58. Arranged marriages are more optional than before: sometimes the person is given the choice of marrying for love or having their marriage arranged. Instead of relying on professional matchmakers, families opt for computerized marriage bureaus and pages of highly specific matrimonial ads in the Sunday papers. Young people are allowed to meet alone a few times before the final decision is made.

59. *Single people:* tendency to postpone marriage; choosing to remain single; elderly who are divorced or surviving spouses. *Single-parent families:* increasing divorce rate; greater tolerance of single parenting by choice. *Cohabitation:* greater cultural tolerance; more people leery of marriage due to divorce. *Homosexual and lesbian couples:* greater cultural tolerance of homosexuality; increased political demands for equal treatment of gays and lesbians.

60. Nationwide, people are spending more time at work than ever. Many companies have family-friendly policies, yet few workers choose these options. New management techniques have helped transform the workplace into a more appreciative, personal sort of social world, while at home, the divorce rate has risen and the emotional demands have become more baffling and complex. Job mobility has taken families farther from relatives who might lend a hand. As women have acquired more education and have joined men at work, they have adopted the view that the work world is more honorable and valuable than the world of home and children.

CHAPTER ELEVEN

1. t	12. t	22. t	32. t	42. a
2. f	13. t	23. t	33. b	43. b
3. t	14. f	24. t	34. a	44. e
4. f	15. f	25. a	35. e	45. c
5. d	16. t	26. b	36. t	46. a
6. c	17. d	27. b	37. f	47. a
7. e	18. a	28. d	38. t	48. t
8. d	19. c	29. e	39. f	49. d
9. d	20. d	30. a	40. f	50. a
10. t	21. a	31. b	41. t	51. t
11. e				

52. *Ritual:* pattern of behavior or practices that are related to the sacred. *Prayer:* a means for individuals to address or communicate with supernatural beings or forces. *Emotion:* one of the functions of ritual and prayer is to produce an appropriate state of emotion. *Belief:* all religions endorse a belief system that usually includes a supernatural order and also often a set of values to be applied to daily life. *Organization:* many religions have an organizational structure through which specialists can be recruited and trained, religious meetings conducted, and interaction facilitated; also promotes interaction among the members of a religion.

53. Demonstrate how your chosen religion (1) satisfies individual needs, (2) promotes social cohesion, (3) provides a worldview, and (4) provides adaptations to society.

54. Find basic characteristics of Protestantism, Catholicism, Judaism, and Islam in the section "Major Religions of the United States." Show how these characteristics are similar or different.

55. In the same manner as the printing press permitted publication and the distribution of The Bible, thereby spreading the message of Christianity across Europe, virtually every religion now maintains its own home page through which it delivers its message. These messages reach members as well as nonmembers.

56. God means something different to everyone. The God of today is not the same as the God of the patriarchs, the God of the prophets, the God of the philosophers, the God of the mystics, or the God of the eighteenth-century deists. As soon as the idea of God ceases to be effective it will be changed.

CHAPTER TWELVE

1. f	12. d	23. e	34. f	44. d
2. t	13. c	24. d	35. t	45. a
3. t	14. d	25. b	36. f	46. t
4. t	15. f	26. d	37. t	47. c
5. t	16. t	27. t	38. b	48. e
6. a	17. c	28. d	39. a	49. a
7. d	18. t	29. f	40. a	50. e
8. c	19. f	30. t	41. b	51. a
9. t	20. f	31. f	42. c	52. b
10. b	21. d	32. t	43. e	53. b
11. d	22. a	33. f		

54. *Functionalist view:* Education functions to support and maintain the system through socialization, cultural transmission, teaching academic skills, providing the basis for innovation, and, in a latent fashion, providing child care and postponing job hunting. *Conflict theory view:* The educational system is a means of maintaining the status quo in terms of inequality of wealth and power in society. Schools teach a hidden curriculum of social control that prepares children to internalize the values, attitudes, and behaviors required by the existing hierarchical system. Educational success is determined more by possession of appropriate personality traits and conformity to norms than by intelligence or merit. Schools also screen and track students into statuses required by the existing stratification system.

55. (opinion)

56. (1) African-Americans first attended schools segregated de jure, then, since *Brown vs. Board of Education,* have attended schools segregated de facto. In both cases, the schools have generally been inferior. Thousands of qualified minority students either never attend college or fail to graduate. (2) About 14% of school-age children in the United States speak English as a second language. There is often not enough programming available in the schools to meet their needs, and they are at higher risk of dropping out. (3) Students whose parents have low educational attainment often do not get the necessary support and encouragement at home to do well and are at higher risk of dropping out. (4) Students who attend schools where a climate of violence prevails are more focused on survival than on education. (5) Low-income students are faced with middle-class bias in standardized testing and are at higher risk of dropping out. (6) Due to ambivalent attitudes toward the gifted, these students often do not have access to programs that appropriately challenge them.

57. The Internet has enabled students to continue their learning at home beyond the formal period of schooling during the daytime. Students have become intensely involved in the tasks of sharing files, downloading materials, creating music, playing video games and so on. They are also learning materials beyond that which many of their parents know.

58. Nearly 3 million thefts and violent crimes occur on or near school campus every year. Many urban school systems screen students with metal detectors when entering school grounds and nearly one-fifth of all students in grades 9–12 reported they had carried a weapon at least once during the previous month. Increasingly students are missing school because they feel unsafe. Finally, during the last couple of years there have been more than one hundred cases of school associated violent deaths at more than 101 different schools.

59. Home schooling is almost always a matter of choice and not a necessity because of the unavailability of schools. In the last decade, there has been an explosion in the growth of this type of schooling. On testing, these students typically score at or above average. The quality of home schooling depends on the level of parental involvement, and usually involve the mother, who spends her time instructing the children. Normally children involved in home schooling are from religiously conservative households and their school performance on standardized testing may be a matter of self-selection. Research which goes beyond case studies of these families or of self reported studies is needed to provide an accurate assessment of this form of education.

CHAPTER THIRTEEN

1. t	13. b	24. t	35. a	46. f
2. f	14. t	25. f	36. t	47. a
3. t	15. t	26. t	37. t	48. c
4. f	16. a	27. f	38. d	49. c
5. t	17. c	28. f	39. d	50. e
6. b	18. a	29. f	40. d	51. e
7. b	19. d	30. c	41. d	52. d
8. a	20. c	31. b	42. b	53. a
9. a	21. b	32. b	43. f	54. c
10. c	22. b	33. a	44. t	55. e
11. d	23. e	34. b	45. t	56. t
12. d				

57. *Autocracy:* could be either legitimate or coercive; authority is likely charismatic, though it could be traditional. *Totalitarianism:* could be legitimate or coercive; authority is likely legal-rational, though possibly traditional (e.g., Nazi evocations of German history, Soviet appeals to Marx, Engels, and Lenin). *Democracy:* legitimate power, based on legal-rational authority. *Democratic socialism:* legitimate power, based on legal-rational authority; charismatic authority probably important in transition to implementation of this form of state.

58. Democracy and capitalism both emphasize individual rights and decision making. Socialists argue that democracy is not really possible in capitalism due to the disparities of wealth and power. Ideally, by ensuring the basic needs of all citizens, socialism provides a firm foundation for democracy. This has not been true in practice, say capitalists. Democratic socialism is a convergence of capitalism and socialism that maintains democracy.

59. *Forms of media influence:* Journalists decide (1) how much coverage to give campaigns and candidates; (2) what interpretation to give to campaign events; (3) how favorably to present candidates in the news; (4) whether to endorse a candidate. *Regarding an African-American or Hispanic candidate:* (1) could be dismissed as a fringe candidate and given little coverage; (2) could be portrayed as candidate for minority groups only, and not someone who could represent all the people; (3) could be held to different, higher standards than white candidates; (4) unlikely to be endorsed by major mainstream media if the candidate is perceived as too controversial for his/her target advertising audience.

60. More than 1 in 5 say they could not take the time off from work or were too busy. Another reason increasingly cited for not voting is apathy about the political process.

61. African-American married couples have closed the gap with white incomes, but the proportionate share of married couples in the total black population is smaller than a decade ago. More African-Americans are in poverty today than a decade ago. Part of the reason is the fragmentation of the African-American family. Forty-seven percent of all African-American families were married couples, 45% were maintained by women without a husband present and 8% were maintained by men without a wife. When we look at married couple African-American families we find that only about 7% are below the poverty level.

62. Today the political-moral values are more conservative than the 1960s that were a period of massive social change ranging from the civil rights legislations, to youth culture that challenged the moral authority of government in many areas. Today there appears to be greater tolerance for different groups in society, for example interracial marriages and greater support for equal job opportunities. On the other hand, Americans are far more fearful of big government and far fewer want the government to do things for them than in the '60s. Overall, one might say we are a more tolerant society today than in the '60s.

CHAPTER FOURTEEN

1. f	12. a	23. f	34. b	45. t
2. t	13. d	24. t	35. d	46. a
3. t	14. a	25. t	36. d	47. c
4. f	15. b	26. f	37. d	48. c
5. t	16. c	27. t	38. c	49. e
6. t	17. d	28. t	39. b	50. f
7. t	18. e	29. f	40. e	51. f
8. b	19. c	30. t	41. f	52. t
9. e	20. c	31. t	42. t	53. c
10. a	21. e	32. t	43. c	54. d
11. c	22. d	33. a	44. e	55. c

56. *First demographic transition:* Stage 1: high fertility, high mortality, high fecundity due to relative youth of the population, no overall population growth; local population growth through migration. Stage 2: continued high fertility, decreasing mortality, slightly decreasing fecundity due to increased life expectancy; migration less of a factor in population growth. Stage 3: decreasing mortality, fertility and fecundity; migration more of a factor. Stage 4: Low fertility and mortality; decreasing fecundity as life expectancy continues to rise; migration a major factor in local population increase or decrease. *Second demographic transition:* fertility below replacement; low mortality; fecundity declines as population ages; migration is major factor in local population change.

57. Factors associated with lower fertility rates: (1) raise the average age of first marriage; (2) encourage breast feeding; (3) lower and stabilize infant mortality rates; (4) discourage strong preference for male offspring; (5) increase availability of effective, affordable contraception; (6) provide good-paying jobs and reliable old-age support; (7) educate the population, particularly women.

58. *Malthus:* Overpopulation is taking place because of a basic, unchanging principle: Population inevitably tends to outrun resources. When the preventive checks have failed—as it appears they have done today—then the positive checks will slow population growth. *Marx:* Industrial capitalists exploitation requires a large class of the poor who are easily exploitable. Lacking security in the existing system, these people turn to their families and have large numbers of children. In addition, capitalists require expanding product and labor markets. Population growth can only be harnessed by eliminating capitalism and implementing socialism, a system in which meeting people's basis needs comes first.

59. Children work all over the world. They are difficult to protect not only from wage exploitation but also from physical and sexual abuse. Their working conditions are often bad, and it is difficult to gather accurate statistics on how many are working and where.

60. As a woman's education increases, the number of children she has decreases. The effects of female education on fertility are much greater than male education. Educated women are more likely to use contraception. Women with 10 years of education desire fewer children than women with less education.

61. Overpopulation affects the environment through the changeover of land from natural states to agricultural and urban usage. Overfarming increases erosion and river pollution, both of which have negative environmental effects. Overpopulation may also have a negative effect on climate.

62. Moore, an economists, believes that population growth will level off. He notes that although naysayers are warning about overpopulation, in fact birth rates around the world are lower rather than higher with the birthrate in developing countries moving from 6 per couple in the 1960s to just more than 3 today. EPA statistics indicate that air and water pollution is decreasing rather than increasing. He suggests that values, for example, the planting of new trees rather the birth of a baby, is regarded as progress by those advocating population control.

CHAPTER FIFTEEN

1. f	12. b	22. d	32. a	42. c
2. f	13. a	23. d	33. f	43. d
3. f	14. e.	24. t	34. e	44. a
4. t	15. t	25. t	35. t	45. f
5. f	16. f	26. t	36. b	46. f
6. f	17. d	27. f	37. b	47. b
7. t	18. c	28. f	38. e	48. d
8. c	19. d	29. t	39. a	49. c
9. d	20. c	30. t	40. c	50. t
10. b	21. b	31. t	41. e	51. d
11. a				

52. See Table 15–1 for a comparison of the preindustrial and industrial city. See Figures 15–1, 15–2, and 15–3 for models of city structure.

53. (1) *Tonnies: Gemeinschaft* (intimate, cooperative, personal relationships) in rural communities; *Gesellschaft* (impersonal, independent relationships) in the city. (2) *Durkheim:* mechanical solidarity (strong commitment to the collective conscience) in rural society; organic solidarity based on division of labor in the city. (3) *Simmel:* rich, multilayered relationships based on multiple role relationships at once in rural society; limited relationships based on one role set at a time in the city. (4) *Wirth:* city is large dense, heterogeneous, alienating; rural areas the opposite. (5) *Gans and Suttles:* people recreate the social solidarity of small towns within urban neighborhoods.

54. Three social problems are (1) urban decline—which involves the movement of middle-class residents to the suburbs and beyond, resulting in continued shrinking revenues; (2) the homeless—who are increasing in central-city areas and are often the result of a combination of failed social policies, one of which was the deinstitutionalization of the mentally ill; and (3) serious social problems—the city centers are still magnets for crime, drug usage, and illegal sexual activities.

55. *Mechanical and organic solidarity* were developed by Durkheim to describe two different forms of the collective conscience in society. In a *mechanically integrated* society, the collective conscience is strong and there is great commitment to that collective conscience. In an *organically integrated* society, social solidarity depends on the cooperation of individuals in many different positions who perform specialized tasks. The mechanically organized society is more likely to be characterized by *gemeinschaft* relationships while the organically organized society is more likely to be characterized by *gesellschaft* relationships.

56. Three main factors were important in suburban growth: (1) the increase in rapid transportation was a major factor, as was (2) the desire for the single family home and (3) the development of savings and loan businesses after the 1920s, which made it easier to purchase homes.

57. (1) Urban decline caused by out-migration of middle-class residents and businesses, reducing the tax base and tax revenues; (2) gentrification and real-estate speculation deplete the housing supply for the poor; (3) increased numbers of homeless people, particularly those deinstitutionalized from mental

health agencies and institutions (4) aging suburbs lack central-city amenities and resources and must overcome poor planning in their original design; (5) exurbs are less dense, and, therefore, it is difficult to deliver services there.

58. One of the major factors contributing to the homeless in urban America is the displacement of mentally ill persons from state and community hospitals. In addition, many of the homeless are HIV infected without places to live. Many more are homeless due to the loss of affordable housing due to the gentrification of urban areas as well as the lack of minimum wage job opportunities for many persons with low skill levels. Most homeless also have a history of illiteracy, crime, poor family organization, and welfare dependency.

CHAPTER SIXTEEN

1. t	13. f	25. a	37. d	49. d
2. f	14. f	26. f	38. d	50. f
3. t	15. e	27. t	39. a	51. t
4. e	16. c	28. t	40. c	52. f
5. a	17. c	29. t	41. d	53. t
6. e	18. e	30. t	42. a	54. d
7. f	19. b	31. f	43. c	55. d
8. t	20. c	32. f	44. a	56. a
9. f	21. d	33. f	45. e	57. a
10. t	22. d	34. b	46. d	58. e
11. f	23. t	35. b	47. c	
12. t	24. e	36. c	48. e	

59. The focus of the U.S. system is on curing disease, not preventing it, and doing so with high technology in complex hospital settings. Care is delivered by highly differentiated (and highly paid) medical specialists who treat the disease, but not the person. Comprehensive, family physicians are the smallest in number and the lowest paid of doctors. While medical doctors are predominantly white, male, and upper-middle class, the bulk of the rest of medical workers are female, disproportionately minority, and low-paid. Health insurance exists primarily to ensure that doctors and hospitals get paid. People without health insurance have difficulty obtaining medical services.

60. *Gender:* Males have a lower life expectancy, due to differing gender roles and socialization; women appear to be sick more often; while the overall rate does not differ, the types of psychiatric illnesses do differ by gender. *Race:* The infant mortality rate is twice as high for African-Americans as whites; life expectancy is six to eight years lower for African-Americans. The health situation is particularly bad for black males. *Social class:* On every measure, social class influences longevity. *Age:* Longevity has increased, and the top causes of death have changed in this century, resulting in more hospitalization and institutionalization of the elderly.

61. *Medical prevention* is directed at the individual's body, *behavioral prevention* is directed at changing people's behavior, and *structural prevention* is directed at changing the society or environments within which people work or live.

62. We tend to think of the elderly as a homogeneous group, which is not the case as there are at least three groups; the young old (65–74); the middle-old (75–84) and the old-old (85 and older). Thinking stereotypically about the elderly reduces their opportunities and deprives society of important resources. In a recent survey, more than half of the elderly believe money, poor health, and fear of crime are serious problems for other persons. Close to half (46%) believed loneliness was a serious problem for other elderly. Interestingly, most of the elderly do not regard these issues as serious problems for themselves. In a sense, when the elderly hold stereotypes about other elderly, it produces higher levels of life satisfactions for themselves when they believe they themselves, are not seriously affected by these problems. The elderly hold many of the same stereotypes about their age group as does the younger public at large.

63. A few of the outcomes are: reduced life expectancy; the breakdown of extended families, the effects on families when parents die and children are left without care; the faltering educational systems; the loss of productive workers for business and government, and the overall inability of the health care systems to care for the number of infected patients. The data suggest the numbers of newly infected persons are increasing and the situation throughout many nations in Africa is likely to get much worse before it begins improving.

CHAPTER SEVENTEEN

1. t	11. t	21. t	31. c	41. c
2. t	12. f	22. f	32. d	42. d
3. a	13. f	23. t	33. e	43. b
4. a	14. t	24. t	34. t	44. c
5. d	15. b	25. f	35. t	45. c
6. a	16. b	26. t	36. t	46. b
7. c	17. a	27. f	37. b	47. d
8. c	18. c	28. c	38. a	48. b
9. d	19. e	29. d	39. c	49. c
10. d	20. t	30. b	40. b	

50. *Contagion theory:* Individuals are transformed by the anonymity of the crowd and become highly receptive to group sentiments. (1) This explains why once it gets started, an acting crowd can become very emotional, active, and even violent. (2) Likewise, people in an expressive crowd give in to the emotions of the moment. (3) In a conventional crowd, peer pressure creates a contagious conformity. (4) A causal crowd may form through contagious curiosity. Weakness: Cannot explain what gets the contagion started in any type of crowd and which people are and are not vulnerable to it. *Emergent norm theory:* Crowds acquire standards by watching and listening to each other. (1) Members of an acting crowd come to believe that action or violence is justified by the situation. (2) Members of an expressive crowd exert interpersonal control over the nature and extent of emotional expression. (3) Peer pressure reinforces conventional conformity. (4) A casual crowd is on the verge of emerging norms. Weakness: has difficulty explaining what norms will emerge and why they spread. *Convergence theory:* Crowds behave they way they do because of the predispositions of the people who are drawn to them. (1) Acting crowds draw people predisposed to action, (2) expressive crowds are those predisposed to emotional expression, (3) conventional crowds are those predisposed to conformity, and (4) casual crowds are those predisposed to only casual, if any, joint interaction. *Weakness:* Cannot account for changes in crowd dynamics. *Value-added theory:* Six necessary, but not sufficient conditions, must occur for collective behavior to happen; these conditions also shape that behavior. In each case, there must be structural conduciveness, structural strain, growth, and spread of a generalized belief, precipitating factors, mobilization for action, and mechanisms of social control. *Weakness:* Cannot explain conventional or causal crowds very well.

51. *Fad*—Transitory social changes which are widely dispersed.

Craze—Especially short-lived changes.

Fashion—Standards of dress at a particular point in time.

Rumor—Information that is shared informally that spreads quickly through a crowd

Propaganda—Advertisements of a political nature seeking to mobilize public support behind one specific party, candidate, or point of view.

Mass Hysteria—When large numbers of people are overwhelmed with emotion and frenzied activity or become convinced they have experienced something for which investigators can find no discernable evidence.

Panic—An uncoordinated group flight from perceived danger.

52. *Incipiency:* Relative deprivation (RD) theory says that a movement begins because people perceive that they lack something possessed by an important reference group; Resource mobilization theory (RM) theory says that leaders strike a responsive chord in people. *Coalescence:* RD says that. as awareness of deprivation increases, people are more motivated to come together. RM says it is because resources continue to be mobilized in larger and more visible amounts., Institutionalization: RD says that the sense of deprivation is so great that people insist on institutional changes. RM says that the sense of deprivation wanes or becomes defined in a variety of diverse ways., RM says that the sense of relative deprivation is no longer present. RM says that the leadership can no longer mobilize sufficient resources to sustain the movement.

53. **Reactionary**—embraces the aims of the past and seeks to return general society to yesterday's values. Based in history (examples: KKK, Neo-Nazis).

    Conservative—seek to maintain society's current values by reacting to change/threats of change that they think will undermine the status quo.

    Revisionary—seek partial or slight changes within the existing order, but do not threaten the order itself.

    Revolutionary—seek to overthrow all or nearly all of the existing social order and replace it with something they think is more suitable.

54. **Vegetarian beliefs** have existed throughout history, so it could be considered a *reactionary social movement*; however, given the extent to which it challenges the fundamental beliefs about eating in a wide variety of countries where animals are views as legitimate sources of food, vegetarianism can be considered a *revolutionary social movement.* It also has a belief system and a growing number of adherents as well as increased support from the medical profession.

CHAPTER EIGHTEEN

1. f	12. f	23. t	34. c	45. t
2. t	13. t	24. t	35. a	46. f
3. f	14. a	25. f	36. d	47. f
4. f	15. t	26. t	37. b	48. t
5. t	16. b	27. f	38. c	49. t
6. t	17. c	28. d	39. c	50. t
7. t	18. b	29. b	40. t	51. e
8. b	19. b	30. c	41. f	52. d
9. e	20. a	31. a	42. t	53. e
10. c	21. f	32. c	43. d	54. a
11. f	22. t	33. e	44. e	55. t

56. *Internal social change* example: caused by technological innovation, ideology, or reactions to institutionalized inequality. *External social example:* caused by diffusion or forced acculturation.

57. You can employ one of the following theories: *Evolutionary Theory, Functionalist Theory, or Cyclical Theory.* The details of these theories are presented in the section on "Theories of Social Change" in your text.

58. **Clifford Stoll** argues that we have always lived in the information age; however, today we find technocrats proclaiming themselves the high priests of a new order. Stoll argues that information itself is not power; nor is it money per se. Cyber communication tends to focus primarily on securing information as quickly as possible thereby shifting our cognitive processes away from contemplative thought. While the Internet has positive features, Stoll argues that it is addictive and isolating for individuals. He suggests that more information is not necessarily improved information. Finally, he states that this global network of information has not brought us any closer to solving our most pressing problems.

59. Ritzer believes the fast-food chain McDonald's is a model for other institutions in modern society. McDonald's affects not only the restaurant business, but also education, work, travel, leisure-time

activities, dieting, political, religion, and the family. Children are bombarded by its commercials so often that the restaurant's Ronald McDonald has name recognition second only to Santa Claus. McDonald's is successful because it offers customers efficiency and predictability and it presents the diner a good value. Unfortunately its production, assembly line format is often a dehumanizing setting in which to work and eat.

60. Improved health and longevity along with greater affluence and more leisure time will be outcomes of modernization as it impacts the individual. Computers will enable individuals to do amazing things; however, as Weber noted, many of these effects of modernization will be damaging to the spirit of the individual. Some persons will experience severe psychological dislocation and there will be a collapse of many traditional cultures that have provided meaning for individuals in the past. Those unable to adjust may find themselves emotionally adrift in a world they do not understand and cannot control. Turnbull's study of the **Ik** presents a possible picture of this disintegration. The critical factors will be the conditions under which modernization is introduced and the degree to which the masses are allowed to share in the material benefits of technological change without losing all of their traditional cultural values.

61. In terms of work, only 12% of the workforce will be engaged in blue-collar manufacturing work due to increased efficiency and greater automation. There will be more women in the labor force as well as minorities; for example Hispanics and African-Americans. Jobs requiring higher levels of education will be increasing and people who cannot read will have a very difficult time finding employment. Computer engineers, scientists, home health care workers, medical assistants, and corrections officers are occupations that will grow faster than the rest of the employment market.